PROGRESS IN COLLOID & POLYMER SCIENCE

Editors: F. Kremer (Leipzig) and G. Lagaly (Kiel)

Volume 103 (1997)

Amphiphiles at Interfaces

Guest Editor:

J. Texter (Rochester)

ISBN 978-3-662-15700-8
ISSN 0340-255 X

Die Deutsche Bibliothek –
CIP-Einheitsaufnahme

Amphiphiles at interfaces / guest ed.:
J. Texter.
  (Progress in colloid & polymer science ;
  Vol. 103)
    ISBN 978-3-662-15700-8
    ISBN 978-3-7985-1662-5 (eBook)
    DOI 10.1007/978-3-7985-1662-5

© 1997 by Springer-Verlag Berlin Heidelberg
Originally published by Dr. Dietrich Steinkopff Verlag
GmbH & Co. KG, Darmstadt in 1995
Softcover reprint of the hardcover 1st edition 1995

Chemistry editor: Dr. Maria Magdalene
Nabbe; English editor: James C. Willis;
Production: Holger Frey, Bärbel Flauaus.

Typesetting: Macmillan Ltd.,
Bangalore, India

Progr Colloid Polym Sci (1997) V
© Steinkopff Verlag 1997

PREFACE

This volume is loosely based upon papers presented and topics discussed at several recent international meetings. The first of these meetings was the international symposium on *Interfacial Structure* held in conjunction with the 210th National Meeting of the American Chemical Society on August 20–25, 1995 in Chicago. The second of these was the 1996 Chemistry at Interfaces Gordon Research Conference held July 21–26, 1996 in Meriden, New Hampshire. Additional contributors to this volume were selected because of work they presented at the 11th Surfactants in Solution conference held June 9–13, 1996 in Jerusalem or because their work was known to the Editor. All of the papers included in this volume, however, are of more recent vintage, and were written in the fall and early winter of 1996. The subject of this volume is limited in that it addresses amphiphiles at liquid/air, liquid/liquid, and liquid/solid interfaces, with little attention paid to vapor/solid interfaces. This subject is an important focal point for examining physical phenomena that occur in colloidal systems such as micellar solutions, microemulsions, emulsions, dispersions, slurries, etc., as well as flows at and wetting of macroscopic interfaces. This volume hopefully will serve to summarize our current understanding of interfacial structure at the molecular level in these systems, and the relation of this structure to chemical and physicochemical phenomena. Simulations and experiments are becoming complementary approaches to investigating the structure of these interfacial systems, and a balanced presentation of these various approaches has been attempted.

Invaluable support for these meetings was provided by the Division of Colloid and Surface Chemistry of the American Chemical Society, Eastman Kodak Company, 3M Specialty Chemicals Division, the Petroleum Research Fund of the American Chemical Society, Chemical Transport & Separations Division of the National Science Foundation, and the Chemistry Division of the Army Research Office. This support was very helpful in assisting with the travel expenses of distinguished scientists from Asia, the middle east, and Europe, and in making it possible for a significant number of graduate students and postdoctoral research associates to attend these meetings.

The contents of this volume have been distributed among six sections. The first of these sections, Liquid/Liquid Interfaces, is the least well characterized interface in colloid science. The first two papers by Brevet and Girault and by Richmond and co-workers give a comprehensive review of the great progress made recently in the application of two-photon processes in the analysis of molecular orientation and conformation at oil-water interfaces. The paper by Volkov and Deamer reviews theoretical aspects of charge transfer at oil-water interfaces, and discusses experimental interfacial catalytic systems. Pohorille, Wilson, and Chipot provide a comprehensive review of simulation studies of water-oil interfaces and present some exciting new results on surface segregation, high-lighting the importance of electric dipoles in the genesis of interfacial activity. Murray presents some novel Langmuir trough studies applied both to oil-water and to air-water interfaces. Andelman and Diamant present an elegant variational study of the derivation of kinetics at liquid-liquid interfaces, and their work will interestingly be compared to the experimental study of Pitt at the end of this volume and the lattice-gas simulations of Khan and Shnidman. Stauffer alludes to the inclusion of amphiphiles in an Ising-like treatment, and this allusion is illustrated by Khan and Shnidman later in this volume.

The second section on Vesicles, Bilayers, and Membranes addresses the most biologically relevant area for amphiphiles. Chaimovich introduces the section with an interesting survey of chemical factors affecting transport through liposomes and bilayers. Beyer introduces a mechanism for bilayer formation from micelles, and Klopfer and Vanderlick present a molecular dynamics study illustrating how small nonionic amphiphilic structure can lead to the selection of bilayer structure over micellar structure. McIntosh and Simon present a comprehensive review of work done on the experimental characterization of fluctuations in lipid bilayers, and Berkowitz and coworkers articulate various approaches to defining the role of hydration forces in the interactions of such bilayers. Stouch reviews the key physical features affecting transport and partitioning in bilayers, and atomic-level molecular dynamics simulations of such bilayers. Green and Lu review studies applied to trying to model transport through membrane pores.

Micellar Aggregation is the focal point of the third section. Care and coworkers lead off with a paper modeling micellar aggregation on a lattice. They examine relatively short-chain surfactants, and calculate the thermodynamics, shapes, and aggregate-size distributions. Mattice and coworkers present a comprehensive review of their work modeling the formation of association colloids of triblock copolymers. They also use a lattice approach. Shelley, Sprik, and Klein outline key issues to consider in molecular dynamics modeling of micellization and other structured surfactant-water phases. They find that inclusion of polarizability results in counter ions being heavily solvated, but that the water-hydrocarbon interface and the electrical double layer at the surfactant water interface is clearly distinguishable irrespective of polarizabilty assumptions. Mavelli presents an exciting new application of stochastic dynamical modeling in illustrating surfactant aggregation. Texter and coworkers present a review of experimental studies directed at examining the role of cosurfactants in modifying reverse micelle aggregation and the onset of percolation processes therein.

The fourth and fifth sections deal with amphiphiles at solid-liquid interfaces. Section four on Amphiphiles at Electrode Surfaces contains a collection of important applications. The first by Rusling illustrates how surfactant films on electrode surfaces may be fabricated to electrocatalyze a variety of reactions. Koglin and coworkers provide some exciting new surface enhanced Raman spectroscopy results that contravene some of the reigning dogma on how cationic surfactants assemble on anionic surfaces. Kaifer presents a concise review of self-assembled monolayers on electrodes derivatized with thiolates and containing preformed binding sites. Somewhat similarly, the paper by Baszkin and coworkers in the last section investigates the incorporation of cyclodextrins into phospholipid monolayers. Bizzotto and Lipkowski comprehensively review nonionic amphiphile adsorption on electrodes and show how hemimicellar and micellar formation, adsorption, and desorption processes may be studied experimentally. The fifth section on Adsorption at Solid/Liquid Interfaces commences with an important review by Thomas on the resolution of surfactant structure by neutron reflectivity. Manne follows with a review of recent, exciting force microscopy results that go further to upset some of the dogma on surfactant hemimicelle formation, and yield some direct morphological evidence for hemimicellar structure. Balazs and coworkers present a self-consistent field treatment of polymers tethered to solid surfaces, and explore association structures formed by these polymers in various solvents. Grainger reviews polymeric thin film formation by adsorption from solution using chemisorption and using stratified polyelectrolyte layers. Khan and Shnidman present a lattice gas treatment of capillary dynamics, and include the effects of amphiphiles. Some formal similarity with the treatments of Daimant and Andelman and of Stauffer in the first section exist in the formalism adopted to attack these flow problems. Cohen and coworkers present a review of small angle x-ray scattering in the analysis of adsorbed layers surrounding colloidal particles.

Section six is devoted to Amphiphiles at Vapor-Liquid Interfaces. Zasadzinski and coworkers present a review of lung surfactant protein and the effects of such protein on altering phase structures in palmitic acid monolayers. They articulate some of the minimal functional requirements for such protein, with a view to expediting the design and synthesis of less expensive replacements for human lung surfactant. Siepmann reviews vapor-liquid phase equilibria and examines by Monte Carlo methods the suitability of two different force fields in modeling such equilibria. Goedel shows how hydrophobic polymers may be assembled at the air-water interface to prepare models of polymer "melt brushes". The experimental approach is viable for preparing nanometer thick coatings. Baszkin and coworkers examine the miscibility of two phospholipids with lipid-derivatized β-cyclodextrin and with lipid-derivatized poly(ethylene oxide). While Murray's paper on protein dynamics appears in the first section, it is equally directed to air-water interfaces, and it will be interesting to see if these protein dynamics can meaningfully be treated by models developed for more classical surfactants. Abbott reviews recent applications of electrochemically active surfactants in modifying interfacial tension, and illustrates some exciting electrochemical engineering in deducing structural features that facilitate control of surface adsorption. This work also defines a mechanism whereby localized release of surfactants may be utilized in the localized permeabilization of membranes. The paper by Pitt presents exciting experimental results for different classes of surfactants in the development of an understanding of key features that control dynamic interfacial tension and the factors important in modifying the efficiency of dynamical surface tension lowering. Correlation of these data with various models put forth by Diamant and Andelman earlier in the volume will help refine our perspective of the modeling of these systems.

John Texter

CONTENTS

VIII

Adsorption at Solid/Liquid Interfaces

Amphiphiles at Vapor-Liquid Interfaces

Progr Colloid Polym Sci (1997) 103:1–9
© Steinkopff Verlag 1997

P.F. Brevet
H.H. Girault

Optical SHG measurements of amphiphiles at liquid/liquid interfaces

Received: 6 December 1996
Accepted: 12 December 1996

Dr. P.F. Brevet (✉) · H.H. Girault
Laboratoire d'Electrochimie
Ecole Polytechnique
Fédérale de Lausanne
1015 Lausanne, Switzerland

Abstract Surface second harmonic generation (SSHG) is described as a powerful surface tool to investigate amphiphiles at liquid/liquid interfaces. In particular, the molecular picture of the interface which is retrieved from these measurements is emphasized. In a first part, a theoretical analysis of the origin of the SH signal from these systems is discussed. The dipole electric contribution, which is highly surface specific, is shown to be the dominant one at liquid/liquid interfaces, the volume contributions of higher orders in the multiple expansion of the nonlinear polarization being overwhelming at pure solvent air/liquid interfaces only. In the second part, the experimental results available to date are reported. The orientation and solvation properties of the amphiphile monolayers are shown to principally depend on the hydrophobic–hydrophilic forces. Photoisomerization experiments are reported and they show that the interfacial molecular friction is dramatically different from the friction found in bulk aqueous solution. Chemical reactions, and in particular acid/base equilibria, are also discussed and the role of the external potential at polarized interfaces in the apparent pKa value measured is underlined. Finally, charge-transfer reactions at polarized liquid/liquid interfaces are presented.

Key words ITIES – SSHG – amphiphiles – photochemistry – surface chemistry

Introduction

The study of surfaces and interfaces has always been hampered by the search of appropriate experimental tools. Optical surface second-harmonic generation (SSHG) has been shown to be a powerful method and has thus been widely applied over the last decades in surface science. Phenomena like adsorption processes, surface reconstruction or interfacial chemical reactivity have been investigated at both solid and liquid surfaces and have already been extensively reviewed in the past [1–5]. At liquid/liquid interfaces experimental results are far less numerous, despite the growing interest for these interfaces which may be used as mimetic models for biological cell membranes [6].

Second-harmonic generation is a nonlinear optical process through which two photons at a fundamental frequency are converted to one photon at twice the fundamental frequency. In the electric dipole approximation, this process is forbidden in media with inversion symmetry and therefore only occurs at interfaces where the inversion symmetry is broken. As a main drawback to such a surface specificity, the process is very weak, its efficiency being in the range of 10^{-12}%. However, with the availability of high intensity lasers and highly sensitive detectors, signal levels of few photons per pulse are routinely detected in laboratories. The suitability of the technique to the study of liquid surfaces has been recognized for a long time, since the experiments of C.C. Wang at the air/water interface [7] but its use to liquid/liquid interfaces, and in particular at interfaces between two immiscible electrolyte solutions (ITIES), has only developed in the last ten years [6].

In this review, we focus on the recent advances of SSHG at liquid/liquid interfaces. We first present the theoretical background with the fundamental equations necessary to analyse the data. Then, we discuss the origin of the SH response at liquid/liquid interfaces before presenting a review on experimental results at liquid/liquid interfaces. We first describe the structure of amphiphile monolayers at free interfaces and ITIES, in particular, the interfacial orientation and solvation properties of the monolayers. We then present dynamics studies like isomerization processes, chemical equilibria and finally charge-transfer reactions across liquid/liquid interfaces.

The SH response from liquid/liquid interfaces

The question of the origin of the nonlinear optical response of an interface has been described with two models in the past: the model of the thick slab and the model of the nonlinear polarization sheet. The two models yield similar results although the model of the nonlinear polarization sheet, owing to its phenomenological aspect, is usually preferred.

The first model, which we call the model of the thick slab, appeared with the early developments of nonlinear optics. Indeed, in the paper of Bloembergen and Pershan in 1968 [8], a section was already devoted to the problem of SH response from a thick nonlinear slab embedded in a linear optical medium. Here it is understood that the thickness of the slab is larger than the optical wavelength. From the equations derived for the SH field amplitude transmitted in the linear media, the limiting case of a vanishing thickness yielded results applicable to the SH response at the boundary between two centrosymmetric media. The problem was soon revisited from another point of view. Since the SH generation is expected to arise from a very thin strata of material, thin as compared to the wavelength of light, the nonlinear polarization induced by the fundamental wave may be taken as a sheet of vanishing thickness. In this case, the mathematical description for the nonlinear polarization is a Dirac delta function in the direction normal to the boundary surface. Heinz and co-workers first used this approach and their results compared well with the previous model [9]. One of the major issues which quickly arises is then the value of the optical dielectric constant of the nonlinear layer. Guyot-Sionnest et al. [10] have stressed the need for an independent

measurement of this dielectric constant and this question has been discussed in detail by Zhang et al. [11]. The phenomenological approach, first suggested by Mizrahi and Sipe [12, 13] and recently reformulated for a three-layer model [14], has been derived from Heinz's model, the sheet of nonlinear polarization being embedded in a thin strata of linear material which is in turn taken between two-half spaces of linear optical dielectric media. In this model, the SH waves generated by the sheet of nonlinear polarization within the thin linear slab are transmitted to either half space through Fresnel coefficients. Hence, the SH intensity I_i^{Ω} is given by [14, 15]

$$I_i^{\Omega} = \frac{\omega^2}{4\varepsilon_0 c^3} \frac{\mathrm{Re}(\sqrt{\varepsilon_i^{\Omega}})}{\mathrm{Re}^2(\sqrt{\varepsilon_1^{\omega}})|\sqrt{\varepsilon^{\Omega}}\cos\theta^{\Omega}|^2} |\mathbf{e}_i^{\Omega}\cdot\boldsymbol{\chi}_{\mathrm{eff}}^{(2)}:\mathbf{e}^{\omega}\mathbf{e}^{\omega}|^2 (I_1^{\omega})^2 ,$$

(1)

where I_1^{ω} is the intensity of the incoming fundamental wave, ε_i^{Ω} the optical dielectric constant of medium i at frequency $\Omega = 2\omega$ and θ^{Ω} the angle of refraction of the harmonic wave within the thin slab containing the nonlinear polarization sheet and of optical dielectric constant ε^{Ω}. Equation (1) is given in SI units where the surface susceptibility tensor has the units of $\mathrm{m\,V^{-2}}$. Also, we have

$$\mathbf{e}_i^{\Omega}\cdot\boldsymbol{\chi}_{\mathrm{eff}}^{(2)}:\mathbf{e}^{\omega}\mathbf{e}^{\omega} = a_{i1}\chi_{\mathrm{eff},XXZ}^{(2)}\sin 2\gamma \sin\Gamma$$

$$+ (a_{i2}\chi_{\mathrm{eff},XXZ}^{(2)} + a_{i3}\chi_{\mathrm{eff},ZXX}^{(2)} + a_{i4}\chi_{\mathrm{eff},ZZZ}^{(2)})\cos^2\gamma\cos\Gamma$$

$$+ a_{i5}\chi_{\mathrm{eff},ZXX}^{(2)}\sin^2\gamma\cos\Gamma$$

(2)

where the a_{ij}, $i = 1, 2$ and $j = 1 \ldots 5$ are the geometrical factors embedding both the Fresnel coefficients and the angular dependence [15]. The angles γ and Γ are, respectively, the angle of polarization of the incoming fundamental and outgoing harmonic waves. Finally, we add that $i = 1$ in reflection and $i = 2$ in transmission. The SH response of a liquid interface, owing to its in-plane isotropy, has thus a characteristic shape as shown in Fig. 1. From the curve obtained for the harmonic light S-polarized, one usually extracts the component $\chi_{\mathrm{eff},XXZ}^{(2)}$, whereas from the curve obtained for the harmonic light P-polarized, one extracts the components $\chi_{\mathrm{eff},ZXX}^{(2)}$ and $\chi_{\mathrm{eff},ZZZ}^{(2)}$. If the S-polarized curve has always a similar shape owing to the factor $\sin 2\gamma$, the P-polarized one may take a different shape depending on the relative weight of the three components of the tensor [16]. In Eq. (2), we have used an effective tensor. Indeed, it has been shown that at air/liquid interfaces, and the air/water interface in particular, contributions from the next order in the multipole expansion of the nonlinear polarization could overwhelm the electric dipole surface sensitive contribution [17, 18]. These contributions arise from the strong field gradients present at the

Progr Colloid Polym Sci (1997) 103:1–9
© Steinkopff Verlag 1997

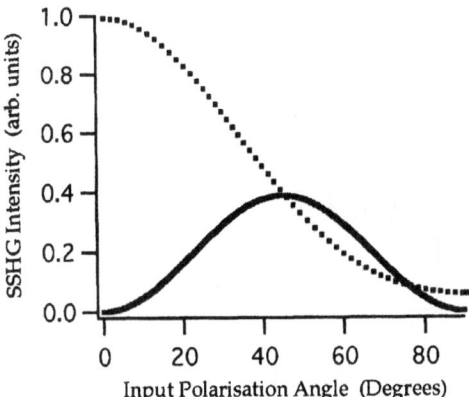

Fig. 1 Theoretical curve, as calculated from Eq. (2), for an air/liquid or a liquid/liquid interface for the S-Polarized (solid curve) and the P-polarized (dotted curve) SH wave in reflection as a function of the input polarization angle

interface and the electric polarization takes the following form:

$$\mathbf{P}^{(\Omega)} = \tfrac{1}{2}\varepsilon_0(\boldsymbol{\chi}^{(2)} : \mathbf{E}^\omega \mathbf{E}^\omega - \boldsymbol{\nabla}\cdot\mathbf{Q}^{(2)} + \frac{\mu_0}{i\omega}\boldsymbol{\nabla}\times\mathbf{M}^{(2)})\,, \qquad (3)$$

where $\mathbf{Q}^{(2)}$ is the second-order electric quadrupole tensor and $\mathbf{M}^{(2)}$ the second-order magnetization tensor. At liquid/liquid interfaces however, the volume contributions cancel owing to the better matching of the optical dielectric constants of the two media and the SSHG technique is again surface specific. This is an interesting feature which allows very low surface coverages to be studied. The complete expression for the three nonvanishing independent tensor elements are thus [15]:

$$\chi^{(2)}_{\text{eff},XXZ} = \chi^{(2)}_{s,XXZ} + \chi^{(2)}_{Q2,XZXZ} - \chi^{(2)}_{Q1,XZXZ}\,,$$

$$\chi^{(2)}_{\text{eff},ZXX} = \chi^{(2)}_{s,ZXX} + \frac{b_{i1}}{a_{I5}}(\gamma^{\text{equiv.}}_{i1} + \gamma^{\text{equiv.}}_{i2})$$

$$+ \chi^{(2)}_{Q2,ZXX} - \chi^{(2)}_{Q1,ZXX}\,,$$

$$\chi^{(2)}_{\text{eff},ZZZ} = \chi^{(2)}_{s,ZZZ} + \left(\frac{b_{i2}}{a_{i5}} - \frac{b_{i1}a_{i3}}{a_{i4}a_{i5}}\right)(\gamma^{\text{equiv.}}_{i1} + \gamma^{\text{equiv.}}_{i2})$$

$$+ \chi^{(2)}_{Q2,ZZZZ} - \chi^{(2)}_{Q1,ZZZZ}\,, \qquad (4)$$

where b_{ij}, $i = 1, 2$ and $j = 1, 2$ are geometrical factors and $\gamma^{\text{equiv.}}_{i1}$, $\gamma^{\text{equiv.}}_{i2}$ two parameters related to the quadrupole electric susceptibility tensor.

Several experiments have been devoted to the study of the structure of bare liquid/liquid interfaces. As shown above, different contributions may interfere in the SH generation and a careful analysis is required in order to assess the surface specificity of the technique. Such a work has been carried out at the air/water, air/1,2-dichloro-

ethane and air/hexane interfaces and the values of the three components of the macroscopic susceptibility tensor compared to their counterparts at the water/1,2-dichloroethane and water/hexane interfaces [17]. It is then demonstrated that the SH signal from air/liquid interfaces is overwhelmingly of electric quadrupole origin, and therefore nonspecific to the surface, whereas the response from liquid/liquid interfaces is mainly attributed to the surface specific dipole electric contribution. The main reason lies in the weakness of the hyperpolarisability tensor of the solvent molecules, and subsequently, in the weakness of the surface electric dipole susceptibility tensor. Hence, terms from higher orders in the multipole expansion have to be included. The two main volume contributions are thus arising from the electric field gradient present at the interface owing to the strong mismatch of the optical dielectric constants between air and the liquid phase and to the gradient of the electric quadrupole susceptibility tensor across the air/liquid interface. At liquid/liquid interfaces, the mismatch of the optical dielectric constants and of the electric quadrupole susceptibility tensors is dramatically reduced and the SH signal is again surface specific. However, in this case, the signal levels are rather weak. A second survey of bare liquid/liquid interfaces has been performed at the n-alkane/water interfaces, for the alkanes ranging from heptane to decane [19]. In order to avoid the problem of weak signal levels, the experiment has been conducted in the total internal reflection mode (TIR) thus dramatically enhancing the signal. In this study, based on the magnitude of the ratio between the two components $\chi^{(2)}_{\text{eff},XXZ}$ and $\chi^{(2)}_{\text{eff},ZXX}$, a close correlation is made between the violation of Kleinman symmetry rule, i.e. the loss of the equality between $\chi^{(2)}_{\text{eff},XXZ}$ and $\chi^{(2)}_{\text{eff},ZXX}$, and the molecular order at the interface. It is thus suggested that the odd n-alkanes/water interfaces are less ordered than the even n-alkanes/water ones, the ratio between $\chi^{(2)}_{\text{eff},XXZ}$ and $\chi^{(2)}_{\text{eff},ZXX}$ being around 1.0 for octane and decane and 1.3–1.5 for heptane and nonane. It is interesting to note here that the value for this ratio is also found to be 1.0 for hexane [17]. In this work, the alternation between the odd and even alkanes for the ratio between the elements $\chi^{(2)}_{\text{eff},XXZ}$ and $\chi^{(2)}_{\text{eff},ZXX}$ is correlated with a similar alternation found in the heats of fusion.

Before closing this section, the case of polarized interfaces has to be introduced since SSHG at interfaces between two immiscible electrolyte solutions (ITIES) constitues one of the main trends of nonlinear optical applications at liquid/liquid interfaces. It has been shown for metals, that upon polarization by an externally applied electric potential, a specific SH response was generated from the coupling between the static dc-field established across the interface and the fundamental electromagnetic wave [20]. The main property of this contribution is that it evolves with the square of the static applied dc-field. At metal/electrolyte interfaces, this dc-field may be quite large since on the metal side, the potential drop is restricted to a very thin layer of a few atomic units at the surface. This property has been used to study adsorption processes, for example [21]. On the contrary, in electrolytes the space charge region may be much wider, in the range of tens of nanometers, owing to the reduced number of charges available in the solution. To date, this contribution to the total SH response has not been directly observed at polarized liquid/liquid interfaces yet [22], only at the air/water interface in presence of charged monolayers [23]. This question is of great importance to SSHG studies since the inversion symmetry may be broken within the diffuse layer, hence the region probed with the technique may expand further towards the bulk solutions.

Orientation and solvation of amphiphiles at liquid/liquid interfaces

The study of amphiphile ordering at interfaces is necessary to understand many phenomena, like microemulsions, foams or interfacial reactivity. It is expected that the preferential orientation taken by these compounds at interfaces is entirely determined by their interactions with the two solvents forming the interface and the intermolecular repulsion or attraction within the monolayer. As mentioned above, the SH response at liquid/liquid interfaces is dominated by electric dipole contributions and is therefore surface specific. Neglecting the contribution from the solvent molecules, which usually only have a weak nonlinear optical activity, the passage from the macroscopic susceptibility tensor $\chi_s^{(2)}$ to the microscopic molecular hyperpolarizability β of the adsorbate is obtained by merely taking the SHG response of the amphiphile monolayer as the superposition of the contribution from each single moiety. Hence, it yields

$$\chi_s^{(2)} = \frac{N_s}{\varepsilon_0} L^{(2)} \langle \mathbf{T} \rangle \beta , \qquad (5)$$

where N_s is the number of molecules per unit surface and $\mathbf{T}$ the tensor referencing the adsorbate in the fixed laboratory frame with the three Euler angles ϕ, θ and ψ, see Fig. 2. Here, the brackets emphasize the averaging over the different molecular orientations. $L^{(2)}$ is the local field correction factor and its expression is

$$L^{(2)} = \left(\frac{\varepsilon^\Omega + 2}{3} \right) \left(\frac{\varepsilon^\omega + 2}{3} \right) \left(\frac{\varepsilon^\omega + 2}{3} \right) \qquad (6)$$

in the Lorentz model. In liquids, where optical dielectric constants are not too large and isotropic, this correction

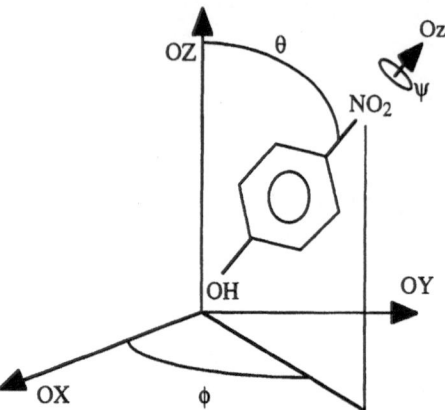

Fig. 2 Euler angles (ϕ, θ and ψ) referencing the molecular frame, with the Oz axis taken along the molecular axis, to the laboratory fixed frame where the OZ axis is along the surface normal

factor is often omitted. It has also been demonstrated that it would only lead to a rescaling of the susceptibility tensor $\chi_s^{(2)}$ in isotropic media. Equation (5) is only tractable when the molecular symmetry strongly reduces the number of independent nonvanishing components of the molecular hyperpolarizability tensor. In many instances, the symmetry point group of the amphiphile may be reduced to the C_{2v} point group, hence reducing the number of independent nonvanishing tensor elements to only three, namely β_{zzz}, β_{zxx} and β_{xzx}. Assuming a random distribution for the angle ϕ then yields:

$$\chi_{s,XXZ}^{(2)} = \frac{N_s}{2\varepsilon_0} \left[\langle \sin^2 \theta \cos \theta \rangle \beta_{zzz} - \langle \sin^2 \psi \cos \theta \sin^2 \theta \rangle \right.$$
$$\left. \times (\beta_{zxx} + 2\beta_{xzx}) + \langle \cos \theta \rangle \beta_{xxz} \right] , \qquad (7a)$$

$$\chi_{s,ZXX}^{(2)} = \frac{N_s}{2\varepsilon_0} \left[\langle \sin^2 \theta \cos \theta \rangle \beta_{zzz} - \langle \sin^2 \psi \cos \theta \sin^2 \theta \rangle \right.$$
$$\left. \times (\beta_{zxx} + 2\beta_{xzx}) + \langle \cos \theta \rangle \beta_{zxx} \right] , \qquad (7b)$$

$$\chi_{s,ZZZ}^{(2)} = \frac{N_s}{\varepsilon_0} \left[\langle \cos^3 \theta \rangle \beta_{zzz} + \langle \sin^2 \psi \cos \theta \sin^2 \theta \rangle \right.$$
$$\left. \times (\beta_{zxx} + 2\beta_{xzx}) \right] . \qquad (7c)$$

The extraction of the molecular parameters requires an assumption on the orientation distribution for the ψ angle and on the relative magnitude of the different elements of the molecular hyperpolarizability. This is achieved with the knowledge of the symmetry of the molecular electronic transitions lying in the vicinity of the fundamental and the harmonic wavelengths. Finally, working with relative intensities on the macroscopic susceptibility tensor, one can extract the orientation angle θ, usually determined for a narrow distribution, and the ratio of the two dominant hyperpolarizability tensor elements.

Progr Colloid Polym Sci (1997) 103:1–9
© Steinkopff Verlag 1997

The first determination of a molecular orientation appeared in 1983 for p-nitrobenzoic acid at the air/silica and ethanol/silica interfaces [24] but only in 1988 at the liquid/liquid interface for sodium 1-dodecylnaphtalene-4-sulfonate (SDNS) at the octane/water and the carbon tetrachloride/water interfaces [25]. The fundamental wavelength of 532 nm was used and the harmonic light collected at 266 nm. At these wavelengths, it was noted that only one molecular tensor element was dominant, namely β_{zzz}, the element along the molecular C_2-axis. The orientation angle of SDNS was measured at 21° at the decane/water interface and at 38° at the water/carbon tetrachloride interface, the hydrophilic sulfonate head pointing into the aqueous phase. Such a considerable change was attributed to the interactions between the organic solvent and SDNS. The solvent is screening the tail–tail interactions within the monolayer and consequently relaxing the constraints on the orientation of SDNS. Also, the larger angle found at the water/carbon tetrachloride interface was correlated with the decrease in the surface tension as compared with the water/decane interface and the question of the influence of the sharpness of the interface on the orientation of the amphiphile raised. The screening effect of the intermolecuar interactions within the monolayer by the solvent has been studied in greater detail for several phenol derivatives [16]. In this series of experiments, p-nitrophenol, phenol and p-propylphenol were studied at the hexane/water interface and the results compared with similar data obtained at the air/water interface. All experiments were performed at full monolayer coverage and showed that the hexane phase was screening the interaction between the aqueous phase and the substituent group placed in the para position: the nitro group in p-nitrophenol, the hydrogen atom in phenol and the propyl group in p-propylphenol. Indeed, at the air/water interface, the three moieties take an angle determined by the hydrophilicity strength of the substituent group, namely 48° for the p-nitrophenol, 43° for phenol and 39° for the p-propylphenol. At the hexane/water interface, the organic phase was shown to completely screen this interaction, the angle of orientation being measured at 43° for the three compounds. It was also noted that the hexane phase allowed more flexibility for the most hydrophobic group, as seen from the larger angle taken by p-propylphenol at the hexane/water interface as compared to the air/water interface. These results were in complete agreement with the ones reported for SDNS. Comparison with surface tension measurements then led to the conclusion that the surface coverage at full monolayer was completely determined by the angle taken by the compounds at the interface. For these small compounds, the angle has been found to be invariant with the surface coverage [26]. This behavior is thus completely different than the one observed in monolayers where tail–tail interactions dominate. In particular, no two-dimensional phase transitions can occur, like in monolayers of pentadecanoic acid at the air/water interface [27]. Similar studies have been conducted at the polarized water/1,2-dichloroethane interface. The orientation angle of 4-octyl-oxybenzoic acid (OBA), n-octyl-4-hydroxybenzoate (OHB) and 4-(4'-dodecyl-oxyazobenzene)-benzoic acid (DBA) were thus reported in a series of experiments by Corn's group [28–30] and the measured values were 34°, 40° and 29° for OBA, OHB and DBA, respectively. The difference in the orientation angle of these compounds stems from the different hydrophilicity and hydrophobicity of the groups on both sides of the nonlinear active chromophore, the benzene ring for OBA and OHB and the azobenzene group for DBA, as already mentioned, at free interfaces. Also, no reorientation was observed as the applied potential between the two phases was swept, indicating that the strength of these adsorbate–solvent interactions are much larger that the electrostatic interactions.

SSHG may also be used as a tool to study adsorption processes. From Eq. (5), we note that the susceptibility tensor is proportional to the number of molecules per unit surface. A simple analysis of the SH intensity as a function of the bulk solution concentration thus leads to adsorption isotherms. This has been performed at the air/water interface for phenol or SO_2, for example [31, 32], but the SH intensity may be monitored as a function of the applied potential across the interface. These measurements have been performed for both ONS and DBA, yielding the surface coverage as a function of potential. In the case of ONS, the increase in the surface coverage has also been correlated with the drop in the surface tension [28]. An expression for the surface coverage may be found using the Frumkin isotherm. It yields:

$$\ln\left(\frac{\Theta}{1-\Theta}\right) = \ln\left(\frac{a_{ONS}}{a_{DCE}}\right) - \frac{\Delta G^0}{RT} + \frac{bF}{RT}(\phi_w - \phi_0) - \frac{c\Theta}{RT}, \quad (8)$$

where Θ is the surface coverage, $\phi_w - \phi_0$ is the potential drop across the interface, ΔG^0 is the free energy of adsorption when $\phi_w = \phi_0$, F is Faraday's constant, a_{ONS} and a_{DCE} are, respectively, the activities of ONS in the organic phase and the organic solvent itself. The parameter b gives the portion of the electric potential felt by ONS and c is a Frumkin interaction parameter. Hence, in Eq. (8), b provides information on the position of ONS relative to the interface. In these measurements, the value of $b = 0.67$ was determined indicating an intermediate position for ONS, neither completely in the aqueous phase nor in the organic phase.

Solvation of amphiphiles is of great interest since it directly gives insight into the molecular environment of the molecules. Depending on the position of the moiety along the direction normal to the interface, the molecule will undergo a transition from solvation in one bulk solution to the other. One way of determining the molecular environment is to use the potential-dependent adsorption properties of amphiphiles as explained above for the case of ONS. However, a more direct method consists in recording the absorption spectrum of the molecules. Molecules with large solvatochromic shifts present different absorption maxima, depending on the solvent polarity. At the interface, it is known that the absorption maximum takes a different value from that in the bulk, owing to the modified solvation experienced by adsorbates in this region [16]. Such measurements can be conducted at liquid/liquid interfaces where a full spectrum is recorded as a function of the fundamental wavelength. Another way to obtain the information is to compare the ratio of the two dominant elements of the hyperpolarizability tensor at different interfaces. This latter method was performed at the heptane/water interface for p-nitrophenol and at the hexane/water interface for p-nitrophenol, phenol and p-propylphenol, the data being compared with the air/water interface in both cases. A change in the ratio of the two dominant elements is attributed to a change in solvation. At the heptane/water interface, it was found that the solvation of p-nitrophenol was changing whereas at the hexane/water interface this was surprisingly not the case. Owing to the hydrophilicity of the nitro group, the moiety lies flat at the air/aqueous interface and therefore the benzene ring is highly solvated. The presence of the organic phase does not perturb the hydration at the hexane/water interface and the change observed at the heptane/water interface may stem from the structure of the interface itself [19, 33]. On the contrary, for the aromatic ring of phenol and p-propylphenol, it is found to clearly lie out of the water phase and thus to undergo a change of solvation when going from the air/water to the hexane/water interface [16]. Molecular dynamics simulations of phenol and p-pentylphenol at the liquid/liquid interface have yielded similar results, both for the orientation angle and the solvation properties [34, 35]. It is important to note here, that SHG is sensitive to electronic transitions and therefore only to a restricted portion of the nonlinear active molecules. In the case of long alkyl chain molecules, this implies that the carbon chains are not probed and therefore may take a different angle from the one measured for the nonlinear chromophore. An illustration of this question may be found in molecular dynamics calculations performed for p-pentylphenol [35].

Dynamics at liquid/liquid interfaces

We have described above the orientation and the solvation of amphiphiles at liquid/liquid interfaces. However, in order to understand the processes ocurring at these interfaces, an insight into the solvent properties like surface roughness, capillary waves or molecular friction are necessary. Different types of motion may then be studied, like orientation relaxation or isomerization. The experimental procedure is rather similar in both cases. For photoisomerization studies, a first laser pulse photoexcites the probe molecule from its ground state to one of its excited state. The molecule then undergoes motion along the reaction coordinate, involving torsion of part of the molecule, before relaxing to the ground state through internal conversion. A second pulse, delayed from the first one with an optical delay path, then probes the molecule by SSHG as it relaxes.

The first experiments on photoisomerization at liquid/liquid interfaces were reported by A. Shi et al. [36] for Malachite Green. In this work, a femtosecond laser with pulses of 130 fs duration is used to sample the Malachite Green isomerization at the different aqueous interfaces. The relaxation of Malachite Green after photoexcitation is measured to be 2.0 ps at the air/water interface, a significantly shorter time scale than at liquid/liquid interfaces. At the octane/water interface, the relaxation time is found to be 3.0 and 3.6 ps at the pentadecane/water interface. In bulk water, where the shortest time scale is found, a value of 0.7 ps is measured. These results suggest that the Malachite Green photoisomerization process occurs through twisting of the two dimethylaniline groups projecting into the water phase whereas the rotation around the axis going through the central carbon and the phenyl ring is not significant (see Fig. 3). The slower isomerization time obtained at liquid/liquid interfaces is therefore consistent with an increased structuring of the water phase at the interface as compared to the bulk water phase. However, one must be cautious in generalizing these results. It is indeed known that the photoisomerization of the dye 3,3'-diethyloxadicarbocyanine iodide (DODCI) is faster at

Fig. 3 Schematic of the izomerization process in Malachite Green at the alkane/water interface. The phenyl moiety protrudes into the alkane phase whereas the nitrogen containing groups are on the aqueous side. From the experimental results, the torsional motion is undergone along the two arrows

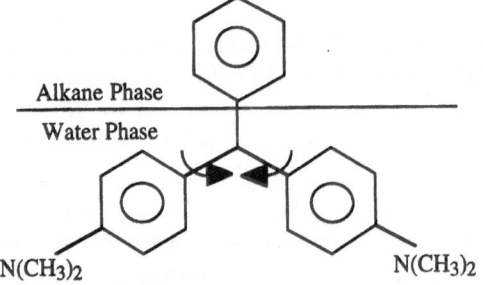

the air/water interface than in the bulk aqueous phase. This result, correlated to the one on Malachite Green, points towards a strong influence of the position of the dye at the interface and the groups within the molecule involved in the process of photoisomerization. In the case of DOCI, it is assumed that the rotation is achieved around the methylene bond in a region of low solvent coupling, i.e. out of the aqueous phase [37].

At the polarized water/DCE interface, the photoisomerization of DBA has been studied by Naujok et al. [38]. In this study, it is observed that nearly a full monolayer of *trans*-DBA may be converted to a full monolayer of *cis*-DBA, with no appreciable thermal conversion back to the *trans*-DBA form. Also, the achievement of the conversion of the complete monolayer on a short time scale was attributed to surface mixing and surface tension driven convection effects. Interestingly, the conversion of the monolayer back to the *trans*-DBA form could easily be performed by proper illumination of the interface.

Acid–base equilibria at liquid/liquid interfaces

The interfacial properties of liquid/liquid interfaces are of great interest in the understanding of transfer mechanisms across an oil/water interface. Neutral species easily partition between the two phases. On the contrary, ions usually require an external work to be transferred into the organic phase but in some cases the transfer is facilitated by the presence of an ionophore. This is indeed the case for alkali ions for example, as seen in section below on transfer reaction and complex formation across liquid/liquid interfaces. Another possible process responsible for a facilitated transfer is protonation or deprotonation. It is indeed known that protonation or de-protonation facilitates the transfer of ionic drugs across the oil/water interface [39]. The interfacial pH is thus a determining parameter in the transfer mechanism as will be the interfacial equilibrium constant between the different forms of the transferring species. We have seen above different structural properties of the interface, like orientation or solvation, and we focus now our attention to chemical equilibria and acid/base equilibria in particular. These equilibria have principally been studied at charged air/water interfaces [40, 41] but recent investigations at the free air/water interface with the use of highly water soluble nonlinear chromophores have also yielded apparent shifts for interfacial equilibrium constants as compared to their bulk solution values [42]. The polarized water/1,2-dichloroethane has received much less attention. The study of the surfactant DBA at the interface as a function of the pH of the bulk aqueous phase has been monitored by SSHG [30]. The nonlinear optical signal from the DBA was found to vary like a titration curve,

from a value close to zero at low pH, up to the maximum observed around a bulk pH value of 10. An apparent interfacial pKa could be deduced from these results, the signal being attributed to the presence of the deprotonated DBA anion. However, this pKa value was also found to vary with the DBA concentration in bulk DCE solution. This system was thus described with a model coupling an adsorption process for the neutral form of DBA between the bulk DCE phase and the interface and an acid/base equilibrium between the neutral and the anionic form of DBA at the interface, this latter step being therefore responsible for the dependence on the bulk aqueous pH of the nonlinear response. For polarized interfaces, it is known that the interfacial pH_s is shifted relatively to the bulk aqueous pH according to [43, 44]:

$$pH_s = pH - \frac{F\Delta\phi_{ws}}{2.3RT}, \qquad (9)$$

where $\Delta\phi_{ws}$ is the potential drop from the bulk aqueous phase to the position of the surfactant. This part of the potential drop has been shown to be only 20% of the overall drop across the interface at a bulk pH of 8.0 for DBA, suggesting that DBA principally resides on the water side of the interface. In the absence of an externally applied potential, Eq. (9) is still valid and suggests equality between surface and bulk pH. However, the problem may be further complicated if adsorption takes place since the formation of a diffuse layer of counterions may break down the equality between bulk and surface pH.

Photo-induced electron transfer at liquid/liquid interfaces

Photo-induced electron transfer reactions are very well studied since they are a way to convert light energy into chemical energy. However, one of the main problems in homogeneous solution reactions has always been the photoproducts separation. The liquid/liquid interface has thus been used as a simple form of a molecular architecture designed to achieve such a separation. The scheme may be the following: an electron donor is placed in one phase, the aqueous phase, for example, respectively, the organic phase, and an electron acceptor in the organic phase, respectively, in the aqueous phase. The separation of the reactants is usually achieved with a careful choice of the hydrophilic–hydrophobic properties of the compounds. Also, the same phase separation must hold for the products. The electron transfer reaction then proceeds as follows:

$$D(\text{aq.}) + A(\text{org.}) \xrightarrow{h\nu} D^+(\text{aq.}) + A^-(\text{org.}) \qquad (10)$$

and the charge required to maintain charge electroneutrality can be supplied by a counterion or an external potential. Photocurrents associated with this reaction have already been observed [45]. However, no direct spectroscopic observation of the interfacial process has been achieved before the work of Kott et al. [46] by SSHG. Since then, optical evidence has been achieved for an electron transfer across the liquid/liquid interface by other techniques like total internal reflection luminescence decay [47]. In the work of Kott et al., the acceptor is $Ru(bpy)_3^{2+}$ dissolved in the aqueous phase. Upon photoexcitation by light at 488 nm, the donor *trans*-1-ferrocene-2-[4-(trimethylammonio)phenyl]ethylene (1^+) is oxidized by $Ru(bpy)_3^{2+}$. The acceptor 1^+ was used as the SSHG active chromophore and the formation of its oxidized ferrocenium form, more optically nonlinear active than the ferrocene form, was monitored as the photoexcitation light was turned on and off. It was then deduced from the observation of a change in the SH response from the interface, that upon photoexcitation of $Ru(bpy)_3^{2+}$, an electron transfer occurred between $Ru(bpy)_3^{2+}$ and the donor 1^+. The slow rise time of the SSHG signal, in the range of 10 s, was then attributed to the diffusion of the $Ru(bpy)_3^{2+}$ excited species to the interface where they underwent electron transfer. These results clearly demonstrate the possibility of monitoring by SSHG an interfacial electron transfer reaction. It is to be noted though, that the mechanism described above is essentially an interfacial electron transfer and not a true heterogeneous electron transfer. With a pump–probe experimental arrangement, using short pulse duration lasers, a full characterization and time resolution of the mechanism for the electron transfer reaction, and ultimately the backtransfer reaction as well as the intermediate complex species formation, should be available.

Ion transfer reaction and complex formation across liquid/interfaces

Another class of charge transfer reactions at liquid/liquid interfaces is ion transfer across liquid/liquid boundaries. This field covers a wide range of applications, from the recovery of heavy metal ions from waste effluents to the transfer of ionic drugs across biological cell membranes. It is here of importance to study the transfer mechanism step by step since several stages may be involved like desolvation, resolvation, reorientation, isomerization, protonation, and deprotonation, etc. for example. This field is very innovative and very few results have appeared in the literature so far. This is explained by the large number of questions arising on the origin of the nonlinear signal upon the perturbations produced by the flow of charges

crossing the interface. Also, since resonant SSHG experiments are usually preferred to nonresonant ones, the SH response may disappear altogether owing to reabsorption due to the presence of the molecular probes in both phases.

The work done at the polarized water/DCE interface by Crawford et al. [48] consisted of the study of the complexation and transfer of the crown ether 4'-nitrobenzo-5-crown-5 with the cation Na^+. When no cation is present in the aqueous phase, no transport across the interface is observed. Nevertheless, the SSHG signal is still exhibiting strong variations when the potential across the interface is swept. This signal evolution is ascribed to the reorientation of the crown ether in order to align with the interfacial electric field. Indeed, since the crown ether possesses a dipole moment, the interfacial electric field attempts to orient the crown ether present at the interface in the direction along the normal to the interface. In the absence of an applied potential, without any electrolyte present, crown ether species which are adsorbed at the interface take a definite orientation, most probably with the nitro group in the aqueous phase and the crown ether ring in the organic phase. Hence, hydrophobic–hydrophilic interactions do play a role in the orientation of the crown ether at the polarized interface, as observed with other compounds (see above). The SSHG response when both the crown ether and the cation are present in the electrochemical cell has also been recorded and strong variations are observed as the potential across the interface is swept. These variations were attributed to the transfer of the cation–ionophore complexes. Also, since the results are dependent on the light polarization, it is expected that the complexes and the uncomplexed crown ethers have a different orientation and therefore that a reorientation occurs before the complexation and the transfer reaction take place.

SSHG measurements have also been performed at ion-selective membrane electrodes (ISE) in presence of the cations Li^+, K^+ and Na^+ dissolved in the aqueous phase and several ionophores like dibenzo-24-crown-8 in a poly(vinyl chloride) (PVC) membrane [49]. In these experiments, the nonlinear optical signal has been observed as a function of the analyte concentration in the aqueous phase and has been ascribed to the oriented cation–ionophore complexes at the aqueous phase/liquid membrane. Langmuir isotherms were built from the data and the results suggested that a saturation of the interface was achieved at high analyte concentration.

Conclusions

SSHG is a powerful tool to study liquid/liquid interfaces at a molecular level. When structural properties of the interface are investigated, orientation and solvation of the

Progr Colloid Polym Sci (1997) 103:1–9
© Steinkopff Verlag 1997

amphiphiles at the interface are determined. It has to be pointed out here though, that a careful analysis of the experimental conditions has to be performed beforehand to assess the surface specificity of the configuration. At free interfaces, that is in the absence of an interfacial applied potential, and at polarized interfaces when surfactants are considered, the signal is however surface specific although the question of the contribution from the two noncentrosymmetric diffuse layers is still open. Dynamics studies can be performed with a time resolution only limited by the laser pulse duration. This allows the study of the dynamics of reorientation processes or photoisomerization reactions, leading to an insight in the molecular friction which is shown to be higher at the interface on the aqueous side. Chemical equilibria have also been studied by SSHG at polarized liquid/liquid interfaces for an azobenzene surfactant. From this study, information on the position of the surfactant relative to the interface has been retrieved and a surface acid/base equilibrium constant has been deduced. Finally, charge-transfer reactions can be investigated, either interfacial electron-transfer or ion-transfer reactions, and clearly demonstrate the possibilities of the technique as a laboratory analytical tool.

Acknowledgments The authors wish to thank A.A. Tamburello-Luca and Ph. Hébert for fruitful discussions. This work is supported by the Fonds National Suisse pour la Recherche Scientifique.

References

1. Shen YR (1989) Ann Rev Phys Chem 40:327
2. Richmond GL (1991) In: Ed Bard AJ (ed) Electroanalytical Chemistry. Marcel Dekker Inc, New York, Vol 17
3. Eisenthal KB (1992) Ann Rev Phys Chem 43:627
4. Higgins DA, Corn RM (1994) Chem Rev 94:107
5. Eisenthal KB (1996) Chem Rev 96:1343
6. Brevet PF, Girault HH (1996) In: Volkov AG, Deamer DW (eds) Liquid–Liquid Interfaces, Theory and Methods. CRC Press, Boca Raton
7. Wang CC (1996) Phys Rev 178:1457
8. Bloembergen N, Pershan PS (1962) Phys Rev 128:606
9. Heinz TF (1991) In: Agranovich VM, Maraduddin AA (eds) Modern Problems of Condensed Matter Science. North-Holland, Amsterdam, Vol 29
10. Guyot-Sionnest P, Shen YR, Heinz TF (1987) Appl Phys B 42:237
11. Zhang TG, Zhang CH, Wong GK (1990) J Opt Soc Am B 7:902
12. Sipe JE (1987) J Opt Soc Am B 4:481
13. Mizrahi V, Sipe JE (1988) J Opt Soc Am B 5:660
14. Brevet PF (1996) J Chem Soc Faraday Trans 92:4541
15. Brevet PF (1997) In: Surface Second Harmonic Generation. Presses Polytechniques Universitaires Romandes, Lausanne
16. Tamburello-Luca AA, Hébert Ph, Brevet PF, Girault HH (1996) J Chem Soc Faraday Trans 92:3079
17. Tamburello-Luca AA, Hébert Ph, Brevet PF, Girault HH (1995) J Chem Soc Faraday Trans 91:1763

18. Goh MC, Hicks JM, Kemnitz K, Pinto GR, Bhattacharyya K, Eisenthal KB, Heinz TF (1988) J Phys Chem 92:5074
19. Conboy JC, Daschbach JL, Richmond GL (1994) J Phys Chem 98:9688
20. Corn RM, Romagnoli M, Levenson MD, Philpott MR (1984) Chem Phys Lett 30:106
21. Tamburello-Luca AA, Hébert Ph, Brevet PF, Girault HH (1996) J Electroanal Chem 409:123
22. Conboy JC, Richmond GL (1995) Electrochimica Acta 40:2881
23. Zhao X, Ong S, Eisenthal KB (1993) Chem Phys Lett 20:513
24. Heinz TF, Tom HWK, Shen YR (1983) Phys Rev A 28:1883
25. Grubb SG, Kim MW, Rasing Th, Shen YR (1988) Langmuir 4:452
26. Vogel V, Mullin CS, Shen YR (1988) Langmuir 7:1222
27. Rasing Th, Shen YR, Kim MW, Grubb S (1985) Phys Rev Lett 55:2903
28. Higgins DA, Corn RM (1993) J Phys Chem 97:489
29. Higgins DA, Naujok RR, Corn RM (1993) Chem Phys Lett 213:485
30. Naujok RR, Higgins DA, Hanken DG, Corn RM (1995) J Chem Soc Faraday Trans 91:1411
31. Hicks JM, Kemnitz K, Eisenthal KB, Heinz TF (1990) J Phys Chem 90:560
32. Donaldson DJ, Guest JA, Goh MC (1995) J Phys Chem 99:9313
33. Bell AJ, Frey JG, VanderNoot TJ (1992) J Chem Soc Faraday Trans 88:2027
34. Pohorille A, Benjamin I (1991) J Chem Phys 94:5599

35. Pohorille A, Benjamin I (1993) J Phys Chem 97:2664
36. Shi A, Borguet E, Tarnovsky AN, Eisenthal KB (1996) Chem Phys 205:167
37. Sitzmann EV, Eisenthal KB (1989) J Chem Phys 90:2831
38. Naujok RR, Paul HJ, Corn RM (1996) J Phys Chem 100:10497
39. Reymond F, Staeyert G, Carrupt PA, Testa B, Girault HH (199■) J Am Chem Soc
40. Xiao XD, Vogel V, Shen YR (1989) Chem Phys Lett 163:555
41. Marowsky G, Chi LF, Möbius D, Steinhoff R, Shen YR, Dorsch D, Rieger B (1988) Chem Phys Lett 147:420
42. Tamburello-Luca AA, Hébert Ph, Brevet PF, Girault HH, submitted
43. Marecek V, Koryta J, Samec Z (1988) Science 29:1
44. Zhao X, Subrahmanyan S, Eisenthal KB (1990) Chem Phys Lett 171:558
45. Samec Z, Brown AR, Yellowless LJ, Girault HH (1998) J Electroanal Chem 288:245
46. Kott KL, Higgns DA, McMahon J, Corn RM (1993) J Am Chem Soc 115:5342
47. Dryfe RWA, Ding Z, Wellington RG, Brevet PF, Kuznetzov AM, Girault HH (submitted)
48. Crawford MJ, Frey JG, VanderNoot TJ, Zhao Y (1996) J Chem Soc Faraday Trans 92:1369
49. Tohda K, Umezawa Y, Yoshiyagawa S, Hashimoto S, Kawasaki M (1995) Anal Chem 67:570

Progr Colloid Polym Sci (1997) 103:10–20
© Steinkopff Verlag 1997

J.C. Conboy
M.C. Messmer
R.A. Walker
G.L. Richmond

An investigation of surfactant behavior at the liquid/liquid interface with sum-frequency vibrational spectroscopy

Received: 28 October 1996
Accepted: 6 November 1996

J.C. Conboy[1] · M.C. Messmer[2]
R.A. Walker · Prof. G.L. Richmond (✉)
Department of Chemistry
University of Oregon
Eugene, Oregon 97403, USA

[1] *Present address*:
Department of Chemistry
University of Minnesota
Minneapolis, Minnesota 55455, USA

[2] *Present address*:
Department of Chemistry
Lehigh University
6 East Packer Avenue
Bethlehem, Pennsylvania 18015-3172, USA

Abstract Vibrational sum-frequency spectroscopy in conjunction with interfacial pressure measurements are used to provide the first direct spectroscopic information about the structure of amphiphillic molecules adsorbed to the interface between two immiscible liquids by total internal reflection sum-frequency vibrational spectroscopy (TIR SFVS). The effect of the ionic head group on the conformational order of sodium dodecyl sulfate (SDS), sodium dodecylsulfonate (DDS), dodecyltrimethylammonium chloride (DTAC), and dodecylamine hydrochloride (DAC) adsorbed at the D_2O/CCl_4 interface has been examined. In addition, the effect of the length of the alkyl chain on the conformation and orientation of sodium hexylsulfonate (HS), sodium undecylsulfonate (UDS), and sodium dodecylsulfonate (DDS) is also presented. SF vibrational spectra indicate the presence of gauche conformations in the hydrocarbon chains of all the surfactants examined. An increase in the surface coverage results in the reduction of gauche defects in the hydrocarbon chains as determined from the intensity ratio of the methyl to methylene symmetric stretch vibrational modes. Significantly different alkyl chain conformations are observed for the various head groups and chain lengths examined. A series of saturated symmetric dialkyl phosphocholines adsorbed at the D_2O/CCl_4 interface are also examined. Temperature controlled experiments carried out with aqueous solutions of dilauroylphosphocholine (DLPC), dimyristoylphosphocholine (DMPC), dipalmitoylphosphocholine (DPPC), and distearoylphosphocholine (DSPC) show the lipid bilayer gel to liquid crystalline phase transition temperature to play a pivotal role in determining interfacial monolayer concentration and alkyl chain structure. Even at equivalent interfacial concentrations longer chain phosphocholine species form more disordered monolayers with a greater number of gauche defects than shorter chain phosphocholine species, as determined from relative intensities of vibrational bands in the CH stretching region.

Key words Surfactants – interfaces – vibrational sum frequency generation – monolayers

Progr Colloid Polym Sci (1997) 103: 10–20
© Steinkopff Verlag 1997

Introduction

Whereas much has been learned in recent years on a molecular level about how amphiphillic molecules adsorb at air/liquid interfaces, much less is known about their behavior at an insoluble organic/water interface. This lack of information is primarily due to the absence of experimental methods for studying the liquid/liquid interface on a molecular level. Studies have been done to investigate the orientation and adsorption of the head group of surface active dyes at the oil/water interface using fluorescence [1, 2], resonance Raman scattering [3–5], and second harmonic generation SHG [6–8]. However, little is known about the conformation of the alkyl chains of amphiphiles adsorbed at a liquid/liquid interface and how factors such as head group functionality and alkyl chain length play a role in conformational order. This is particularly true for simple water soluble surfactants which are commonly used in commercial products [9–11].

Studies of monolayers on solid surfaces and at the air/water interface show that the van der Waals interactions between the alkyl chains of amphiphiles play a very important role in how they assemble at the air/water interface [10, 12]. These attractive forces can work to counter the repulsive electrostatic forces between the charged head groups, leading to a relatively ordered assembly of molecules. At a liquid/liquid interface, the van der Waals attractions between adjacent alkyl chains are diminished as solvent is introduced [13]. Under such conditions the chemical and electrostatic environment of the hydrophilic head group plays an even greater role in the adsorption behavior of the amphiphile [13].

The work presented here focuses on providing information about how different head groups and chain lengths affect the conformational order of the alkyl chain of surfactants at the liquid/liquid interface. An understanding of these systems will be invaluable to such areas of research as enhanced oil recovery and the technologically important field of emulsions [10, 14]. In addition, the investigation of phospholipid monolayers at the liquid/liquid interface are presented. Such systems have been used as model systems for lipid bilayers and biomembranes [15–19]. The technique used to gather this information is sum-frequency vibrational spectroscopy (SFVS). With SFVS, the chemical sensitivity of vibrational spectroscopy is coupled with the interface specificity of this nonlinear technique to greatly expanding its capabilities for examining buried interfaces [20–31]. Information about molecular conformation of the hydrocarbon chain is obtained specifically from the C–H symmetric stretch region of the vibrational spectra.

In this paper, the effect of the ionic head group on alkyl chain conformation is examined by studying a series of anionic and cation surfactants, sodium dodecyl sulfate (SDS), sodium dodecylsulfonate (DDS), dodecyltrimethylammonium chloride (DTAC) and dodecylammonium chloride (DAC) adsorbed at the D_2O/CCl$_4$. A series of alkyl sulfonates sodium hexylsulfonate (HS), sodium undecylsulfonate (UDS) and sodium dodecylsulfonate (DDS), at the water/carbon tetrachloride interface have been studied to determine the effect of chain length on the relative conformational order and orientation. In addition a number of phospholipids are also examined to gain insight into the interfacial behavior of this important class of biological surface active agents. The phospholipids used in this study belong to a family of saturated, symmetric, dialkyl phosphocholines (PC) having alkyl chain lengths of 12 carbon atoms (dilauroyl-PC or DLPC), 14 carbon atoms (dimyristoyl-PC or DMPC), 16 carbon atoms (dipalmitoyl-PC or DPPC) and 18 carbon atoms (distearoyl-PC or DSPC).

Experimental

Materials

D_2O (99%) and HPLC grade CCl$_4$ were purchased from Aldrich. The CCl$_4$ was distilled in order to remove any residual hydrocarbon compounds, and its purity confirmed by transmission FTIR. D_2O was shaken with purified CCl$_4$ prior to use and decanted. Sodium dodecyl sulfate (SDS) (Aldrich, 99.8%), Sodium dodecyl sulfonate (DDS) (TCI America, 99%), dodecyltrimethylammonium chloride (DTAC) (TCI America, 99%) and dodecylamine hydrochloride (DAC) (Kodak, 98%) were used as received. Sodium hexylsulfonate (HS) (TCI America, 99%), sodium undodecylsulfonate (UDS) (TCI America, 99%), and sodium dodecylsulfonate (DDS) (TCI America, 99%), were used as received. Dilauroylphosphocholine (DLPC), dimyristoylphosphocholine (DMPC), dipalmitoylphosphocholine (DPC), and distearoylphosphocholine (DSPC) were purchased in powder form from Avanti Polar Lipids and used without any further purification.

Interfacial tension measurements

Interfacial tension measurements of DTAC, DAC, SDS, DDS, UDS and HS at the H_2O/CCl$_4$ interface were obtained by means of the drop-volume method [32]. A Gilmont micrometer syringe was used for drop delivery of the CCl$_4$. Measurements were made at room temperature with aqueous surfactant concentrations in the range

0.1–7.5 mM, with the exception of HS for which the concentration range was 0.1–75.0 mM. The interfacial tension was obtained from the drop volume by means of the method of Wilkinson [33]. Interfacial tension measurements of the phospholipids, DLPC, DMPC, DPPC and DSPC, were performed using the Wilhelmy plate method. The equipment consisted of an electronic balance equipped with a platinum plate with a resolution of 4 μN/m. Measurements were performed at the interface of an aqueous pH 7.0 buffer solution containing the phospholipid and CCl_4. Measurements were obtained after the rate of change in the interfacial pressure had drop to less than ~2% per hour. The interfacial tension of the neat interface between a buffer solution and CCl_4 was used as a reference in determining the interfacial pressure.

Sum-frequency experiments

In IR-Vis SFG, two coherent laser beams, one visible and the other from a tunable IR laser source (ω_1 and ω_2) impinge on a surface. The induced nonlinear polarization at the surface results in the coherent generation of light at the sum-frequency ($\omega_3 = \omega_1 + \omega_2$) [34]. The intensity of the sum-frequency light is given by

$$I(\omega_{SF}) = |\tilde{f}_{SF} f_{vis} f_{IR} \chi^{(2)}|^2, \qquad (1)$$

where $\tilde{f}_{sf}$ is the nonlinear Fresnel factor for the generated sum-frequency light [35, 36], $\chi^{(2)}$ is the second-order susceptibility tensor and f_{vis} and f_{IR} are the geometric Fresnel factors of the incident fields. The symmetry constraints on $\chi^{(2)}$ prohibit nonlinear interactions in the bulk of centrosymmetric medium. As a result, the spectroscopy of the molecules residing in the interfacial region can be probed selectively without any contributions from the molecules present in the more pervasive bulk liquids.

The second-order susceptibility $\chi^{(2)}$ can be separated into a nonresonant contribution $\chi_{NR}^{(2)}$ arising from the bare interface and a resonant contribution $\chi_R^{(2)}$ arising from the vibrational resonances of the molecules at the interface:

$$\chi^{(2)} = \chi_{NR}^{(2)} + \chi_R^{(2)}(\omega_{IR}). \qquad (2)$$

For an IR vibrational mode, the resonant component of the susceptibility is given by

$$(\chi_R^{(2)}(\omega_{IR}))_{lmn} = \sum_v \frac{N A_{n_v} M_{lm_v}}{(\omega_v - \omega_{IR} - i\Gamma_v)}, \qquad (3)$$

where N is the adsorbate surface density, A_{n_v} is the IR transition moment, M_{lm_v} is the Raman transition strength, ω_v is the transition frequency with a damping constant of Γ_v for a specific transition, v, and ω_{IR} is the frequency of the incident infrared beam. In order for a transition to be SFG

active it must satisfy the constraint of Eq. (3) which requires that the vibrational mode be both infrared and Raman allowed.

For the SFVS experiments, tunable IR light was generated using a $LiNbO_3$ optical parametric oscillator (OPO), described elsewhere [37]. The OPO was pumped with the fundamental output of a Q-switched Nd:YAG laser generating 1064 nm pulses at 10 Hz with a pulse duration of 12 ns. Tunability throughout the desired wavelength region was achieved by angle tuning the $LiNBO_3$ crystal. IR light pulses in the 2700–3100 cm^{-1} region with a bandwidth of 6 cm^{-1} and energies of 2–3 mJ were obtained over the entire spectral region. Calibration of the OPO was performed using a polystyrene sample. The remainder of the 1064 nm YAG fundamental was frequency doubled in a KDP crystal to generate the visible 532 nm.

In the case of linear spectroscopic methods such as FTIR, attenuated total internal reflection (ATR) has been used to enhance the sensitivity of these methods [38–40]. In a similar fashion, a total internal reflection geometry has been used here to enhance the otherwise weak sum-frequency (SF) response from the liquid/liquid interface. In addition to a strong resonant contribution from the C–H stretching vibrations within the molecules, the intensity is also dependent upon the Fresnel factors for the input fields, f_i, and the outgoing SF, $\tilde{f}_{SF}$, as shown in Eq. (1). When the incident beams are directed on the interface at their respective critical angles, an enhancement of several orders of magnitude in the SF response is achievable over that of an external reflection geometry [25, 35, 36, 41–43].

For this reason, the sum-frequency experiments were performed in a cylindrical quartz cell, describe elsewhere [44]. In order to achieve the desired optical geometry for total internal reflection, both the visible and IR beams were directed onto the D_2O/CCl_4 interface through the CCl_4, the high index phase, in a copropagating manner. D_2O was used due to the weak but significant absorption by H_2O in the spectral region of interest which resulted in notable heating of the interface. The IR was focused on the interface at an angle of 70°. The visible 532 nm was collimated to a diameter of 1–2 mm and incident on the interface at an angle of 66°. Laser powers used were typically 1–2 mJ/pulse between 2800 and 3100 cm^{-1} and 5 mJ/pulse at 532 nm.

The sum-frequency signal was collected in reflection at an angle of 66°. The resulting sum-frequency light was polarization selected with a broad band Glann–Taylor polarizer. The residual 532 nm light was removed by a combination of absorptive, interference and holographic notch filters. The resulting signal was detected using a PMT and gated electronics. Variation of the input IR polarization was accomplished with a Soleil-Babinet compensator and an IR polarizer. Polarization of the visible

Progr Colloid Polym Sci (1997) 103: 10–20
© Steinkopff Verlag 1997

light was selected with a half-wave plate. Data points, collected every 2 cm^{-1}, were an average of 200 pulses. The SF spectra were corrected for Fresnel contributions for each polarization and for the intensity variation of the infrared beam throughout the spectral region.

Results and discussion

SF vibrational spectra

Representative sum-frequency vibrational spectra of DTAC, DAC, SDS, UDS, DDS and HS adsorbed at the D$_2$O/CCl$_4$ interface are displayed in Fig. 1. All spectra were collected at a bulk aqueous concentration of 5.0 mM except for HS which was obtained at a bulk concentration of 50 mM. The spectra in Fig. 1 were collected with the polarization combination of ssp (s polarized SF, s polarized visible, p polarized IR). The vibrational spectra for the various surfactants are similar with differences apparent only in the relative peak intensities.

Spectral assignments have been made as follows. The spectra for p polarized IR for all the surfactants studied (Fig. 1(a–f)) exhibit strong intensities for the methylene asymmetric stretch (d$^-$) at 2930 cm^{-1}, in agreement with the value observed in the IR spectrum, (2925 cm^{-1}) [45]. Peaks of moderate intensity are observed for the methylene symmetric (d$^+$) and methyl symmetric (r$^+$) stretches at 2848 and 2872 cm^{-1}, respectively. A weak methylene Fermi resonance (d$^+_{FR}$) at 2900 cm^{-1}, resulting from interaction of an overtone of the methylene bending mode with the methylene symmetric stretch, is observed as a shoulder of the methylene asymmetric stretch. This can be compared to the methylene Fermi resonance in polymethylene appearing in the IR (d$^+(\pi)$FR) at 2898–2904 cm^{-1} and in the Raman (d$^+(0)$FR) at 2890 cm^{-1} [46, 47].

Conformation of the alkyl chain

Effect of head group

The effect of the ionic head group on the conformation of the alkyl chains of DTAC, DAC, SDS and DDS adsorbed at the D$_2$O/CCl$_4$ interface have been determined by measuring the vibrational spectra. Information on the conformation of the alkyl chain can be obtained specifically from the intensities of the spectral features seen in Fig. 1. To understand the relative intensities of the peaks in these spectra, it is necessary to consider the local symmetry of the CH$_2$ hydrocarbon backbone. An all-*trans* hydrocarbon chain is locally centrosymmetric with respect to methylene groups. Therefore, under the dipole approximation for sum-frequency vibrational spectroscopy, little contribution from methylene resonances should be observed for a system of well ordered (all-*trans*), hydrocarbon chains [20, 48]. This makes the methyl and methylene region of the SF spectrum an especially sensitive indicator of alkyl chain conformation. The presence of a strong methylene peak at 2850 cm^{-1} in the spectra of Fig. 1 suggests that the alkyl chains contain a number of gauche conformations which relax the local symmetry. A schematic of the transition moments for the symmetric methyl and methylene modes for both gauche and trans conformations is illustrated in Fig. 2.

To understand how the conformational order of the alkyl chains varies with surface concentration, the ratio of the intensities of the symmetric methyl and symmetric methylene stretch modes is used in conjunction with the interfacial tension data. Previous studies have demonstrated that the ratio of the intensities of the symmetric

Fig. 1 Sum-frequency vibrational spectra of (a) DAC, (b) DTAC, (c) SDS, (d) DDS, (e) UDS and (f) HS acquired with the polarization combination of ssp (s polarized SF, s polarized visible, p polarized IR). The solid lines represent a fit to the spectra using a combination of Gaussian Lorentzian functions for each peak

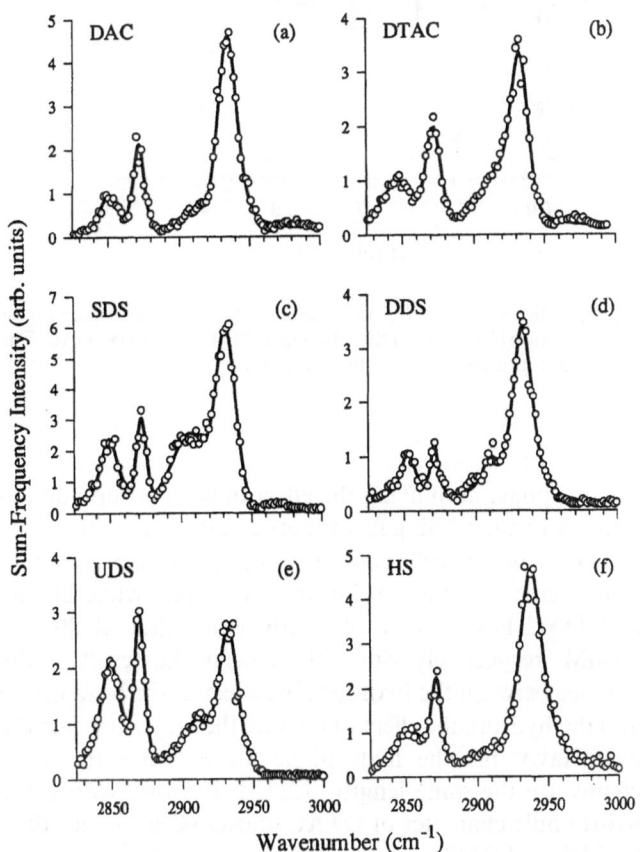

Fig. 2 Schematic representation of the net transition dipole moment for the symmetric methyl and methylene vibrational modes for an all-trans and gauch conformation

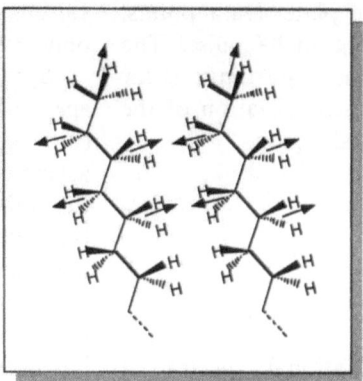
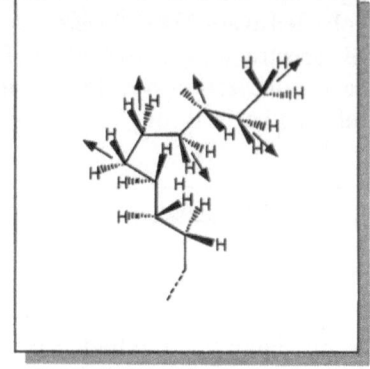

All-*trans* conformation   Gauche conformation

methyl and methylene modes, I_{r^+}/I_{d^+}, can be used as an indicator of the relative order within the hydrocarbon chains [44, 49]. A low methyl to methylene ratio reflects a relatively large number of gauche defects in the alkyl chain.

Plotting the surface concentration obtained from interfacial tension measurements against the I_{r^+}/I_{d^+} ratio derived from the SF vibrational spectra (Fig. 3) reveals the correlation between the surface concentration of surfactant and chain order. First, a linear relationship is observed between the I_{r^+}/I_{d^+} ratio and the surface density for all the surfactants. These results suggest that there is a good correlation between alkyl chain order and surface concentration. This correlation is consistent with the argument that as the surface density increases the degree of conformational mobility within the alkyl chains decreases, leading to more alkyl chain ordering. Secondly, the data points for SDS and DDS are seen to fall on nearly the same line whereas the data for DTAC and DAC are distinctly different. A much higher degree of alkyl chain order is observed for DTAC and DAC over SDS and DDS at comparable surface concentrations. However, if the surface density alone were determining the number of gauche defects within the hydrocarbon chain then the I_{r^+}/I_{d^+} ratio for DTAC and DAC should coincide with that of SDS and DDS. Instead, DTAC and DAC show a consistently higher degree of ordering, in the form of few gauche conformations, than SDS and DDS at all surface concentrations.

The fact that the surfactants, SDS, DDS, DTAC and DAC are completely soluble in the aqueous phase, unlike insoluble monolayers, means that the head group interaction with the water plays an important role in determining the orientation and conformation of the molecules. In addition, at the oil/water interface chain–chain interactions are minimal due to solubility of the chains in the

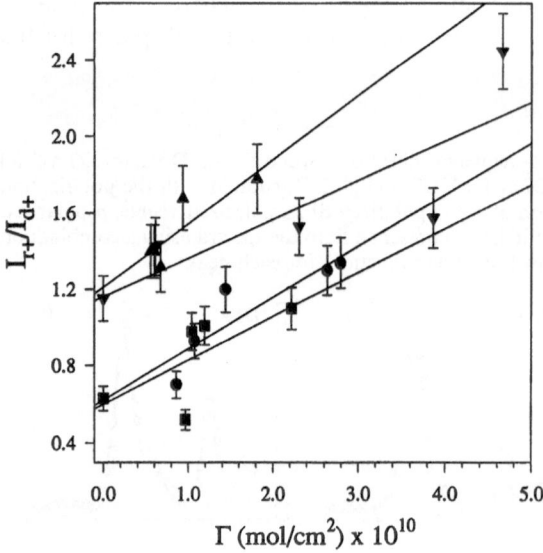

Fig. 3 A plot of the I_{r^+}/I_{d^+} ratio as a function of the surface excess concentration (Γ) for (●) SDS, (■) DDS, (▲) DTAC, and (▼) DAC. The solid line represents a linear fit to the data

organic phase amplifying the effect of head group interactions. A distinguishing factor between the four surfactants is there critical micelle concentration (cmc). SDS and DDS have a cmc of 8.2 and 9.7 mM, respectively while DTAC and DAC have a cmc of nearly twice that at 20 and 18 mM, respectively. Since the cmc is the result of the balance between the hydrophobic effect of the head group and the hydrophilic effect of the tail, the larger the cmc the more favorable the hydrophilic forces, since the alkyl chains are the same length [12, 13]. In other words, the hydrophilic character of DTAC and DAC is greater than for SDS and DDS since the chain length is identical for the series. The similarities of the trends observed in our own data with the micelle formation suggests that similar energetics are occurring in the two cases. It is important to point out that all spectra where recorded well below the cmc for all the surfactants.

Solvation of the head group has two effects on the adsorbed surfactant molecules; one is the penetration depth of the ionic head group into the aqueous phase and the other surface roughness. For example, neutron reflection studies have found a difference in the penetration depth of the ionic head groups of SDS and hexadecyltrimethylammonium bromide (HDTAB) at the air/water interface [50]. The measured separation between the position of the ionic head group and that of the mean position of the aqueous interface is (7.5 ± 0.1) Å for SDS and (8.0 ± 0.1) Å for HDTAB. These results are found to be independent of the structure of the adsorbed layer.

By drawing the alkyl chain further into the aqueous phase the conformational fluidity of the alkyl chain could be reduced. Recent neutron reflection studies of dodecyltrimethylammonium bromide layers at the air/water interface have found that the majority of conformational defects in the hydrocarbon tail resides within the portion of the hydrocarbon chain farthest from the ionic head group [51]. The portion of the alkyl chain closest to the head group is restricted due to its proximity to the head group and solvation of the hydrocarbon chain decreasing the conformational mobility. By drawing the head group further into the aqueous phase the effective chain length in the organic phase is then also reduced making it more difficult for gauche defects to be introduced. This may account for the SF results which show that the alkyl chain of DTAC and DAC are found to contain fewer gauche defects than SDS and DDS.

Mixed ionic surfactants

The conformational changes in the hydrocarbon tail of the surfactants induced by the ionic head groups can also be examined by studying the SFG spectra of a mixed anionic and cationic surfactant layer. The strong electrostatic interactions between oppositely charged head groups results in unique properties of aqueous mixtures of anionic and cationic surfactants. An example of their unique properties can be seen in the increased surface activity of such a mixture in comparison to that of the pure surfactants [52, 53]. Upon introduction of the oppositely charged surfactant a dramatic increase in the methyl peak at 2873 cm^{-1} is observed with an accompanying decrease in the CH$_2$ symmetric stretch at 2850 cm^{-1}. While forming a mixed surfactant monolayer, the repulsive electrostatic force responsible for limiting the packing density of the film is

diminished. The result is that an extremely closely packed ensemble is formed and the tighter packing of the alkyl chains is manifested in the relatively high I_{r^+}/I_{d^+} ratio. DDS and DAC at 0.1 mM bulk aqueous concentration have a methyl/methylene ratio of 1.3 and 0.8, respectively. However, the mixed surfactant film of the same cumulative bulk surfactant concentration has a ratio of 2.4.

Effect of chain length

Spectra were collected for a series alkylsulfonates of varying chain length in order to ascertain the effect of the alkyl length chain on alkyl chain order. The SF vibrational spectra of the three surfactants examined, DDS, UDS, and HS, are shown in Fig. 1(d, e, and f). The peak intensities of the symmetric methyl and methylene vibrational modes were obtained from spectral fits, as done above. The surface excess concentration (Γ_i) of surfactant at the interface is obtained from the bulk aqueous concentration by way of the interfacial pressure isotherms.

The ratio of the methyl to methylene peak intensity, for the range of concentrations examined for HS, DDS, and UDS, are plotted as a function of the surface concentration in Fig. 4. These results suggest a correlation between alkyl chain order and surface concentration as seen previously for DTAC, DAC, and SDS. This trend is consistent with the argument that as the surface density increases, the degree of conformational mobility within the alkyl chains

Fig. 4 A plot of the I_{r^+}/I_{d^+} ratio as a function of the surface excess concentration (Γ) for (▲) HS, (■) UDS, and (●) DDS. The solid line represents a linear fit to the data

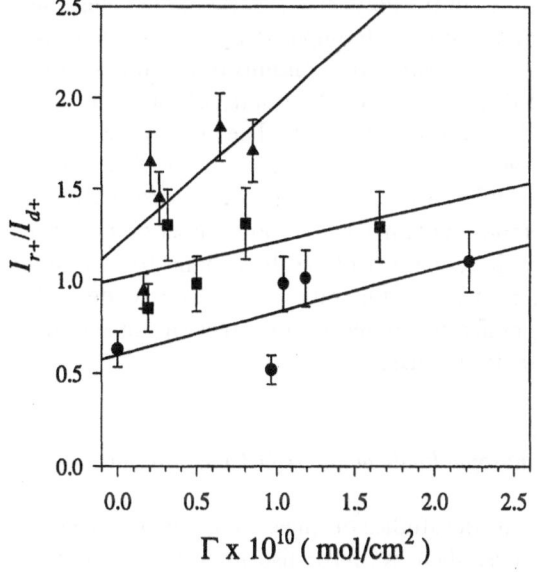

decreases, leading to greater alkyl chain ordering. It is clear from the data in Fig. 4 that the length of the alkyl chain also plays a role in the conformational mobility of the alkyl chain. The C_6 alkyl chain of hexylsulfonate possesses the largest I_{r^+}/I_{d^+} ratio even with the lowest surface concentration. In contrast, UDS and DDS show much smaller increases in the I_{r^+}/I_{d^+} ratio and exhibit nearly identical behavior as a function of surface coverage.

The data presented above suggest that there are fewer gauche defects in the alkyl chains of HS than for UDS and DDS. This observation is not surprising, as the hexyl chain of HS is much shorter than the 11- and 12-membered alkyl chains of UDS and DDS and will contain fewer geometrically allowed gauche conformations. Since the technique of SFG is not capable of distinguishing between the position and proximity of the gauche conformations in the alkyl chain, we have weighted all such conformations equally. That is, a gauche defect in the portion of the chain near the head group is assumed to have the same effect as one further down the chain. Although the portion of the chain adjacent to the gauche site is reoriented, this part of the chain will still posses an all-*trans* conformation and will not contribute to the SF spectra. This leaves the symmetric methylene resonance SF inactive due to symmetry constraints. It is only the two methylene units forming the dihedral angle that will experience a disruption in the local symmetry and result in the appearance of a CH_2 resonance in the SF spectra. The intensity of the symmetric methylene resonance with *ssp* polarization is therefore proportional to the number average of such defects oriented along the *z*-axis within the ensemble of alkyl chains which are being sampled.

The data therefore suggest that the larger I_{r^+}/I_{d^+} ratios observed for HS can be seen as a result of the reduction in the possible number of gauche conformations for the shorter alkyl chain compared to UDS and DDS. The conformational order of the alkyl chains is therefore determined by the spatial constraints imposed by neighboring alkyl chains, as can be seen in the concentration studies and by the degree of conformational mobility of the chain itself, determined by the chain length. The diminished chain–chain interaction may also account for the fact that no discernible difference is found in the conformation of a chain composed of an even or odd number of methylene units. Since the proximity of the alkyl chains to each other is reduced by the introduction of solvent, the tendency for a packing preference based on the structure of the alkyl chain is greatly reduced.

SFVS of phosopholipids at the D_2O/CCl_4 interface

In addition to the studies of single chain alkyl surfactants previously described, we have also undertaken an inves-

tigation of the phospholipid structure at the water/CCl$_4$ interface. We have studied a series of symmetric, saturated, dialkyl phosphocholines differing only in the length of their alkyl chains, 12 carbon atoms (dilauroylphosphocholine or DLPC), 14 carbon atoms (dimyristoylphosphocholine or DMPC), 16 carbon atoms (dipalmitoylphosphocholine or DPPC) and 18 carbon atoms (distearoylphosphocholine or DSPC). The molecular structures are shown in the insert of Fig. 5. When fully hydrated, phospholipids spontaneously form vesicles, uni or multilamellar bilayer structures [54]. At an interface (air/water, liquid/liquid), the vesicles decompose to form a monolayer with headgroups solvated in the aqueous phase and the alkyl tails in the hydrophobic phase [55]. As with the alkyl surfactants studied, vibrational sum-frequency generation is used in conjunction with interfacial pressure measurements to provide information about the structure of these phospholipid monolayers.

The interfacial activities of the four different phosphocholines show a clear dependence on alkyl chain length. Isotherms for the four phospholipids examined are displayed in Fig. 5. The DLPC pressure vs. bulk concentration isotherm rises steeply with increasing concentration

Fig. 5 Adsorption isotherms for DLPC, DMPC, DPPC, and DSPC taken at the D_2O/CCl_4 interface. Lines are shown as guides for the eye. (Insert) Molecular structure of dialkyl phosphocholines used in this study. R = $C_{n-1}H_2CH_3$ where $n = 12$ is dilauroylphosphocholine (DLPC), $n = 14$ is dimyristoylphosphocholine (DMPC), $n = 16$ is dipalmitoylphosphocholine (DPPC), and $n = 18$ is distearoylphosphocholine (DSPC)

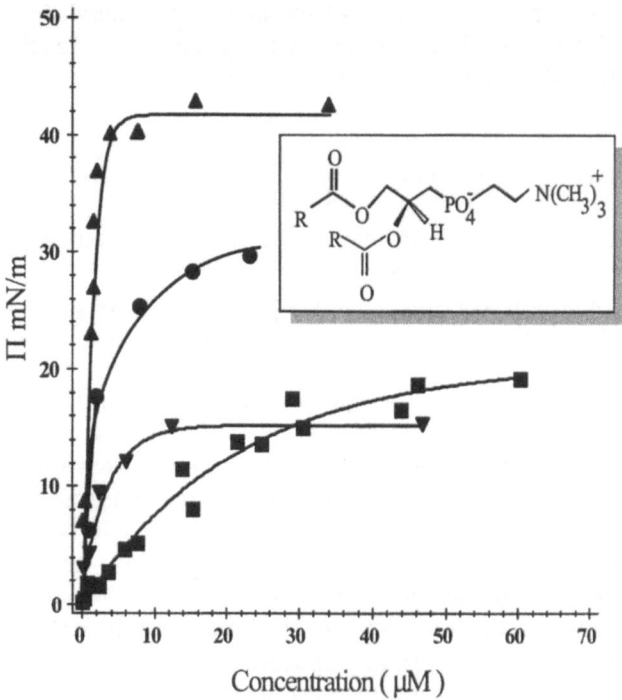

before leveling off at a terminal interfacial pressure of 39 mN/m. The interfacial pressure of DLPC remains constant at bulk concentrations above 8.0 μM. The DMPC isotherm does not rise as sharply as that of DLPC and reaches a terminal pressure of 30 mN/m at bulk concentrations in excess of 20 μM. Both DPPC, and DSPC have isotherms which climb more gradually with bulk concentration; both achieve terminal interfacial pressures of less than 20 mN/m.

The molecular areas of the different phosphocholines at their terminal surface pressures have been calculated using the Gibbs equation and the isotherm data shown in Fig. 5. Consistent with the interfacial pressure measurements, DLPC monolayers show the tightest packing with molecular areas of (50 ± 15) Å^2/molecule. DMPC monolayers pack more loosely achieving molecular areas of 70 Å^2/molecule. DPPC and DSPC monolayers form the most expanded monolayers with areas greater than 100 Å^2/molecule.

The vibrational spectra of DLPC, DMPC, DPPC and DSPC with the polarization combination of *ssp* (*s* polarized SF, *s* polarized visible, *p* polarized IR) have been obtained in order to ascertain the conformation of the alkyl chains of the various phospholipids. Representative spectra of DLPC and DSPC in the symmetric stretch region are shown in the inset of Fig. 6. A clear correlation between the adsorption isotherms and the I_{r^+}/I_{d^+} ratio obtained from spectroscopic measurements is observed. The methyl/methylene symmetric stretch ratio data for DLPC when plotted against bulk concentration (Fig. 6) resembles the DLPC pressure–concentration isotherm plotted in Fig. 6. Starting out with a value I_{r^+}/I_{d^+} ratio of ~1 at sun-μM concentration, the DLPC methyl/methylene ratio rises to a terminal value of 2.8 at bulk concentrations above 8 μM. DLPC forms the most ordered monolayers of any amphiphile studied thus far. Earlier in this report terminal ratios for SDS and DAC were shown to be 1.2 and 2.4, respectively. This high degree of order may arise from a greater density of alkyl chains per headgroup or tighter packing of zwitterionic headgroups. Alkyl chain order in DMPC monolayers depends less upon bulk concentration with the I_{r^+}/I_{d^+} ratio, rising from, 0.8 at sub-μM concentration up to a terminal value of 1.3 at concentrations above 20 μM (Fig. 6). Both DPPC and DSPC form disordered monolayers with small ratios (0.8) that appear insensitive to bulk phospholipid concentration.

Interfacial pressure measurements in conjunction with vibrational SF spectra of phosphocholines adsorbed to the water/CCl$_4$ interface provide a clear picture of how alkyl chain structure depends upon interfacial concentration. Namely, as interfacial concentration increases, alkyl chains adopt more of an all *trans* conformation as reflected

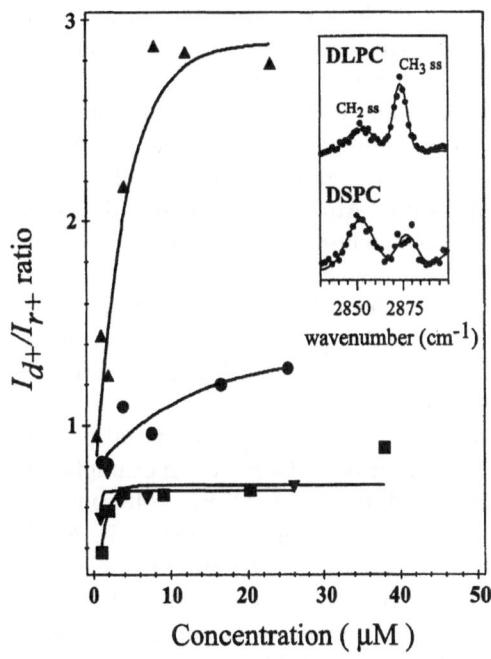

Fig. 6 Intensity ratios of the CH$_3$ symmetric stretch to the CH$_2$ symmetric stretch for DLPC, DMPC, DPPC, and DSPC versus bulk aqueous phospholipid concentration. Lines are shows as guides for the eye. (Insert) SF spectra of DLPC and DSPC at saturated monolayers coverage under ambient conditions ($T = 22\,°C$). Spectra contain both the methylene and methyl symmetric stretches (CH$_2$ ss at 2850 cm^{-1} and CH$_3$ ss at 2872 cm^{-1}). Solid lines represent fits to the spectra using Voight profiles for the different bands

by a larger I_{r^+}/I_d ratio. Still unanswered, though, is the question of why the different phospholipids, which differ only by the length of the alkyl chains, exhibit such desparate behavior. The solution to this riddle lies in the physical characteristics of the vesicles formed by the phospholipids in the aqueous phase. Differential scanning calorimetry [56], infrared [57] and Raman [58] measurements have all characterized two distinct thermodynamic phases of the lipid bilayers which constitute vesicle walls, gel and liquid crystalline. Above a characteristic phase transition temperature (T_c), the well ordered, solid gel phase melts into a more disordered, fluid-like liquid crystalline state. This phase transition temperature depends quite sensitively upon the hydrocarbon chain length of the phospholipid. DLPC has a T_c of $-1\,°C$, DMPC is at $23\,°C$ and DPPC and DSPC have transition temperatures at $41\,°C$ and $55\,°C$, respectively. For the experiments described above performed at room temperature, DLPC vesicles are in their liquid crystalline state. The phase transition temperature of DMPC lies at $23\,°C$, essentially the same temperature at which experiments discussed above were carried out. DMPC vesicles, then, were composed of lipid bilayers in coexistence between gel and liquid crystalline states.

Both DPPC and DSPC have transition temperatures well above ambient which means that these vesicles were held together tightly by the bilayers in their gel state.

The spectroscopic and thermodynamic data presented above suggest a correlation between molecular order, interfacial packing and the gel to liquid crystalline phase transition temperature. Aqueous phospholipid vesicles above their transition temperature readily form tightly packed, well-ordered monolayers at a water/CCl$_4$ interface (e.g. DLPC). Vesicles below their transition temperature possess greater stability, forming monolayers which are considerably expanded and which show greater disorder (i.e. DPPC and DSPC).

The relationship between T_c and interfacial packing and order can be seen by using DSPC as an example. The interfacial pressure of a 20 μM DSPC solutions was monitored as a function of temperature (Fig. 7). As the temperature is increased to T_c (55 °C), the interfacial pressure rises slightly. The interfacial pressure steps up sharply in the vicinity of T_c reaching a terminal pressure of ~40 mN/m before leveling out. At temperatures above T_c the interfacial pressure does not change, indicating formation of a tightly packed monolayer. The calculated molecular area of DSPC above 55 °C is ~50 Å^2/molecule, an area equal to that of DLPC at room temperature.

After forming a tightly packed DSPC monolayer above T_c, reducing the temperature below T_c does not lead to a corresponding reduction in the interfacial pressure that would accompany removal of DSPC molecules from the interface. The irreversible nature of monolayer formation is not surprising. With an enthalpy of solvation in CCl$_4$ of ~4.5 kJ/mol per CH$_2$ unit [59], alkyl chains once adsorbed to the water/CCl$_4$ interface face a very steep barrier to desorption and vesicle re-formation. The irreversible nature of the adsorption process can be used to

form tightly packed DSPC monolayers above T_c and then cool the interface down to experimentally accessible temperatures. We describe this process as annealing the interface.

Annealing a DSPC monolayer at the water/CCl$_4$ interface allows us to record spectra of tightly packed DSPC monolayers without the need for mechanical compression. Increasing the interfacial concentration of DSPC affects alkyl chain structure within the monolayer, but the magnitude of the change is surprisingly slight. Accompanying the abrupt climb in interfacial pressure at 55 °C is a reproducible rise in the I_{r^+}/I_d ratio from 0.8 ($T < T_c$) up to 1.1 ($T > T_c$). We observe similar behavior for both DPPC (I_{r^+}/I_d ratio rises from 0.8 to 1.2) and DMPC (I_{r^+}/I_d ratio rises from 1.3 to 1.7). Transition temperature of DPPC (41 °C) and DMPC (23 °C) are experimentally accessible and spectra taken with $T > T_c$ are identical to those taken with annealed interfaces. In no instance do the terminal ratios of tightly packed (~50 Å^2/molecule) DMPC, DPPC or DPPC monolayers approach the value 2.8 found in DLPC monolayers. At the equivalent interfacial concentrations longer alkyl chains show a greater susceptibility to sustain gauche defects than shorter alkyl chains, unlike at an air–water interface where longer chain alkyl chains demonstrate a greater propensity to adopt an all-*trans* conformation.

Conclusions

The effect of the ionic head group on the conformation of the hydrocarbon chain has been studied. By examining the SF vibrational spectra the conformational ordering of SDS, DDS, DTAC and DAC was investigated. The CH$_2$ backbone displays pronounced resonances due to the symmetric and asymmetric methylene modes at 2850 and 2925 cm^{-1}, respectively. The large intensity observed for these resonances is the result of the relaxation of the local symmetry constraint for the CH$_2$ vibrations by the introduction of gauche defects as the surface density increases. The ratio of the I_{r^+}/I_{d^+} methyl/methylene intensity as a function of bulk concentration is used as a measure of chain conformation. An increase in the I_{r^+}/I_{d^+} ratio is observed with increasing surface coverage suggesting a reduction in the number of gauche defects. The I_{r^+}/I_{d^+} correlates well with the surface concentration obtained from interfacial tension measurements. From the spectral data DTAC and DAC are found to possess the least number of gauche defects relative to SDS and DDS at similar surface concentrations. For DTAC and DAC versus SDS and DDS, the greater penetration depth of the ionic head groups of DTAC and DAC into the aqueous phase may account for these observations.

Fig. 7 Interfacial pressure of a 20 μM aqueous solution of DSPC versus temperature. The gel to liquid crystalline phase transition temperature (T_c) for DSPC and the terminal interfacial pressure (40 mN/m) of DLPC at room temperature are shown

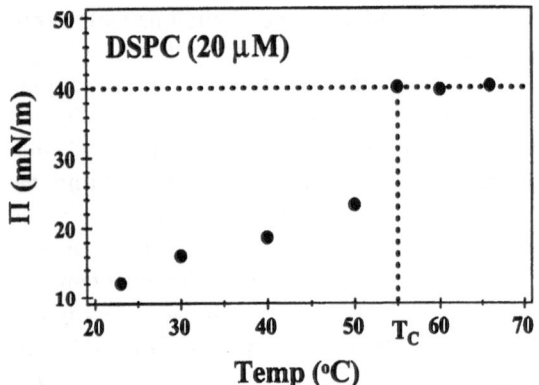

Progr Colloid Polym Sci (1997) 103:10–20
© Steinkopff Verlag 1997

In addition to contribution from the ionic head group, the length of the alkyl chain and its effect on the relative conformational order and orientation of a series of alkyl sulfonates at the water/carbon tetrachloride interface was investigated. Specifically, the conformational order of sodium hexylsulfonate (HS), sodium undecylsulfonate (UDS) and sodium dodecylsulfonate (DDS), adsorbed at the D_2O/CCl_4 interface were examined with vibrational sum-frequency generation. From the spectral data, HS is found to display the least number of gauche defects relative to UDS and DDS at similar surface concentrations, as would be expected due to its shorter chain length. The number of gauche conformation within the alkyl chain is found to be dependent upon both the packing density, as determined from interfacial tension measurements, and the length of the alkyl chain.

Vibrational sum frequency spectroscopy in conjunction with interfacial pressure measurements provide the first direct information about the structure of phospholipid monolayers composed of DLPC, DMPC, DPPC, and DSPC, adsorbed to the interface between two immiscible liquids. Temperature controlled experiments carried out with aqueous solutions of DSPC show the lipid bilayer gel to liquid crystalline phase transition temperature to play a pivotal role in determining interfacial monolayer concentration and alkyl chain structure. Even at equivalent interfacial concentrations longer chain phosphocholine species form more disordered monolayers with a greater number of gauche defects than shorter chain phosphocholine species, as determined from relative intensities of vibrational bands in the CH stretching region.

These studies demonstrate the promise of TIR SFVS for studying adsorption of amphiphillic species at the liquid/liquid interface. An investigation of molecular species adsorbed at a liquid/liquid phase boundary by conventional vibrational spectroscopy has been inaccessible until now. Complications arising from distinguishing between the spectral contributions of interfacial molecules from those in the more pervasive bulk liquids has presented a formidable experimental challenge. These limitations have been overcome by the use of the surface selective technique of SFVS. By coupling a total internal reflection (TIR) optical geometry with this nonlinear technique the sensitivity is greatly increased making possible the investigation of sub-monolayer coverages. TIR SFVS can provide molecular level information with the use of conventional optical techniques compared to other facilities oriented experiments such as neutron scattering and X-ray scattering which have only recently been applied to the liquid/liquid interface. Many studies are in progress in this laboratory to ascertain the effect of the head group on the solvent structure at the interface and the effect of the organic phase composition on hydrocarbon tail conformation. The results reported here have far reaching implications for the investigation of adsorption and transport properties at the interface between two immiscible liquids by vibrational spectroscopy.

Acknowledgments Funding from NSF (CHE 9416856), the Petroleum Research Fund (30557) of the American Chemical Society and the DOE for the alkylsulfonate studies is gratefully acknowledged.

References

1. Wirth MJ, Burbage JD (1992) J Phys Chem 96:9022
2. Piasecki DA, Wirth MJ (1993) J Phys Chem 97:7700
3. Takenaka TT, Nakanaga TJ (1976) J Phys Chem 80:475
4. Takenaka T (1978) Chem Phys Lett 55:515
5. Tian Y, Umemura J, Takenaka T (1988) Langmuir 4:1064
6. Grubb SG, Kim MW, Rasing T, Shen YR (1988) Langmuir 4:452
7. Higgins DA, Corn RM (1993) J Phys Chem 97:489
8. Higgins DA, Naujok RR, Corn RM (1993) Chem Phys Lett 213:485
9. The Royal Society of Chemistry (1992) In: Karsa DR (ed) Vol. 3. Industrial Applications of Surfactants
10. Myers D (1991) Surface, Interfaces and Colloids: Principles and Applications. VCH New York
11. Myers D (1992) Surfactant Science and Technology, 2nd ed. VCH New York
12. Rosen MJ (1987) Surfactants and Interfacial Phenomena. Wiley, New York
13. Tanford C (1973) The Hydrophobic Effect: Formation of Micelles and Biological Membranes. New York
14. Heinz TF, Himpsel FJ, Palange E, Burstein E (1989) Phys Rev Lett 63:644
15. Bayerl TM, Thomas RK, Penfold J, Rennie A, Sackmann E (1990) Biophys J 57:1095
16. Pallas NR, Pethica BA (1985) Langmuir 1:509
17. Phillips MC, Chapman D (1968) Biochim Biophys Acta 163:301
18. Dluhy RA, Wright NA, Griffiths PR (1988) Appl Spectrosc 42:138
19. Hwang J, Tamm LK, Böhm C, Ramalingam TS, Betzig E, Edidin M (1995) Science 270:610
20. Guyot-Sionnest P, Hunt JH, Shen YR (1987) Phys Rev Lett 59(14):1597–1600
21. Hunt JH, Guyot-Sionnest P, Shen YR (1987) Chem Phys Lett 133(3):189
22. Bain CD, J Chem Soc Faraday Trans 91:1281
23. Miragliotta J, Polizzotti RS, Rabinowitz P, Cameron SD, Hall RB (1990) Chem Phys 143(1):123–130
24. Hall RB, Russell JN, Miragliotta J, Rabinowitz PR (1990) Springer Ser Surf Sci 22 (Chem Phys Solid Surf 8):87–132
25. Hatch SR, Polizzotti RS, Dougal S, Rabinowitz P (1993) J Vac Sci Tech 11(4):2232
26. Akamatsu N, Domen K, Hirose C, Onishi T, Shimizu H, Masutani K (1991) Chem Phys Lett 181(2–3):175–178
27. Hirose C, Yamamoto H, Akamatsu N, Domen K (1993) J Phys Chem 97:10064
28. Wolfrum K, Graener H, Laubereau A (1993) Chem Phys Lett 213:41–46

29. Wolfrum K, Lobau J, Laubereau A (1994) Appl Phys A 59:605
30. Zhang D, Gutow J, Eisenthal KB (1994) J Phys Chem 98:13 729
31. Harris AL, Rothberg L, Dhar L, Levinos NJ, Dubois LH (1991) J Chem Phys 94(4):2438–2448
32. Adam NK (1941) The Physics and Chemistry of Surfaces. 3rd ed, Oxford University Press, London
33. Wilkinson (1972) J Colloid Interface Sci 40:14
34. Shen YR (1984) The Principles of Nonlinear Optics. Wiley, New York
35. Guyot-Sionnest P, Shen YR, Heinz TF (1987) Appl Phys B 42:237
36. Dick B, Gierulski A, Marowsky G (1987) Appl Phys B 42:237
37. Wong EKL (1992) In: Wong EKL (ed) Comparative Studies of Optical Second Harmonic Generation of Single Crystal Noble Metal Electrodes Under Resonant and Nonresonant Conditions. University of Oregon, Eugene, p 180
38. Haller GL, Rice RW (1970) J Phys Chem 74:4386

39. Harrick NJ, du Pre FK (1966) Appl Opt 5:1739
40. Harrick NJ (1967) Internal Reflection Spectroscopy. Wiley-Interscience, New York
41. Bloembergen N, Pershan PS (1962) Phys Rev 128:606
42. Bloembergen N (1966) Opt Acta 13:311
43. Bloembergen N, Simmon HJ (1969) Phys Rev 181:1261
44. Conboy JC, Messmer MC, Richmond GL (1995) J Phys Chem 100:7617–7622
45. Snyder RG, Hsu SL, Krimm S (1978) Spectrochim Acta 34A:395
46. MacPhail RA, Strauss HL, Snyder RG, Elliger CA (1984) J Phys Chem 88:334–341
47. Snyder RG, Strauss HL, Elliger CA (1982) J Phys Chem 86:5145
48. Aljibury AL, Snyder RG, Strauss HL, Raghavachari K (1986) J Phys Chem 84:6873
49. Messmer MC, Conboy JC, Richmond GL (1995) J Am Chem Soc 117:8039
50. Lu JR, Simister EA, Lee EM, Thomas RK, Rennie AR, Penfold J (1992) Langmuir 8:1837–1844

51. Lyttle DJ, Lu JR, Su TJ, Thomas RK, Penfold J (1995) Langmuir 11: 1001–1008
52. Kaler EW, Herrington KL, Miller DD, Zasadzinski JAN (1992) In: Kaler EW, Herrington KL, Miller DD, Zasadzinski JAN (eds) Phase Behavior of Aqueous Mixtures of Anionic and Cationic Surfactants Along a Dilution Path; Kluwer Academic Publishers, Dordrecht, pp 571–577
53. Lucassen-Reynders EH, Lucassen J, Giles D (1981) J Colloid Interface Sci 81:150
54. Szoka F, Papahadjopoulos D (1980) Ann Rev of Biophys Bioeng 9:467–508
55. MacDonald RC, Simon SA (1987) Proc Natl Acad of Sci, USA 84:4089–4093
56. Hinz HJ, Sturtevant JM (1972) Journal of Biological Chemistry 247: 6071–6075
57. Spiker RC, Levin IW (1976) Biochim Biophys Acta 433:457–468
58. Asher IM, Levin IW (1977) Biophys Acta 468:63–72
59. Fuchs R, Chambers EJ, Stephenson WK (1987) Can J Chem 65:2624–2627

Progr Colloid Polym Sci (1997) 103:21–28
© Steinkopff Verlag 1997

A.G. Volkov
D.W. Deamer

Redox chemistry at liquid/liquid interfaces

Received: 25 October 1996
Accepted: 7 November 1996

Dr. A.G. Volkov (✉) · D.W. Deamer
Department of Chemistry and Biochemistry
University of California
Santa Cruz, California 95064, USA

Abstract The interface between two immiscible liquids with immobilized photosynthetic pigments can serve as the simplest model of a biological membrane convenient for the investigation of photoprocesses accompained by spatial separation of charges. As it follows from thermodynamics, if the resolvation energies of substrates and products are very different, the interface between two immiscible liquids may act as a catalyst. Theoretical aspects of charge transfer reactions at oil/water interfaces are discussed. Conditions under which the free energy of activation of the interfacial reaction of electron transfer decreases are established. The activation energy of electron transfer depends on the charges of the reactants and dielectric permittivity of the non-aqueous phase. This can be useful when choosing a pair of immiscible solvents to decrease the activation energy of the reaction in question or to inhibit an undesired process. Experimental interfacial catalytic systems are discussed. Amphiphilic molecules such as chlorophyll or porphyrins were studied as catalysts of electron transfer reactions at the oil/water interface.

Key words Liquid/liquid interfaces – catalysis – amphiphilic molecules – chlorophyll – charge transfer – oil/water interface

Liquid–liquid interfaces as a biomembrane model

Vectorial charge transfer at the interface between two dielectric media is an important stage in bioelectrochemical processes such as those mediated by energy transducing membranes [1, 2]. Boundary membranes play a key role in the cells of all contemporary organisms, and simple models of membrane function are therefore of considerable interest. The interface of two immiscible liquids has been widely used for this purpose. For example, the fundamental processes of photosynthesis [3], membrane fusion [4], ion pumping [5] and electron transport [6, 7] have all been investigated in such interfacial systems.

Multielectron redox reactions at the interface between two immiscible liquids were first investigated by Bell [8]. This approach was later extended to redox and hydrolysis reactions catalyzed by enzymes [2, 9–11], photosynthetic pigments [12], metal complexes of porphyrins [13–15], and submitochondrial particles [16], as well as in systems with an extended surface such as microemulsions [12], vesicles and reversed micelles [17]. Enzymes and pigments embedded in a hydrophilic–hydrophobic interface have properties similar to their functional state in a membrane. For instance, certain enzymes can be highly active at the interface, but virtually inactive in a homogeneous medium. The interface between two immiscible liquids with immobilized photosynthetic pigments can also serve as a simple model for investigating photoprocesses accompanied by spatial separation of charges across a membrane [18]. Such light-dependent redox reactions at the oil/water interface have been discussed in recent reviews and books [12, 19, 20]. Here we will first present some theoretical aspects of interfacial redox reactions, then show how the theory has been applied to experimental results.

Catalytic properties of liquid interfaces: Theory

It is possible to shift the redox potential scale in a desired direction by selecting appropriate solvents, thereby permitting reactions to occur that are highly unfavorable in a homogeneous phase. It follows from thermodynamics, that if the resolution energies of substrates and products are very different, the interface between two immiscible liquids can act as a catalyst. The kinetic mechanism underlying the catalytic properties of the liquid/liquid interface was developed by Kharkats and Volkov [21–25]. The quantum theory of chemical reactions in polar media [26] can be used to describe charge transfer at the interface between two dielectric media such as the oil/water interface, allowing the electron transfer rate to be expressed in terms of the dielectric properties of the medium and the electronic properties of reactants.

Kharkats and Volkov first calculated the energy of activation and solvent reorganization of change transfer across the interface between two immiscible liquids [21–24]. The expression for the probability of electron transfer can be written as

$$W = A \exp \left\{ -\frac{U_i}{kT} - \frac{[E_s + \Delta G_c + U_f - U_i]^2}{4E_s kT} \right\}, \quad (1)$$

where U_i is the work that must be performed upon the system to place the reactants at distances h_1 and h_2 from the interface (Fig. 1), U_f is the corresponding work for the reaction products, ΔG_c is the configurational Gibbs free energy, E_s is the solvent reorganization energy, and A is the pre-exponential factor, which is proportional to the transmission coefficient. The transmission coefficient, κ, of the reaction for a non-adiabatic process is proportional to the square of the electronic matrix element. Theoretical analysis shows that the most effective electron transfer takes place at the closest disposition of reaction centers [27]. If κ is smaller than unity, the process may be considered to be nonadiabatic. The transmission coefficient for multi-electron transfer also corresponds to a nonadiabatic process and may be substantially lower with respect to κ for a one-electron reaction. At the same time the activation factors for one- and multi-electron processes may differ considerably.

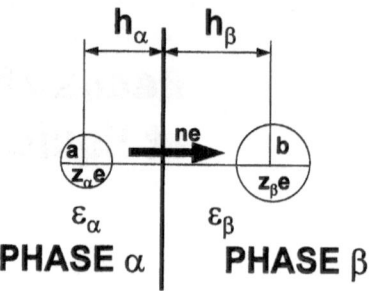

Fig. 1 Schema of locations of charge donors and acceptors at the interface between two immiscible electrolyte solutions and vectorial electron transport

The difference between Gibbs free energies for substrates and products can be found using corrected on a solvophobic effect the Born equation [22]:

$$\Delta G_c = \Delta G_c(\text{svp}) + \frac{e^2}{4\pi\varepsilon_0} \left(\frac{n^2 + 2z_\alpha n}{2\varepsilon_\alpha a} + \frac{n^2 - 2z_\beta n}{2\varepsilon_\beta b} \right), \quad (2)$$

where the solvophobic component of Gibbs free energy $\Delta G_c(\text{svp})$ does not depend on the dielectric properties of media α and β. The calculation of the Born electrostatic contribution to the solvation Gibbs free energy is not very precise. More accurate calculations can be performed using non-local electrostatics [28, 29].

Marcus [30–32] estimated the rate constant k_r for an electron transfer reaction between two redox components dissolved in two different liquid phases (Fig. 1):

$$-\frac{dN_1}{dt} = -k_r c_1 c_2 S, \quad (3)$$

where N_1 is the number of molecules of type 1 in the phase α, c_1 is the mean concentration of reactant 1 in phase α, c_2 is the mean concentration of reactant 2 in phase β, and S is the interfacial area. The rate constant k_r can be approximately determined as

$$k_r = \kappa v v \exp \left(-\frac{E_a}{RT} \right), \quad (4)$$

where v is some relevant frequency for the molecular motion, and E_a is the activation energy. If the liquid/liquid interface is a sharp boundary and if $h_\alpha \geq a$ and $h_\beta \geq b$, Marcus's expression for v [30–32] is

$$v = 2\pi(a + b)(\Delta R)^3, \quad (5)$$

where ΔR is the center-to-center distance between reagents (Fig. 1). In the system shown in Fig. 1, $\Delta R = h_\alpha + h_\beta$. If the ions penetrate to a second contacting phase but the reactants do not overlap, Marcus [30–32] derived another equation:

$$v \approx \pi(a_1 + a_2)^3 \Delta R. \quad (6)$$

The solvent reorganization energy is an important parameter in the quantum theory describing charge transfer in polar media. In the case of homogeneous reactions that occur in one phase it can be estimated by the relation:

$$E_s = \frac{1}{2\varepsilon_0}\left(\frac{1}{\varepsilon_{op}} - \frac{1}{\varepsilon_s}\right) \int\limits_{\infty - V_a - V_b} (\mathbf{D}_i - \mathbf{D}_f)^2 \, dV \,, \tag{7}$$

where ε_{op} and ε_s are the optical and the static dielectric permittivities of the medium, and $\mathbf{D}_i$ and $\mathbf{D}_f$ are the inductions of the electric fields which are created in the solvent during the initial and final state of charge transfer. Integration in Eq. (7) is carried out over the entire volume of the medium except the reactant volumes.

An approximate calculation of the solvent reorganization energy during charge transfer was performed by Marcus [33]. Assuming that the distance h_{12} between the reactant centers of the reactant is much larger than their radii a and b, and that the reactants can be described as non-polarizable spheres with charges rigidly and uniformly distributed over the surfaces, the expression for the reorganization energy is

$$E_s = \frac{e^2 n^2}{4\pi\varepsilon_0}\left(\frac{1}{\varepsilon_{op}} - \frac{1}{\varepsilon_s}\right)\left(\frac{1}{2a} + \frac{1}{2b} - \frac{1}{h_{12}}\right). \tag{8}$$

If the charge transfer occurs between reactants in two different dielectric media, Eq. (7) can be written as

$$E_s = \frac{1}{2\varepsilon_0}\left[\int\limits_{\infty - V_a - V_b} \frac{1}{\varepsilon_s(r)} (\mathbf{D}_f - \mathbf{D}_i)^2 \, dV\right]_{static}$$
$$- \frac{1}{2\varepsilon_0}\left[\int\limits_{\infty - V_a - V_b} \frac{1}{\varepsilon_{op}(r)} (\mathbf{D}_f - \mathbf{D}_i)^2 \, dV\right]_{optical}. \tag{9}$$

Girault [34] stressed the need for heterogeneous electron transfer reactions to differentiate static from optical integrals. For a sharp interface between two immiscible liquids the solvent reorganization energy can be written as

$$E_s = \frac{(ne)^2}{8\pi\varepsilon_0 a}\left(\frac{1}{\varepsilon_{op\alpha}} - \frac{1}{\varepsilon_\alpha}\right) + \frac{(ne)^2}{8\pi\varepsilon_0 b}\left(\frac{1}{\varepsilon_{op\beta}} - \frac{1}{\varepsilon_\beta}\right)$$
$$+ \frac{(ne)^2}{16\pi\varepsilon_0 h_\alpha}\left(\frac{\varepsilon_{op\alpha} - \varepsilon_{op\beta}}{\varepsilon_{op\alpha}(\varepsilon_{op\alpha} + \varepsilon_{op\beta})} - \frac{\varepsilon_\alpha - \varepsilon_\beta}{\varepsilon_\alpha(\varepsilon_\alpha + \varepsilon_\beta)}\right)$$
$$- \frac{(ne)^2}{16\pi\varepsilon_0 h_\beta}\left(\frac{\varepsilon_{op\alpha} - \varepsilon_{op\beta}}{\varepsilon_{op\beta}(\varepsilon_{op\alpha} + \varepsilon_{op\beta})} - \frac{\varepsilon_\alpha - \varepsilon_\beta}{\varepsilon_\beta(\varepsilon_\alpha + \varepsilon_\beta)}\right)$$
$$- \frac{(ne)^2}{2\pi\varepsilon_0(h_\alpha + h_\beta)}\left(\frac{1}{\varepsilon_{op\alpha} + \varepsilon_{op\beta}} - \frac{1}{\varepsilon_\alpha + \varepsilon_\beta}\right), \tag{10}$$

where ne is the charge transferred in the reaction, subscripts α and β denote the dielectric permittivities in media α and β, and the reactants are spheres of radii a and b which are located at distances h_1 and h_2 from the interface with charges $z_1 e$ and $z_2 e$, respectively (Fig. 1).

Similarly, U_i and U_f can be expressed in terms of integrals of inductions $\mathbf{D}_i$ and $\mathbf{D}_f$.

$$U_i = \frac{1}{8\pi\varepsilon_1}\int\limits_I D_i^2 \, dV + \frac{1}{8\pi\varepsilon_2}\int\limits_{II} D_i^2 \, dV - \frac{z_1^2 e^2}{2a\varepsilon_1} - \frac{z_2^2 e^2}{2b\varepsilon_2} \,, \tag{11}$$

$$U_f = \frac{1}{8\pi\varepsilon_1}\int\limits_I D_f^2 \, dV + \frac{1}{8\pi\varepsilon_2}\int\limits_{II} D_f^2 \, dV$$
$$- \frac{(z_1 + n)^2 e^2}{2a\varepsilon_1} - \frac{(z_2 - n)^2 e^2}{2b\varepsilon_2} \,, \tag{12}$$

where the integration ranges I and II represent the two half-spaces of media α and β excluding the volume of reactants.

Calculation of the integrals in (11) and (12) is conveniently carried out by changing to surface integrals. For instance, Kharkats [23] calculated for $\int_I D_i^2 \, dV$ with an accuracy of $(a/h)^3$, $(b/h)^3$:

$$\int\limits_I D_i^2 \, dV = \frac{4\pi e^2 z_1^2}{a} - \frac{2\pi e^2 z_1^2}{h_1}$$
$$+ \pi e^2 \left(\frac{2\varepsilon_1}{\varepsilon_1 + \varepsilon_2}\right)^2 \left(\frac{z_1^2}{h_1} + \frac{z_2^2}{h_2} + \frac{4z_1^2 z_2^2}{h_1 + h_2}\right). \tag{13}$$

The expression for $\int_{II} D_i^2 \, dV$ is obtained from Eq. (13) by making the substitutions $a \rightarrow b$, $\varepsilon_1 \rightarrow \varepsilon_2$, $h_1 \rightarrow h_2$ and $z_1 \rightarrow z_2$. The corresponding expression for $\int_{I,2} D_f^2 \, dV$ can be obtained by making substitutions $z_1 \rightarrow (z_1 + n)$ and $z_2 \rightarrow (z_2 - n)$.

The simplest expressions for E_s, U_i and U_f are obtained when the reactions take place at equal distances from the interface, $h_1 = h_2 = h$:

$$E_s = \frac{(ne)^2}{8\pi\varepsilon_0 a}\left(\frac{1}{\varepsilon_{op\alpha}} - \frac{1}{\varepsilon_\alpha}\right) + \frac{(ne)^2}{8\pi\varepsilon_0 b}\left(\frac{1}{\varepsilon_{op\beta}} - \frac{1}{\varepsilon_\beta}\right)$$
$$- \frac{(ne)^2}{4\pi\varepsilon_0 h}\left(\frac{1}{\varepsilon_{op\alpha} + \varepsilon_{op\beta}} - \frac{1}{\varepsilon_\alpha + \varepsilon_\beta}\right), \tag{14}$$

$$U_i = \frac{z_\alpha z_\beta e^2}{(\varepsilon_\alpha + \varepsilon_\beta)h} + \frac{z_\alpha^2 e^2(\varepsilon_\alpha - \varepsilon_\beta)}{4\varepsilon_\alpha(\varepsilon_\alpha + \varepsilon_\beta)h} - \frac{z_\beta^2 e^2(\varepsilon_\alpha - \varepsilon_\beta)}{4\varepsilon_\beta(\varepsilon_\alpha + \varepsilon_\beta)h} \,, \tag{15}$$

$$U_f = \frac{(z_\alpha + n)(z_\beta - n)e^2}{(\varepsilon_\alpha + \varepsilon_\beta)h} + \frac{(z_\alpha + n)^2 e^2(\varepsilon_\alpha - \varepsilon_\beta)}{4\varepsilon_\alpha(\varepsilon_\alpha + \varepsilon_\beta)h}$$
$$- \frac{(z_\beta - n)e^2(\varepsilon_\alpha - \varepsilon_\beta)}{4\varepsilon_\beta(\varepsilon_\alpha + \varepsilon_\beta)h} \,, \tag{16}$$

$$E_a = U_i + \frac{(E_s + \Delta G_c + U_f - U_i)^2}{4E_s} \,. \tag{17}$$

In the case of homogeneous electron transfer in a dielectric medium, the work required to bring the reactants or reaction products together approaches zero when one of the reactants or products is electrically neutral, whereas in the process discussed here, U_i values are never zero because of the interactions with image charges.

The activation energy of electron transfer depends on the charges of the reactants and dielectric permittivity of the non-aqueous phases. This can be useful when choosing a pair of immiscible solvents to decrease the activation energy of the reaction in question or to inhibit an undesired process [18]. For example, suppose that an electron is transferred from a donor in the aqueous phase α to an acceptor in organic phase β (Fig. 1). Assuming that ΔG_c is negligible compared to E_s, the activation energy E_a depends on the dielectric permittivity of non-aqueous phase ε_2:

$$E_a \cong U_i + \frac{(E_s + U_f - U_i)^2}{4E_s}. \qquad (18)$$

The reorganization energy increases with ε_α to a maximum asymptotic value at $\varepsilon_\beta \gg \varepsilon_{op\beta}$:

$$E_s = \frac{(ne)^2}{8\pi\varepsilon_0 a}\left(\frac{1}{\varepsilon_{op\alpha}} - \frac{1}{\varepsilon_\alpha}\right) + \frac{(ne)^2}{8\pi\varepsilon_0 b}\left(\frac{1}{\varepsilon_{op\beta}}\right)$$
$$- \frac{(ne)^2}{4\pi\varepsilon_0 h}\left(\frac{1}{\varepsilon_{op\alpha} + \varepsilon_{op\beta}} - \frac{1}{\varepsilon_\alpha + \varepsilon_\beta}\right) \qquad (19)$$

and at $\varepsilon_\beta = \varepsilon_{op\beta}$ reorganization energy is minimal and equal to

$$E_s = \frac{(ne)^2}{8\pi\varepsilon_0 a}\left(\frac{1}{\varepsilon_{op\alpha}} - \frac{1}{\varepsilon_\alpha}\right) - \frac{(ne)^2}{4\pi\varepsilon_0 h}\left(\frac{1}{\varepsilon_{op\alpha} + \varepsilon_{op\beta}} - \frac{1}{\varepsilon_\alpha + \varepsilon_\beta}\right). \qquad (20)$$

Examples of the dependencies of E_s, E_a and U_i for different sets of parameters z_1, z_2 and h/a are plotted in Figs. 2–4.

Figure 2 shows how the dielectric constant of the organic phase ε_2, and the distance from the interface or between reagents affect the medium reorganization energy. A decrease of h or ε dramatically decreases E_s (Fig. 2), which reaches a minimum value as ε approaches 2.

Equation (18) is plotted in Figs. 3 and 4, which show that the activation energy of the process decreases (or increases) greatly at small ε_β and, accordingly, the rate constant of charge transfer across the interface increases (or decreases) sharply at relatively small ε_β. As is clearly shown in Fig. 4, the liquid–liquid interface has selectivity properties and can catalysis or inhibit interfacial charge transfer reactions due to effects caused by the electrostatics.

To summarize, the kinetic parameters of interfacial charge transfer depend on the charge being transferred, the charges of reactants, their location in relation to the interface, as well as the dielectric properties of the media forming the liquid/liquid interface. The charge transfer

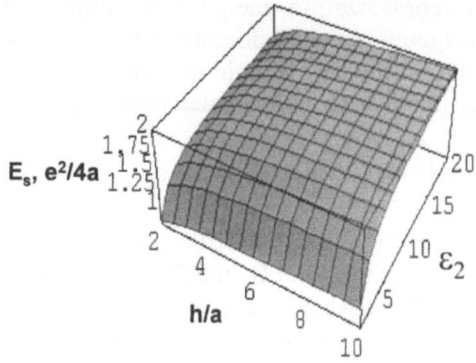

Fig. 2 Dependence of solvent reorganization energy on dielectric permittivity of nonaqueous phase ε_2 and distance $h = h_1 = h_2$ from the interface at parameters: $\varepsilon_{op\alpha} = \varepsilon_{op\beta} = 1.8$, $\varepsilon_\alpha = 78$, $a = b$, $n = 1$

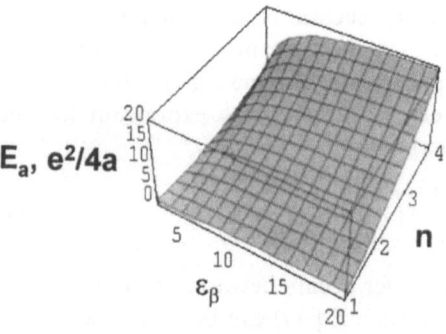

Fig. 3 Dependence of activation energy E_a on dielectric permittivity of nonaqueous phase ε_β and number of transferred electrons: $\varepsilon_{op\alpha} = \varepsilon_{op\beta} = 1.8$, $\varepsilon_\alpha = 78$, $a = b = h/2$, $z_\alpha = -1$, $z_\beta = +1$, $\Delta G_c = e^2/2a$

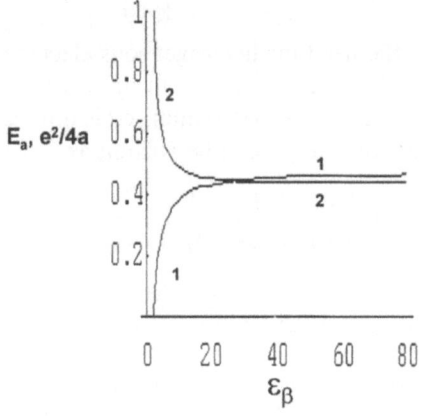

Fig. 4 Dependencies of the activation energy, E_a, on dielectric permittivity of a non-aqueous phase ε_β, calculated by Eq. (13) with parameters: $a = b = h$, $n = 1$, $\Delta G_c = 0.01e^2/a$ and (1) $z_\alpha = 0$, $z_\beta = +1$; (2) $z_\alpha = 0$, $z_\beta = -1$

processes in simple models are described by equations that can in turn be extended to more complicated processes, such as phase transfer and micellar catalysis as well as bioenergetic processes taking place in biomembranes.

Experimental systems

Redox reactions at the interface between immiscible liquids fall into two classes. The first class includes spontaneous processes that occur in the absence of external electromagnetic fields. This type of redox transformation has been investigated in bioenergetics [2], model membrane systems [20] and at oil/water interfaces [1]. Redox reactions in the second class occur at the interface between immiscible electrolytes when external electrical fields are applied to the interface, and under these conditions interfacial charge transfer reactions take place at controlled interfacial potentials [11, 35, 36]. Such electrochemical interfacial reactions are usually multi-stage processes that proceed through five stages: (i) diffusion of reactants to the interface; (ii) adsorption of reactants onto the interface; (iii) electrochemical reaction at the interface; (iv) desorption of products from the interface; (v) diffusion of products from the interface.

Long before interfacial charge transfer reactions began to be studied systematically, Bell [8] observed a multielectron transfer reaction across a benzene/water interface. This involved permanganate oxidation of benzoyl-o-toluidine to benzoylanthranilic acid:

$$\tag{21}$$

This reaction occurred at the water/benzene interface and its kinetics did not depend on the rate of stirring of the contacting phases.

Electrochemical metallization at the water/1,2-dichloroethane and water/dichloromethane interfaces was studied by Guainazzi et al. [37]. The electrical current passing through the liquid/liquid interface caused the growth of a metallic layer of copper or silver at the phase boundary:

$$[V(CO)_6]^-_{\{in\ oil\}} + 2Cu^{2+} \rightarrow V^{3+}_{(aq)} + 6CO + 2Cu \,, \tag{22}$$

$$[V(CO)_6]^-_{\{in\ oil\}} + 4Ag^+ \rightarrow V^{3+}_{(aq)} + 6CO + 2Ag \,. \tag{23}$$

Redox reactions at bilayer lipid membrane/water interfaces have been studied by many authors [2, 7, 38, 39]. Usually the BLM, doped by ubiquinone, tetracyano-p-quinodimethane (TCNQ), or by ferrocene, has the properties of a bipolar electrode [39]. A variety of redox couples in the aqueous phases have been used, including ascorbic acid/dehydroascorbic acid, KI/I_2, $[PtCl_4]^{2-}$, Sn^{4+}/Sn^{2+},

Cd^{2+}, vitamin K_3, benzoquinone/hydrobenzoquinone, HADH, and O_2 [40, 41]. Tien used traditional electrochemical techniques such as cyclic voltammetry to detect redox reaction products [39] and electrometric titration of Fe^{2+} with $KMnO_4$ using a pigmented BLM as the indicator electrode. According to Tien, "a modified BLM can be considered either as a reversible electrode, or as an inert electrode, analogous to a platinum electrode that merely serves as a conductor for facilitating electron transfer". Tien introduced the term "*electrostenolysis*" which means that a reduction occurs on one side of the BLM and the accompanying oxidation reaction takes place on the other side.

An entirely novel heterogeneous process, called "*Phase transfer catalysis*" or PTC [42], has been widely used by chemists for preparative purposes. The basis of the method is to create a two-phase system (usually with an organic and an aqueous phase), in which non-polar and ionic reactants are present in the different phases, and catalysts that are sources of lipophilic cations. The role of the catalysts is to form lipophilic ion pairs between the cation of the catalyst and the reacting anion which are then capable of migrating within the organic phase. PTC is one of the simplest and most economical methods for intensifying the production of a wide range of organic materials. The main advantage of the PTC method is that it is general, mild, and catalytic. PTC also has disadvantages: the complexity of separating the catalyst from the reaction medium, the tendency to form stable emulsions, and the impossibility of performing repeated or continuous processes. These disadvantages can be eliminated by three-phase catalysis (TPC) in which a catalyst is immobilized on a polymeric support. The insoluble catalysts are easily separated from the reaction medium by simple filtration and can be used repeatedly.

If a catalytic reaction of electron or ion transfer takes place at the oil/water interface between reagents located in two different contacting phases, the result is "*interfacial catalysis*" discovered by Volkov and Kharkats [24, 25]. The interface itself can serve as a catalyst for such heterogeneous charge transfer reactions [18]. If the interfacial catalysis requires an electrical field, the reaction can take place at the interface between two immiscible electrolyte solutions having a fixed interfacial potential, a process called *electrocatalysis* [36].

A number of other reactions have been demonstrated in which an oxidative transformation occurs at an interface, but the factors affecting the rates have not been examined [2]. Some progress in this direction was made in a simple redox reaction between hydrophobic porphyrin dissolved in octane and a hydrophilic donor – sodium dithionite dissolved in water [43].

Chlorophyll is an amphiphilic catalyst of electron transfer reactions at the oil/water interface

The photosynthetic pigment chlorophyll mediates the primary act of photosynthesis in higher plants. However, chlorophyll can also catalyze electron transfer reactions at oil/water interfaces and on bilayer lipid membranes without illumination [2, 7, 38]. The reversible chemical reduction of chlorophyll by zinc was first studied by Timiriazeff [44], and the role of chlorophyll in oxidation-reduction reactions has been the subject of numerous later investigations [45].

In early work, Rabinowitch and Weiss [46] observed a homogeneous redox reaction of chlorophyll (Chl) with ferric chloride:

$$Chl + Fe^{3+} \Leftrightarrow oChl^+ + Fe^{2+} , \qquad (24)$$

where $oChl$ is oxychlorophyll, a yellow oxidized form of the pigment. If ferric salts are added to chlorophyll dissolved in CH_3OH, the solution changes color from green to yellow, and spectroscopic analysis of the products shows a nearly complete disappearance of the chlorophyll adsorption spectrum. The oxychlorophyll can be returned to chlorophyll by quickly adding ferrous chloride or other reducing agents, but with time the oxychlorophyll forms allomerized pigments in an irreversible reaction [46]. The standard redox potentials and polarographic half-wave potentials of chlorophylls have been reviewed by Seely [45].

Chlorophyll a adsorbed at the oil/water interface catalyzes the electron transport between donors and acceptors of electrons located in the two phases [2, 7]. The chemical structure of chlorophyll accounts for its high catalytic activity at the oil/water interface. The asymmetric amphiphilic chlorophyll molecule consists of a hydrophilic "head" formed by four pyrrole rings located around magnesium and a long "tail" – the hydrophobic chain of phytol. The hydrophilic head faces the water, while the hydrophobic tail attaches chlorophyll to the non-aqueous phase.

Electron transfer reactions involving catalytic chlorophyll were studied by the vibrating plate method [3, 7, 47, 48] in the electrochemical circuit:

Au	Air	Octane	Water	Water	Hg₂Cl₂, Hg
		Acceptor	Substrate	sat. KCl	
		Catalyst	Tris-HCl		

$$(I)$$

Figure 5 shows that if a hydrophobic acceptor of electrons such as $2N$-methylamino-1,4-naphtoquinone, vitamin K_3 is added to the octane and a donor of electrons (NADH, NADPH, potassium ascorbate) to the aqueous

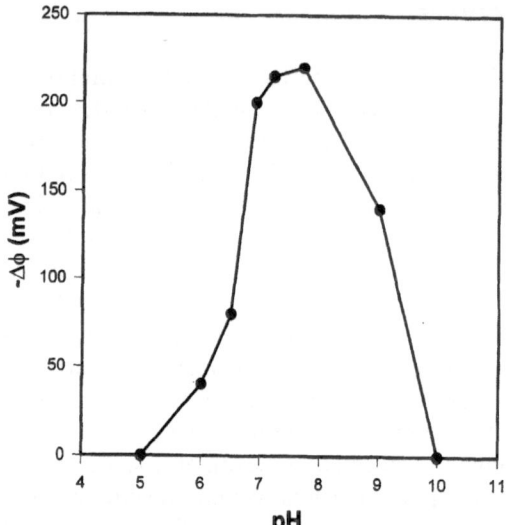

Fig. 5 Volta-potential, measured in the chain (I), as a function of NADH concentration. Medium: 10^{-5} M 2-N-methylamino-1,4-naphtoquinone + 10^{-2} M tris-HCl + 5 μg/ml chlorophyll [7]

phase, the Volta-potential measured in the chain (I) shifts in the negative direction (minus in the octane) due to the electron-exchange reaction at the oil/water interface catalyzed by chlorophyll:

$$NADH_{aq} + Q_{oil} \xrightarrow{\text{chlorophyll}} NAD_{aq}^+ + H^+ + Q_{oil}^{2-} \qquad (25)$$

where Q is the 2N-methylamino-1,4-naphthoquinone. The magnitude of the potential does not depend on how the redox reaction is initiated, that is by addition of substrate, chlorophyll or a charge acceptor. No change in the Volta-potential was observed in the absence of any component of the reaction [7]. The interfacial transfer of electrons catalyzed by chlorophyll has on optimum at pH 6.5–9.1 and is inhibited in the acid and alkaline regions (Fig. 5). Inhibition of the electron transfer reaction in the acidic region is probably due to conversion of chlorophyll to pheophytin. Inhibition in the alkaline region may be due to the fact that at pH ≥ 10 the carboxylic ring breaks and salts of chlorophilins are obtained since the complex ester bonds are simultaneously hydrolyzed [7].

The observed Volta potential does not change at a sufficiently large buffer capacity of the solution (from 5 to 100 mM Tris-HCl) [47, 48]. This indicates that the potential shift at the octane/water interface results from electron transfer across the interface and not from the pH change in the boundary layer. During a redox reaction on BLM containing chlorophyll, a layer adjacent to the membrane is formed with a proton concentration different from that in the bulk phase [7, 38]. Boguslavsky et al. [47, 48]

Progr Colloid Polym Sci (1997) 103: 21–28
© Steinkopff Verlag 1997

27

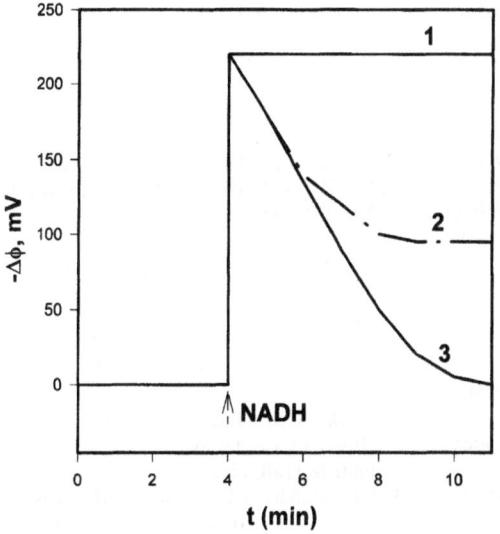

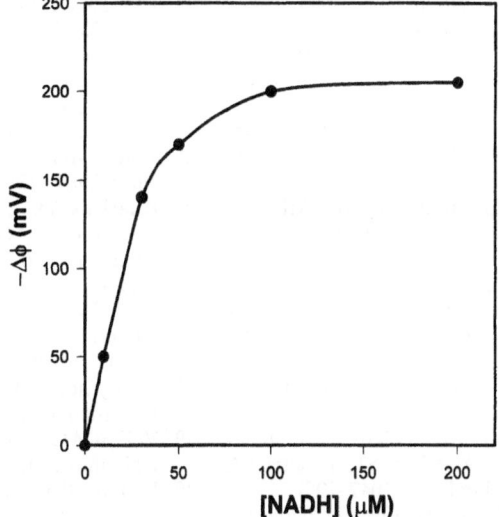

Fig. 6 Volta-potential as a function of pH of an aqueous solution. Composition of reaction mixture: 20 mM tris-HCl + 1 mM NADH + 10^{-5} M 2-N-methylamino-1,4-naphtoquinone + 5 μg/ml chlorophyll [48]

Fig. 7 Volta-potential in the chain (I) as a function of incubation time of reaction mixture. Medium: 20 μM NADH + 10 μM vitamin K$_3$ and tris-HCl: 1–10 mM, 2–1 mM, 3–0 mM, pH 6.5 [47]

investigated such boundary layers in the octane/water system. It follows from Eq. (25), that a charge transfer reaction must involve proton ejection into the aqueous phase. Figure 6 shows the dependence of the Volta potential on the incubation time of the reaction mixture at different Tris-HCl buffer concentrations. The aqueous phase pH was chosen so that acidification of the boundary layer inactivated the chlorophyll due to its pheophytinization. As shown in Fig. 7, at low Tris-HCl concentrations $\Delta\phi$ decreases at pH < 6.5, while in the buffered solution this effect is absent.

From these results, it is clear that chlorophyll adsorbed at an interface can catalyze redox reactions between solutes in the two liquid phases, and these reactions are accompanied by injection of negative charges into the low dielectric membrane interior. Metallo-complexes of porphyrins also have similar catalytic properties at the oil/water interface without illumination [13, 14, 43]. An advantage of porphyrins as catalysts in such experiments is their chemical stability in different media.

In conclusion, the utilization and storage of energy during photosynthesis and respiration are accompanied by redox reactions at the membrane/electrolyte interface, catalyzed by chlorophyll, metalloporphyrins, or membrane enzyme systems, the prosthetic group of which is usually a metalloporphyrin. Experimental models of such processes include primarily various membranes, where the transmembrane current or potential is measured directly, as well as liposomes and microemulsions, where the process is recorded according to the accumulation of the reaction products. In this review, we have described the theory underlying the fundamental reactions, as well as several experimental approaches. Future research in this area will provide further insight into both biological and chemical applications of redox chemistry at liquid/liquid interfaces.

Acknowledgment This work was supported by NASA Grant NAGW 4235.

References

1. Volkov AG, Deamer DW (eds) (1996) Liquid–Liquid Interfaces: Theory and Methods. CRC-Press, Boca Raton, London, Tokyo
2. Boguslavsky LI, Volkov AG (1987) In: Kazarinov VE (ed) The Interface Structure and Electrochemical Processes at the Boundary Between Two Immiscible Liquids. Springer, Berlin, pp 143–178
3. Volkov AG (1989) Bioelectrochem Bioenerg 21:3–24
4. Gingell D, Todd I, Parsegian VA (1977) Nature 268:767–769
5. Yaguzhinsky LS, Boguslavsky LI, Volkov AG, Rakhmaninova AB (1975) Nature 259:494–496
6. Faraday M (1857) Philos Trans R Soc London 147:145–181
7. Volkov AG, Lozhkin BT, Boguslavsky LI (1975) Doklady Akad Nauk SSSR 220:1207–1210
8. Bell RP (1928) J Phys Chem 32:882–893
9. Boguslavsky LI, Kondrashin AA, Kozlov IA, Metelsky ST, Skulachev VP, Volkov AG (1975) FEBS Lett 50: 223–226
10. Volkov AG (1986) J Electroanal Chem 205:245–257

11. Kakiuchi T (1996) In: Volkov AG, Deamer DW (eds) Liquid–Liquid Interfaces. Theory and Methods. CRC-Press, Boca Raton, New York, London, Tokyo, pp 317–331

12. Volkov AG, Gugeshashvili MI, Deamer DW (1995) Electrochim Acta 40: 2849–2868

13. Volkov AG, Bibikova MA, Mironov AF, Boguslavsky LI (1983) Bioelectrochem Bioenerg 10:477–483

14. Volkov AG, Gugeshashvili MI, Mironov AF, Boguslavsky LI (1983) Bioelectrochem Bioenerg 10:485–491

15. Volkov AG, Mironov AF, Boguslavsky LI (1976) Elektrokhimiya 12:1326–1329

16. Boguslavsky LI, Volkov AG, Kondrashin AA, Metelsky ST, Yasaitis AA (1976) Biokhimiya 41:1047–1051

17. Garcia E, Texter JJ (1994) J Colloid Interface Sci 162:262–264

18. Volkov AG (1984) J Electroanal Chem 173:15–24

19. Kotov NA, Kuzmin MG (1996) In: Volkov AG, Deamer DW (eds) Liquid–Liquid Interfaces. Theory and Methods. CRC-Press, Boca Raton, New York, Tokyo, pp 375–400

20. Volkov AG, Deamer DW, Tanelian DI, Markin VS (1997) Liquid Interfaces in Chemistry and Biology. J Wiley, New York

21. Kharkats Yu I, Volkov AG (1985) J Electroanal Chem 184:435–439

22. Kharkats Yu I, Volkov AG (1987) Biochim Biophys Acta 891:56–67

23. Kharkats Yu I (1976) Sov Electrochem 12:1370–1377

24. Volkov AG, Kharkats Yu I (1985) Kinetica Kataliz 26:1322–1326

25. Volkov AG, Kharkats Yu I (1986) Chem Phys 5:964–971

26. Kharkats Yu I, Kuznetsov AM (1986) In: Volkov AG, Deamer DW (eds) Liquid–Liquid Interfaces. Theory and Methods. CRC-Press, Boca Raton, New York, Tokyo, pp 139–154

27. Kharkats Yu I (1990) Sov Electrochem 26:1032–1039

28. Kornyshev AA, Volkov AG (1984) J Electroanal Chem 180:363–381

29. Volkov AG, Kornyshev AA (1985) Elektrokhimiya 21:814–817

30. Marcus RA (1990) J Phys Chem 94: 1050–1055

31. Marcus RA (1990) J Phys Chem 94: 4152–4155

32. Marcus RA (1991) J Phys Chem 95: 2010–2013

33. Marcus RA (1956) J Chem Phys 24: 966–978

34. Girault HHJ (1995) J Electroanal Chem 388:93–100

35. Samec Z (1996) In: Volkov AG, Deamer DW (eds) Liquid–Liquid Interfaces. Theory and Methods. CRC-Press, Boca Raton, New York, Tokyo pp 155–178

36. Cunnane V, Murtomaki L (1996) In: Volkov AG, Deamer DW (eds) Liquid–Liquid Interfaces. Theory and Methods. CRC-Press, Boca Raton, New York, Tokoyo, pp 401–416

37. Guainazzi M, Silvestri G, Serravalle G (1975) J Chem Soc Chem Commun 200–201

38. Boguslavsky LI, Lozhkin BT, Kiselev BA (1975) Doklady Akad Nauk SSSR 222:228–231

39. Tien H Ti (1986) Bioelectrochem Bioenergy 15:19–38

40. Tien H Ti (1989) Progress Surf Sci 30:1–199

41. Tien H Ti (1974) Bilayer Lipid Membranes. Theory and Practice. M Dekker, New York

42. Starks CM, Liotta CL, Halper NM (1994) Phase Transfer Catalysis. Chapman & Hall, New York

43. Gugeshashvili MI, Volkov AG, Yaguzhinsky LS, Mironov AF, Boguslavsky LI (1983) Bioelectrochem Bioenerg 10:493–498

44. Timiriazeff C (1885) Nature 32:342

45. Seely GR (1977) In: Barber J (ed) Primary Processes of Photosynthesis Elsevier, Amsterdam, pp 1–53

46. Rabinowitch E, Weiss I (1937) Proc R Soc (London) A 162:251–267

47. Boguslavsky LI, Volkov AG, Kandelaki MD (1976) FEBS Lett 65:155–158

48. Boguslavsky LI, Volkov AG, Kandelaki MD (1977) Bioelectrochem Bioenerg 4:68–72

Progr Colloid Polym Sci (1997) 103:29–40
© Steinkopff Verlag 1997

A. Pohorille
M.A. Wilson
C. Chipot

Interaction of alcohols and anesthetics with the water–hexane interface: a molecular dynamics study

Received: 13 December 1996
Accepted: 19 December 1996

Dr. A. Pohorille (✉) · M.A. Wilson
Department of Pharmaceutical Chemistry
University of California, San Francisco
San Francisco, California 94143, USA

A. Pohorille · M.A. Wilson · Chr. Chipot[1]
NASA-Ames Research Center
MS 239-4, Moffett Field
California 94035-1000, USA

[1]*On leave from*:
Laboratoire de Chimie Théorique
Unité de Recherche Associée
au CNRS no. 510
Université Henri Poincaré–Nancy 1, BP 239
54506 Vandoeuvre-lès-Nancy Cedex, France

Abstract The transfer of eight solutes across the water–hexane interface is studied using molecular dynamics computer simulations. Four of these solutes are model amphiphiles, straight chain alcohols – methanol, ethanol, butanol and hexanol. The remaining four molecules – cyclopropane, nitrous oxide, isoflurane and desflurane – are non-amphiphilic and polar or weakly polar. All of them are clinical anesthetics. All eight molecules exhibit free energy minima at the interface, indicating that they are interfacially active. Whereas interfacial activity of amphiphiles has been well known, it is shown here that a similar, although somewhat weaker behavior is also characteristic of a wide range of polar solutes. This can be explained as a balance between electrostatic and non-electrostatic contributions to the free energy, that change monotonically, but oppositely near the interface. Qualitatively, similar results are expected for solutes at interfaces between water and other non-polar liquids or lipid bilayers. Based on the results showing a very good correlation between anesthetic potencies and interfacial concentrations of 20 anesthetic compounds, it is proposed that the site of anesthetic action is located near the interface between water and the neuronal membrane.

Key words Water – hexane interface – straight chain alcohols – anesthetics – computer simulations

Introduction

Interfaces between water and non-polar phases are of considerable interest as environments for heterogeneous catalysis, as well as many important electrochemical and photochemical processes [1–3]. Emerging new industrial applications involve the building up of layers of mono-molecular films at interfaces [4, 5]. In biological systems, interactions of small molecules and peptides with water–membrane interfaces and protein receptor sites near these interfaces are highly relevant for modulation of receptor action, signal transduction [6, 7] and membrane fusion [8]. Studies in all these areas require knowledge of the principles determining the behavior of small molecules in interfacial environments. These principles, however, are not very well understood. Recently, considerable progress in this direction has been achieved by applying computer simulation methods to interfacial systems [9, 10]. In this paper, this line of research is continued.

A compound is interfacially active when its concentration in the interfacial region exceeds those in the two adjacent bulk phases. In the limit of infinite dilution, interfacial activity can be identified by the presence of an interfacial minimum in the free energy of the solute, as a function of its position along the direction perpendicular to the interface. Traditionally, activity at the interface between water and a non-polar liquid has been associated with the concept of amphiphilicity. In an interfacial environment, the polar parts of amphiphilic solutes are immersed in water while their non-polar parts are buried in the non-polar phase. Straight chain alcohols and fatty acids are examples of amphiphilic molecules. Although amphiphilicity is sufficient to ensure interfacial activity, it is not a necessary condition. In recent theoretical studies [11–13], a wide range of polar, but not amphiphilic molecules, have been predicted to accumulate at the interface.

In this paper, we discuss the behavior of eight solutes at the interface between two immiscible liquids – water and hexane. Four of the solutes, methanol, ethanol, butanol and hexanol, are amphiphilic, straight chain alcohols. This choice allows us to examine systematically the interfacial activity of amphiphiles as a function of the hydrophobic chain length. The remaining four molecules, nitrous oxide, cyclopropane, isoflurane and desflurane are not amphiphilic. Even though nitrous oxide, cyclopropane and the two halogenated ethers are structurally unrelated, they, nonetheless, share an important property – all of them are clinical anesthetics.

We have two main objectives. The first is to understand the general principles that determine the activities, orientations and conformational equilibria of small solutes at interfaces between water and non-polar media. Once these principles are established, they can be applied to predict the interfacial behavior of different solutes, not only at simple liquid–liquid interfaces, but also in more complex systems, such as water–membrane interfaces.

The second objective is to apply our knowledge of interfacial systems to a concrete problem of considerable medical interest – the determination of a relationship between the interfacial behavior of anesthetics and their anesthetic activity. This represents an alternative view to the century-old Meyer–Overton hypothesis [14, 15], which relates anesthetic activity of anesthetics to their solubility in the bulk oily phase.

Method

Description of the system

The system consisted of 480 water molecules in a lamellar arrangement located between two lamellae of hexane, each containing 80 molecules. As can be seen in Fig. 1, two water–hexane interfaces were present in this system. Liquid hexane was in equilibrium with its gas phase, yielding two liquid–vapor interfaces. The x,y-dimensions of the simulation box, parallel to the interfaces, were 24×24 Å and the z-dimension, perpendicular to the interfaces, was 150 Å. The approximate widths of the water, and each of the hexane lamellae, were 25 and 32 Å, respectively.

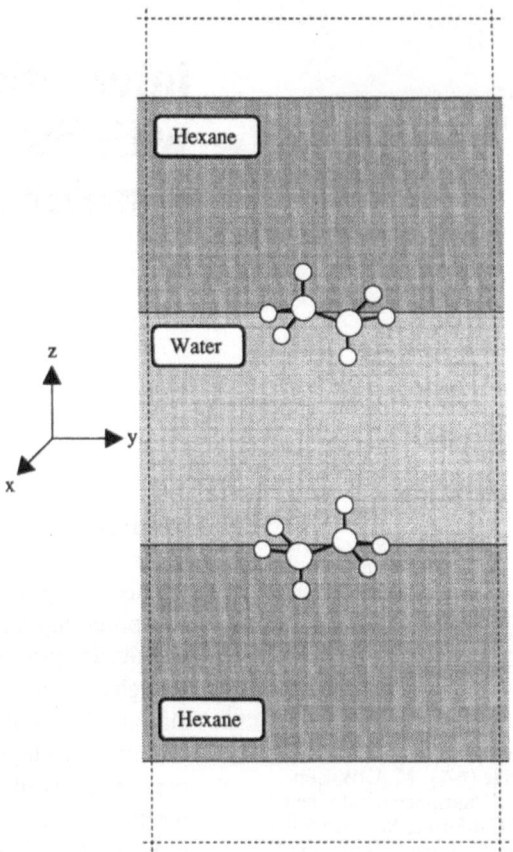

Fig. 1 A schematic of the geometry of the systems used in the present calculations. There are two water–hexane interfaces, and a solute molecule is located at each interface

In addition to the solvent, the system also contained two solutes. These molecules were located in the vicinity of each hexane–water interface. Such an arrangement is computationally more efficient than a system containing one solute molecule at a single water–hexane interface, because information provided from one molecular dynamics trajectory would otherwise require two separate trajectories.

Potential energy functions

In the molecular dynamics simulations presented here, the energies of the solute molecules were evaluated using an empirically based Hamiltonian, H_{total}, which consisted of a sum of individual contributions arising from bond stretching, deformation of valence and torsional angles, as well as Lennard–Jones and Coulomb interactions between non-bonded atoms:

$$H_{\text{total}}(\mathbf{r}) = \sum_{\text{bonds}} k_r(r - r_0)^2 + \sum_{\text{angles}} k_\theta(\theta - \theta_0)^2$$

$$+ \sum_{\text{dihedrals}} \sum_{n=1}^{n=3} V_n[1 + \eta \cos(n\phi - \phi_0)]$$

$$+ \sum_{i<j} \left[\frac{A_{ij}}{R_{ij}^{12}} - \frac{B_{ij}}{R_{ij}^6} + \frac{q_i q_j}{R_{ij}} \right], \qquad (1)$$

where $\mathbf{r}$ represents the full set of intramolecular coordinates of the solute molecule, k_r, r, and r_0 are the bond stretching constant, the bond length and the equilibrium bond length, respectively; k_θ, θ and θ_0 are the valence angle stretching constant, the valence angle and the equilibrium valence angle, respectively. V_n is the coefficient for dihedral angle ϕ with phase ϕ_0, n is the periodicity of the torsional term, and $\eta = 1$, -1, and 1 for $n = 1$, 2, and 3, respectively. R_{ij} is the interatomic distance between atoms i and j, A_{ij} and B_{ij} are the Lennard–Jones non-bonded repulsion and attraction coefficients, respectively, and q_i is the net atomic charge on atom i.

Water molecules were described using the TIP4P model [16]. Hexane molecules were represented by the OPLS potential functions [17, 18]. In this approach, the aliphatic–CH_n groups were treated as electrically neutral united atoms. The water–hexane Lennard–Jones parameters were derived from the standard OPLS combination rules. The bond lengths and the valence angles of both water and hexane were constrained to their equilibrium values, using the SHAKE algorithm [19, 20].

The intramolecular parameters k_r, r_0, k_θ, and θ_0 from the AMBER force field [21, 22] were used. The oxygen atoms of alcohols and ethers were described by the OPLS parameters [17, 18]. The Lennard–Jones parameters for halogens and the atoms of the aliphatic groups were optimized to reproduce the experimentally measured solubilities of methane and its halogenated derivatives in water and hexane, separately [23]. The resulting parameters do not follow the standard combination rules because we were unable to find a set of parameters consistent with the rules which would yield sufficiently accurate solubilities of the solutes in both phases. A similar difficulty was encountered in other studies of interfacial systems, in which the united atom representation was used to describe non-polar molecules [24, 25].

Molecular charge distributions of the solute molecules were represented as sets of partial charges centered on the nuclei. For methanol, the ESP charges were used [26]. For the other three alcohol molecules and the anesthetics, the charges were fitted to reproduce the quantum-mechanically computed electrostatic potentials around the molecules. For flexible solutes, a "multi-conformational" fit of the atomic charges was carried out, whereby these

charges were chosen to reproduce simultaneously the electrostatic potentials of two to four distinct conformers [27, 28]. The relative energies of these conformers were utilized in the optimization of the coefficients, V_n, in the torsional potential. The various conformations in the family of alcohols were investigated at the restricted Hartree–Fock level of approximation, using the split-valence 6-31G** basis set. The second-order Møller–Plesset level of theory was employed for nitrous oxide, cyclopropane, desflurane and isoflurane in conjunction with the 6-31G** basis set. In all cases, the geometries of the solutes were fully optimized prior to determining their molecular charge distributions.

Once a set of charges was obtained for a molecule, the free energy of solvation in water was determined using molecular dynamics and compared with experimental results. It should be, however, borne in mind that using a simple potential function in Eq. (1) with the charges derived from electrostatic potentials does not guarantee that solubilities in water are well-reproduced. One difficulty is that atomic charges should effectively account for the polarization effects not explicitly included in the Hamiltonian [27, 29]. Another difficulty is that multi-conformational fits of atomic charges should be properly weighted toward conformations most populated in aqueous solution. To deal with these difficulties the weights of different conformations in fitting the charges were appropriately changed or the charges on atoms were simply uniformly scaled whenever necessary to improve the agreement between the calculated and experimental values. Obtaining such an agreement was required to have confidence that interfacial properties of the solutes were correctly described in computer simulations.

The solvation free energies of cyclopropane and nitrous oxide were found to be in good agreement with experimental results. For the clinical anesthetics, isoflurane and desflurane, and the alcohols, ethanol, butanol and hexanol, the agreement was not as good. For the clinical anesthetics, perturbation theory was used to determine a scaling factor to increase the electrostatic moments of the molecules. A scale factor of 1.2 was found to give good results. For butanol and hexanol, a different weighting of conformations was able to reproduce the measured solubilities. For ethanol, the charges of the alcohol group were scaled by 1.15. The atomic charges found to give correct solubilities in water are shown in Fig. 2.

Molecular dynamics simulations

Molecular dynamics simulations were performed in the microcanonical ensemble. The equations of motion were

Fig. 2 The structure and partial charges of the solutes that were investigated in this study: (a) nitrous oxide, (b) ethanol, (c) butanol, (d) hexanol, (e) cyclopropane, (f) desflurane, and (g) isoflurane

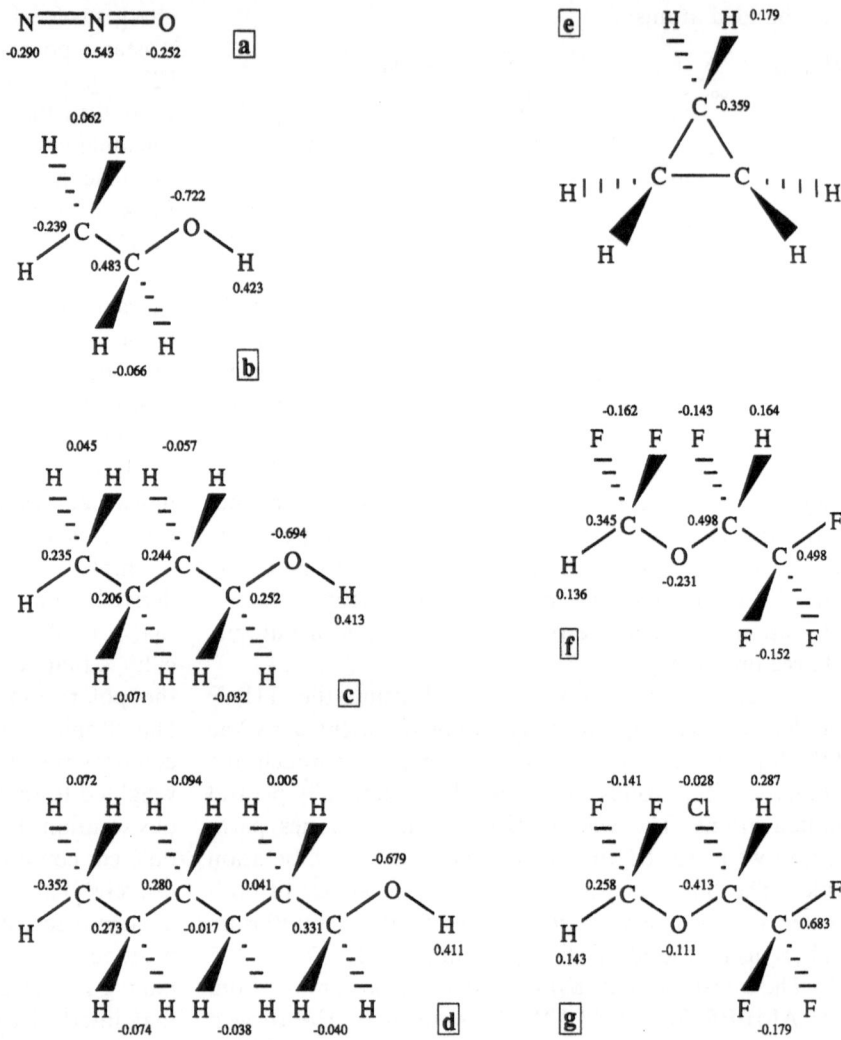

integrated using the Verlet algorithm [30, 31] with a time step of 2 fs. The temperature of the system was maintained at 310 K by occasionally rescaling the velocities of all particles in the system. All the simulations presented here were carried out at 310 K, because the solubilities in water and in hexane, as well as the anesthetic potencies of all eight compounds were measured at this temperature.

The free energies along the z-direction were calculated using the umbrella sampling method [32]. The distance along z was divided into a sequence of "windows" approximately 5 Å wide. Two consecutive windows overlapped by at least 1.5 Å. Approximately, six windows were needed to obtain a complete free energy profile. As mentioned above, each molecular dynamics trajectory yielded results for two windows. Thus three trajectories were needed per free energy profile. For each window, a molecular dynamics

trajectory 0.5–2.0 ns long was obtained. A biasing potential $U_b(z)$ was added in some windows to ensure a more uniform sampling of all positions within these windows, and, therefore, to improve the statistical accuracy of the results. In each window, the free energy along z, $\Delta A_{w-h}(z)$, was calculated from the probability $P(z)$ of finding the center of mass of the solute at position z within the window:

$$A_{w-h}(z) = -k_B T \ln P(z) + U_b(z) \,, \qquad (2)$$

where k_B is the Boltzmann constant and T is the temperature. The free energy profile in the full range of z was obtained from the one-dimensional Weighted Histograms Analysis Method [33], exploiting the requirement that $\Delta A_{w-h}(z)$ be a continuous function of z.

Results and discussion

Alcohols at the water–hexane interface

The main quantity of interest is $\Delta A_{w-h}(z)$, the free energy profile characterizing the transfer of a solute molecule along the z-coordinate. As may be seen in Fig. 3, $\Delta A_{w-h}(z)$ are qualitatively similar for all four alcohols. The profiles are flat on the water and hexane sides and exhibit minima at the interface, approximately 12 to 15 Å wide. In the flat regions $\Delta A_{w-h}(z)$ reaches the values characteristic of the bulk media. The differences between these values yield the free energies of transfer from water to hexane, ΔA_{w-h}^{transf}. They are listed in Table 1 and compared with the free energies of transfer obtained from the measured Ostwald

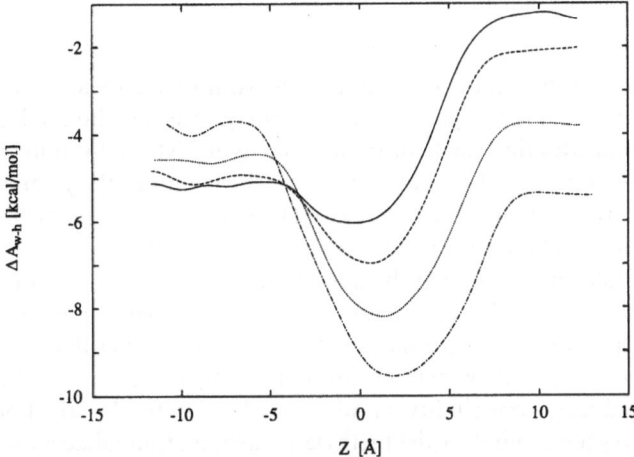

Fig. 3 The free energy profiles for transferring methanol (solid line), ethanol (dashed line), butanol (dotted line), and hexanol (dot-dashed line) across the water–hexane interface. The aqueous phase is to the left, the hexane phase is to the right, and the interface is located at $z = 0$

solubilities of the alcohols in water and hexane [34]. In all four cases, the agreement is quite good – the differences between these two values do not exceed 0.3 kcal/mol.

Another approach to calculate ΔA_{w-h}^{transf} consists of considering the transfer of a solute along the cyclic pathway: gas phase → water → hexane → gas phase. Since the initial and final states along this pathway are identical, the total free energy should be equal to zero. Then $\Delta A_{w-h}^{transf} = \Delta A_{g-w}^{transf} - \Delta A_{g-h}^{transf}$, where ΔA_{g-w}^{transf} and ΔA_{g-h}^{transf} are the free energies of dissolving the solute in water and hexane, respectively. They were obtained from the calculated free energy profiles across the vapor–liquid interfaces of water [13] and hexane, in the same way as ΔA_{w-h}^{transf}. The experimental and calculated free energies of dissolving the alcohols in water and hexane, and the corresponding free energies of transfer are given in Table 1. The excellent agreement between ΔA_{w-h}^{transf} obtained from simulating direct transfer of the solutes across the water–hexane interface, and from the cyclic pathway, provides confidence in the accuracy of the calculations.

The existence of an interfacial minimum in $\Delta A_{w-h}(z)$ indicates that the alcohols are interfacially active. The free energy of transferring a solute molecule from the gas phase to the interface decreases with the chain length and is equal to −6.0, −6.9, −8.2 and −9.6 kcal/mol for methanol, ethanol, butanol and hexanol, respectively. The free energy of transferring the four alcohols from hexane to the interface, however, is almost constant, and equals to 4.5 ± 0.3 kcal/mol. This, in turn, implies that when the alcohols are in the minimum free energy positions, their hydrocarbon portions are buried in hexane.

The free energy profiles across the interface are consistent with the amphiphilic nature of alcohols. More information about the behavior of amphiphilic solutes can be obtained by examining the statistical properties that characterize the positions and orientations of the alcohols at the interface. The probability distribution functions of finding the hydroxyl oxygen and terminal methyl carbon

Table 1 Free Energies of solvation of the anesthetic solutes in water and hexane

Molecule[a]	Water		Hexane		Water/hexane		
	Calc.	Exp.[b]	Calc.	Exp.[b]	Calc.	Exp.	Calc.(w/h)[c]
Methanol	− 5.2	− 4.94	− 1.3	− 1.35	− 3.9	− 3.59	− 3.9
Ethanol (b)	− 5.2	− 4.96	− 2.1	− 2.23	− 3.0	− 2.73	− 3.1
Butanol (c)	− 4.9	− 4.74	− 3.8	− 3.87	− 0.8	− 0.87	− 0.9
Hexanol (d)	− 3.9	− 3.74	− 5.4	− 5.46	1.5	1.72	1.5
Nitrous oxide (a)	0.4	0.51	− 0.7	− 0.76	− 1.2	− 1.27	− 1.2
Cyclopropane (e)	0.6	0.95	− 2.2	− 1.95	2.7	2.80	2.8
(R)-desflurane (f)	0.7	0.92	− 2.1	− 1.75	− 3.4	− 2.67	− 2.8
(R)-isoflurane (g)	0.5	0.31	− 3.2	− 2.67	− 3.9	− 2.98	− 3.7

[a] See Fig. 2.

[b] Unpublished results provided by Eger, E.I.

[c] Calculated from the individual simulations at the water–vapor and hexane–vapor interfaces.

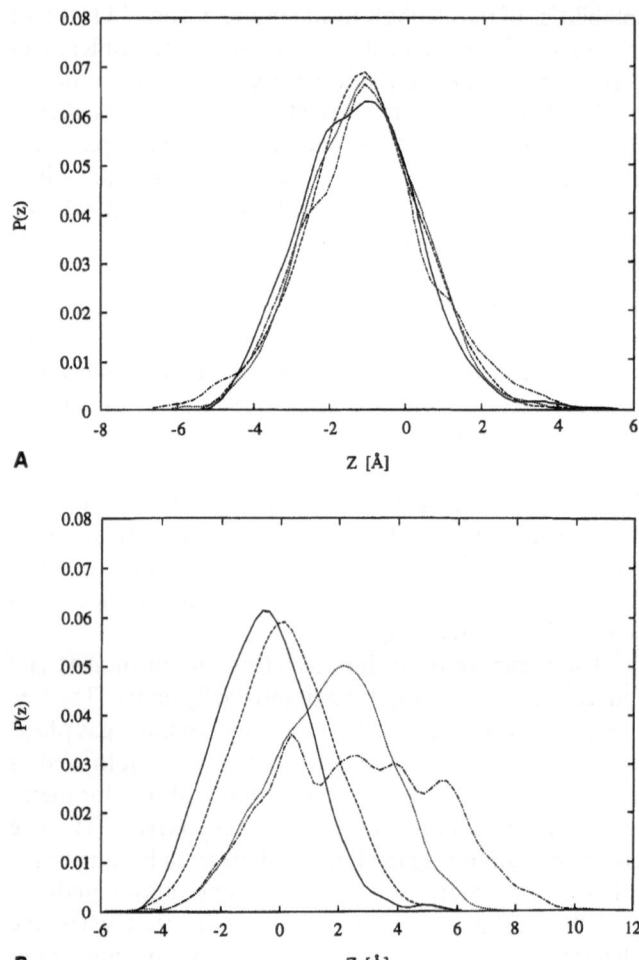

A

B

Fig. 4A The probability density of the location of the oxygen atom of methanol (solid line), ethanol (dashed line), butanol (dotted line), and hexanol (dot-dashed line) at the water–hexane interface; **B** the same as in (a) but for the terminal CH$_3$ group of the molecules. The geometry of the system is the same as in Fig. 3

atoms along the z-direction are shown in Fig. 4. The most probable positions of the oxygen atoms with respect to the interface are almost identical for all four alcohols, approximately 2 Å from the interface towards the aqueous phase. In contrast, the positions of the methyl groups differ. As the length of the hydrocarbon chain increases, they are shifted deeper into hexane. This means that the polar head group of the alcohols is immersed in water, whereas the non-polar chain extends towards the non-polar phase – a behavior expected for amphiphilic molecules.

The hydration of individual atoms in the alcohols can be characterized by the radial distribution functions, g_{X-O}, of finding water oxygen atoms around atom X of the solute. These functions, for the oxygen atom of hexanol, the adjacent methylene carbon atom and the terminal methyl carbon atom, are shown in Fig. 5. For the hydroxyl oxygen atom, g_{O-W} exhibits a sharp peak at 2.75 Å, and

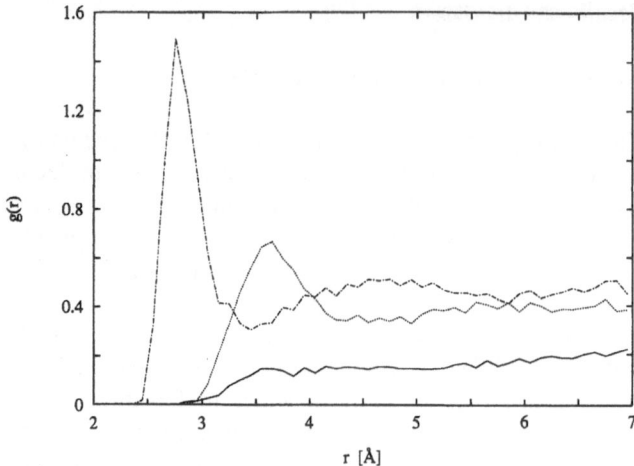

Fig. 5 The alcohol oxygen–water oxygen, $g_{O-O}(r)$ (dot-dashed line), alcohol C1 methylene – water oxygen, $g_{CH_2-O}(r)$ (dotted line), and alcohol C6 methyl – water oxygen $g_{CH_3-O}(r)$ (solid line), radial distribution functions for hexanol at the water–hexane interface. The geometry of the system is the same as in Fig. 3

a very broad peak around 4.6 Å. Almost identical peaks are observed in g_{O-O} of the remaining three alcohols. This indicates that the hydration of the hydroxyl head group at the interface does not depend on the length of the hydrocarbon chain, which is removed from the aqueous environment. The positions and shapes of the peaks in g_{O-W} for molecules located at the interface and in the bulk aqueous solution [35] are very similar. The only difference is that the peak in the interfacial environment is smaller than the peak in bulk water by approximately 25%, due to the reduced accessibility of water molecules to the alcohol oxygen atom. For the methylene carbon atom adjacent to the hydroxyl group, g_{C-O} also exhibits a maximum. Its position is again the same as in the aqueous solution, albeit its height is reduced by nearly a factor of three. This is because the carbon atom is partially removed from water. For the carbon atoms further away from the hydroxyl group, g_{C-O} becomes structureless and progressively smaller. As can be seen in Fig. 5, the methyl carbon atom is practically inaccessible to water.

To describe directly the orientation of the hydrocarbon chain with respect to the interface we define an end-to-end vector pointing from the methylene carbon atom adjacent to the hydroxyl group, to the methyl carbon atom. Then, we calculate the probability distribution function, $P(\theta)$, of finding the angle θ, formed between this vector and the normal to the interface directed from hexane to water. $P(\theta)$ is normalized to describe the probability in the same solid angle for all values of θ.

$$P(\theta) = \left\langle \frac{N(\theta)}{\sin \theta} \right\rangle, \tag{3}$$

Progr Colloid Polym Sci (1997) 103: 29–40
© Steinkopff Verlag 1997

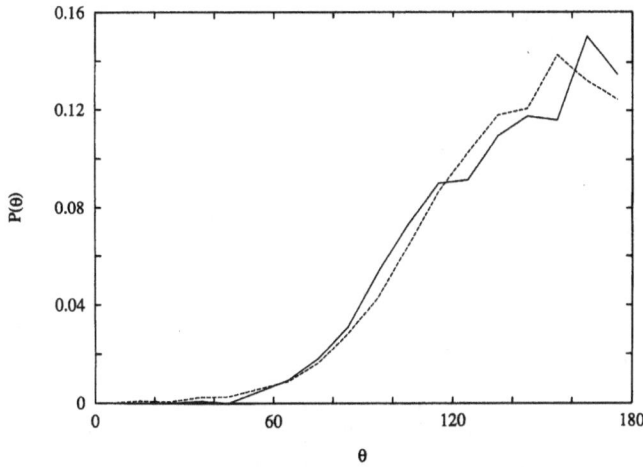

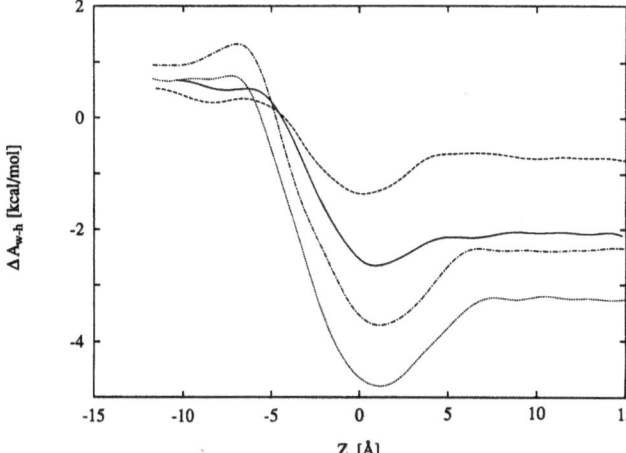

Fig. 6 The probability distribution of the angle formed between the surface normal pointing from water towards hexane and the vector pointing from the C4 methyl to the C1 methylene in butanol (solid line) and from the C6 methyl to the C1 methylene in hexanol (dashed line) when the alcohols are located at the water–hexane interface

Fig. 7 The free energy profiles for transferring the anesthetic molecules cyclopropane (solid line), nitrous oxide (dashed line), isoflurane (dotted line), and desflurane (dot-dashed line) across the water–hexane interface. The aqueous phase is to the left, the hexane phase is to the right, and the interface is located at $z = 0$

where $N(\theta)$ is the fraction of the total number of vectors forming an angle θ with the normal to the interface, $\sin \theta$ corrects for the Jacobian, and $\langle \cdots \rangle$ represents a statistical average. $P(\theta)$ for butanol and hexanol are shown in Fig. 6. Both distributions are strongly non-uniform, and peak near $\theta = 180°$, which corresponds to the end-to-end vector pointing towards hexane. The probability of finding this vector pointing towards water is quite small – an expected effect, considering the hydrophobic nature of the hydrocarbon chain.

Anesthetics at the water–hexane interface

The free energy profiles for transferring the four clinical anesthetics, cyclopropane, nitrous oxide, desflurane and isoflurane, across the water–hexane interface are shown in Fig. 7. Again, the free energies of transfer across the two bulk phases have been estimated as the free energy differences between the end points of the profiles. As may be seen from Table 1, the agreement with the measured values of $\Delta A_{\text{w-h}}^{\text{transf}}$ is very good for cyclopropane and nitrous oxide. For desflurane and isoflurane, the calculated free energies of transfer are not as accurate and deviate from the respective experimental values by 0.6 and 0.9 kcal/mol. This is a sum of small inaccuracies in describing the free energies of dissolving the solutes in both water and hexane. In the case of isoflurane, there is also a 0.6 kcal/mol difference between $\Delta A_{\text{w-h}}^{\text{transf}}$ obtained from the direct simulations across the water–hexane interface, and along the cyclic pathway. Why this error persists,

despite using long molecular dynamics trajectories to calculate the contributing free energy values, is not clear.

For the four clinical anesthetics, $\Delta A_{\text{w-h}}(z)$ exhibits minima at the interface. Thus, all these molecules are interfacially active. However, the degree to which they tend to accumulate at the interface differs. For cyclopropane and nitrous oxide, the depths of the interfacial minima in $\Delta A_{\text{w-h}}^{\text{transf}}$, measured with respect to the free energy in hexane, is smaller than 1 kcal/mol. The minima for isoflurane and desflurane are considerably deeper, approximately 2–2.5 kcal/mol. The interfacial activity of the solutes can be related to their polarity. Isoflurane and desflurane have an average permanent dipole moment of approximately 2 D. In contrast, cyclopropane is a symmetrical molecule with no permanent dipole moment, and the dipole moment of nitrous oxide is only 0.2 D. In spite of their small dipole moments, however, these molecules should be considered as weakly polar rather than non-polar. They are involved in non-negligible electrostatic interactions with water, that are primarily due to higher molecular multipole moments arising from substantial partial charges on the atoms.

In addition to influencing the free energy profiles, the presence of the interface affects the orientational distributions of polar solutes. This is particularly clear in the case of isoflurane and desflurane, both bearing a considerable dipole moment. To describe this effect, we consider the probability distribution function, $P(\theta)$, of finding the angle θ between the molecular dipole moment of the solute and the normal to the interface, pointing from hexane to water. $P(\theta)$ is defined as in Eq. (3). In the isotropic environment of

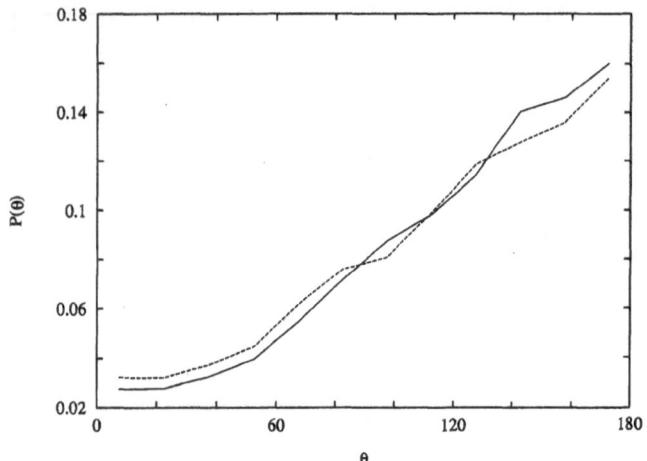

Fig. 8 The probability distribution of the angle formed between the surface normal pointing from water towards hexane and the molecular dipole vector for the anesthetic molecules desflurane (solid line) and isoflurane (dashed line) when the molecules are located near the free energy minimum at the water–hexane interface

the bulk solvents, $P(\theta)$ is constant. This is not the case, however, at the interface. In this region, water molecules are in an anisotropic environment and, therefore, exhibit preferred orientations. This, in turn, yields an excess electric field perpendicular to the interface, which points towards the water phase. The charge distributions of the solutes interact with the interfacial electric field. In the first-order, perturbation theory approximation, these interactions are favorable if the molecular dipole moments are approximately aligned with the excess field. As can be seen in Fig. 8, this is indeed the case for both isoflurane and desflurane.

Relationship between interfacial activity and the mechanism of anesthetic action

According to the Meyer–Overton hypothesis [14, 15], the anesthetic potency of anesthetic compounds is proportional to their solubilities in a non-polar phase, similar to the interior of the neuronal membrane. The remarkable accuracy of this relationship for all clinical and many other anesthetics led to its broader interpretation that anesthetics act inside the neuronal membrane [36]. Then, the Meyer–Overton hypothesis implies that the same concentration is required for all anesthetics to exert anesthetic action, irrespective of their molecular structure. Their action may be either nonspecific or directed at selected membrane receptors.

Once a large database of anesthetic compounds is examined the correlation predicted by the Meyer–Overton relationship is less convincing. In Fig. 9, we show the

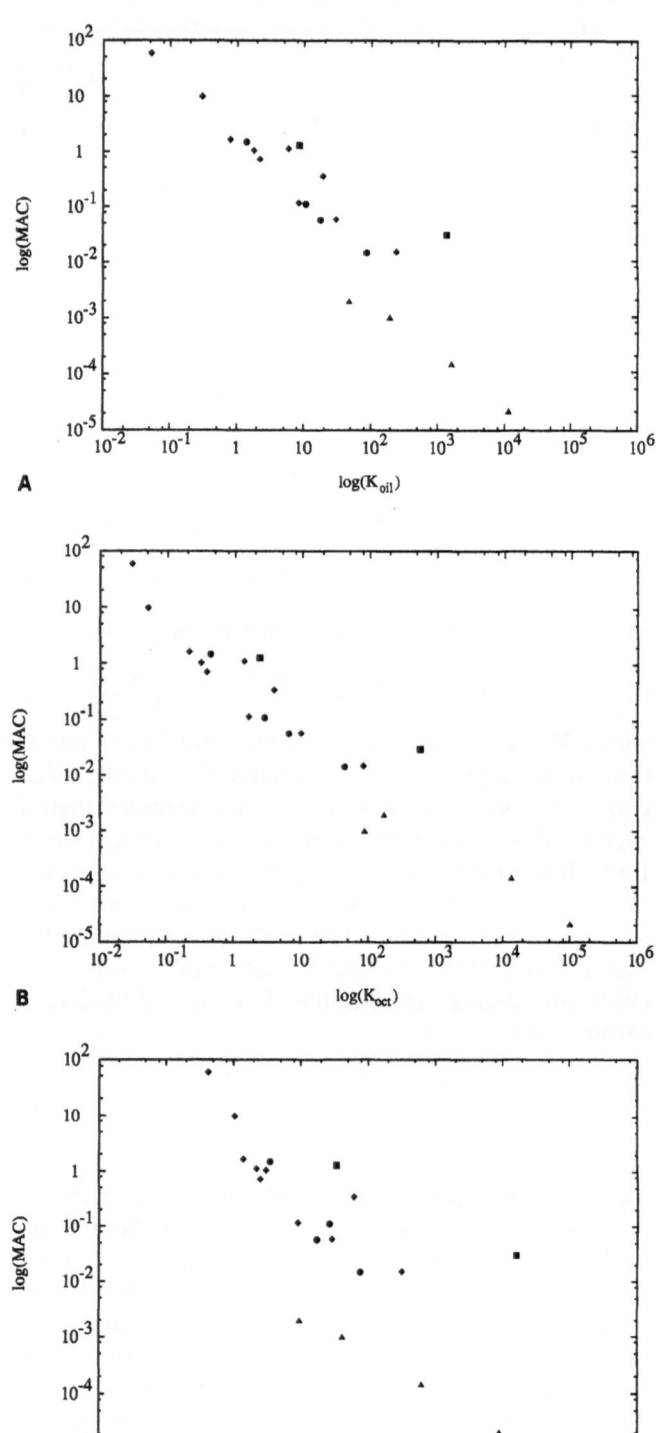

Fig. 9 Log–log of MAC vs. solubility in **A** olive oil, **B** octanol, and **C** hexane for clinical anesthetics (circles), n-alkanols (triangles), transition compounds (squares) and other anesthetic compounds (diamonds)

Progr Colloid Polym Sci (1997) 103:29–40
© Steinkopff Verlag 1997

relation between a measure of the anesthetic potency, the maximum alveolar concentration (MAC), and the solubilities in hexane, octanol and olive oil for 20 anesthetics. MAC is defined as the average applied partial pressure that suppresses the response to painful stimuli, i.e., the lower the MAC, the higher the potency. The solubilities (Ostwald solubilities) are defined as

$$\frac{c(z)}{c_{\text{bulk}}} = e^{-\Delta A(z)/k_{\text{B}}T} , \qquad (4)$$

where $c(z)$ and c_{bulk} are the concentrations of the solute at z and in the bulk phase, respectively. The selected anesthetics include eight molecules discussed in this paper and 12 other compounds, transfer of which across the water–hexane interface has been previously studied in molecular dynamics simulations [13]. The correlation between MAC and the solubilities in hexane is not very good, even though the dielectric constants of hexane and the interior of the membrane bilayer [37–39] are approximately the same. A better correlation is obtained if olive oil is used as a model of a non-polar phase, albeit it has a higher dielectric constant. The correlation holds particularly well for clinical anesthetics. However, even in this liquid, there are considerable deviations from a linear relationship. Two compounds (transition compounds) are markedly less potent, whereas all four alcohols are more potent than predicted by the Meyer–Overton hypothesis. If the solubilities in octanol are used, deviations for the alcohols are reduced. The solubilities of the two transitional compounds, however, still deviate from the Meyer–Overton relationship, and the scatter of the data points, from the correlation line, remains considerable. A similar conclusion was reached by Fang et al., on the basis of their experimental work [34]. Furthermore, bulk octanol cannot be considered as a realistic model of the membrane interior since it contains small, polar clusters of hydroxyl groups [40], absent in the core of the bilayer.

In an alternative hypothesis, the anesthetic potency correlates with interfacial solubility, defined as

$$s_{\text{int}} = \frac{1}{\Delta z}\int_{z_1}^{z_2} e^{-\Delta A_{\text{exc}}(z)/kT}\, dz , \qquad (5)$$

where $\Delta z = z_2 - z_1$. This correlation for solubilities at the water–hexane interface is shown in Fig. 10. As may be observed, the correlation holds for all the compounds. Neither the transitional compounds nor the alcohols deviate from that relationship. Equally important, clinical anesthetics obey this relationship as well as they obey the Meyer–Overton hypothesis. An equally good correlation is found for solubilities at the interface between water and glycerol 1-monooleate (GMO) [41].

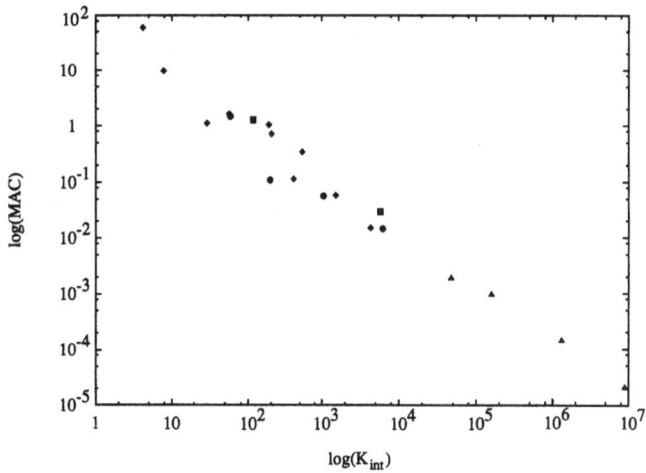

Fig. 10 Log–log plot of MAC vs. interfacial solubility for clinical anesthetics (circles), n-alkanols (triangles), transition compounds (squares) and other anesthetic compound (diamonds)

Within the narrow interpretation, our results suggest that solubilities at interfaces between water and a non-polar liquid allow the prediction of potencies for a broad range of anesthetic compounds better than solubilities in environments that model the interior of the membrane. In a broader view, a very good correlation between MAC and interfacial solubilities suggests that the sites of anesthetic action are located near an interface. A natural candidate site is the head group region of the neuronal membrane. Another possibility is an interface between the aqueous solvent and a hydrophobic patch in a water-exposed portion of a membrane receptor.

A general view of interfacial activity

All eight molecules discussed in this work are interfacially active. This does not, however, mean that *all* neutral solutes tend to concentrate at the interface. To illustrate this point we show in Fig. 11, the free energies of transfer, $\Delta A(z)$, across the water–hexane and the water–GMO bilayer interfaces for two simple molecules, methane and fluoromethane [11]. Whereas fluoromethane has a pronounced free energy minimum at the interface, $\Delta A(z)$ characteristic of methane decreases monotonically when the solute is transferred from water to the interior of the membrane. At the water–hexane interface, the $\Delta A(z)$ profile exhibits only a very shallow minimum, smaller than the thermal energy. Calculations of the free energy profiles for several molecules very poorly soluble in water [12, 13] – perfluoroethane, 1,2-dichlorohexafluorocyclobutane, octafluorocyclobutane and 2,3-dichlorooctafluorobutane

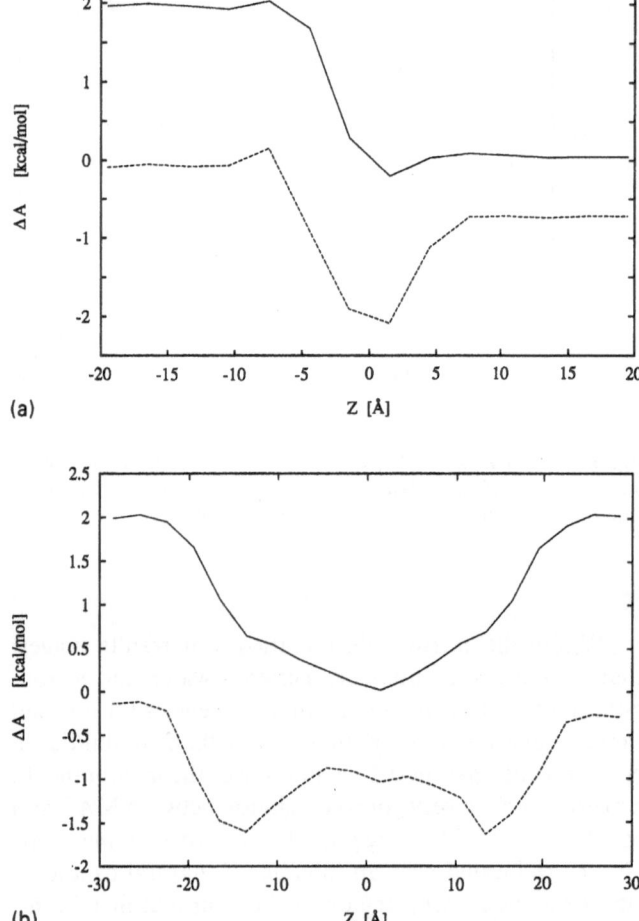

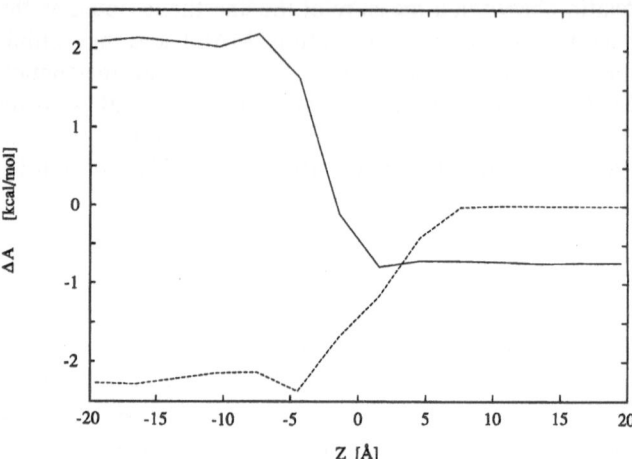

Fig. 12 Non-electrostatic (solid line) and electrostatic (dashed line) components of the free energy of transferring fluoromethane across the water–hexane interface. The geometry is the same as in Fig. 11a

Fig. 11 Free energy of transferring methane (solid line) and fluoromethane (dashed line) across (a) the water–hexane interface and (b) the water–GMO membrane interface. In (a) the aqueous phase is to the left, the hexane phase is to the right and the interface is at $z = 0$. In (b) the interior of the membrane is located at $z = 0$, the head group region is around $z = \pm 15$ and the aqueous phase is $z < 18$ and $z > 18$

– allow a generalization of this observation: non-polar molecules are not interfacially active.

In order to understand the difference in the interfacial behavior of polar and non-polar molecules, we consider the free energy, $\Delta A(z)$, for model solutes that have exactly the same van der Waals parameters as methane and fluoromethane, but do not bear any partial charges on atoms. This free energy will be referred to as the non-electrostatic contribution to $\Delta A(z)$ for the corresponding, fully charged solutes. The difference between the free energies of the molecules with and without atomic partial charges will be called the electrostatic contribution to $\Delta A(z)$. The sum of the two terms, shown in Fig. 12, yields the total $\Delta A(z)$. It should be noted that each term involves

both electrostatic and non-electrostatic contributions from solvent reorganization.

For both solutes, the non-electrostatic contribution, $\Delta A_{nel}(z)$, is positive and decreases when the solute is moved across the interface, from water to a non-polar phase. This term is dominated by the free energy required to create in the solvent a cavity that can accommodate the solute. The free energy of cavity formation is always positive and is larger in water than in a non-polar solvent. This is a manifestation of the hydrophobic hydration – the effect responsible for poor solubility of non-polar species in water [42, 43]. $\Delta A_{nel}(z)$ also includes an attractive contribution from solute–solvent van der Waals interactions, albeit this contribution does not change the shape of the free energy profile.

The electrostatic term, $\Delta A_{el}(z)$, is negative and becomes progressively less attractive as the solute is transferred from water to a non-polar phase. A similar result is well known for solubilities of small molecules in liquids of different polarities – the electrostatic contribution to the free energy of solvation becomes less favorable as the dielectric constant of the solvent decreases. Note that $\Delta A_{el}(z)$ begins to change only when the solute closely approaches the interface from the water side and most of the reduction in this term occurs on the non-polar side.

As can be seen from Fig. 12, the two contributions to $\Delta A(z) - \Delta A_{el}(z)$ and $\Delta A_{nel}(z)$ – change oppositely in the interfacial region. For non-polar solutes, $\Delta A_{el}(z)$ is small whereas $\Delta A_{nel}(z)$ dominates. As a consequence, these solutes preferentially partition into the non-polar phase and their free energy changes monotonically, or nearly monotonically, in the interfacial region. The opposite situation is

Progr Colloid Polym Sci (1997) 103:29–40
© Steinkopff Verlag 1997

also possible, in particular for small, polar or ionic species. Here, changes in $\Delta A_{nel}(z)$ at the interface are markedly smaller than changes in $\Delta A_{el}(z)$, and $\Delta A(z)$ decreases monotonically when the solute moves towards water. In most cases, however, both contributions are important. At the interface, the repulsive non-electrostatic term is considerably reduced, compared to its value in water, whereas the attractive electrostatic contribution does not undergo large changes. This results in the interfacial free energy minimum.

So far we have not discussed free energy profiles that yield interfacial barriers to the transfer of solutes between the two adjacent bulk phases. Not only have we not encountered such a case among the solutes studied here, but also it has been argued, based on qualitative arguments, that this case cannot exist [4]. Our microscopic-level considerations indicate that this conclusion is likely to be correct for liquid–liquid interfaces. In water–membrane systems, the free energy of cavity formation in the densely packed head group region of the membrane could be more unfavorable than in water and in the membrane interior. This, in turn, could yield an interfacial maximum in $\Delta A(z)$. It has not been shown yet, however, that such a possibility is actually encountered in any water–membrane system.

It is interesting to wonder whether it is possible to develop a reliable approach for calculating interfacial free energies without appealing to molecular-level computer simulations. Although this has not yet been accomplished, it appears that all the ingredients for such an approach are in place. Recently, it has been shown that for small solutes the free energy of cavity formation can be obtained from computer simulations of the neat solvent by employing a simple Gaussian model for the probability distribution of the number of water molecules located in the cavity volume [43]. Although the model has not yet been tested for interfacial systems, it is expected that it will be as accurate as in bulk liquids. Complemented by a WCA-type perturbation theory [44] for van der Waals solute–solvent interactions, the model would yield $\Delta A_{nel}(z)$.

The electrostatic contribution can be obtained by solving the Poisson equation in the *continuum* dielectric approximation. Although this approximation has not been systematically tested for interfacial systems, its recent applications to bulk solutions proved to be highly successful [45, 46]. The conventional *continuum*, dielectric model can be considered as an implementation of second-order perturbation theory [47]. The first-order term is assumed to vanish and all terms beyond the second order are neglected. It is not clear, however, how well these approximations hold near an interface. In particular, interfacial solvent molecules have preferred orientations due to the interfacial excess electric field. They will, therefore, not be randomly oriented around the cavity volume of the solute – a requisite for the first-order term to vanish. Furthermore, it has

been shown for fluoromethane at the water–hexane interface, that the second-order approximation is insufficient to reproduce $\Delta A(z)$ accurately [11]. Including the third-order term was also necessary, but higher-order terms were not needed. In conclusion, the *continuum* dielectric model, with some extensions, may provide means for calculating $\Delta A_{el}(z)$ at interfaces.

Conclusions

A unique property of interfaces between water and non-polar phases is that many small solutes exhibit increased concentrations in this region, compared to the adjacent bulk phases. These solutes are characterized by interfacial minima in the free energy of transfer from water to the non-polar medium. One, well-known class of such solutes are amphiphiles, examples of which are four straight-chain alcohols, discussed in this paper. Compared to the free energies in aqueous solution, interfacial minima for these molecules become markedly deeper as the length of the hydrocarbon chain increases. However, the depth of the minima, measured with respect to the non-polar phase, is approximately constant. This is because the hydroxyl groups of all four alcohols are located in the same place, near the interface, whereas their hydrophobic portions remain almost completely removed from water. To generalize this observation, the degree to which the polar head group of an amphiphile penetrates the aqueous solvent depends on the chemical nature of this head group but not on the length of the non-polar chain.

Other examples of interfacially active solutes are molecules that are not amphiphilic but exhibit some degree of polarity. The origin of this activity becomes apparent from the analysis of electrostatic and non-electrostatic contributions to the free energy of transferring these solutes across the interface. These contributions are defined in the previous section. The non-electrostatic term, dominated by the free energy of cavity formation, decreases monotonically, or nearly monotonically, from water to a non-polar environment. The electrostatic term also changes monotonically, but oppositely. For most polar molecules, both terms are important and the resulting free energy profile reflects the balance between these terms. When the absolute values of the rate of change of both contributions are equal the free energy exhibits a minimum. This will occur for many solutes. Exceptions are either strongly polar or charged species, when the electrostatic term dominates, or non-polar molecules, when this term is negligible. For weakly polar solutes, the interfacial minima are often shallow, not much deeper than the thermal energy. For more polar molecules, however, the minima can be substantial, sometimes comparable to those observed for amphiphiles.

Although most of the discussion in this paper concentrated on the water–hexane interface, a qualitative description of interfacial activity applies also to interfaces between water and other non-polar liquids or lipid bilayers. It should be, however, borne in mind that the structure of bilayers is strongly anisotropic and this feature should enter the analysis of the free energy profiles. In particular, the region in which the electrostatic term changes, will be broader and, possibly, somewhat non-monotonic due to the presence of charged or zwitterionic head groups. Another factor to consider is the packing of head groups, which may influence the free energy of cavity formation.

Interfacial activity of small molecules at the interface has several implication of biological and pharmacological importance. For example, interfacial concentrations of a broad range of anesthetic compounds correlate very well with their anesthetic potencies, suggesting that the site of anesthetic action is located near the water–membrane interface. In other computer simulations it was demonstrated that terminally blocked amino acids and dipeptides [48] also belong to the group of interfacially active molecules. Their conformations and orientations at the interface depend on the polarity of the side chains. The principles governing the interfacial behavior of these small peptides are qualitatively the same as those discussed here. Their knowledge helps to determine structural preferences of longer peptides at water–membrane interfaces. Similarly, these principles may be applied to analyze interactions of ligands with interfacial sites of membrane receptors.

Acknowledgments This work was supported by a National Institute of Health grant (GM47818-01) and a grant from the National Aeronautics and Space Administration Exobiology Program. C.C. was supported by a National Research Council Associateship. Computational resources from the National Cancer Institute are gratefully acknowledged. We thank Dr. Michael H. New for helpful suggestions to improve this manuscript.

References

1. Bard AJ, Faulkner LR (1980) Electrochemical Methods: Fundamentals and Applications. Wiley, New York
2. MacRitchie F (1990) Chemistry at Interfaces. Academic Press, San Diego
3. Bockris JOM, Gonzalez-Martin A (1990) In: Spectroscopic and Diffraction Techniques in Interfacial Chemistry, Kluwer Academic Publishers, Dordrecht
4. Israelachvili J (1992) Intermolecular and Surface Forces. Academic Press, New York
5. Starks CM, Liotta CL, Halpern M (1994) Phase Transfer Catalysis. Chapman & Hall, New York
6. Petty HR (1993) Molecular Biology of Membrane Structure and Function. Plenum Press, New York
7. Goodsell D (1993) The Machinery of Life. Springer, New York
8. White J (1992) Science 258:917
9. Pohorille A, Wilson MA (1993) Mol Struct (Theochem) 284:271
10. Benjamin I (1996) Chem Rev (1996) 96:1449
11. Pohorille A, Wilson MA (1996) J Chem Phys 105:3760
12. Pohorille A, Cieplak P, Wilson MA (1996) Chem Phys 204:337
13. Chipot C, Wilson MA, Pohorille A (1997) J Phys Chem (in press)
14. Meyer H (1899) Arch Exptl Pathol Pharmakol 42:109
15. Overton E (1901) Studien über die Narkose zugleich ein Betrag zur Allgemeinen Pharmakologie. Verlag von Gustav Fischer, Jena
16. Jorgensen WL, Chandrasekhar J, Madura JD, Impey RW, Klein ML (1983) J Chem Phys 79:926
17. Jorgensen WL, Madura JD, Swenson CJ (1984) J Am Chem Soc 106:6883
18. Jorgensen WL, Tirado-Rives J (1988) J Am Chem Soc 110:1657
19. Ryckaert J, Cicotti G, Berendsen HJC (1979) J Comput Phys 23:327
20. Van Gunsteren WF, Berendsen HJC (1977) Mol Phys 34:1311
21. Weiner SJ, Kollman PA, Nguyen DT, Case DA (1986) J Comput Chem 7:230
22. Cornell WD, Cieplak P, Bayly CI, Gould IR, Merz Jr. KM, Ferguson DM, Spellmeyer DC, Fox T, Caldwell JC, Kollman PA (1995) J Am Chem Soc 117:5179
23. Pohorille A, Wilson MA (work in progress)
24. Van Buuren AR, Marrink SJ, Berendsen HJC (1993) J Phys Chem 97:9206
25. Marrink SJ, Berendsen HJC (1994) J Phys Chem 98:4155
26. Carlson HA, Nguyen TB, Orozco M, Jorgensen WL (1993) J Comput Chem 14:1240
27. Reynolds CA, Essex JW, Richards WG (1992) J Am Chem Soc 114:9075
28. Chipot C, Ángyán JG (1995) GRID Version 3.0: Point Multipoles Derived From Molecular Electrostatic Properties. QCPE No. 655
29. Stouch TR, Williams DE (1992) J Comput Chem 13:622
30. Allen MP, Tildesley DJ (1987) Computer Simulation of Liquids. Clarendon Press, Oxford
31. Verlet L (1967) Phys Rev 159:98
32. Torrie GM, Valleau JP (1974) Chem Phys Lett 28:578
33. Kumar S, Bouzida D, Swendsen RH, Kollman PA, Rosenberg JM (1992) J Comput Chem 13:1011
34. Fang Z, Ionescu P, Chortkoff BS, Kandel L, Sonner J, Laster M, Eiger II Anesthesia Analgesia (submitted)
35. Wilson M, Pohorille A (1997) J Phys Chem (in press)
36. Janoff AS, Miller KW (1982) In: Biological Membranes, Chapman D (ed) Vol. 4. Academic Press, New York, p 417
37. Fettiplace R, Andrew DM, Haydon DA (1971) J Membr Biol 5:277
38. Ohki S (1968) J Theor Biol 19:97
39. Dilger JP, Fisher LR, Haydon DA (1982) Chem Phys Lipids 30:159
40. DeBolt SE, Kollman PA (1995) J Am Chem Soc 117:5316
41. Pohorille A, Wilson MA (unpublished results)
42. Pratt LR, Pohorille A (1992) Proc Natl Acad Sci USA 89:2999
43. Hummer G, Garde S, Garcia AE, Pohorille A, Pratt LR (1996) Proc Natl Acad Sci USA 93:8951
44. Andersen HC, Chandler D, Weeks JD (1976) Adv Chem Phys 34:105
45. Rashin AA (1990) J Phys Chem 94:1725
46. Honig B, Nicholls A (1995) Science 268:1144
47. Pratt LR, Hummer G, Garcia AE (1994) Biophys Chem 51:147
48. Chipot C, Pohorille A (1997) J Mol Struct (Theochem) (in press)

Progr Colloid Polym Sci (1997) 103:41–50
© Steinkopff Verlag 1997

LIQUID LIQUID INTERFACES

B.S. Murray

Dynamics of proteins at air-water and oil-water interfaces using novel Langmuir trough methods

Received: 28 October 1996
Accepted: 15 November 1996

Abstract Spread monolayers of β-lactoglobulin and bovine serum albumin and adsorbed films of lysozyme and β-lactoglobulin were studied at the oil (n-tetradecane) –water (O–W) and air–water (A–W) interfaces. In general, spread monolayers were more expanded at the O–W interface than at the A–W interface. Desorption rates from monolayers increased greatly with increasing interfacial pressure, π, but were still quite low until typical equilibrium adsorption π were exceeded. Desorption rates as a function of the energy barrier to desorption were similar at both types of interface. Adsorption kinetics of lysozyme at the A–W interface were in agreement with measurements obtained via ellipsometry. Spread and adsorbed films of β-lactoglobulin were subjected to dilatational expansion. The dynamics of the change in π were slower at the O–W interface, but for both interfaces the dynamics of the in-film relaxation processes were quite fast, diffusion from the bulk aqueous phase accounting for the long time-scale of the relaxation back to the equilibrium π for adsorbed films. A new type of Langmuir trough was used for these measurements.

Key words Protein(s) – interface(s) – monolayer(s) – dynamic(s)

Dr. B.S. Murray (✉)
Food Colloids Group
Procter Department of Food Science
University of Leeds
Leeds LS2 9JT, United Kingdom

Introduction

Whether or not emulsions and foams are able to form and whether or not they are stable is largely dependent on how the adsorbed film of surfactant at the interface responds to mechanical deformation [1, 2]. These deformations may be the result of Brownian motion or shear fields arising through convection, gravitational forces or through stirring, for example. Emulsions and foams are metastable states and the key requirement for stability is that a sufficient density of adsorbed surfactant should be maintained at the interface to prevent bridging of the drops or bubbles so that a single volume of the dispersed phase is not formed. Mechanical deformations cause changes in the surface density of surfactant molecules as the total area of the interface is expanded or contracted. Both dilatational and shear deformation of the adsorbed film of molecules will occur. The rate at which surfactant molecules are able to redistribute within the interface will depend upon all the factors affecting interactions between them. The creation of a non-uniform distribution of surfactant leads to surface tension gradients which tend to oppose any such deformation.

The motion of molecules in the interface is not independent of the motion of fluid in the adjacent bulk phases but the two flows are hydrodynamically coupled. For high molecular weight polymeric surfactants, such coupling is even more important since, in general, the majority of the segments of the adsorbed macromolecules are solvated in either of the two bulk phases.

In addition to the motion of molecules within the interface there is always excess surfactant present in one or both of the bulk phases which is capable of moving in and "healing the gaps" in the film produced by the disturbance. When the surface concentration of molecules changes, the local equilibrium with the bulk phase(s) is perturbed, leading to adsorption or desorption in accordance with the adsorption isotherm. If the surfactant molecules have a tendency to aggregate in the bulk then the relative concentrations of aggregates and monomers in the bulk phase immediately adjacent to the interface will also be disturbed, in turn leading to diffusion of aggregates, etc., to and from the interface along concentration gradients.

Proteins and proteoglycans are important stabilisers of emulsions and foams in many food and non-food applications [3]. The interactions between protein molecules, either when adsorbed at interfaces or when present in the bulk, are very complex and are connected with changes in the conformation of the folded polypeptide chains. These changes have a wide range of characteristic time-scales [4]. When these particular features of proteins are superimposed on the above picture of a deformed interfacial film it is seen that the task of understanding the mechanism of action of proteins as surfactants is a daunting task.

Unfortunately, some of the most basic aspects of the dynamics of adsorbed protein films are still unclear. For example, adsorption kinetics predicted by the bulk diffusion coefficient of proteins are usually too slow due to the existence of barriers to adsorption which may be entropic (orientational) or enthalpic (e.g. electrostatic in origin) [5]. This is important since diffusion of proteins to and from interfaces is generally much slower than with low molecular weight surfactants so that the contribution from bulk transport effects to changes in the stability may be enhanced.

One way of trying to gain a better understanding of the different contributions to the adsorbed film dynamics is to study separately spread films (monolayers) of proteins and adsorbed films of proteins under similar conditions. This paper is a summary of some new experiments which should help to gain a better understanding of the dynamic behaviour of adsorbed protein films. A particular type of Langmuir trough was used as the principal tool of investigation, which allowed experiments at both A–W and O–W interfaces. Most of this work was presented at a recent conference – "Surfactants in Solution", Jerusalem, June 1996.

Materials

Imidazole, potassium dihydrogen phosphate, disodium hydrogen phosphate, sodium chloride and hydrochloric acid were all AnalR grade reagents from BDH Merck. Bovine serum albumin (prod. code A-7638, lot no. 14H9348), β-lactoglobulin (prod. code L-0131, lot no. 91H7005) and n-tetradecane (99%) were from Sigma Chemicals. Water used for all the experiments was from a Millipore alpha-Q purification system with a surface tension of 72.0 mN m^{-1} at 25°C.

Methods

The Langmuir trough

The Langmuir trough used in this study has been described in detail recently [6, 7a, 7b]. Figure 1 illustrates the essential features of the apparatus. A solid PTFE frame, or barrier, hinged at the corners, provides a continuous, leak-free enclosure which can be used to reduce the area of an

Fig. 1 Schematic illustration of the apparatus: view from side **A** and above **B**. Key: a = trough walls, b = lower (aqueous) phase, c = barrier, d = upper (oil) phase, e = hydrophobic Wilhelmy plate, f = siphon, g = direction of stepper motor drive for compression

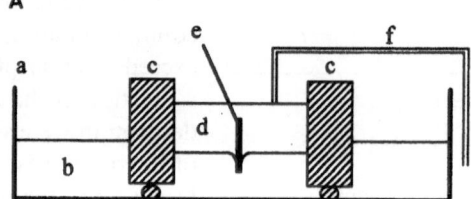

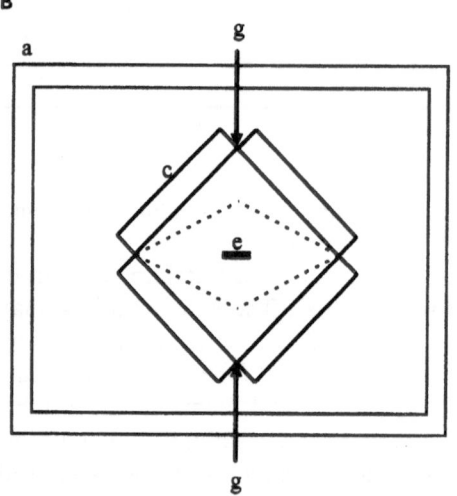

oil–water (O–W) or air–water (A–W) interface. This is achieved by driving together or apart two opposite corners of the barrier via a stepper motor interfaced to a PC. The drive movement is extremely accurate and flexible, allowing slow, rapid or sinusoidal changes of the interfacial area. The interfacial tension/pressure, π, is monitored by measuring the force on a Wilhelmy plate dipping into the interface. The temperature of the subphase was maintained at $25 \pm 0.3°C$, unless stated otherwise. For A–W films Wilhelmy plates of roughened, cleaned mica were used; for O–W films hydrophobic mica plates were used. Further details are given elsewhere [6, 7a, 7b].

Interfacial pressure–area (π–A) isotherms

Before spreading monolayer material, the A–W interface was rapidly reduced to an area lower than that used in the subsequent monolayer experiment. The interface was sucked clean with a vacuum line, the interface expanded and the process repeated until $\pi < 0.1$ was obtained on compression. For the A–W interface, spreading was then performed (see below). For the O–W interface, a 6–7 mm layer of n-tetradecane was gently layered over the top of the aqueous phase and the material spread immediately afterwards. The concentration of protein in spreading solutions, made up in pure α–Q water, was typically 0.2 mg ml^{-1}. For spreading, a drop of spreading solution was slowly formed on the tip of the syringe and then the drop slowly lowered to touch the interface, the syringe tip raised and the process repeated until all the solution had been spread. The material was spread at high area per molecule, A, generally at $A > 25\,000$ Å^2 per molecule. Spreading took 5–10 min and measurement of the π–A isotherm was begun 10 min after spreading. Films were compressed at constant speed, dA_r/dt approximately $= 1.5 \times 10^{-4}$ s^{-1}, where A_r is the fraction of the maximum area (225 cm^2) inside the barrier and t is the time. Below this compression rate no difference in the π–A isotherms was observed – this was therefore considered to be slow enough that the π–A isotherm obtained represented the "equilibrium" isotherm.

Desorption from monolayers

For desorption measurements, monolayers were compressed slowly to the desired π, π_{des}, as for the measurements of the π–A isotherms. When π_{des} was reached, the barrier was then moved only so as to maintain π constant at π_{des} and the total film area was monitored. Computer control of the barrier allowed π_{des} to be maintained at $\pi_{des} \pm 0.1$ mN m^{-1} easily.

Adsorption experiments

For experiments on the kinetics of adsorption the trough was filled with the protein solution and the interfacial film compressed. The interface was sucked clean with a vacuum line for 1–2 min and then the interface was rapidly expanded (in 1–2 s) to the starting area. This point was taken as zero adsorption time. At certain time intervals the interface was rapidly compressed until π reached a certain value, π^*, then the interface was quickly re-expanded to the full, original film area. As the protein surface concentration, Γ, increases with time the trough area at which $\pi = \pi^*$ increases with time. From the area of the interface when $\pi = \pi^*$ and provided the corresponding value of the area per molecule, A^*, is known, Γ may be calculated.

Dilatational experiments

For dilatational experiments on monolayers, films were compressed slowly up to an initial π as for the π–A isotherm measurements and then rapidly expanded. The film area was increased by 10% and the barrier speed adjusted so as to produce this expansion in exactly 1.0 s, unless otherwise stated. For dilatational measurements on adsorbed films, films were expanded in the same way, after waiting for the equilibrium value of π to be reached.

Results

π-A isotherms

Figure 2 shows π–A isotherms for spread films of BSA and β-lactoglobulin at the A–W interface and at the O–W interface. The size of the symbol may be taken as the magnitude of the error in the measurement of the quantity on the ordinate axis in this figure (and also in the other figures in this paper). For BSA the aqueous phase was pH 7.4, 0.3 M NaCl, for comparison with the result of MacRitchie [8], obtained under the same conditions. For β-lactoglobulin the aqueous phase was 0.02 mol dm^{-3} imidazole buffer, adjusted to pH 7.0 with HCl. The relative molecular mass of BSA was taken as 66 500 Da [9] and that of β-lactoglobulin (dimeric form) taken as 36 700 Da [10] for calculation of the π–A isotherms.

The results shown in Fig. 2 are representative of those found by spreading 100–200 μl of protein solution, i.e. isotherms were independent of the amount of solution spread, indicative of reproducible spreading. The A–W isotherm for BSA is in good agreement with the result of MacRitchie [8], particularly at low π. The π–A curves were completely reproducible in that on re-expansion

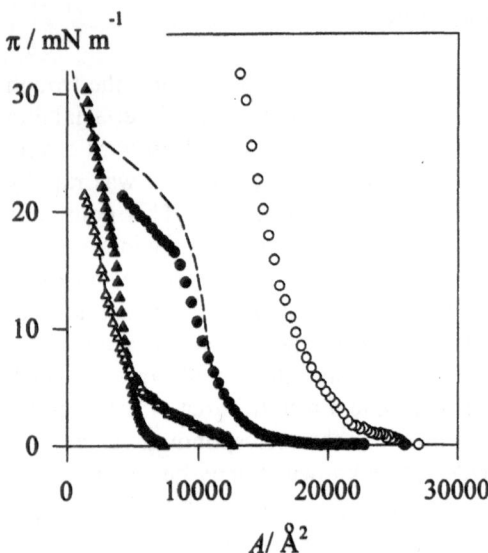

Fig. 2 Interfacial pressure (π) versus area per molecule (A) for BSA (circles) and β-L (triangles). Open symbols are for O–W films, filled symbols for A–W films. Dashed line = isotherm for A–W BSA film due MacRitchie (8)

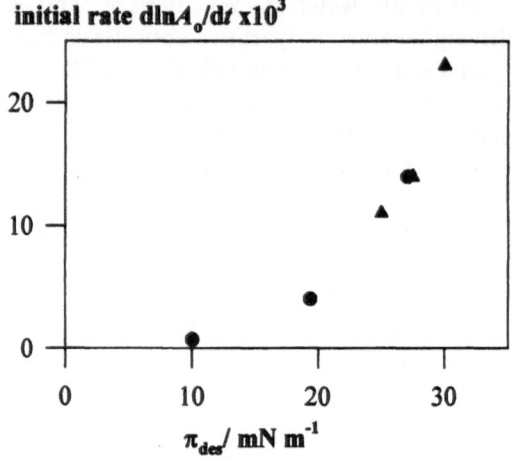

Fig. 3 Initial rate of relative area loss, dln A_0/dt, versus π_{des}: this work ($\bullet$); MacRitchie [11]. ($\blacktriangle$)

followed by immediate compression the same π–A curve was obtained. It was observed, however, that if an A–W protein film was left for 12 h or more and then re-compressed, the π–A curve appeared to be slightly more expanded (e.g. A was increased by ca. 10% at $\pi = 15\ \text{mN m}^{-1}$) on the first compression. But if this was immediately followed by a second and subsequent π–A experiments the original π–A curve (shown in Fig. 2) was recovered.

The results in Fig. 2 show that the π–A isotherms for both BSA and β-lactoglobulin are considerably more expanded at the O–W interface compared to the A–W interface, typically by a factor of about 1.7 in the low π (e.g. $\pi < 2\ \text{mN m}^{-1}$) region. For $\pi > 7\ \text{mN m}^{-1}$, however, β-lactoglobulin is apparently more compressible at the O–W interface, whilst for BSA the film is considerably more expanded up to the highest value of π measured. Interestingly, no ageing effects were observed for the O–W protein isotherms – leaving the films for 12 h or more did not result in an initially more expanded film on subsequent compression. The n-tetradecane that was used contained low levels of weakly surface-active impurities, but these were completely dominated by spread protein and had negligible effect on the O–W isotherms obtained.

Desorption from monolayers

It is useful to know the rate of desorption of protein from an interface because this may be important when films

undergo compression but also because the ease of desorption may provide clues as to the configuration of the protein already adsorbed at the interface. Figure 3 shows the initial rate of decrease in film area, A_0, i.e. the average of dln A_0/dt in the first 2 min, plotted as a function of π_{des}. Comparison with some earlier results obtained by MacRitchie [11] shows reasonable agreement. MacRitchie and co-workers [11–15] established that protein adsorption is indeed reversible, not irreversible as was often previously held, albeit that desorption from protein films can be very slow. For diffusion-controlled desorption the data should obey the equation [13]

$$\frac{dn}{dt} = C_b \left(\frac{D}{\pi}\right)^{1/2} t^{-1/2}, \tag{1}$$

where dn/dt is the rate of desorption, C_b is the bulk concentration of protein, D is the diffusion coefficient of the protein, $\pi = 3.14\ldots$ Thus, a plot of rate versus $t^{1/2}$ should be linear with a negative gradient $= -C_b(D/\pi)^{0.5}$. Figure 4 shows plots of desorption rate, dn/dt versus $t^{1/2}$ for the A–W and O–W interfaces at a range of π_{des} where desorption rates were significant. Values of dn/dt have been expressed as mg m^{-2} s^{-1}, calculated from the area per molecule corresponding to π_{des} and the area of the interface as a function of time. Within experimental error the plots do appear to be linear. These absolute desorption rates at the A–W interface are of the same order of magnitude to those obtained by MacRitchie [11–13] for a variety of other globular proteins. Even the highest desorption rates may be considered as still quite low. For example, typical equilibrium surface concentrations of adsorbed proteins are of the order of 2 mg m^{-2}. In this case

Progr Colloid Polym Sci (1997) 103:41–50
© Steinkopff Verlag 1997

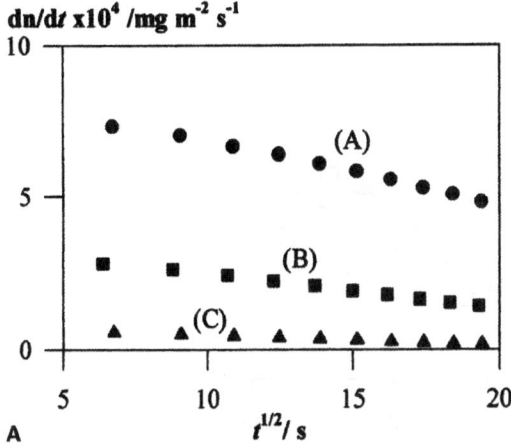

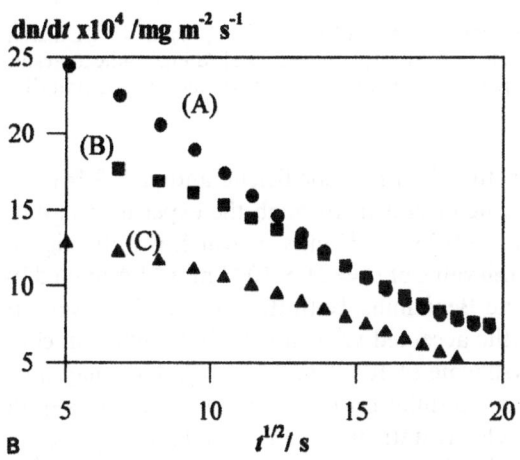

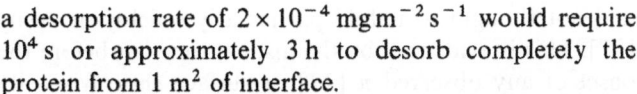

Fig. 4 A Desorption rate, dn/dt, versus time, t, for β-L A–W films for π_{des} of (A) 27.1, (B) 19.4 and (C) 10.1 mN m^{-1}. **B** Desorption rate, dn/dt, versus time, t, for β-L O–W films for π_{des} of (A) 21.0, (B) 17.4 and (C) 14.5 mN m^{-1}

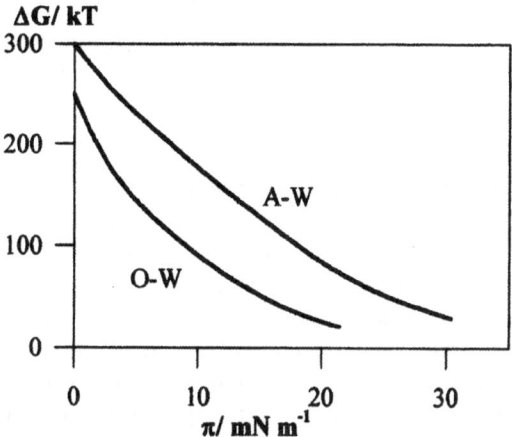

Fig. 5 Free energy barrier to desorption, ΔG, obtained via Eq. (2), versus interfacial pressure, π, for β-L A–W and O–W films

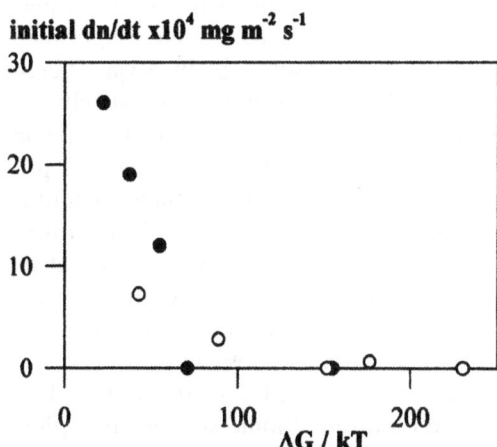

Fig. 6 Initial desorption rate, dn/dt, versus free energy barrier to desorption, ΔG, for: β-L O–W films (●) and β-L A–W films (O)

a desorption rate of 2×10^{-4} mg m^{-2} s^{-1} would require 10^4 s or approximately 3 h to desorb completely the protein from 1 m^2 of interface.

It is seen that desorption rates at the O–W interface are generally higher than at the A–W interface for the same value of π. However, this does not take into account the different energy barriers to desorption, ΔG, as a function of π for the two interfaces. ΔG at any particular π may be obtained by integrating the π–A curve from π_0 to π, i.e.

$$\Delta G = \int_{\pi_0}^{\pi} A\, d\pi, \qquad (2)$$

where π_0 is the interfacial pressure at which ΔG is effectively zero, i.e. protein molecules desorb spontaneously. The value of π_0 is therefore not accessible experimentally, because on approaching π_0 molecules are desorbing rapidly and the π–A curve in this region is not an equilibrium one. However, this may be estimated by extrapolation of

the existing to π–A curve to a very low value of A_0, say 10 Å^2 [14]. Using this procedure, the values of π_0 for the A–W and O–W interfaces are 38.9 and 29.1 mN m^{-1}, respectively. At these values of π, ΔG is therefore set equal to zero and Fig. 5 shows the variation of ΔG with π obtained using Eq. (2). Values of ΔG have been expressed in units of kT. It is seen that the energy barrier to desorption decrease rapidly with increasing π, but starts to level off as π approaches π_0. (For this reason a small error in the estimation of π_0 does not greatly affect the values of ΔG at low π.) It is seen that at values of π less than about 20 mN m^{-1} the energy barrier to desorption is considerably less at the O–W interface compared to the A–W interface. This explains why desorption is faster at the O–W interface at a given value of π and is a consequence of the different shapes of the π–A curves. Figure 6 shows the initial desorption rates plotted as a function of the corresponding ΔG. The dependence of the desorption rate on the energy barrier to desorption is similar for both types of interface.

Adsorption kinetic experiments – lysozyme

The above experiments on monolayers illustrate the strong dependence of desorption rates on π. In real systems stabilised by proteins, π for the film on average does not exceed a particular maximum value at which the rate of adsorption from solution is balanced by the rate of desorption. On perturbation from the equilibrium state of the film, such as a transient (local) expansion or compression a knowledge of both rates is important. Unfortunately, measurements of adsorption rates are not so straightforward since the surface concentration of protein, Γ, must be monitored with time and is not predetermined as in the spread monolayers. There is often disagreement between adsorption kinetic results obtained via different techniques – see below, for example. Relatively few measurements have been made of the adsorption kinetics of β-lactoglobulin at the A–W interface and for all proteins, because of experimental difficulties, there seem to be almost no direct measurements of $\Gamma(t)$ at O–W interfaces.

One protein which has been intensively studied at the A–W interface is lysozyme [16–18]. The results of several repeat experiments using the trough method to obtain $\Gamma(t)$ are shown in Fig. 7. Γ has been plotted against $t^{1/2}$. The bulk concentration of lysozyme, C_b, used was 1.5×10^{-4} wt% at pH 7 in $0.1\ \mathrm{mol\,dm^{-3}}$ ionic strength phosphate buffer. The value of π^*, used to obtain these results, as described in the Methods section above, was $3\ \mathrm{mN\,m^{-1}}$. Surface concentrations were calculated using a value of $A^* = 1870\ \mathring{\mathrm{A}}^2$ ($\Gamma^* = 1.25\ \mathrm{mg\,m^{-2}}$) corresponding to $\pi^* = 3\ \mathrm{mN\,m^{-1}}$, from the study due to de Feijter and Benjamins [17]. The results of these workers, who used ellipsometry to measure Γ, show that A was independent of C_b at $\pi = 3\ \mathrm{mN\,m^{-1}}$ over the concentration range $C_b = 10^{-4}$ to 10^{-1} wt%. It is seen that the results obtained here are quite similar to those of de Feijter and Benjamins, but quite different from those obtained by Xu and Damodaran [18], who used techniques employing radiolabelled proteins but under exactly the same solution conditions as in this study. De Feijter and Benjamins used slightly different solution conditions: $C_b = 1.0 \times 10^{-4}$ wt%, pH 6.7, ionic strength $0.02\ \mathrm{mol\,dm^{-3}}$. At low C_b and at short adsorption times the rate of adsorption should obey the equation [17]

$$\frac{\mathrm{d}\Gamma}{\mathrm{d}t} = C_b \left(\frac{D}{\pi t}\right)^{1/2}, \qquad (3)$$

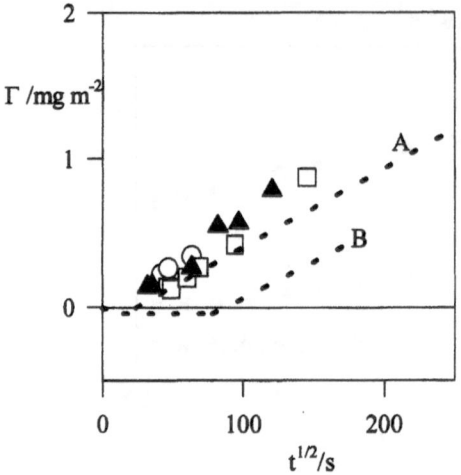

Fig. 7 Surface concentration of lysozyme, Γ, versus $t^{1/2}$. Different symbols indicate separate experiments. Dashed lines indicate results due to de Feijter and Benjamins (A) and Xu and Damodaran (B)

where D is the diffusion coefficient and $\pi = 3.141\ldots$. Taking the line of best fit through the experimental data gave $D = 1.5 \pm 0.2 \times 10^{-11}\ \mathrm{m^2\,s^{-1}}$, which is only slightly lower than the value of $D = 2.0 \times 10^{-1}\ \mathrm{m^2\,s^{-1}}$ obtained by de Feijter and Benjamins. Both these values, however, are lower than the accepted value for the bulk diffusion coefficient of lysozyme of $10^{-10}\ \mathrm{m^2\,s^{-1}}$ [18]. This may indicate a barrier to diffusion to the interface which is quite likely to be electrostatic in origin, the high charge on the lysozyme molecules at pH 7 at the interface preventing the adsorption of further molecules from the bulk. It has recently been recognised that such barriers play an important role in determining the adsorption dynamics of surface-active molecules [19, 20].

For lysozyme at pH 7 the molecule is quite compact and not thought to unfold particularly quickly or easily [21]. This is evidenced by the distinct time lag before the onset of any observed π [17] or surface shear viscosity [1, 22] on adsorption from low bulk concentration. This is because the protein film may be in a "gaseous monolayer state", with large areas of bare interface between isolated adsorbed protein molecules. At high protein surface concentrations/long adsorption times, however, the trough method will be invalid because surface concentrations are known [17] to continue to change after π has reached an equilibrium value, due to continued unfolding and re-arrangements in the interface.

Monolayer dilatation experiments

The above measurements on protein monolayers confirm that rates of protein desorption are quite low until high

π are reached. These π must be considerably higher than the equilibrium π reached by adsorption from solution under similar conditions. What is not clear is how such a film responds to a transient increase or decrease in π due to compression or expansion. The kinetics of this response must reflect the rates of rearrangement of the molecules in the film. The following results illustrate the dilatational response of A–W and O–W β-lactoglobulin monolayer films.

Figure 8 shows results for the expansion of two β-lactoglobulin A–W monolayers, expanded from initial π, π_i, of 26 and 13 mN m^{-1}. Immediately following expansion there is a rapid fall in π but after 2–3 s this is followed by a slower recovery of π, which finally reaches the equilibrium π corresponding to the new point on the π–A isotherm. At these π the desorption rates are negligible, so that the kinetics of π reflect the kinetics of rearrangements within the film and not the expulsion of protein from the film. The change in π, $\Delta\pi$, following expansion is defined as

$$\Delta\pi = (\pi_{eq} - \pi), \qquad (4)$$

where π_{eq} is the final, equilibrium value of π corresponding to the point on the π–A isotherm. Values of $\Delta\pi$ are plotted in Fig. 9, for the A–W monolayer where $\pi_i = 26$ mN m^{-1}

and an O–W monolayer where $\pi_i = 23$ mN m^{-1}. (Data in the region of the initial rapid fall in π after expansion has been ignored.) These values of π_i correspond closely to the equilibrium values of π for β-lactoglobulin adsorption at the A–W interface [23] and the O–W interface [24] of 25 and 22 mN m^{-1}, respectively.

In Fig. 10 the $\Delta\pi$ versus time data have been plotted in terms of the reduced variable R, R being defined by

$$R = \Delta\pi/\Delta\pi_{max}, \qquad (5)$$

where $\Delta\pi_{max}$ is the maximum value of $\Delta\pi$ observed. R versus time has then been fitted to a double exponential decay to characterise the decay kinetics. Table 1 shows the resultant values of the parameters, which give a reasonable fit to the data, as seen in the Fig. 10. The longer of the two relaxation times, τ_2, is seen to be approximately $3 \times$ larger for the O–W monolayer compared with the A–W monolayer.

An alternative method of analysing the data is to treat the monolayer as a mechanical entity. The interfacial stress, i.e. π versus time decay curve may be Fourier-transformed to obtain the dilatational elasticity, ε, and viscosity, κ, of the film as a function of frequency (ω) of deformation. For a step change in area, ΔA, which

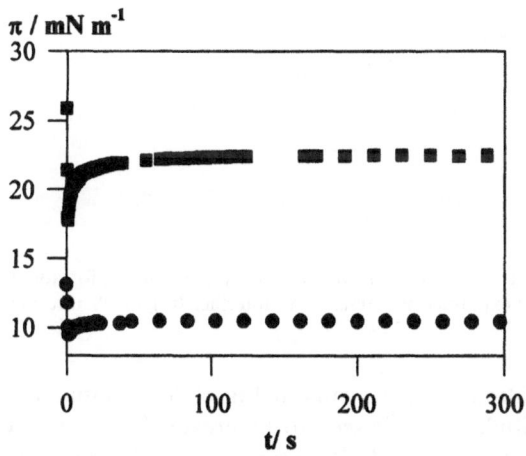

Fig. 8 Interfacial pressure, π, versus time, t, on expanding a β-L A–W film by 10%, of π_i of: 13.0 (●) and 26.0 mN m^{-1} (■)

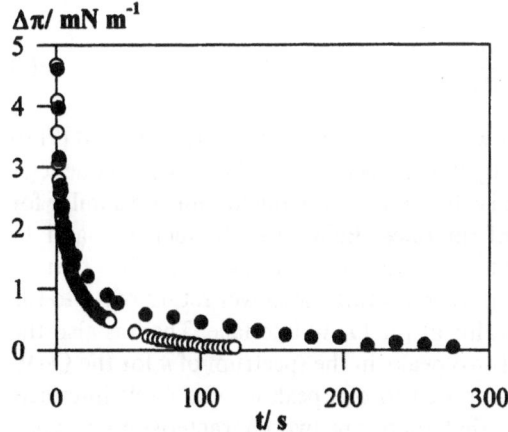

Fig. 9 Change in interfacial pressure, $\Delta\pi$, versus time, t, for β-L films expanded by 10% for: A–W film of $\pi_i = 26.0$ mN m^{-1} (O) and O–W film of $\pi_i = 23.0$ mN m^{-1} (●)

Table 1 Fit of monolayer relaxation to double exponential. Parameters used to fit R versus t data in Fig. 10 to a double exponential decay, i.e. $\Delta\pi = b_1 \exp(t/\tau_1) + b_2 \exp(t/\tau_2)$. SE = standard error, CV% = confidence value (%)

Parameter	O–W			A–W		
	Value	SE	CV%	Value	SE	CV%
$\tau 1$	02.03	0.17	8	02.05	0.07	4
$\tau 2$	64.60	5.80	9	24.50	5.70	2
$b1$	00.68	0.03	4	00.67	0.01	2
$b2$	00.39	0.01	3	00.41	0.01	2

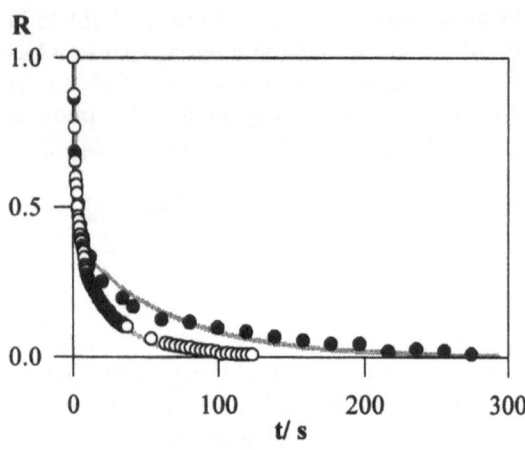

Fig. 10 Normalised change in interfacial pressure, R, versus time, t, for β-L films expanded by 10% for: A–W film of $\pi_i = 26.0$ mN m^{-1} (O) and O–W film of $\pi_i = 23.0$ mN m^{-1} (●). Light curves show best fit of double exponential decay (see text)

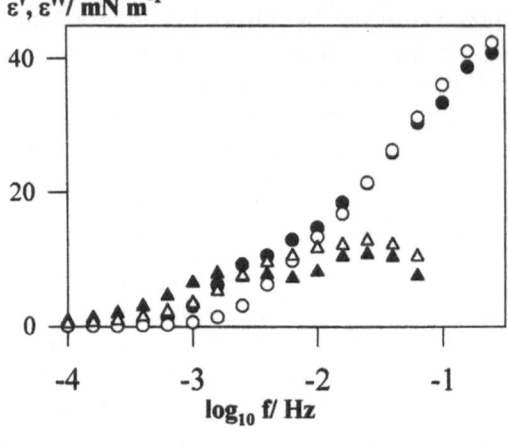

Fig. 11 Dilatational moduli, ε', (circles) and ε'', (triangles) versus frequency, f, obtained by Fourier transformation of data in Fig. 10. Open symbols = β-L A–W film, filled symbols = β-L O–W film

proceeds faster than the time scale (frequency) of the recovery considered [25, 26]:

$$\varepsilon' = \varepsilon = \frac{\omega}{(\Delta A/A)} \int_0^\infty \Delta\pi(t)\sin(\omega t)\,dt \qquad (6)$$

and

$$\varepsilon'' = \omega\kappa = \frac{\omega}{(\Delta A/A)} \int_0^\infty \Delta\pi(t)\cos(\omega t)\,dt. \qquad (7)$$

Figure 11 shows the results of applying Eqs. (6) and (7) to the data in Fig. 9. It is seen that at the higher frequencies (short times), i.e. for $>10^{-2}$ Hz, the moduli are similar for both types of interface. Below this frequency ε and to a lesser extent κ are greater at the O–W interface than at the A–W interface, reflecting the slower return of $\Delta\pi$ to the equilibrium value at the O–W interface. There is also the suggestion of two peaks in the spectrum of κ for the O–W interface, as opposed to one peak for the A–W interface. This signifies that there are two characteristic processes going on in the relaxation at the O–W interface.

Adsorbed films of β-lactoglobulin

Finally, similar measurements on the dilatational response of adsorbed films of a β-lactoglobulin have been made. The β-lactoglobulin solution used had a protein concentration of 10^{-3} wt%, made up in 0.02 mol dm^{-3} imidazole buffer adjusted to pH 7.0 with HCl (i.e. as for the monolayer experiments). Figure 12 shows $\Delta\pi$ versus t on expansion of the A–W and O–W interfaces by 10%. The data have been plotted on the same time scale as for the mono-

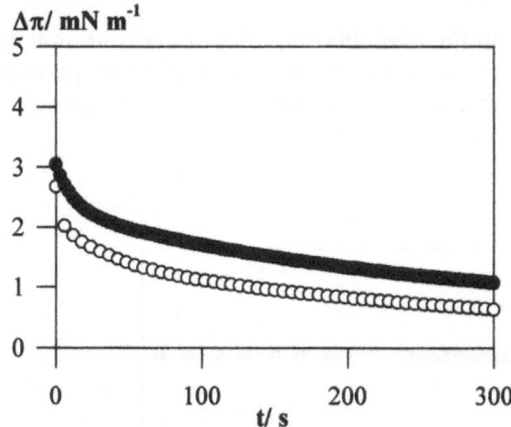

Fig. 12 Change in interfacial pressure, $\Delta\pi$, versus time, t, for adsorbed β-L films expanded by 10%: A–W interface (O), O–W interface (●)

layer dilatational experiments in Fig. 9, for comparison. The magnitude of $\Delta\pi$ is seen to be greater for the O–W interface than for the A–W interface and the decay to $\Delta\pi = 0$ is slower. However, the common feature for both interfaces is that for the adsorbed films it is seen that the recovery to the equilibrium π (i.e. $\Delta\pi = 0$) is much slower than for relaxation of the corresponding monolayers. In Fig. 13, $\Delta\pi$ for the adsorbed films has been subtracted from $\Delta\pi$ for the monolayer films to give $\Delta\pi_{soln}$. Examination of the π–A isotherms (Fig. 2) shows that the difference in the equilibrium π for the monolayer films before and after expansion by 10% is negligible for $\pi = 26$ mN m^{-1} (A–W monolayer) and 23 mN m^{-1} (O–W monolayer). Therefore, since π for the monolayer films was approximately the equilibrium π for the adsorbed films before

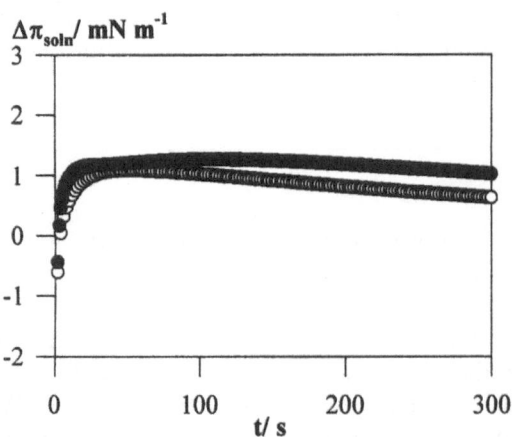

Fig. 13 $\Delta\pi_{\text{soln}}$, versus time, t, for β-L films expanded by 10%: A–W interface (O), O–W interface (●)

expansion, $\Delta\pi_{\text{soln}}$ should represent the contribution from the solution only to the relaxation of the adsorbed film. In Fig. 13 it is seen that the behaviour of $\Delta\pi_{\text{soln}}$ is quite similar for both the A–W and O–W interfaces.

Discussion

The more expanded nature of the π–A isotherms at the O–W interface may be explained by greater solvency of oil than "air" for the more hydrophobic side chains of the amino acid residues in the protein polypeptide chain. This may be expected to lead to a more unfolded state at the O–W interface. The greater compressibility of β-lactoglobulin at low A may also be a consequence of the same effect – because of the greater solvency of the oil for the hydrophobic parts of molecule the protein is more flexible in terms of the configurations which it can adopt at the O–W interface. In the A–W monolayer β-lactoglobulin is not as compressible and the ageing effects on the A–W films indicate that protein may be in a more aggregated, insoluble state – very slow unfolding perhaps continuing long after spreading.

If β-lactoglobulin is more unfolded at the O–W interface one might expect a lower rate of desorption than at the A–W interface. From Fig. 4 it can be seen that the opposite is true when the initial desorption rates are considered at similar values of π. However, when the rate of desorption is considered as a function of surface concentration, Γ (i.e. the inverse of A) desorption is slower for $\Gamma < 2\ \text{mg m}^{-2}$ ($A > 3050\ \text{Å}^2$). Above this Γ the rate is higher than that at the A–W interface. Perhaps the only useful interpretation is that shown in Fig. 6, where it is demonstrated that the dependence of the desorption rates on the energy barrier to desorption is similar for both

types of interface. Thus, the mechanism of desorption appears to be the same for both types of interface. More data are required, particularly for the A–W interface at higher values of π, to confirm any real differences between the two types of interface in terms of the ease of desorption. Clearly, desorption rates are quite low in the region of π typical of adsorbed films of β-lactoglobulin. Thus, for an equilibrium adsorbed film the rate of exchange of molecules between the bulk and the interface is low.

The adsorption experiments with lysozyme indicate that the trough method may be a way of obtaining the adsorption kinetics of other globular proteins, including β-lactoglobulin. This is necessary for the complete analysis of the response of an adsorbed film to a dilatational deformation. In order to use the method, however, reliable values of π^* and A^* area are required.

The dilatational experiments on the A–W and O–W monolayers of β-lactoglobulin seem to confirm the idea that the protein is more unfolded at the O–W interface. Whether or not the actual values of the decay times in Table 1 have any physical meaning is not clear at present, but the longer of the two relaxation times, τ_2, is seen to be approximately $3\times$ larger for the O–W monolayer compared with the A–W monolayer. This might be expected if greater unfolding of the protein occurs at the O–W interface, which may tend to hinder the rapid adoption of a new configuration required to fill gaps in the interfacial film after an expansion. On the other hand, the relaxation of $\Delta\pi$ back to zero for both the A–W and O–W monolayers is surprisingly fast, given that if often appears to take considerable time before proteins adopt their equilibrium configuration when adsorbing at interfaces [3, 4]. This suggests that apparently long equilibration times for adsorbed films may be more indicative of barriers to adsorption and/or diffusion to the interface, rather than barriers to conformational changes within the interface. Comparison with the final results on the adsorbed β-lactoglobulin films seems to confirm this view.

$\Delta\pi_{\text{soln}}$ should represent the contribution from the solution to the relaxation of the adsorbed film. This assumes that the state of the protein in the adsorbed film is the same as in the monolayer at the same value of π. Whilst it may be argued that this may not be the case [8], in Fig. 13 it is seen that the behaviour of $\Delta\pi_{\text{soln}}$ is quite similar for both the A–W and O–W interfaces. One might expect $\Delta\pi_{\text{soln}}$ to be decreasing at all values of t, since adsorption to the interface from the solution and the unfolding of newly adsorbed protein at the interface will decrease the interfacial tension back to its equilibrium value. However, initially, $\Delta\pi_{\text{soln}}$ is negative and increases. The negative values of $\Delta\pi_{\text{soln}}$ can only be explained by the fact that in the adsorbed film experiments $\Delta\pi$ changes too quickly in the first second or so to be measured properly, since the film expansion itself was arranged so as to take place over 1 s. Thereafter the increase in $\Delta\pi_{soln}$ in the first 100 s or so is a consequence of the fact that the in-film relaxation in the protein monolayers is more rapid than the in-film relaxation in the corresponding adsorbed films. This may be taken as evidence that the state of the protein is different in the monolayer and adsorbed films, even though π is the same. On the other hand, it is possible that the adsorption of more protein from the bulk inhibits the relaxation of protein molecules already present in the film. The work referred to earlier [17, 1, 22] shows that there may be a distinct time lag before globular proteins start to unfold after adsorption, even when π is close to zero. Unfolding against an already existing π (of around 20 mN m^{-1} here) is expected to be even more inhibited. Certainly it seems possible, for adsorbed β-lactoglobulin films under these conditions, that the major part of the relaxation of π after an expansion may be ascribed to rearrangements of the molecules already adsorbed, whilst adsorption and unfolding of additional molecules is only responsible for the long time-scale relaxation back to the equilibrium π. An improved device for monitoring the pull on the Wilhelmy plate should allow accurate measurements of π over time scales well below 1 s for more rapid film expansions than employed here and provide more useful information in this short time-scale region.

Conclusions

Using the variety of trough methods described here it is possible to monitor protein adsorption, desorption and rearrangement processes at both A–W and O–W interfaces. The results indicate that β-lactoglobulin is more unfolded at the O–W interface and, consequently, the dynamics of rearrangement are slower than at the A–W interface. For adsorbed films of β-lactoglobulin at both A–W and O–W interfaces subjected to a dilatational expansion, the contribution to the relaxation from the already adsorbed molecules appears to be dominant. Adsorption and unfolding of additional molecules contribute to the minor, long time-scale part of the film relaxation.

Acknowledgment Financial support from the Nuffield Foundation is gratefully acknowledged.

References

1. Murray BS, Dickinson E (in press) Food Sci Technol Int
2. Lucassen-Reynders EH (1993) Food Struct 12:1
3. Dickinson E, McClements DJ (1995) Advances in Food Colloids. Blackie Academic, Glasgow
4. Miller R, Kretzschmar G (1991) Adv Colloid Interface Sci 37:97
5. Dickinson E, Murray BS, Stainsby G (1988) In: Dickinson E, Stainsby G (eds) Advances in Food Emulsions and Foams. Elsevier, London, p 123
6. Murray BS, Nelson PV (1996) UK patent application 96 11936.7
7. a. Murray BS, Nelson PV (in press) Langmuir
   b. Murray BS (in press) Colloids Surf
8. MacRitchie F (1986) Adv Colloid Interface Sci 254:341
9. van Holde KE (1971) Physical Biochemistry. Prentice-Hall, Englewood Clifts
10. Fox PF (1982) Developments in Dairy Chemistry – 1 Proteins. Applied Science, London
11. MacRitchie F (1985) J Colloid Interface Sci 105:119
12. MacRitchie F (1993) Colloids Surf A 76:159
13. Gonzalez G, MacRitchie F (1970) J Colloid Interface Sci 32:55
14. MacRitchie FJ (1977) Colloid Interface Sci 61:223
15. MacRitchie F, Ter-Minassian-Saraga L (1984) Colloids Surf A 10:53
16. Graham DE, Phillips MC (1979) J Colloid Interface Sci 70:403, 415, 427
17. de Feijter JA, Benjamins J (1987) In: Dickinson E (ed) Food Emulsions and Foams. Royal Society of Chemistry, London, p 72
18. Xu SQ, Damodaran S (1994) Langmuir 10:472
19. Sharpe D, Eastoe J (1996) Langmuir 12:2303
20. Bonfillon A, Sicoli F, Langevin D (1994) J Colloid Interface Sci 168:497
21. Izmailova VN, Yampol'skaya GP, Lapina GP, Sorokin MM (1982) Colloid J USSR 44:195
22. Murray BS (1987) Ph D Thesis, University of Leeds
23. Thompson L (1985) M Sc Thesis, University of Leeds
24. Murray BS, Lallement C, Ventura A (in press)
25. Loglio G, Tesei U, Cini R (1984) J Colloid Interface Sci 100:393
26. Cárdenas-Valera AE, Bailey AI (1993) Colloids Surf A 79:115

Progr Colloid Polym Sci (1997) 103 : 51–59
© Steinkopff Verlag 1997

H. Diamant
D. Andelman

Adsorption kinetics of surfactants at fluid–fluid interfaces

Received: 7 November 1996
Accepted: 5 December 1996

H. Diamant · Prof. D. Andelman (✉)
School of Physics and Astronomy
Raymond and Beverly Sackler Faculty
of Exact Sciences
Tel Aviv University
Ramat Aviv
69978 Tel Aviv, Israel

Abstract We review a new theoretical approach to the kinetics of surfactant adsorption at fluid–fluid interfaces. It yields a more complete description of the kinetics both in the aqueous solution and at the interface, deriving all equations from a free-energy functional. It also provides a general method to calculate dynamic surface tensions. For non-ionic surfactants, the results coincide with previous models. Non-ionic surfactants are shown to usually undergo diffusion-limited adsorption, in agreement with the experiments. Strong electrostatic interactions in salt-free ionic surfactant solutions are found to lead to kinetically limited adsorption. In this case, the theory accounts for unusual experimental results which could not be understood using previous approaches. When salt is added, the electrostatic interactions are screened and the ionic surfactant adsorption becomes similar to the non-ionic case. The departure from the non-ionic behavior as the salt concentration is decreased is calculated perturbatively.

Key words Fluid–fluid interfaces – adsorption – adsorption kinetics – interfacial tension

Introduction

The kinetics of surfactant adsorption is a fundamental problem of interfacial science playing a key role in various processes and phenomena, such as wetting, foaming and stabilization of liquid films. Since the pioneering theoretical work of Ward and Tordai in the 1940s [1], it has been the object of thorough experimental and theoretical research [2].

The problem being a non-equilibrium one, a few theoretical questions immediately arise. One question concerns the kinetic adsorption mechanism to be employed by the model. One might assume a sort of an equilibrium adsorption isotherm to hold at the interface (e.g. as in refs. [3–5]), or, alternatively, use a full kinetic equation (e.g. [6–9]). Another important question relates to the definition and calculation of the time-dependent interfacial tension as measured in experiments.

Previous theoretical works have addressed these questions by adding appropriate assumptions to the theory. Such models can be roughly summarized by the following scheme: (i) consider a diffusive transport of surfactant molecules from a semi-infinite bulk solution (following Ward and Tordai); (ii) introduce a certain adsorption equation as a boundary condition at the interface; (iii) solve for the time-dependent surface coverage; (iv) assume that the equilibrium equation of state is valid also out of equilibrium and calculate the dynamic surface tension [10].

In the current paper, we would like to review an alternative approach based on a free-energy formalism [11, 12]. The main advantage is that all the equations are derived from a single functional, thus yielding a more complete and consistent description of the kinetics in the entire system. Results of previous models can be recovered as special cases, and one can check the conditions under which such cases hold. The definition and calculation of the dynamic surface tension results naturally from the formalism itself, and extension to more complicated interactions can follow.

We restrict ourselves in the current paper to a simple, yet rather general case. A sharp, flat interface is assumed to separate an aqueous surfactant solution from another fluid, non-polar phase. The solution is assumed to be below the critical micelle concentration, i.e., it contains only monomers. We start in Section 2 by considering the adsorption of non-ionic surfactants, for which previous theories yield satisfactory results. We then proceed in Section 3 to discuss salt-free ionic surfactant solutions, where strong electrostatic interactions exist and interesting time dependence has been observed in experiments [13]. In Section 4 the effect of added salt to ionic surfactant solutions is examined.

We shall not describe various experimental techniques which have been devised in the context of adsorption kinetics of surfactants. Such information can be found in ref. [2] and in the contribution by A. Pitt included in this volume.

Non-ionic surfactants

We identify the measurable change in interfacial tension, $\Delta\gamma$, with the excess in free energy per unit area due to the adsorption at the interface. This definition is assumed to hold both at equilibrium and out of equilibrium. The free energy excess can be written as a functional of the volume fraction profile of the surfactant, $\phi(x, t)$, x being the distance from the interface and t the time,

$$\Delta\gamma[\phi] = \int\limits_0^\infty \Delta f[\phi(x, t)]\, \mathrm{d}x , \tag{1}$$

where Δf is the local excess in free-energy density over the bulk, uniform solution.

We take the bulk solution to be dilute and assume a contact with a reservoir, where the surfactant has fixed volume fraction and chemical potential, ϕ_b and μ_b, respectively. Steric and other short-range interactions between surfactant molecules are assumed to take place only within a molecular distance from the interface. This is motivated by the observation that the profile of a soluble surfactant monolayer is in practice almost "step-like", the volume fraction at the interface itself being many orders of magnitude larger than that in the solution.

Hence, we write the local free-energy density as

$$\begin{aligned}\Delta f = &\{T[\phi(\ln\phi - 1) - \phi_\mathrm{b}(\ln\phi_\mathrm{b} - 1)] - \mu_\mathrm{b}(\phi - \phi_\mathrm{b})\}/a^3 \\ &+ \{T[\phi\ln\phi + (1 - \phi)\ln(1 - \phi)] \\ &- (\alpha + \mu_1)\phi - (\beta/2)\phi^2\}\delta(x)/a^2, \end{aligned} \tag{2}$$

where a denotes the surfactant molecular dimension and T the temperature (taking the Boltzmann constant as 1). Note that this functional divides the system into two distinct, coupled subsystems – the bulk solution and the interface [14]. As a result, we shall obtain distinct equilibrium and kinetic equations for these two subsystems. The contribution from the bulk contains only the ideal entropy of mixing in the dilute solution limit and contact with the reservoir. In the interfacial contribution, we have included the entropy of mixing accounting for the finite molecular size, a linear term accounting for the surface activity and contact with the adjacent solution [$\mu_1 \equiv \mu(x \to 0)$ being the chemical potential at the adjacent layer], and a quadratic term describing short-range lateral attraction between surfactant molecules at the interface. The surface activity parameter, α, is typically of order $10T$, and the lateral attraction parameter, β, is typically a few T.

Although the functional (2) has a simple form, it yields physically non-trivial results. More complicated cases, e.g., certain surfactants whose adsorption seems to be hindered by a potential barrier, may require additional terms. Such terms, however, can be easily incorporated, as demonstrated in the next section for electrostatic interactions.

Equilibrium relations are readily obtained by setting the variation of the free energy with respect to $\phi(x)$ to zero,

$$\frac{\delta\Delta\gamma}{\delta\phi(x)} = 0, \quad \text{equilibrium} .$$

This yields in the current simple case a uniform profile in the bulk, $\phi(x > 0) \equiv \phi_\mathrm{b}$, and recovers the Frumkin adsorption isotherm (or the Langmuir one, if $\beta = 0$) [15] at the interface,

$$\phi_0 = \frac{\phi_\mathrm{b}}{\phi_\mathrm{b} + \mathrm{e}^{-(\alpha + \beta\phi_0)/T}} , \tag{3}$$

where $\phi_0 \equiv \phi(x = 0)$ denotes the surface coverage. Substituting these results in the free-energy functional recovers also the equilibrium equation of state,

$$\Delta\gamma = [T\ln(1 - \phi_0) + (\beta/2)\phi_0^2]/a^2 . \tag{4}$$

Kinetic equations can also be derived from the variation of the free energy. The conventional scheme in the case of a conserved order parameter is [16]

$$\frac{\partial\phi}{\partial t} = (a^3 D/T)\frac{\partial}{\partial x}\left[\phi\frac{\partial}{\partial x}\left(\frac{\delta\Delta\gamma}{\delta\phi}\right)\right],$$

where D is the surfactant diffusion coefficient. This leads to an ordinary diffusion equation in the bulk,

$$\frac{\partial \phi}{\partial t} = D \frac{\partial^2 \phi}{\partial x^2} , \tag{5}$$

and to a conservation condition at the layer adjacent to the interface,

$$\frac{\partial \phi_1}{\partial t} = (D/a) \left. \frac{\partial \phi}{\partial x} \right|_{x=a} - \frac{\partial \phi_0}{\partial t} , \tag{6}$$

where $\phi_1 \equiv \phi(x \to 0)$ is the local volume fraction. Finally, at the interface itself, we get

$$\frac{\partial \phi_0}{\partial t} = (D/a^2) \phi_1 \left[\ln \frac{\phi_1(1 - \phi_0)}{\phi_0} + \frac{\alpha}{T} + \frac{\beta \phi_0}{T} \right] . \tag{7}$$

We have assumed, for simplicity, that the surfactant diffusion coefficient D is the same in the bulk and near the interface in spite of the different environments. In reality this should not be strictly accurate.

Our formalism has led to a diffusive transport in the bulk [Eqs. (5) and (6)] coupled to an adsorption mechanism at the interface [Eq. (7)]. Yet unlike previous models, all of the equations have been derived from a single functional, and hence, various assumptions employed by previous works can be examined. Treating Eqs. (5) and (6) using the Laplace transform with respect to time, we obtain a relation similar to the Ward and Tordai result [1],

$$\phi_0(t) = (\sqrt{D/\pi}/a) \left[2\phi_b \sqrt{t} - \int_0^t \frac{\phi_1(\tau)}{\sqrt{t - \tau}} d\tau \right]$$
$$+ 2\phi_b - \phi_1 , \tag{8}$$

with a small difference coming from the finite thickness we have assigned to the subsurface layer of solution (vanishing for $a \to 0$).

The diffusive transport from the bulk solution [Eq. (8)] relaxes like

$$\phi_1(t)/\phi_b \simeq 1 - \sqrt{\tau_d/t} \qquad t \to \infty$$
$$\tau_d \equiv (a^2/\pi D)(\phi_{0,eq}/\phi_b)^2 , \tag{9}$$

where $\phi_{0,eq}$ denotes the equilibrium surface coverage. The molecular diffusion time scale, a^2/D, is of order 10^{-9} s, but the factor $\phi_{0,eq}/\phi_b$ in surfactant monolayers is very large (typically 10^5–10^6), so the diffusive transport to the interface may require minutes. The kinetic process at the interface [Eq. (7)] relaxes like

$$\phi_0(t)/\phi_{0,eq} \simeq 1 - e^{-t/\tau_k} , \qquad t \to \infty$$
$$\tau_k \equiv (a^2/D)(\phi_{0,eq}/\phi_b)^2 e^{-(\alpha + \beta \phi_{0,eq})/T} . \tag{10}$$

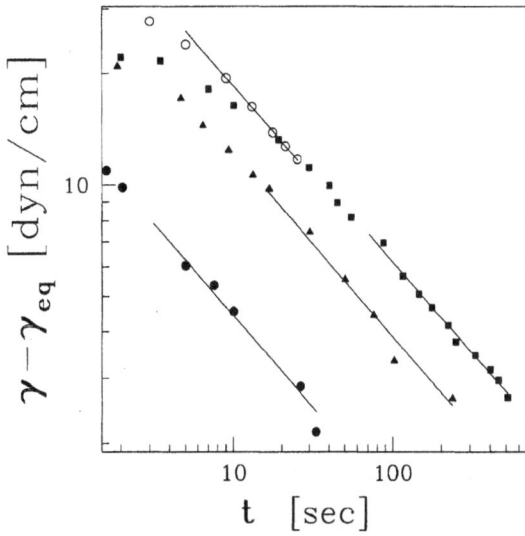

Fig. 1 Diffusion-limited adsorption exhibited by non-ionic surfactants. Four examples for dynamic surface tension measurements are shown: decyl alcohol at concentration 9.49×10^{-5} M (open circles) adapted from ref. [17]; Triton X-100 at concentration 2.32×10^{-5} M (squares) adapted from ref. [8]; $C_{12}EO_8$ at concentration 6×10^{-5} M (triangles) and $C_{10}PY$ at concentration 4.35×10^{-4} M (solid circles), both adapted from ref. [18]. The asymptotic $t^{-1/2}$ dependence shown by the solid fitting lines is a "footprint" of diffusion-limited adsorption

Since α for common surfactants is of order $10T$, we expect τ_k to be much smaller than τ_d. In other words, the adsorption of many non-ionic surfactants, not hindered by any high potential barrier, is expected to be *diffusion-limited*. The asymptotic time dependence (9) yields a distinct "footprint" for diffusion-limited adsorption, as demonstrated in Fig. 1.

In mathematical terms, the adsorption being diffusion-limited means that the variation of the free energy with respect to ϕ_0 can be taken to zero at all times whereas the variation with respect to $\phi(x > 0)$ cannot. This has two consequences. The first is that the relation between ϕ_0 and ϕ_1 is given at all times by the equilibrium adsorption isotherm [(3) in our model]. The solution of the adsorption problem in the non-ionic, diffusion-limited case amounts, therefore, to the simultaneous solution of the Ward–Tordai equation (8) and the adsorption isotherm. Exact analytical solution exists only for the simplest, linear isotherm, $\phi_0 \propto \phi_1$ [19]. For more realistic isotherms such as (3), one has to resort to numerical techniques (useful numerical schemes can be found in refs. [2, 8]). The second consequence of the vanishing of $\delta \Delta \gamma/\delta \phi_0$ is that the dynamic surface tension, $\Delta \gamma(t)$, approximately obeys the equilibrium equation of state (4). These two consequences show that the validity of the schemes employed by previous theories is essentially restricted to diffusion-limited cases.

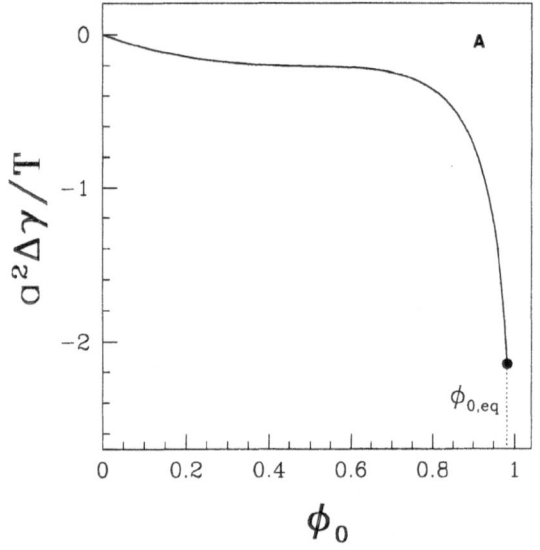

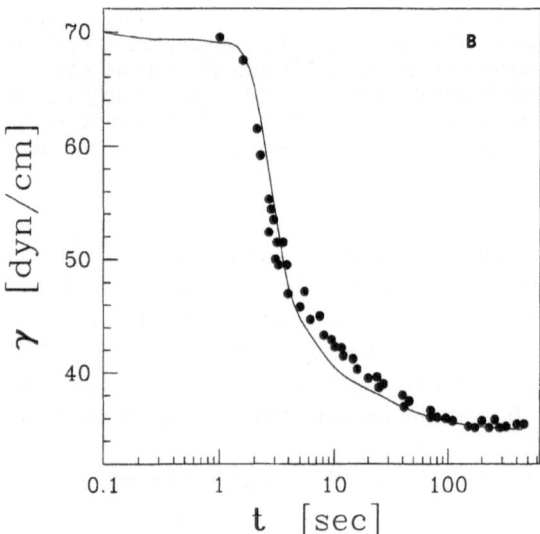

Fig. 2A Dependence between surface tension and surface coverage in diffusion-limited adsorption [Eq. (4)]. The values taken for the parameters match the example in (b). **B** Typical dynamic surface tension curve in diffusion-limited adsorption (reproduced from ref. [20]). The solution contains 1.586×10^{-4} M decanol. The solid line is a theoretical fit using the following parameters: $a = 4.86$ Å, $\alpha = 11.6T$, $\beta = 3.90T$ (all three fitted from independent equilibrium measurements), and $D = 6.75 \times 10^{-6}$ cm²/s

The dependence defined by the equilibrium equation of state (4) is depicted in Fig. 2A. As a result of the competition between the entropy and interaction terms in Eq. (4) the surface tension changes very little for small surface coverages. As the coverage increases beyond about $1 - (\beta/T)^{-1/2}$, the surface tension starts decreasing until reaching equilibrium. This qualitatively explains the shape of dynamic surface tension curves found in experiments for non-ionic surfactants (e.g. [8, 20]). We have reproduced in

Fig. 2B, one such curve published by Lin et al. [8]. The theoretical solid curve was obtained by these authors using a scheme similar to the one just described – solution of the Ward–Tordai equation together with the Frumkin isotherm and substitution in the equation of state to calculate the surface tension. Note that the parameters α, β and a can be fitted from independent equilibrium measurements, so the dynamic surface tension curve has only one fitting parameter, namely the diffusion coefficient, D. As can be seen, the agreement with experiment is quite satisfactory. However, when the adsorption is not diffusion-limited, such a theoretical approach is no longer applicable, as will be demonstrated in the next section.

Salt-free ionic surfactant solutions

We turn to the more complicated but important problem of ionic surfactant adsorption, and start with the salt-free case where strong electrostatic interactions are present. In Fig. 3 we have reproduced experimental results published by Bonfillon-Colin et al. for SDS solutions with (open circles) and without (full circles) added salt [13]. The salt-free ionic case exhibits a much longer process with a peculiar intermediate plateau. Similar results were presented by Hua and Rosen for DESS solutions [21]. A few theoretical models were suggested for the problem of ionic surfactant adsorption [22–24], yet none of them could produce such dynamic surface tension curves. It is also rather clear that a theoretical scheme such as the one discussed in the previous section cannot fit these experimental results. On the other hand, addition of salt to the solution screens the electrostatic interactions and leads to a behavior very similar to the non-ionic one. We shall return to this issue in Section 4. We thus infer that strong electrostatic interactions affect drastically the adsorption kinetics. Let us now study this effect in more detail. We follow the same lines presented in the previous section while adding appropriate terms to account for the additional interactions.

Our free-energy functional in the salt-free ionic case is divided into three contributions: a contribution from the surfactant, one from the counterions and one from the electrostatic field. It depends on three degrees of freedom: the surfactant profile, $\phi^+(x, t)$ (we take the surfactant ion to be the positive one), the counterion profile, $\phi^-(x, t)$, and an electric potential, $\psi(x, t)$.

$$\Delta\gamma[\phi^+, \phi^-, \psi] = \int_0^\infty [\Delta f^+(\phi^+) + \Delta f^-(\phi^-)$$

$$+ f_{el}(\phi^+, \phi^-, \psi)]\, dx \ . \qquad (11)$$

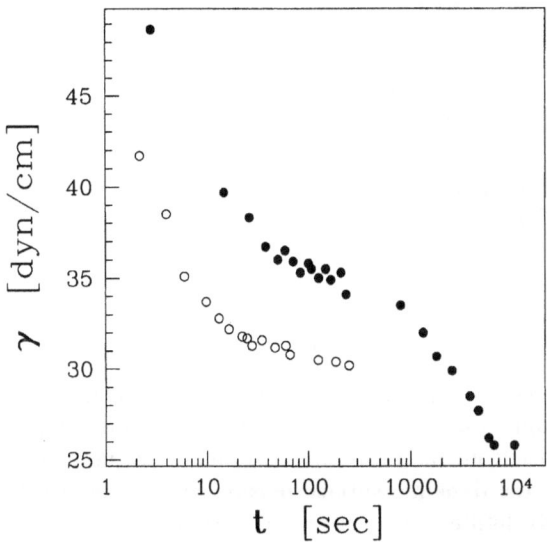

Fig. 3 Dynamic interfacial tension between SDS aqueous solution and dodecane, adapted from ref. [13]: 3.5×10^{-4} M SDS without salt (filled circles); 4.86×10^{-5} M SDS with 0.1 M NaCl (open circles)

The surfactant contribution, Δf^+, is identical to that of the non-ionic case [Eq. (2)]. In the counterion contribution, Δf^-, we include only the bulk part of Eq. (2), taking the counterions at this stage to be completely surface-inactive. The electrostatic contribution contains interactions between the ions and the electric field and the energy stored in the field itself,

$$f_{\mathrm{el}} = e\left(\frac{\phi^+}{(a^+)^3} - \frac{\phi^-}{(a^-)^3}\right)\psi - \frac{\varepsilon}{8\pi}\left(\frac{\partial \psi}{\partial x}\right)^2 + \frac{e}{(a^+)^2}\,\phi^+\psi\delta(x)\,,\tag{12}$$

where $a^\pm$ are the molecular sizes of the two ions, e the electronic charge and ε the dielectric constant of water. For simplicity, we have restricted ourselves to fully ionized, monovalent ions [which implies that $\phi_{\mathrm{b}}^+/(a^+)^3 = \phi_{\mathrm{b}}^-/(a^-)^3 = c_{\mathrm{b}}$, the bulk concentration].

Ions in solution, apart from interacting with each other, also feel repulsion from the interface due to "image-charge" effects, as discussed by Onsager and Samaras [25]. It can be shown, however, that these effects become negligible as soon as the surface coverage exceeds about 2% [12].

Equilibrium equations are readily obtained, as in the previous section, by setting the variation of the free energy with respect to the various degrees of freedom to zero,

$$\frac{\delta \Delta \gamma}{\delta \phi^\pm(x)} = \frac{\delta \Delta \gamma}{\delta \psi(x)} = 0, \quad \text{equilibrium}\,.$$

These equations yield the Boltzmann ion profiles,

$$\phi^\pm(x > 0) = \phi_{\mathrm{b}}^\pm\, e^{\mp\, e\psi(x)/T}\,,$$

the Poisson equation,

$$\frac{\partial^2 \psi}{\partial x^2} = -\frac{4\pi e}{\varepsilon}\left(\frac{\phi^+}{(a^+)^3} - \frac{\phi^-}{(a^-)^3}\right),\tag{13}$$

the electrostatic boundary condition,

$$\left.\frac{\partial \psi}{\partial x}\right|_{x=0} = -\frac{4\pi e}{\varepsilon(a^+)^2}\,\phi_0^+\,,\tag{14}$$

and, finally, recovers the Davies adsorption isotherm [26],

$$\phi_0^+ = \frac{\phi_{\mathrm{b}}^+}{\phi_{\mathrm{b}}^+ + e^{-(\alpha + \beta\phi_0^+ - e\psi_0)/T}}\,.\tag{15}$$

Combining Eq. (13) with the Boltzmann profiles leads to the Poisson–Boltzmann equation,

$$\frac{\partial^2 \psi}{\partial x^2} = \frac{8\pi e c_{\mathrm{b}}}{\varepsilon}\sinh\frac{e\psi}{T}\tag{16}$$

for the equilibrium double-layer potential [27, 28]. By means of the Poisson–Boltzmann equation, the Davies isotherm can be expressed as

$$\phi_0^+ = \frac{\phi_{\mathrm{b}}^+}{\phi_{\mathrm{b}}^+ + [b\phi_0^+ + \sqrt{(b\phi_0^+)^2 + 1}]^2 e^{-(\alpha + \beta\phi_0^+)/T}}\,,\tag{17}$$

where $b \equiv a^+/(4\phi_{\mathrm{b}}^+ \lambda)$, and $\lambda \equiv (8\pi c_{\mathrm{b}} e^2/\varepsilon T)^{-1/2}$ is the De-bye–Hückel screening length [29]. Similar to Section 2, one can calculate the equilibrium equation of state,

$$\Delta\gamma = [T\ln(1 - \phi_0^+) + (\beta/2)(\phi_0^+)^2$$
$$- (2T/b)(\sqrt{(b\phi_0^+)^2 + 1} - 1)]/(a^+)^2\,.\tag{18}$$

For weak fields the electrostatic correction to the equation of state is quadratic in the coverage, thus merely modifying the lateral interaction term, and for strong fields it becomes linear in the coverage.

Kinetic equations are derived using the same scheme as before,

$$\frac{\partial \phi^\pm}{\partial t} = \frac{(a^\pm)^3 D^\pm}{T}\frac{\partial}{\partial x}\left[\phi^\pm \frac{\partial}{\partial x}\left(\frac{\delta \Delta\gamma}{\delta \phi^\pm}\right)\right],$$

where $D^\pm$ are the diffusion coefficients of the two ions. This variational scheme yields in the bulk solution the Smoluchowski diffusion equations,

$$\frac{\partial \phi^\pm}{\partial t} = D^\pm \frac{\partial}{\partial x}\left(\frac{\partial \phi^\pm}{\partial x} \pm \frac{e}{T}\phi^\pm \frac{\partial \psi}{\partial x}\right),\tag{19}$$

at the layer adjacent to the interface

$$\frac{\partial \phi_1^\pm}{\partial t} = \frac{D^\pm}{a^\pm}\left(\left.\frac{\partial \phi^\pm}{\partial x}\right|_{x=a^\pm} \pm \frac{e}{T}\phi_1^\pm \left.\frac{\partial \psi}{\partial x}\right|_{x=a^\pm}\right) - \frac{\partial \phi_0^\pm}{\partial t}\,,\tag{20}$$

and finally, at the interface itself

$$\frac{\partial \phi_0^+}{\partial t} = \frac{D^+}{(a^+)^2}\, \phi_1^+ \left[\ln \frac{\phi_1^+(1-\phi_0^+)}{\phi_0^+} + \frac{\alpha}{T} \right.$$

$$\left. + \left(\frac{\beta \phi_0^+}{T} - \frac{4\pi l}{a^+} \right) \phi_0^+ \right]. \tag{21}$$

We have made use of the electrostatic boundary condition (14) in order to replace an electrostatic barrier term, $e(\psi_0 - \psi_1)/T$, with the approximate term $(4\pi l/a^+)\phi_0^+$, where $l \equiv e^2/\varepsilon T$ is the Bjerrum length (about 7 Å for water at room temperature).

We neglect electrodynamic effects, so the Poisson equation continues to hold. The kinetic equations just derived, along with the Poisson equation and the necessary boundary and initial conditions, can be solved numerically (a similar set of equations is solved in ref. [24]).

The relaxation in the bulk solution, accounted for by the Smoluchowski equations (19), has the time scale

$$\tau_e = \lambda^2/D\,,$$

where D is an effective ambipolar diffusion coefficient. This time scale is typically very short (microseconds), i.e. the bulk relaxation is by orders of magnitude faster than in the non-ionic case. The relaxation at the interface [Eq. (21)], by contrast, is slowed down by the electrostatic repulsion, and has a time scale of

$$\tau_k = \tau_k^{(0)} \exp[e(\psi_0 + \psi_1)/T]$$

$$\simeq \tau_k^{(0)} [(a^+/2\lambda)(\phi_{0,\,eq}^+/\phi_b^+)]^4 \exp[-(4\pi l/a^+)\phi_{0,\,eq}^+]\,,$$

where $\tau_k^{(0)}$ denotes the kinetic time scale in the absence of electrostatics [Eq. (10)]. In salt-free surfactant solutions the surface potential reaches values significantly larger than T/e, and hence, the interfacial relaxation is by several orders of magnitude slower than in the non-ionic case.

This analysis leads us to the conclusion that ionic surfactants in salt-free solutions undergo *kinetically limited adsorption*. Indeed, dynamic surface tension curves of such solutions do not exhibit the diffusive asymptotic time dependence of non-ionic surfactants, depicted in Fig. 1. The scheme of Section 2, focusing on the diffusive transport inside the solution, is no longer valid. Instead, the diffusive relaxation in the bulk solution is practically immediate and we should concentrate on the interfacial kinetics, Eq. (21). In this case the subsurface volume fraction, ϕ_1^+, obeys the Boltzmann distribution, not the Davies adsorption isotherm (15), and the electric potential is given by the Poisson–Boltzmann theory. By these observations Eq. (21) can be expressed as a function of the surface

coverage alone,

$$\frac{\partial \phi_0^+}{\partial t} = \left(\frac{D^+ \phi_b^+}{(a^+)^2} \right) \frac{\exp[(4\pi l/a^+)\phi_0^+]}{[b\phi_0^+ + \sqrt{(b\phi_0^+)^2 + 1}]^2}$$

$$\times \left\{ \ln \left[\frac{\phi_b^+(1 - \phi_0^+)}{\phi_0^+} \right] + \frac{\alpha}{T} + \frac{\beta \phi_0^+}{T} \right.$$

$$\left. - 2 \sinh^{-1}(b\phi_0^+) \right\}, \tag{22}$$

thus reducing the problem to a single integration.

Not only does the scheme for solving the kinetic equations differ from the non-ionic case, but also the way to calculate the dynamic surface tension has to change. In kinetically limited adsorption the variation of the free energy with respect to the surface coverage does not vanish, and, therefore, the equation of state (18) is strictly invalid out of equilibrium. We derive the expression for the dynamic surface tension in the kinetically limited case from the general functional (11) by assuming quasi-equilibrium inside the bulk solution (i.e. using Boltzmann profiles and the Poisson–Boltzmann equation). This gives

$$\Delta\gamma[\phi_0^+(t)] = \{T[\phi_0^+ \ln(\phi_0^+/\phi_b^+) + (1 - \phi_0^+)\ln(1 - \phi_0^+)]$$

$$- \alpha\phi_0^+ - (\beta/2)(\phi_0^+)^2 + 2T[\phi_0^+ \sinh^{-1}(b\phi_0^+)$$

$$- (\sqrt{(b\phi_0^+)^2 + 1} - 1)/b]\}/(a^+)^2\,. \tag{23}$$

Assuming high surface potentials ($b\phi_0^+ \gg 1$), the function defined in Eq. (23) becomes non-convex for $\beta/T > 2(2 + \sqrt{3}) \simeq 7.5$, as demonstrated in Fig. 4. If that is indeed the case, our model predicts an unusual time dependence for the dynamic surface tension, as observed in experiments (Fig. 3). We thus infer that the shape of the experimental dynamic surface tension curves is a consequence of a kinetically limited adsorption brought about by strong electrostatic interactions. Physically, the non-convexity results from a competition between short- and long-range interactions. It suggests the following scenario: As the surface coverage increases, the system reaches a local free-energy minimum leading to a pause in the adsorption (the intermediate plateau of the experimental curves). This metastable state lasts until domains of the denser, global-minimum phase are nucleated, resulting in further increase in coverage and decrease in surface tension. A complete, quantitative treatment of such a scenario cannot be presented within this framework, since our current formalism inevitably leads to a monotonically decreasing free energy as a function of time, and hence, cannot account for nucleation [16].

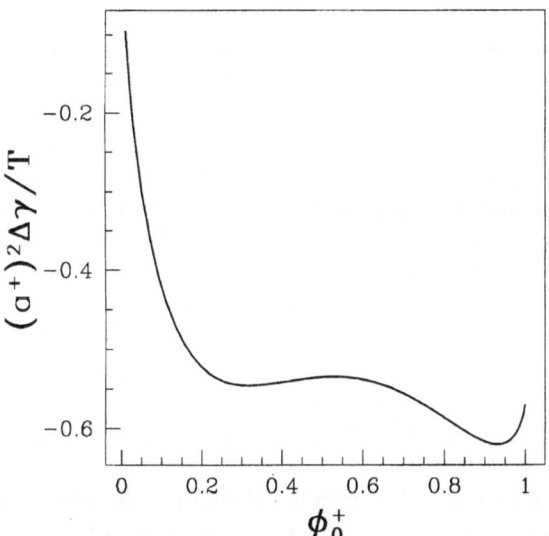

Fig. 4 Dependence between surface tension and surface coverage in kinetically limited adsorption [Eq. (23)]. The values taken for the parameters are: $a^+ = 8.1$ Å, $\phi_b^+ = 1.1 \times 10^{-4}$, $\alpha = 14T$ and $\beta = 9T$. Such a curve should lead to the qualitative time dependence found in certain salt-free cases (see Fig. 3)

A value of $\beta > 7.5T$ is somewhat large for the lateral attraction between surfactant molecules. Experimental estimation of this parameter for common non-ionic surfactants yields a few T [20]. Throughout the above calculations we have assumed, to a sort of a zeroth approximation, that no counterions are adsorbed at the interface. It can be shown that the presence of a small amount of counterions at the interface introduces a correction to the free energy which is, to a first approximation, quadratic in the surfactant coverage, i.e. leading to an effective increase in lateral attraction [12]. The addition to β due to the counterions turns out to be $[2\pi la^-/(a^+)^2]T$, which may amount to a few T. This addition accounts for the larger value of β required for non-convexity.

Ionic surfactants with added salt

Finally, we consider the effect of adding salt to an ionic surfactant solution. For simplicity, and in accordance with practical conditions, we assume that the salt ions are much more mobile than the surfactant and their concentration exceeds that of the surfactant. In addition, we take the salt ions to be monovalent and surface inactive.

Under these assumptions, we can neglect the kinetics of the salt ions and reduce their role to the formation of a thin electric double layer near the interface, maintaining quasi-equilibrium with the adsorbed surface charge. We take the

double-layer potential to be in the linear, Debye–Hückel regime [28, 29],

$$\psi(x, t) = \frac{4\pi e\lambda}{\varepsilon a^2} \phi_0(t) e^{-x/\lambda} ,$$

with a modified definition of the Debye–Hückel screening length, $\lambda \equiv (8\pi c_s l)^{-1/2}$, $c_s \gg c_b$ being the salt concentration (the superscript $+$ is omitted from the surfactant symbols in this section).

Substituting this double-layer potential in Eqs. (19) and (20), we obtain the kinetic equations in the bulk and at the layer adjacent to the interface,

$$\frac{\partial \phi}{\partial t} = D \frac{\partial}{\partial x} \left(\frac{\partial \phi}{\partial x} - \frac{\phi_0 e^{-x/\lambda}}{2a^2\lambda^2 c_s} \phi \right), \tag{24}$$

$$\frac{\partial \phi_1}{\partial t} = (D/a) \left(\frac{\partial \phi}{\partial x}\bigg|_{x=a} - \frac{\phi_0}{2a^2\lambda^2 c_s} \phi_1 \right) - \frac{\partial \phi_0}{\partial t} . \tag{25}$$

The kinetic equation at the interface itself remains the same as (21).

Considering the electric potential as a small perturbation, Eqs. (24) and (25) lead to the asymptotic expression

$$\phi_1(t)/\phi_b \simeq 1 - \phi_{0,eq}/(2a^2\lambda c_s) - \sqrt{\tau_d/t}, \quad t \to \infty$$

$$\tau_d \equiv \tau_d^{(0)} \left[1 - \frac{c_b}{2c_s} - \frac{\phi_{0,eq}}{2a^2\lambda c_s} \left(1 - \frac{3c_b}{2c_s} \right) \right]^2 , \tag{26}$$

where $\tau_d^{(0)}$ denotes the diffusion time scale in the non-ionic case [Eq. (9)]. As expected, the screened electrostatic interactions introduce a small correction to the diffusion time scale. This correction decreases with increasing salt concentration.

Since the kinetic equation at the interface is identical to the one in the absence of salt, so is the expression for the corresponding time scale. However, in the case of added salt the electrostatic interactions are screened, the surface potential is much smaller than T/e, and, therefore, the kinetic time scale, τ_k, is only slightly larger than the non-ionic one [Eq. (10)].

We infer that ionic surfactants with added salt behave much like non-ionic surfactants, i.e. undergo diffusion-limited adsorption provided that no additional barriers to adsorption exist. The departure from the non-ionic behavior depends on the salt concentration and is described to first approximation by Eq. (26). The "footprint" of diffusion-limited adsorption, i.e. the $t^{-1/2}$ asymptotic time dependence, is observed in experiments, as demonstrated in Fig. 5. Consequently, the scheme described in Section 2 for solving the adsorption problem and calculating the dynamic surface tension in the non-ionic case is applicable also to ionic surfactants in the presence of salt, and good fitting to experimental measurements can be obtained [13].

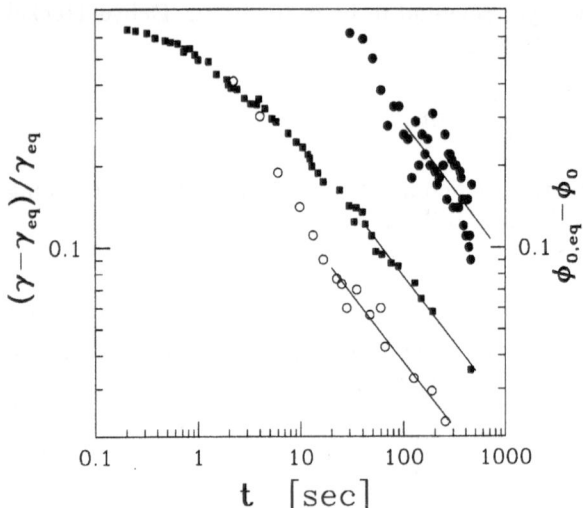

Fig. 5 Diffusion-limited adsorption exhibited by ionic surfactants with added salt: Dynamic interfacial tension between an aqueous solution of 4.86×10^{-5} M SDS with 0.1 M NaCl and dodecane (open circles and left ordinate), adapted from ref. [13]; Dynamic surface tension of an aqueous solution of 2.0×10^{-4} M SDS with 0.5 M NaCl (squares and left ordinate), adapted from ref. [30]; Surface coverage deduced from second-harmonic-generation measurements on a saturated aqueous solution of SDNS with 2% NaCl (filled circles and right ordinate), adapted from ref. [31]. The asymptotic $t^{-1/2}$ dependence shown by the solid fitting lines is a "footprint" of diffusion-limited adsorption

limited adsorption, in agreement with experiments. In the non-ionic case our general formalism coincides with previous ones and helps clarify the validity of their assumptions. Strong electrostatic interactions in salt-free ionic surfactant solutions are found to have a dramatic effect. The adsorption becomes kinetically limited, which may lead to an unusual time dependence, as observed in certain dynamic surface tension measurements. Such a scenario cannot be accounted for by previous models. Addition of salt to ionic surfactant solutions leads to screening of the electrostatic interactions, and the adsorption becomes similar to the non-ionic one, i.e. diffusion-limited. The departure from the non-ionic behavior as the salt concentration is lowered has been described by a perturbative expansion.

A general method to calculate dynamic surface tension is obtained from our formalism. In the diffusion-limited case, it coincides with previous results which used the equilibrium equation of state, but in the kinetically limited case it produces different expressions leading to novel conclusions.

Our kinetic model is restricted to simple relaxation processes, where the free energy monotonically decreases with time. In order to provide a quantitative treatment of more complicated situations, such as the ones described in Section 3 for certain ionic surfactants, a more accurate theory is required.

Finally, the approach presented here may be easily extended to more complicated systems. This flexibility has been demonstrated in Section 3 by introducing electrostatic interactions. Solutions above the critical micelle concentration and adsorption accompanied by lateral diffusion [32] are just two examples for other interesting extensions.

Summary

We have reviewed an alternative theoretical approach to the fundamental problem of the adsorption kinetics of surfactants. The formalism we present is more complete and general than previous ones as it yields the kinetics in the entire system, both in the bulk solution and at the interface, relying on a single functional and reducing the number of externally inserted assumptions previously employed.

Non-ionic surfactants, not hindered by any high barrier to adsorption, are shown to usually undergo diffusion-

Acknowledgments We are indebted to D. Langevin and A. Bonfillon-Colin for introducing us to the problem and for further cooperation. We benefited from discussions with M.-W. Kim. Support from the German-Israeli Foundation (G.I.F.) under grant No. I-0197 and the US–Israel Binational Foundation (B.S.F.) under grant No. 94-00291 is gratefully acknowledged.

References

1. Ward AFH, Tordai L (1946) J Chem Phys 14:453
2. For reviews of both experiments and theory see: Borwankar RP, Wasan DT (1988) Chem Eng Sci 43:1323; Miller R, Kretzschmar G (1991) Adv Colloid Interface Sci 37:97; Dukhin SS, Kretzschmar G, Miller R (1995) In: Möbius D, Miller R (eds) Dynamics of Adsorption at Liquid Interfaces: Theory, Experiment, Application. Studies in Interface Science Series. Elsevier Amsterdam
3. Delahay P, Fike CT (1958) J Am Chem Soc 80:2628
4. Hansen RS (1960) J Phys Chem 64:637
5. van den Bogaert R, Joos P (1980) J Phys Chem 84:190
6. Miller R, Kretzschmar G (1980) Colloid Polym Sci 258:85
7. Borwankar RP, Wasan DT (1983) Chem Eng Sci 38:1637
8. Lin SY, McKeigue K, Maldarelli C (1990) AIChE J 36:1785
9. Chang CH, Franses EI (1992) Colloids Surf 69:189
10. An earlier discussion of this assumption is found in: Fordham S (1954) Trans Faraday Soc 54:593

Progr Colloid Polym Sci (1997) 103:51–59
© Steinkopff Verlag 1997

11. Diamant H, Andelman D (1996) Europhys Lett 34:575
12. Diamant H, Andelman D (1996) J Phys Chem 100:13732
13. Bonfillon A, Langevin D (1993) Langmuir 9:2172; Bonfillon A, Sicoli F, Langevin D (1994) J Colloid Interface Sci 168:497
14. For an earlier discussion of such a distinction, see: Tsonopoulos C, Newman J, Prausnitz JM (1971) Chem Eng Sci 26:817
15. Adamson AW (1990) Physical Chemistry of Surfaces, 5th ed. Wiley, New York, Chapters XI, XVI
16. See, for example, Langer JS (1991) In: Godrèche C (ed) Solids Far From Equilibrium. Cambridge University Press, Cambridge
17. Addison CC, Hutchinson SK (1949) J Chem Soc (London): 3387
18. Hua XY, Rosen MJ (1991) J Colloid Interface Sci 141:180
19. Sutherland KL (1952) Austral J Sci Res A 5:683
20. Lin SY, McKeigue K, Maldarelli C (1991) Langmuir 7:1055
21. Hua XY, Rosen MJ (1988) J Colloid Interface Sci 124:652
22. Dukhin SS, Miller R, Kretzschmar G (1983) Colloid Polym Sci 261:335; Miller R, Dukhin SS, Kretzschmar G (1985) Colloid Polym Sci 263:420
23. Borwankar RP, Wasan DT (1986) Chem Eng Sci 41:199
24. MacLeod CA, Radke CJ (1994) Langmuir 10:3555
25. Onsager L, Samaras NNT (1934) J Chem Phys 2:528
26. Davis JT (1958) Proc Roy Soc A 245:417
27. Verwey EJW, Overbeek JThG (1948) Theory of the Stability of Lyophobic Colloids, Elsevier, New York
28. Andelman D (1995) In: Lipowsky R, Sackmann E (eds) Handbook of Biological Physics, Vol 1B. Elsevier Amsterdam
29. Debye P, Hückel E (1923) Phyzik 24:185; Debye P, Hückel E (1924) Phyzik 25:97
30. Fainerman VB (1978) Colloid J USSR 40:769
31. Rasing Th, Stehlin T, Shen YR, Kim MW, Valint Jr P (1988) J Chem Phys 89:3386
32. There are cases encountered in practice where lateral diffusion seems to play an important role. See Joos P, Fang JP, Serrien G (1992) Colloid Interface Sci 151:144; Menger FM, Littau CA (1993) J Am Chem Soc 115:10083

Progr Colloid Polym Sci (1997) 103:60–66
© Steinkopff Verlag 1997

D. Stauffer

Oil–water interfaces in the Ising-model

Received: 7 November 1996
Accepted: 13 December 1996

Dr. D. Stauffer (✉)
Institute for Theoretical Physics
Cologne University
50923 Köln, Germany

Abstract We review here computer simulations for determining the interfacial structure and interfacial tension of liquid–liquid interfaces using the three-dimensional Ising model.

Key words Ising model – oil–water interfaces – computer simulations

Introduction

The theory of interface structures goes back to Van der Waals, about a century ago. Today, with the help of computers, we can both confirm some of his assumptions, as well as go beyond them. This is better done with a simple model than with a complicated one. On the other hand, we do not want to use a model which describes only one of the two phases that are separated by an interface. Thus we ignore solid-on-solid approximations and concentrate here on the three-dimensional Ising model, that describes well the behavior near the liquid–gas critical point, or near the demixing temperature of binary fluids. First, the shortcomings of square-gradient theories are explained, and the 76 year old Ising model is introduced as an alternative. Then we summarize what is known about the surface or interfacial tension of this model, and finally, we discuss roughening of the interface.

Van der Waals square-gradient theory

Traditional square-gradient theories [1], as introduced by van der Waals, assume a free energy density f depending in a unique way on the local density ρ and added to a free energy density proportional to the gradient $\nabla\rho$. Thus

integration over the whole volume V gives the total free energy F:

$$F = \int_V \left(f(\rho(\mathbf{r})) + K(\nabla\rho)^2 \right) d^3 r \ . \tag{1}$$

Minimization of this free energy functional gives what is often denoted as a Ginzburg–Landau equation, using the derivative f' of the function $f(\rho)$:

$$f'(\rho) - 2K\nabla^2\rho = 0 \ . \tag{2}$$

If we look at a binary fluid mixture instead of a liquid–vapor equilibrium, the variable ρ is the concentration or mixing ratio instead of the density. In both cases, for temperatures T below the critical or demixing critical temperature T_c, two bulk phases with $\rho = \rho_1$ and $\rho = \rho_2 > \rho_1$ can coexist with each other, and this is the situation we assume now.

For this phase coexistence, the free energy density $f(\rho)$ has two equally low minima at $f(\rho_1) = f(\rho_2)$ corresponding to the two coexisting phases. Often one expands

$$f(\rho) = f(\rho_c) + \Delta\mu(\rho - \rho_c) - A(\rho - \rho_c)^2 + B(\rho - \rho_c)^4 + \cdots \ , \tag{3}$$

where A, B and $\rho_c = (\rho_1 + \rho_2)/2$ depend on temperature; then $(\rho_{1,2} - \rho_c)^2 = A/2B$. The chemical potential minus its value on the coexistence curve is $\Delta\mu$ and is set to zero from now on.

Progr Colloid Polym Sci (1997) 103:60–66
© Steinkopff Verlag 1997

For a horizontal interface at $z = 0$ with density ρ_1 for strongly positive heights z and ρ_2 for strongly negative z in the vessel, a smooth density profile follows from Eq. (2) near $z = 0$, with the asymptotic behavior $\rho(z) - \rho_1 \propto \exp(-z/\xi)$. Here ξ is called the correlation length and is basically the thickness of the interface, since Eqs. (2) and (3) are solved by

$$\rho(z) - \rho_c \propto \tanh(z/2\xi) \ . \tag{4}$$

These approximations (1–3) give the correct "critical exponents" near the critical demixing temperature, like $\rho_{1,2} - \rho_c \propto (T_c - T)^{0.32}$, $\xi \propto |T_c - T|^{-0.63}$ or interface tension $\sigma \propto (T_c - T)^{1.26}$, provided the coefficients A, B, K follow proper power laws in $T - T_c$ and $|\xi \nabla \rho| < \rho_2 - \rho_1$, $|\rho - \rho_c| < \text{const}(T_c - T)^{0.32}$. Nevertheless, these approximations are not satisfactory for two reasons: The function $f(\rho)$ is undefined for $\rho_1 < \rho < \rho_2$, and Eqs. (1–3) neglect capillary waves leading to interface roughening.

The function $f(\rho)$ is supposed to describe the bulk equilibrium free energy. For $\rho_1 < \rho < \rho_2$ the system is not in equilibrium; instead it wants to separate into two phases $\rho = \rho_1$ and $\rho = \rho_2$ via nucleation or spinodal decomposition (except when the range of interaction is infinite, an unrealistic limit ignored here). In general, the same density difference $\rho - \rho_1$ can be achieved by many small or a few large droplets (heterophase fluctuations) of the second phase within the first phase (see Chapter 7 of Ref. [2] for a droplet description of phase transitions). The free energy is basically the number of different droplets, and thus will be different in the two cases of few large or many small droplets: f is not a unique function of ρ. Only in an equilibrium situation, where also the average number of droplets is fixed through this thermodynamic equilibrium, does a well-defined function $f(\rho)$ exist, similar to Eq. (3). Similarly the height of a mountain should be defined by its peak (= equilibrium) and not by the numerous paths (= nonequilibrium) leading to this peak.

Interface roughening comes from long-range capillary waves by which the interface fluctuates. For a wave $A_q \sin(qx)$ with a wave vector q in x-direction, the (free) energy varies as $q^2 A_q^2 \sigma$ and is about $k_B T$ in thermal equilibrium; here σ is the interfacial tension and quantum effects are neglected. Thus, the thermal average of the wave amplitude is $\langle A_q^2 \rangle \propto k_B T q^{-2}/\sigma$. Integration over all two-dimensional wavevectors (q_x, q_y) parallel to the interface gives the squared total amplitude or squared interface width:

$$W^2 = \sum_q A_q^2 \propto \int_{1/a}^{1/L} k_B T q^{-2} \sigma^{-1} 2\pi q \, dq$$

$$\propto \int_{1/a}^{1/L} \frac{1}{q} \, dq = \ln(L/a) \ . \tag{5}$$

Thus, the interface width W diverges logarithmically if the linear extent L of the interface becomes much larger than the intermolecular distance a. Since resulting interfacial thicknesses are typically in the nanometer range (except near a critical point), this logarithmic divergence presumably has thus far been seen only in computer simulations [3] and not in laboratory experiments. Also, a finite thickness of the fluid film hampers interfacial fluctuations and leads to experimentally observed corrections [18]. But ideally, for an interface in an infinite medium without gravity, the thickness is infinite and is not equal to ξ, and this effect is not described by the square gradient theory. Thus, we need a better way to approach the structure of the interface, though *approximately* with a suitably renormalized ξ the above hyperbolic tangent of Eq. (4) is a good overall fit.

Ising model

The Ising model was suggested by Lenz in 1920 for magnets; his student Ern (e)st Ising solved it for one dimension in 1925, Nobel laureate Onsager solved it for two dimensions in 1944, and for more than four dimensions its critical exponents are described by the van der Waals equation. (Some of the statements below on interfacial tensions are no longer valid above four dimensions.) But for the experimentally relevant case of three dimensions only numerical approximations, no exact solutions, are available. These techniques, however, are good enough to show that the critical exponents of this model are within the experimental error bars in agreement with those of real fluids near their liquid–vapor critical point, or real binary fluid mixtures near their demixing temperature.

We take a large three-dimensional lattice and place a "spin" S_i on each lattice site i. This spin, a term physicists like, is for our purposes just a binary variable which is either $+1$ or -1: instead we could also choose it as 1 or 0, up or down, occupied or empty, black or white, water or isobutyric acid, etc. Regions with most of the sites empty correspond to vapor, those with most of the sites occupied correspond to liquid, if we use the Ising model for liquid–vapor transitions. Neighboring spins tend to have the same status, that means they are coupled by an interaction energy $-JS_iS_j$ when lattice site j is a nearest neighbor to i. Thus, $2J$ is the energy needed to break a "bond" between two neighbors. For a binary fluid A–B mixture, this energy is related to the interaction energies between A–A, A–B, and B–B molecules. The chemical potential leads to an energy contribution proportional to $\sum_i S_i$ which for magnets would be the magnetization. The total Ising energy then is

$$E = -J \sum_{\langle ij \rangle} S_i S_j - H \sum_i S_i + \text{const} \ , \tag{6}$$

where the first sum goes over all pairs of nearest neighbors, and where the "magnetic field" H is half of the chemical potential difference $\Delta\mu$ (since an energy $\pm 2H$ is needed to flip a spin, that means to transform an A molecule into a B molecule or back).

In thermal equilibrium at temperature T, different spin configurations appear with Boltzmann's probability proportional to $\exp(-E/k_B T)$. Numerically, one finds a critical temperature T_c with $J/k_B T_c \simeq 0.221654$ such that for $T < T_c$, $H = 0$ phase separation ("ferromagnetism") occurs with one phase predominantly $S_i = 1$ and the other one predominantly $S_i = -1$. For binary fluids of components A and B this means a demixing into an A-rich and a B-rich phase. For $T > T_c$, on the other hand, a magnet becomes paramagnetic, and the two fluid components A and B can mix; for $H = 0$ we have half the sites occupied by A-molecules $S_i = +1$ and half by B-molecules $S_i = -1$. The normalized magnetization $m = \sum_i S_i/N$ for N spins varies between -1 and 1, the concentration or density $\rho = (m+1)/2$ between 0 and 1. Only at $H = 0$ can the two phases coexist; and these two phases are completely symmetric to each other: $\rho_2 - \rho_c = \rho_c - \rho_1 = m/2$. Thus, the critical density $\rho_c = 1/2$ is independent of temperature in this Ising model. This symmetry greatly simplifies the analysis compared to more realistic fluid models in a continuum [4].

Interface structure

The fluid–fluid interface, or the interface separating a liquid from its vapor, is in the Ising model a domain wall between a region of predominantly up spins and a region of predominantly down spins. Leamy et al. [5] presumably published the first interface profile, and it looks like Fig. 1 (except that our figure was made for the purposes of this review in less than 150 min on an Intel Paragon computer, using 1088 million spins distributed among 136 processors). In contrast to laboratory experiment, this computer simulation is not plagued by impure materials or by gravity. Similar to real experiments, the computer experiment is made for a finite sample during a finite time (though our sizes and times are much smaller than those of typical experiments in the laboratory). Thus, in spite of the surface roughening mentioned above we see a well-defined interface separating the two phases on the left and right of the figure. (A difference between real and computer experiments is that computer simulators imagine the height to vary from left to right, as for mathematical functions.) Qualitatively, van der Waals is fully justified with his idea of a smooth density profile interpolating between liquid and vapor.

The appendix describes the Monte Carlo method used to get such figures. To study interfaces, one can put adjacent to the bottom (or left) boundary a layer of up spins, and to the top a layer of down spins. Then one starts with all spins up in the lower half, and all spins down in the upper half of the lattice. Such a program is given in the appendix. However, for simulations of large lattices and moderate times (as measured by the number of Monte Carlo steps per spin; 1000 in the above figure) the inertia of the interface is so large, that the interface barely moves and also shows no intention to change its orientation. Thus one does not really need the two boundary conditions: The lower neighbor plane of the lowest lattice plane is identified with the uppermost lattice plane, except that an up spin there is treated as a down neighbor, and the other way round. Analogous tricks are used for the upper boundary of the lattice. In this way we avoid the deviations seen in Fig. 1a on the left and right of the density profile arising from the boundary effects. Thus with antiperiodic boundary conditions we get the same quality of data with less lattice planes to be simulated.

Figure 1b shows the central part of Fig. 1a, and we see that Eq. (3), in spite of its fundamental shortcomings, is still a good fit to the data if one fits freely not only the "spontaneous magnetization" m but also effective correlation length ξ to the data. Closer inspection of this and other figures shows that m was still not in equilibrium (Fig. 1a has it at 0.372 while its equilibrium value is near 0.367), and also the steepness of the profile still diminished with time. To my knowledge, the asymptotic tail of the profile, when $|\rho - \rho_{1,2}| \ll \rho$, has been investigated only experimentally [6] and not yet by computer simulation.

To measure the width W of the liquid–vapor interface, it seems practical [7] to let all the molecules fall down. That means for each column (x, y) of the lattice we look at all the sites along the z-axis having the same x and y coordinates. Then we evaluate $Z(x, y) = \sum S_i$ along this column. This sum would correspond to the height of the interface if all the internal holes would be filled by molecules moving down along the column, and if all the liquid droplets above the interface would fall down onto the interface, without changing their x and y coordinates. Then we simply calculate

$$W^2 = \langle Z^2 \rangle - \langle Z \rangle^2$$

by averaging over all x and y coordinates. In this way we avoid some fluctuation effects inherent in other methods, and we can observe how the width W increases towards a finite limit for time $\to \infty$, and then increases weakly to infinity if $L \to \infty$. There are also many other models for which the behavior of W as a function of time and system size have been studied [8].

In summary, therefore, a well-defined interface profile of the van der Waals type does not exist in principle for infinitely large systems in complete equilibrium, but it

Fig. 1 Interface profile from computer simulations; part a is an overview and part b expands the central part for comparison with Eq. (4)

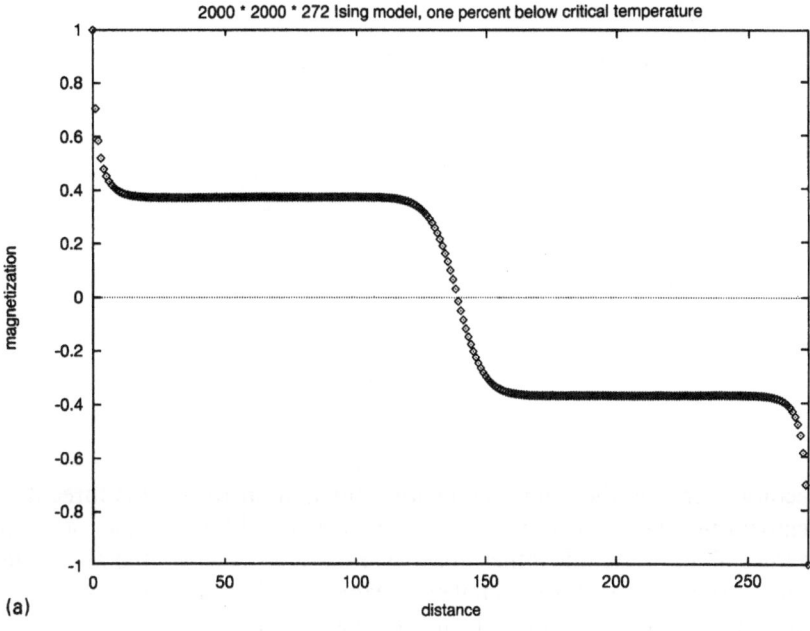

(a)

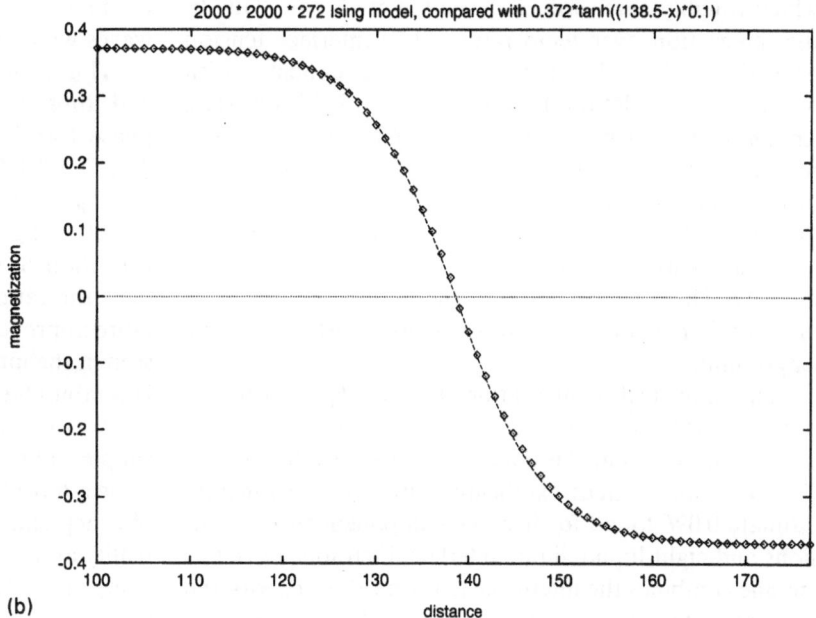

(b)

exists in practice. Careful computer simulations have shown [3] that the equilibrium width, which is constant in the square-gradient theory, increases with the square-root of the logarithm of the system size.

Interfacial tension and roughening

At zero temperature, the interfacial tension is the excess energy per square centimeter stored in the interface; at finite temperatures it is the free energy. This interfacial free energy σ goes to zero at the critical temperature T_c, roughly proportional to $(T_c - T)^{1.26}$. Expressed in units of $k_B T$, the interfacial free energy in the correlation area ξ^2 approaches a finite value R for $T \to T_c$, and R is material independent or "universal" [9].

Computer simulations can find the interfacial energy by simulating the same lattice first with antiperiodic boundary conditions as described above, and then with periodic boundary conditions and with initially all spins up. Then an interface exists only in the first and not in the second case, and the difference between the first and the

Fig. 2 Example of experimental versus theoretical interface tension for solid helium [17]

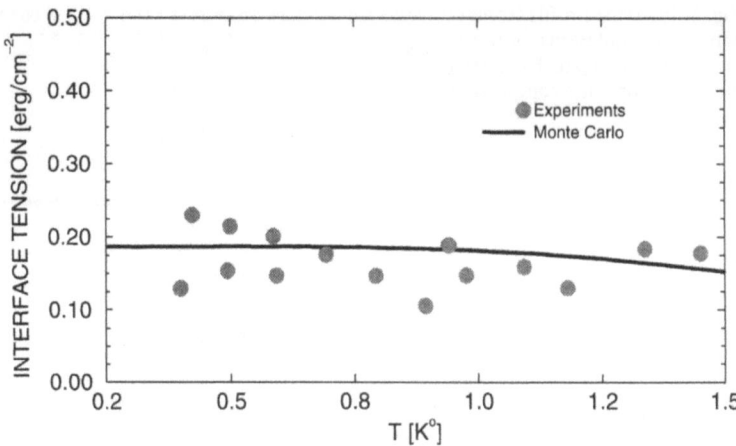

second energy is the interfacial energy. Integration over temperature then yields the interfacial free energy. Many studies along these and other lines were made in recent years, and in terms of total number of Monte Carlo steps the last simulations [10] are the most extensive ones. Ito [10] also summarized the earlier high-quality studies which roughly agree within the stated error bars. Due to the integration over temperature, Ito's interface tension depends on the value assumed for the exponent ν with which the correlation lenght $\xi \propto (T_c - T)^{-\nu}$ diverges, since $\sigma \propto (T_c - T)^{2\nu}$. Assuming $\nu = 0.63$ as found in more recent simulations [11], Fig. 13 in Ref. [10] gives the proportionality factor $\sigma(1 - T/T_c)^{-2\nu} = 1.48$ for the Ising model on the simple cubic lattice, which is in excellent agreement with later "series expansion" methods [12] giving 1.49. These Ising interfacial energies are expressed in units of $k_B T$ with the nearest-neighbor distance as the length unit.

The universal combination $R = \sigma \xi^2 / k_B T$ then approaches 0.096 to 0.106 in the Ising model, when both σ and ξ are taken at the same temperature slightly below T_c. This value agrees excellently with the experimental estimate 0.096 found for five one-component surface tensions and eight liquid–liquid interfacial tensions [13]. Often one combines the interfacial tension below T_c with the correlation length above T_c at the same distance from the critical temperature; since this correlation length is nearly twice as high, the universal combination R then takes the value 0.37. Thus near the critical point Ising computer simulations and laboratory experiments agree very well for the interfacial or surface tension.

This is no longer true at temperatures far below T_c where the correlation length is small, of the order of the nearest-neighbor distance, and where the lattice structure of the Ising model, which seems negligible near T_c, plays a crucial role. Now the interfacial tension depends on the direction with respect to the lattice axes, an effect important if one would make precise nucleation theories [16]. In

the three-dimensional Ising model, not necessarily in real fluids, a roughening temperature T_R is found, below which the above statements of the width W increasing logarithmically with size no longer are valid: For $T < T_R$ the width remains finite even in the limit of an infinitely large lattice. The thermal energy is then no longer large enough to create on a lattice the small long-wavelength capillary waves which dominate in Eq. (5).

This roughening transition happens in the interface and thus might at first be identified with Onsager's phase transition in the two-dimensional Ising model at $J/k_B T = 0.44069$. However, it actually occurs [14] at $J/k_B T_R = 0.40758$ and is not described by the two-dimensional Ising model, but by the Kosterlitz–Thouless transition of the two-dimensional XY-model which predicts the ratio $W^2/\ln(L)$ to approach $1/\pi^2$ if the temperature approaches T_R from above. At T_R, no anomaly was seen in the interfacial tension. For more details we refer to Hasenbusch et al. [14].

All these simulations refer to the Ising model on the simple cubic lattice. Of course, also more complicated lattices have been studied, and Fig. 2 shows a comparison of a hcp simulation with solid helium. Again model and reality agree nicely. The Ising model has also been used to study oil–water systems where amphiphilic molecules may form membranes, micelles, and vesicles [15].

We thank the German–Israeli Foundation and SFB 341 for partial support and A. Hashibon for Fig. 2.

Appendix: A simple Monte Carlo program

For the novice in Ising model simulations, we present here a short program (Fig. 3) and its description, in old-fashioned Fortran; it looks very similar in BASIC. It treats an interface by keeping the upper and the lower border of an $L \times L \times L_z$ lattice all spin up and all spin down, respecively. It should convince the reader that the basics of Monte

Progr Colloid Polym Sci (1997) 103:60–66
© Steinkopff Verlag 1997

```
c       3D Ising Model with interface between fixed walls, primitive technique
        parameter(Lz=20,L=100,L2=L*L,Lmax=(Lz+2)*L2)
        dimension iex(-6:6),is(Lmax)
        data  t,max,iseed/0.9,20,1  /
        t=t/0.2216544
        ibm=2*iseed-1
        print *, L,t,max,iseed
        do 2 ien=-6,6
          ex=exp(2*ien/T)
2         iex(ien)=(2.0/(1.0+ex) - 1.0)*2147483648.0
        do 1 i=1,Lmax/2
          is(Lmax+1-i)=-1
1         is(i)=1
c       end of initialization (Glauber dynamics), start of simulation
        do 3 itime=1,max
        do 3 i=L2+1,L2*(Lz+1)
          ien=is(i)*(is(i-1)+is(i+1)+is(i-L)+is(i+L)+is(i-L2)+is(i+L2))
          ibm=ibm*16807
3         if(ibm.lt.iex(ien)) is(i)=-is(i)
        do 4 iplane=0,Lz+1
          m=0
          j=L*L*iplane
          do  5 k=1,L2
5           m=m+is(j+k)
4         print *, iplane,m*1.0/L2
        stop
        end
```

Fig. 3 Monte Carlo program

Carlo simulations are quite simple; this test program finishes within seconds.

Random odd integers, ibm, on a 32-bit workstation are produced by multiplication with 16 807, where most computers simply throw away the leading bits if the product requires more than 32 bits. As a result, ibm lies between -2^{31} and $+2^{31}$, and a command is executed with probability p if the random number ibm is smaller than $(2p-1)2^{31}$. This we do in the line with label 3 where a spin is(i) is flipped to $-$ is(i) if the random number ibm is less than the limit, iex, which depends on the interaction energy, ien, and is basically Boltzmann's probability.

Quite generally, Monte Carlo simulations go numerous times through the following steps: (1) Select randomly a new atom to be investigated; (2) Calculate the energy E needed to change the status of that atom (e.g. to flip it spin); (3) Calculate a random number and compare it with the properly normalized Boltzmann probability $\exp(-E/k_B T)$ or $1/(1 + \exp(E/k_B T))$; (4) Make the change if and only if the random number is smaller than this probability; (5) Calculate the desired averages independently of whether this change was made or rejected.

Completing these five steps once for every atom in the system corresponds to one time unit, called one Monte Carlo step per site.

The innermost loop 3 therefore calculates the interaction energy of the newly selected spin is(i) with its six neighbors, then the random number, and finally flips the spin if this random number is small enough. The appropriate Boltzmann factors, normalized to the interval $-2^{31} \cdots +2^{31}$, are evaluated once in advance in loop 3, using the normalized (Glauber) probability $1/(1 + \exp(E/k_B T))$. Loop 1 fixes the boundary planes to all spins up and all spin down; loop 4 evaluates at the end the density profile and prints it out. The temperature is given in the data statement and is measured in units of the critical temperature, since $J/k_B T_c = 0.221654 \dots$. We make max Monte Carlo steps per spin and start our random numbers with an arbitrary iseed number.

More sophisticated "multi-spin coding" programs store spins in single computer bits and simulate up to 5888 spins [19]. More details on Monte Carlo research can be found in the book series "Monte Carlo Methods..." edited by Binder [20].

References

1. Widom B (1972) Domb Green MS (eds) Phase Transitions and Critical Phenomena, Vol. 2. Academic, London, p 79; Amit DJ, Luban M (1989) Phys Lett 27A:487

2. Stauffer D, Aharony A (1994) Introduction to Percolation Theory. Taylor & Francis, London

3. Evertz HG, Hasenbusch M, Marcu M, Pinn K, Solomon S (1991) J Physique I 1:1669

4. Wilding NB (1996) In: Annual Reviews of Computational Physics 1V, Stauffer D (ed.) World Scientific, Singapore
5. Leamy HJ, Gilmer GH, Jackson KA, Bennema P (1973) Phys Rev Lett 30:601
6. Huang JS, Webb WW (1969) J Chem Phys 50:3677
7. Kremer S, Wolf DW (1992) Physica A 182:542
8. Barabasi A-L, Stanley HE (1995) Fractal Concepts in Surface Growth. (Cambridge University Press, Cambridge)
9. Privman V, Hohenberg PC, Aharony A (1991) In: Phase Transitions and Critical Phenomena, vol. 14. Domb C, Lebowitz JL (eds) Academic, London
10. Ito N (1993) Physica A 196:591
11. Blöte HWJ, Luijten E, Heringa JR (1995) J Phys A 28:6289
12. Zinn S-Y, Fisher ME (1996) Physica A 226:168
13. Mainzer T, Woermann D (1996) Physica A 225:312
14. Hasenbusch M, Meyer S, Pütz M (1996) J Stat Phys 85:383; Hasenbusch M, Pinn K (1997) J Phys A 30:63
15. Liverpool TB, page 317 in ref. 4
16. Stauffer D (1992) Int J Mod Phys C 3:1059
17. Hashibon A, Adler J, Lipson S, in preparation
18. Kerle T, Klein J, Binder K (1996) Phys Rev Lett 77:1318
19. Stauffer D, Knecht R (1996) Int J Mod Phys C 7:893
20. Binder K (ed) (1978) Topics in Current Physics, Vols. 7, 36, and 71. Springer, Berlin

Progr Colloid Polym Sci (1997) 103:67–77
© Steinkopff Verlag 1997

H. Chaimovich
I.M. Cuccovia

Quantitative analysis of reagent distribution and reaction rates in vesicles

Received: 26 November 1996
Accepted: 5 December 1996

Dr. H. Chaimovich (✉) · I.M. Cuccovia
Departamento de Bioquímica
Instituto de Química
Universidade de Sao Paulo
CP 26077 Sao Paulo, SP, Brazil

Abstract Vesicles prepared with synthetic amphiphiles constitute useful microreactors, where reaction rates can be delicately controlled. Here we review our work on quantitative analysis of reaction rates in vesicles and show that reaction at several vesicular sites can be probed and controlled. Vesicles prepared with dialkyldimethylammonium halides, (DODA)X, can accelerate bimolecular reactions by more than a million fold. Quantitative analysis of the vesicular effect on ester thiolysis, using a pseudophase ion exchange formalism, suggests that the rate increase is primarily due to reagent concentration in the bilayer and interfacial effects on ion distribution, as well as contributions from enhanced nucleophile reactivity. Vesicle-containing solutions exhibit a variety of potential reaction sites: the inner and outer surfaces, bilayer and internal aqueous compartment.

Site dissection and reagent distribution has been accomplished, in several cases. For this purpose we have prepared vesicles of various sizes, determined some of their physical properties and developed theoretical and experimental tools for probing vesicular sites. The reactivity of OH^- in the internal compartment is identical to that in bulk solution. Moreover, the reaction rates of OH^- at the inner and outer vesicular interface are also comparable. However, since OH^- permeation through the bilayer can limit reaction rate, the relative inner/outer rate ratios can be controlled by changing the composition of the external and/or internal medium.

Key words Vesicles – reaction sites in vesicles – kinetic analysis in vesicles – vesicular catalysis – reaction rate – control with vesicles

Introduction

Understanding chemistry in organized assemblies is essential to learn and interpret, at a molecular level, biological chemistry. Menger and Gabrielson, in a recent review, presenting fascinating properties of micron size vesicles, expressed this idea concisely: "Biology is, in effect, organized organic chemistry" [1]. Fundamental features of organized systems derive from the organization itself and are hard to predict from the properties of individual molecules [2, 3]. Chemistry in these systems, for similar motives, is also complicated to forecast from knowledge obtained in dilute solution. Organized systems can bring together, from dilute solutions, reactants with contrasting chemical properties, at high local concentrations [2]. In a dilute primordial soup, several interfaces, such as clays and self-organized assemblies of primordial lipids, may have

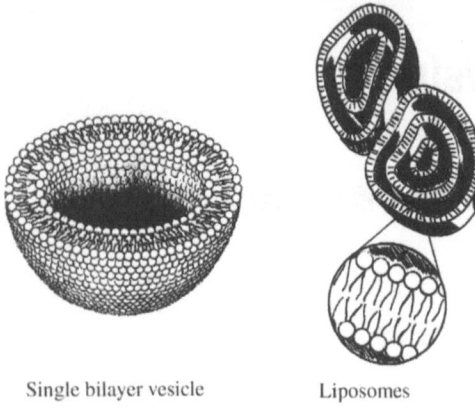

Single bilayer vesicle   Liposomes

Scheme 1

played a central prebiotic role [4, 5]. Life without an in-side—outside separation is unimaginable and primordial vesicles are one of such possible forms of organization [6]. Biomimetic chemistry, however, is useful not only to understand biology or to simulate prebiotic conditions, but also to create new chemistry, based on biological paradigms.

Organized supramolecular systems are abundant and one of the most investigated models results from the spontaneous aggregation of amphiphiles in excess water [1–3]. Closed bilayer vesicles, or liposomes, first described by Bangham and Horne [7], are intensely explored supra-molecular structures (Scheme 1). Liposomes, simple models for biological membranes, are widely used as vehicles in drug and DNA delivery as well as in cosmetics [8]. Totally synthetic amphipatic monomers can also form liposomes and vesicles [2]. The work of Kunitake, showing that dioctadecyldimethylammonium bromide, (DODA)B, forms vesicles, marked the beginning of membrane mimetic chemistry [9]. This pioneering finding was soon followed by the description of vesicles prepared with a simple phosphate diester (sodium di-hexadecyl phosphate, DCP) [10]. Thereafter, an increasing amount of basic and applied information concerning vesicles prepared with a wide variety of synthetic amphiphiles has emerged [11].

Here we will review the limited amount of information regarding the quantitative analysis of the effect of vesicles on the rate of chemical reactions in the ground state. We will concentrate on reactions occurring in vesicles of synthetic amphiphiles, emphasizing the work done in our laboratory. Excellent reviews, focusing other aspects of properties and applications of synthetic vesicles, have appeared recently [11].

Vesicular catalysis, reagent binding and kinetic analysis

The choice of a well known reaction is a key element for understanding the effect of an aggregate on the resultant

chemistry since aggregate-induced changes of mechanism and rate limiting steps are known [12]. The thiolysis of p-nitrophenyl esters by alkyl mercaptans is rate limited by the attack of the dissociated form of the nucleophile [13], and the rate limiting step is maintained upon transferring the reaction site from bulk solution to an amphiphile aggregate [14]. Positively charged hexadecyltrimethylam-monium bromide ((CTA)B) micelles increase the reaction rate essentially by reagent concentration and only marginally affect nucleophilic reactivity. This reaction was chosen for analyzing the effect of a more organized bilayer, that of (DODA)X vesicles, on chemical reactivity.

In our initial attempts to study the effect of (DODA) Chloride, (DODA)C, sonicated vesicles on the thiolysis of p-nitrophenylacetate (NPA) by n-heptyl mercaptan (HM) we found that metal from the probe catalyzed the oxidation of the mercaptan [Cuccovia, I.M., unpublished]. In order to avoid mercaptan oxidation we developed an alcohol injection method for the preparation of (DODA)C vesicles, that were characterized by gel filtration and by the incorporation of lipid soluble probes [15]. It is not clear, today, if the aggregates prepared by ethanol injection are highly permeable closed vesicles or other types of bilayer assemblies [16].

The rate of thiolysis of NPA by HM was substantially accelerated by (DODA)C vesicles, the maximum rate enhancement reaching 7×10^4 fold [15]. By changing ester and mercaptan hydrophobicity, using p-nitrophenyl octanoate (NPO) and hexadecylcysteinamide (HCys) as a nucleophile, we obtained rate accelerations reaching several millions over that in water [17]. Selected data on the effects of several amphiphiles on the rate of thiolysis of p-nitrophenyl octanoate are presented in Table 1. It is striking to find rate accelerations of the magnitudes shown in Table 1 in model systems [18]. The most interesting feature of the results in Table 1 is, however, the variation of the effect of transferring the reaction site from water to the interfacial site at a particular aggregate. The thiolysis of NPO by HM is inhibited by sodium dodecyl sulfate (SDS) micelles, accelerated by zwitterionic detergents (N-hexa-decyl-N,N-dimethyl-3-ammonium-1-propanesulphonate, HPS) and positively charged (CTA)B micelles. Changing the structure of the interface, to a (DODA)C positively charged vesicular surface, results in a rate acceleration of more than a million fold. Although the aggregated systems are different and, therefore, not strictly comparable, the results presented in Table 1 suggest that charge and structural changes at an amphiphile interface can produce rate control in excess of 10^8. This level of rate control is an essential and common feature of living systems. To make practical use of this idea in interfacial chemistry, it is critical to dissect quantitatively the factors that lead to the observed effects. Only then the simple models these

Table 1 Effect of aggregate type, and charge on the rate of NPO thiolysis [17, 20]

Thiol	Amphiphile	Type of aggregate	k_{max}/k_w	k_{2m}/k_{2w}
n-Heptyl mercaptan	SDS	Micelle	0.1	0.27
n-Heptyl mercaptan	HPS	Micelle	266	1.0
n-Heptyl mercaptan	(CTA)B	Micelle	1.6×10^4	1.2
n-Heptyl mercaptan	(DODA)C	Vesicle	7.7×10^6	15
HCys	(CTA)B	Micelle	1×10^6	15
HCys	(DODA)C	Vesicle	9×10^6	50

k_{max} is the maximum rate constant produced by the aggregate, k_w the rate constant in the absence of aggregate; k_{2m} and k_{2w} the second order rate constants in the vesicle and in water, respectively.

systems represent will be useful to understand both the complex biological realities intended to represent and the underlying interfacial chemistry.

The rates of vesicle-modified reactions were first quantitatively analyzed using a pseudophase model with explicit consideration of ion exchange [19, 20].

Pseudophase models, as well as other forms of quantitatively analyzing the effects of supramolecular aggregates on reaction rates, were derived for micellar solutions [21]. The full description, as well as critical analysis of the models, has been reviewed [22]. Pseudophase models start by considering that a system can be treated as if it were composed of two separate pseudophases, micelles and an intermicellar aqueous phase. It is further assumed that monomers, substrates and ions equilibrate much faster than reaction rates. This last assumption renders these treatments only applicable to some ground state reactions, since excited state processes and some electron transfer reactions, can be faster than equilibration rates [21, 23]. Further assuming that the reaction rate in the two pseudophases can be characterized by two separate and independent rate constants (k_m, in the micelle and k_w in the intermicellar aqueous phase) the rate constant ($k\psi$) of a spontaneous monomolecular decomposition of a substrate **S** is described by

$$k_\psi = k_w \frac{S_w}{S_t} + k_m \frac{S_m}{S_t} ,$$ (1)

where S_t is the total concentration of **S**, S_w and S_m represent the stoichiometric concentrations of S in the intermicellar and micellar pseudophases, respectively. A bimolecular reaction can be described by

$$k_\psi = k_m \frac{A_m}{A_t} B_m + k_w \frac{A_w}{A_t} B_w ,$$ (2)

where **A** and **B** are the reagents and the subscripts have the same meaning as above. Micelle–water reagent distribu-

tion, for neutral substrates, can be represented by a simple association constant K_s [24]

$$K_s = \frac{A_m}{A_w C_d}$$ (3)

where C_d is the concentration of micellized detergent ($C_d = C_t - $ cmc, where cmc is the critical micelle concentration). The supposition that in a multi-reagent reaction the association constant of a particular substrate is independent of the relative occupation of the micelle by other reagent(s) need not be true and has to be evaluated, specially if the relative occupation number (mole fraction of one of the reagents in the micelle) is large [25]. This caution may be more important when dealing with vesicles (see below).

The distribution of ions between the micellar interface and bulk solution cannot be treated using an association constant [26]. Ion binding to ionic micelles is determined by ion–interface interactions, characterized by a combination of coulombic (point charge) contributions and a specific interaction term containing all the binding contributions in excess of the coulombic interaction [26]. The relative free energy contribution of the specificity term to ion binding has been described for homologue series of carboxylates and dicarboxylates [27]. Ions also bind specifically to zwitterionic micelles and can be selectively displaced from the interface by added salt [28]. The need to differentiate between neutral substrate binding and ion association was clear in very early descriptions of quantitative analysis of micellar effects on reaction rates [29]. One of the most useful forms of describing ion binding to charged micelles is the concept of ion exchange [30]. Considering that the interface of an ionic micelle is ion-saturated, the binding of an ion produces the release of a similarly charged ion, maintaining the relative charge neutralization at the surface [30]. Using the pseudophase ion exchange (PPIE) formalism, the binding of a univalent ion, X, to a micelle containing counterions, Y, can be described by [30]

$$K_{X/Y} = \frac{[X_b][Y_f]}{[X_f][Y_b]}, \qquad (4)$$

where the subscripts b and f represent bound and free species, respectively.

Ion distribution around a spherical micelle can also be described with models that consider that polarizable and not very hydrophilic species (such as Br^-) interact both coulombically and by a specific, noncoulombic, interaction [26]. This latter interaction allows the ion to intercalate at the micellar surface and to neutralize an equivalent number of head groups. Ion distribution around a micelle is then calculated by solving the Poisson–Boltzmann equation (PBE) in the spherical symmetry with allowance for specific interactions via a Langmuir or Volmer isotherm [31]. The original kinetic treatment for a micelle of radius a, aggregation number N in a cell of radius R yields [31]:

$$\frac{1}{r^2}\frac{d(r^2\,d\phi/dr)}{dr} = -\frac{4\pi e^2}{\varepsilon RT}\left(\sum Z_i N_i^z \exp(Z_i\phi) + n_d g(r)\right), \quad (5)$$

where ϕ, the reduced potential, equals $e\psi/kT$; ψ is the electrical potential, T the absolute temperature, e the electrostatic charge, ε the dielectric constant, k the Boltzmann constant, z_i the valency of ion i and n_i and n_d are the number concentration of ions and head groups, respectively and r is the distance from the micelle center. Roughness of the surface can be corrected through the term $n_d g(r)$. In general, the boundary conditions are

$$\phi = d\phi/dr \quad \text{at } r = R$$

and

$$d\phi/dr = (1-f)Ne^2/\varepsilon^2 kT, \qquad (6)$$

where f is the fractional specific interfacial coverage by counterions which neutralize an equivalent number of headgroups. Headgroup neutralization by specifically interacting ions can be calculated from:

$$f = \frac{\delta_x \exp(-f/(1-f))[X_w^-]}{1 + \delta_x \exp(-f/(1-f))[X_w^-]} \qquad (7)$$

and the micellar surface charge density is modified accordingly. Langmuir or Volmer isotherms can also be used when two specifically interacting ions are present in solution using ion-specific δ's [31].

Within the limitations of present theory the distribution of neutral substrates and reactive ions can, consequently, be determined. Thus, the rate constants in the aggregate and in bulk solution can be deduced from the experimentally determined (bulk) $k\psi$ (see Eqs. (1) and (2)).

In some cases the distribution constants, as well as ion selectivity coefficients, have been measured independently, making the subsequent analysis of kinetic data less dependent on adjustable parameters.

Quantitative analysis of the effect of (CTA)B micelles and (DODA)C vesicles on the rate of decomposition of 5,5'-dithiobis(2-nitrobenzoic acid) (DTNB), using pseudophase models, has been described [19]. The maximum acceleration by vesicles is 1500 fold and was ascribed, essentially, to reagent concentration at the vesicle bilayer [19].

Cationic surfactant vesicles accelerate the rate of thiolysis several million fold, permitting the observation of mercaptan reactivity more than 6 pH units below the pK of the SH group [17]. A full kinetic analysis, including the effect of the vesicle' interface on the pK_a of a hydrophobic derivative of cysteine, HCys, was done using the following equation [17]:

$$k_\psi =$$

$$\frac{k_2^w + (k_2^v/\bar{V})K_S K_{HC} C_d}{(1 + K_S C_d)\left[(1 + K_{HC}C_d) + \dfrac{Hf}{K_a}\left(1 + \dfrac{K_{HC}K_w^0 K_a C_d}{K_{OH/Br}K_b}\dfrac{Cl_f}{Cl_b}\right)\right]}, \qquad (8)$$

where K_S and K_{HC} are the distribution constants of p-nitrophenyl octanoate and undissociated mercaptan, Hf is calculated from the buffer' pH, K_a is the acid dissociation constant of the mercaptan in water, $K_b = K_a/K_w$ where K_w is the ionic product of water, K_w^0 the ratio between the ionic product of water in the micellar pseudophase and water phase and $K_{OH/Br}$ the hydroxide/bromide selectivity coefficient, Cl_f, Cl_b refer to the analytical concentrations of free and bound chloride, respectively. The second order rate constants in water and in the vesicle are k_2^w and k_2^v, respectively, and $\bar{V}$ the molar volume of the amphiphile. The substrate distribution constants as well as the ion exchange parameters were determined independently. Curves in panel A were fitted using $k_2^v/\bar{V}$ as the only adjustable parameter and the resultant value was used to predict the experimental values shown in panel B (Fig. 1). Taking the value of $\bar{V}$ as 0.37 l/mole, the second order rate constant in the vesicle is 50 fold larger than that in water (Table 1). Quantitative analysis, using PPIE, suggests that the rate enhancement is due primarily to reagent concentration in the dimensionally restricted environment provided by the vesicle, coupled with contributions from enhanced dissociation and reactivity of the nucleophile at the vesicle surface [17].

One of the conditions for the application of PPIE models is the exchange of ions at the interface. Vesicle gel filtration in the presence of buffer shows that some of the initially added counterion is displaced by the buffer anion,

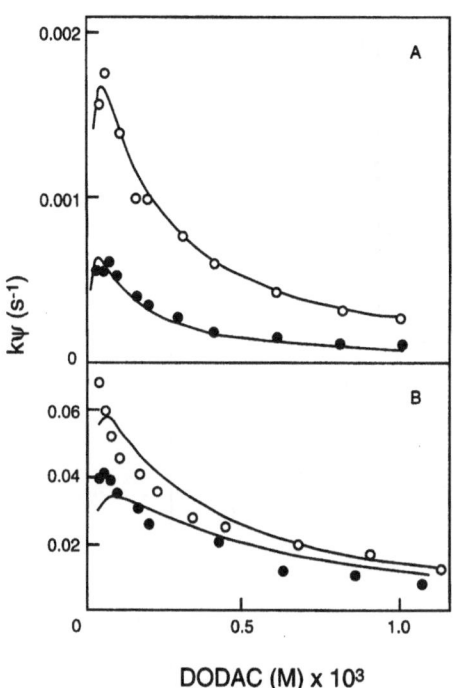

Fig. 1 Effect of (DODA)C vesicles on the rate of thiolysis of *p*-nitrophenyl octanoate (NPO) by HCys. [HCys] = 4.0×10^{-5} M, [NPO] = 4×10^{-6} M   **A**:   [HCl] = 2.65 mM,   pH = 2.65   (○),   [HCl] = 5 mM, pH = 2.35 (●) **B**: sodium acetate buffer (0.005 M) pH = 4.50 (○), pH = 3.75 (●) [17]

strongly indicating ion exchange at the vesicular interface [15].

Direct evidences for exchange and determination of selectivity constants for ion binding were obtained directly making use of fluorescence quenching methods [32]. The fluorescence of 1-pyrenenonanoic acid (1-Py), incorporated in large vesicles of (DODA)C and (DODA)B, is quenched by iodide addition [33]. Below the gel–liquid crystalline phase transition temperature (T_C), temporal changes of the fluorescence of 1-Py can be interpreted as a fast ion exchange at the outer vesicular surface followed by a slower iodide permeation through the bilayer. In the final state ion exchange occurs at both inner and outer interfaces. The selectivity constants for I^-/Br^- exchange ($K_{I/Br}$) at the outer and inner surfaces are 8 and 7, respectively. The average value of $K_{I/Cl}$ at both interfaces is 31 ± 5. An exchange constant of ca. 4 for the Br^-/Cl^- exchange was calculated from the experimentally determined selectivity coefficients for exchange of the halides with I^- [33]. The value of $K_{Br/Cl}$ obtained for the exchange at the surface of (DODA)C vesicles is very similar to data obtained with positively charged micelles of comparable headgroup, i.e. hexadecyltrimethylammonium (CTA) halide [34]. This similarity suggests that the noncoulombic (specific) component of ion binding at the interface of

aggregates of different architecture, but similar headgroup structure, is a local phenomenon, depending on close molecular contact between the counterion and the charged headgroup.

Reaction sites and vesicle size

The PPIE model fits the kinetic data representing the effects of vesicles on thiolysis and hydrolysis. However, several of the parameters used in the fitting procedure were derived from the effect of micelles on the same reaction or depended on assumptions regarding solubilization sites or rates of substrate permeation. In addition, use of a single rate constant to represent reactivity in the vesicle ignores the fact that a solution containing vesicles has, in principle, several potential reaction sites, and no a priori theory predicts that reactivity at all the sites is equal or comparable.

In a solution containing vesicles one can distinguish, at least, five reaction sites (Scheme 2): (1) the inner compartment, (2) the internal interface, (3) the hydrophobic bilayer itself, (4) the external interface and (5) the aqueous phase [35]. We decided to probe each site with appropriate test reactions, elaborate theoretical and experimental approaches to describe reagent distribution and reactivity in vesicle-containing solutions.

Small unilamellar vesicles (SUV)(hydrodynamic diameter, $D_h < 50$ nm) are not adequate for the purpose of site dissection. With total internal volumes seldom exceeding 1 l/mole of amphiphile in the bilayer, SUV's do not permit the entrapment of analytically convenient amounts of substrate. Larger (DODA)C and DCP vesicles ($D_h > 300$ nm), LUV, were first obtained by chloroform vaporization at 70 °C [36]. (DODA)C LUV have a sharp gel–liquid crystalline phase transition temperature (T_C), an internal volume of 9 ± 1 l/mole and exhibit ideal osmotic behavior towards KCl (0–0.050 M) and sucrose [36b]. LUV of comparable properties, obtained by dichloromethane vaporization, at 40 °C and membrane extrusion,

Scheme 2

Potential Reaction sites in a vesicle-containing solution

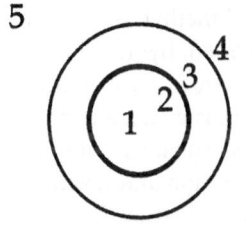

- 1 Inner Compartment
- 2 Inner Interface
- 3 Membrane
- 4 Outer Interface
- 5 Intervesicle Compartment

Table 2 Selected properties of (DODA)C vesicles prepared by different methods [38]

Preparation method and conditions	D_h [nm]	α	ζ [mV]
EtOH injection			
No salt	23	0.29	96
5 mM NaCl	25	0.23	47
Sonication			
No salt	53	0.21	85
5 mM NaCl	38	0.13	58
CHCl$_3$ injection			
No salt	285	0.04	64
4.4 mM NaCl	350	—	—

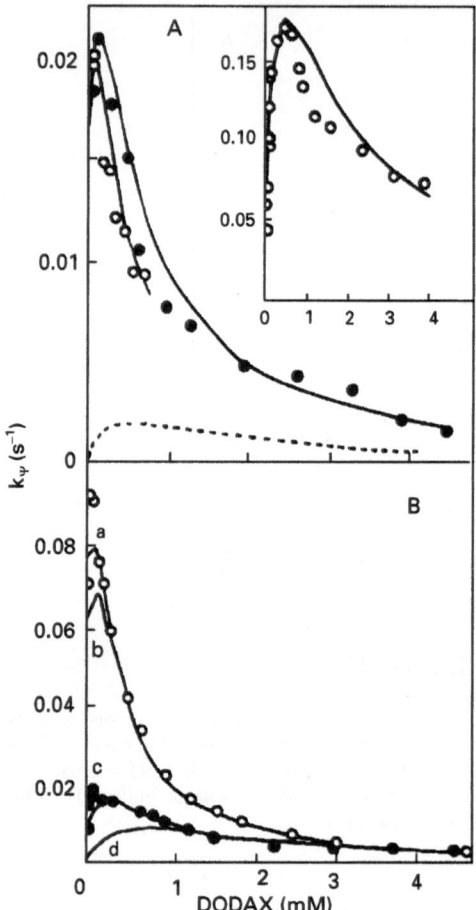

Fig. 2 Effect of vesicles on the thiolysis of NPO by heptyl mercaptan (HM). [NPO] 5×10^{-6} M **A** (DODA)C: (○) ethanolic (Tris-acetate buffer (0.05 M) pH = 5.35, [HM] = 5×10^{-5} M); (●) sonicated (sodium acetate buffer (0.05 M), pH = 5.35, [HM] = 5×10^{-5} M); (-----) large vesicles (calculated). Inset shows large vesicles (Tris-acetate buffer (0.006 M), pH = 7.17, [HM] = 8.2×10^{-5} M). **B** (DODA)B: (○) sonicated (piperazine-acetate buffer (0.0066 M) pH = 6.46, [HM] = 8.18×10^{-5} M); (●) large vesicles (piperazine-acetate buffer (0.0066 M) pH 6.57, [HM] = 8.4×10^{-5} M). Lines were calculated [41]

permit the incorporation of probes that are stable below ca. 45 °C in the internal aqueous compartment [37].

Several vesicular properties are size sensitive. D_h's of sonicated, alcohol injected, chloroform or dichloromethane vaporized, and extruded (DODA)C vesicles, determined by light scattering, range from 23 to 500 nm, depending on salt concentration (Table 2) [37, 38]. Electrophoretic mobility, measured by free flow electrophoresis, and conductivity data, yield size dependent values of zeta potentials (ζ) and degrees of ion dissociation (α) from the outer monolayer (Table 2). The decrease in α with size is probably related to decreasing headgroup area and the increasing counterion association needed to relax the surface electrostatic potential [38]. The size-dependent structural differences of the vesicles can modulate aggregate' effects on chemical reactions by affecting reagent distribution or stability as well as energy barriers along the reaction pathway.

The effect of monomer packing in reaction rates was first determined decades ago studying reactions in monolayers [39]. The rate of lactonization of γ-hydroxystearic acid (HSA) was determined in substrate monolayers as a function of applied lateral pressure (π). As π increases the film of HSA passes from an expanded to a condensed state and at the transition the rate of acid catalyzed lactonization of HSA decrease sharply. The rate decrease was attributed to a change in the "accessibility factor" related to the alteration of the fluidity of the monolayer, preventing the approach of hydroxyl and carboxylate groups [39].

Catalysis of 6-nitrobenzisoxazole-3-carboxylate (NBOC) decarboxylation by N,N dimethyl dialkylammonium bromide vesicles is modulated by the bilayer fluidity [40]. Catalytic efficiency increases with temperature, but at the phase transition there is a sharp increase in the catalysis. Several parameters, such as substrate binding constants, extent of ion dissociation and reactivity of NBOC, may change simultaneously at the T_C, com-

plicating the understanding of the temperature effect on reaction rate.

The kinetic effects of vesicles in reaction rates are sensitive to the structural consequences of size variation. Even small differences in bilayer packing, with no changes in medium or aggregate composition, modulate the rates of chemical reactions occurring at vesicular interfaces. The rate/[amphiphile] dependence obtained by studying the effect of small and large (DODA)C vesicles on the thiolysis of esters are quite different (Fig. 2) [41]. Smaller vesicles are 2–5 fold more efficient as reaction catalysts. The kinetic results were analyzed quantitatively utilizing PPIE, using substrate binding constants and selectivity coefficients for ion exchange obtained independently from fluorescence quenching and vesicle' effects on pK_a of the mercaptan. The fit of the PPIE model to the experimental data, under several conditions, and with vesicles of varying size, is excellent (Fig. 2). This analysis demonstrated that the size-dependent differences in kinetic efficiencies is attributable to differences in ion dissociation, substrate binding constants and small changes in nucleophilic reactivity due to different packing of amphiphile in the bilayer of vesicles of different sizes [41].

The inner aqueous compartment

Reactions at the inner aqueous compartment of unilamellar vesicles were investigated using water soluble probes, likely charged with respect to the vesicular interface, reacting with hydrophilic ions [42]. The necessary condition for using a reaction to probe, exclusively, the inner aqueous compartment is the choice of a particular substrate that will be essentially excluded from the interface. Even relatively minor contributions from reaction at the interface, because of the difference in the properties between interface and bulk solution [2, 3], can lead to erroneous interpretations. This condition can be best understood making reference to a very much more explored interfacial model: that of micelles. The local counterion concentration at the micelles interface can be orders of magnitude higher than that in bulk solution [2]. Hence reaction of a small proportion of the substrate with a counterion, at a site where the local concentration is high, can dominate the reaction rate (Eq. (2)).

These conditions were met using as a reaction probe of the inner aqueous compartment of (DODA)X vesicles the alkaline hydrolysis of N-methyl-4-cyanopyridinium ion (MCP). The alkaline hydrolysis of N-alkyl-4-cyanopyridinium ions (RCP) produces two products: N-alkyl-4-pyridone (P) and N-alkyl-4-carboxamido pyridinium ion (A) (Scheme 3) [43]. The limiting P/A ratio in water is

ca. 2 [43]. Hydrolysis rates and product distribution of alkaline hydrolysis of RCP for $R > C_4$ significantly increase by the addition of (CTA)X micelles [44]. The reaction rate increases upon RCP and OH^- concentration in the micelle and pyridone is basically the only product formed at the interface [44]. On the other hand, (CTA)X micelles affect neither rate or product composition of MCP hydrolysis, demonstrating that this substrate is essentially excluded from the positively charged micellar interface [45]. Micellar exclusion of MCP is also suggested by independent fluorescence quenching experiments [45]. The alkaline hydrolysis of MCP was, therefore, selected as a test reaction to explore the internal aqueous compartment of positively charged vesicles, since this substrate does not bind to the bilayer to any significant extent.

The relative impermeability of the bilayer of (DODA)B vesicles allows control of reaction rates. Choosing a vesicle-retained substrate permits the isolation of the internal compartment. Changes in the composition of the external media can lead to slow or fast change in the internal media, depending on the permeabilities of the bilayer. This system was explored studying the rate of hydrolysis of MCP in the internal compartment of (DODA)B and lecithin vesicles as a function of external pH [42]. Large (DODA)B vesicles containing MCP chloride only in the aqueous compartment were added to buffers at several pHs. In the intravesicular aqueous phase the rate constant of MCP hydrolysis is lower than that in aqueous phase over a wide pH range (Fig. 3A). Similar results were obtained, using MCP-containing large egg lecithin/DCP (10/1 mole/mole) reverse phase vesicles (Fig. 3B). These data could be rationalized by assuming that: (a) the rate of MCP hydrolysis is limited by OH^- permeation across the vesicle bilayer; (b) a slow permeation of the MCP through the bilayer reacting in the external aqueous compartment or (c) a change in OH^- reactivity in the aqueous internal compartment.

The diffusion of OH^-/H^+ through the membrane of (DODA)B and Lecithin vesicles was determined incorporating a water soluble positively charged pH probe, 2-hydroxy-5-(2-trimethyl-ammonium) acetyl benzoate (2-HTAB, $pK_a = 10.5$) (Scheme 3) in both preparations. If OH^- diffusion in, or probe leakage out, were fast, no kinetics of probe dissociation could be observed in the time-scale of these experiments (several seconds). With both (DODA)B and lecithin vesicles the half life for deprotonation was approximately 7 min, in the range of the half-life observed for MCP hydrolysis in the internal aqueous compartment of the same vesicles (Fig. 3C). These result indicate that slow OH^- diffusion through the membrane is responsible for the difference observed in the rate of MCP hydrolysis in the outer and inner aqueous compartments. The vesicle-entrapped dissociated probe can be

Scheme 3

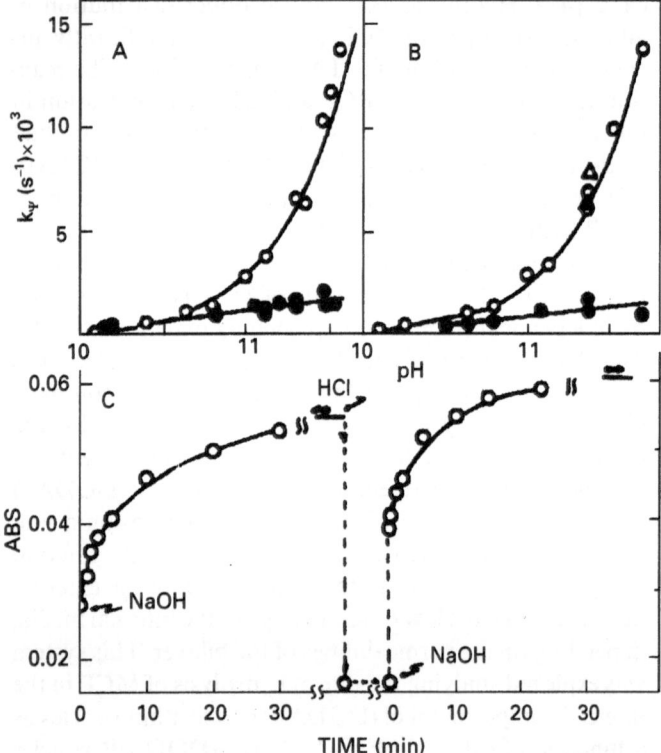

Fig. 3 Effect of external pH on internal vesicle pH and MCP hydrolysis. In all experiments, a vesicle sample of 1 ml, was eluted from a Sephadex G-25 column (1 × 50 cm) with the appropriate solution. The column was previously saturated with sonicated vesicles of (DODA)B or lecithin to avoid vesicle disruption due to adsorption on the Sephadex. The vesicles eluted in the void volume were pooled and a 0.05 ml aliquot was added to 1.5 ml buffer. Borate buffer (5 mM) was used at pH 10.2 and NaOH for all other pHs. The samples were maintained at 30 °C and the hydrolysis of MCP was followed at 265 nm. **A** Effect of pH on the hydrolysis of free (o) and (DODA)B-entrapped MCP (●). Chloroformic (DODA)B vesicles were prepared in 5 mM MCP chloride and 0.95 M erythritol. The vesicles were eluted from the column with erythritol 0.95 M and NaCl 5 mM. Erythritol 0.95 M was used in all buffers and the ionic strength was maintained at 5 mM with NaCl in order to avoid osmotic and ionic stress. Final concentrations of MCP and (DODA)B were 5×10^{-6} M and 7×10^{-5} M, respectively. **B** Effect of pH on hydrolysis of free (o) and Lecithin entrapped MCP (●). Lecithin vesicles were prepared by reverse phase evaporation (0.01 M, with 10% DCP) in 0.16 M KCl and 5 mM MCP, and the sample was eluted from Sephadex with KCl (0.165 M). All buffers contained KCl (0.16 M) and the kinetics were obtained with lecithin (5.6×10^{-5} M) and MCP (5×10^{-6} M). (▲) Lecithin vesicles eluted from the column were incubated with 5×10^{-7} M Val for 15 min. After pH change the rate of MCP hydrolysis was the same as in aqueous phase. In a separate experiment, lecithin vesicles containing MCP only in the internal compartment were incubated with Val for 90 min and refiltered in Sephadex. No free MCP was found. An aliquot of the refiltered vesicles was added to NaOH and the rate constant was measured (△). **C** Kinetics of the effect of pH on the absorbance of intravesicular 2-HTAB. (DODA)B vesicles (5 mM) prepared with 5 mM 2-HTAB, 5 mM NaCl and 0.95 M erythritol were eluted from Sephadex G-25 with NaCl 5 mM and erythritol 0.95 M. Aliquots (0.05 ml) of (DODA)B vesicles were added to 1.5 ml of a solution containing erythritol (0.95 M), NaOH, pH 11.59 and the absorbance was measured at 340 nm. At the end of the reaction, HCl was added. A new addition of NaOH promotes a similar increase of absorbance [42]

rapidly protonated by addition of HCl and slowly deprotonated with NaOH, indicating that 2-HTAB does not leak from the internal compartment to the external aqueous phase [42].

These results were confirmed using lecithin vesicles containing MCP and Valinomycin (Val). This peptide, a known K^+ transporter should compensate the electrogenic inward OH^- diffusion, facilitating OH^- transport. The rate of MCP hydrolysis in lecithin vesicles in the presence of Val reaches the same value as that observed in aqueous phase, indicating that the OH^- inward diffusion can limit the reaction rate of a substrate contained in the inner vesicular compartment (Fig. 3B). The incubation of Lecithin vesicles containing only internal MCP with Val, followed by refiltration on Sephadex and external alkalinization yields the same rate constant. This result confirms that the external OH diffuses through the vesicle bilayer, that MCP reacts in the internal compartment and there is no leakage of the probe to the external aqueous medium. These experiments were not done with (DODA)B vesicles since Val precipitated. Hence, at least for reactions with small, hydrophilic and positive or zwitterionic probes, the reactivity in the aqueous compartment of large (DODA)X and lecithin vesicles is identical to that in the external aqueous phase. In lecithin vesicles, the presence of DCP does not affect reaction rate probably because with the high ionic strength used (0.96 M KCl) MCP does not extensively bind to the interface and is restricted to the aqueous phase (see below).

The demonstration that the reactivity of substrates in the large internal aqueous compartment of vesicles with $D_h > 200$ nm is identical to that in free solution is consistent with results in reverse micelles, showing that chemical reactivity in water/detergent ratios higher than 20–30 is comparable to that in the continuous solvent [46].

The results obtained probing reactivity in the internal aqueous compartment raised several questions concerning ion distribution at the inner outer surfaces and about chemical reactivity in vesicles. It was of interest to theoretically understand ion distribution in the different compartments of a vesicle-containing solution and to determine whether inward reagent flux could be used to control reactivity. Other experimental data demonstrating slow OH^- permeation have been reported [47].

Ion distribution in a solution containing vesicles

In a solution containing vesicles ions, can reside in the continuous solution around the vesicles or in the internal aqueous compartment. The ion distribution has been described using PBE under several limiting conditions such as (a) no added salt, (b) low amphiphile and salt, (c) assuming electroneutrality in the inner compartment, (d) vesicles with no curvature, and (e) hard shells [48]. The

Progr Colloid Polym Sci (1997) 103:67–77
© Steinkopff Verlag 1997

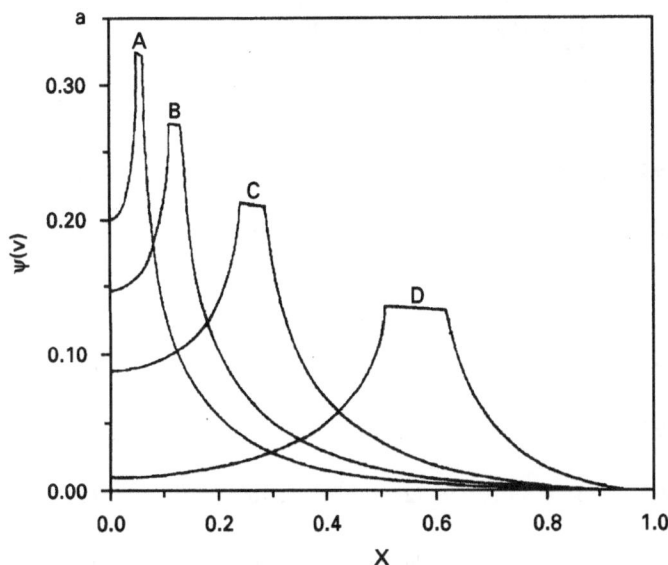

Fig. 4 Electrical potential (ψ, V) as a function of reduced distance (x = distance from the center of the vesicle to the diameter of the aspherical cell containing one micelle/diameter of the cell) from the vesicle center for different amphiphile concentrations in the absence of added salt. Curves A through D correspond to the following amphiphile concentrations ($\times 10^3$): 0.1, 1.0, 10.0, 100 M, respectively. The corresponding cell diameters (nm) are: 428.7, 199.0, 92.4 and 42.9, respectively. The vesicle diameter was 265 nm and the area per head group was taken as 75 Å^2 [49]

PBE equation was also solved numerically assuming water and ion-permeable hollow spheres treating specific ion adsorption using a Volmer isotherm [49]. The calculations suggest that the distribution of ions in the internal aqueous of (DODA)X vesicles is measurable and that the value of the electrical potential, ψ, at the vesicle center is not negligible at moderately low salt concentration. As could be expected the value of ψ at the vesicle center decreases with vesicle size and vesicle concentration (Fig. 4). Using realistic parameters, for a 265 nm (DODA)X vesicle, the value of the potential at the vesicle center can reach 100 mV [49]. The calculations yielded results that are consistent with measured values for external ion dissociation and zeta potentials of vesicles of synthetic amphiphiles.

The PBE calculations showing appreciable concentrations of counterions in the inner aqueous compartment of the vesicle is consistent with the results showing that the reactivity of OH⁻ in (DODA)C is comparable in the inner and outer compartments. Direct measurement of ion concentrations in the internal aqueous of synthetic amphiphile vesicles has not been reported. We have recently measured the concentration of counterions in the intermicellar aqueous phase using the dediazionation method first described by Romsted [50]. Basically, this method consists in determining the products from the fast reaction of a substituted

phenyl cation, formed by spontaneous dediazoniation of a diazo derivative in mildly acid solutions, with the surrounding nucleophiles. The choice of a very water soluble diazo compound (2, 4, 6 trimethylbenzene diazonium), which is essentially excluded from the micellar interface, allowed the determination of the extent of ion dissociation from (CTA)B and (CTA)C, as well as the determination of the ion selectivity coefficient for exchange [51]. We have recently communicated that this method can be applied for the determination of Cl⁻ in the aqueous compartment of (DODA)C vesicles [52]. Our preliminary data suggest that the internal chloride ion concentration is consistent with the previous calculation of ion distribution obtained by integration of the PBE equation [52].

Vesicular interfaces

Reactivity at the vesicular interfaces can be probed with substrates that bind and/or react preferentially in the inner or outer surfaces. The negatively charged DTNB and OH⁻ bind to the positive surfaces of (DODA)B [53]. The rate of alkaline decomposition of DTNB increases with [(DODA)B] since both reagents are concentrated at both internal and external surfaces. DTNB was selectively incorporated (at pH 5.3) in the inner and/or outer surfaces of positively charged (DODA)B vesicles in order to probe reactivity with OH⁻ at both surfaces. A pH jump, obtained by NaOH addition, leads to a slow OH⁻ permeation through the vesicle membrane and slow increase in the pH of the internal vesicle compartment. Under these reaction conditions the rate constant for DTNB hydrolysis, at the internal surface of (DODA)B vesicles (k_{in}) is 90 fold slower than that at the external surface (k_{out}) ($k_{out}/k_{in} = 90$) (Table 3). Alkalinization with a buffer containing rapidly permeating species, such as triethylammine at pH 11, where a high fraction of triethylamine is neutral, yields equal values for reaction rates at both surfaces ($k_{out}/k_{in} = 1$). This latter results shows that the reactivity of OH⁻ at both surfaces is identical (at least in this reaction) and that the lower k_{in} measured upon alkalinization with NaOH is due to a slow OH⁻ permeation through the membrane [53].

The reaction rate at the external surface can be modulated by changing the nature of added salt. (DODA)B vesicles are unstable at moderately high salt and are prepared in 0.005 M NaCl and 5×10^{-5} M DTNB. The external DTNB is removed by Dowex 21 K filtration and the vesicles are eluted with an isoosmolar NaCl solution. The ion composition of the external solution can be changed by dilution of the vesicles with, for example, NaBr 0.005 M, to maintain the total ionic strength. The addition of NaBr at the external aqueous

Table 3 Effect of buffer composition on the rate constants of DTNB hydrolysis at the internal (k_1) and external (k_2) interface of (DODA)C vesicles [53]

Buffer composition	pH	[(DODA)C] [M] $\times 10^5$	k_1 [s^{-1}] $\times 10^3$	k_2 [s^{-1}] $\times 10^3$	k_2/k_1
Borate (5 mM), NaCl (2.5 mM)	9.2	5.3	2.0	5.6	2.8
Borate (5 mM), NaBr (2.5 mM)	9.2	5.3	1.8	3.7	2.1
NaOH (1 mM), NaCl (4 mM)	10.9	5.3	2	180	90
NaOH (1 mM), NaBr (4 mM)	10.8	5.3	3.3	44	13.3
Triethylamine HCl (5 mM)	9.7	5.3	7.8	13	1.7
Triethylamine HBr (5 mM)	9.7	5.3	6.2	3.9	0.6
Triethylamine HCl (10 mM)	10.8	13	72	160	2.2
Triethylamine HBr (10 mM)	10.9	13	98	63	0.6

compartment of (DODA)B vesicles in NaOH changes the ratio k_{out}/k_{in} to 13, indicating the externally bound OH$^-$ and DTNB are exchanged by Br$^-$. At the inner interface, because of the low rate of Br$^-$ permeation, the local concentration of OH$^-$ and DTNB remain unchanged. This ratio can be even lower when the NaBr is added in triethylamine/HCl buffer. Since triethylamine rapidly permeates the bilayer the internal pH equilibrates fast upon external alkalinization. Moreover the nonpermeant bromide ion exchanges with OH$^-$ exclusively at the outer interface. As a result the rate at the internal interface is higher than that at the outer interface and the k_{out}/k_{in} ratio reaches 0.6 (Table 3) [53]. These results, exemplifying selective reaction site control by surface composition, nicely demonstrate that differential ion binding at the surface can promote rate modulation of compartmentalized substrates. This in/out selectivity can be magnified in asymetric vesicles, where the composition of the external leaflet is different from that in the internal interface [54]. In fact, natural selection has provided cells with a variety of bilayer bordered compartments, where the lipid composition of outer and inner leaflets are highly asymmetric, affording added pathways for reaction site control [55].

Conclusions and perspectives

Vesicles are multicompartimentalized microreactors capable of concentrating reactants or maintaining reactants separated in solution. The complexity of vesicular systems offer multiple applications for reaction control. Vesicles can catalyze reactions by significant factors. Quantitative analysis of vesicle-modified reaction rates, made possible by using models initially developed for analyzing micellar-modified reaction rates, permits dissection of factors leading to catalysis or inhibition. Further developments of these systems as microreactors depend on deeper understanding of reagent distribution in the vesicular compartments, specially in the inner aqueous compartments, and the development of more stable bilayers.

Acknowledgments This work was partially supported by the following brazilian granting Agencies: FAPESP, CNPq, PADCT-FINEP.

References

1. Menger F, Gabrielson KD (1995) Angew Chem Int Ed Engl 34:2091–2106
2. Fendler JH (1982) Membrane Mimetic Chemistry. Wiley, New York
3. Israelachvili J (1991) Intermolecular and Surface Forces, 2nd ed. Academic Press, San Diego
4. Ferris JP, Hill AR, Liv RH, Orgel LE (1996) Nature 381:59–61
5. Pozzi G, Birault V, Werner B, Dannenmuller O, Nakatani Y, Ourisson G, Terakawa S (1996) Angew Chem Int Ed Engl 35:177–180
6. Luisi P, Varela FJ (1986) Origin Life Evol Biosphere 19:633–643
7. Bangham AD, Horne RW (1964) J Mol Biol 8:660–668
8. Lasic DD (1989) La Recherche 200:904–913

9. Kunitake T, Okahata Y, Tamaki K, Kumamuro F, Yakayanagi M (1977) Chem Lett 387–390

10. Mortara RA, Quina FH, Chaimovich H (1978) Biochem Biophys Res Comm 81:1080–1086

11. (a) Lasic DD, Barenholz Y (eds) (1966) Handbook of Nonmedical Applications of Liposomes. CRC Press, New York (b) Engberts JBFN (1995) Biochim Biophys Acta 1241:232–240

12. Jager DA, Clennan MW, Jamrozik J (1990) J Am Chem Soc 112:1171–1176

13. Hupe DJ, Jencks WP (1977) J Am Chem Soc 99:451–464

14. Correia VR, Cuccovia IM, Chaimovich H (1993) J Phys Org Chem 6:7–14

15. Cuccovia IM, Aleixo RMV, Mortara RA, Berci-Filho P, Bonilha JBS, Quina FH, Chaimovich H (1979) Tetrahedron Lett 3065–3068

16. Cuccovia IM, Chaimovich H (1966) In: Lasic DD, Barenholz Y (eds) Handbook of NonMedical Applications of Liposomes. CRC Press, New York, pp 239–256

17. Chaimovich H, Bonilha JBS, Zanette D, Cuccovia IM (1984) In: Mittal KL, Lindmann B (eds) Surfactants in Solution, Vol. 2. Plenum Press, New York, pp 1121–1138

18. Jencks WP (1969) Catalysis in Chemistry and Enzymology, McGraw-Hill, New York

19. Fendler JH, Hinze WL (1981) J Am Chem Soc 103:5439–5447

20. Cuccovia IM, Quina FH, Chaimovich H (1982) Tetrahedron 38:917–920

21. Chaimovich H, Aleixo RMV, Cuccovia IM, Zanette D, Quina FH (1984) In: Mittal KL, Fendler EJ (eds) Solution Behavior of Surfactants, Vol 2: Plenum Press, New York, pp 949–973

22. Bunton CA (1991) In: Mittal KL (ed) Surfactants in Solution, Vol 11. Plenum Press, New York, pp 17–40

23. Singer LA (1982) In: Mittal KL, Fendler EJ (eds) Solution Behaviour of Surfactants, Vol 1. Plenum Press, New York, pp 73–112

24. Quina FH, Alonso EO, Farah JPS (1995) J Phys Chem 99:11 708–11 714

25. Abuin EB, Lissi EA (1983) J Colloid Interface Sci 95:198–203

26. Bunton CA, Nome F, Quina FH, Romsted LS (1991) Acc Chem Res 24:357–364

27. (a) Lissi EA, Abuin EB, Cuccovia IM, Chaimovich H (1986) J Colloid Interface Sci 112:513–520; (b) Lissi E, Abuin E, Ribot G, Valenzuela E, Chaimovich H, Araujo P, Aleixo RMV, Cuccovia IM (1985) J Colloid Interface Sci 103:139–144

28. Brochsztein S, Berci-Filho P, Toscano VG, Chaimovich H, Politi MJ (1990) J Phys Chem 94:6781–6785

29. Kolthoff IM, Johonson WF (1951) J Am Chem Soc 73:4563–4568

30. (a) Romsted LS (1977) In: Mittal KL (ed) Micellization, Solubilization and Microemulsions, Vol 2. pp 509–530 (b) Quina FH, Chaimovich H (1979) J Phys Chem 83:1844–1850

31. Bunton CA, Savelli G (1986) Adv Phys Org Chem 22:213–309

32. Abuin EA, Lissi EB, Backer E, Zanocco A, Whitten D (1989) J Phys Chem 93:4886–4890

33. Cuccovia IM, Chaimovich H, Lissi EB, Abuin EA (1990) Langmuir 6:1601–1604

34. Abuin EA, Lissi EB, Araujo PS, Aleixo RMV, Chaimovich H, Bianchi N, Miola L, Quina FH (1983) J Colloid Interface Sci 96:293–295

35. Furhop J, Mathieu J (1984) Angew Chem Int Engl 23:100–113

36. (a) Carmona-Ribeiro AM, Chaimovich H (1983) Biochim Biophys Acta 733:172–179; (b) Carmona-Ribeiro AM, Yoshida LS, Chaimovich H (1985) J Phys Chem 89:2928–2933

37. Cuccovia IM, Sesso A, Abuin E, Okino PF, Tavares PG, Campos JFS, Florenzano FH, Chaimovich H, J Mol Liquids, accepted

38. Cuccovia IM, Feitosa ES, Chaimovich H, Reed W (1990) J Phys Chem 94:3722–3725

39. Davies JT (1949) Trans Faraday Soc 45:448–460

40. Kunitake T, Okahata Y, Ando R, Shinkai S, Hirakawa S (1980) J Am Chem Soc 102:7877–7881

41. Kawamuro MK, Chaimovich H, Abuin EB, Lissi EA, Cuccovia IM (1991) J Phys Chem 95:1458–1463

42. Cuccovia IM, Chaimovich H (1990) Bol Soc Chil Quim 35:39–42

43. Kosower EM, Patton JW (1966) Tetrahedron 22:2081–2093

44. Politi MJ, Chaimovich H (1991) J Phys Org Chem 4:207–216

45. Chaimovich H, Bonilha JBS, Politi MJ, Quina FH (1979) J Phys Chem 86:1951–1953

46. Wong M, Thomas TK, Nowak T (1977) J Am Chem Soc 99:4730–4734

47. Moss RA, Ganguli S, Okumura Y, Fujita T (1990) J Am Chem Soc 112:6391–6392

48. (a) Mille M, Vanderkooi G (1977) J Colloid Interface Sci 61:455–474; (b) Bentz J (1982) J Colloid Interface Sci 90:164–182

49. Feitosa ES, Agostinho A, Chaimovich H (1993) Langmuir 9:702–707

50. Chaudhuri A, Loughlin JA, Romsted LS, Yao J (1993) J Am Chem Soc 115:8351–8361

51. Cuccovia IM, da Silva IN, Chaimovich H, Romsted LS Langmuir, in press

52. da Silva IN, Chaimovich H, Cuccovia IM (1995) Resumos da XXV Reunião Anual da Sociedade Brasileira de Bioquímica e Biologia Molecular, Caxambú, MG, Brasil, p 155

53. Cuccovia IM, Kawamuro MK, Krutman MAK, Chaimovich H (1989) J Am Chem Soc 111:365–366

54. (a) Moss RA, Bhattacharya S, Scrimin P, Swarup S (1987) J Am Chem Soc 109:5740–5744; (b) Moss RA, Bhattacharya S, Chatterjee S (1989) J Am Chem Soc 111:3680–3687

55. Jain M, Wagner RC (1980) Introduction to Biological Membranes. Wiley, New York

Progr Colloid Polym Sci (1997) 103:78–86
© Steinkopff Verlag 1997

K. Beyer

Packing and bilayer–micelle transitions in mixed surfactant–lipid systems as studied by solid state NMR

Received: 15 November 1996
Accepted: 30 November 1996

Dr. K. Beyer (✉)
Institut für Physikalische Biochemie
der Ludwig-Maximilians-Universität
Schillerstraße 44
80336 München, Germany

Abstract Chain packing and phase transitions in liquid crystalline mixtures of phospholipids and fatty acids with the akyl poly(oxyethylene) surfactant $C_{12}E_8$ were studied by ^{2}H- and ^{31}P-NMR. The packing constraints that the lipids encounter in different phase structures are reflected by order parameter profiles and by the average chain length, average interfacial area and isobaric expansion coefficients that can be obtained from the ordering profiles. The lamellar–hexagonal or lamellar–micellar transition is shown to result from the lateral packing stress as a consequence of the temperature dependence of headgroup hydration in the phospholipid/$C_{12}E_8$ system. Micelle formation in saturated phospholipid/$C_{12}E_8$ mixtures probably involves a disorder/order transition in the phospholipids as shown by X-ray scattering.

Key words Phospholipids – alkyl poly(oxyethylene) – surfactants – mixed micelles – lamellar – hexagonal – NMR

Introduction

The solubility of phospholipid membranes by non-ionic detergents is a basic technique in membrane biochemistry. Non-ionic rather than ionic surfactants are usually the first choice in biomembrane work when the preservation of structure and function of membrane proteins is desired. Two classes of non-ionic surfactants have found broad application, namely the n-alkylglycosides and the poly(oxyethylene)-n-alkyl- and acyl ethers. The former compounds are sometimes preferable for the reconstitution of proteins into closed membrane vesicles, due to their large critical micelle concentrations (cmc), while better protection against protein denaturation is usually achieved by the latter class of surfactants [1, 2].

The present contribution is focussed on the relation between chain packing and mesophase structure in liquid crystalline mixtures of phospholipids with alkyl poly(oxyethylene) surfactants. These compounds are frequently referred to as C_mE_n where m and n denote the number of carbons in the alkyl chain and the number of ethylene oxide units, respectively. Complete phase diagrams for a series of binary C_mE_n/H_2O systems with $m = 8, 12, 16$ and n ranging from 3 to 8 have been obtained by Mitchell et al. [3]. The sequence of mesomorphic phase structures observed upon changing temperature, composition and chain length was interpreted on the basis of molecular shape considerations introduced by Israelachvili et al. [4]. This model assumes that the hydrocarbon chains in the interior of the mesomorphic structures behave like an incompressible fluid and, as a corollary, that the average area per molecule determines the phase structure. Using the notation of Israelachvili et al. the condition for lamellar (L_α) packing is given by $\frac{1}{2} < V/A_0L_C < 1$ and that for normal hexagonal (H_I) packing by $\frac{1}{3} < V/A_0L_C < \frac{1}{2}$, where V, A_0 and L_C are the volume, the interfacial area and the critical length of the hydrophobic chain of the surfactant monomers, respectively. The length L_C which is defined as the cutoff value beyond which the chains cannot stretch is assumed to be equal or slightly less than the *all trans* length of the chains and the value V is given as the sum of the segment (CH_2 and CH_3) volumes. Therefore, as V and L_C are fixed, the phase geometry is a function of the average interfacial area A_0.

These considerations do not account explicitly for the volume of the hydrated headgroups. Headgroup hydration plays a major role in the mesomorphism of the C_mE_n surfactants, as it contributes to the average interfacial area and thereby affects the aggregate geometry. The hydration of the oxyethylene groups in the series $C_{12}E_n$ has been studied recently using the isopiestic method [5]. At room temperature the water binding capacity was approximately 2 mol and the average free energy of water binding 7 kJ per mol of oxyethylene residues, in good agreement with earlier reports on water binding in the $C_{12}E_n$ series [6] and in the isooctylphenol poly(oxyethylene) surfactant Triton-X100 [7].

The problem of chain packing in lamellar (L_α) and hexagonal (H_I) phospholipid–surfactant *mixtures* will be addressed here with emphasis on the technically important surfactant $C_{12}E_8$. Among the $C_{12}E_n$ detergents for which complete binary phase diagrams have been published, $C_{12}E_8$ exhibits the greatest propensity for the spontaneous formation of liquid crystalline structures with positive surface curvature[1] (i.e. normal hexagonal, H_I or micellar, L_1), as a result of its cone-like average molecular shape. Thus, the incorporation of $C_{12}E_8$ into phospholipid liquid crystalline bilayers results in a force imbalance (usually referred to as "frustration") that eventually leads to the L_α–L_1 or equivalently, to the L_α–H_I transition. The infinite rods of the normal hexagonal phase can be considered as a paradigm of the spherical or non-spherical mixed micelles that form as a result of membrane solubilization [8].

Relevant NMR theory

Solid state NMR spectroscopy is especially suited for the determination of local order and segmental dynamics, providing that local and whole aggregate motions take place on different time scales. Thus, tensorial quantities such as quadrupolar or dipolar couplings and anisotropic Zeeman interactions can be directly observed. These interactions are averaged in small aggregate structures such as sonicated vesicles or small mixed micelles. The desired information can be obtained, however, in mesomorphic

phases with long-range order by solid state ^{31}P- and ^2H-NMR spectroscopy. The concepts pertinent to the application of these techniques to liquid crystalline aggregates will be only briefly introduced here. Comprehensive treatments can be found in refs. [9–11].

Lipids in a liquid crystalline aggregate undergo rapid axial motion where the normal to the aggregate surface represents an axis of motional averaging. In a planar bilayer axial symmetric motion about the normal to the bilayer plane results in the experimentally observed quadrupolar splitting

$$\Delta v_Q^{(i)} = \tfrac{3}{2}\chi P_2(\cos \Theta)S_{CD}^{(i)} , \tag{1}$$

where P_2 denotes the second rank Legendre polynomial, Θ the angle between the bilayer normal and the magnetic field and χ the quadrupolar coupling constant (170 kHz for C–^2H bonds). The local bond order parameter $S_{CD}^{(i)}$ is given by

$$S_{CD}^{(i)} \equiv \langle P_2(\cos \alpha_i) \rangle = \tfrac{1}{2}\langle 3 \cos^2 \alpha_i - 1 \rangle , \tag{2}$$

where α_i denotes the time-dependent angle between the ith C–^2H-bond and the bilayer normal. The angular brackets indicate a time or ensemble average. After selective deuteration, replacing C–H by C–^2H vectors, local ordering information can be obtained by ^2H-NMR. Bond order parameters can be defined for any C–^2H bond vector in a lipid molecule.

In a cylindrical aggregate there is an additional symmetry axis due to rapid motion of the molecules over the cylinder surface. Thus,

$$\Delta v_Q^{(i)} = \tfrac{3}{2}\chi P_2(\cos \zeta) P_2(\cos \Theta)S_{CD}^{(i)} , \tag{3}$$

where ζ denotes the angle between the average molecular long axis (which is identical with the bilayer normal in the lamellar case) and the cylinder axis. The angle ζ is 90° which results in scaling of the quadrupolar splittings in the hexagonal phase by a factor of $-\tfrac{1}{2}$.

The segmental order parameters in a phospholipid bilayer as determined from the quadrupolar splittings can be translated into the average projected length of the hydrocarbon chains [12, 13]

$$\langle L_{chain} \rangle = l_0 \left[\frac{b-a+1}{2} + \sum_{i=a}^{b-1} |S_{CD}^{(i)}| + 3|S_{CD}^{(b)}| \right] , \tag{4}$$

where the index i refers to the numbering of the acyl chains starting with the carbonyl carbon ($i = 1$) and ending with the methyl carbon ($i = b$), e.g. $b = 14$ for DMPC. The effective length of the sn-1 acyl chain is taken as extending from the C_1 carbon to the methyl carbon ($a = 2$) while in the case of the sn-2 chain the contribution of C-2 is neglected ($a = 3$). The length of a single C–C-bond projected on to the long axis of the *all trans* reference state is

---

[1] Positive surface curvature means convex with respect to the bulk water phase

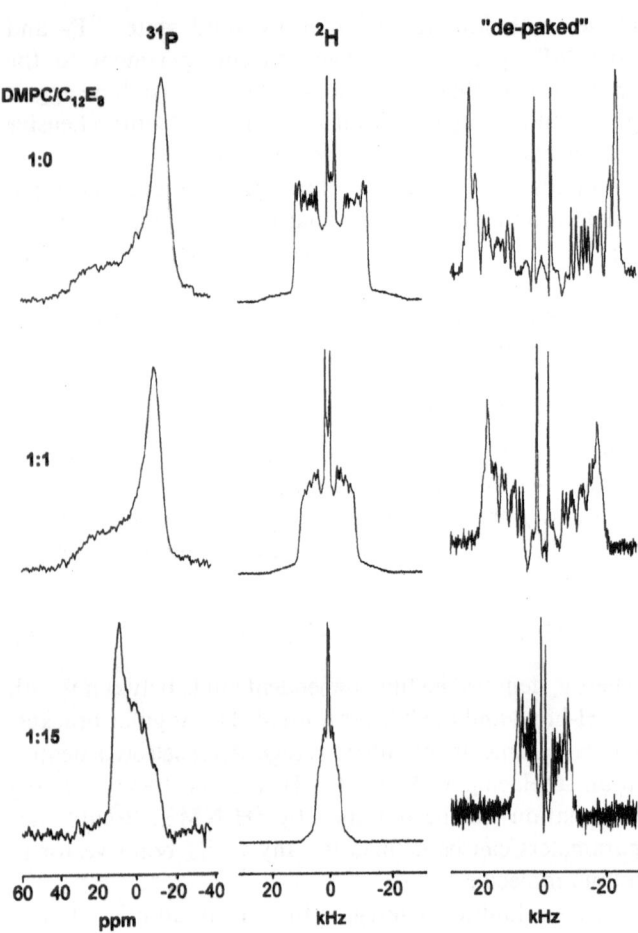

Fig. 1 ^{31}P- and ^{2}H-NMR spectra of liquid crystalline lamellar DMPC-d$_{54}$ (first row) and of lamellar (second row) and hexagonal (third row) DMPC-d$_{54}$/C$_{12}$E$_{8}$ mixtures. Powder type (center column) and corresponding dePaked (right column) ^{2}H spectra are shown. Water content in all samples, 50 wt%. Temperature, 45 °C

The temperature dependence of $\langle L \rangle$ yields an estimate of the isobaric thermal expansion coefficient parallel to the director axis

$$\alpha_{\parallel} \equiv \frac{1}{\langle L \rangle}\left[\frac{\partial \langle L \rangle}{\partial T}\right]_{p} = \left[\frac{\partial \ln \langle L \rangle}{\partial T}\right]_{p}. \tag{7}$$

The ^{31}P-NMR line shape is dominated by the chemical shielding tensor [10]. In a bilayer the orientation dependence is again given by the second Legendre polynomial, due to the axial motion of the phosphate around the bilayer normal. For randomly distributed bilayers this results in a spectrum with a high-frequency shoulder for $\Theta = 0°$ and a low-frequency peak for $\Theta = 90°$. In a structure with cylindrical symmetry, diffusion about the cylinder axis leads to a scaling of the spectrum by a factor of $-\frac{1}{2}$ (cf. Fig. 1).

Hydrocarbon chain packing in mixed lamellar and hexagonal phospholipid/detergent systems

Phospholipid bilayers take up small quantities of non-ionic surfactants without disruption of the bilayer structure. The spatial relations in oriented POPC bilayers[2] incorporating non-lytic amounts of C$_{12}$E$_n$ surfactants with n ranging from 2 to 6 have been recently studied in detail. Combining X-ray and neutron scattering results it was shown that the α-methylene groups of the surfactants are strongly anchored near the hydrocarbon–water interface of the bilayers. These data also suggest that the first oxyethylene group penetrates beyond the carbonyl region of the phospholipid. A decrease of the hydrocarbon layer thickness as a result of surfactant incorporation into the bilayer was deduced from the reduction of the repeat distance and from the reduced distance of the phosphates with respect to the bilayer center. These effects which became more pronounced with increasing number of oxyethylene units were attributed to the increasing motional freedom of the phospholipid acyl chains [15].

The diffraction methods provide information on the long-range order of liquid crystalline aggregates, i.e. thermal motions of individual atoms as well as the absence of perfect long-range order may create problems that must be taken into account when analyzing diffraction data [15]. By contrast, ^{2}H- and ^{31}P-NMR give useful results even in the absence of perfect long-range order. ^{2}H-NMR provides a convenient and conclusive means for a study of order and mobility in phospholipid–surfactant mixtures, while the local geometry of the aggregates is immediately obvious from the line shape of the ^{31}P-NMR spectrum [9, 19]. Thus solid state NMR, employing the anisotropy of quadrupolar, dipolar and chemical shift interactions is

$l_0 = 1.25$ Å [13]. It must be noted that the average projected length $\langle L_{\text{chain}} \rangle$ is smaller than the bilayer half thickness $\langle t \rangle$ as obtained by diffraction techniques. This is a consequence of the so-called "chain end effects" which are also reflected by a "plateau" usually observed in order parameter profiles (cf. Fig. 2).

An estimate of the average interfacial area may be obtained from $\langle L_{\text{chain}} \rangle$, i.e.

$$\langle A_{\text{chain}} \rangle = V_{\text{chain}}/\langle L_{\text{chain}} \rangle, \tag{5}$$

where V_{chain} is given by $n_{\text{CH}_2} V_{\text{CH}_2} + V_{\text{CH}_3}$ with $V_{\text{CH}_2} = 28.0$ Å^3 and $V_{\text{CH}_3} = 2V_{\text{CH}_2}$ [14]. This certainly represents an upper limit value, while considering only the plateau values of the order parameter profile

$$\langle A \rangle' = V_{\text{plat}}/\langle L_{\text{plat}} \rangle \tag{6}$$

yields a lower limit, i.e. $V_{\text{plat}}/\langle L_{\text{plat}} \rangle \leq \langle A \rangle \leq V_{\text{chain}}/\langle L_{\text{chain}} \rangle$.

Progr Colloid Polym Sci (1997) 103:78–86
© Steinkopff Verlag 1997

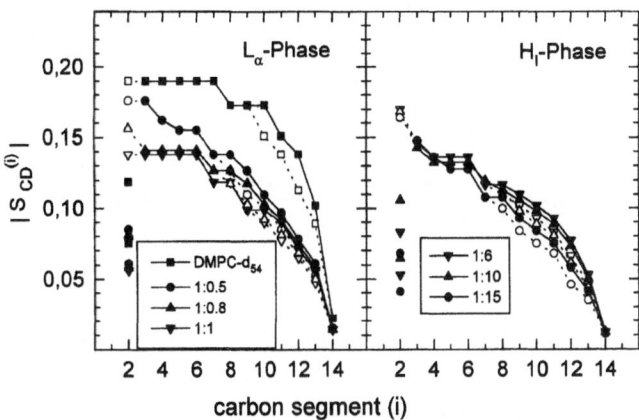

Fig. 2 Deuterium ordering profiles of the sn-1 and sn-2 chains of DMPC-d$_{54}$ in lamellar (left panel) and hexagonal (right panel) liquid crystalline mixtures with C$_{12}$E$_8$. Water content 50 wt%. Temperature, 30 °C. The molar ratios DMPC-d$_{54}$/C$_{12}$E$_8$ are given in the insets. The profile for pure DMPC-d$_{54}$ is included for comparison

complementary to X-ray and neutron scattering in liquid crystalline media.

The way ^{2}H- and ^{31}P-NMR can be used for an analysis of a phospholipid–surfactant mixture is illustrated here for the ternary system DMPC-C12E8-water[2] with the hydrogens in the phospholipid acyl chains being replaced by deuterons (DMPC-d$_{54}$). The samples in this study were made up with a water content of 50% by weight. This degree of hydration is close to the maximal swelling capacity of the oxyethylene headgroup [16] which ensures that whole body motion and surface undulations are restricted, due to strong interbilayer forces.

The spectra in Fig. 1 reflect the changes in segmental order and phase geometry arising from increasing surfactant concentration in this mixture. Numerical deconvolution of the ^{2}H-spectra obtained from the macroscopically non-oriented samples, using the so-called de-Paking routine [17], results in a considerable simplification as shown in the right column of Fig. 1. Thus, profiles of segmental order parameters $S_{CD}^{(i)}$ (Eq. (2)) can be obtained by integration of individual signals in the de-Paked spectra. The resolved ^{2}H-signals can be easily assigned by assuming that the quadrupolar splittings decrease monotonously with increasing distance of the respective carbon segments from the bilayer surface, as observed previously using selectively deuterated lipids [18].

It can be recognized from the profiles collected in Fig. 2 that addition of C$_{12}$E$_8$ leads to an abrupt decrease of the

order parameters of all chain segments. The reduction is more pronounced for segments close to the interface which also results in reduction of the curvature of the profiles. In order to obtain quantitative information the ordering profiles can be translated into the average acyl chain length projected onto the director axis, according to Eq. (4) (Table 1). The average length $\langle L_{chain} \rangle$ as obtained from the entire sn-1 acyl chain of the phospholipid decreases with increasing surfactant concentration in the membrane, in agreement with the membrane thinning observed by X-ray and neutron diffraction [15]. Unfortunately, the available diffraction and NMR data are not directly comparable due to differences in temperature and phospholipid chain length [15, 8].

Also included in Table 1 are the average areas $\langle A_{chain} \rangle$ and $\langle A \rangle'$ obtained from the entire chains (i.e. $\langle L_{chain} \rangle$) or from the plateau, according to Eqs. (5) and (6). As noted above, the calculation of $\langle A \rangle'$ from the plateau order parameters accounts for the fact that $\langle L_{chain} \rangle$ is shorter than the average bilayer half thickness $\langle t \rangle$ [19, 20, 8]. The problems encountered when correlating chain order parameters and the average interfacial areas per chain in liquid crystalline aggregates have been recently discussed by Nagle [21].

The temperature dependence of the order parameter profiles yields the isobaric thermal expansion coefficient $\alpha_{\parallel}$, according to Eq. (7). The data from the L_α mixed phase collected in the last column of Table 1 are in broad agreement with thermal expansion coefficients obtained from X-ray diffraction experiments in pure phospholipid bilayers [22, 23] indicating that in the L_α-phase the phospholipid acyl chain motion is not constrained by an increasing proportion of the surfactant.

The order parameters of the phospholipid acyl chains probably do not reflect the corresponding ordering of the surfactant in the lamellar phase. Recently, complementary order parameter profiles of the phospholipid *and* of the surfactant alkyl chains were obtained in the lamellar system DMPC/C$_{12}$E$_4$ [24]. The order parameters of the surfactant alkyl chain segments close to the hydrocarbon–water interface where by approximately 30% lower than the order parameters of the corresponding phospholipid acyl chain positions. Similarly, an order parameter of approximately 0.1 has been reported for the α-methylene group of pure liquid crystalline C$_{12}$E$_4$ in the lamellar state at 25 °C [25] which is significantly lower than the corresponding value in the DMPC/C$_{12}$E$_8$ mixtures (0.18 for the C$_2$-segment at 30 °C, see Fig. 2). It remains to be shown how the "averaging" of local order in such mixtures of single and double chain amphiphiles correlates with macroscopic (thermodynamic) properties.

The system DMPC/C$_{12}$E$_8$/50 wt% water forms a normal hexagonal (H$_1$) phase at DMPC/C$_{12}$E$_8$ molar ratios

---

[2] The following abbreviations will be used for the phospholipids. POPC, 1-palmitoyl-2-oleyl-sn-glycero-3-phosphocholine; DLPC, 1,2-dilauroyl-sn-glycero-3-phosphocholine; DMPC, 1,2-dimyristoyl-sn-glycero-3-phosphocholine; DPPC, 1,2-dipalmitoyl-sn-glycero-3-phosphocholine

Table 1 Average length $\langle L \rangle$, area $\langle A \rangle$, interfacial area $\langle A \rangle'$, and thermal expansion coefficient $\alpha_{\parallel}$ of the sn-1 chain of DMPC-d_{54} in $C_{12}E_8$/DMPC-d_{54} liquid–crystalline mixtures (water content 50 wt%)

$C_{12}E_8$/DMPC-d_5 mol:mol (45 °C)	Phase	$\langle L \rangle^a$ [Å]	$\langle A_{\text{chain}} \rangle^b$ [Å²]	$\langle A \rangle'$ [Å²]	$\alpha_{\parallel}^d$ [10^{-3} K^{-1}]
0.5	L_α	10.0	39.2	33.1^c	− 2.25^e
0.6	L_α	9.96	39.4	34.5^c	− 2.36^e
0.8	L_α	9.88	39.7	35.0^c	− 2.11^e
1.0	L_α	9.81	40.0	35.1^c	− 1.88^e
6	H_I	9.85	–	–	− 1.30^f
10	H_I	9.81	–	–	− 1.27^f
15	H_I	9.69	–	–	− 1.05^f

^{a} Calculated according to Eq. (4).

^{b} Calculated according to $\langle A \rangle = V_{\text{chain}}/\langle L_{\text{chain}} \rangle$ with $V_{\text{chain}} = n_{CH_2}V_{CH_2} + V_{CH_3}$, where $V_{CH_3} \approx 2V_{CH_2}$ and $V_{CH_2} = 28.0$ Å^3.

^{c} Calculated according to Eq. (6).

^{d} Calculated according to Eq. (7), involving the entire acyl chain length $\langle L_{\text{chain}} \rangle$.

^{e} Determined in the temperature range 30–50 °C.

^{f} Determined in the temperature range 15–50 °C.

< 1:6 which corresponds with the extensive H_I phase region in the binary $C_{12}E_8$/water system [3]. The presence of a hexagonal phase can be easily detected by ^{31}P-NMR as noted above (see Fig. 1). The topology of the structure (H_I or H_{II}) is not directly available from the NMR spectra, however. A suitable criterion is the behaviour upon addition of excess water. The H_I-structure breaks down into spherical micelles (as observed in the present case), while H_{II}-structures coexist with an excess of water [26].

The extreme curvature of the H_I-cylinder surface inevitably results in restricted motion of the hydrocarbon chains *within* the cylinder core, as shown by the ordering profiles (Fig. 2b). The "plateau" is much shorter here and is shifted away from the interfacial segments. The entire profile is almost linear, in contrast to the curved profiles encountered in the mixed L_α-phase. The effect of increasing surfactant concentration in the L_α- and H_I-phases can be recognized when selected order parameters are shown as a function of the phospholipid–surfactant molar ratio over a broad range of compositions (Fig. 3). The surfactant leads to rapidly decreasing chain order in the lamellar phase while the order parameters are almost concentration independent in the hexagonal state where the phospholipid represents a dilute solute.

A somewhat unexpected feature of this representation is the increasing rather than decreasing order parameter of the interfacial segments (C_2–C_4) when going from the L_α to the H_I phase which is probably a result of the motional restriction of the phospholipid chains in the normal hexagonal phase. Thus, a neutral or pivotal plane, approximately at the level of C_5, may be assumed about which the molecules turn in the lamellar–hexagonal phase transition [8]. A similar relation between order parameters and spontaneous curvature has been observed for liquid crystalline sodium laurate where an L_α–H_I phase transition can be induced by changing the water content

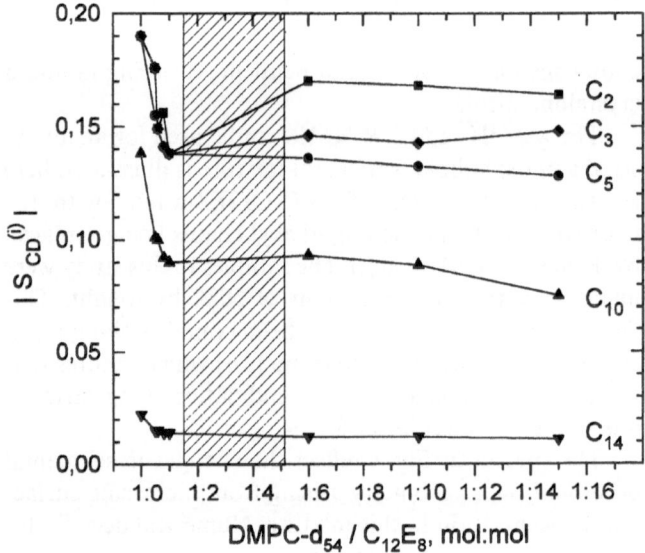

Fig. 3 Selected orientational order parameters as a function of the DMPC-d_{54}/$C_{12}E_8$ molar ratio. The hatched area indicates lamellar–hexagonal phase separation and the respective boundaries of the pure lamellar and hexagonal mixtures (cf. Fig. 2)

[20]. Likewise, the acyl chain order parameters of phosphatidylethanolamines collectively *decrease* when going from the L_α to the H_{II} phase, due to the greater motional freedom of the chains in the inverted hexagonal structure [27, 28, 20]. Geometrical models have been recently proposed that relate chain order parameters with the cylinder curvature in a H_{II}-phase [20, 29]. It has been noted that similar considerations may not be valid for the H_I phase due to the possibility of chain backbending [20].

Analogous to the lamellar phase, the temperature dependence of the order parameter profile can be used for an estimate of the thermal expansion coefficient $\alpha_{\parallel}$ (Table 1). The thermal expansion coefficients obviously reflect the packing constraints within the cylinder core. Moreover, the dependence of $\alpha_\parallel$ on the composition of the hexagonal phase indicates, that phospholipid chain motions become even more restricted at lower phospholipid concentration. It is important to realize, however, that in the normal hexagonal phase backbending conformations of the chains cannot be definitely excluded which makes the calculation of a projected chain length somewhat uncertain.

Another structural feature of the phospholipid packing in the hexagonal detergent host phase is worth mentioning. The sn-1 and sn-2 chains in pure L_α-state phospholipids are inequivalent as a consequence of the orientation of the first few segments of the sn-2 chain parallel to the membrane surface [30–32]. The resulting orientation of the α-methylene deuterons leads to reduced quadrupolar splittings. Distinct splittings from these deuterons were found in both mesomorphic phospholipid/detergent mixtures, indicating that the phospholipid backbone conformation is similar in the hexagonal and in the lamellar state [8].

The solubilization of lipids with an extreme range of chain lengths in dilute micellar solutions of non-ionic detergents is not well understood regardless of the broad application of this property. This problem has been addressed by incorporation of a series of perdeuterated fatty acids into the hexagonal phase of the $C_{12}E_8$ [33]. Again, $C_{12}E_8$ was chosen as it forms a stable H_I-phase within a broad temperature range at a fatty acid/surfactant molar ratio of 1:8 (total water content 50 wt%), thus providing a suitable model for the chain packing in mixed micelles. The fatty acids, ranging from decanoic to octadecanoic acid, were added in free (protonated) form so as to avoid electrostatic interactions in the hydrocarbon–water interface. The individual ^{2}H-NMR ordering profiles were similar to those obtained from DMPC-d_{54} in the hexagonal mixture (cf. Fig. 2), considering the short plateau and the continuously decreasing order parameter values towards the chain ends. The whole series, however, reflects the packing constraints with increasing fatty acid chain length (Fig. 4A). The entire profiles are shifted with respect to each other, i.e. the average order of the interfacial chain segments increases with increasing chain length.

An alternative representation of the same data exhibits the differential effects of chain packing on the interfacial and terminal segment. In Figs. 4B and C, order parameter values from chain segments at different distances from the interface and from the chain termini are correlated with the total fatty acid chain lengths. The order parameter values from the chain positions C_2–C_9 increase with chain length, while those from C_{n-4} to C_n decrease with the total number of carbons in the fatty acids. Notably, the latter set of order parameters tends to a common small value suggesting that the terminal chain segments sample an

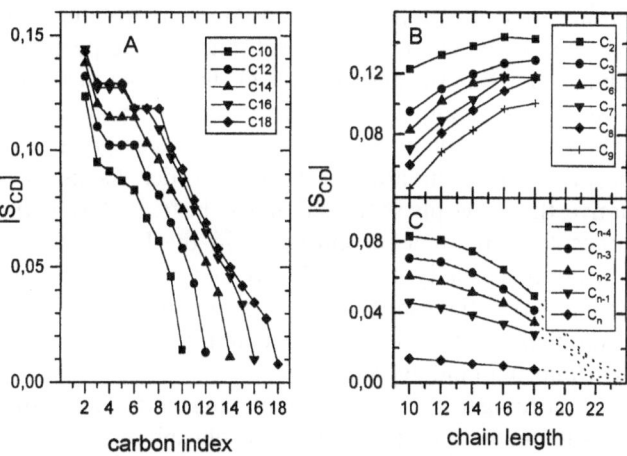

Fig. 4 ^{2}H-NMR ordering profiles of fatty acids ranging from decanoic (C10) to octadecanoic acid (C18) in a hexagonal liquid crystalline $C_{12}E_8$-host phase (A) and selected order parameters of chain segments with increasing distance from the carboxyl group (B) or from the terminal methyl group (C) as a function of the total fatty acid chain length. Temperature, 35 °C. Water content, 50 wt%

increasing configurational space, probably including backbending or hairpin conformations as noted above [34].

The shape of the order parameter profile for the longest fatty acid, octadecanoic acid, is noteworthy as it exhibits a small change in curvature beyond chain position C_{13} (Fig. 4A). This kind of deviation from the monotonous behaviour of the ordering profile has been previously observed in a bilayer membrane containing a glycosphingolipid with a deuterated N-lignoceric acid, which was by six carbon segments longer than the adjacent sphingosin chain [35]. The upward bending of the profile was attributed to finite probability of the unusually long hydrocarbon chain to bump into the opposite monolayer of the membrane where it encounters a region of increasing average order. Analogously, it was assumed that the packing stress of the long chain fatty acid in the hexagonal $C_{12}E_8$ host phase can be relieved by traversing the central rod-axis [33].

Average molecular properties and the bilayer micelle transition

The transition from planar aggregates without surface curvature into aggregates with positive surface curvature is a consequence of an imbalance of repulsive forces among the headgroups and hydrocarbon chains in the lamellar mixture [36]. The average interfacial area per molecule in liquid crystalline aggregates of a single chain amphiphile is related to the total hydrocarbon chain volume V and to the fully extended length of the chain L_C. Based on the simple packing considerations mentioned above a minimal surface area of V/L_C, $2V/L_C$ and $3V/L_C$ is obtained for bilayers, cylinders and spherical micelles, respectively. Thus, increasing the average area at the level of the headgroup in an amphiphile with a given hydrocarbon chain is expected to result in the appearance of aggregates with positive surface curvature in the above order. As an alternative, the above phase transitions can be treated using the concept of spontaneous curvature in the framework of a bending elasticity model [37]. It may therefore be anticipated that the spontaneous curvature of the detergent is the major determinant of the stability of mixed detergent/phospholipid aggregates.

According to a somewhat simplifying model [38] a surfactant can be incorporated into a phospholipid bilayer membrane up to a critical molar fraction X_S^{sat} where mixed spherical or cylindrical micelles of composition X_S^{sol} start to appear, in equilibrium with the surfactant saturated membrane. Thus, increasing the average *headgroup* area of the surfactant causes the mixed bilayers to undergo the lamellar–micellar transition at lower values of X_S^{sat}. This has been shown experimentally for the $C_{12}E_n$-series with n ranging from 5 to 8 where X_S^{sat} decreases approximately from 0.65 to 0.35 [39]. It may be noted, that the $C_{12}E_n$ surfactants with $n < 5$ fail to solubilize phospholipid bilayers at all.

Apart from changing the number of oxyethylene units a variation of the equilibrium interfacial area can be achieved by changing the headgroup hydration. The hydration of the oxyethylene headgroup is temperature-dependent, i.e. the average number of water molecules associated with the headgroup decreases with increasing temperature. Effective binding constants for D_2O in lamellar surfactant/water mixtures of $C_{12}E_3$, $C_{12}E_5$ and $C_{12}E_6$ were found to decrease by two orders of magnitude between 5 °C and 35 °C [6].

The temperature dependence of intraaggregate packing can be demonstrated in mesomorphic (low water content) and in pseudobinary (excess water) mixtures. As an example, a mesomorphic $DMPC/C_{12}E_8$/water mixture (molar ratio 1.5:1, 50 wt% of water; cf. Fig. 3) undergoes a L_α–H_I phase transition with decreasing temperature according to the ^{31}P- and ^2H-NMR line shapes [8]. The distinction of spectral features that are attributable to lamellar and hexagonal domains indicates that lipid diffusion across the domain boundaries is slow on the NMR time scale. An ^{31}P-NMR saturation transfer experiment in a similar system showed that the average lifetime of a lipid molecule in one of the coexisting subphases is ≥ 40 ms [40].

In the pseudobinary $DMPC/C_{12}E_8$-system (i.e. in the presence of an excess of water) a L_α phase exists at a surfactant mole fraction $X_S = 0.33$ above 30 °C which probably consists of large unilamellar vesicles [41, 42].

Unilamellar vesicles were observed by cryo-electron microscopy in egg lecithin/$C_{12}E_8$-mixtures at the same surfactant concentration [43]. The salient feature of the lamellar $DMPC/C_{12}E_8$ system at this particular phospholipid–surfactant molar ratio is the almost perfect orientation of the lamellae in a sufficiently strong magnetic field with the normal to the bilayer plane being perpendicular to the field vector [42]. Field orientation is a consequence of the anisotropy of the molecular diamagnetic susceptibility, i.e. $\Delta\chi = \chi_\parallel - \chi_\perp$ (assuming an axially symmetric susceptibility tensor). Thus, in a liquid crystalline aggregate containing on average N monomers at temperature T and field strength H the degree of orientation will be proportional to $\exp(-N\Delta\chi H^2/kT)$. The cooperative contribution of the N monomers to the magnetic interaction energy requires that the lipid molecules are parallel correlated in a flat bilayer. The free-energy gain resulting from field alignment thereby compensates for the bending rigidity of the curved surface of the bilayer vesicles [44]. The spontaneous field orientation is an advantage here as ^2H quadrupolar splittings and signal integrals can be directly obtained without the necessity of numerical deconvolution [17]. The orientation with respect to the magnetic field can be deduced at the same time from the chemical shift and from the line width of the ^{31}P signal (Fig. 5).

As noted above, water is more weakly bound with increasing temperature, i.e. the surfactant headgroup will be dehydrated. Subtle effects of the headgroup dehydration on the interfacial region of the phospholipid can then be studied using selectively deuterated DMPC, e.g. the order parameters of chain and headgroup segments which are close to the interface increase with temperature, in contrast to what is observed with pure DMPC [42]. Thus, the motional restriction of the interfacial segments of the phospholipid can be attributed to the "shrinkage" of the surfactant headgroup in this system. This is shown schematically in Fig. 6.

These trends are reversed with decreasing temperature, analogous to what has been observed at low water content. The average interfacial area is expected to increase due to the headgroup becoming more strongly hydrated while the average area per hydrocarbon chain decreases due to progressive chain ordering (considering the whole chain, Eq. (6)). The lateral force imbalance thus created eventually causes the membrane to undergo a transition into a structure with positive surface curvature, i.e. rods or micelles, depending on the total amount of water present in the system.

The lamellar–micellar phase transition can be recognized in the ^{31}P- and ^2H-NMR spectra (Fig. 5). The variations of chemical shift and line width in the ^{31}P spectra of the homologous system $DLPC/C_{12}E_8$ and $DMPC/C_{12}E_8$ are summarized in Fig. 7. With decreasing temperature,

Progr Colloid Polym Sci (1997) 103:78–86
© Steinkopff Verlag 1997

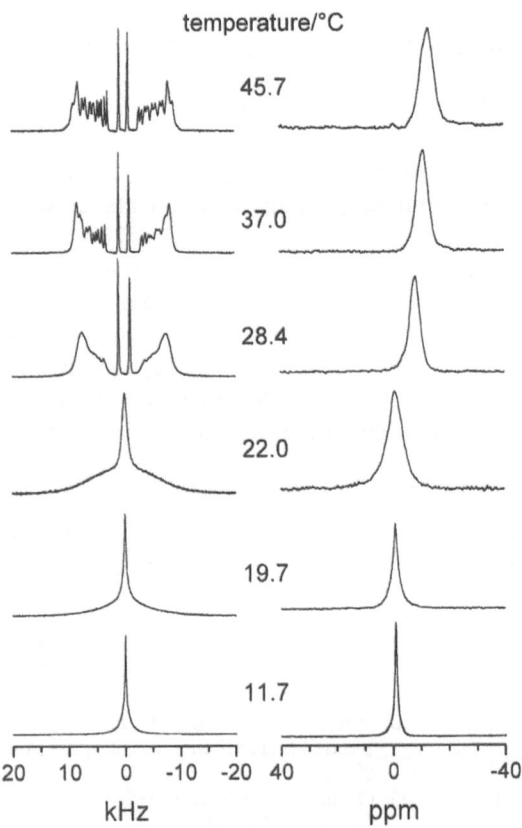

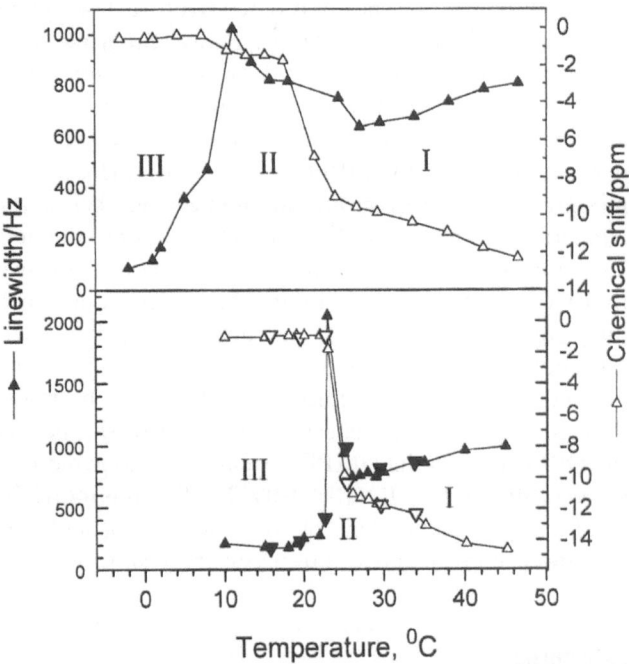

Fig. 7 ^{31}P-NMR chemical shifts and line broadenings as a function of temperature in mixtures of DLPC (upper panel) and DMPC (lower panel) with $C_{12}E_8$. Molar ratio phospholipid/surfactant, 2:1. Water content, approximately 90 wt%

Fig. 5 Temperature dependence of ^{2}H-NMR (left column) and ^{31}P-NMR spectra (right column) of a DMPC-d$_{54}$/$C_{12}E_8$ mixture at a molar ratio 2:1. Water content, approximately 90 wt%

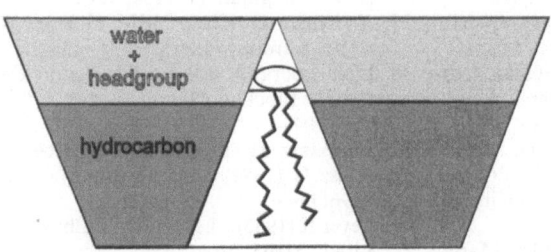

Fig. 6 Schematic representation of the geometric restraints in a lamellar phospholipid/$C_{12}E_8$ mixture. The $C_{12}E_8$ molecules are shown as inverted truncated cones [8]

the ^{31}P signal moves downfield while the line width decreases (region I). This behaviour has been attributed to a gradual loss of field orientation of the vesicles [42] as a result of increasing bending rigidity [44]. In the transition region (region II), the line width increases drastically and the samples become turbid, which may be due to size and shape fluctuations. Finally, on the low-temperature side of the transition, the narrow and homogeneous ^{31}P line at -1 ppm, the singlet signal in the ^{2}H spectrum (Fig. 5) and the absence of sample turbidity signify the

presence of small mixed micelles. It may be noted that the temperature dependence of the solubilization capacity has also been demonstrated by isothermal titration calorimetry in the system POPC/C12E8 [45].

The transition region (region II in Fig. 7) is much broader for DLPC than for DMPC. The DMPC/$C_{12}E_8$ mixture also differs from the DLPC mixture in that it involves a complex calorimetric transition (not shown) which can be attributed to chain crystallization and partial demixing of the components (see below). The total calorimetrically detectable transition ranges from 26 °C to 15 °C, i.e. it is broader than the transition range detected by ^{31}P-NMR (Fig. 7). An analogous transition was observed in the DPPC/$C_{12}E_8$ mixture. In the latter case the total enthalpy was three fold larger and the transition was by approximately 50% narrower than in the DMPC/$C_{12}E_8$ system. The enthalpy and the onset temperatures in the mixtures containing DMPC and DPPC suggest that the acyl chain order/disorder phase transitions of these phospholipids is involved in the overall transformation from bilayers to micelles and vice versa. Hence, it seems justified to assume that the hydrocarbon chains of the phospholipids (DMPC, DPPC) are in an ordered state below the micellization temperature where particles with the typical size of small spherical micelles (approximately 7 nm) were observed by different techniques [42].

Recently, the presence of ordered hydrocarbon chains could indeed be demonstrated by X-ray scattering using synchrontron radiation. A single wide angle reflection, corresponding to an ordered structure with a spacing of approximately 4.2 Å was observed in the low-temperature micellar phases of the DPPC/$C_{12}E_8$ and DMPC/$C_{12}E_8$ mixtures. This reflection appears immediately below the temperature where micellization sets in (approximately 26 °C for DMPC and 38 °C for DPPC) and persists down to 10 °C, the lowest temperature that could be attained for technical reasons. There was definitely no low angle reflection, indicating that the chain ordering is a feature of randomly distributed micelles [46]. It may be noted that the wide angle reflection was much sharper in the DPPC than in the DMPC system at the same reduced temperature ($(T - T_C)/T_C = 0.0439$; T_C, thermotropic $P_{\beta'}-L_{\alpha}$ phase transition of the pure phospholipid). Additional experiments will be necessary to show whether this difference is due to the domain (micelle) size or to the size distribution. The formation of ordered micelles has been assumed previously, when it was observed that mixtures of DPPC with the short chain phospholipid 1,2-diheptanoyl-sn-glycero-3-phosphorylcholine (DHPC) undergoes a lamellar–micellar transition and phase separation of short and long chain components close to the transition temperature of pure DPPC [47]. Interestingly, these micelles are most stable at a mole fraction of the short chain lipid component of 0.3, which is similar to the surfactant molar fraction in the present system ($x_S = 0.33$). Our preliminary X-ray study, for the first time directly reveals the existence of ordered bilayer patches in such phase separated pseudobinary systems.

Acknowledgment This research was supported by the Deutsche Forschungsgemeinschaft, SFB 266.

References

1. Helenius A, McCaslin DR, Fries E, Tanford C (1979) Methods Enzymol 63:734–749
2. Møller JV, le Maire M, Andersen JP (1986) In: Watts A, De Pont JJHHM (eds) Progress in Protein Lipid Interactions. Vol. 2, Elsevier/North-Holland, Amsterdam, pp 147–196
3. Mitchell DJ, Tiddy GJT, Waring L, Bostock T, McDonald MP (1983) J Chem Soc Faraday Trans 79:975–1000
4. Israelachvili JN, Marcelja S, Horn RG (1980) Q Rev Biophys 13:121–200
5. Klose G, Eisenblätter S, Galle J, Islamov A, Dietrich U (1995) Langmuir 11:2889–2892
6. Rendall K, Tiddy GJT (1984) J Chem Soc Faraday Trans 80:3339–3357
7. Beyer K (1981) J Colloid Interface Sci 86:73–89
8. Thurmond RL, Otten D, Brown MF, Beyer K (1994) J Phys Chem 98:972–983
9. Seelig J (1977) Q Rev Biophys 10:353–418
10. Seelig J (1978) Biochim Biophys Acta 515:105–140
11. Davis JH (1983) Biochim Biophys Acta 737:117–171
12. Schindler H, Seelig J (1975) Biochemistry 14:5893–5903
13. Salmon A, Dodd SW, Williams GD, Beach JM, Brown MF (1987) J Am Chem Soc 109:2600–2609
14. Nagle JF, Wilkinson DA (1978) Biophys J 23:159–175
15. Klose G, Islamov A, König B, Cherezov V (1996) Langmuir 12:409–415
16. Sadaghiani AS, Khan A, Lindman B (1989) J Colloid Interface Sci 132:352–362
17. Sternin E, Bloom M, MacKay AL (1983) J Magn Reson 55:274–282
18. Seelig A, Seelig J (1974) Biochemistry 13:4839
19. Thurmond RL, Dodd SW, Brown MF (1991) Biophys J 59:108–113
20. Thurmond RL, Lindblom G, Brown MF (1993) Biochemistry 32:5394–5410
21. Nagle JF (1993) Biophys J 64:1476–1481
22. Luzzati V (1968) In: Champman (ed) Biological Membranes. Academic Press, London–New York, pp 71–123
23. Seddon JM, Cevc G, Kaye RD, Marsh D (1984) Biochemistry 23:2634–2644
24. Mädler B (1995) Thesis, University of Leipzig
25. Ward AJI, Ku H, Phillippi MA, Marie C (1988) Mol Cryst Liq Cryst 154:55–60
26. Seddon JM (1990) Biochim Biophys Acta 1031:1–69
27. Thurmond RL, Lindblom G, Brown MF (1990) Biochem Biophys Res Commun 173:1231–1238
28. Lafleur M, Cullis PR, Fine B, Bloom M (1990) Biochemistry 29:8325–8333
29. Lafleur M, Bloom M, Eikenberry EF, Gruner SM, Han Y, Cullis PR (1996) Biophys J 70:2747–2757
30. Seelig A, Seelig J (1975) Biochem Biophys Acta 406:1–5
31. Engel AK, Cowburn D (1981) FEBS Lett 126 (2):169–171
32. Hauser H, Pascher I, Pearson RH, Sundell S (1981) Biochim Biophys Acta 650:21–51
33. Otten D, Beyer K (1995) Chem Phys Lipids 77:203–215
34. Gruen DWR (1985) J Phys Chem 89:146–153
35. Morrow MR, Singh D, Lu D, Grant CWM (1995) Biophys J 68:179–186
36. Fattal DR, Andelman D, Ben-shaul A (1995) Langmuir 11:1154–1161
37. Andelman D, Kozlov MM, Helfrich W (1994) Europhys Lett 25:231–236
38. Lichtenberg D, Robson RJ, Dennis EA (1983) Biochim Biophys Acta 737:285–304
39. Heerklotz H, Binder H, Lantzsch G, Klose G (1994) Biochim Biophys Acta 1196:114–122
40. Beyer K (1983) Chemistry and Physics of Lipids 34:65–80
41. Heerklotz H, Lantzsch G, Binder H, Klose G, Blume A (1995) Chem Phys Lett 235:517–520
42. Otten D, Löbbecke L, Beyer K (1995) Biophys J 68:584–597
43. Edwards K, Almgren M (1991) J Colloid Interface Sci 147:1–21
44. Jansson M, Thurmond RL, Trouard TP, Brown MF (1990) Chem Phys Lipids 54:157–170
45. Heerklotz H, Binder H, Lantzsch G, Klose G, Blume A (1996) J Phys Chem, in press
46. Rapp G, Funari S, Beyer K, in preparation
47. Bian J, Roberts MF (1990) Biochemistry 29:7928–7935

Progr Colloid Polym Sci (1997) 103:87–94
© Steinkopff Verlag 1997

Self-assembly of volatile amphiphiles

K.J. Klopfer
T.K. Vanderlick

Received: 29 December 1996
Accepted: 3 January 1997

K.J. Klopfer · Dr. T.K. Vanderlick (✉)
University of Pennsylvania
Department of Chemical Engineering
220 South 33rd Street
Philadelphia, Pennsylvania 19104-6393
USA

Abstract Computer simulations are used to study the aggregation phenomena of volatile amphiphiles in a system displaying liquid/vapor coexistence. These molecular dynamics simulations are based on a simple, yet versatile, model used previously to study oil/water/amphiphile systems: amphiphiles are nothing more than water and oil particles connected together by stiff springs. We observe a highly regulated self-assembly process wherein amphiphiles form bilayers within the liquid phase. The density of amphiphiles in a bilayer varies from a well-defined lower to upper limit as the overall concentration of amphiphiles in the system is changed; we examine how these limits vary for amphiphiles having different hydrophobic chain lengths. The vapor phase plays an important role in this self-assembly process, serving as a reservoir for amphiphiles not included in the bilayers. Finally, we show that bilayer formation can be completely suppressed if the geometry of the amphiphiles is altered, as is predicted by simple packing arguments.

Key words Computer simulations – amphiphiles – bilayers – packing parameter

Introduction

Many molecules are composed of segments that if not bonded would phase separate. Amphiphiles made up of hydrophilic and hydrophobic moieties, and block copolymers formed of immiscible monomers are important examples. The interface that forms between oil and water is a natural habitat for amphiphiles; similarly, block copolymers are preferentially adsorbed at the interface between phase separated monomers.

As with the water/oil interface, amphiphiles can specifically adsorb to the interface between water and air. One limiting case is complete adsorption resulting in the so-called insoluble monolayer. Experimental studies of insoluble monolayers at the water/air interface are routinely carried out using a film balance developed by Langmuir [1] over 70 years ago; with this, the average area per molecule can be varied and the resulting change in surface tension can be measured. Amphiphiles which exhibit significant solubility in water form Gibbs monolayers; here, partitioning of amphiphiles between the interface and the bulk phase establishes the surface pressure, which is a function of the overall concentration of amphiphiles.

The interfacial activity of amphiphiles is but one manifestation of their discordant intramolecular makeup. These molecules can also self-assemble in solution to form a variety of microstructures, such as micelles and vesicles to name a few. Of course structures based on a bilayer motif, such as vesicles, have direct biological relevance as they serve as models for cell membranes.

Insight into the fundamental behavior of amphiphiles can be provided through computer simulations; however, the number of degrees of freedom is inherently large. Hence, in practice, various restrictions or approximations are often made, such as using external potentials (as opposed to explicit molecular models) to represent bulk fluid phases [2, 3]. Furthermore, different schemes have been developed to deliberately maintain, or at least coax, the amphiphiles into specified microstructures. Simulations of phospholipid membranes, for example, often start with amphiphiles placed in the bilayer configuration [4]. Consequently, the initial set points of molecular area and bilayer thickness may influence simulation results [5].

An alternative simulation strategy is to relinquish exacting molecular models in exchange for a simple particulate representation of all components in the system. In one of the simplest scenarios, all particles in the simulation interact with Lennard–Jones-type potentials; each is identified as one of two species, "oil" or "water"; and amphiphiles are nothing more than oil and water particles bonded together by way of stiff harmonic potentials. Simple simulations of this sort have been recently used to investigate the behavior of amphiphiles at the oil/water interface. In particular, Smit and co-workers [6–10] have employed this methodology to study the behavior of linear and branched chained amphiphiles. We extended this analysis further and showed how the observed tension reducing properties of various amphiphiles can be predicted with simple equations of state [11]. Moreover, simulations of this type are now being used to study membrane dynamics [12] and self-assembly in amphiphilic systems [13–15]. Beyond the specific results of all these studies, it is clear that this simple model of amphiphiles is sufficient to capture many basic elements of their behavior. Hence simulations of this nature can be quite useful, especially for exploratory work where more realistic molecular models become prohibitive.

In this study, we use the same simulation scheme to study the behavior of these model amphiphiles in the presence of a liquid/vapor interface. This is quite feasible since Lennard–Jones particles are known to exhibit liquid/vapor coexistence [16]. In short, we observe a highly regulated self-assembly process leading to the formation of one or more well-formed bilayers in the liquid phase; amphiphiles not included in the bilayers are dispersed in the vapor phase. In this paper, we report on this phenomenon, and examine how the size and overall concentration of amphiphiles govern the development and characteristics of the bilayers formed. We also show that bilayer formation can be suppressed if the geometry of the amphiphiles is altered, as is predicted by simple packing arguments.

Model and computational details

The model used was inspired by the free energy density functional theory developed by Telo da Gama and Gubbins [17] and put into practice by Smit and co-workers [6]. Each particle in the simulation can be identified as one of two species: oil or water. Amphiphiles are nothing more

than water and oil particles connected via Hookean springs; this construction captures, at a basic level, the dual hydrophilic/hydrophobic character of these molecules. By connecting different amounts of particles in different configurations, this simple approach can be used to create molecules with varying architectures.

The two species of particles in the simulation interact with truncated Lennard–Jones potentials with energy parameter ε_{ij}, distance parameter σ_{ij}, and cutoff radius R_{ij}^{cut}.

$$\phi_{ij} = \begin{cases} \phi_{ij}(r) - \phi_{ij}(R_{ij}^{\mathrm{cut}}), & r \leq R_{ij}^{\mathrm{cut}}, \\ 0, & r \geq R_{ij}^{\mathrm{cut}}, \end{cases}$$

$$\phi_{ij}(r) = 4\varepsilon_{ij}\left[\left(\frac{\sigma_{ij}}{r}\right)^{12} - \left(\frac{\sigma_{ij}}{r}\right)^{6}\right].$$

We have assumed that all interactions are governed by the same ε. Therefore, the only difference between oil–oil, water–water, and oil–water interactions is determined by the choice of R_{ij}^{cut}. Intra-species interactions have $R_{ij}^{\mathrm{cut}} = 2.5\sigma_{ij}$, and so attractive forces are present. On the other hand, the inter-species interactions have $R_{ij}^{\mathrm{cut}} = 2^{1/6}\sigma_{ij}$, restricting the interactions to be repulsive.

Amphiphiles are constructed by adding to the potential above (which continues to apply to all particles) harmonic potentials between designated pairs of their constituent particles. For this we use

$$\phi_{ij} = \tfrac{1}{2}k(r_{ij} - \sigma_{ij})^2,$$

where k is the spring constant, and σ_{ij} is the separation at zero force. The value of k employed in the simulations should not affect thermodynamic properties; evidence of this is given in a previous publication [11]. To avoid confusion with other particles in the system, we shall at times refer to water and oil particles which make up amphiphiles as head and tail particles, respectively.

Unless otherwise noted, we used a system of 512 particles in a rectangular box of size $6.84\sigma \times 6.84\sigma \times 27.36\sigma$. Periodic boundary conditions were kept in all three directions. A constant temperature of $T = 1\varepsilon/k_{\mathrm{b}}$ was maintained by incorporating a constraint method into our leap-frog verlet algorithm as described in Allen and Tildesley [18]. We employed a time step of 0.005τ, where $\tau = \sigma(m/\varepsilon)^{1/2}$ (m is defined as particle mass).

We launched our simulations in the following way. We first started with a single component system of water particles in a homogeneous liquid phase. From this a vapor phase was formed using the method described by Harris [19], after which the system was equilibrated for approximately 1×10^5 time steps. At this point, sets of particles were randomly chosen to make up the amphiphiles; the requisite number were changed from water to oil species; and harmonic potentials were added to

Progr Colloid Polym Sci (1997) 103:87–94
© Steinkopff Verlag 1997

connect head and tail particles in the desired fashion. Since the particles which make up an amphiphile were not necessarily in close proximity to one another, the spring constant was initially set at a small value (ca. $1.0\varepsilon/\sigma^{-2}$) and then slowly ramped to its final value (ca. $1 \times 10^5 \varepsilon/\sigma^{-2}$) over 5×10^3 time steps. We note that creation of amphiphiles frequently disturbed the liquid/vapor phase coexistence, often causing cavitation in the liquid phase. This disruption subsided, and thermodynamic quantities were usually stabilized after 5×10^4 time steps; we allowed, however, an additional 2×10^4 steps to pass before equilibrium was assumed.

On average, data were collected for 1.2×10^5 time steps after equilibration. However, some simulations were run for an additional 4×10^5 time steps with no appreciable changes in thermodynamic quantities. The accuracy of our simulations was estimated by dividing the data collection steps into sub-runs of 1×10^4 time steps; determining the average quantities within each sub-run; and calculating standard deviations from sub-run averages. Density profiles were calculated at every time step within a sub-run. Density profiles shown herein represent a sub-run averaged profile.

Results and discussion

We first consider amphiphiles which have a cylindrical geometry, composed of one water particle and numerous oil particles – all of the same size–linked together in series as shown in Fig. 1. Furthermore, the size of the particles which comprise the amphiphiles is also equal to that of the free water particles which make up the liquid and vapor coexisting phases. The general behavior of these amphiphiles as seen in our simulations can be described as follows. At low overall concentrations, the amphiphiles are present exclusively in the vapor phase, showing only slight preferential adsorption to the vapor/liquid interface. At higher concentrations, the molecules form a distinct bilayer (or more than one) solubilized within the liquid water phase. Under conditions where bilayers do form, the vapor phase plays an important role, serving as a sink for excess amphiphiles so that the bilayers can exist with a preferred density (i.e. number of amphiphiles per unit area). Figure 2 shows the density profile of amphiphiles in a system displaying bilayer formation. In this particular simulation (consisting of 1300 total particles), the amphiphile was composed of four-tail particles attached to one head particle; henceforth denoted (4:1).

Before describing this self-assembly process in more detail, it is worth comparing this overall behavior to that which we observed for the same amphiphiles in the presence of a water/oil interface. In that case, we found [11]

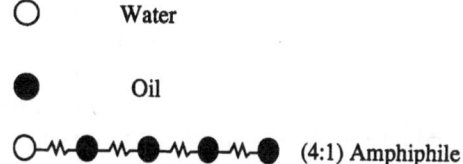

Fig. 1 Schematic drawing of water/oil/amphiphile model

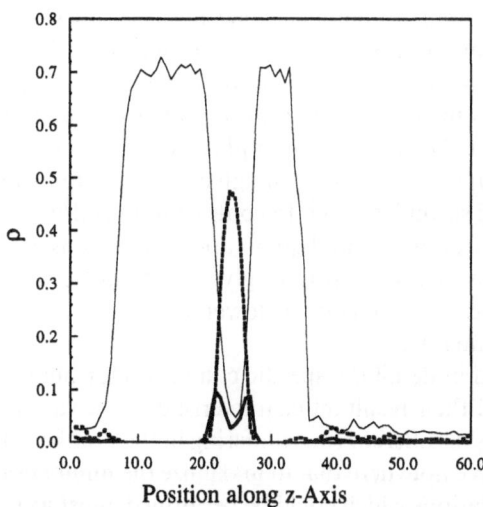

Fig. 2 Density (σ^{-3}) vs. position along z-axis (σ) for: (——) free water particles; (■■) head particles; ■■■ (tail particles)

that the amphiphiles behaved as an insoluble monolayer when present at low enough concentrations. At higher concentrations, amphiphiles exhibited solubility in the oil phase but showed no tendency to self-organize therein. In our small-scale simulations we also saw no evidence of micelle formation in the water phase; however, this was seen in the larger-scale simulations of Smit et al. [8, 13]. Interestingly, over large simulation times, these investigators noted that micelles tended to cluster until a single bilayer structure was formed [10]. The simulations reported in this paper demonstrate that these mock amphiphiles are in fact quite volatile, and that the presence of a vapor phase plays an important role in the self-assembly process which occurs in the denser liquid phase. This same self-assembly process is not observed when the vapor phase is exchanged for an oil phase.

Returning the to self-assembly process that occurs in our liquid/vapor system, we now describe in general terms the evolution of bilayers as the concentration of amphiphiles in the system is increased. At some point a concentration is reached where amphiphiles become soluble in the liquid phase. When this first happens the amphiphiles aggregate into featureless clusters. However, soon thereafter, as the concentration continues to increase, the amphiphiles in the liquid all come together to form a bilayer, displaying a characteristic width given by the distance between the exposed head groups. As more amphiphiles are added to the system, these molecules are directly incorporated into the bilayer making it more dense (in amphiphiles) and also decreasing the degree of chain interdigitation. Concomitantly, water trapped within the bilayer is gradually squeezed out, as the bilayer width increases to some asymptotic value. Thereafter, any further increase in overall amphiphile concentration does not change the properties of the first bilayer, but rather leads ultimately to the formation of a second bilayer. The new bilayer evolves in structure and form in the same way as the first. As stated before, the vapor phase serves as a reservoir for amphiphiles when not enough are present to form the first, or the second, bilayer. In contrast to the oil/water interface, adsorption at the liquid/vapor interface is not significant; this is not surprising given that the tension [20] is an order or magnitude lower than that of the liquid/liquid interface.

We studied in detail the specific conditions for bilayer formation and their resultant characteristics for a variety of amphiphiles of different sizes (differing in the number of oil particles). We note here that to maximize the number of different simulations which could be performed, most were conducted with 512 total particles. A few larger simulations (such as the one reported in Fig. 2) were conducted to confirm that the properties of the bilayers, as reported below, were not affected by system size. We did notice, however, that simulations containing only 512 total particles did not usually contain enough free water particles to display bulk liquid behavior on both sides of the bilayer; instead one side of the bilayer was usually contacted by a thin water film.

As discussed qualitatively above, the properties of the bilayered microstructures depend on the number of amphiphiles present in the system; it is, in fact, the nature of this dependence that we wish to report. The overall number density of amphiphiles in this two-phase system is not, however, a useful independent variable; in particular, differently shaped simulation boxes of the same volume (containing the same total number of particles) would not exhibit the same behavior. A better choice, which allows us to compare directly simulations of different sizes, is a two-dimensional number density, χ, defined to be the total number of amphiphiles in the system divided by the area of the simulation box which lies parallel to the liquid–vapor interface. It is also convenient to define a similar number density χ^B, associated with the number of amphiphiles in the simulation that are incorporated in one or more (if they exist) bilayers. Thus, $\chi^B \le \chi$, with the equality holding only when all the amphiphiles in the system are assimilated in bilayers.

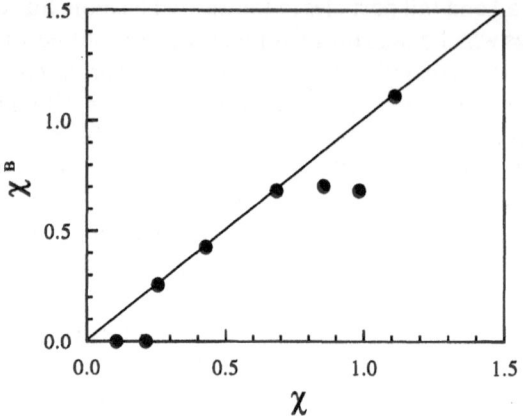

Fig. 3 χ^B (σ^{-2}) vs. χ (σ^{-2}). The 45° line corresponds to the situation where all amphiphiles are incorporated in bilayer(s)

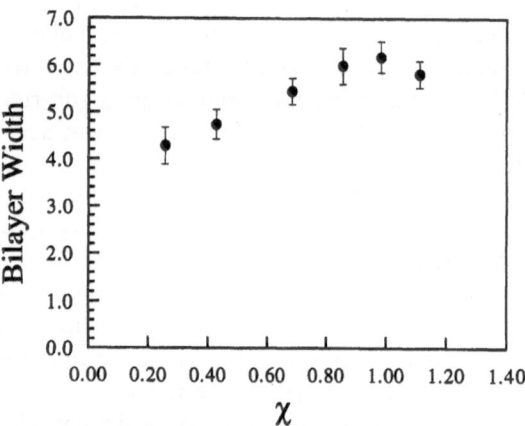

Fig. 4 Bilayer width (σ) vs. χ (σ^{-2})

As a basis for comparison, it is worthwhile to present first the results for the (4:1) amphiphiles; these, in fact, display the most sensitivity to concentration. The development of bilayers as a function of concentration is shown in Fig. 3 (χ^B versus χ). As can be seen therein, a bilayer is first formed when χ reaches 0.26 (e.g. 12 amphiphile molecules in a simulation with interfacial area $46.8\sigma^2$). At this point, the effective area per amphiphile, determined by dividing the area of the bilayer (equal to the areal size of the simulation box) by the number of amphiphiles in the bilayer, is $3.9\sigma^2$. As χ is increased up to 0.68, the added amphiphiles are incorporated directly into the bilayer (hence the associated points in Fig. 3 fall on the 45° line). The bilayer widens as it takes on more amphiphiles, as can be seen in Fig. 4 which plots the width of the bilayer, as measured by the distance between the centers of the exposed head particles, as a function of χ. At its maximum width, the effective area per amphiphile in the bilayer diminishes to $1.5\sigma^2$. Further increase in overall amphiphile

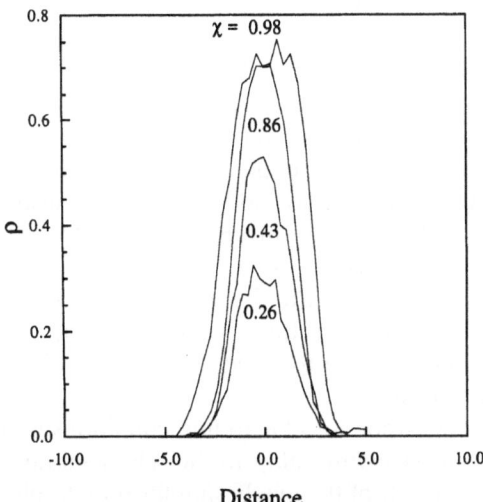

Fig. 5 The density profile of amphiphilic tail particles (σ^{-3}) vs. position along z-axis (σ) for various χ values

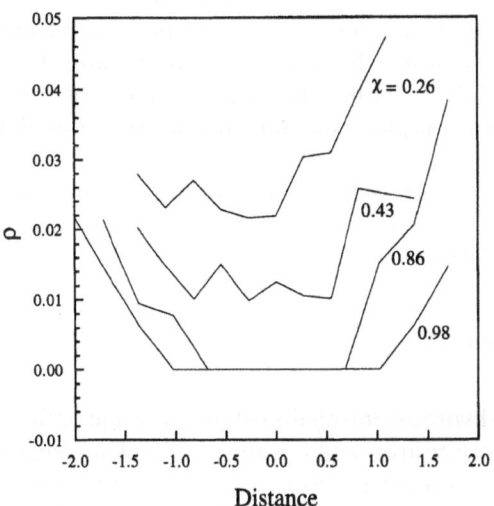

Fig. 6 The density profile of free water particles (σ^{-3}) vs. position along z-axis (σ) for various χ values

concentration beyond this point does not affect the properties of this bilayer; excess amphiphiles simply build up in the vapor phase until enough are present to form a second bilayer. As can be seen in Fig. 3, this occurs at $\chi = 1.1$ (as all the amphiphiles now make up two bilayers, this point then lies on the 45° line).

In summary, we see that a surprisingly limited number of amphiphiles is required to form a stable bilayer. And although more amphiphiles are readily accepted into the microstructure, an upper limit is eventually reached. The evolution of bilayer structure from one limit to the other can be further appreciated by examining the density profiles of tail particles and free water particles within the bilayer; these are shown in Figs. 5 and 6, respectively. As is

readily seen, as the bilayer takes on more amphiphiles, the density of tail groups in the interior region rises and broadens, reaching an asymptotic limit. Correspondingly, water in the interior of the bilayer is reduced, until the region is completely void of water. It appears that the maximum number of amphiphiles included in the bilayers is that which yields a interior tail particle density of about 0.7, which corresponds to the density of bulk oil that could exist in equilibrium with bulk water at this temperature [16].

It is worthy to examine more fully the role of the vapor phase in the bilayer formation process. To do this, we examined the behavior of (4:1) amphiphiles in a one phase liquid system, composed of water particles. A constant volume simulation was carried out with the concentration of amphiphiles set to yield an effective χ equal to 0.86; at this same concentration, the two phase liquid–vapor system shows one bilayer at the asymptotic width and density, and excess amphiphiles are hosted in the vapor phase. In the single-phase system, we found that a bilayer did form, but all amphiphiles in the simulation were included in it. Although the width of this bilayer (5.6σ) was not significantly different from that formed in the two-phase system (6σ), the density of tail particles in the interior was considerably greater. We did find, however, that if the normal pressure was lowered (by increasing the length of the simulation box along the axis perpendicular to the bilayer) the bilayer stretched out (to 7.1σ) and the density in the interior relaxed to the same value found for bilayers in the two-phase system. The main point is, however, that the presence of the vapor reservoir allows self-regulation of the intrinsic properties of the bilayers formed.

In examining similar amphiphiles of different lengths, the general features of bilayer formation exhibited by the (4:1) amphiphiles were followed, as long as the chain of the amphiphile contained at least 3 oil particles. On the other hand, we found that (1:1) and (2:1) amphiphiles showed no tendency to self-assemble into layers. The (1:1) amphiphile was, in fact, soluble in both the vapor and liquid phases at all concentrations, and did not demonstrate any form of aggregation. The (2:1) was slightly soluble in the liquid phase at higher concentrations. Clearly, inter-chain interactions are very important in stabilizing the lamellar structures. In fact, for the longer chained amphiphiles, stable bilayers are formed sooner, i.e., with less amphiphiles, the longer the chain. Figure 7 shows this dependence (circular symbols) for amphiphiles containing 3-, 4- and 6-tail particles. Chain length also affects the asymptotic density of the bilayers: bilayers formed of longer chained molecules contain less molecules per unit area. This is also illustrated in Fig. 7 (square symbols). As expected, longer chains also form bilayers with larger asymptotic widths.

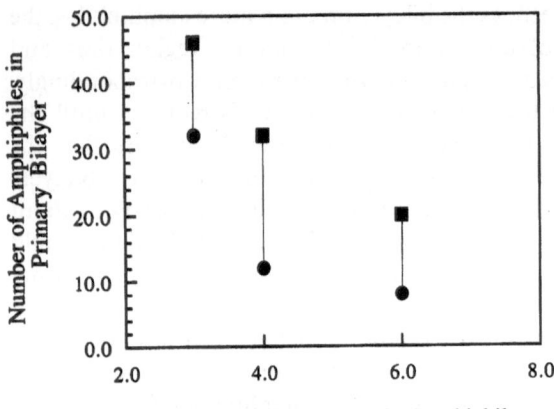

Fig. 7 The number of amphiphiles in primary bilayer vs. number of tail segments in amphiphile. (●) first appearance of stable bilayer; (■) packing limit of bilayer

One common trait shared by all bilayer forming systems we examined is that every bilayer experiences a large degree of interdigitation, having widths which are much smaller than the sum length of two extended amphiphiles. Notably this is not the characteristic of cell membranes, which display widths nearly twice the thickness of a well-packed monolayer; however, interdigitated bilayers are not entirely uncommon, particularly in systems of mixed phosphatidylcholines [21]. Within the gross limitations of the simple molecular model we are employing, we probed the influence of two factors that might seem intuitively to influence bilayer structure: the rigidity of amphiphiles; and the relative strength of inter- and intra-molecular interactions. These are now discussed in turn.

In the simulations reported above, the amphiphiles were composed of a train of particles, each linked by springs to its immediate neighbors. The resulting molecules are quite flexible, bearing no constraints other than a preferred separation between each successive pair of particles (as set by the length of interconnecting spring). It is easy, however, to create more rigid molecules by additionally connecting together, with a spring of appropriate length, the first and last particles of the train. We did this for simulations of (4:1) amphiphiles and found that the bilayer width increased by only about 10% (to achieve no inter-digitation would require an approximately 60% increase). We also noticed that in comparison to the flexible amphiphiles, larger overall concentrations of rigid amphiphiles were required to achieve the first stable bilayer: $\chi = 0.68$ as opposed to $\chi = 0.26$.

One seeming advantage of tightly packed, non-inter-digitated, bilayer is the screening of interactions between chain particles and bulk water particles. However, in the simulations reported above, oil–water interactions are

very short-ranged: the oil particles do not interact with water particles until they are within 1.12σ of one another, at which point they experience repulsive interactions given by the Lennard–Jones potential. We attempted, in an *ad hoc* fashion, to increase the range and strength of repulsive forces exerted between free water particles and amphiphile tail particles; in particular, we added to the repulsive inverse-12 potential a repulsive inverse-6 potential which acts out to a distance of 2σ. We found that this alteration had no effect on the bilayer width. However, because of the longer-ranged repulsive interactions between water and oil particles, no water was able to penetrate into any of the bilayers which formed.

In summary, we were not able to find conditions that lead to "bilayered" structures akin to the cell membrane; this is arguably a result of the overly simplified molecular model used in these simulations. On the other hand, the ease of the model facilitates investigations which examine the role of general molecular characteristics, such as size and shape. We take advantage of this possibility to study the role of molecular geometry, and compare our results with predictions using the simple packing arguments of Israelachvili [22, 23]. Like the simulation model itself, these arguments neglect the fine details of molecular structure.

According to this analysis, the so-called packing parameter dictates the type of microstructures that an amphiphile will form:

$$\text{Packing parameter} = \frac{v}{a_o \cdot l_c}.$$

Here, v is the hydrocarbon chain volume, a_o is the optimal head group surface area, and l_c is the critical chain length. For packing parameters between $\frac{1}{2}$ and 1, amphiphiles should form flexible bilayers (vesicles), or planar bilayers as the packing parameter approaches 1. On the contrary, a packing paramter less than $\frac{1}{2}$ should hinder the formation of bilayers; in particular, the theory predicts micelle formation.

Esselink and co-workers [10] showed that the shapes of micelles formed in amphiphile/oil/water systems were indeed consistent with packing parameter arguments; they changed the geometry of the amphiphiles by connecting together different numbers of particles in different arrangements. Interestingly, they found that packing parameters were not always predictable *a priori*. For example, a linear train of particles of equal size, composed of four tails and four heads, formed highly spherical micelles stabilized by the large area conformation taken on by the chain-like headgroup. In our study, we changed the packing parameter by simply increasing the size of the one water particle which forms the head group of the amphiphiles. We

Progr Colloid Polym Sci (1997) 103:87–94
© Steinkopff Verlag 1997

used three values for the diameter of the head particle, 1.0σ, 1.25σ, and 1.5σ; these led to packing parameter values of 1.0, 0.64, and 0.44, respectively. Following the theoretical predictions, the first two cases should self-assemble into bilayers while the third geometry should not. Indeed, we found this to be the case. In the third case (where bilayers are not predicted to form), we saw significant but not well-defined aggregation of amphiphiles within the bulk liquid. The small system size employed in our simulations probably prohibits the formation of micelles [8, 11, 13]. Nevertheless, these simulations reveal that this simple molecular model used to construct amphiphiles is sufficient to capture many important elements of their behavior, and provide a valuable tool for studying the role of molecular architecture in self-assembling systems.

Granted, the simulations reported herein may have limited practical relevance. Most real surfactants demonstrate negligible volatility; the high surface tension of water induces strong adsorption to the interface; and applications based on closed liquid–vapor systems are not prevalent. We are, in fact, unaware of any systems that demonstrate the reported vapor-regulated self-assembly process. Of course there is always the possibility that some or all aspects of this phenomenon may be ultimately realized; short-chained alcohols, or mixed amphiphilic systems containing them, may be candidate systems. We do believe, however, that the lack of direct relevance does not totally diminish the utility of these simulations. In particular, these simulations might be ideal for fundamental investigations of the effects of perturbants on self-assembly. For example, one might be able to determine the size and/or shape of inclusions that serve to break apart a bilayer. The distinct formation of bilayers in the liquid phase, combined with their self-regulating properties associated with having the vapor reservoir, makes these simulations useful platforms for these investigations.

Conclusions

We have used a simple molecular model employed in simulations of oil/water/amphiphiles to investigate the self-assembly of amphiphiles in systems displaying liquid/vapor equilibrium. Studies of oil/water/amphiphile systems rarely concentrate on the role of the oil phase. Although it is rarely emphasized, many (if not most)

amphiphiles in such systems are dissolved in the high-density oil phase. Exchanging the oil phase for one of water vapor significantly modifies the behavior of the amphiphiles. As we have shown herein, this creates a high driving force to self-assemble into well-formed bilayers within the dense water phase. Curiously, we are unaware of any oil/water/amphiphile simulations that exhibit similar well-formed layers. (The closest is perhaps the work Smit et al. [13], who report seeing "bilayer micelles".) The enhanced tendency of these amphiphiles to self-assemble in the water/vapor environment seems clearly related to the lack of any other oil particles in the system.

We probed the formation and structure of bilayers as a function of amphiphile chain length (i.e. number of oil particles). We found that amphiphiles must have at least three oil particles to form stable bilayers. For those that do, we found that the resulting bilayer had a characteristic width that increases with increasing amphiphile concentration until reaching an asymptotic value. At this point, amphiphiles added to the system are hosted in the vapor phase until enough are available to form a new bilayer. The minimum number of amphiphiles needed to form a bilayer, as well as the maximum number that one bilayer can have, depends on the chain length of the amphiphiles: both are less the longer the chain. We did not, however, find conditions leading to closed-packed bilayers with small or no chain interdigitation.

Finally, the architecture dependence of bilayer formation was also examined by altering the size of the amphiphiles' head particle and thus varying its packing parameter. Results of these simulations were in agreement with conditions for bilayer formation as predicted by simple geometric arguments. Thus there is increasing evidence that this simple model of amphiphiles – one that can be readily implemented on stand-alone workstations – is sufficient to capture many elements of their inherent behavior. In future work, we plan to take advantage of the building block nature of this model to study the behavior of amphiphiles with more complex architectures, such as those containing rings. In addition, we plan to use the simulations reported herein to study the effects of inclusions on bilayer formation.

Acknowledgments We gratefully acknowledge support for this work provided by the David and Lucile Packard Foundation and the National Science Foundation (CTS-89-57051).

References

1. Langmuir I (1917) J Am Chem Soc 39:1848
2. Karaborni S, Toxvaerd S (1992) J Chem Phys 96:5505
3. Karaborni S, Toxvaerd S, Olsen OH (1992) J Phys Chem 96:4965
4. Taga T, Kazuhumi M (1995) J Comp Chem 16:235
5. Nagle JF (1993) Biophys J 64:1476
6. Smit B, Hilbers PAJ, Esselink K, Rupert LAM, van Os NM, Schlijper AG (1990) Nature 348:624

7. Smit B, Schlijper AG, Rupert LAM, van Os NM (1990) J Phys Chem 94:6933
8. Smit B, Hilbers PAJ, Esselink K, Rupert LAM, van Os NM, Schlijper AG (1991) J Phys Chem 95:6361
9. van Os NM, Rupert LAM, Smit B, Hilbers PAJ, Esselink K, Bohmer MR, Koopal LK (1993) Colloids Surfaces A: Physicochem Eng Aspects 81:217
10. Esselink K, Hilbers PAJ, van Os NM, Smit B, Karaborni S (1994) Colloids Surfaces A: Physicochem Eng Aspects 91:155–167
11. Klopfer KJ, Vanderlick TK (1995) Colloids Surfaces A: Physicochem Eng Aspects 96:171
12. Lasic DD, Papahadjopoulos D (1995) Science 267:1275
13. Smit B, Hilbers PAJ, Esselink K (1993) Tenside Surf Det 30:287
14. Karaborni S, Esselink K, Hilbers PAJ, Smit B, Karthauser J, van Os NM, Zana R (1994) Science 266:254
15. Karaborni S, Esselink K, Hilbers PAJ, Smith B (1994) J Phys Condens Matter 6:A351
16. Nicolas JJ, Gubbins KE, Street WB, Tildesley DJ (1979) Molecular Physics 37:1429
17. Telo Da Gama MM, Gubbins KE (1986) Molecular Phys 59:227
18. Allen MP, Tildesley DJ (1987) Computer Simulation of Liquids. Oxford University Press, Oxford
19. Harris JG (1992) J Phys Chem 96:5077
20. Nijmeijer MJP, Bakker AF, Bruin C, Sikkenk JH (1988) J Chem Phys 89:3789
21. Adachi T, Takahashi H, Ohki K, Hatta I (1995) Biophys J 68:1850
22. Israelachvili JN, Mitcdhell DJ, Ninham BW (1976) J Chem Soc Farad Trans 2 72:1525
23. Israelachvili Jacob N (1991) Intermolecular and Surface Forces. Academic Press, London

Progr Colloid Polym Sci (1997) 103:95–106
© Steinkopff Verlag 1997

T.J. McIntosh
S.A. Simon

Experimental tests for thermally-induced fluctuations in lipid bilayers

Received: 25 November 1996
Accepted: 2 December 1996

Abstract For several years researchers have been investigating the interactions between solvated lipid bilayers. Presently there is some disagreement regarding the range, magnitude, and origin of the long- and short-range repulsive and attractive interactions between bilayers. To address this issue experimentally, we have used the osmotic stress/X-ray diffraction method to measure the total repulsive pressure as a function of interbilayer distance for both fluid and solid bilayers. These bilayers were primarily composed of the most common phospholipids found in biological membranes, the zwitterionic lipids phosphatidyl-choline (PC) and phosphatidyl-ethanolamine (PE). For PC bilayers the pressure–distance data can be explained by the presence of an attractive van der Waals pressure and short- and long-range repulsive pressures. The short-range underlying pressure, which extends about 4 Å into the fluid space from each monolayer, is due to an enthalpically driven, exponentially decaying pressure arising from the work to remove water from the hydrophilic polar head groups, and an excluded volume contribution arising from the non-ideal interactions between head groups from apposing bilayers. This review focusses on experiments designed to determine quantitatively the effects of a long-range repulsive pressure, due to undulations of the entire bilayer arising from thermally induced bending moments. In these experiments pressure–distance relations were measured for bilayers with a range of bending moduli (measured independently) obtained as a function of temperature, the number of double bonds in the lipid acyl chains, and the presence of exogenous compounds such as lysophosphatidyl-choline. For PC bilayers the equilibrium fluid spacings can be manipulated in a manner predictable by a theory of continuous unbinding. However, for PE bilayers there is an additional attractive pressure arising from interactions between the PE head groups. This additional interaction, responsible for the large adhesion energy of PE bilayers, gives rise to discontinuous disjoining of PE bilayers.

Key words Bilayer – hydration – adhesion – bending – X-ray diffraction

T.J. McIntosh
Department of Cell Biology
Duke University Medical Center
Durham, North Carolina 27710, USA

Dr. S.A. Simon (✉)
Department of Neurobiology
Duke University Medical Center
Durham, North Carolina 27710, USA

Introduction

The types of long- and short-range interactions that are present between amphiphiles are of interest to surface chemists and also to biophysicists investigating the structure and interactions between biological membranes and lipid bilayers. These interactions are important in processes involving adsorption, adhesion, and cell fusion [1–5]. Classically two forces were used to describe the stability of colloids, electrostatic double layer repulsion, arising from fixed or adsorbed charges, and van der Waals attraction, arising from dispersion forces between colloidal particles that interact through a water phase [6, 7]. For both these interactions the range, magnitude, and physical origin are well understood. In this classical analysis, in the absence of applied external pressures, particles are in a bound state when the attractive van der Waals energy is greater than the repulsive electrostatic energy by $k_B T$, where T is temperature and k_B is Boltzmann's constant.

For several years investigators have been studying the origin, range, and magnitude of several "non-classical" short- and long-range repulsive interactions that have been shown to exist between neutral lipid bilayers [2–10], DNA molecules [11, 12], polysaccharides [13], and proteins [14]. We have chosen to investigate lipid bilayers since their composition, phase behavior, fluid spacing, and chemical potential of the aqueous phase can be experimentally manipulated to test theoretical models of these repulsive pressures. Hydrated lipid bilayers, composed of amphiphilic molecules containing solvated polar head groups and acyl chains, can exist in crystalline, gel, and liquid crystalline phases. Bilayers can be found as single closed vesicles of various sizes or as laminated stacks of closed vesicles. The adhesion energy between apposing bilayers, that arises from the balance between the repulsive and attractive interactions, has been measured by the surface force apparatus [15–17], the osmotic stress/X-ray diffraction method [8, 18, 19], and by the joining of two vesicles held in micropipettes [20].

The short-range interbilayer interactions which are important at fluid spacings less than 8 Å [21, 22] include the repulsive hydration pressure, that arises from the reorganization of the solvent by the molecularly rough lipid polar head groups [2, 8, 21, 23], and pressures arising from the excluded volume of the polar head groups from apposing bilayers [24]. Given that these interactions decay very rapidly, it is difficult to provide a completely separate theoretical and experimental description. Although we have made some progress towards this end [21, 22, 25, 26], for the purpose of this report we lump these two classes of interactions together and consider them to represent the "bare" or underlying repulsive pressure. Experimentally,

a component of this pressure has been shown to decay exponentially from the physical edge of the bilayer into the fluid space with a decay length of 1 to 2 Å [2, 8, 19, 21, 24]. In this review we present evidence that part of this underlying, or bare repulsive potential, is enthalpic in origin, meaning that it has a comparatively small temperature dependence.

For fluid bilayers that are not adsorbed on mica (or other molecularly smooth surfaces) there is another repulsive pressure, named the undulation pressure, that arises from thermally induced density fluctuations in the two monolayers of the bilayer [27–30]. These density fluctuations give rise to bending modes or undulations of the entire bilayer. Clearly, bilayers that undulate have a larger entropy (more degrees of freedom) than those that are flat. Therefore the undulation pressure (P_u) is entropic in origin and thus should have a large temperature dependence. Helfrich and colleagues [28, 31] originally showed that for large fluid spacings (d_f), $P_u \propto (k_B T)^2 / B d_f^3$ where B is the bilayer bending modulus ($B \approx 10–25 \, k_B T$). Although other forms of P_u have been developed for bilayers with different adhesion energies [29, 30, 32–34], they all predict that P_u should increase with increasing temperature and with decreasing bending modulus.

The existence of the undulation pressure has been postulated to have several consequences. Theoretically it means that the bare repulsive potential will have to be "re-normalized" to reflect this additional repulsion [30, 32–34]. At fluid spacings where the adhesion energy is small, it is predicted theoretically [29], and observed experimentally [35], that the presence of an undulation pressure reveals itself in plots of log P (where P is the applied osmotic pressure) versus d_f by increasing the decay length of the bare potential. This should be true whether the bare potential is electrostatic in nature, and thus characterized by the Debye length, or if the bare potential arises principally from solvent re-orientation, as for uncharged bilayers. If the magnitude of the undulation pressure becomes sufficiently large the bilayers may completely disjoin. For phosphatidylcholine (PC) bilayers the attractive interaction arises entirely from the van der Waals pressure (P_V). For lipid bilayers, P_V has been shown to be proportional to A_H / d_f^3, where the Hamaker constant $A_H \propto k_B T$. Since both P_V and P_u decrease as d_f^{-3} at large fluid spacings and have numerators of the order of $k_B T$, increasing the temperature or decreasing the bending modulus should cause an increase in d_f (unbinding) and consequently a decrease in adhesion energy. This crossover between binding and unbinding has been treated by theoreticians as a phase transition.

The focus of this review is an experimental analysis of the factors that modulate the range and magnitude of the undulation pressure between bilayers composed of the two most common phospholipids found in biological membranes, PC and phosphatidylethanolamine (PE). For these bilayers we present evidence, obtained using the osmotic stress/X-ray diffraction method, of the consequences of systematically modulating the undulation pressure. For PC bilayers we find that increasing P_u, either by increasing the temperature or changing the bilayer composition to decrease the bilayer bending modulus, causes PC bilayers to unbind or disjoin. Theoretical analysis [33] of the unbinding transitions provides predictions of the extent of unbinding given values of B, A_H, P_u, the equilibrium fluid spacing, and the magnitude and decay length of the bare potential. We find that these PC bilayers unbind continuously in a manner consistent with the theory. In contrast, PE bilayers disjoin discontinuously as a consequence of a strong short-range attractive interaction between apposing head groups.

Methods

We have examined by the X-ray diffraction/osmotic stress method two types of multibilayer systems, unoriented suspensions of multiwalled vesicles (liposomes) and oriented multilayers. Known osmotic pressures were applied to each of these systems by published procedures [8, 10, 19, 24]. In the case of liposomes, osmotic pressures (P) from 0 to 5.7×10^7 dyn/cm^2 (log $P = 7.8$) were applied by incubating the vesicles in aqueous solutions of large neutral polymers such as dextran or polyvinylpyrrolidone (PVP). Since these polymers are too large to enter the lipid lattice, they compete for water with the lipid multilayers, thereby applying a known osmotic pressure [8, 10, 18, 36]. Higher pressures, between 3.0×10^7 dyn/cm^2 (log $P = 7.5$) and 2.3×10^9 dyn/cm^2 (log $P = 9.4$) were applied to oriented multibilayers by incubating them in constant relative vapor pressures (p/p_0) maintained with saturated salt solutions [37, 38]. In this case the applied pressure is given by $P = -(RT/V_W)\cdot\ln(p/p_0)$ where R is the molar gas constant, T the temperature in degrees Kelvin, and V_W the partial molar volume of water [10].

For both oriented and unoriented specimens, X-ray diffraction patterns were recorded on X-ray film and integrated intensities were obtained for each diffraction order h by measuring the area of the diffraction peak. Structure amplitudes $F(h)$ were obtained from the measured intensities by applying standard correction factors [24, 39, 40]. Electron density profiles, $\rho(x)$, on a relative electron density scale were calculated from

$$\rho(x) = (2/d)\sum \exp\{i\phi(h)\} \cdot F(h) \cdot \cos(2\pi xh/d) , \qquad (1)$$

where x is the distance from the center of the bilayer, d the lamellar repeat period, $\phi(h)$ the phase angle for other h,

and the sum is over h. Phase angles were determined by the use of the sampling theorem [41] as described in detail previously [19, 42, 43]. Electron density profiles described in this paper are at a resolution of $d/2h_{max} \approx 7$ Å.

Results

PC bilayers

In the first series of experiments we measured pressure–distance relations for bilayer systems having a wide range of bending moduli (B), and therefore a wide range of predicted undulation pressures. These systems include egg phosphatidylcholine (EPC) in the liquid-crystalline phase ($B = 5.1 \times 10^{-13}$ ergs [25]), diarchidonylphosphatidyl choline (DAPC) in the liquid-crystalline phase ($B = 2.8 \times 10^{-13}$ ergs [25]), an equimolar mixture of EPC and monooleoylphosphatidylcholine (MOPC) in the liquid-crystalline phase ($B = 1.3 \times 10^{-13}$ ergs [25]), and the subgel (crystalline) phase of dipalmitoylphosphatidylcholine (DPPCsg) (B can not be measured for the subgel phase but would be expected to be much larger than for any liquid-crystalline bilayer). For all applied pressures all specimens gave X-ray diffraction patterns containing a series of low-angle reflections that indexed as orders of a single lamellar repeat period. A plot of the logarithm of applied pressure (log P) versus the lamellar repeat period is shown in Fig. 1 for DPPC in the subgel phase, and EPC, DAPC, and EPC:MOPC and in the liquid-crystalline phase. The repeat periods in excess buffer, with no applied pressure, were 59.4 Å for DPPCsg, 63.2 Å for EPC, and 64.7 Å for DAPC, and 74.6 Å for EPC:MOPC (shown on the x-axis in Fig. 1). For each system the repeat period decreased with increasing applied pressure.

The lamellar repeat period includes the total thickness of the lipid bilayer plus the water space between apposing bilayers. To obtain information on the bilayer and fluid layer thicknesses we [19, 21, 24, 25] have calculated electron density profiles for each osmotic stress experiment shown in Fig. 1. Figure 2 shows typical profiles for EPC multilayers both in the absence of applied pressure and at an applied pressure of log $P = 7.8$. Each profile contains two unit cells, with two adjacent bilayers and the intervening fluid space. For each profile the bilayer on the left is centered at the origin and the high electron density peaks centered at ± 19 Å correspond to the lipid polar head groups of the bilayer. The trough at the origin corresponds to the low density terminal methyl groups, and the medium density regions between the terminal methyl trough and the head group peaks correspond to the methylene groups of the lipid hydrocarbon chains. In the case of the profile for EPC in water (solid line), the fluid space

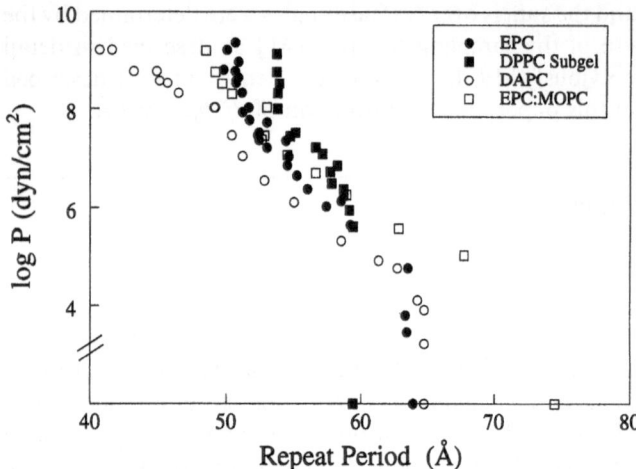

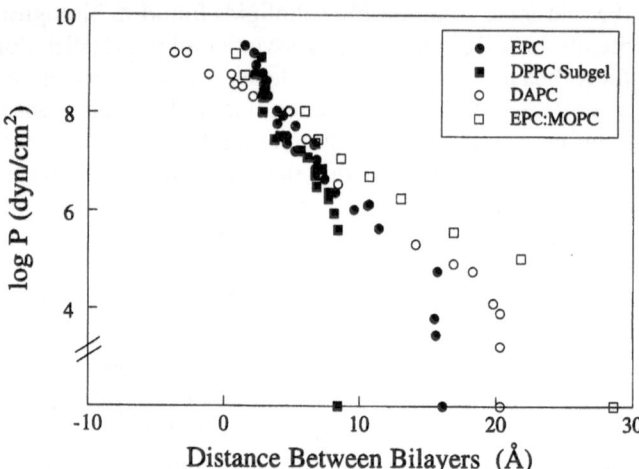

Fig. 1 Logarithm of applied pressure (log P) plotted versus the lamellar repeat period (d) for subgel bilayers of DPPC and liquid-crystalline bilayers of EPC, DAPC, and 1:1 EPC:MOPC. The repeat periods in excess water (with no applied pressure) are shown on the x-axis. Data are taken from [19, 21, 24, 25]

Fig. 3 Logarithm of applied pressure (log P) plotted versus the distance between bilayers (d_f) for subgel bilayers of DPPC and liquid-crystalline bilayers of EPC, DAPC, and 1:1 EPC:MOPC. The fluid spaces in excess water (with no applied pressure) are shown on the x-axis. Data are taken from [19, 21, 24, 25]

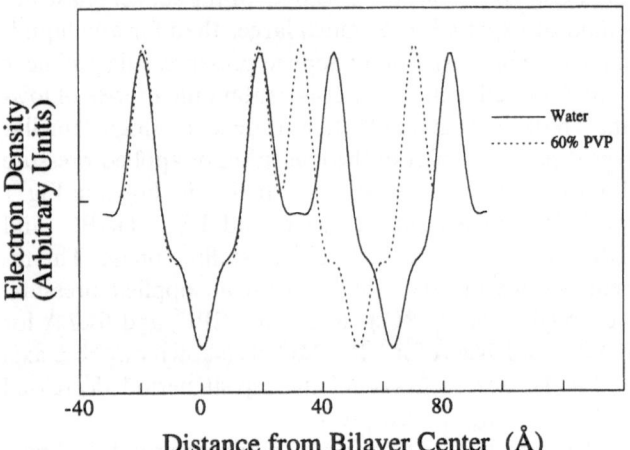

Fig. 2 Electron density profiles of egg phosphatidylcholine (EPC) bilayers in water (solid line) and in 60% PVP (dashed line). Each profile contains two unit cells, with two apposing bilayers and the intervening fluid space

between adjacent bilayers is centered at 32 Å, and the head group peaks from the adjacent bilayer are centered at 44 and 82 Å. When EPC was incubated in 60% PVP (dashed line), water was removed from the lipid lattice, resulting in a much narrower fluid space between bilayers as evidenced in the profiles by the close apposition of the head group peaks from adjacent bilayers. However, the profile across the bilayer on the left was very similar for EPC in water and in 60% PVP, indicating that this amount of osmotic stress did not appreciably change the bilayer structure [19].

As detailed in [19, 21, 24, 36, 42, 44], electron density profiles, such as those shown in Fig. 2, can be used to estimate the fluid space between apposing bilayers for each value of applied pressure. The definition of the lipid/water interface is somewhat arbitrary, because the bilayer surface is not smooth and water penetrates into the head group region of the bilayer [45–47]. We operationally define the bilayer width as the total physical thickness of the bilayer assuming that the conformation of the phosphorylcholine head group in bilayers is the same as it is in single crystals of phosphatidylcholine [48]. In that case the high density head group peak in the electron density profiles would be located between the phosphate group and the glycerol backbone, so that the edge of the bilayer lies about 5 Å outward from the center of the high density peaks in the electron density profiles [19, 24, 42, 44].

Using this definition of the lipid/water interface, we plot in Fig. 3 the logarithm of applied osmotic pressure (log P) versus the distance between bilayer surfaces (d_f) for subgel DPPC and liquid crystalline EPC, DAPC, and EPC:MOPC. The equilibrium values for d_f for each lipid in the absence of applied pressure are shown on the x-axis. Comparisons of the pressure–distance data shown for the various lipids in Fig. 3 indicate the following. First, for $6 < \log P < 8$ the pressure–distance data are similar for all lipids. In this region, the plots of log P versus d_f can be fit with straight lines, indicating that the total repulsive pressure decays exponentially with a decay constant of about 1.5 Å for all of these lipid systems [19, 21, 22, 25]. However, for $\log P > 8$ and $\log P < 6$ the pressure–distance relations are quite different for the various lipids.

In particular, for log $P > 8$ there is a sharp upward break in the log P versus d_f curve at $d_f \approx 3$ Å for DPPCsg, but no such upward break for DAPC. Thus, at the highest applied pressure (log $P = 9.2$) the fluid space for DAPC is about 7 Å smaller than d_f for subgel DPPC. For log $P > 8$ the values of d_f for EPC and EPC:MOPC are intermediate between those of DAPC and DPPCsg. For log $P < 6$ the pressure–distance curves extends to much larger fluid spaces for EPC:MOPC than for DPPCsg. The differences in the range of the total repulsive pressure can best be appreciated by noting that the equilibrium fluid separation at zero applied pressure is about 8 Å for DPPCsg, 16 Å for EPC, 20 Å for DAPC, and 28 Å for EPC:MOPC (Fig. 3).

In order to get more information on the contribution of entropic pressures to the total repulsive pressure between bilayers, we performed osmotic stress experiments as a function of temperature [49]. These experiments were performed with bilayers in both the gel phase (large bending modulus so small undulation pressure) and liquid crystalline phase (small bending modulus so larger undulation pressure). The gel phase lipid was dibehenoylphosphatidylcholine (DBPC), a saturated PC which has a melting temperature of about 70 °C, whereas the liquid crystalline lipid was EPC. Figure 4 shows pressure–distance data for gel phase DBPC and liquid crystalline phase EPC for temperatures ranging from 5 to 50 °C. For DBPC the pressure–distance data were nearly independent of temperature over the entire pressure range (Fig. 4), and d_f in excess buffer was only 1 Å larger at 50 °C than at 5 °C. For EPC the pressure–distance relations were nearly independent of temperature for $6 < \log P < 9$, but were strongly temperature dependent for log $P < 6$ (Fig. 4). In particular, the distance between bilayers in excess buffer in the absence of applied pressure was significantly larger at 50 °C ($d_f \approx 21$ Å) than at 5 °C ($d_f \approx 14$ Å).

PE bilayers

We have also analyzed the interactions between gel and liquid crystalline bilayers composed of the second most common membrane phospholipid PE. Figure 5 shows electron density profiles of bacterial PE (BPE), which is in the liquid crystalline phase at room temperature. The figure shows profiles of BPE at zero applied pressure and in 60% PVP, the same conditions shown for the profiles of EPC (Fig. 2). Each profile shows two bilayers, with the intervening fluid space. The bilayer on the left is centered at the origin with the high density head group peaks located at ± 20 Å. Note that the bilayers for BPE in water and in 60% PVP nearly superimpose, indicating that, as was the case for EPC (Fig. 2), the amount of osmotic stress

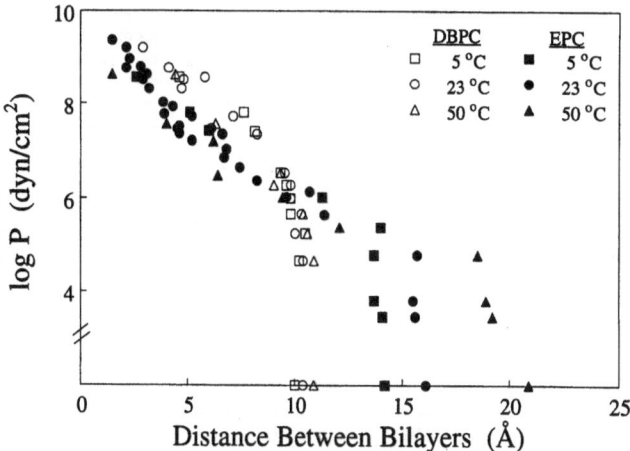

Fig. 4 Logarithm of applied pressure (log P) plotted versus the distance between bilayers (d_f) obtained as a function of temperature for DBPC bilayers in the gel phase and EPC bilayers in the liquid crystalline phase. The fluid spacings in excess water (with no applied pressure) are shown on the x-axis. Data are taken from [49]

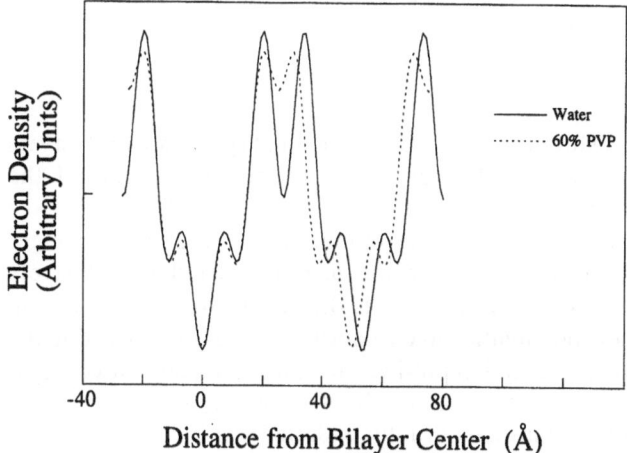

Fig. 5 Electron density profiles of bacterial phosphatidylethanolamine (BPE) bilayers in water (solid line) and in 60% PVP (dashed line). Each profile contains two unit cells, with two apposing bilayers and the intervening fluid space

applied by 60% PVP does not appreciably change the width of the bilayer. The most important difference between EPC (Fig. 2) and BPE (Fig. 6) is that the fluid spacing between apposing bilayers is much smaller in the case of BPE bilayers. From profiles such as these we have estimated that the fluid space between bilayers is 5–6 Å for PE bilayers in *both* the liquid crystalline [19, 50] and gel phases [50, 51]. These results are consistent with many studies which have found that PE bilayers imbibe less water than PC bilayers [2, 9, 52, 53].

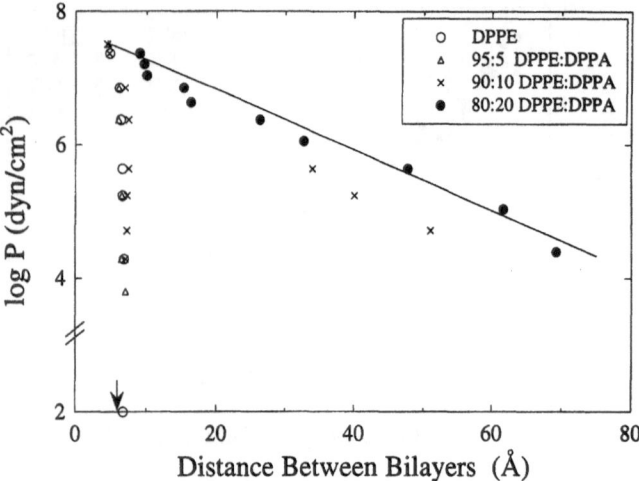

Fig. 6 Logarithm of applied pressure (log *P*) plotted versus the distance between bilayers (d_t) for gel phase bilayers composed of mixtures of the zwitterionic lipid DPPE and the negatively charged lipid DPPA. The circle on the *x*-axis represents the fluid spacing of gel phase DPPE bilayers in water and the arrow indicates the fluid spacing for liquid-crystalline phase BPE bilayers. The line represents the repulsive electrostatic pressure calculated from double-layer theory for 80:20 DPPE:DPPA bilayers. Data were taken from [19, 51]

This observation of smaller fluid spaces for PE bilayers than for PC bilayers has implications in terms of the interbilayer interactions, as it means that PE bilayers have either weaker interbilayer repulsive pressure(s) or stronger interbilayer attractive pressure(s) compared to PC bilayers [51]. Experimentally it is difficult to distinguish among these possibilities, in part due to problems in extracting the attractive and repulsive pressures from measured pressure–distance relations because of the extremely small fluid separations in fully hydrated PE bilayers. We [51] have used the osmotic stress/X-ray diffraction method to measure pressure–distance relations between PE and PC bilayers over an expanded distance regime. The interbilayer distance was extended by incorporating into both gel phase dipalmitoylphosphatidylethanolamine (DPPE) and dipalmitoylphosphatidylcholine (DPPC) bilayers various concentrations of the negatively charged lipid dipalmitoylphosphatidic acid (DPPA), which adds a known electrostatic repulsive pressure to the bilayer surface. The experiments were designed to: (1) provide an estimate of the adhesion energy between gel phase PE bilayers, (2) test the hypothesis that there is an additional attractive interaction that operates between PE but not PC bilayers, and (3) to determine whether the disjoining transition occurs continuously or discontinuously. The rationale was that if an additional attractive interaction exists, then it should take a larger repulsive electrostatic

pressure (the addition of more negatively charged lipid) to swell PE bilayers than PC bilayers.

Figure 6 shows pressure–distance relations for gel phase DPPE bilayers containing 0, 5, 10 and 20 mol% of the negatively charged lipid DPPA. DPPA was chosen because it is negatively charged (at neutral pH in 0.1 M NaCl) and has the same acyl chain length as DPPC or DPPE, so it should mix ideally at small mole fractions. Several points are worth noting from this figure. First, the pressure–distance data from DPPE and 95:5 DPPE:DPPA superimpose, meaning that the electrostatic repulsion provided by 5 mol% charged lipid was not sufficient to increase the fluid space between apposing bilayers. In contrast, 5 mol% DPPA markedly increased the fluid spacing between DPPC bilayers [51]. Second, the pressure–distance relation for 10 mol% DPPA gave two repeat periods (indicating phase separation). The first repeat period was the same as DPPE, and the second gave a fluid space that could be predicted by electrostatic double layer theory [51]. As will be elaborated on in the discussion, this result is good evidence for a discontinuous disjoining of the bilayers. The pressure–fluid distance data for the 80:20 DPPE:DPPA bilayers were close to the values expected for bilayers with a repulsive pressure due to the electrostatic repulsive pressure predicted for bilayers containing this amount of charge lipid (solid line, Fig. 6). Third, at zero applied pressure the equilibrium fluid separations for both gel phase DPPE and liquid-crystalline phase BPE (arrow in Fig. 6) are about 6 Å, much smaller than the equilibrium fluid separations for typical PC bilayers in both the gel and liquid-crystalline phases (Figs. 3 and 4).

Discussion

The experiments in this paper were obtained by analysis of X-ray diffraction patterns of osmotically stressed liposomes and from direct measurements of the bending modulus (*B*) of single-walled bilayers. As described below, these data provide evidence, in the case of electrically neutral PC bilayers, for several distinct "non-classical" repulsive interactions and, in the case of PE bilayers, for an attractive interaction that is larger than can be accounted for by van der Waals interactions between the hydrocarbon regions of apposing PE bilayers. Initially we discuss the origin of the "bare" or underlying interactions between PC bilayers. This is followed by a comparison of the experimental data with a theory developed for the continuous unbinding of lipid bilayers. Finally, the nature of the unbinding transition seen between PE bilayers is analyzed and compared with that observed with PC bilayers.

PC bilayers

Origin of underlying pressure

For bilayers containing phospholipids with small polar head groups, such as PC and PE, the underlying repulsive pressure contains contributions from the energy to desolvate the head groups (the hydration pressure) and the pressure arising from the non-ideal interactions between head groups from apposing bilayers (the excluded volume pressure). These two energies are coupled, because decreasing the fluid space will increase the excluded volume and re-organize the solvent. To obtain an estimate of the range of the underlying pressure, the interactions between DPPCsg bilayers are especially useful, since the van der Waals pressure is the only attractive interaction, the acyl chains are crystalline and consequently all out-of plane fluctuations, whatever their origin, are markedly quenched. For DPPCsg bilayers the underlying pressure is indeed short-ranged, since the equilibrium fluid spacing in the absence of applied pressure is only 8 Å (Fig. 3), meaning the repulsive pressure extends only about 4 Å from each bilayer surface.

Hydration pressure: The data presented in Figs. 3 and 4 show that for PC bilayers containing one (EPC, DBPC, DAPC, DPPCsg) or two lipids (EPC:MOPC), in the pressure range of $6 < \log P < 8$, there is a large short-range pressure that decays exponentially with a decay length of about 1.5 Å. There are two reasons that the pressure in this region has a relatively small contribution from the undulation pressure. First, in this region the magnitude of the pressures are independent of the bending modulus (B), because they are about the same in fluid (EPC, EPC:MOPC, DAPC), gel (DBPC) and crystalline (DPPCsg) bilayers. Second, as detailed below, these pressures are independent of temperature, at least from 5 to 50 °C (Fig. 4).

We argue that the major contributing pressure in the region $6 < \log P < 8$ is the hydration pressure due to solvent re-organization. This is consistent with several theoretical treatments [23, 54, 55] showing that the hydration pressure should decay exponentially with increasing separation. As outlined below, there are several lines of experimental evidence indicating the existence of a hydration pressure.

For both gel phase DBPC and liquid crystalline phase EPC bilayers, the exponentially decaying short range pressure in the region $6 < \log P < 8$ is, within experimental error, independent of temperature (Fig. 4). Entropic pressures such as excluded volume interactions and the undulation pressure are temperature dependent. That is, the excluded volume pressure can, to a first

approximation, be modeled as a van der Waals gas such that $P \approx k_B T/(V - V_0)$ where V_0 is the molecular volume of the head group. The simplest model of the undulation pressure, an ideal gas of N uncorrelated humps confined by an external potential (with each hump having an area determined by the longest permitted wavelength), yields $P_u \approx (k_B T)^2/B d_f^3$. Stated differently, the origin of the observed short-range (for $d_f < 8$ Å) exponentially decaying pressure (Figs. 3 and 4) is not one simply dominated by entropy, but rather must be based on enthalpic considerations. The work to dehydrate the bilayer (ΔG) has two main components, work arising from changes in bilayer structure (ΔG_{bs}) and work required to remove water from between apposing bilayers (ΔG_{deh}). In the case of DBPC, electron density profiles [49] indicate that there is little change in bilayer structure with dehydration, so that $\Delta G \approx \Delta G_{deh}$. Therefore, ΔG_{deh} can be obtained from integration of the pressure–distance curves [10]. The free energy of dehydration can be expressed as $\Delta G_{deh} = \Delta H_{deh} - T \Delta S_{deh}$, where ΔH_{deh} and ΔS_{deh} represent the enthalpy and entropy of dehydration, respectively. The entropy can be calculated from $\Delta S_{deh} = - \partial \Delta G_{deh}/\partial T$. Since the pressure–distance data for DBPC are nearly temperature independent from 5 to 50 °C (Fig. 4), it follows that the energy of dehydration arises primarily from changes in the enthalpy of dehydration. The near equivalence of energy and enthalpy of dehydration is also found in measurements of the work to remove water from ions or dipoles [56, 57]. The main contribution to the hydration energy of ions arises from solvent immobilization in a primary hydration shell, electrostriction, and further effects of water surrounding the shell [57]. The energy to remove one water molecule from the hydration shell of a tetramethylammonium ion (which resembles choline) is 8.7 kcal/mol [58], and the energy to completely dehydrate this ion is 35.1–38.2 kcal/mol [57, 58]. These numbers can be compared to values calculated for dehydrating DBPC obtained by integration of the pressure–fluid-spacing curve (Fig. 4). The energy to decrease the fluid spacing at constant area from $d_f = 10$ Å to $d_f = 1$ Å is about 9.3 kcal/mol. It follows that the energy of dehydration of DBPC to $d_f = 1$ Å can be entirely accounted for by partial dehydration of the choline moiety.

For EPC, EPC:MOPC, DAPC or DBPC bilayers Figs. 3 and 4 show that, for $6 < \log P < 8$, the pressure decays with increasing fluid spacing in a manner that can be described by $P = P_0 \exp(-d_f/\lambda)$, where P_0 is the magnitude of the pressure at $d_f = 0$ and λ is the decay length. A serious issue is to determine the physical basis of P_0 and λ. To this end we have shown that the decay length (λ) depends on the size of the solvent [36], which is consistent with the hypothesis that this pressure arises from head group solvation. Moreover, for a variety of lipids and solvents we [59–61] found a good correlation between P_0 as obtained from the osmotic stress/X-ray measurements and the value of P_0 obtained from a theoretical model based on solvent polarization [55]. That is, the theoretical value of P_0 can be calculated from dipole potential measurements and compared to values obtained using the osmotic stress/X-ray method [59–61]. Since the dipole potential contains a significant contribution from the solvent dipoles [62–64], we take this correlation as evidence for the role of the hydration pressure in the bare pressure. For example, we note that the dipole potential is larger for gel than liquid crystalline bilayers [59]. Therefore, from dipole potential measurements the hydration pressure of the gel phase is expected to be larger than liquid crystalline bilayers. That this is indeed the case can be seen in Fig. 4 where the repulsive pressure of the gel phase DBPC bilayers is greater than the repulsive pressure for liquid crystalline phase EPC bilayers.

Excluded volume interactions: As noted above, short-range repulsive hydration interactions for amphiphiles with small head groups are extremely difficult to separate from excluded volume interactions. There is, however, strong evidence that the excluded volume interactions also contribute to the short-range repulsive pressure. For lipid bilayers with head groups markedly larger than PC, such as glycolipids or lipids with covalently bound polymers, we have shown that the magnitude and range of the repulsive pressure arising from excluded volume considerations depends on both the head group size and surface density [65, 66]. Obviously for lipids with small polar head groups, such as PC and PE, excluded volume terms are also present but are more difficult to unequivocally attribute to this phenomenon. However, the repulsion arising from excluded volume considerations for PC bilayers should, to a first approximation, be inversely proportional to the number of head groups per unit area at the interface. Therefore decreasing the density of head groups should "soften" this repulsion in a manner similar to that found with glycolipids or lipids with covalently attached polymers [65, 66]. Indeed comparisons of the log P versus d_f plots of log $P > 8$ (Fig. 3) are consistent with this prediction. Specifically, the sharp upward break seen for DPPCsg (molecular area $\approx 48 \text{ Å}^2$) at $d_f \approx 3 \text{ Å}$, "softens" for EPC (molecular area $\approx 64 \text{ Å}^2$), and essentially disappears for DAPC (molecular area $\approx 73 \text{ Å}^2$). It should also be noted that protrusions of individual lipid molecules from the plane of the bilayer [67, 68] could contribute to this short-range repulsion.

Thus, we argue that the underlying short-range pressure for PC bilayers contains at least two contributions: an enthalpically driven, exponentially decaying hydration pressure arising from the work to remove water from the hydrophilic polar head groups, and excluded volume contributions arising from non-ideal interactions between head groups from apposing bilayers.

Undulation pressure

Pressure–distance data for liquid crystalline EPC obtained as a function (Fig. 4) are nearly independent of temperature for log $P > 6$. However, for EPC for log $P < 6$, the range of the total repulsive pressure dramatically increases with increasing temperature, so that the equilibrium fluid separation is about 7 Å larger at 50 °C than at 5 °C. This temperature dependence implies that for liquid crystalline bilayers near the equilibrium fluid spacing there is a strong entropic component that is increasing (re-normalizing) the range of the underlying bare pressure [49].

If undulations truly increase (renormalize) the range of the underlying pressure, then, at constant temperature, decreasing the bending modulus should also increase the magnitude of undulation pressure and extend the range of the pressure–distance plots. Measurements of the bending moduli of these lipid bilayers have the order EPC > DAPC > EPC:MOPC [25]. The pressure–distance data for liquid crystalline bilayers of EPC, DAPC, and 1:1 EPC:MOPC (Fig. 3) show that the equilibrium fluid separation increases as the bending modulus decreases. We now show that these increases in the equilibrium fluid separation can be explained quantitatively by increases in the undulation pressure as a result of changes in the bilayer bending modulus.

Theory of continuous unbinding: Continuous unbinding between bilayers occurs when the potential energy well that characterizes the bilayer adhesion energy continuously decreases in magnitude as the undulation pressure increases and/or the van der Waals pressure decreases. Evans [33] derived a self-consistent-field theory to predict the continuous unbinding of multilamellar vesicles. His analysis provides a method of predicting the changes in equilibrium fluid spacing (d_{eq}) between multilamellar bilayers as functions of the magnitude (P_0) and decay length (λ) of the underlying exponential hydration repulsion, a power-law attraction characterized by the Hamaker constant (A_H), the bilayer thickness, and the bilayer bending modulus (B) (Fig. 7). Thus the theory makes specific predictions for the value of d_{eq} which depend on a number of parameters that must be obtained by methods based on entirely different physical principles. In this context any agreement between theory and experiment is not likely to be fortuitous. When the Hamaker constant is very large, the bilayers will be in a bound state, and when it is small,

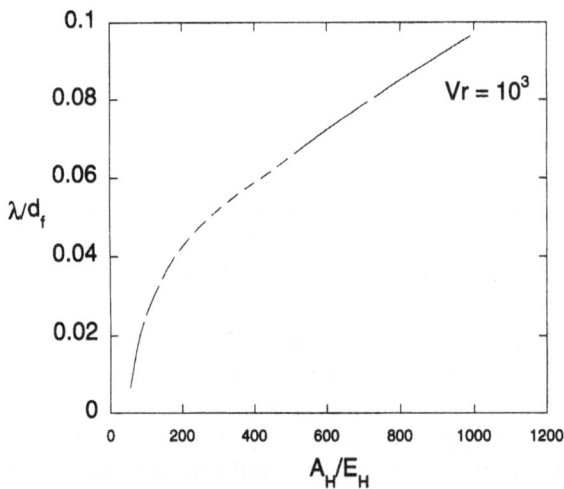

Fig. 7 Predictions for continuous unbinding of multilamellar vesicles that have an underlying exponential repulsion $V_r = P_0\lambda^2/E_H = 10^3$. See text for definition of other variables. Adapted from Evans [33]

Table 1 Theoretical and experimental equilibrium fluid spacings

Lipid	Temperature (°C)	d_{eq} (experimental) (Å)	d_{eq} (undulation theory) (Å)
EPC	5	22.1	22.1
EPC	23	23.3	23.1
EPC	50	28.8	28.0
DAPC	23	28.2	30.1
EPC:MOPC	23	36.5	48.3

and the bending modulus is small, the bilayers will completely unbind. That is, they will swell to the extent determined by the water in the system [69].

We initially consider the temperature-induced increase in fluid spacing between liquid-crystalline EPC membranes as functions of the underlying pressure ($P_h = P_0 \exp(-d_f/\lambda)$), the attractive van der Waals pressure (P_v), and the bilayer undulation pressure (P_u). For our calculations, we use for P_h the repulsive pressure measured for DPPCsg bilayers (Fig. 3), where entropic pressures are small [22]. Evans [33] asssumed that the van der Waals pressure decays as d_f^{-3} and has a magnitude characterized by the Hamaker constant (A_H), and that the undulation pressure is characterized by the "Helfrich energy scale", $E_H = (k_B T)^2/B \cdot 1.6\pi^2$. Evans plotted λ/d_{eq} versus A_H/E_H for different values of the parameter, $V_r = P_0\lambda^3/E_H$. Evans used a different definition than ours for the plane of origin of the attractive and repulsive pressures, as he used the plane of origin obtained from gravimetric analysis of X-ray diffraction data [8, 70]. From our previous comparisons of electron density profiles and gravimetric analysis of EPC bilayers [36], we have found that the fluid spaces calculated using these different planes of origin differ by 7.9 Å. This means that to use our data with the Evans' plane of origin we must: (1) increase the magnitude of P_0 that we obtained for DPPCsg (1.1×10^9 dyn/cm²) by the factor $\{\exp(7.9 \text{ Å}/1.4 \text{ Å})\}$, where 1.4 Å is λ for the subgel phase [21], and (2) add 7.9 Å to our measured value of the equilibrium fluid separation (Fig. 3). Using these modified values, we calculate $V_r = 3 \times 10^3$ for EPC bilayers. In subsequent calculations we use the closest published curve, which is the one where $V_r = 10^3$ (shown in Fig. 7).

To predict d_{eq} for EPC bilayers at different temperatures it is necessary to know the value of E_H for EPC bilyers at 23 °C. This can be estimated by assuming $A_H = 3 \times 10^{-14}$ erg for PC bilayers [71] and selecting the value of A_H/E_H that corresponds to our value of $\lambda/d_{eq} = 1.4 \text{ Å}/23.3 \text{ Å} = 0.06$. This gives $A_H/E_H = 400$, so that $E_H = 7.5 \times 10^{-17}$ erg, a reasonable value [33]. We make the simplest assumptions and determine whether varying only one parameter, the bending modulus (B), is sufficient to explain the observed increase in d_{eq} with increasing temperature (Fig. 4). The bending modulus depends on the bilayer hydrocarbon chain thickness (h) by $B = bk_T h^2$ [72], where b is a constant and k_T is the isothermal compressibility modulus [49]. As a first approximation, we assume that k_T is nearly constant for 5–50 °C [49], use electron density profiles to calculate h and values of E_H for each temperature, and use the relation in Fig. 7 to determine the equilibrium fluid spacings at 5 °C and 50 °C. The results (Table 1) indicate excellent agreement between theory and experiment. Thus, at zero or small externally applied pressures these calculations show that from 5 to 50 °C the undulation pressure is the dominant repulsive pressure between electrically neutral liquid-crystalline EPC bilayers. The presence of the undulation pressure explains why liquid-crystalline bilayers adsorb more water (i.e., have larger fluid spacings) than gel phase bilayers (Figs. 3 and 4). As expected from theoretical treatments [29, 33], we find that the undulation pressure increases with increasing temperature primarily due to decreases in the bilayer bending modulus.

For DAPC and EPC:MOPC bilayers which have lower bending moduli than EPC bilayers (at the same temperature), we use the same theory [33] to determine the predicted changes in the equilibrium fluid spacing. After substituting in the appropriate values of P_0, λ, B, A_H [25], we find that the agreement between experiment and theory is quite good for EPC, DAPC, and for 1:1 EPC:MOPC (Table 1) especially considering the large experimental uncertainty in measurements of B and the assumptions involved in these calculations. Thus, thermally-induced undulations can explain the large differences in fluid spacing observed between gel and liquid crystalline

PC bilayers (Fig. 4), the large fluid separations observed for DAPC and 1:1 EPC:MOPC bilayers (Fig. 3), as well as the increases in fluid spacing for EPC bilayers as a function of temperature (Fig. 4).

Thus, for uncharged, fluid PC bilayers we argue that the undulation pressure, arising from thermally induced bending modes, increases the range of the total pressure for fluid PC bilayers. When the magnitude of the undulation pressure becomes sufficiently large, PC bilayers disjoin continuously.

PE bilayers

Several studies have shown that, compared to PC bilayers, PE bilayers have smaller equilibrium fluid spaces [2, 9, 52, 53] and larger adhesion energies [73]. These small fluid spacings and large adhesion energies have several experimental and theoretical consequences. Experimentally PE bilayers differ from PC bilayers in that the equilibrium fluid separation is about the same below and above the phase transition temperature [50]. This means that renormalization of the underlying pressure with the undulation pressure does not produce a sufficiently large repulsive energy to cause PE bilayers to unbind. Since both the underlying hydration pressure (as estimated independently from dipole potential measurements), and the bending moduli of PC and PE bilayers are about the same [51], the fact that PE bilayers do not unbind at their phase transition can not be attributed to an abnormally small hydration or undulation pressure. Moreover, the large adhesion energy between PE bilayers does not arise from in-plane correlations among the PE head groups since separating the head groups with spacer molecules, like cholesterol, does not change d_{eq} [47, 74].

We had two main goals in investigating the interaction energy between PE bilayers. The first was to obtain an estimate of the adhesion energy between PE bilayers that did not include contributions from the undulation pressure or from the mica to which PE bilayers are adsorbed in surface force apparatus experiments. The second was to compare the unbinding relationships between DPPE and DPPC. To accomplish these goals a given mole fraction of charged lipid DPPA was added to each lipid for the purpose of putting a known amount of repulsive energy in the system and unbind the bilayers. From a knowledge of the surface charge density the repulsive pressure arising from double layer repulsion were calculated and thus the energy to separate or disjoin the bilayers was obtained [51].

For DPPC bilayers, increasing the surface charge density increases d_{eq} in proportion to the amount of DPPA present [51]. That is, DPPC bilayers containing only 5 mol% DPPA disjoin and increasing the DPPA mol% increases the magnitude of the repulsive pressure in a manner completely consistent with the range and magnitude predicted from double layer theory [51].

In contrast, DPPE bilayers exhibit a very different behavior upon the addition of DPPA. The addition of 5 mol% does not alter the fluid spacing (Fig. 6). The addition of 10 mol% DPPA results in multilayers with two distinct fluid spacings – one being the same as the control DPPE and the other considerably larger [51]. At 20 mol% DPPA, only one phase is present, the bilayers are disjoined, and the pressure–distance plots can be accurately predicted from double layer repulsion (Fig. 6). These data suggest that a potential energy barrier has to be surmounted for the bilayers to disjoin and get into a regime that can be described by double layer theory. The existence of such a barrier indicates that the unbinding transition between DPPE bilayers is discontinuous. From the observed pressure–distance data (Fig. 6) the adhesion energy between DPPE bilayers is estimated to be -0.7 erg/cm^2, which is significantly larger than the adhesion energy measured between gel phase DPPC bilayers [73]. It follows that the potential energy well characterizing the interactions between DPPE bilayers is both deeper and narrower than for DPPC bilayers.

The adhesion energy (E) between bilayers can be written

$$E = E_a + E_r,   \tag{2}$$

where E_a and E_r represent the total attractive and repulsive energies, respectively, at the equilibrium fluid separation. It is well established that for gel phase PC bilayers the adhesion energy is determined by a balance between a short-range repulsive pressure and an attractive van der Waals pressure [73]. However, the situation is more complex with PE bilayers. The attractive van der Waals energy can be estimated from $E_v = -(A_H/12\pi)(1/d_v)^2$ where d_v represents the distance from the plane of origin of the van der Waals pressure. Given the similarity in composition of DPPE and DPPC, we assume that the Hamaker constant should be approximately the same for DPPE and DPPC bilayers and set $A_H = 3 \times 10^{-14}$ erg. We also assume that the plane of origin for E_v is at the center of the DPPE head group region, near the peak in the electron density profiles. With these assumptions we calculate that at the equilibrium fluid spacing, $E_v \approx -0.03$ erg/cm^2. Thus, for DPPE bilayers our estimated adhesion energy ($E \approx -0.7$ erg/cm^2) is much larger in magnitude than the van der Waals energy calculated with this standard formalism. Since the repulsive component of the energy (E_r) must be positive, the only way that Eq. (2) can be balanced for DPPE is for there to be an additional attractive pressure besides the usual representation of the van der Waals pressure.

Attractive interactions that have been proposed for PE bilayers involve interbilayer interactions between an amine group in one bilayer and a phosphate group in the apposing bilayers, either through electrostatic interactions, direct hydrogen-bonding, or hydrogen-bonded water bridges [50–52]. Recent molecular dynamics simulations show that water molecules are hydrogen-bonded to PE head groups, whereas water forms a 'clathrate-like' structure around the PC head group [75, 76], suggesting that PE bilayers may be stabilized by water bridges across the small fluid space [51, 77].

Summary

The water spacings between uncharged, fluid PC bilayers are determined by a balance between hydration and steric repulsive pressures and the attractive van der Waals pressure. The equilibrium fluid spacings can be manipulated in a predictable manner by changing the bilayer's bending modulus and thereby changing the repulsive undulation pressure. For PE bilayers another attractive pressure, arising from interactions between the PE head groups, is larger than the repulsion arising from undulations, and is responsible for the large adhesion energy. These interactions are reflected in the discontinuous manner that PE bilayers disjoin.

Acknowledgments This work was supported by a grant from the National Institutes of Health (GM-27278).

References

1. Parsegian VA, Rau DC (1984) J Cell Biol 99:196s–200s
2. Rand RP, Parsegian VA (1989) Biochim Biophys Acta 988:351–376
3. Chernomordik LV, Kozlov MM, Zimmerberg J (1995) J Membr Biol 146:3
4. Cafiso DS (1995) In: Disalvo EA, Simon SA (eds) Permeability and Stability of Lipid Bilayers. CRC Press, Boca Raton, Florida
5. Simon SA, McIntosh TJ (1996) Cold Spring Harbor Symposia on Quantitastive Biology LX. Cold Spring Harbor Laboratory Press, pp 601–608
6. Derjaguin BV, Landau L (1941) Acta Physiochim USSR 14:633–662
7. Verwey EJW, Overbeek JTG (1948) Theory of the Stability of Lyophobic Colloids. Elsevier, Amsterdam
8. LeNeveu DM, Rand RP, Parsegian VA, Gingell D (1977) Biophys J 18:209–230
9. Lis LJ, McAlister M, Fuller N, Rand RP, Parsegian VA (1982) Biophys J 37:657–666
10. Parsegian VA, Fuller N, Rand RP (1979) Proc Nat Acad Sci USA 76:2750–2754
11. Rau DC, Lee B, Parsegian VA (1984) Proc Nat Acad Sci USA 81:2612–2625
12. Rau DC, Parsegian VA (1992) Biophys J 61:246–259
13. Rau DC, Parsegian VA (1990) Science 249:1278–1281
14. Leikin S, Rau DC, Parsegian VA (1994) Proc Nat Acad Sci USA 91:276–280
15. Israelachvili JN, Adams GE (1976) Nature 262:774–776
16. Israelachvili JN, Tandon RK, White LR (1979) Nature 277:120–121
17. Israelachivili J, Marra J (1986) Methods Enzymol 127:353–361
18. Parsegian VA, Rand RP, Fuller NL, Rau RC (1986) Methods in Enzymology 127:400–416
19. McIntosh TJ, Simon SA (1986) Biochemistry 25:4058–4066
20. Evans E, Metcalfe M (1984) Biophys J 46:423–426
21. McIntosh TJ, Simon SA (1993) Biochemistry 32:8374–8384
22. McIntosh TJ, Simon SA (1994) Annu Rev Biophys Biomol Struct 23:27–51
23. Marcelja S, Radic N (1976) Chem Phys Lett 42:129–130
24. McIntosh TJ, Magid AD, Simon SA (1987) Biochemistry 26:7325–7332
25. McIntosh TJ, Advani S, Burton RE, Zhelev DV, Needham D, Simon SA (1995) Biochemistry 34:8520–8532
26. McIntosh TJ, Simon SA (1996) Colloids and Surfaces A: Physiochemical and Engineering Aspects 116:251–268
27. Harbich W, Helfrich W (1984) Chem Phys Lipids 36:39–63
28. Helfrich W, Servuss R-M (1984) Il Nuovo Climento 3:137–151
29. Evans EA, Parsegian VA (1986) Proc Nat Acad Sci USA 83:7132–7136
30. Lipowsky R (1995) In: Lipowsky R, Sackman E (eds) Handbook on Physics of Biological Systems. Elsevier Press, New York
31. Helfrich W (1973) Z Naturforsch 28C: 693–703
32. Evans E, Needham D (1987) J Phys Chem 91:4219–4228
33. Evans E (1991) Langmuir 7:1900–1908
34. Podgornik R, Parsegian VA (1992) Langmuir 8:557–562
35. Podgornik R, Rau DC, Parsegian VA (1994) Biophys J 66:962–971
36. McIntosh TJ, Magid AD, Simon SA (1989) Biochemistry 28:7904–7912
37. O'Brien FEM (1948) J Sci Instrum 25:73–76
38. Weast RC (1984) Handbook of Chemistry and Physics. CRC Press, Boca Raton, Florida
39. Blaurock AE, Worthington CR (1966) Biophys J 6:305–312
40. Herbette L, Marquardt J, Scarpa A, Blasie JK (1977) Biophys J 20:245–272
41. Shannon CE (1949) Proc Inst Radio Engrs NY 37:10–21
42. McIntosh TJ, Magid AD, Simon SA (1989) Biochemistry 28:17–25
43. McIntosh TJ, Holloway PW (1987) Biochemistry 26:1783–1788
44. McIntosh TJ, Simon SA, Needham D, Huang C-h (1992) Biochemistry 31: 2020–2024
45. Griffith OH, Dehlinger PJ, Van SP (1974) J Membr Biol 15:159–192
46. Worcester DL, Franks NP (1976) J Mol Biol 100:359–378
47. Simon SA, McIntosh TJ, Latorre R (1982) Science 216:65–67
48. Pearson RH, Pascher I (1979) Nature 281:499–501
49. Simon SA, Advani S, McIntosh TJ (1995) Biophys J 69:1473–1483
50. McIntosh TJ, Simon SA (1986) Biochemistry 25:4948–4952
51. McIntosh TJ, Simon SA (1996) Langmuir 12:1622–1630
52. Rand RP, Fuller N, Parsegian VA, Rau DC (1988) Biochemistry 27:7711–7722
53. Nagle JF, Wiener MC (1988) Biochim Biophys Acta 942:1–10
54. Gruen DWR, Marcelja S (1983) J Chem Soc Faraday Trans 2 79:225–242

55. Cevc G, Marsh D (1985) Biophys J 47:21–32
56. Israelachvili JN (1991) Intermolecular and Surface Forces. Academic Press, London
57. Marcus YJ (1991) Chem Soc Faraday Trans 87:2995–2999
58. Cevc G, Marsh D (1988) Phospholipid Bilayers, Physical Principles and Models. Wiley, New York
59. Simon SA, McIntosh TJ (1989) Proc Nat Acad USA 86:9263–9267
60. Simon SA, Fink CA, Kenworthy AK, McIntosh TJ (1991) Biophys J 59:538–546
61. Simon SA, McIntosh TJ, Magid AD, Needham D (1992) Biophys J 61:786–799
62. Smaby JM, Brockman HL (1990) Biophys J 58:195–204
63. Brockman HL (1994) Chem Phys Lipids 73:57–79
64. Ebmann U, Perera L, Berkowitz ML (1995) Langmuir 11:4519–4531
65. McIntosh TJ, Simon SA (1994) Biochemistry 33:10477–10486
66. Kenworthy AK, Hristova K, Needham D, McIntosh TJ (1995) Biophys J 68:1921–1936
67. Israelachvili JN, Wennerstrom H (1990) Langmuir 6:873–876
68. Israelachvili JN, Wennerstrom H (1992) J Phys Chem 96:520–531
69. McIntosh TJ, Magid AD, Simon SA (1989) Biophys J 55:897–904
70. Tardieu A, Luzzati V, Reman FC (1973) J Mol Biol 75:711–733
71. Gingell D, Parsegian VA (1972) J Theor Biol 36:41–51
72. Needham D (1995) In: Disalvo EA, Simon SA (eds) Permeability and Stability of Lipid Bilayers. CRC Press, Boca Raton, Florida
73. Marra J, Israelachvili J (1985) Biochemistry 24:4608–4618
74. Simon SA, McIntosh TJ (1986) Methods in Enzymology 127:511–521
75. Damodaran KV, Merz KM (1993) Langmuir 9:1179–1183
76. Damodaran KV, Merz KM, Jr (1994) Biophys J 66:1076–1087
77. Perera L, Essmann U, Berkowitz ML (1996) Langmuir 12:2625–2629

Progr Colloid Polym Sci (1997) 103:107–115
© Steinkopff Verlag 1997

VESICLES. BILAYERS. AND MEMBRANES

L. Perera
U. Essmann
M.L. Berkowitz

The role of water in the hydration force – molecular dynamics simulations

Received: 10 January 1997
Accepted: 17 January 1997

L. Perera · U. Essmann
Prof. M.L. Berkowitz (✉)
Department of Chemistry CB 3290
University of North Carolina
Chapel Hill, North Carolina 27599, USA

Abstract To understand the contribution of water to the repulsive force acting between phospholipid membrane molecules we performed molecular dynamics computer simulations on lamellar systems of phospholipid bilayers in water. Four simulations were performed. Two simulations were done on dilauroyl-phosphatidylethanolamine (DLPE) in water systems and two on dipalmitoylphosphatidylcholine (DPPC) in water systems. The simulations differed by the amount of water per phospholipid headgroup. From the simulations we concluded that even at the hydration limit the headgroups of the opposing membranes come in to close proximity, so that they are separated by only one or two water layers. Since the water structure in the first solvation shell is distinctly different from bulk water, the hydration force is likely to be due to the solvation layer of water around the headgroups, which are separated by up to two water layers, rather than due to a long range perturbation of the water structure. We also observed that different solvation patterns exist for water around the phosphatidylcholine (PC) and phosphatidylethanolamine (PE) headgroups. We propose that this solvation pattern is connected to the difference in the hydration of DPPC and DLPE membranes.

Key words Water – phospholipid bilayers – molecular dynamics computer simulations

Introduction

The strong exponentially decaying repulsive force acting between electrically neutral lecithin bilayers was first measured twenty years ago [1]. According to the measurements the pressure due to this force has a form

$$P = P_0 \exp(-d_w/\lambda) \qquad (1)$$

where d_w is the separation between membrane surfaces occupied by water, λ is the characteristic decay length of the pressure and P_0 is the value of the pressure when d_w is zero. From the initial measurements it was concluded that d_w was in the range of 1.4 to 2.6 nm, and $\lambda = 0.193$ nm.

Since the observed pressure cannot be due to the electrostatic interaction and since the value of λ was close to the size of a water molecule, it was suggested that the origin of the force is due to the water. This is the reason that the force was named "hydration" force. Later it was shown that the values of λ are distributed in a rather wide region (from 0.1 to more than 0.3 nm) [2]; it was also realized that the range of d_w values depends on the definition of the membrane water boundary. For example, according to the definition adopted by McIntosh and Simon [3] the values of d_w are somewhere in the region of 0.5 to 1.5 nm.

After the initial discovery of the hydration force acting between membrane surfaces the same type of force was observed between DNA molecules [4] and between

108
L. Perera et al.
Origin of the hydration force

polysaccharides [5]. As we can see the force is ubiquitous and therefore it is important to have a clear understanding of its nature.

Immediately after the publication of the first measurements on the hydration force Marcelja and Radic (MR) [6] proposed a phenomenological Landau type theory that explained the origin of such a force. In this theory it was assumed that water is ordered between membrane surfaces and therefore an order parameter can be assigned to describe its state. According to the MR theory the decay of the order parameter was characteristic of the solvent and therefore the theory predicted that the value of λ was supposed to be the same when different surfaces were involved. This prediction of the MR theory did not materialize, since the experiments indicated that the measured decay length of the repulsive force was strongly dependent on the physical state and type of the surface [2]. To explain this dependence, Kornyshev and Leikin [7] (KL) extended the MR theory by including the inhomogeneous boundary conditions into the minimization problem. Both MR and KL theories emphasized the role of water in the hydration force. A very different point of view related to the nature of the hydration force was taken by Israelachvili and Wennerstrom (IW) [8]. In the IW theory it was assumed that the hydration force is mostly due to the roughness of the interface. Based on a simple model of protruding lipids IW showed that an exponentially decaying repulsive force can result just from the entropic confinement of lipids at opposing membranes. IW theory allowed a simple interpretation of some experimental data, while the interpretation of other data became less obvious [9].

Simulations

In an attempt to study the molecular nature behind the hydration force we performed a series of simulations on dipalmitoylphosphatidylcholine (DPPC) [10] and dilauroylphosphatidylethanolamine (DLPE) [11] membranes in their gel and liquid crystalline (l.c.) states. We will concentrate here on four simulations. All four of them were done on membranes in their l.c. states. Two of the simulations were performed on DLPE membranes and they differed by the amount of water per headgroup. In the first simulation we considered the case of 10.3 waters per headgroup, which is about the experimentally estimated limit for full hydration [12–14]. In the second simulation we considered 20.5 waters per headgroup. This is well above the limit of full hydration, but allows us to perform a comparison with the simulation where the same number of waters per headgroup is used to study the full hydration of DPPC. We call the first and the second simulations

small and large DLPE simulations respectively. The third and the forth simulations were done on the DPPC system and again the simulations differed by the amount of water per headgroup. In the third simulation we considered a case of 11 waters per headgroup, while in the fourth simulation we had 20.5 waters per headgroup. We call the third and fourth simulations small and large DPPC simulations, respectively. There were 64 phospholipid molecules in every simulation. The detailed description of the simulations was given in our previous work [10, 11]; here we briefly summarize the most important points. The simulations were done using a constant volume constant temperature ensemble. For the small DLPE system the box dimensions were determined by using the experimental data on bilayer repeat distance and the area per headgroup [13,14]. In the large DLPE system the area per headgroup was kept at the same value as the area in the small DLPE system, and the repeat distance was evaluated by adding water with a density of 1 g/cm^3. A different strategy was used to find the box dimensions in the water/DPPC systems. Since the number of molecules was given, we needed the density to determine the volume of the system. The system density we used in our simulations was obtained from experiment [15]. We also used the experimental bilayer repeat period [16] to determine the size of the simulation box along the direction perpendicular to the bilayer/water interface (z direction). These data allowed us to determine the area per headgroup and therefore the lengths of the simulation box in the directions parallel to the interface. We used periodic boundary conditions in all three dimensions, these conditions are very appropriate for simulations of lamellar systems. To take into account the long-ranged Coulomb forces we used a particle mesh Ewald method [17]. For the representation of the molecules we used the united atom approach from the program AMBER [18]. All parameters including the parameters for the water molecules were taken from the STUB section of the PARM91 file, except for the alkane chains of the lipid tails. Intramolecular 1–4 interactions within alkane chains were represented by the Ryckaert–Bellemans potential [19]. The values of the charges were determined using Gaussian programs and CHELPG module [20]. All intramolecular bond lengths were fixed using the SHAKE algorithm [21]. The time step was 2 fs. Each simulation was performed for 300 ps after initial equilibration. Conditions for the four simulations are summarized in Table 1.

Results

To compare the results from our simulations with some of the results from the experiments, we calculated the

Progr Colloid Polym Sci (1997) 103:107–115
© Steinkopff Verlag 1997

Table 1 Conditions of the four MD simulations		Small DLPE	Large DLPE	Small DPPC	Large DPPC
Area/headgroup (nm²)		0.504	0.504	0.613	0.658
Repeat distance (nm)		4.610	5.815	5.150	5.680
Density (g/cm³)		1.09	1.07	0.98	0.98
Temperature (K)		310	310	333	333
No. of waters/lipid		10.3	20.5	11.0	20.5
Thickness of water slab (nm)		0.6	1.79	0.49	1.21
Width of water interface (nm)		0.92	1.01	1.09	1.22

Fig. 1 Experimental and simulated electron density profiles of DLPE in the liquid crystalline phase

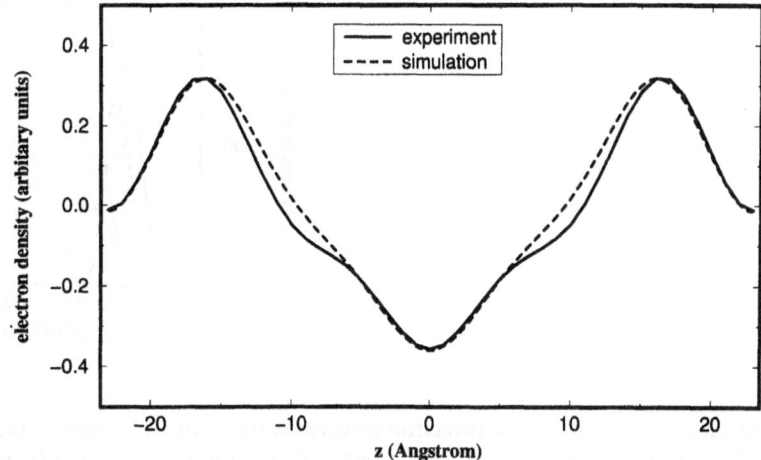

Fig. 2 Order parameters of DPPC in the liquid crystalline phase. Closed circles are experimental data at $T = 333$ K; open squares are data from the small liquid crystalline simulation and open circles are data from the large liquid crystalline simulation

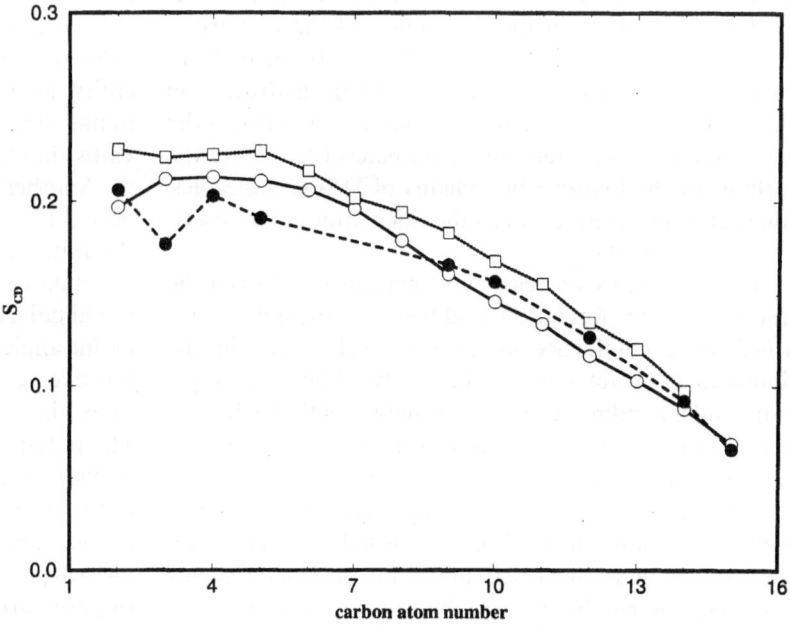

electron density profiles of the systems. The calculated density profiles for the small DLPE system are given in the Fig. 1 along with the corresponding profiles from the experiment [3]. As we can see the agreement is reasonably good. A reasonable agreement with the available experiments was also obtained for the simulations with the DPPC membranes (not shown here). The electron density profiles can be used to establish the boundaries between the bilayer and water. Following McIntosh and Simon [3] we defined them to be located R nm away from the peak of

Fig. 3 Distributions of atom (or group) densities along the bilayer normal. **A** small DLPE system, **B** large DLPE system, **C** small DPPC system, and **D** large DPPC system. Dotted–dashed line: water, dotted line: P, short dashed line: N, long dashed line: carbonyl C, and solid line: methyl groups on N in the DPPC systems. Vertical lines represent the lipid/water boundaries (see text)

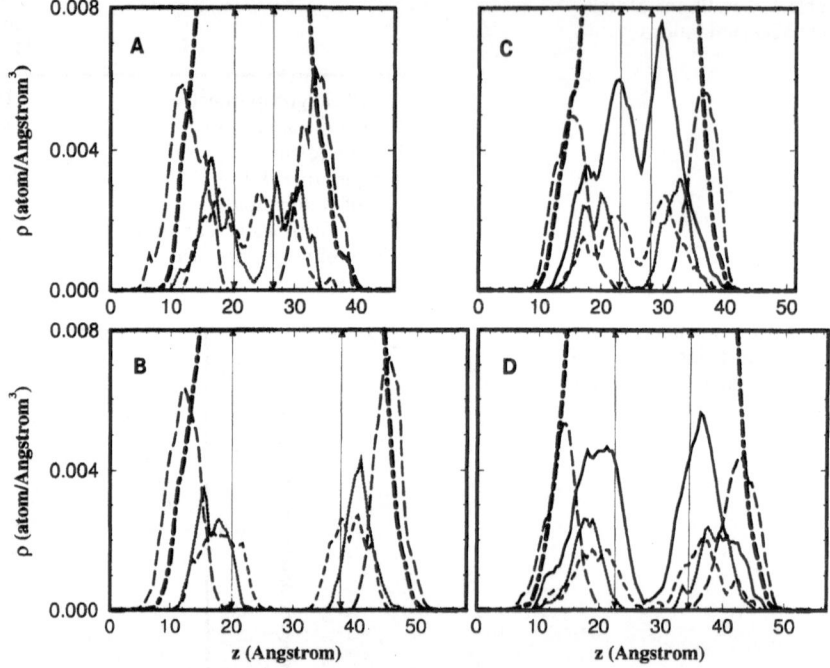

the electron density in the direction towards water. For the DLPE/water systems we used the value $R = 0.4$ nm and for the DPPC/water systems $R = 0.5$ nm. These values were also used by McIntosh and Simon. To check how our force field performs in the description of the conformational changes along the chains of the membrane molecules we calculated the order parameter for the hydrocarbon chains. In Fig. 2 we present the comparison between the calculated and measured order parameters [22, 23] for the carbons in the hydrocarbon chains of DPPC molecules. Again the agreement between the calculations and experiment is reasonably good.

Having achieved a reasonable agreement between the results from our simulations and the experiment we established some confidence in the force fields used in the simulations. Therefore we further analyzed our data to get more understanding related to the nature of the hydration force. The distributions of atomic density along the surface normal have been commonly used to characterize the spatial arrangement of the water/lipid system. We display such distributions for our four systems in Fig. 3. As we can see from the figure the distributions for the small systems look very similar. In all cases the water distributions are extended up to carbonyl positions as has been found in the experiment [24]. Also in all cases the distributions are broad indicating a raggedness of the interface. To get an estimate of the size of the interface we consider a distance for which the water density decreases from 90% of its maximum value to 10%. Such water densities in g/cm^3 are displayed in Fig. 4. From this figure we obtained inter-

facial thicknesses of 0.92 and 1.01 nm for small and large DLPE systems respectively. For the DPPC/water systems with similar amounts of water the interfacial thicknesses are 1.09 and 1.22 nm. These numbers mean that the interface constitutes $\sim 40\%$ of the system's volume. That indicates that one should not treat the interface as a sharp entity, as it was done in simplified models including the initial MR model. This was also emphasized in the previous simulations [25].

Another issue related to the initial MR model is connected to the order parameter. Although not specified in the initial paper by MR, the order parameter was later specified by Gruen and Marcelja [26] to be the orientational polarization of water, which is defined as an average cosine angle between the water dipole and the vector in the positive z direction. The profiles of orientational polarizations obtained from our simulations are shown in Fig. 5. The polarizations show no oscillations, contrary to some previous results from simulations where restrictions on headgroup dynamics were utilized [27]. The smoothness of the water polarization profile is another indication of the roughness of the membrane surface. Next to the smooth surfaces (e.g. metal or mica) one can see oscillations in the density of water that propagate 2–3 layers from the surface to bulk [28]. The rugged surface of the membrane smoothes such oscillations. Orientational polarization profiles from Fig. 5 also do not show substantial polarization of water beyond one water layer from the furthest extention of nitrogen atoms. This is clearly seen in the larger DLPE system with excess water. The extra

Progr Colloid Polym Sci (1997) 103:107–115
© Steinkopff Verlag 1997

Fig. 4 Distributions of water densities along the bilayer normal. **A** small DLPE system, **B** large DLPE system, **C** small DPPC system, and **D** large DPPC system. Vertical lines represent the boundaries of lipid/water interface

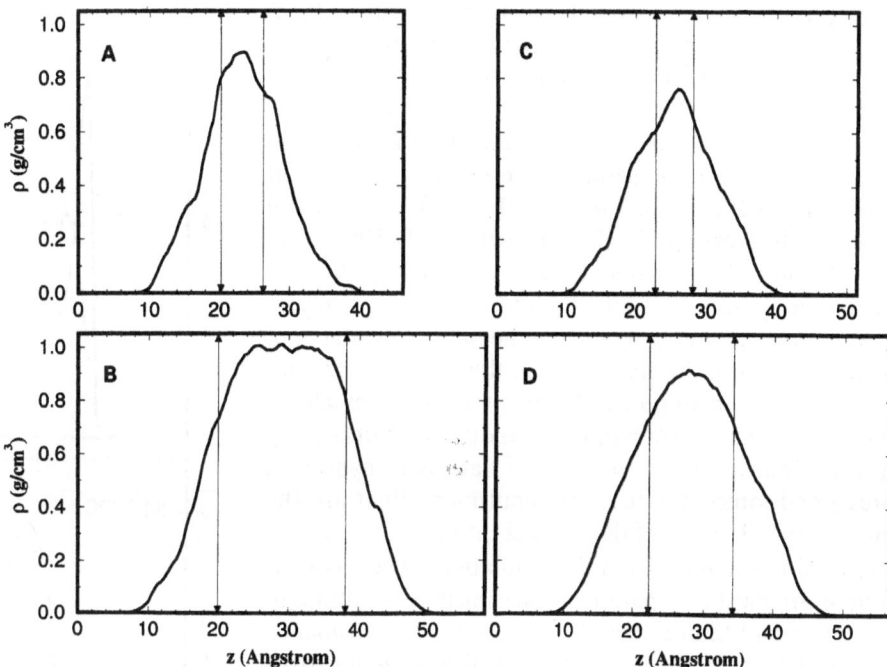

Fig. 5 Orientational polarization profiles: **A** small DLPE system, **B** large DLPE system, **C** small DPPC system, and **D** large DPPC systems. Vertical dotted lines represent the boundaries of lipid/water interface

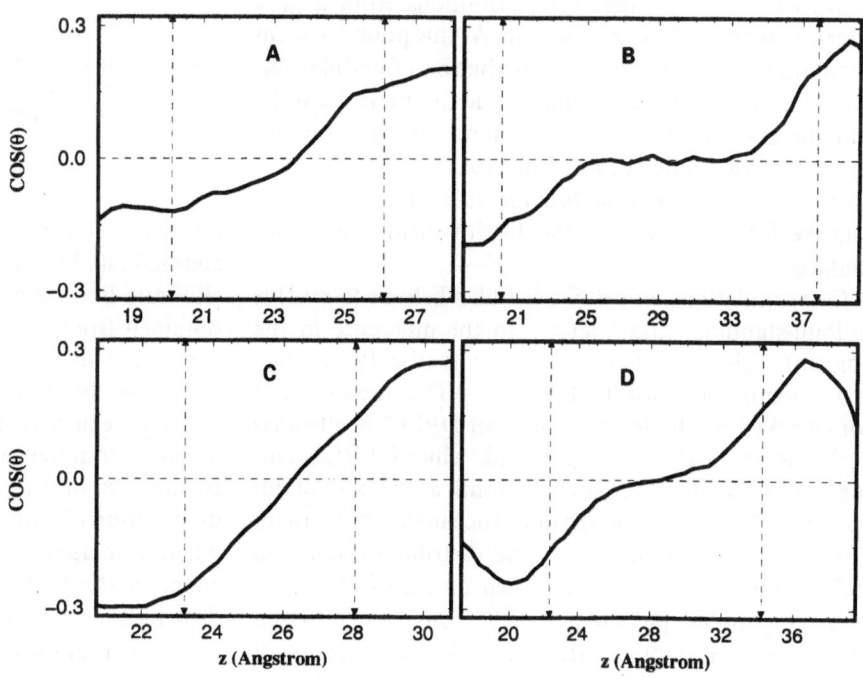

water region in the center of the water slab displays bulk-like properties and therefore attractive forces will expel this water and maintain the waters that are needed for hydration.

So far our data lead us to conclude that the surfaces of the membranes are rough on the molecular scale. The interface occupies ∼ 40% of the bilayer/water system vol-

ume and the orientational properties of the water are only very slightly perturbed beyond one hydration layer. From the data on the thickness of the water slab given in Table 1 and from the good agreement between our electron density distribution and the experimental density distribution, we conclude that not too many water layers on average are involved in the solvation of each side of the bilayer. As we can see from Table 1, in case of the large DPPC system with 20.5 waters per headgroup, the distance d_w is only 1.2 nm indicating that there are on the average about four layers of water between membranes. The estimated distance d_w in the simulation is in a good agreement with the experimental estimate [24] that at maximum hydration of DPPC with ~23 molecules of water per headgroup, the distance between DPPC layers is ~1.5 nm. From Table 1 we also conclude that at maximum hydration of DLPE molecules only two layers of water (on the average) separate the opposite surfaces of membranes, since the distance d_w in this case is ~0.6 nm.

To bring the opposing layers of membranes closer together one has to overcome the undulation forces [29], the forces due to the restructuring of water (true hydration forces) and forces due to steric repulsions. What are the relative contributions of these forces into the total force is very hard to establish from the simulations. These contributions are hard to separate in experiment also, although McIntosh and Simon, based on a series of experiments, proposed that the repulsive force which is acting over distances 0.8–1.5 nm is mostly determined by the undulation force [30]. The force that is observed to act at distances below 0.8 nm is due to contributions from a true hydration force and steric repulsion. At this point we want to remind the reader that the contribution of undulations into membrane dynamics is impossible to obtain from the simulations of the type described here, since the present simulations are limited in time and size. The present simulations were done to establish the role of water in the repulsive force and in the stability of membrane/water interface.

Our simulations, hopefully, can also help us to resolve the long-standing puzzle related to the difference in the hydration behavior of membranes with the PC groups versus membranes with the PE groups. The question that is often asked is why for example, can DPPC membranes imbibe up to ~23 waters per lipid, while DLPE membranes take up only less than half of that at the limit of full hydration? To understand the difference in the solvation of PE versus PC we studied the radial distribution function (r.d.f.) of the water around nitrogen atoms of the headgroups. Figure 6 depicts these r.d.f.'s around the nitrogen atoms of NH_3 groups of DLPE and $N(CH_3)_3$ groups of DPPC. Since the NH_3 groups of the DLPE molecules are positively charged and the size of the groups is relatively small, the water oxygen centers are closer to the amine group of the membrane molecules than the hydrogen centers. This behavior is different from that observed in the solvation of the choline headgroups of DPPC molecules. In that case we found that the distance from the N atoms to the first peak of water O atoms is nearly the same as to the first peak of water H atoms. This indicates that water

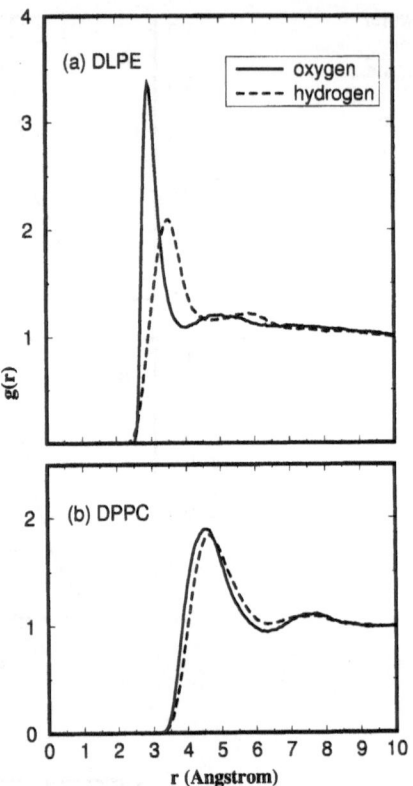

Fig. 6 Radial distribution functions of the head group N atom and water oxygens (solid line) and the head group N atom and water hydrogens (dashed line)

creates a clathrate-like structure around the positively charged $N(CH_3)_3$ groups [31,32]. Another indication that clathrate-like structures around choline groups exist is obtained from the analysis of the water pair interaction energies. This is shown in Fig. 7. As we can see from this figure the distribution for the pair interaction energy of water molecules in the first solvation shell of headgroups is different from the pair interaction energy of the bulk water. In the case of a membrane with the PE headgroups the distribution of water pair energies is shifted towards more positive energies compared to the bulk distribution. For water in the first solvation shell of PC headgroups the distribution is shifted towards the more negative energies. This observation agrees with our speculation about the clathrate structure around the PC headgroups and water bridging across the PE headgroups. Moreover, from Fig. 7 we can observe that the energetics in the second shell of water around the headgroups are close to that of bulk water, again indicating that water is slightly perturbed beyond the first hydration layer.

We have calculated the number of waters per lipid required to solvate the polar head groups in each system (solvation number) by counting the number of water

Progr Colloid Polym Sci (1997) 103:107–115
© Steinkopff Verlag 1997

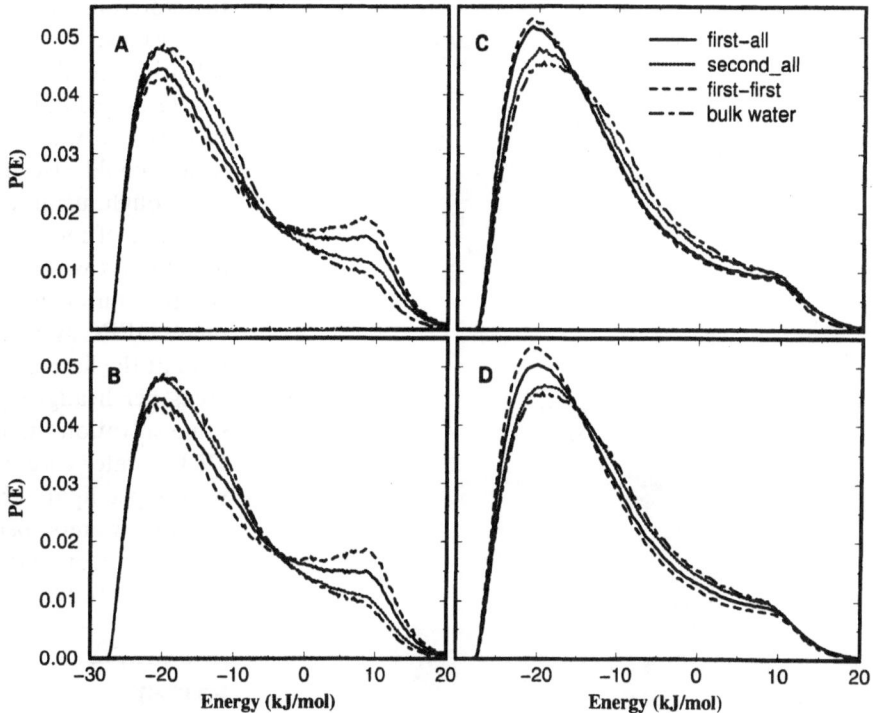

Fig. 7 Distributions of water pair energies: (i) solid line: one molecule belongs to the first hydration shell; (ii) dotted line: one molecule belongs to the second shell; and (iii) dashed line: both molecules belong to the first shell: **A** small DLPE system; **B** large DLPE system; **C** small DPPC system; and **D** large DPPC system. For comparison, the distributions of pair energies for bulk water are also shown (dotted–dashed line)

molecules that do not fall into the first peak of the r.d.f. of any lipid atom and subtracted that from the total number of water molecules. In the calculations for the DPPC membranes we considered each $N(CH_3)_3$ group as one entity. On average, 9 waters per lipid were found in the first solvation shell of both DLPE systems. This number corresponds closely to the hydration limit of 8.8–10.3 for the DLPE system [12–14]. For the large DPPC system which is close to its full hydration limit of 23 waters per lipid [12] we found 15.5 waters per lipid in the first solvation shell. The difference in the hydration number is mostly due to the difference in the size of the headgroups as well as the larger area per headgroup in DPPC membranes. In the limit of hydration, DLPE membranes contain around 9–10 waters per lipid [12–14], while DPPC membranes contain around 23 waters per lipid [12], showing that both DLPE and DPPC membranes have their hydration shells completed. While DPPC membranes at the hydration limit contain ∼ 7 more waters per headgroup than needed for full hydration of the headgroup, the DLPE membranes contain only one extra water per headgroup. That means that in the lamellar phase PE membranes are separated on the average by just two water layers, whereas PC membranes on the average are separated by more than two water layers, in agreement with the previously drawn conclusions. A possible explanation that accounts for this difference in the amount of water has its origin in the different solvent structures around PE and PC headgroups. In PE/water systems, water molecules can create

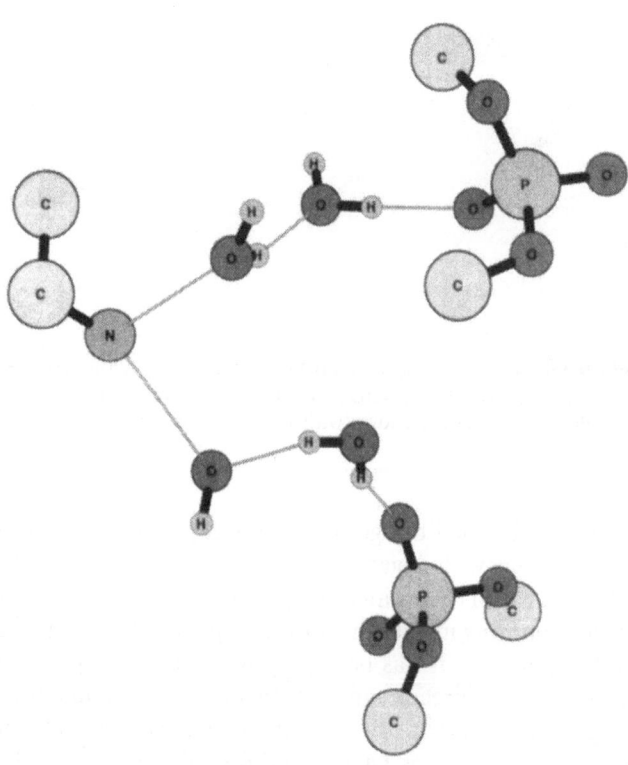

Fig. 8 Snapshot showing water bridges between opposing headgroups (taken from the simulation of the small DLPE system

114
L. Perera et al.
Origin of the hydration force

therefore one needs extra water molecules between them. This is illustrated in Fig. 9. Since the difference between the solvation number and the true hydration limit is about 7 waters/lipid in the DPPC case this roughly corresponds to only two water layers. A substantial fraction of those waters could also be accommodated in voids within the very rough membrane surfaces. Another evidence for the existence of more than two water layers between DPPC membranes when they are fully hydrated stems from the comparison of the number of water molecules in the gel phase DPPC system [10] and the small l.c. DLPE system, both at the limit of hydration. These systems have similar areas per headgroup (0.47 versus 0.50 nm^2 and also the same solvation numbers (9 waters/lipid). However, in the DLPE/water case, the hydration limit is reached with 9 waters per lipid, whereas DPPC membrane can imbibe 4 more waters per lipid showing that the clathrate structures around the DPPC headgroups are separated by additional waters.

Summary

In summary, from our simulations we conclude that the DLPE membranes are separated by two layers of water on average at the full hydration limit. Removal of these water layers contributes to the repulsive force. Due to the ragged character of the membrane surfaces the steric interactions between them also contribute to the force. The DPPC membranes are separated by four layers of water on average (two from each side of the membrane) at the limit of full hydration. Our simulations indicate that only one layer of water next to the membrane is substantially perturbed. This conclusion was based on the analysis of the orientational profiles and the energetics of water. Therefore, we think that the contribution of water to the total repulsive force comes into play only when we remove this hydration layer, which means that the true hydration force is probably appearing to act at distances d_w below 0.8 nm in agreement with the experimental assessment given by McIntosh and Simon. As in the case of DLPE membranes, the raggedness of the surface can contribute to the repulsive force and again we cannot quantitate the different contributions. We also allow for the possibility of the protrusion mechanism to contribute into the total force, although again the time and length scale of our simulations does not allow us to make concrete predictions. Our simulations also reveal the difference in the water structure around the membranes with PE versus PC headgroups. We connect this difference to the difference in the hydration limit of such membranes.

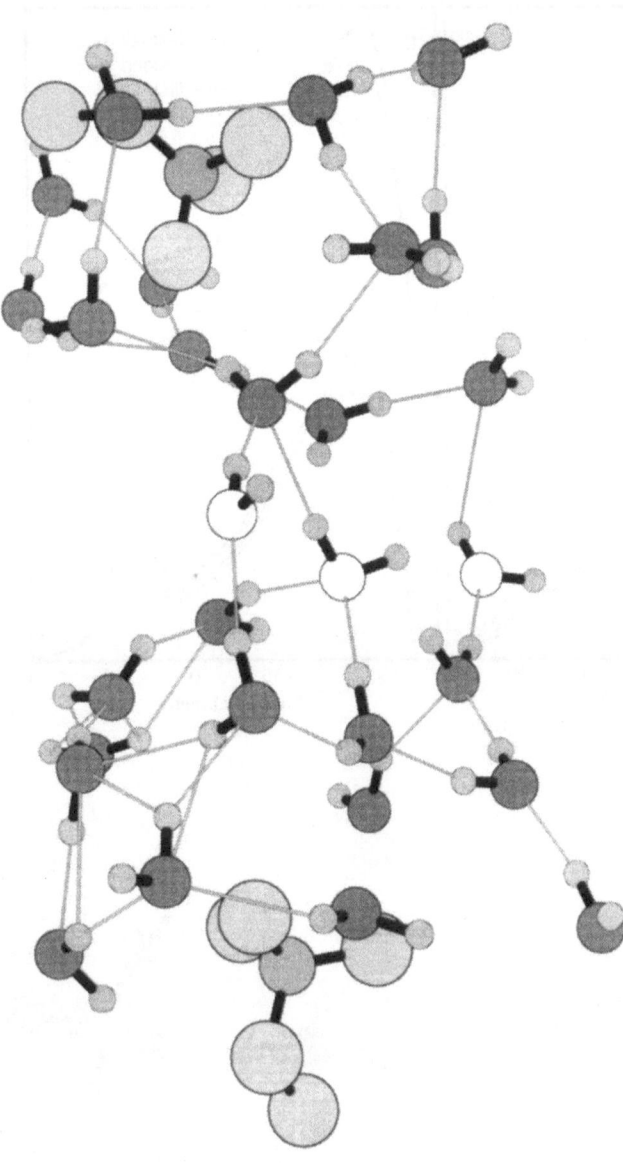

Fig. 9 Snapshot from the large DPPC system depicting the clathrate cages around N(CH$_3$)$_3$ groups. Water molecules which bridge opposing clathrate cages are shown with white oxygen centers

hydrogen bonded bridges that connect positively charged NH$_3$ groups with negatively charged oxygens attached to phosphorus. These water bridges serve as a "glue" that keeps opposing membrane surfaces together. The existence of bridging waters has been proposed previously [13, 33] and we can directly observe them in the simulations (Fig. 8). In PC/water systems, the first solvation shells of the N(CH$_3$)$_3$ groups have a clathrate-like structure. The opposing clathrates will not hydrogen bond effectively,

References

1. Leneveu DM, Rand RP, Parsegian VA (1976) Nature 259:601
2. Rand RP, Parsegian VA (1989) Biochim Biophys Acta 988:351
3. McIntosh TJ, Simon SA (1986) Biochemistry 25:4058
4. Rau DC, Lee B, Parsegian VA (1984) Proc Natl Acad Sci USA 81:2612
5. Rau DC, Parsegian VA (1990) Science 249:1278
6. Marcelja S, Radic N (1976) Chem Phys Lett 42:129
7. Kornyshev AA, Leikin S (1989) Phys Rev 40:6431
8. Israelachvili JN, Wennerstrom HJ (1990) Langmuir 6:873
9. Parsegian VA, Rand RP (1991) Langmuir 7:1299
10. Essmann U, Perera L, Berkowitz ML (1995) Langmuir 11:4519
11. Perera L, Essmann U, Berkowitz ML (1996) Langmuir 12:2625
12. Nagle JF, Wiener MC (1988) Biochim Biophys Acta 942:1
13. McIntosh TJ, Simon SA (1986) Biochemistry 25:4948
14. McIntosh TJ, Simon SA (1986) Biochemistry 25:8474
15. Tristram-Nagle S, Wiener MC, Yang C, Nagle JF (1987) Biochemistry 26:4288
16. Ruocco MJ, Shipley GG (1982) Biochim Biophys Acta 691:309
17. Darden TA, York D, Pedersen LG (1993) J Chem Phys 98:10089
18. Singh UC, Weiner PK, Caldwell JW, Kollman PA (1988) AMBER (Version 3.1), Dept Pharmaceutical Chemistry, University of California, San Francisco
19. Ryckaert J-P, Bellemans A (1975) Chem Phys Lett 30:123
20. Frisch MJ, Trucks GW, Head-Gordon M, Gill PMW, Wong MW, Foresman JB, Johnson BG, Schlegel HB, Robb MA, Replogle ES, Gomperts R, Andres JL, Raghavachari K, Binkley JS, Gonzalez C, Martin RL, Fox DJ, Defrees DJ, Baker J, Stewart JP, Pople JA (1992) Gaussian 92, Revision E.2. Gaussian Inc, Pittsburgh, PA
21. Ryckaert J-P, Ciccotti G, Berendsen HJC (1977) J Comp Phys 23:327
22. Seelig A, Seelig J (1974) Biochemistry 13:4839
23. Seelig A, Seelig J (1977) Biochemistry 16:45
24. McIntosh TJ, Simon SA (1994) Ann Rev Biomol Struct 23:27
25. Marrink SJ, Berkowitz ML, Berendsen HJC (1993) Langmuir 9:3122
26. Gruen DWR, Marcelja S (1983) J Chem Soc Faraday Trans 2, 79:211
27. Raghavan K, Reddy MR, Berkowitz ML (1992) Langmuir 8:233
28. Christenson HK, Horn RG, Israelachvili JN (1982) J Colloid Interface Sci 88:79
29. Helfrich W (1978) Z Naturforsch 33a:305
30. McIntosh TJ, Simon SA (1993) Biochemistry 32:8374
31. Damodaran KV, Merz KM (1993) Langmuir 9:1179
32. Damodaran KV, Merz KM (1994) Biophys J 66:1076
33. Rand RP, Fuller N, Parsegian VA, Rau DC (1988) Biochemistry 27:7711

Progr Colloid Polym Sci (1997) 103:116–120
© Steinkopff Verlag 1997

T.R. Stouch

Solute transport and partitioning in lipid bilayers: molecular dynamics simulations

Received: 9 January 1997
Accepted: 16 January 1997

Dr. T.R. Stouch (✉)
Mail Stop H23-07
Bristol-Myers Squibb Pharmaceutical
Research Institute
Princeton, New Jersey 08543-4000
USA

Abstract Results of atomic-level molecular dynamics computer simulations of lipid bilayers are reviewed, including studies of the transport and partitioning of small molecules (solutes) within them. The structure and timescales of the motions of the bilayer are discussed. These are contrasted with the organic/water interface, to which bilayers are often compared. It is shown that the interactions of solutes with the bilayer are governed by the fine aspects of bilayer structure and dynamics and not simply by the hydrophobicity of the solutes. The simulations were shown to support and further refine theories of the mechanism of bilayer permeation and help to rationalize experimental data.

Key words Lipid bilayer – molecular dynamics simulations – computer simulation – permeability – transport

Life is driven by water. This is evident in the importance of the delicate balance between the hydrophilic and hydrophobic forces responsible for protein structure and folding. It is nowhere as obvious as its role in the formation of biomembranes whose structures are principally determined by the amphipathic properties of the class of small biological molecules known as lipids. Membranes exist throughout organisms and many, perhaps most, important biological processes occur within them. In fact, the primordial formation of biomembranes might have been the first step in the emergence of life.

One of the important roles of biomembranes is to moderate the passage of substances between compartments of organisms. In particular, the passage of ions and proteins are actively controlled. The passage of some small nutrients are also regulated. Many small, neutral molecules can passively diffuse (without the assistance of protein channels) across the membrane, however, and this is the principal means of transport of drug molecules into cells. It is this passive transport of small molecules that will be the focus of this paper. We will show how the details of the structure and dynamics of membranes are responsible for the observed rates of transport.

Although the focus of this paper is biomembranes, the work and concepts discussed here have a broader reach, impinging on the structure of lipid-based amphiphilic assemblies in general. This is especially true since we will primarily address homogeneous lipid bilayers as membrane models. Although biological membranes are often complex, being composed of a mixture of lipids and a substantial fraction of protein, simple homogeneous bilayers have been successfully used experimentally as membrane models for decades. In fact, little concrete physical data exists for anything more complex than these systems. In particular, the dialkylphosphatidylcholines (PC), among the most common lipids, have been most widely studied experimentally and will be the subject of most of the investigations discussed here.

It is unfortunate that, for all their importance, membranes and even simple bilayers can be very difficult to study by experiment. There exist many unanswered questions about their structure, dynamics, properties, and

Progr Colloid Polym Sci (1997) 103:116–120
© Steinkopff Verlag 1997

functions. Further, many of the experiments that can be conducted effectively can be interpreted only using models. It is becoming clear that in most cases detailed atomic-level models are required. Further, static models are not appropriate since experiments measure time-averaged dynamic properties. At physiological temperatures lipid bilayers and membranes are very fluid, as will be shown. This fluidity is important to life processes and is highly regulated by cells.

In response, we have been performing atomic-level molecular dynamics (MD) simulations of PC bilayers to provide dynamical models with which to study bilayers and membranes [1–7]. MD simulations have become a standard tool in biophysical research. Although most commonly applied to studies of proteins and nucleic acids its application to lipids and membranes has a recent, growing literature. Despite the growing pains of this young field, such computational studies provide insight not available otherwise. MD simulations of lipid bilayers have done a reasonable job at duplicating and maintaining lipid bilayer structure and dynamics in a number of laboratories. At least, this is true within the resolution of experiment, which is not always as precise as desired.

Much can be said about the details of these simulations, including issues such as potential energy function form and parameterization [8–14], simulation conditions, treatment of long-range interactions [1, 2], the correct ensemble to simulate, and the proper timescales of the simulations. However, much of this is published already and here we will concentrate on the results of these studies which illustrate how small molecules move in and through the bilayer.

Despite extensive experimental research, the process, energetics, and kinetics of permeation of membranes by small solutes is unclear. The prediction of permeation rates, a quantity intensely desired by the pharmaceutical community, lags even further behind. This might be surprising considering that permeation rates have been estimated by correlations with the partition coefficients between water and an organic phase for almost a century [15]. These correlations have led to the deceptively simple picture of a bilayer as a simple one-component organic phase. Unfortunately, such correlations do not necessarily span different families of molecules. Even within homologous series of molecules there is a size dependence to permeation which ruins the correlations with partition coefficients by altering the rate of diffusion within the membrane for small molecules [4, 5, 16–18]. In those instances where these correlations do exist, they extend only to the total free energy of partitioning. Often, the entropic and enthalpic contributions to partitioning from water are quite different between membrane and a neat organic phase [19]. This indicates that, despite the correlation,

the actual process of permeation is different between the systems.

Just as the organic nature of the membrane interior helps to rationalize why correlations are often found, a consideration of the details of the internal structure of membranes helps in understanding why they fail. It is quite clear, both from experiment and simulation, that the membrane exhibits a structure that would not be expected at an interface between an organic solvent and water. Further, at no place in its structure can a membrane be considered to be a bulk phase. Although the gross features of a bilayer are clear, the polar interface between water and the charged regions of the lipids and the hydrocarbon interior and the finer structure of these regions are also important to their permeation. Here we will consider this structure and its effect on bilayer/solute interactions.

Adjustments to account for structural features have improved the correlations. For example, Lieb and Stein [17, 18] related the interior of bilayers to that of soft polymers (still considering bulk properties, however). They suggested that the passage of small solutes (but only to a certain size) could be enhanced by their ability to move within voids within the bilayer, much as happens within these polymers. Subsequently, by adjusting for the molecular weight of solutes, Walter and Gutknecht [16] were able to improve the correlations. This was done without a detailed knowledge of biomembrane structure, guided solely by analogy with soft polymer properties.

Attempts to relate diffusion rates, measured permeation rates, and partition coefficients led Miller to consider the distinct regions of the membrane as independent compartments (polar headgroup, hydrocarbon interior) [20]. Although these adjustments were advances, they still were not universally applicable. A more detailed understanding of lipid bilayer structure, obtained by a consideration of data from both MD simulations and experiment, helps us to understand why this is so.

While the experimental results are suggestive, simulation results are conclusive in their support for Lieb and Stein's hypothesis. Further, the simulations suggest refinement of their hypothesis. MD simulations of the passive diffusion within the bilayer of molecules of a range of sizes, including methane, benzene, adamantane, and nifedipine clearly show that small molecule (methane and benzene) movement includes discrete rapid jumps between voids within the bilayer. These voids are often as much as the volume of benzene, but seldom much larger. These jumps have been observed to be moderated by the torsional isomerization of the lipid hydrocarbon chains, whose motion creates passages between existing voids [4, 5, 7, 21]. These jumps are occasional for benzene, however they are very frequent for the much smaller molecule, methane. The movement of adamantane and nifedipine is

devoid of these rapid, large movements (for benzene a contiguous movement of 6–8 Å within 2–3 ps).

However, the simulations go beyond simple confirmation of the Lieb and Stein hypothesis. They further show
that these jumps are dependent on the location of the
solute within the bilayer. Jumps are smaller and less frequent high in the hydrocarbon chains, near the headgroup
regions. The jumps become larger and more frequent as
the bilayer center is approached.

This provides a gradient across the bilayer to the rate
of diffusion of the solutes. Diffusion is slowest close to the
headgroups and becomes more rapid (by a factor of 4 for
benzene) as the center is approached. The larger the solute,
lesser is the gradient. It is very large for methane (a factor
of 10 difference between the rates at various locations,
a similar difference has been seen for water [22] but is
nonexistent for the much larger nifedipine.

As noted above, Miller hypothesized that the bilayer
was composed of different "compartments" and suggested
they should be considered individually when describing
solute movement and partitioning. He suggested that permeation rate was the summation over these compartments
of the products of the partition coefficients and diffusion
constants in each.

Just as Lieb and Stein advanced the field by their
hypothesis, so did Miller. But also like Lieb and Stein's
hypothesis, the reality of bilayers is more detailed and
complex. This is illustrated by the gradient of the diffusion
coefficients mentioned above. No sharp boundaries are
seen, but rather a smooth progression between what Miller
had considered to be discrete compartments.

This is further illustrated by the free-energy profile for
bilayer permeation by a simple hydrophobe, methane,
determined from MD simulation (it is impossible to obtain
such profiles experimentally) (Fig. 1). Distinct barriers for
the penetration of headgroups, and a clear preference for
localization in the hydrophobic hydrocarbon center are
observed. Although one can easily determine general regions of the bilayer, distinct, discrete compartments
cannot be observed. Rather, the bilayer shows a smoothly
changing profile.

This behavior of both the diffusion and free-energy
profile of solute permeation is easy to understand if the
atomic-level structure of the bilayer is considered.

Our consideration of bilayer structure starts with one
of its least considered but most important constituents,
water. Bilayers are bounded by water layers and simulation indicates that, at about 12 Å away from lipid molecules, water has the properties of bulk water. At closer than
12 Å, however, these properties are substantially perturbed: the water movement slows down and individual water
molecules become oriented. At the true interface, a region
composed of both waters and lipid molecules (their polar

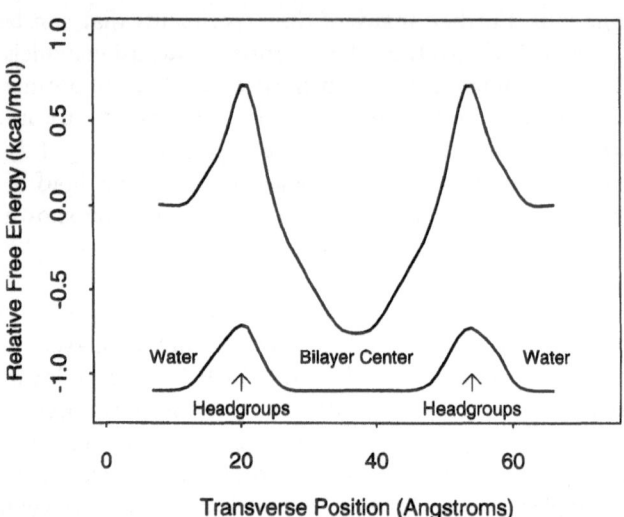

Fig. 1 Potential of mean force (relative free energy) for methane
across a lipid (dimyristoylphosphatidylcholine) bilayer. The areas
corresponding to the regions of the bilayer are labeled. The lower
curve shows the probability distribution of the atoms in the PC
headgroups

headgroups), the water molecules hydrogen bond with the
lipid and have even slower translational and rotational
diffusion rates. Even more important, on average, these
molecules experience a net orientation such that the positive ends of their dipoles point slightly toward the bilayer
center. This, we believe, contributes to what has become
known as the membrane dipole potential, a positive potential located inside the bilayer [1, 2, 23].

This potential is manifested in the permeation rates of
hydrophobic ions such as tetraphenylborate and tet-
raphenylphosphonium. It has been known for some time
that the positive ion passes more slowly than the negative
ion and hence has a higher free-energy barrier for passage.
This is despite their similar sizes and hydrophobicities.
The long-accepted explanation is that this is due to an
interior positive potential at the center of the bilayer resulting from the orientation of the dipoles of the lipid
molecules. Until recently, it was speculated that this was
due to the orientation of lipid dipoles, the PC headgroup
or the carbonyl groups of the ester linkage. However,
NMR experiments show that the headgroup dipole is
oriented parallel to the bilayer plane and so can contribute
nothing to a net dipole potential. Additionally, recent
permeability experiments on ether-linked (rather than ester-
linked) phospholipids continue to show a potential of
several hundred millivolts. The only explanation can be
water orientation [24]. Although the experimental results
are indirect, simulation shows this orientation directly and
clearly.

However, it is significant and interesting that these
waters, although oriented, are not bound. Note that we

Progr Colloid Polym Sci (1997) 103:116–120
© Steinkopff Verlag 1997

indicated that water diffusion slows (by as much as a factor of 2–5), but does not stop, even relative to that of the lipid molecules. Rather, the water molecules are in rapid exchange about the lipid molecules. Waters in the first shell of the PC headgroup were found by simulation to reside for about 20 ps, in good agreement with NMR studies. When one water molecule leaves a headgroup another not only takes its place but does so, on average, in approximately the same orientation as the water molecule it replaced [1].

This headgroup/water interface is quite varied and dynamic. Contrary to popular opinion it is quite broad, about 15 Å, and together the two interfacial regions make up about $\frac{2}{3}$ of the bilayer width. It is populated by mobile, but oriented, waters, the highly charged lipid headgroups, and interfaces with the nonpolar hydrocarbon chains of the lipid molecules. White has called it a region of tumultuous chemical heterogeneity [25]. The presence of water exhibits a dramatic and sharp decline in this region. Across the 15 Å span, its probability drops from a certain occurrence in bulk water to an essentially zero probability by the time the hydrocarbon region is reached. The hydrocarbons are not impenetrable to water, however, just as organic liquids and water, have measurable miscibilities. Water is thought to exist at a millimolar concentration even within the hydrocarbon core. However, at our atomic level consideration, this represents a very low concentration. Water completely solvates the headgroups and penetrates to the ester carbonyl groups with which it hydrogen bonds at least 50% of the time. The strong polar headgroup/water interactions make this region highly packed and of high density.

A solute entering this region will find tight quarters, slow diffusion rates, polar interactions, and many opportunities to hydrogen bond. The density makes for favorable van der Waals interactions. The headgroups, carbonyl groups, and water serve as a sink for charged and polar molecules and those that hydrogen bond. These hydrogen bonds can be strong and long-lived. We have observed instances of hydrogen-bonding between solutes and lipid molecules strong enough such that the two exhibited correlated translational diffusion (this has been observed also for a transmembrane alpha-helix) [3, 26]. However, despite these favorable interactions, this region is also populated by the large, bulky headgroups which limit the volume available to small molecules and, hence, present an entropic barrier to permeation.

As solutes progress from this interfacial region into the region populated by the lipid hydrocarbon chains one might expect a fairly homogeneous environment. However, this is not so. To be sure, chemically it is not very interesting; the ester groups were the last hydrogen bond acceptors to be found and water, the last hydrogen bond donor, remains largely above the ester groups. However, the hydrocarbon region, unlike the bulk alkanes to which it is often compared, has substantial structure.

This has been known for quite some time, based on the order parameters of individual methylene groups calculated from the quadrapolar splittings seen in NMR spectra of specifically deuterated lipid molecules [27]. Rather than randomly oriented chains, the degree of disorder of the hydrocarbon varies with position in the bilayer. Toward the lipid headgroups, to which the chains are covalently attached, the chains are quite ordered. The amphiphilic nature of the lipid molecule as a whole orients the molecule with the headgroup in the water. The hydrophobic chains orient away from the water. This upper region of the chains is also thought to be fairly tightly packed, relative to bulk alkanes. Some feel that this "waxy", closely packed hydrocarbon region is the primary barrier to membrane permeation [22, 28]. Deeper in the bilayer, the portions of the hydrocarbon chain closer to its terminal methyl end are found and approximately halfway down the chain a distinct decrease in order is seen. It is at this point that the dynamic properties of the bilayer appear to resemble liquid alkanes most. Progressing finally to the very center of the bilayer, the chains are very mobile and fluid, having a fluidity perhaps even greater than that of liquid alkanes.

Although the headgroup region is quite dense, the hydrocarbon region has a low density throughout, becoming quite low at the bilayer center. X-ray scattering studies of the L-alpha (physiologically relevant) phase of bilayers show a characteristic low-density "dip" at the juncture between the two monolayers [29]. At the atomic level, this low density is manifested by voids, i.e. unoccupied volume. It is this volume that is so accommodating to smaller molecules, as we discussed above.

Just as the structural properties of the hydrocarbon "core" vary, so too do the dynamic properties. As mentioned, although the water has a rapid exchange rate, the headgroups reorient slowly, in the order of tens of nanoseconds (individual torsions within the headgroups interconvert between rotamers on the order of hundreds of picoseconds). The torsions of upper, ordered region of the hydrocarbon chains interconvert between rotamers at a rate in the order of hundreds of picoseconds, also, in good agreement with NMR measurements [30]. Progressing down the chain, toward the terminal methyl group, this rate increases gradually until near the end, it is only about 10 ps.

A solute progressing through the bilayer will find increasing fluidity and more and more free volume as it approaches the center. In the upper hydrocarbon region it will experience some constriction and orientation due to the order and packing of the hydrocarbon chains. However, closer to the bilayer center, the increased free volume will allow a solute a wider range of orientations. The increased rate of movement of the hydrocarbon chains breaks-up and moves the voids, allowing more frequent jumps by small solutes.

The above description of the atomic-level structure and dynamics of lipid bilayers demonstrates that they have a structure unlike bulk organic liquids and that their interface with water is unlike those of bulk organics. Fur-

ther, unlike a bulk organic liquid, bilayers might be considered as a continuous interface in that their properties vary continuously from the bulk water layer through the ordered water layer, the headgroup region, the upper hydrocarbon chains and the bilayer center.

Understanding this, it is now easier to rationalize the problems with prediction of permeation rates. Hopefully, this increased understanding will improve predictive approaches such as proposed by Overton, Miller, and Walter and Gutknecht.

References

1. Alper HE, Bassolino D, Stouch TR (1993) J Chem Phys 99:5547–5559
2. Alper HE, Bassolino DA, Stouch TR (1993) J Chem Phys 98:9798–9807
3. Alper HE, Stouch TR (1995) J Phys Chem 99:5724–5731
4. Bassolino D, Alper HE, Stouch TR (1993) Biochem 32:12624–12637
5. Bassolino D, Alper HE, Stouch TR (1995) J Amer Chem Soc 117:4118–4129
6. Stouch TR (1993) Molecular Simulations 10:317–345
7. Stouch TR, Alper HE, Bassolino D (1995) In: Holloway K, Reynolds C, Cox H (eds) Computer-Aided Molecular Design: Applications in Agrochemicals, Materials, and Pharmaceuticals. American Chemical Society, Washington, DC, 589:127–138
8. Liang C, Ewig CS, Stouch TR, Hagler AT (1993) J Amer Chem Soc 115:1537–1545
9. Liang C, Ewig CS, Stouch TR, Hagler AT (1994) J Amer Chem Soc 116:3904–3911
10. Liang C, Yan L, Hill J-R, Ewig CS, Stouch TR, Hagler AT (1995) Journal of Computational Chemistry 16:883–897
11. Stouch TR, Williams DE (1993) J Comput Chem 14:858–866
12. Stouch TR, Williams DE (1992) J Comput Chem 13:622–632
13. Stouch TR, Ward KB, Altieri A, Hagler AT (1991) J Comput Chem 12:1033–1046
14. Williams DE, Stouch TR (1993) J Comput Chem 14:1066–1076
15. Overton E (1899) Vierteljahrsschr Naturforsch Ges Zuerich 44:88–135
16. Walter A, Gutknecht J (1986) J Membrane Biol 90:207–217
17. Lieb WR, Stein WD (1969) Nature 224:240–243
18. Lieb WR, Stein WD (1971) Curr Top Membr Transp 2
19. Diamond JM, Katz Y (1974) J Membrane Biol 17:121–154
20. Miller DM (1986) Biochimica et Biophys Acta 856:27–35
21. Stouch TR, Bassolino D (1996) In: Merz JK, Roux B (eds) Biological Membranes. Birkhauser, Boston, pp 255–277
22. Marrink SJ, Berendsen HJC (1994) J Phys Chem 98:4155
23. Stouch TR, Alper HE, Bassolino D (1994) Int J Supercomput Appl 8:6–23
24. Gawrisch K et al (1992) Biophys J 61:1213
25. White S, Wimley W (1994) Current Opinion in Structural Biology 4:79–86
26. Shen L, Bassolino D, Stouch TR (in press) Biophys J
27. Seelig A, Seelig J (1974) Biochem 13:4839–4845
28. Xiang T, Anderson B (1994) Biophys J 66:561
29. Franks NP (1976) J Mol Biol 100:345–358
30. Seelig J, Seelig A (1980) Quart Rev Biophys 13:19–61

Progr Colloid Polym Sci (1997) 103:121–129
© Steinkopff Verlag 1997

J. Lu
M.E. Green

Simulation of water in a pore with charges: application to a gating mechanism for ion channels

Received: 8 October 1996
Accepted: 6 December 1996

J. Lu · Dr. M.E. Green (✉)
Department of Chemistry
City College of the City University
of New York
138th St. and Convent Avenue
New York, New York 10031, USA

Abstract The electric fields and potential in a pore filled with water are calculated, without using the Poisson–Boltzmann equation. No assumption of macroscopic dielectric behavior is made for the interior of the pore. The field and potential at any position in the pore are calculated for a charge in any other position in the pore, or the dielectric boundary of the pore. The water, represented by the polarizable PSPC model, is then placed in the pore, using a Monte Carlo simulation to obtain an equilibrium distribution. The water, charges, and dielectric boundary, together determine the field and potential distribution in the channel. The effect on an ion in the channel is then dependent on both the field, and the position and orientation of the water. The channel can exist in two major configurations: open or closed, in which the open channel allows ions to pass. In addition, there may be intermediate states. The channel has a water filled pore, and a wall consisting of protein. The open or closed condition of the channel is determined without major conformational changes in the wall protein. Examples of the potential distribution in three dimensions, and the positions of the water molecules, are given for several charge configurations. It is suggested that the pK values of the amino acids in the protein are shifted by several units by the large potentials resulting from the charges which are present. The consequence is that many of the amino acids in the protein, on a particular segment (S4) of Na^+ and K^+ channels, which could bear a positive charge, are not charged. The protons may move from one amino acid to another by tunneling under the influence of the membrane potential, or upon depolarization of the membrane, which is the normal requirement for opening the channel.

Key words Pore water – electric fields – ion channels – gating

Introduction

The role of water in pores of nanometer size is of critical importance in some systems, of which biological ion channels are surely among the most interesting. It is in the hope of understanding these systems that we have studied a model for an ion channel, in which the charges nominally correspond to those in a real channel. We are interested in a type of channel which is embedded in a cell membrane, is *voltage gated*, and which will transmit either Na^+ or K^+ across the membrane in one state ("open"), but allows no ions to pass in the other state ("closed"). *Voltage gated* implies that a change of potential across the membrane in which the channel is embedded is the condition for switching from closed to open. The channels are rectifying, with

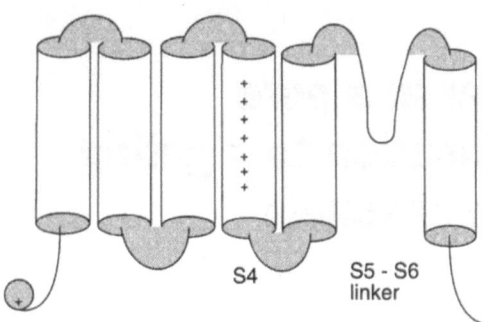

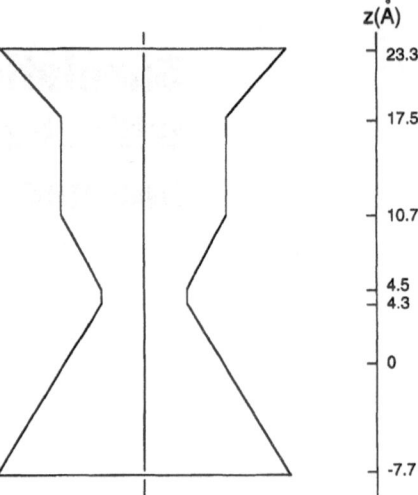

Fig. 1 Cartoon of one-quarter of a sodium or potassium channel; there are 6 trans-membrane segments of slightly more than 20 amino acids. The segments are represented as cylinders since at least most are believed to be α helices. The complete channel has therefore 24 such transmembrane segments, numbered S1–S6, and repeated four times. There are up to eight basic amino acids in each segment S4, either arginine or lysine, in every third position. There are three acidic amino acids in S2 and S3 (total), which salt bridge to three of the basic residues in S4. The S5–S6 linker generally has one or more acidic residues, at least in K^+ channels, but this is too far from S4 to form a salt bridge. It is the linker which lines the pore, and the S4 segment which is primarily involved in gating. As shown, the ball on the left, bearing positive charge, is the N-terminus of the channel; the ball is needed for *inactivation*, a set of states of the channel different from the closed state, but in which the channel is blocked. The tail on the S6 (right-hand) segment is the C terminus. Both are in the cytoplasmic domain (i.e. inside the cell)

Fig. 2 A model of the channel structure as a cylindrically symmetric pore, approximately maintaining the slope of the central part of the channel. This structure, shown here in two-dimensional projection, is used in the simulation. Charges are placed in rings holding up to 4 charges each, at 90° spacing, roughly corresponding to the charges on the protein segments; the rings are at values of $z = 17.4, 13.5, 7.2,$ and 4.5 Å. The radius of the figure at the top ≈ 8.5 Å, at the constriction at $z = 3.5$–4.5 Å, ≈ 1.9 Å. The number of charges can be chosen to correspond to various possible configurations of the charge, so that one can simulate the effects of partial neutralization of the charges, or pK shifts causing them to not all be ionized. In the central pore, there are explicitly represented water molecules; to the side of the pore, the protein is represented by a medium of dielectric constant 4. Above and below the pore, the water behaves approximately as does bulk water, and enters the calculation as a medium of dielectric constant 80

K^+ leaving the cell, Na^+ entering. There are a number of complications, such as *inactivation*, which we will ignore in this discussion. A very thorough review of the field, including essentially all significant experimental background as of 1992, has been given by Hille [1].

It should be stated at the beginning that the model is somewhat abstract, in that there exists no good model of the location of charges in such a channel, although attempts have been made, especially by Guy and coworkers [2]. However, the general structure of these channels is known, and is shown in Fig. 1, with description in the figure caption. Briefly, the channel is composed of a tetramer of a bundle of six transmembrane segments (in the sodium channel, it is a true tetramer, with the sets of segments covalently linked, while in the potassium channel, four very similar units associate to form the channel). Between the fifth and sixth segments of each unit there is a long linking segment which lines the pore, producing a tapering internal structure. It is of central importance to channel gating that there is one segment, S4, on which every third amino acid is basic, and would take a positive charge in solution at physiological pH [3]. We have simplified the model to a cylindrically symmetrical, but tapering, pore. Our pore model is shown in Fig. 2. Above and below the pore are regions of dielectric constant 80, since the pore is lodged between two aqueous environments.

However, the center of the pore in the model has explicitly represented water molecules; it is this which makes this model particularly well suited to study the effects of water on the channel gating, and the interaction between charges and water. We will examine the effects of charges on the potential in the entire region, and calculate the field, in order to understand the qualitative behavior of the water, and to get some hint as to the mechanism of gating of the channel. We will also see that the water itself has a strong effect on the potential. All this will be done with one or two ions in the channel. The K^+ ions are represented by spherical particles with the values of van der Waals parameters appropriate to these ions. A conical model which resembles ours has been simulated by Sansom and coworkers, but without the charges [4]. The same group has simulated ion channels composed of amphipathic α-helices, as well as model channels composed of hydrophobic α-helices and hydrophobic β-barrel. The amphipathic helices modulated the orientation and self-diffusion of the water molecules, slowing the latter by nearly an order of magnitude [5]. A cylindrical model, with ions but no fixed charges in the wall, has been studied by molecular dynamics simulation by Lynden-Bell and Rasaiah [6], who found water structure ordered by the wall of the narrow pore, with reordering by ions.

The calculation

Electric potential and field from a charge
in an arbitrary position

The field calculation is done for charges which may be placed each 2 Å in the x and z directions, and the field and potential found for all positions on a three-dimensional 2 A lattice. *The calculation treats the boundaries by finding the induced charges created by charges in any of the regions of the model.* The central pore has dielectric constant $\varepsilon = 1$, with explicit water molecules; the protein is replaced by a dielectric medium with dielectric constant 4 [7]. The wall of the protein has water molecules fixed to it with density approximately that of normal water, as prepared by the position of the water molecules in an ice lattice, relative to the wall. This allows the wall to be less hydrophobic, as would be expected for a channel protein. A Monte Carlo simulation for the water, representing the water molecules by the PSPC model [8], and using the field and potential values, was carried out. With the water molecules represented explicitly, there is no need to assume $\varepsilon = 80$ for water. However, above and below the region, it is assumed that the water is normal bulk water, and can be so represented. The reason the pore is treated differently is that the water is in a field large enough, and between boundaries small enough, to possibly affect the orientation of the water (and, in fact, affecting the orientation in the conical region near the center, at least). Therefore this region must not be treated as simply $\varepsilon = 80$ water. We have outlined the electrostatic calculation previously, in the context of a greatly truncated model of the pore [9].

The calculation begins (Eq. (1)) with what is essentially Coulomb's Law, in which the first term on the right gives the potential of explicit charges, the second term that from the charges induced on the boundary:

$$\phi = (1/4\pi\varepsilon_0\varepsilon_1)[q/|r - r_q|]$$

$$+ (1/4\pi\varepsilon_0)\int_s \omega(r'')/|r - r''|\,\mathrm{d}s \qquad (1)$$

where ε_0 and ε_1 are, respectively, the dielectric permittivity of free space and the dielectric constant of the medium in which the charge is placed; in the region of explicit water, $\varepsilon_1 = 1$, q the charge (real, not induced), $\omega(r'')$ the charge density at position r'' (for continuous surface), r the posi-

tion; r'' the position on boundary surface, r_q the position of real charge and ds the infinitesimal element of surface area.

The induced charges on the boundary are all that is needed to take the boundaries into account. For the sake of being definite, take the medium in which the real charge resides to be medium 1, with dielectric constant ε_1. Medium 2 is a dielectric continuum on the other side of the boundary. In order to obtain the induced charges, use the boundary condition:

$$\omega(r'') = \varepsilon_0(\chi_2 \nabla\phi_{n2} - \chi_1 \nabla\phi_{n1})$$

$$= \varepsilon_0\mathbf{n}\cdot(\chi_2 \nabla\phi_2 - \chi_1 \nabla\phi_1) \qquad (2)$$

where $\mathbf{n}$ is a normal vector directed from 1–2, $\nabla\phi_{n\alpha} \equiv \mathbf{n}\cdot\nabla\phi_\alpha$, α the medium, $\chi_\alpha = \varepsilon_\alpha - 1$ the dielectric coefficient in region α. Note that this form of the standard boundary condition $D_1 = D_2$, where D_i are the normal components of the displacement vectors in the two media, is valid when no real charges are present in the dielectric boundary, as is the case here. We next discretize the equation in order to solve it, treating the surface as a set of surface elements, where r_k is the position of center of surface element k; ω_k the charge density on surface element k, at position r_k; s_k the area of surface element k; k the index of surface element: if $k = k''$, it is the element being computed; any other element $k = i \neq k''$.

This condition is to be applied at each surface element k'', as it is computed. Each surface element is small but finite; the charge is taken as constant over the element. Real charges can exist either within the dielectric medium (protein), because of the amino acids which bear charge, or in the channel proper (charges on the water in the PSPC model, ions):

$$\nabla\phi = \nabla\phi(q) + \sum \nabla\phi(\omega_k) . \qquad (3)$$

In Eq. (3), the first term is for each real charge, and if the potential is to be found in the presence of more than one charge, this potential would be a sum over real charges. The second term is a sum over all surface elements. For a given surface element k'', satisfy the boundary condition by splitting the gradient of the potential into three terms, two of which pertain to the contribution of charges at the location of the surface element k''.

$$\nabla\phi_{n2}(q) = \nabla\phi_{n1}(q) \quad \{\text{contribution of real charge}\} , \qquad (4a)$$

$$\nabla\phi_{n2}(\omega_{i \neq k}) = \nabla\phi_{n1}(\omega_{i \neq k})$$

{contribution of other surface charge elements} .

One pertains to the element itself:

$$\nabla\phi_{n2}(\omega_{k''}) = -(\tfrac{1}{2}\varepsilon_0)(\omega_{k''}) , \qquad (4b)$$

$$\nabla\phi_{n1}(\omega_{k''}) = +(\tfrac{1}{2}\varepsilon_0)(\omega_{k''}) .$$

The contributions of the terms in Eq. (4a) are continuous across the boundary at k''. When added to the sum of the Eq. (4b) terms, the boundary condition, Eqs. (2) and (3), is satisfied. This is most easily seen by looking at Eq. (3): the first term of Eq. (4a) is the first term of Eq. (3), and the second term of Eq. (4a), plus both of Eq. (4b) add to give the second term of Eq. (3). Equation (4b) terms are the discontinuity. This leads to equations satisfying the boundary conditions. Uniqueness insures that these are the only possible equations. Return to Eq. (3), writing for the field at the surface element k'':

$$\nabla \phi_{n1}(r_{k''}) = \left(\frac{1}{4\pi \varepsilon_0 \varepsilon_1}\right) \mathbf{n} \cdot \nabla |q/(r_{k''} - r_q)|$$

$$+ \left(\frac{1}{4\pi \varepsilon_0}\right) \mathbf{n} \cdot \nabla \sum_{i \neq k''} \omega_{k''} s_{i''}/|r_{k''} - r_i|$$

$$+ \omega_{k''}/2\varepsilon_0 , \tag{5a}$$

$$\nabla \phi_{n2}(r_{k''}) = \left(\frac{1}{4\pi \varepsilon_0 \varepsilon_1}\right) \mathbf{n} \cdot \nabla |q/(r_{k''} - r_q)|$$

$$+ \left(\frac{1}{4\pi \varepsilon_0}\right) \mathbf{n} \cdot \nabla \sum_{i \neq k''} \omega_{k''} s_{i''}/|r_{k''} - r_i|$$

$$+ \omega_{k''}/2\varepsilon_0 , \tag{5b}$$

where the last terms come from Eq. (4b) all the rest from Eq. (4a). $\omega(r_i)s_i$ is the charge on the surface element i. The first two terms are continuous at the boundary, and the last satisfies the discontinuity in the boundary condition. We can insert Eqs. (5) into Eq. (2)

$$\omega_{k''} = q/\varepsilon_1 \mathbf{n} \cdot \nabla |1/(r_{k''} - r_q)|(\chi_2 - \chi_1)/4\pi$$

$$+ \frac{1}{4\pi}(\chi_2 - \chi_1) \sum_{i \neq k''} \omega(r_i) s_i \mathbf{n} \cdot \nabla |1/(r_{k''} - r_i)|$$

$$- (\tfrac{1}{2})(\chi_2 + \chi_1)\omega_{k''} . \tag{6}$$

We can move all the $\omega_{k''}$ terms to the left-hand side.

$$(1 + (\tfrac{1}{2})(\chi_2 + \chi_1))\omega_{k''} - 1/4\pi(\chi_2 - \chi_1) \sum_{i \neq k''} s_i \mathbf{n} \cdot$$

$$\nabla |1/(r_{k''} - r_i)|\omega_i = q\mathbf{n} \cdot \nabla |1/(r_{k''} - r_q)|(\chi_2 - \chi_1)/4\pi . \tag{7}$$

On the left-hand side there are a set of terms which are linear in the $\omega_{k''}$; the terms in the sum have the form of an ω_i times a geometric term; the terms on the right-hand side also depend only on the geometry of the system. For a set of linear equations, we can rewrite Eq. (7) as a simple matrix equation:

$$\mathbf{A}\omega = B , \tag{8}$$

where $\mathbf{A}$ matrix depends only on the geometry of the system, not the arrangement of the real charges, while the vector B depends on the geometry and location of the real charges, but not the surface elements, so that it is a vector.

As a consequence, it is possible to solve for the components of the ω vector, which are the charges on the surface elements, by simply inverting the $\mathbf{A}$ matrix:

$$\omega = \mathbf{A}^{-1} B . \tag{9}$$

From Eq. (9), we have the charges on all surface elements for any one charge, and therefore also for all possible real charges, by summation. From this we can get the potential and field at any location in the volume for which the solution has been carried out, from Eq. (1). This calculation depends on the surface elements being small. We found that the accuracy was fairly good (test described below) with $1 \text{ Å} \times 1 \text{ Å}$ elements for most of the volume, but with the second K^+ ion in the conical region near the narrow neck, the accuracy degenerated. However, we were able to restore adequate results by choosing elements $0.5 \text{ Å} \times 0.5 \text{ Å}$ in this region. The final $\mathbf{A}$ matrix which had to be inverted was 1630×1630; that is, there were 1630 total surface elements in the entire model. Inversion time was roughly 45 min CPU time on a Decalpha 2100, and the entire calculation took less than 2 h. The final result was a set of four (potential plus 3 field components) five-dimensional arrays, each slightly more than 6 Mb.

Test of validity: The interaction energy of the K^+ ion with the water molecules should be the same, regardless of whether the water molecules have equilibrated or not, and regardless of whether one counts the ion as inducing the surface charge and the molecules reacting, or vice versa. When this is checked, the two results match within 10% for K^+ in most of the region, but can be as much as 25% different in the most extreme cases with K^+ in the tapered region just above the constriction, where the approximation, and the 2 Å scale of the lattice, are least accurate. Prior to adjusting the size of the surface elements in the tapered region, the results were not this good.

Addition of water molecules: determination
of local fields and potentials for the entire volume

After completing the electric field calculation, the field and potential are known at any point in the volume for a charge at any other point in the volume. In order to get the total fields and potentials, it is necessary to include the water molecules; in order to orient the water molecules properly, a Monte Carlo simulation was carried out. This requires a model for the water molecules, and the PSPC model [7] was chosen, as the high fields suggest the importance of allowing the molecules to be polarized. In this model the molecules are treated as having three-point charges and an induced dipole, in addition to a Lennard–Jones potential. Each water molecule therefore provides a set of charges which can react in the field of all other water molecules, plus the fixed charges in the "protein", and the K^+ ion(s). In addition to the electric field and potential, we obtain the equilibrium properties of the molecules in the pore.

Monte Carlo simulation

In addition to the PSPC model for water, the following conditions were used for the simulations:

1) The Metropolis criterion was used for the acceptance of moves [10].

2) In most simulations, there were 6000 Monte Carlo Steps per molecule; in some cases 12000 were used to insure that the 6000 step equilibration was adequate; it was. The first 4000 steps were used to equilibrate, and the next 2000 for data collection. The energy was constant after 4000 steps, and the system was not near any known phase transition, so that this should have been an adequate criterion for equilibration, even without the 12000 step check.

3) The ensemble was defined as follows: The volume was fixed, and the temperature set to 300 K. The upper section of the simulation volume, $z = 17.5$ to 23.3 Å in Fig. 2, was set to have a constant number of molecules, at approximately the density of bulk water. Molecules were allowed to disappear in the other sections by dropping out through the bottom of the section. The upper section therefore had an *average* chemical potential for the water which was set by the value of NVT in the upper section (obviously, the chemical potential must fluctuate, as it is not possible to fix N, V, T, and μ simultaneously). The remainder of the system could exchange water molecules

Fig. 3A, B Potential at the end of simulation, "high charge" configuration: 12 charges plus the K^+ ion. The 12 charges are arranged as follows: The two lowest rings have three negative, one positive, charge. The two upper rings have two positive charges each. In (A), the K^+ ion is at $z = 10$ Å; range: white $= -8.4$ V, black $= 4.9$ V; (B) K^+ ion at $z = 4$, near the lowest, negative, ring, range: white $= -5.1$ V, black $= 4.2$ V (C, D) the "snapshots" of the molecules at the end of the simulation, corresponding to the charge distributions in (A) and (B), respectively

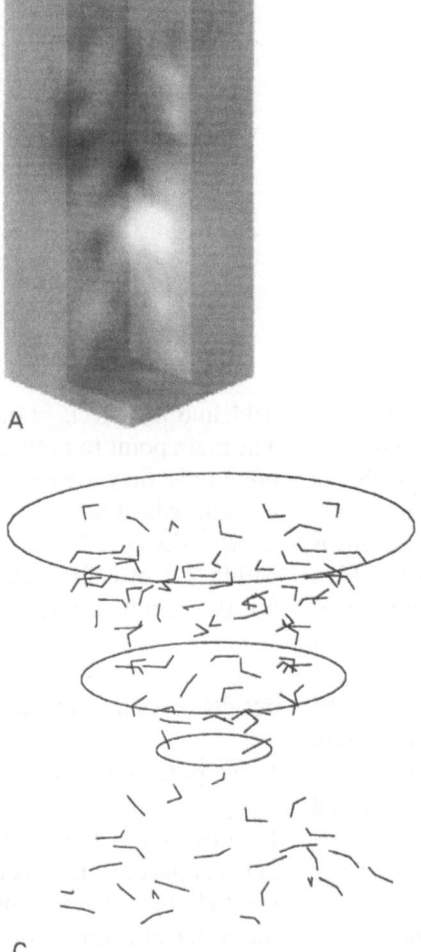

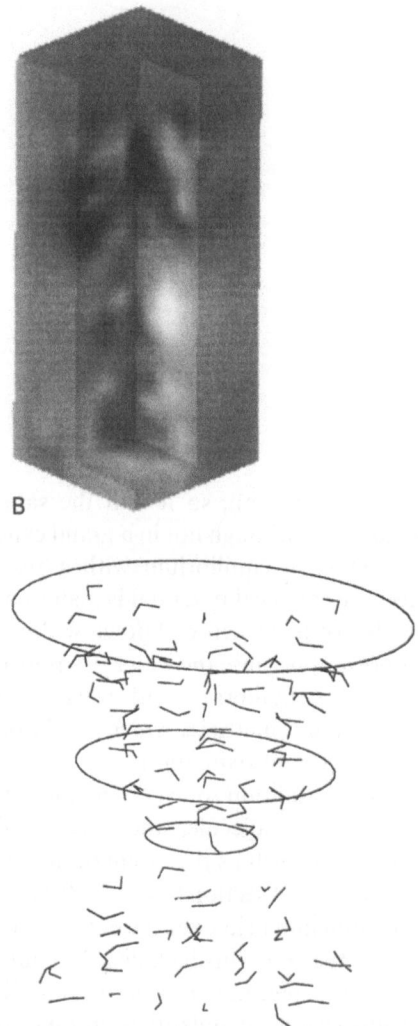

A

B

C

D

Fig. 4 Similar to Fig. 3 (same K⁺ ion positions), except that the charge distribution corresponds to fewer charges: one positive, one negative in each lower ring, one positive in each ring above (A) range: white = − 10.7 V, black = 2.4 V; (B) white = − 4.4 V, black = 4.1 V

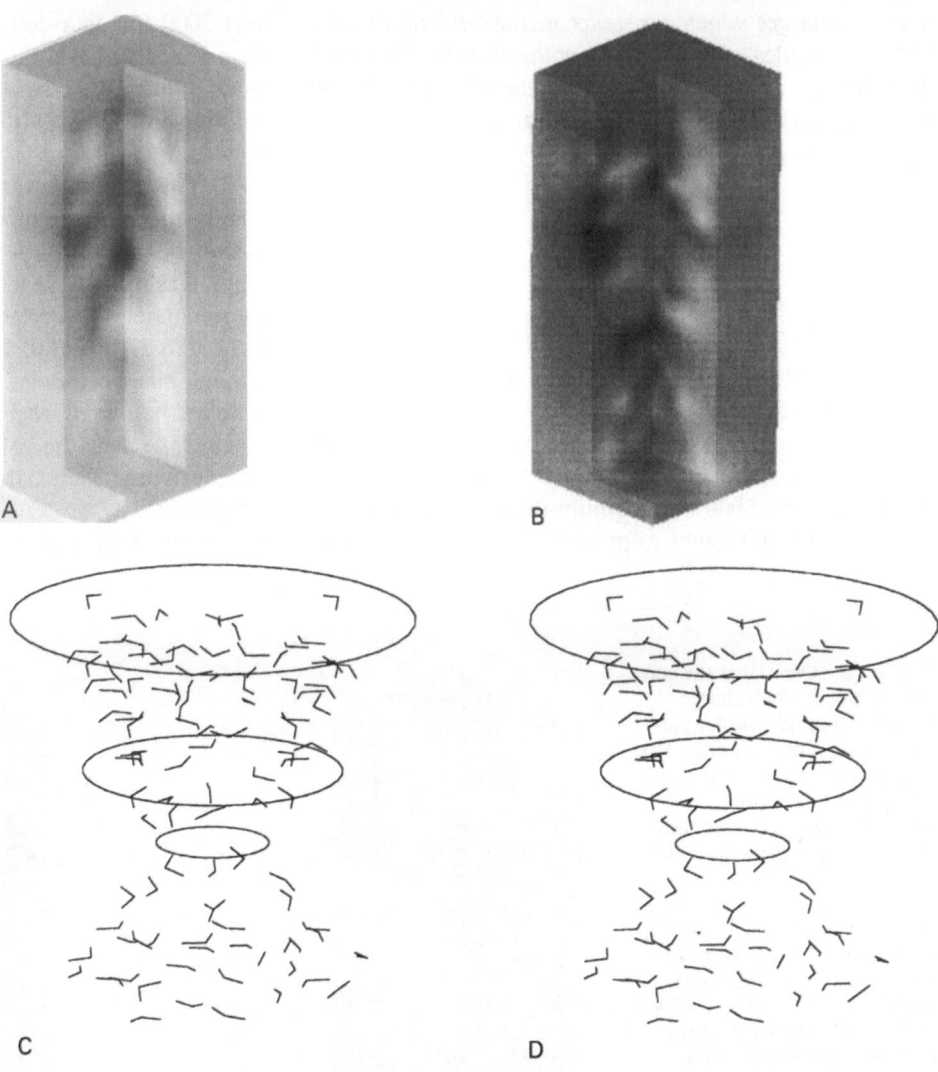

with that section, so it had the same average chemical potential. Although not in a grand canonical ensemble, the system is in equilibrium with a reservoir in which the average chemical potential is maintained at a value which, while not fixed, can be known within fluctuations. This is important because the water in a pore of this type would in fact be in equilibrium with water in the intracellular and extracellular matrices, which would behave as bulk water to a good approximation.

4) In addition to averaging the usual data, such as the energy per molecule, after the simulation, and certain properties, such as the orientation of the molecules, in the lower tapered section ($z = 4.5$–10.7 Å in Fig. 2), the overall potential and field over the entire volume was saved at the end of the run. This included the contribution of all water molecules, plus the fixed charges and K⁺ ion. These are the results shown, as potentials, in Figs. 3 and 4. One can see the positive potentials created by the K⁺ ions, the reaction

field, and the effects of the charges embedded in the walls. The main point to notice, however, is the magnitude of the potentials; they are extremely large. The potentials range to hundreds of mV, in some places exceeding 1 V, even at a few Å distance from the nearest fixed charge (the points within 2 Å of a fixed charge have even larger potentials, but this is not so interesting).

Results of the potential calculations and simulations

Large potentials and fields

Figures 3 and 4 show these very large potentials, hundreds of mV, approaching a volt, even at some distance from the fixed charges. These potentials change over distances of the order of a nm, producing fields in some instances in excess of 10^9 V m⁻¹. While not quite large enough to reliably tear molecules apart, such a field certainly is capable of orienting them.

Orientation of molecules

Molecules in the lower tapered region of the model, just above the constriction, are oriented to a significant extent by the fields. In other regions, the fields depend on the choice of charges, and no single direction is selected, so that no particular orientation is expected, and none is found.

"Snapshots" of the molecules at the end of the run

Accompanying the potential and field diagrams of Figs. 3 and 4, "snapshots" of the molecules at the end of the run are shown. These show the extent to which the molecules are oriented, together with the potentials which produce the fields responsible for the orientation. The results are approximately as one would expect. The K^+ ion does not appear among the water molecules, but its presence is marked by an empty space around which the adjacent water molecules orient.

Other quantities

The quantities which may be obtained from the simulation include the density of the molecules in the lower section, and in the entire volume, the energy per molecule, and the total energy.

The latter is dominated by the fixed charge interactions, and is fairly well determined once these charges have been placed. Some configurations are of appreciably lower energy than others, but are not affected by rearrangements during the simulation to a first approximation. However, they do suggest that the arrangement of charges on the amino acids will change if too many charges of the same sign are placed in proximity.

The densities increase in the lower region above the constriction if the charge increases, or decreases if there are no charges. If the charges are placed very close (0.5 Å) to the boundary, there is an apparent further increase in density in the lower region.

Discussion

Large potentials and pK shifts

A shift of 60 mV in potential should shift pK_a or pK_b by one unit. We are seeing potentials of hundreds of mV.

A shift of approximately 300 mV is sufficient to shift the second basic arginine pK_b below the physiogical value of approximately pH 7.4, and 180 mV would do for lysine. In an ion channel, the arginine and lysine residues on the S4 segment are not all unbalanced charges, since there are corresponding acidic residues on the S2 and S3 segments which have been shown, by site-directed mutagenesis, to be salt bridged to the corresponding S4 basic residues [11]. However, there are three to five unbalanced residues on each of the four sections of the tetramer, sufficient to produce a huge field locally. There is also a negative charge, or possibly two (depending on the channel) in the pore lining segment (S5–S6 linker) [12]; however, there appears to be no evidence that this charge is salt-bridged to the basic S4 residues, or even very close to them. More probably the potentials are arranged as we see them in Figs. 3 and 4, with a large local potential and a huge field over a very short distance (probably less than a nanometer, but larger than the 3 to 4 Å of a salt bridge). We are not the only ones to have found such large fields and potentials. Other groups have found high potentials and large pK shifts in proteins, doing the calculation by rather different techniques. For example, Lancaster et al. [13], using a non-linear Poisson–Boltzmann equation approach, obtained results of the same order of magnitude in the *Rhodopseudomonas viridis* photosynthetic reaction center.

It is not surprising to see large shifts in pK values. There are numerous examples of diprotic acids, including amino acids. It is easiest to understand the point with acids such as oxalic, with two identical groups, as the first ionization occurs with a very different pK from the second, (for oxalic acid, $pK_{a1} = 1.23$, $pK_{a2} = 4.19$, for another example, malonic, pK_as 2.83 and 5.69) [14] although the groups are symmetric. As standard texts point out, the large potentials produced by charging the molecule by its first ionization shifts the pK (although texts are often concerned with intramolecular effects, the underlying cause is electrostatic). Charge shifts from external ions create potentials with similar effects. We therefore suggest that the number of charges on each of the four S4 segments, beyond those needed for the three salt bridges, is probably limited to one or two. If the number of charges on S4 grows beyond this level, potentials locally can exceed one volt, and fields larger than 10^{10} V m^{-1}, or 1 V Å^{-1}, may appear. These fields are too large for the channel to remain stable, as the energy of bonds is of the order of 1 V Å^{-1}. The potentials and fields we discussed in the first part of "Results" section represent the upper limits consistent with a stable system, in which the chemical bonds are not disrupted by the local fields.

Conditions for the channel to be open

The tapered region just above the constriction in Fig. 2 has the highest fields, and includes at least four aspartic acids in the S5–S6 linker which lines the pore [12], so as to produce a substantial negative potential, even if only two or three of the aspartic acids are ionized. This should pull a K^+ into that region. However, the ion, once there, is effectively ionically bound, and cannot easily leave and pass through the membrane. However, a second ion could replace the first, pushing it out through the far (lower, in our diagram) end of the channel, so this is not a bar to transport. Such a "knock-on" mechanism was proposed in 1972 by Bezanilla and Armstrong [15] based on their data on the relatively permeability of channels by several ions.

The high fields produce a statistically significant orientation of the water in the tapered region above the constriction (Fig. 2). In the other regions, there is no preferred average orientation of the water molecules. The orientation in the lower region suggests that these molecules are held in place by the field, at least in some charge configurations in which the orientation has the cosine of the angle between the field and the vector along the bisector of the water molecule is greater than 0.5. In these conditions one would expect that it would be difficult for the ion to pass through the section of the channel, and the channel would be closed. With other configurations, the water is less well held, and the channel would be open. Of course, the potential contributes directly to the open or closed state of the channel, by blocking or facilitating the motion of the ion.

Channel gating

We will consider the results of the calculations with a view to understanding channel gating mechanisms. There are charge arrangements which can allow the ions to pass, and others which do not, so that charge rearrangement, without motion of the protein itself, is adequate to produce gating. It is not necessary to physically move S4, as in models of Horn [16] and Isacoff [17] and their coworkers. Of course, nothing in the calculation rules out such motion, but it does show that it is not necessary.

Such charge motion without physical movement would lead to lower values of the temperature coefficient of gating; our model includes proton tunneling, presumably to some extent assisted by vibration, but still not strongly temperature dependent. The physical motion models should separate charges involved in salt bridges, (pairs of opposite sign), and separated in the original state by no more than 4 Å. There should be three such bridges

for each S4 segment, and so one expects at least three charges would have to be separated to a distance of 12 Å, if the charges are to move 8 Å, as contemplated in those models. Let us assume that only two charges are in fact so separated (for example, perhaps some geometric arrangement allows one charge to replace another, the most which can be reasonably imagined, and thus the most generous case which can be made for these models). With the dielectric constant of the protein taken to be four, the Coulombic energy is approximately 3.8×10^{-19} J. This can be inserted into a Boltzmann factor at two temperatures separated by 10 °C, leading to an estimate of more than 20 for the increase in relative rate of gating with 10 °C rise in temperature (Q_{10}) if the salt bridges must break, as in the physical movement models, while values of 2–4 are found [18]. Since three salt bridges are involved, this is probably as generous as it is possible to be to the physical movement models, and, on its face, is an argument against such models.

To pass from one charge arrangement to another, it must be possible to move at least a proton, if the protein is not to move. We propose that the first step in gating is the motion of a proton from one basic residue to another in one S4 segment, probably followed by a similar motion in another. The first movement must be under control of the membrane potential, which is small compared with local potentials. Therefore, it must be in some sense "resonant", that is, near a potential such that a very small change in potential will allow the proton to transfer. We suggest that this is possible if the proton tunnels, and that the critical tunneling distance is approximately 1.4–1.7 Å. This distance is appreciably longer than the 0.6 Å considered in the high probability transfers in strongly hydrogen bonded systems as studied by Zundel and coworkers [19]. Here, we consider the separation between the groups involved in the hydrogen bond to be approximately 1 Å greater, and thus part of a very weak hydrogen bond. Such a weak bond would not be particularly stable itself, and so would require the species to be held rather rigidly, by external groups. (The tunneling may be through water molecules tightly held, to form a proton chain, but in the absence of evidence, we will not speculate on this possibility.) The longer distance makes the tunneling less probable, and so makes it possible for the membrane potential (much smaller than the local fields due to charges) to control it; a resonant exchange of protons (or electrons) can be described by the splitting of energy levels occasioned by the interaction (consider the umbrella inversion of NH_3 as an example, in which a tunneling frequency of 2.4×10^{10} Hz corresponds to a splitting of 1.57×10^{-23} J [20]). If the frequency of exchange is low, the resonance could be destroyed by a small external field, and restored by removing the field. In the case of the channel, the membrane

Progr Colloid Polym Sci (1997) 103:121–129
© Steinkopff Verlag 1997

potential is here postulated to play this role. Calculations to study the likelihood of such a mechanism are now in progress.

Summary

Electric potential and field calculations, together with Monte Carlo simulation of water in a pore, show very large potentials and fields for reasonable charge configurations. Configurations allowing an ion to pass, and others which would not, have been found. The most significant property of the potential and field results is the huge size of these quantities: potentials on the order of hundreds of mV, and fields of several times $10^9 \, V \, m^{-1}$ are calculated. Fields larger than this would be likely to destroy the channel protein. One can reasonably expect significant pK shifts for the amino acids in the protein which creates such a pore, with a 60 mV potential shift corresponding to one pK unit. Properties of the water in the pore, particularly the orientation and the energy, have been studied in the course of the simulations. Based on the results of the calculations, a mechanism is proposed for gating in these ion channels. The water in the pore plays a critical role; it is controlled by the large fields caused by the charges, and has different properties depending on the charge configuration in the protein. We propose that proton tunneling allows the kind of critical transition which is required for this mechanism.

Acknowledgments This work has been supported in part by the National Science Foundation, and by PSC/CUNY grants.

References

1. Hille B (1992) Ionic Channels of Excitable Membranes. Sinauer Associates, Sunderland, MA
2. Durrell SR, Guy HR (1992) Biophys J 62:238–250
3. Hille, loc cit p 251
4. Sansom MSP, Kerr ID, Breed J, Sankararamakrishnan R (1996) Biophys J 70:693–702
5. Breed J, Sankararamakrishnan R, Kerr ID, Sansom MSP (1996) Biophys J 70:1643–1661
6. Lynden-Bell RM, Rasaiah JC (in press)
7. (a) Beroza P, Fredkin DR, Okamura MY, Feher G (1991) Proc Nat'l Acad Sci (USA) 88:5804–5808 (b) Gunner MR, Honig B (1991) Proc Nat'l Acad Sci (USA) 88:9151–9155
8. Ahlstrom P, Wallqvist A, Engstrom S, Jonsson B (1989) Molecular Phys 68:563–581
9. Green ME, Lu J (1995) J Colloid Interface Sci 171:117–126
10. Metropolis N, Rosenbluth AW, Rosenbluth MN, Teller AH, Teller E (1953) J Chem Phys 21:1087–1092
11. Papazian DM, Shao XM, Seoh SA, Mock AF, Huang Y, Wainstock DH (1995) Neuron 14:1293–1301
12. Lu Q, Miller C (1995) Biophys J 68:A24
13. Lancaster CRD, Michel H, Honig B, Gunner MR (1996) Biophys J 70:2469–2492
14. Bell RP, The Proton in Chemistry, 2nd ed. Cornell Univ Press, Ithaca NY, p 97
15. Bezanilla F, Armstrong CM (1972) J Gen'l Physiol 60:588–608
16. Yang N, Horn R (1995) Neuron 15:213–218
17. Mannuzzo LM, Marrone MM, Isacoff EY (1996) Science 271:213–216
18. Hille, loc cit p 50
19. Zundel G (1992) Hydrogen Bonded Systems with large proton polarizability due to collective proton motion as pathways of protons in biological systems. In: Muller A (ed) Electron and Proton Transfer in Chemistry and Biology. Elsevier, Amsterdam, pp 313–327
20. Atkins PW (1983) Molecular Quantum Mechanics, 2nd ed. Oxford University Press, Oxford p 315

Progr Colloid Polym Sci (1997) 103:130–137
© Steinkopff Verlag 1997

C.M. Care
T. Dalby
J-C. Desplat

Micelle formation in a lattice model of an amphiphile and solvent mixture

Received: 5 November 1996
Accepted: 11 November 1996

Prof. C.M. Care (✉) · T. Dalby
Materials Research Institute
Sheffield Hallam University
Pond Street
Sheffield S1 1WB, United Kingdom

J-C. Desplat
Edinburgh Parallel Computing Centre
The University of Edinburgh
Kings Buildings
Mayfield Road
Edinburgh EH9 3JZ, United Kingdom

Abstract Results are presented from Monte Carlo simulations of a three-dimensional lattice model of a binary mixture of amphiphile and solvent. The amphiphiles are represented by connected chains on a simple cubic lattice and free self-assembly is allowed within the simulations. Earlier work on this model, for chains of length four, has shown that it exhibits a critical micelle concentration and a cluster size distribution which is consistent with those observed experimentally. The results presented in this paper use chains of length six, two of which are head segments and also include the effect of chain rigidity. It is found that the mean aggregation number is greater than that achieved with shorter chains and the cluster size distribution has a significantly enhanced minimum, with the micelles more spherical in shape. The excess chemical potential is derived and its enthalpic and entropic components calculated. It is shown that the process of micellization depends critically upon the excess internal free energy per monomer falling more rapidly than the excess entropy per monomer, as the clusters grow. The reduction in entropy per monomer is found to be associated with the entropy of packing and not from a loss of configurational freedom. An analysis of the internal energy and principal moments of inertia shows that the preferred cluster shape changes from a sphere to a cylinder at cluster sizes greater than approximately 45 monomers.

Key words Micelle – lattice model – Monte Carlo simulation – entropy

Introduction

The behaviour of amphiphiles has been the subject of extensive study both experimentally and theoretically [1–4]. However, the complexity of their molecular structure renders detailed statistical mechanics difficult and the theory of these systems is largely empirical.

The computer simulation of amphiphilic systems is now well established [5] although in the majority of the simulations the molecules are constrained to lie within a particular geometry such as a bilayer or micelle. In recent years there has been increasing interest in computer simulations which allow free self-assembly in amphiphilic systems and results from both lattice and off-lattice models have been reported by several groups [6–14]. The results from the various models are essentially consistent and demonstrate that it is possible to simulate the self-assembly of micelles and lyotropic liquid crystal phases. The work of Care et al. [6–8], Larson [9], Rodrigues and Mattice [10], Bernardes et al. [11] and Wijmans and Linse [12] is based on lattice models and that of Smit et al. [13]

Progr Colloid Polym Sci (1997) 103:130–137
© Steinkopff Verlag 1997

and Rector et al. [14] is based on off-lattice molecular dynamics simulations using chains of truncated Lennard–Jones potentials.

In this paper we present, for the first time, detailed results for the enthalpic and entropic contributions to the process of micellization and establish that the entropy of chain packing plays a central role in the formation of micelles for the model considered in this work. It is expected that this result is most relevant to the understanding of the behaviour of non-ionic, rather than ionic, surfactants since the model does not include any long-range forces.

The model

The amphiphiles are represented as N chains, each of s sites, on a three-dimensional simple cubic lattice [15]. The chains are connected only through nearest-neighbour bonds and, in this work, have two-head segments and four-tail segments. It should be emphasized that the segments within this Flory-type model are assumed to represent several repeat units in an alkyl chain. If only nearest-neighbour interactions are allowed, it can be shown that only three independent bond parameters are needed to specify the energy of the system and these are taken to be the head–head, head–solvent and tail–solvent interactions [15]. This gives the following equation for the total potential energy of the system:

$$\frac{U}{k_B T} = \beta \left(n_{TS} + \gamma n_{HS} + \eta n_{HH} + \varepsilon \sum_{i=1}^{N} r_i \right), \qquad (1)$$

where n_{TS} is the number of tail–solvent interactions, n_{HS} is the number of head–solvent interactions, n_{HH} is the number of head–head interactions, β is the ratio of the tail–solvent interaction energy to $k_B T$, γ is the ratio of head–solvent interaction energy to the tail–solvent interaction energy and η is the ratio of the head–head interaction energy to the tail–solvent energy. Chain stiffness is introduced through the parameter ε which represents the scaled energy associated with a single right-angle bond in the chain and r_i is the number of right-angle bonds in the ith amphiphile chain. In the current work, the head–head interactions are ignored ($\eta = 0$). This choice is a reasonable approximation since it is observed that there are only a small number of nearest-neighbour head–head interactions in a typical micellar cluster at the concentrations of interest here.

Monte Carlo simulations have previously been reported for this model in two [6] and three [7, 8] dimensions and the model exhibits lamellar and micellar-like phases as well as other features consistent with those of amphiphiles including a critical micelle concentration

(cmc) and the growth of cylindrical micelles as the concentration is increased. All these previous results were obtained using chains of total length four with a single-head segment. In order to explore the effect of chain geometry on the phase behaviour, the simulations described below use longer chains, our main interest being to elucidate more clearly the mechanisms involved in micellization.

Simulation details

The model has been studied using the Metropolis Monte Carlo method in the NVT ensemble with periodic boundary conditions. The chains are moved using both reptation and configurational bias schemes [16]; N attempted moves are counted as one Monte Carlo step. Simulations were first run using 512 amphiphile chains from a reduced temperature of $\beta^{-1} = 1.50$ to $\beta^{-1} = 1.18$ in steps of $\delta\beta^{-1} = 0.02$ with 1.2×10^4 Monte Carlo steps at each temperature. The final configuration from the $\beta^{-1} = 1.18$ run was then replicated to provide a starting point for the larger simulations. This process was repeated at a number of amphiphile concentrations, X_A, as summarized in Table 1. The larger simulations were undertaken for 10^6 Monte Carlo steps (2×10^4 for thermalization) with data collected every 10^3 Monte Carlo steps. The reduced temperature of $\beta^{-1} = 1.18$ was chosen because previous simulations of the model have shown the clearest micellar behaviour for temperatures around this value [8]; at lower temperatures, the time to reach equilibrium becomes too large. The principal quantity measured in each simulation is the cluster size distribution, which is expressed as the volume fraction of monomers, X_n, in clusters containing n monomers.

In order to undertake a preliminary study of the entropic and enthalpic contributions to the process of micellization, further simulations were carried out at 3 vol% of surfactant. Data were collected to determine the average total number of head–solvent ($\bar{h}^n$), tail–solvent interactions

Table 1 Final micelle simulation parameters. In all these simulations $\beta^{-1} = 1.18$, $\gamma = -2.0$ and $\varepsilon = 1.0$

Concentration [vol%]	Lattice size	Chains
0.25	$214 \times 107 \times 107$	1024
0.50	$170 \times 85 \times 85$	1024
0.99	$136 \times 68 \times 67$	1024
1.99	$108 \times 54 \times 53$	1024
3.02	$94 \times 94 \times 92$	4096
3.96	$86 \times 86 \times 84$	4096
4.92	$80 \times 80 \times 78$	4096
10.66	$62 \times 62 \times 60$	4096
0.125	$135 \times 135 \times 135$	512
0.374	$93 \times 94 \times 94$	512

$(\bar{t}^n)$ and right-angle bends $(\bar{r}^n)$ present in a cluster of size n. The data presented below for these results are derived from 10^7 Monte Carlo steps with data collected every 10^3 steps. Data are also obtained for the average moments of inertia of the clusters as a function of cluster size.

Results

Results are reported here for a set of simulations undertaken over a range of total amphiphile concentrations up to $X_A = 10.66$ vol% with the reduced temperature $\beta^{-1} = 1.18$, the head–solvent parameter $\gamma = -2.0$ and the chain stiffness energy $\varepsilon = 1.0$.

Critical micelle concentration

The volume fraction of monomers in clusters of size n, X_n, is shown as a function of n in the main section of Fig. 1 for various values of the total amphiphile concentration. The growth of the micellar clusters above the cmc is clearly evident and it is interesting to note that the minimum in the cluster size distribution is significantly deeper for this model than for the earlier work [8], which used chains of length four with one-head segment and completely flexible chains. Preliminary investigations indicate that the chain stiffness is more important than chain length in controlling the depth of the minimum in the cluster size distribution. This is a topic of current study.

The monomer concentration as a function of total amphiphile concentration is shown as an inset in Fig. 1 and it can be seen that the model shows a cmc at approximately 1 vol%. The decrease in monomer concentration above the cmc arises because of non-ideality, as discussed below.

In order to relate the results to the usual theoretical analysis we write X_n, the volume fraction of monomers in clusters of size n, as [2, 8]

$$X_n = n \frac{(f_1 X_1)^n}{f_n} \exp\left(- n \frac{(\mu_n^0 - \mu_1^0)}{kT} \right), \qquad (2)$$

where f_1 is the activity coefficient for a monomer, f_n is the activity coefficient for a chain belonging to a cluster of size n, $(\mu_n^0 - \mu_1^0)$ is the excess chemical potential per monomer at infinite dilution of chains belonging to clusters of size n and X_1 is the monomer concentration. We follow [8] and approximate the activity coefficients by $f_n = \exp(aX_A)$ for all n and rewrite Eq. (2) in the form

$$\log X_1 - \frac{1}{n} \log\left(\frac{X_n}{n} \right) = h(n, X_A)$$

$$= \underbrace{\frac{(\mu_n^0 - \mu_1^0)}{kT}}_{\text{intercept}} + \underbrace{\left(-a + \frac{a}{n} \right)}_{\text{gradient}} X_a . \qquad (3)$$

The set of functions $h(n, X_A)$ defined by Eq. (3) are measurable from simulations undertaken at different concentrations, X_A and yield a set of lines, the y-axis intercepts of which give the infinite dilution excess chemical potential as

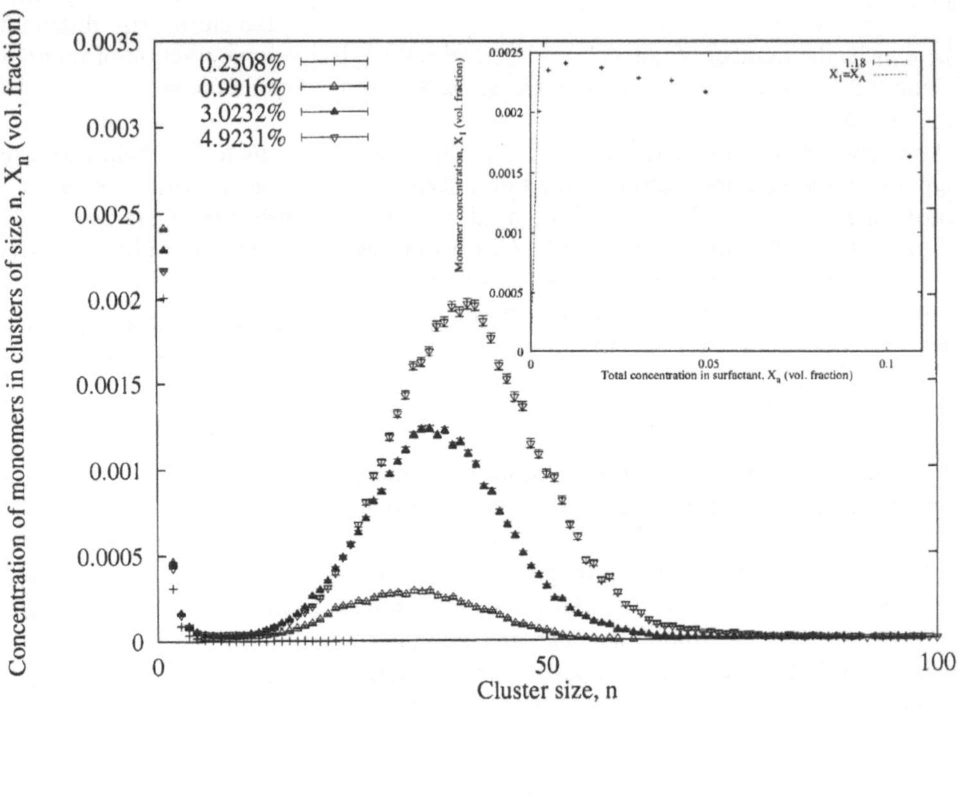

Fig. 1 Concentration of monomers in clusters of size n, X_n as a function of n. The inset figure shows the concentration of monomers, X_1, as a function of total amphiphile concentration X_A. The line $X_A = X_1$ is also marked

Progr Colloid Polym Sci (1997) 103:130–137
© Steinkopff Verlag 1997

a function of n whilst the gradients give an estimate of the activity parameter (a). The results for these quantities are shown in Fig. 2 from which it can be seen that the excess chemical potential is a monotonically decreasing function of n. The data from the gradient of these lines yields a value for the activity parameter $a = 4.86$.

Enthalpic and entropic contributions to the free energy

It is possible to make further progress with the analysis of the results by writing $\bar{N}_n$, the equilibrium number of clusters of size n, in terms of a cluster partition function, Q_n [17, 18]

$$\bar{N}_n = (\bar{N}_1)^n \frac{Q_n}{(Q_1)^n} , \tag{4}$$

where

$$Q_n(V, T) = \sum_{\langle i \rangle} \exp(- U_i^n/kT) , \tag{5}$$

Q_n is the cluster partition function for a *single* physical cluster of size n on a lattice with a total of M sites and at temperature T. The summation $\langle i \rangle$ is over all possible connected physical clusters and U_i^n is the energy of each such cluster as given by Eq. (1). It should be noted that the summation in Q_n includes all the M placings of each distinct cluster on the lattice. The result (4) only applies at sufficiently low concentrations that the clusters do not interact (i.e., the system is ideal) and we assume for the purposes of this further analysis that this is an acceptable approximation up to concentrations of the order of

3 vol%. Combining Eqs. (2) and (4) and remembering that X_n is defined in this work as a volume fraction, it may be shown that

$$\frac{1}{kT}(\mu_n^0 - \mu_1^0) = \frac{1}{kT}\left(\frac{\bar{U}^n}{n} - \bar{U}^1 \right) - \frac{1}{k}\left(\frac{S^n}{n} - S^1 \right)$$
$$+ \left(1 - \frac{1}{n} \right)\ln(s) , \tag{6}$$

where s is the number of segments in an amphiphile chain and we have defined the average internal energy per n-cluster, $\bar{U}^n$ by

$$\bar{U}^n = \sum_{\{i\}} p_i^{cn} U_i^n \tag{7}$$

and the average entropy per n-cluster as

$$\frac{\bar{S}^n}{k} = - \sum_{\{i\}} p_i^{cn} \ln p_i^{cn} \tag{8}$$

with

$$p_i^{cn} = \frac{\exp(- U_i^n/kT))}{Q_n^c} , \tag{9}$$

where p_i^{cn} is the probability of a particular n-cluster and Q_n^c is the associated cluster partition function. The summations $\{i\}$ are over all distinguishable n-clusters, assuming that the amphiphile chains are indistinguishable, the superscript c indicating that p_i^{cn} and Q_n^c *omit* the summation over the M lattice sites. We note that

$$- \ln(Q_n^c) = \frac{\bar{U}^n}{kT} - \frac{S^n}{k} = \frac{\bar{U}^n}{kT} + \sum_{\{i\}} p_i^{cn} \ln p_i^{cn} . \tag{10}$$

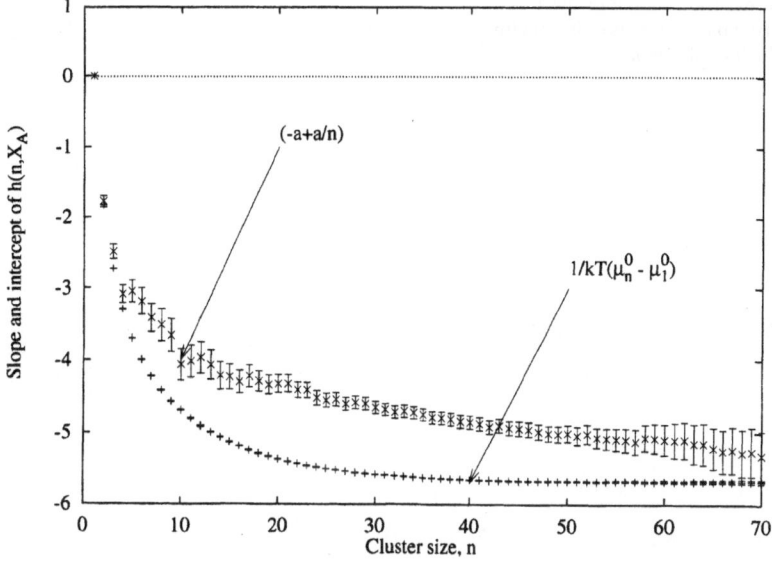

Fig. 2 Excess chemical potential and $(- a + a/n)$ derived from the y-axis intercept and slope of the functions $h(n, X_A)$ defined in Eq. (3)

Using these definitions we identify the *enthalpic contribution* to the excess chemical potential of a cluster of size n as

$$\frac{1}{kT}\left(\frac{\bar{U}^n}{n} - \bar{U}^1\right) = \beta\left(\left(\frac{\bar{t}^n}{n} - \bar{t}^1\right) + \gamma\left(\frac{\bar{h}^n}{n} - \bar{h}^1\right)\right.$$
$$\left. + \varepsilon\left(\frac{\bar{r}^n}{n} - \bar{r}^1\right)\right), \tag{11}$$

where $\bar{t}^n$, $\bar{h}^n$ and $\bar{r}^n$ are the average number of tail–solvent interactions, head–solvent interactions and right-angle bonds for clusters of size n defined, for example, by relations of the form

$$\bar{t}^n = \sum_{\{i\}} p_i^{cn} t_i^n . \tag{12}$$

We also identify the *entropic contribution* as

$$\frac{1}{k}\left(\frac{S^n}{n} - S_1\right) = -\frac{1}{n}\sum_{\{i\}} p_i^{cn} \ln p_i^{cn} + \sum_{\{i\}} p_i^{c1} \ln p_i^{c1} . \tag{13}$$

In order to evaluate the enthalpic and entropic contributions to micellization, a single simulation was undertaken at an amphiphile concentration of 3 vol% and the results for $\bar{t}^n$, $\bar{h}^n$ and $\bar{r}^n$ are shown in Fig. 3, together with an inset showing the average principal moments of inertia as a function of n. In the latter diagram, the moments for each cluster are arranged in the descending order before taking the average. Implicit in the analysis given below is the assumption that the values of $\bar{t}^n$, $\bar{h}^n$ and $\bar{r}^n$ are independent of X_A for small X_A. It is necessary to make this assumption since data for the large cluster sizes can only be collected

Fig. 3 Average total number of tail–solvent, head–solvent and right-angle bonds in clusters of n monomers. The inset figure shows the average of the three principal moments of inertia for clusters of size n

above the cmc where there are a reasonable number of large clusters. It is also possible that Eq. (6) may fail at large n because of cluster–cluster interactions. However, the authors do not believe that either of these effects will lead to errors which will significantly affect the conclusions drawn in this paper.

The average number of head–solvent in an n-cluster obeys the relation $\bar{h}^n = 8.9n$ and this is consistent with the clusters having nearly all the heads fully in contact with the solvent. The number of right-angle bonds behaves as $\bar{r}^n = 2.6n$ and this suggests, perhaps surprisingly, that the conformational freedom of the chains is not inhibited by the requirement that the heads remain in contact with the solvent as the clusters grow. It must be remembered that one of the right-angle bonds may be associated with the head segments which are free within the solvent.

The number of tail–solvent interactions shows two regions as a function of n. Below $n \simeq 45$, $t^n \simeq 16.3n^{2/3}$ which is consistent with the growth of compact, approximately spherical, clusters. Above $n = 45$, $t^n \simeq 5.1n$ which is consistent with the growth of cylindrical micelles, although the statistics become poor for these large clusters. These observations suggest a sphere to cylinder transition for the preferred cluster shape at $n \simeq 45$ and this hypothesis is supported by the moments of inertia data which show a transition from moments consistent with a sphere for $n < 45$ to those consistent with a cylinder for $n > 45$. It should also be noted that the clusters observed in this work for $n < 45$ are more spherical than those observed with chains of length four; this may be related to a reduction in the influence of the underlying lattice as the chain

Progr Colloid Polym Sci (1997) 103:130–137
© Steinkopff Verlag 1997

length increases. The data for $\bar{t}^n$, $\bar{h}^n$ and $\bar{r}^n$ can be combined to yield $(\bar{U}^n/n - \bar{U}^1)/kT$ shown in Fig. 4.

In order to determine S^n, it was first necessary to calculate the monomer partition function Q_1^c and this was achieved by exact enumeration. As a consistency check the quantities $\bar{t}^1$, $\bar{h}^1$ and $\bar{r}^1$ were calculated from Q_1^c using equations of the form (12) and found to agree with the values from the simulation to within 0.6%. This good agreement supports the assumption that the corrections due to non-ideality are reasonably small at 3 vol%. The quantity S^1 was evaluated using (13) and found to be 7.678 for $\beta^{-1} = 1.18$. Equation (6) was then used to evaluate S^n since all the other terms are known. The results for the excess entropy per monomer is shown in Fig. 4 and it can be seen that this quantity initially grows for the smallest cluster sizes and then monotonically decreases as the clusters grow. The initial growth arises because of the rapid increase in the number of distinguishable clusters as the number of monomers in a cluster increases. The subsequent reduction in entropy per monomer, as the clusters grow, must be related to the loss of freedom associated with packing the monomers in such a way as to maintain the head–solvent interactions whilst minimizing the tail–solvent interactions. It is also important to recall from above that the loss of entropy per monomer does not arise from the loss of conformational freedom.

Mechanisms of micellization

It is of interest to speculate about the mechanisms which control the position of the peak in the cluster size distribu-

tion. If we assume X_n to be given by an equation of the form (2), with the activities set to unity, the turning points in the cluster size distribution will occur at the zeros of the function $d(n)$ defined by

$$d(n) = \frac{d \ln(X_n)}{dn} = \frac{1}{n} + \ln(X_1) - \frac{(\mu_n^0 - \mu_1^0)}{kT} - n \frac{1}{kT} \frac{d\mu_n^0}{dn}.$$

(14)

It is common in the literature [2] to identify the maximum in the cluster size distribution with a minimum in the excess chemical potential. However, it is necessary to consider the full expression (14) in order to understand the micellar behaviour exhibited by models of the type discussed in this paper. The function $d(n)$ is plotted in Fig. 5 together with the associated cluster size distribution calculated from Eq. (2), setting all the activity coefficients to unity. For small X_1, $d(n)$ is always negative and since it therefore has no zeros, the cluster size distribution is monotonically decreasing. The cmc is given by the value of X_1 at which $d(n)$ becomes tangential to the x-axis and for values of X_1 just above the cmc, the micellar clusters grow rapidly. For large n, $d(n)$ undergoes a shallow decay to a constant and hence the average cluster size is a rapidly changing function of X_1. It must be noted that the monomer concentration will be a monotonically increasing function of X_A if the activity coefficients are set to unity.

The maximum in $d(n)$, and hence the presence of the micellar phase, arises essentially because the internal energy per monomer falls more rapidly than the entropy per monomer as the cluster size grows (see Fig. 4). This conclusion is in agreement with the arguments of Wijmans

Fig. 4 Excess internal energy per monomer, $(U^n/n - U_1)/kT$, and excess entropy per monomer, $(S_n/n - S_1)/k$

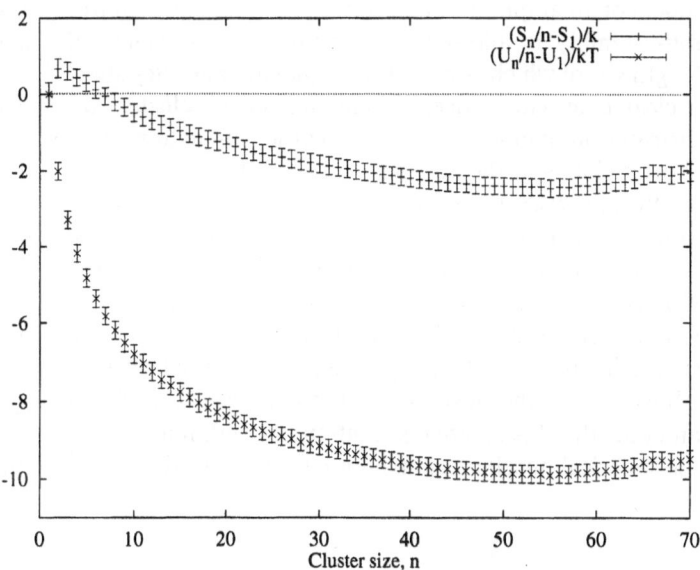

Fig. 5 Function $d(n)$ (Eq. (14)) evaluated using numerical differentiation of the excess chemical potential (Fig. 2) with $X_1 = 2.2 \times 10^{-3}$. The dotted line shows the position of the x-axis for $X_1 = 2.7 \times 10^{-3}$. The inset figure shows the cluster size distribution for these two values of X_1

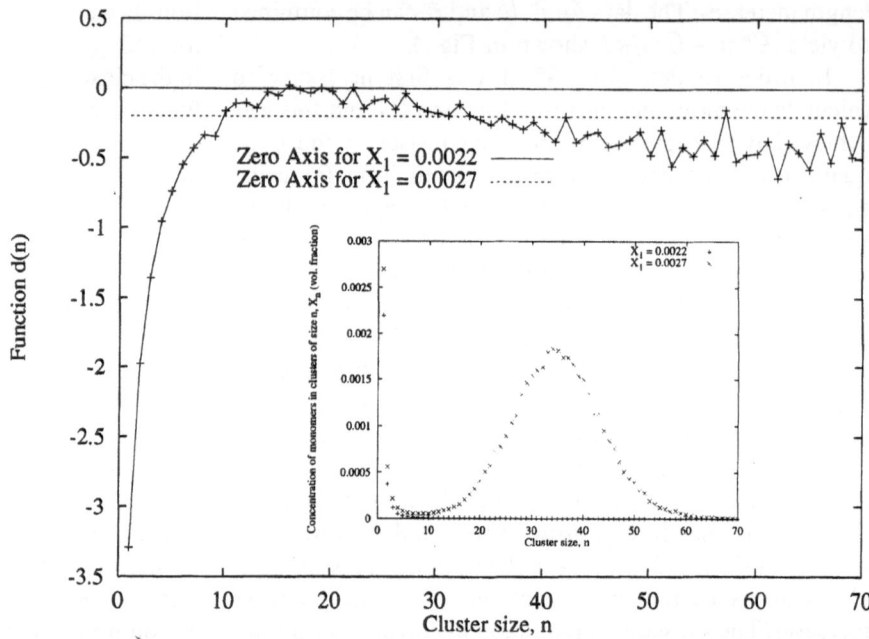

and Linse [12]. In the absence of this differential effect, it seems likely that the system would merely undergo a transition from monomers to very large clusters, possibly cylinders. Thus, we conclude that the n-dependence of the entropy per monomer is the key to determining the presence, and position of, a maximum in the cluster size distribution for this model.

Conclusions

We have presented results from simulations of a lattice model of an amphiphile and solvent mixture in which free self-assembly is allowed. The results are for chains of length six, of which two are head segments, and they show a clear cmc with a deeper minimum in the cluster size distribution than was observed in earlier simulations with shorter chains in the absence of chain stiffness.

Results were obtained which showed that both the number of head–solvent interactions and right-angle bonds per monomer were independent of the cluster size. However, the tail–solvent interactions per monomer are found to decrease as $n^{-1/3}$ for clusters up to size $n \simeq 45$ suggesting the growth of approximately spherical clusters. Above $n \simeq 45$, the cluster shape becomes more cylindrical and thus the cluster size distribution has within it, even in the spherical micelle region, information about the sphere

to cylinder transition which would occur at higher volume fractions of amphiphile.

The data were analysed to separate the enthalpic and entropic contributions to the excess free energy. For clusters greater than a few monomers, the entropy per monomer was found to decrease with increasing cluster size, although this loss of entropy was *not* associated with loss of conformational freedom. The competition between the decreasing internal energy per monomer and entropy per monomer was found to be the source of the micellar behaviour. Thus, in order for the micellar phase to occur, it is necessary for the internal energy per monomer to fall more rapidly than the entropy per monomer as the cluster size increases.

The results obtained establish that the entropy of chain packing plays a central role in the formation of micelles in the class of models used in this work. It seems possible that similar arguments should apply to micellization in nonionic surfactants although electrostatic effects are likely to be important in ionic systems. Work is now being undertaken to study the relative importance of chain length and chain stiffness on the entropy per monomer.

Acknowledgments The authors wish to acknowledge financial support from Albright and Wilson Ltd and thank Dr. D Cleaver for his helpful comments on the manuscript.

Progr Colloid Polym Sci (1997) 103:130–137
© Steinkopff Verlag 1997

References

1. Israelachvili JN, Mitchell DJ, Ninham BW (1976) J Chem Faraday Trans 2, 22:1525
2. Wennerstrom H, Lindman B (1979) Phys Reps 52:1
3. Tanford C (1980) The Hydrophobic Effect. Wiley, New York
4. Puvvada S, Blankschtein D (1990) J Chem Phys 92:3710
5. van Os NM, Karaborni S (eds) (1993) Computer Simulation Studies of Surfactant Systems, Tenside Surf Det 30
6. Care CM (1987) J Chem Soc Faraday Trans 83:2905
7. Brindle D, Care CM (1992) J Chem Soc Faraday Trans 88:2163
8. Desplat JC, Care CM (1996) Mol Phys 87:441
9. Larson RG (1992) J Chem Phys 96:7904
10. Rodrigues K, Mattice WL (1991) J Chem Phys 95:5341
11. Bernardes AT, Henriques VB, Bisch PM (1994) J Chem Phys 101:645
12. Wijmans CM, Linse P (1995) Langmuir 11:3748
13. Smit B, Hilbers PAJ, Esselink K, Rupert LAM, van Os NM (1991) J Phys Chem 95:6361
14. Rector DR, van Swol F, Henderson JR (1994) Molecu Phys 82:1009
15. Care CM (1987) J Phys Chem, Solid State Phys 20:689
16. Siepmann JI, Frenkel D (1992) Mol Phys 75:59
17. Hill TL (1956) Statistical Mechanics: Principles and Selected Applications. Dover, New York
18. Care CM (1989) J Phys, Condens Matter 1:8583

Progr Colloid Polym Sci (1997) 103:138–145
© Steinkopff Verlag 1997

Simulation of self-assembly in solution by triblock copolymers with sticky blocks at their ends

M. Nguyen-Misra
S. Misra
Y. Wang
K. Rodrigues
W.L. Mattice

Received: 15 October 1996
Accepted: 29 October 1996

M. Nguyen-Misra[1] · S. Misra[2] · Y. Wang[3]
K. Rodrigues[4] · Prof. W.L. Mattice (✉)
Department of Polymer Science
The University of Akron
Akron, Ohio 44325-3909, USA

[1] Present address:
H.B. Fuller Co.
Research and Development Laboratory
Vadnais Heights, Minnesota 55110
USA

[2] Present address:
Osmonics Inc.
Minnetonka, Minnesota 55343
USA

[3] Present address:
Department of Chemistry
North Carolina AT&T State University
Greensboro, North Carolina 27411
USA

[4] Current address: Alco Chemical
Chattanooga, Tennessee 37377
USA

Abstract Recent simulations of the self-assembly of ABA triblock copolymers in dilute solution in a selective solvent are reviewed. The medium is a good solvent for the internal block (B) and a poor solvent for the terminal blocks (A). The simulations detect both free chains, which sometimes form intramolecular 'hairpins", and interchain aggregates. Analysis of the equilibrium between the open and intramolecularly aggregated states among the free chains permits a quantitative assessment of the importance of loop entropy in the self-assembly. When the energetic affects are relatively weak, multichain aggregates are loose, with a small number of chains and a preponderance of dangling ends. As the energetic effects become more important, the system forms larger aggregates with a closer approximation to a core-shell model. Most of the chains in these larger aggregates form loops, but a few dangling ends are also present. At sufficiently high concentration, the dangling ends and loops can form bridges that connect micelles (or smaller aggregates) into clusters. Eventually, a concentration is reached at which a single cluster permeates throughout the entire system, resulting in gelation. This gel point can be determined quantitatively in the simulations. The gel is a thermo-reversible network, because the bridges have finite lifetimes. The lifetime distribution of the bridges, which plays an important role in the stress relaxation by the thermo reversible network, can be evaluated in the simulation.

Key words Gels – micelles – networks – thickeners – triblock copolymers

Introduction

Block copolymers spontaneously self-assemble into interesting microstructures in dilute solution, in the bulk, and in the presence of surfaces and interfaces. Monte Carlo simulations of coarse-grained block copolymers on simple lattices have been used in the last few years to investigate the generic properties of block copolymers [1, 2]. Since the coarse-grained chains are not atomistic, it is difficult to make a one-to-one comparison of a simulation and a spe-

cific experiment with a real block copolymer. Instead the simulation investigates the properties shared by a class of block copolymers. The simulation is performed on a lattice in order to gain access to the necessary scales of size and time, which would not be accessible if the simulation employed fully atomistic models for the chains. The simulations are performed with energies of short range, and therefore apply to nonionic block copolymers.

Much of our effort has focused on AB diblock copolymers, where each chain contains N_A beads of A and N_B beads of B. We studied the equilibrium [3–6] and

Progr Colloid Polym Sci (1997) 103:138–145
© Steinkopff Verlag 1997

dynamic [7–9] properties of the micelles formed by di-block copolymers in dilute solution in a selective solvent. Diblock copolymers form other types of ordered structures at higher concentration [10–12]. They segregate at interfaces [13, 14] and can be adsorbed onto surfaces from dilute solutions in nonselective [15, 16] and selective [17, 18] solvents.

A richer behavior in dilute solution is exhibited by ABA triblock copolymers when the medium is a poor solvent for the terminal A blocks and a good solvent for the internal B block. If water is the solvent, nonionic chains of this type are obtained when poly(ethylene oxide) has been blocked at both ends with hydrophobic groups. The hydrophobic blocks can be alkyl groups [19], which might be coupled to the poly(ethylene oxide) via a urethane [20], or they can be blocks of a more hydrophobic polymer, such as poly(propylene oxide) [21, 22] or poly(butylene oxide) [23, 24]. Of course, ABA triblock copolymers in which all of the blocks are insoluble in water can be studied appropriately selected organic solvents [25–27]. Our recent simulations of these ABA triblock copolymers in dilute solution in a medium that is selective for the middle block are reviewed here, and comparisons are made with several recent experimental [19–27] and theoretical [28] studies.

Methods

For details, the reader is referred to the original studies that will be cited later in this section. A brief description is given here. The simulation box of volume $L_x L_y L_z$ is constructed on a cubic lattice. Periodic boundary conditions are applied in all three directions. The box contains N_C chains of ABA triblock copolymers that are self-and mutually avoiding. The triblock copolymers are monodisperse and symmetric, with each terminal A block containing N_A beads of A, and each internal B block containing N_B beads of B. The beads should be identified with statistical segments, each of which is composed of several monomer units for most flexible polymers. The concentration is expressed as the volume fraction, ϕ,

$$\phi = (2N_A + N_B)N_C(L_x L_y L_z)^{-1}. \tag{1}$$

The empty sites are treated as solvent, S. The system is equilibrated with the desired values of the pairwise nearest-neighbor energies, ε_{ij}, where i and j are selected from A, B, and S. In a given simulation, most of the $\varepsilon_{ij} = 0$. The nonzero ε_{ij} are expressed in units of kT. In most of the simulations, $\varepsilon_{AB} = \varepsilon_{AS} = \varepsilon > 0$ and all other pairwise interactions are zero. The Θ solvent for a homopolymer of A (or B) is obtained when $\varepsilon_{AS} = 0.154$ (or $\varepsilon_{BS} = 0.154$) and the remaining ε_{ij} are zero [29]. The values for the dimen-

sionless energies in the simulation place A in a poor solvent ($\varepsilon_{AS} > 0.154$) and B in a good solvent ($\varepsilon_{BS} = 0$).

Open Chain ↔ Hairpin Equilibria for Isolated Chains

Linear chains experience a decrease in conformational entropy when they are restricted to those conformations that bring the two ends into contact [30]. If $N_A = 1$ and N_B is sufficiently large so that Gaussian statistics apply, and excluded volume is ignored, the loop entropy is given by

$$S_{\text{loop}} = -(3/2)k \ln N_B \tag{2}$$

which becomes more severe as the chain becomes longer. This loop entropy is important for the ABA triblock copolymers, because it opposes the placement of both of the terminal hydrophobic blocks in the core of the same micelle. The size of the loop entropy will determine whether or not the ABA triblock copolymer can form a stable micelle [31, 32].

Isolated chains are detected in the simulations. They may be devoid of intramolecular contacts, as sketched in Fig. 1A, or they may form intramolecular aggregates, or hairpins, [33] as sketched in Fig. 1B. When $N_A > 1$ and excluded volume is present, the structural and energetic variables that influence the loop entropy are conveniently identified by examination of the isolated chains in the simulations [34]. The important quantity is the ratio, n_a/n_u, of the number of isolated chains that have at least one contact between the two terminal blocks (n_a) and the

Fig. 1 Sketches of A a free chain that does not form an intramolecular aggregate, B a free chain that forms an intramolecular aggregates, or hairpin, C a small intermolecular aggregate in which each of the three chains has a dangling end, D a larger micelle in which most of the ten chains form loops, and E two aggregates connected by a bridge. Insoluble terminal blocks are drawn with the thicker lines, and the soluble B blocks are drawn with the lighter lines

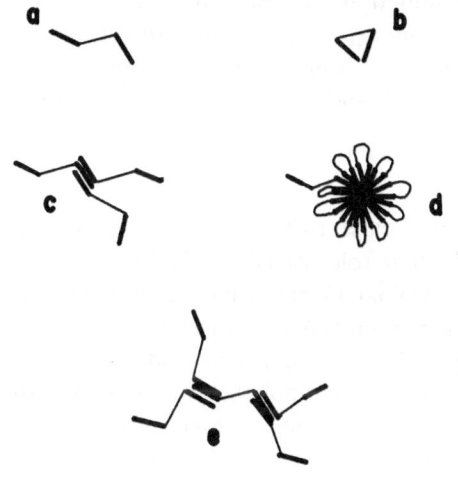

number of isolated chains that have no such contacts (n_u). This ratio is independent of whether any other chains in the system self-assemble into multichain structures. In the limit where $N_A \to 1$, the value of n_a/n_u is close to the expectation for the macrocyclization of a Gaussian chain. The value of n_a/n_u increases as N_A becomes larger (equivalently, the loop entropy effect becomes less severe), because a greater number of conformations of the middle block will bring the larger terminal blocks into direct contact. In simulations where the only nonzero energies are $\varepsilon_{AB} = \varepsilon_{AS} > 0$, a sufficiently large value of N_A, given by

$$N_A = K\chi^{-1}\ln N_B \qquad (3)$$

will cause n_a/n_u to reach the value of one. Here χ is the Flory–Huggins parameter, and K is close to 0.5 when $N_A = 1$, and is a slowly decreasing function of N_A. The simulations show that Eq. (3) provides an upper limit for the value of N_A at which the loops, or hairpins, become the dominant conformation of the isolated chains [34].

Critical micelle concentration (CMC)

A critical micelle concentration can be defined in different ways in the simulations. One approach uses the concentration of the free chains. Their concentration increases initially as the total concentration, ϕ, increases. Eventually, the concentration of free chains becomes nearly constant, at the ϕ where all of the new chains added to the system are incorporated in aggregates, and do not affect the concentration of free chains. This plateau in the concentration of the free chains defines a critical micelle concentration [34].

Alternatively, one can measure the volume fraction of the chains that participate in aggregates, using a set of simulations performed at various ϕ. This volume fraction of the aggregates is then extrapolated to zero, and the value of ϕ at this point defines a critical micelle concentration [35]. When simulations are performed with the only nonzero energies being $\varepsilon_{AB} = \varepsilon_{AS} = \varepsilon$, the critical micelle concentration of the ABA triblock copolymers defined by this latter approach is found to obey a power law of the form

$$\text{CMC} \sim (N_A\varepsilon)^{-b}, \qquad (4)$$

where b has a value close to 0.4 [35]. This expression identifies the important role played by the hydrophobic character of the terminal blocks, which is seen experimentally in the comparison of ABA triblock copolymers in which B is composed of poly(ethylene oxide) and A is either poly(propylene oxide) or poly(butylene oxide), with the latter type of A block giving a much lower CMC [21, 24]. However, the experimental CMC is much more sensitive to the size of the poly(butylene oxide) block than is suggested by Eq. (4) [23]. This discrepancy might be related to our use of $\varepsilon_{BS} = 0$, which may not adequately represent the solubility of poly(ethylene oxide) in water, or it might be a consequence of the small range for N_B in our simulations.

In the simulations, the CMC is much less sensitive to variation in N_B than to variation in N_A.

Small aggregates

Under conditions of weak segregation (small values of εN_A), the system contains free chains in equilibrium with a few small aggregates, sketched in Fig. 1C, that form and dissociate during the simulation [33]. Sometimes an ABA chain will place both of its terminal blocks in a single small aggregate, but more frequently only one of these terminal blocks is in an aggregate, and the other terminal block is dangling free in solution [33]. At concentrations slightly above the CMC, the aggregates contain a much smaller number of chains than would be found with the diblock copolymer of the same overall size and composition under the same conditions [34]. The internal structure of these aggregates is different from that of the classic core–shell micelle. The center of the core is rich in A, but it also contains a large amount of solvent and beads from the B blocks [34]. The radial profiles reveal that the volume fractions of both A and B decline over the same range of distances from the center of mass of the aggregate, showing that the interface is extremely diffuse [34].

The extension of the individual chains is measured by their dimensionless characteristic ratio, defined as

$$C = \langle r^2 \rangle / (2N_A + N_B - 1), \qquad (5)$$

where $\langle r^2 \rangle$ denotes the mean square end-to-end distance, expressed in squared lattice units. The value of this ratio is 1.45 for chains in these small aggregates, which is nearly as large as the value of 1.5 expected for an unperturbed chain with the bond angles and torsional angles of the chains on this lattice [34]. This result shows that the conformations of the chains are not perturbed when they form a small aggregate. The loops that place both terminal blocks in the same core would markedly lower the values of $\langle r^2 \rangle$ if they were present. The absence of any appreciable effect on $\langle r^2 \rangle$ shows that the loops are not the dominant conformation [34]. The preference for dangling ends, rather than loops, in these small aggregates is verified by counting the number of instances of each type of conformation [35]. This structure in the simulations is similar to the one described for polystyrene–polyisoprene–polystyrene in n-heptane [25].

At ϕ much above the critical micelle concentration, there is a second transition to a much larger aggregate that contains an enormous number of chains [34]. Here the radial distribution functions of A beads and B beads are nearly indistinguishable from one another. The aggregate is so large that the volume fractions of A and B do not fall to half of their values at the center of mass until the radial distance exceeds the contour length of the individual blocks. There is a large amount of solvent throughout the entire structure. The values of $\langle r^2 \rangle$ are slightly larger than 1.5, showing that the chains are slightly extended in this very large aggregate. The appearance of this enormous aggregate in the simulations signifies the onset of macrophase separation [34].

Larger micelles

Aggregates with a closer approximation to the classic core–shell model, as sketched in Fig. 1D, are obtained at stronger segregation (larger values of εN_A) [33, 34]. Recent experiments have reported micelles with 15–20 insoluble terminal blocks, which varies little with concentration in the range in which these micelles are the dominant species [19], or about 20 insoluble terminal blocks [26, 36]. When simulations are performed with an absolute constraint which requires that AB diblock copolymers and ABA triblock copolymers must form micelles with exactly 20 chains, each of the same mass and composition, differences in the structures of the micelles at AB and ABA are readily apparent [37]. Several of these differences are detected by calculation of mean square radii of gyration, expressed in Table 1 in units of r_{hs}^2, defined as

$$r_{hs}^2 = [60(2N_A + N_B)/4\pi]^{2/3} \qquad (5)$$

for the ABA triblock copolymers, where $20(2N_A + N_B)$ is the volume occupied by all of the beads in the block copolymers in the micelle. Thus, r_{hs}^2 is the minimum squared radius of gyration for the micelle. A similar definition is employed with the AB diblock copolymers. The

Table 1 Mean square radii of gyration for micelles (in units of r_{hs}^2) of AB diblock copolymers and ABA triblock copolymers with the same number of chains and overall composition[a]

Portion of the micelle	AB diblock	ABA triblock
All insoluble (A) blocks	0.68	0.89
Entire micelle	1.43	1.21
All soluble (B) blocks	2.18	1.53

[a] Data for the AB diblocks and ABA triblocks are from refs. [4, 35], respectively. Each micelle contains 20 chains of $A_{10}B_{10}$ or 20 chains of $A_5B_{10}A_5$.

micelle composed of the ABA triblock copolymers has a mean square radius of gyration that is about 15% smaller than the value obtained for the micelle composed of the AB diblock copolymers. This difference arises because the soluble corona is composed of the dangling ends of the AB diblock copolymers, but it is primarily made up of loops from the ABA triblock copolymers. For this same reason, the radii of gyration for the soluble blocks are larger in AB diblock copolymers than in the ABA triblock copolymers. The reverse situation is seen with the radii of gyration for the insoluble blocks. When the micelle is composed of ABA triblock copolymers, the core is less compact, and contains more solvent, than does the micelle composed of AB diblock copolymers [37].

The characteristic ratio, C, is much smaller (1.08) for the ABA chains in the large micelles than in the small aggregates formed at weak segregation [34], since most of the chains form loops. Loops appear to be dominant in the micelle formed in N,N-dimethyl acetamide by a large triblock copolymer with polystyrene as the soluble internal block and poly(tert-butylstyrene) as the insoluble terminal blocks [26].

Dangling ends vs. loops

Dangling ends are the dominant species in the small aggregates that are formed at weak segregation, as shown in Fig. 1C. As the value of εN_A increases and the micelles become larger, the free terminal blocks of the dangling ends in the simulation are incorporated into the micelle, and the dangling ends are thereby converted into loops [35], as shown in Fig. 1D. This effect has been detected in copolymers in which poly(ethylene oxide) is the soluble middle block, and poly(butylene oxide) is the hydrophobic terminal block, with the size of the hydrophobic block being varied while the size of the middle soluble block is held nearly constant [23]. An increase in N_B tends to convert the loops back into dangling ends in the simulations [35], as expected from the dependence of the loop entropy on the size of the soluble middle block. The value of the characteristic ratio, C, is sensitive to the transition between dangling ends and loops [34], being larger for the former conformation.

Bridges

When two micelles are close together, a single ABA triblock copolymer can place one of its A blocks in the core of one of the micelles, and the other A block in the core of the other micelle, as sketched in Fig. 1E. This copolymer forms a bridge between the two micelles. We shall use the

142

M. Nguyen-Misra et al.
Simulation of the self-assembly in solution by triblock copolymers

word "cluster" to denote two or more micelles that are connected by bridges. In the simulations, bridges are readily seen between the small aggregates that form when εN_A is not too large [33–35]. When the aggregates are isolated from one another, these εN_A are not strong enough to force both of the insoluble blocks of each chain into the core of the small aggregate. Instead there is a strong population of dangling ends. With a sufficient increase in concentration, leading to an increase in the number density of the small aggregates and a corresponding decrease in their separation, these dangling ends produce bridges between the small aggregates. In the larger micelles that can be formed at stronger segregation, loops may be favored over dangling end, but there are still sufficient dangling ends so that bridging can occur [35, 37].

Bridges are obviously favored by an increase in concentration, as expected from the law of mass action, both in simulation [35] and experiment [19, 20]. They are also favored by an increase in the size of the soluble middle block, both in the simulation [35] and in experiments [24].

The bridges produce an apparent attraction between the micelles, which may manifest itself experimentally as a smaller excluded volume than might be expected from the size of the individual micelles [23].

Gelation

As the number of bridges between micelles increases, the sizes of the clusters also increase. One eventually attains the formation of an enormous cluster, which is a very large physical network that can extend throughout the entire sample, both in simulation [35] and in experiment [22, 27]. The gel point for this physical network can be obtained from the simulations. At first sight, this achievement seems unlikely, because usually one only evaluates from a simulation those phenomena that take place on a length scale that is smaller (typical no larger than $\frac{1}{2}$) than the length of the smallest side of the periodic box. However, if one measures the apparent gel point for a series of simulations of the same system, performed in a series of cubic periodic boxes of increasing volume, L^3, one can demonstrate that this apparent gel point converges to a limiting value as L increases [35]. This limiting value can be taken as the gel point for the macroscopic system.

The apparent gel point in any particular periodic box can be determined from a series of simulations at varying ϕ, using well-established methods. In one method, the weight-average number of insoluble blocks in the clusters in the simulation, p_w, is monitored as a function of ϕ. For an infinitely large system, this number will diverge at the concentration that corresponds to the gel point [38]. In our simulations of a system capable of gelation, this number increases gradually with ϕ up to a particular concentration, $\phi_{gel, L}$, and then increases at a much larger rate at higher ϕ. When simulations are performed in boxes of increasing size, the slope $\partial p_w/\partial \phi$ below $\phi_{gel, L}$ is nearly unchanged, but the slope above $\phi_{gel, L}$ increases dramatically as L increases, and approaches its limit of ∞. The value of $\phi_{gel, L}$, where the change in slope occurs, exhibits a weak dependence on L for the largest boxes used. Therefore, we can identify the limiting value of $\phi_{gel, L}$ with the critical gel point for a macroscopic system [35]. We denote this gel point by $\phi_{gel, MW}$. The behavior is sketched diagrammatically in Fig. 2. Actual data is presented in the original reference [35].

The gel point can also be defined using an analogy with the polymerization of multifunctional monomers [39]. For this purpose we need the number of micelles (n_m) and the number of bridges, or elastically effective chains (n_e), in the largest cluster in the system. If this cluster is extremely large, its functionality, Φ, is defined as

$$\Phi = 2n_e/n_m. \tag{7}$$

For an infinitely larger system, gelation occurs at the concentration that produces $\Phi = 2$. In our simulations, n_m can be no larger than $N_C/2$. Since our clusters are of finite size, we prefer the alternative definition,

$$\Phi = [2(n_e - 1) + 6]/n_e \tag{8}$$

which approaches the definition in Eq. (7) as the number of micelles increases. For our systems, gelation occurs when Φ from Eq. (8) is about 2.1. Gel points evaluated by this method, which are denoted by $\phi_{gel, \Phi}$, are always slightly larger than $\phi_{gel, MW}$ in the simulations [35].

There is a third method for detecting gelation that is unique to the simulation. Here we take advantage of the fact that we can inspect in detail the arrangement of every chain at each step in the simulation, and therefore

Fig. 2 Schematic behavior of the weight average number of insoluble blocks in the clusters as a function of concentration, for simulation boxes of increasing size, L

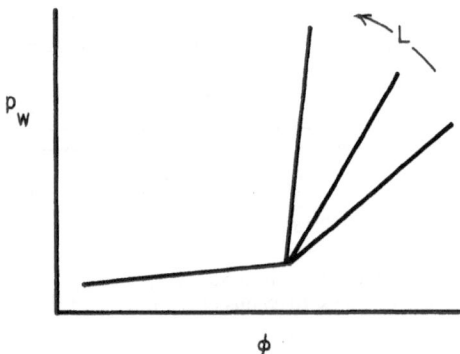

Progr Colloid Polym Sci (1997) 103:138–145
© Steinkopff Verlag 1997

establish the connectivity of any and all clusters that might be present. This direct visualization of the system can unambiguously place every replica in one of two categories. The first category contains a cluster that extends all of the way across the periodic simulation cell in every dimension, as sketched in Fig. 3A. The second category includes all other situations, including replicas in which one cluster spans the box in one dimension, and other independent clusters span the box in the other dimensions (Fig. 3b). Note that it is not necessary that all of the chains must be members of the large cluster in the first category. Free chains, independent micelles, and small clusters might also be present, as suggested in Fig. 3A. This visual analysis of the connectivity of the clusters in the simulation box identifies the gel point with the concentration at which the system switches from the second category to the first [35]. This concentration is denoted by $\phi_{gel,CON}$.

In the simulation, the three gel points are close to one another, and usually fall in the sequence $\phi_{gel,MW} < \phi_{gel,CON} < \phi_{gel,\phi}$ [35]. The functionality is usually slightly less than 2 at $\phi_{gel,MW}$, and therefore $\phi_{gel,MW}$ slightly underestimates the true gel point in the simulation.

The gel point is insensitive to changes in the size of the middle soluble block, but it depends on the size of the insoluble blocks and the strength of their repulsive interaction with the environment [35]. This dependence can be summarized as

$$\phi_{gel} \sim (\varepsilon N_A)^{-a}, \tag{9}$$

where the best values of a are 0.17 for $\phi_{gel,\phi}$ and 0.14 for $\phi_{gel,MW}$. We conclude that the exponent a is in the range 0.1–0.2 in the simulations [35].

The form of Eq. (9) can be compared with the result obtained recently for the gelation of a homopolymer in a solvent [28]. For a particular choice of polymer and solvent at isothermal conditions (which corresponds to constant ε in our simulations) Tanaka finds [28]

$$\phi_{gel} \sim M^{-a}, \tag{10}$$

where M is the molecular weight of the homopolymer. This expression is of the same forms as Eq. (9) is we assume N_A for the triblock copolymer is proportional to M for the homopolymer. The theory extracts the multiplicity of the junctions, s, in the thermoreversible physical network from the value of the exponent [28].

$$a = -(s-1)^{-1}. \tag{11}$$

Our estimate from simulation of a in the range 0.1–0.2, when interpreted by this equation, yields 6–11 for the multiplicity of the junctions. This number is comparable with the number of insoluble blocks in the micelles that provide the junction points in the network that is formed at ϕ_{gel} in the simulations. Of course, it is this number of insoluble blocks that determines the stability of the micelle, which is the junction in our gel.

Dynamics of the network

The physical network formed by the end-associated triblock copolymers contains bridges that are attached to the junction sites (micelles) by non covalent interactions. When one of the terminal blocks of a bridge extracts itself from the core of a micelle, this bridge is converted to a dangling end. Coalescence or fusion of the two micelles will also destroy the bridge, by converting it to a loop. Both of these processes are depicted schematically in Fig. 4. A third process, in which both terminal blocks of a bridge are simultaneously expelled from the cores of the micelles, would produce a free chain. This process is so rare in the simulation that it can be ignored in comparison with the other two processes [40]. The two more common processes, which we will call breaking and fusion, cause a bridge to have a finite lifetime. The distribution of these lifetimes will affect the stress relaxation by the network.

Breaking and fusion are both observed in the simulations, and either one can be dominant under appropriate conditions [40]. At constant ϕ, N_A, and N_B, an increase in the ε causes a continuous decrease in the overall transition rate, as expected, and the transition rate for breaking shows similar behavior. Fusion, however, shows a qualitatively different behavior. The fusion transition rate initially increases as ε increases, passes through a maximum, and then decreases, but this decrease is slower than the decrease in the transition rate for breaking. For this reason, the transition rates for breaking and fusion cross one another as ε increases, with breaking being the dominant process at small ε, and fusion becoming dominant at larger ε. The qualitative behavior of the transition rate for breaking is easily understood, because the amount of energy required to extract a terminal block from a core increases as ε increases. The more complicated behavior of the

Fig. 3 Connectivity of the chains in a two-dimensional rendition of the three-dimensional system when **A** gelation has occurred, and **B** gelation has not occurred

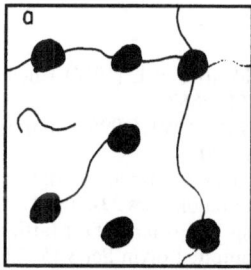
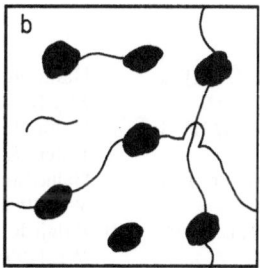

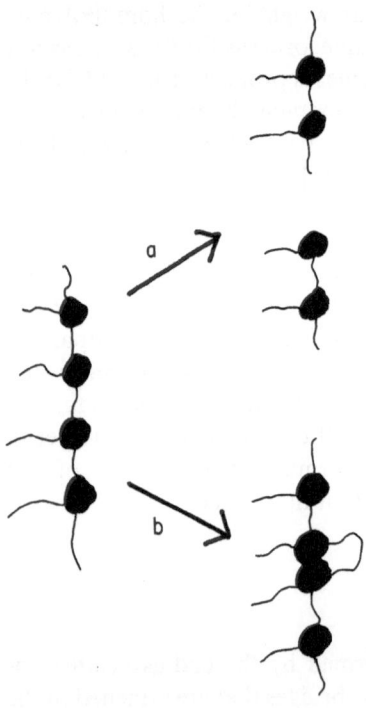

Fig. 4 Destruction of a cross-link in the physical gel by **A** breaking, which converts a bridge to a dangling end, or **B** fusion, which converts a bridge to a loop

transition rate for fusion increases because of increases in the micelle aggregation number and the radius of the core of the micelle. The transition rate for breaking increases because some of the terminal blocks do not make extensive contacts with the core, and hence they are easily expelled [40].

An increase in N_B, at constant ε, ϕ and N_A, produces opposing effects on the transition rates for breaking and fusion, so that there is a much smaller change in the total transition rate [40]. The transition rate for fusion decreases as N_B increases, due to decreases in the aggregation number and number of bridges, as well as an increase in the intermicellar distance. This change is partially compensated by an increase in the transition rate for breaking, apparently because the loop entropy for the loops in the micelle becomes more severe, and therefore the cores are somewhat looser. The looseness of the core facilitates the extraction of one of the terminal blocks of the bridge.

Analysis of the distribution of lifetimes for the bridges can be used to deduce their affect on the shear stress relaxation after a unit shear strain [40]. A similar approach has been used to study the dynamic response of triblock copolymers, adsorbed via their terminal blocks between two parallel plates, when they are subjected to step and sinusoidal shear [41].

transition rate for fusion has a more obscure origin. An increase in ε (in the region of low ε) produces an increase in the aggregation number for the micelles and in the average number of bridges, and for these reasons the transition rate for fusion also increases. At larger ε, the average number of bridges remains nearly constant, and the aggregation number of the micelles decreases somewhat, causing the transition rate for fusion to slowly decrease as ε increases.

An increase in concentration at constant ε, N_A and N_B has little affect on the transition rate for breaking, but produces an increase in the transition rate for fusion, as expected, and hence leads to an increase in the overall transition rate [40].

An increase in N_A, at constant ε, ϕ and N_B, causes an increase in the transition rates for both processes. The

Current work

Current work on triblock copolymers with Prof. Haliloglu examines the structure and dynamics of the layers formed by their adsorption at a surface. With Ms. Xing, we are studying the solubilization of slightly soluble small molecules by triblock copolymers, in order to understand the structure of the complexes and the variables that most strongly affect the strength of the adsorption and the solubilization capacity.

Acknowledgments Our investigation of the self-assembly of triblock copolymers has been supported by grants from the National Science Foundation and the Australian Department of Industry, Technology, and Commerce.

References

1. Binder K (1995) In: Binder K (ed) Monte Carlo and Molecular Dynamics Simulations in Polymer Science. Oxford, New York, pp 356–432
2. Haliloglu T, Mattice WL (1996) In: Webber SW (ed) Solvents and Self-Organization of Polymers. Kluwer, Dordrecht, pp 167–196
3. Rodrigues K, Mattice WL (1991) J Chem Phys 94:761–766
4. Rodrigues K, Mattice WL (1991) J Chem Phys 95:5341–5347
5. Wang Y, Mattice WL, Napper DH (1993) Langmuir 9:66–70
6. Adriani P, Wang Y, Mattice WL (1994) J Chem Phys 100:7718–7721
7. Haliloglu T, Mattice WL (1994) Chem Eng Sci 49:2851–2857
8. Haliloglu T, Mattice WL (1995) Comput Polym Sci 5:65–70
9. Haliloglu T, Bahar I, Erman B, Mattice WL (1996) Macromolecules 29:4764–4771
10. Balaji R, Wang Y, Foster MD, Mattice WL (1993) Comput Polym Sci 3:15–22

Progr Colloid Polym Sci (1997) 103:138–145
© Steinkopff Verlag 1997

11. Haliloglu T, Balaji R, Mattice WL (1994) Macromolecules 27:1473–1476
12. Ko MB, Mattice WL (1995) Macromolecules 28:6871–6877
13. Wang Y, Mattice WL (1993) J Chem Phys 98:9881–9887
14. Wang Y, Li Y, Mattice WL (1993) J Chem Phys 99:4068–4075
15. Zhan Y, Mattice WL, Napper DH (1993) J Chem Phys 98:7502–7507
16. Zhan Y, Mattice WL, Napper DH (1993) J Chem Phys 98:7508–7514
17. Zhan Y, Mattice WL (1994) Macromolecules 27:677–682
18. Zhan Y, Mattice WL (1994) Macromolecules 27:683–688
19. Alami E, Almgren M, Brown W, François J (1996) Macromolecules 29:2229–2243
20. Yekta A, Xu B, Duhamel J, Adiwidjaja H, Winnik MA (1995) Macromolecules 28:956–966
21. Zhou Z, Chu B (1994) Macromolecules 27:2025–2033
22. Mortensen K, Brown W, Jorgensen E (1994) Macromolecules 27:5654–5666
23. Yang YW, Yang Z, Zhou ZK, Attwood D, Booth C (1996) Macromolecules 29:670–680
24. Zhou Z, Chu B, Nace VM, Yang YW, Booth C (1996) Macromolecules 29:3663–3664
25. Raspaud E, Lairez D, Adam M, Carton JP (1994) Macromolecules 27:2956–2964
26. Zhou A, Chu B, Peiffer DG (1995) Langmuir 11:1956–1965
27. Raspaud E, Lairez D, Adam M, Carton JP (1996) Macromolecules 29:1269–1277
28. Tanaka F, Nishinari K (1996) Macromolecules 29:3625–3628
29. Tanaka G, Mattice WL (1996) Macromol Theory Simul 5:499–520
30. Jacobson H, Stockmayer WH (1950) J Chem Phys 18:1600–1606
31. ten Brinke G, Hadziioannou G (1987) Macromolecules 20:486–489
32. Balsara NP, Tirrell M, Lodge TP (1991) Macromolecules 24:1975–1986
33. Rodrigues K, Mattice WL (1991) Polym Bull 25:239–243
34. Wang Y, Mattice WL, Napper DH (1992) Macromolecules 25:4073–4077
35. Nguyen-Misra M, Mattice WL (1995) Macromolecules 28:1444–1457
36. Yekta A, Duhamel J, Adiwidjaja H, Brochard P, Winnik MA (1993) Langmuir 9:881–883
37. Rodrigues K, Mattice WL (1992) Langmuir 8:456–459
38. Stauffer D, Coniglio A, Adam M (1982) Adv Polym Sci 44:103–158
39. Flory PJ (1946) Chem Rev 39:137–197
40. Nguyen-Misra M, Mattice WL (1995) Macromolecules 28:6976–6985
41. Nguyen-Misra M, Misra S, Mattice WL (1996) Macromolecules 29:1407–1415

Progr Colloid Polym Sci (1997) 103:146–154
© Steinkopff Verlag 1997

J.C. Shelley
M. Sprik
M.L. Klein

Structure and electrostatics of the surfactant–water interface

Received: 9 December 1996
Accepted: 12 December 1996

J.C. Shelley[1]
Department of Chemistry
University of British Columbia
Vancouver, British Columbia
Canada V6T 1Z1

[1] Current address:
Dr. J.C. Shelley (✉)
Corporate Research Division
The Procter & Gamble Company
Miami Valley Laboratories
P.O. Box 538 707
Cincinnati, Ohio 45253-8707, USA

M. Sprik
Zürich Research Laboratory
IBM Research Division
8803 Rüschlikon, Switzerland

M.L. Klein
Center for Molecular Modeling
and Department of Chemistry
University of Pennsylvania
Philadelphia, Pennsylvania 19104-6323
USA

Abstract Molecular dynamics simulations using nonpolarizable and polarizable models have been conducted on columnar micelles of sodium octanoate in the liquid crystalline (E) mesophase. In the simulation using a nonpolarizable model, most of the Na^+ ions are in direct contact with the carboxylate headgroups while for the simulations using our polarizable models the Na^+ ions are mostly solvent separated, in better agreement with experiment. For all models the hydrocarbon/water interface is only about 4 Å wide and the electric double layer is clearly distinguishable. In a detailed demonstration of the collective nature of dielectric screening, significant cancellation occurs between the contributions to the electrostatic potential from the different components in the system resulting in a relatively small net electrostatic potential.

Key words Surfactant – interface – electrostatic – potential drop – molecular dynamics – polarizability

Introduction

Interfaces play a central role in electrochemistry, biophysics, drug delivery, and surfactant science. The study of interfacial phenomena at the molecular or atomic level is an active field with significant contributions from both experiment [1, 2] and theory [3]. In this article, we will present some initial results concerning the structure and electrostatics of the surfactant–water interface. In particular, we will focus on the sodium octanoate–water system in the lyotropic liquid crystalline mesophase (E). In this phase the surfactant forms very long micelles which pack in a hexagonal pattern. This is a good system to study since the columnar structures have geometrically simple interfaces and a number of simulations have already been done on aqueous sodium octanoate systems [4–10].

Most classical simulations of molecular systems such as aqueous solutions employ pair-wise additive nonbonded potentials, which are often a combination of Lennard–Jones and coulombic interactions. When molecules possess internal degrees of freedom, additional intramolecular potentials such as bond, bond angle and dihedral angle potentials are typically employed. In aqueous ionic solutions, like sodium octanoate in water, there are strong many-body effects arising from the electrostatic polarization of the components in the system. A common

Progr Colloid Polym Sci (1997) 103:146–154
© Steinkopff Verlag 1997

approach is to employ effective potentials which are parametrized to include at least some of these contributions in the parameters used in their two-body potentials. However, it is not clear that many-body effects can be adequately modeled in this manner. One of the properties that one might expect to be the most sensitive to such contributions is the electrostatic potential drop through an interface. In this paper, we will compare the results for a columnar micelle obtained using a model based on effective two-body potentials, a model that incorporates polarizability in the surfactant, and a fully polarizable model. More detailed results for the current system along with those from simulations of a globular micelle from the isotropic L_1 phase will be published elsewhere [11].

Techniques

These simulations were conducted in much the same manner as has been previously employed [5–9]. In the current paper, in addition to a nonpolarizable (NP) model [5–8] and a similar model incorporating polarizability in just the surfactant ions (PS) [9], we employ a fully polarizable (FP) model for all components in the system (with the exception of the sodium ions which are not very polarizable). We start by describing these models and then outline how the simulations were performed.

Simulation models

The nonbonded interactions in the NP model are pairwise additive Lennard–Jones and Coulombic interactions between some of the atomic sites within the molecules in the system. The model employed for water in these studies was the SPC/E effective pair potential [12] which has an enhanced dipole moment (2.35 debye) relative to the gasphase value (1.85 debye) to imitate the polarization of water molecules in bulk water. Na^+ was represented using a potential taken from the literature [13] in which a full charge of 1 e is used. Each CH_2 and CH_3 group in the octanoate ion were represented by one force and mass center, a pseudoatom [14]. The carbon atom and both oxygen atoms in the carboxylate headgroup were explicitly modeled and the headgroup carried a net charge of -1 e. For clarity, the headgroup carbon atom and pseudoatoms were numbered starting from the headgroup.

The C–C and O–C bond lengths in the octanoate ion were held fixed at 1.53 and 1.257 Å, respectively, using the SHAKE algorithm [15]. This method was also used to fix the geometry of the headgroup end of the octanoate ion (the $O–C_1–C_2$ and $O–C_1–O$ angles were 118.9° and 122.2°, respectively). Intramolecular bond bending [16],

torsional [16], and nonbonded potentials (between sites further apart than three bonds) were used. No torsional potential was employed for the rotation of the headgroup about the $C_1–C_2$ bond. The nonbonded potential parameters and charges for this model are given in Table 1. Lorentz–Berthelot combining rules [17] were used to define the Lennard–Jones potential parameters between force centers of different type.

The extended Lagrangian scheme was employed to model polarization in the PS and FP models [18–21]. This scheme involves introducing additional degrees of freedom into the simulation. Within our approach polarization was modeled by introducing additional charge sites in the models for the molecules. The charges that these sites carry were treated as additional classical degrees of freedom which evolve in response to the time-dependent local electric field.

The PS model employed the same models for the water and sodium ions as the NP model. In the PS model for octanoate each site in the NP model has two additional sites located equal distances (0.15 Å) from and on opposite sides of the original site. These sites carried charges of opposite sign and equal magnitude. The orientation of this triplet of sites and the magnitude of charges associated with the additional sites respond to the local electric field in a dynamic manner. Since the size of this triplet of atoms is small (0.3 Å) compared to the overall size of the groups (at least 3.0 Å) this arrangement mimics the polarization of the site. This model is depicted in Fig. 1A. Interactions between permanent charges and polarization centers on adjacent atoms were excluded since the polarizations along bonds in the model were represented by the permanent point charges and including such interactions would introduce additional polarizations along such bonds. The parameters for the PS model were the same as for the NP model except the parameters for the Lennard–Jones potentials between the headgroup and the sodium ions (see Table 2). These parameters were adjusted to maintain the

Table 1 Potential parameters for the non-polarizable and polarizable surfactant models

Site	σ[Å]	ε[K]	q[e]	α^a[Å^3]
$O_{SPC/E}$	3.166	78.2	-0.8476	
$H_{SPC/E}$			0.4238	
Na^+	1.897	808.73	1	
O	3.083	87.94	-0.7	1.5
C_1	4.437	13.2	0.4	1.16
CH_2	3.983	57.48	0.0	1.87
CH_3	3.861	91.15	0.0	2.24

[a] α is used in the PS model only.
In the PS model, ζ for the permanent charge and polarization sites was 0.0 and 0.7 Å, respectively.

148
J.C. Shelley et al.
Structure and electrostatics of surfactant–water interface

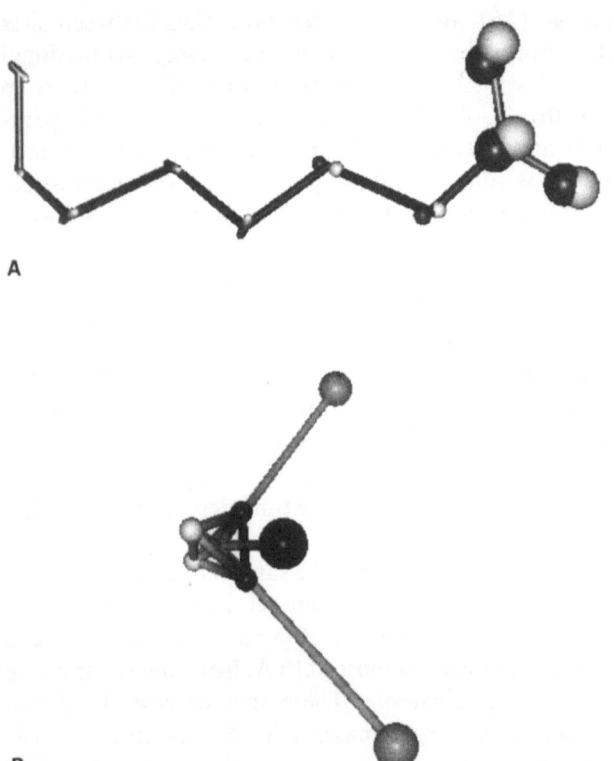

Fig. 1 A depiction of the polarizable models used for the octanoate ion **A** and the PW water molecule **B**. Light and darks spheres represent negatively and positively charged polarization sites, respectively. The size of the spheres represents the magnitude of the charge associated with them. For reference, the dipole moment on the oxygen atoms in the octanoate ion (for an individual monomer in solution) is roughly 1.5 Debye

Lennard–Jones interaction center. Polarizability is represented in this model by a rigid tetrahedron of variable charges centered on the oxygen atom. The orientation of the tetrahedron is fixed in the molecular frame with two of the vertices of the tetrahedron lying in the molecular plane close to the O–H bonds. The geometries of the SPC/E and PW water models were held rigid using the method of constraints [22]. The potential parameters for the FP model are given in Tables 3 and 4. As for the PS model, Lorentz–Berthelot combining rules were used for all cross interactions except the Na^+-octanoate interactions listed in Table 4.

The additional polarization sites within the octanoate anions, i.e., the two sites at the ends of the triplets pivoting about the middle site, have a real physical mass. In both the PS and FP models, this mass was assigned the value 1 AMU and the mass for the original (central) site in each triplet was correspondingly reduced by 2 AMU. The (fictitious) conjugate masses for the magnitudes of the charges on the polarization sites in the octanoate ion in both polarizable models have a value of 0.5μ a.u. where μ is the mass ratio of the proton and the electron ($\mu = 1822$). The fictitious mass for the polarization charges in the PW model is 0.25μ a.u.

Experience has convinced us that some form of control over the induction, particularly in aqueous media is desirable. In our approach this was accomplished by replacing the point charges by Gaussian charge distributions for the polarization sites and for selected permanent charge sites [9, 18–20]. The width of these Gaussian charge distributions, ξ, was one of the parameters adjusted to obtain

Table 2 Na^+-surfactant Lennard–Jones parameters for the polarizable surfactant model

Site	$\sigma[\text{Å}]$	$\varepsilon[\text{K}]$
Na^+–O	3.107	27.78
Na^+–C_1	3.784	10.762
Na^+–CH_2	3.557	22.458
Na^+–CH_3	3.496	28.28

same geometry and overall strength of interaction for a Na^+ interacting with an octanoate ion in the gas phase as was found for the NP model.

In the FP model, the overall framework for modeling the octanoate ion was the same as in the PS model although the potential parameters differed somewhat. An existing polarizable water (PW) model [20] (see Fig. 1B) was used. This model employs permanent fractional positive charges on each of the H atoms and a balancing negative charge located at a site (M) 0.26 Å away from the oxygen atom along the bisector of the H–O–H bond angle. As in the SPC/E model the oxygen atom is a

Table 3 Potential parameters for the fully polarizable surfactant model

Site	$\sigma[\text{Å}]$	$\varepsilon[\text{K}]$	$q[e]$	$\alpha[\text{Å}^3]$	$\xi[\text{Å}]$
O_{PW}	3.1925	101.0	0.0	1.444	0.775
M_{PW}			−1.2		0.0
H_{PW}			0.6	0.0	0.0
Na^+	1.900	2.0	1	0.0	0.7
O	3.085	71.56	−0.6	1.36	0.65
C_1	3.75	24.76	0.2	1.16	0.65
CH_2	3.983	57.48	0.0	1.87	0.65
CH_3	3.861	91.15	0.0	2.24	0.65

Table 4 Na^+ Lennard–Jones parameters for the fully polarizable model

Site	$\sigma[\text{Å}]$	$\varepsilon[\text{K}]$
Na^+–O(PW)	3.488	3.5
Na^+–O	3.107	4.03
Na^+–C_1	3.849	1.55
Na^+–CH_2	3.971	2.00
Na^+–CH_3	3.915	2.62

matches for properties such as the gas-phase dipole moments of related molecules and gas-phase association energetics for the ion pairs. Such Gaussian distributions, with $\xi = 0.7$ Å, were employed for only the polarization charges in the PS model. In the FP model the situation was somewhat more complex with ξ being assigned values of 0.775 Å for the polarization sites in the PW water model, 0.7 Å for the permanent charge on the Na$^+$, and 0.65 Å for both the permanent and polarization sites in the octanoate anion.

Simulation methods

All of these simulations consist of a single columnar micelle containing 54 octanoate ions, 54 Na$^+$ counterions, and 503 water molecules as depicted in Fig. 2. This configuration was based on that obtained from the end of a previous simulation [8] but only half of the system in the z direction (along the column) has been used in the current studies. The actual box shape used in the simulations was monoclinic. In the NP and PS simulations $a = b = 29.9$ Å, $c = 36.3$ Å, $\alpha = \beta = 90°$ and $\gamma = 60°$. In the FP simulation, the box dimensions were reduced by 2.2% to maintain a positive pressure. In order to eliminate surface effects and to appropriately take into account the long-range character of the electrostatic interactions Ewald boundary conditions were employed [23]. A cut-off distance of 12 Å was employed for both the Lennard–Jones potentials and the real space portion of the Ewald sums. The equations of motion for these systems were integrated using the Verlet

Fig. 2 The final configuration for the FP system. **A** is the entire system looking along the surfactant column. In **B** (along the column) and **C** (perpendicular to the column) the water has not been displayed to better illustrate the structure. **D** is a view looking along the column with only the water visible. Note the complete absence of water from the center of the column. Van der Waals radii are used for the atoms in these images

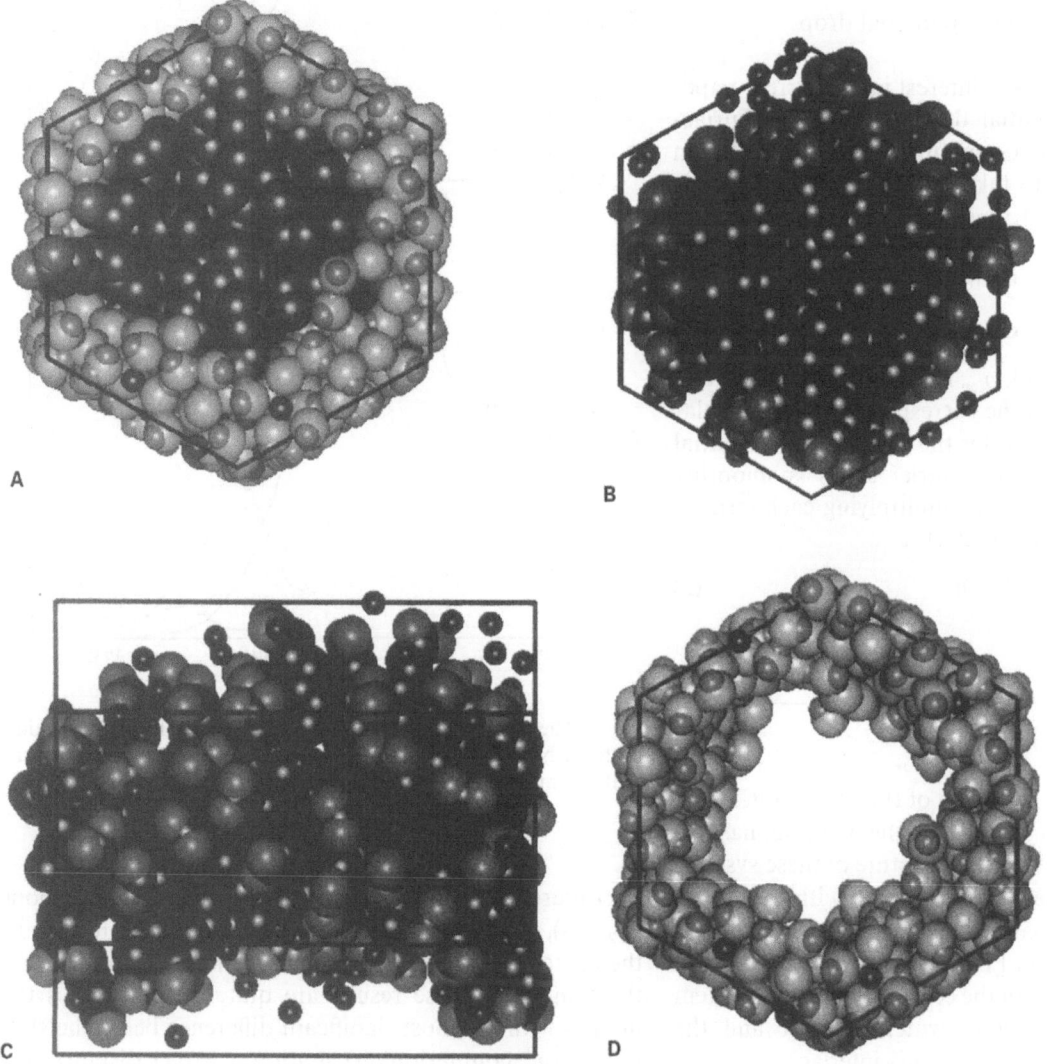

150

J.C. Shelley et al.
Structure and electrostatics of surfactant–water interface

algorithm [23] with a time step of 2.5 fs for the NP and PS models, and 1.0 fs for the FP model. These simulations were carried out in the canonical (N, V, T) ensemble at a temperature of 298.15 K using Nosé–Hoover thermostats [24]. We employ a nonequilibrium adiabatic scheme for both the size and orientation of the induced dipoles in which the temperature of these degrees of freedom was maintained at 5 K using Nosé–Hoover thermostats [9].

Prior to the production phase of the simulations these systems were simulated until their properties remained unchanged. This amounted to 75, 120, and 37 ps for the NP, PS and FP systems, respectively. The production phase of these simulations lasted 121, 80 and 66.5 ps for the NP, PS and FP systems, respectively. Since these simulations were somewhat cpu intensive, they were carried out using a prallelized code on 8 processors of an IBM scalable parallel computer at the Cornell Supercomputing center.

Calculation of the electrostatic potential drop

One of the main properties of interest in the current paper is the electrostatic potential drop through the surfactant–water interface. In estimating this property from a simulation one needs to take into account the periodic boundary conditions and typically employ some sort of filtering process to reduce the noise in this property. Previously, we have presented a method [25] for doing so:

$$\text{div}\,\mathbf{E} = 4\pi(\rho_{\text{ion}} - \text{div}\,\mathbf{P} + \text{div}(\text{div}\,\mathbf{Q})) \qquad (1)$$

where E is the electric field, ρ_{ion} is the charge density, $\mathbf{P}$ is the polarization and $\mathbf{Q}$ is the corresponding quadrupolar tensor. Eq. (1) can be solved for the electrostatic potential using Fourier methods. The Fourier series is smoothly truncated at $k_{\text{max}} = 2.5\ \text{Å}^{-1}$, by multiplying each term in the Fourier series, k by the prefactor

$$c(k) = \exp(\tau/k_{\text{max}} + \tau/(k - k_{\text{max}})) \qquad (2)$$

with τ set to 0.5 Å^{-1}.

Results and discussion

This paper concerns the structure of the surfactant–water interface of a columnar micelle in the E mesophase of sodium octanoate in water. The structure of these systems will be examined in detail and compared with results for globular sodium octanoate micelles in the L$_1$ phase in a forthcoming publication [11]. Our discussion here will focus on the arrangement of the components of the system with respect to the surfactant–water interface and the nature of the electrostatics across this interface.

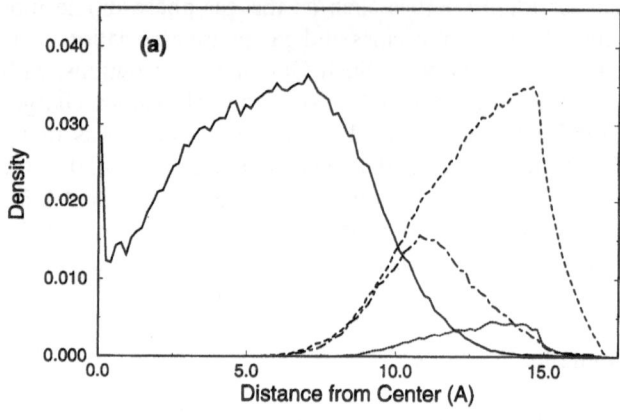

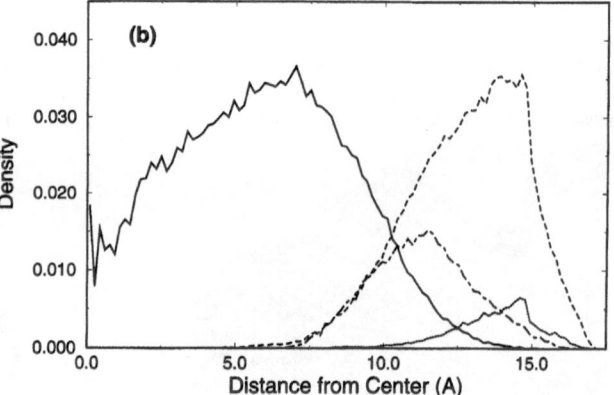

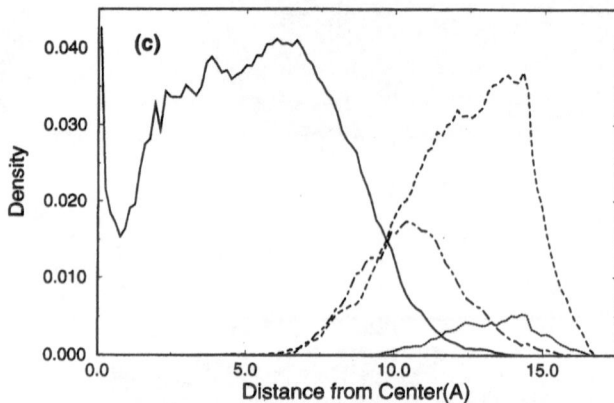

Fig. 3 The densities for the hydrocarbon (solid), head-group (dot-dashed), Na$^+$ ion (dotted) and water (dashed) with respect to the center of mass of the column for the NP (a), PS (b), and FP (c) models

Figure 3 gives the densities of the hydrocarbons (C$_2$–C$_8$), headgroups, Na$^+$ ions and water with respect to the center of mass of the column of the surfactant for all three models. These results are quite similar across the models with the most significant difference being the shift in the Na$^+$ counterion distribution to larger distances for

Progr Colloid Polym Sci (1997) 103:146–154
© Steinkopff Verlag 1997

Table 5 Hydration and polarization of the octanoate ion

Site	Hydration numbers				Polarization			
	Monomer PS	Column		FP	Monomer		Column PS	FP
		NP	PS		Gas PS	Solution PS		
O					0.88	1.47	1.46	1.15
C_1					0.08	0.52	0.55	0.69
C_2	3.6	1.7	1.5	1.4	0.42	0.51	0.48	0.14
C_3	2.9	1.5	1.5	1.6	0.80	0.27	0.28	0.17
C_4	2.7	0.4	0.4	0.4	0.45	0.14	0.09	0.07
C_5	2.5	0.5	0.5	0.6	0.33	0.11	0.07	0.06
C_6	2.7	0.3	0.3	0.3	0.29	0.10	0.05	0.03
C_7	3.0	0.3	0.4	0.4	0.21	0.10	0.04	0.03
C_8	4.8	0.4	0.4	0.4	0.22	0.15	0.05	0.04

The monomer results are taken from ref. [9].

the polarizable models. The lower hydrocarbon density towards the middle of the micelle is attributable to the prevalence of the larger CH_3 groups in this region. Consistent with previous studies [9, 26], the width of the interface between the hydrocarbon portion of the micelle and the water (defined here as the distance between the points where the water and the hydrocarbon densities cross through 10% of their bulk values) is narrow at 4 Å. Water does not penetrate much further than the headgroups and the sodium ions typically lie outside of the inner half of the hydrocarbon–water interface. Figure 2d, in which only the water molecules are displayed, provides a graphic illustration of the limited water penetration into the micelle.

Hydration of the pseudoatoms in the hydrocarbon tail (see Table 5) is consistent across the models and much less than for a solvated monomer. It is also significantly smaller than that found for globular micelles [9]. The level of hydration is slightly higher near the CH_3 end of the surfactant ion as has been found by previous simulations [5–9, 26, 27] of micellar systems. The hydration of the pseudoatoms in the hydrocarbon tail away from the headgroup is not due to the deep penetration of water but rather due to the fact that some of these pseudoatoms spend time near the surface of the surfactant column. Water molecules are considered to be in contact with the headgroups if they lie within 4.25 Å of the C_1 atom or 3.45 Å of either O atom. The number of water molecules near the headgroups in the NP, PS and FP models were 7.2, 7.6, and 7.1, respectively, which is somewhat lower than that for the PS monomer in solution (7.9) [9].

The induced dipole moments of sites within the surfactant ion for a gas phase and solvated monomer for the PS model from a previous study [9] and for the current study are given in Table 5. Clearly the induced moments on the headgroup atoms are enhanced by solvation. The enhancement does not seem to matter if the monomer is part of a micelle or an individual monomer in solution. The difference between the PS and FP results is due to the reparametrization of the model for the octanoate ion. However, the polarization of the carbon pseudoatoms away from the headgroup is reduced by up to a factor of three upon solvation. For instance, for the C_4 pseudoatom the dipole moment was 0.45 debye in the gas phase but dropped to 0.14 debye for the solvated monomer. This reduction happens despite the fact that no water molecules typically lie directly in between the headgroup and this site. The reduction occurs because of the collective nature of dielectric screening in which many molecules, not just those that lie directly between given sites, participate. The polarization of the hydrocarbon pseudoatoms is further reduced for octanoate ions in the columnar micelles. Since the headgroup polarization is unaffected by micellization and also the corresponding addition of Na^+ ions, this reduction is most likely due to the lower level of contact between the hydrocarbon pseudoatoms and water molecules within the columnar micelle.

The NP model gives a significant number (75%) of contact ion pairs (Na^+ ions closer than 4.0 Å to the headgroup C_1 atoms) while the two polarizable models give far fewer (~13%). Experiment indicates that these ions remain solvent-separated until there is too little water to solvate them [28, 29]. Our results should not be taken to indicate that polarizable models are needed to get solvent-separated ions. As has been demonstrated for SDS micelles it is possible to devise nonpolarizable potentials which will yield this result [27]. On the other hand, all publications on simulations of sodium octanoate micelles (with full ionic charges) using nonpolarizable models prior to our PS model gave contact ion pairs. It is instructive to note that the PS model was intended as an improved, polarizable version of the NP model and was not parametrized with the intention of achieving solvent separation of the

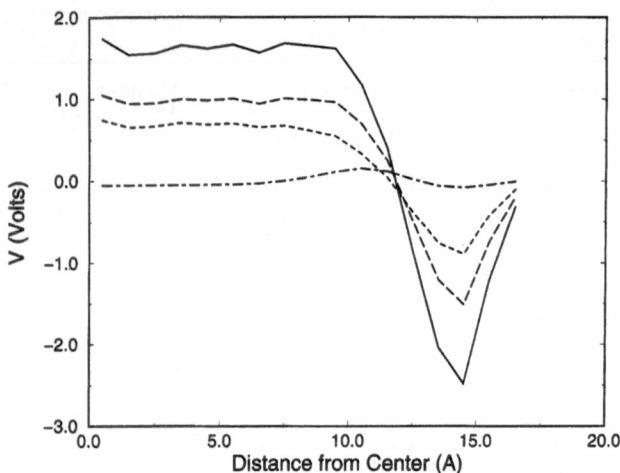

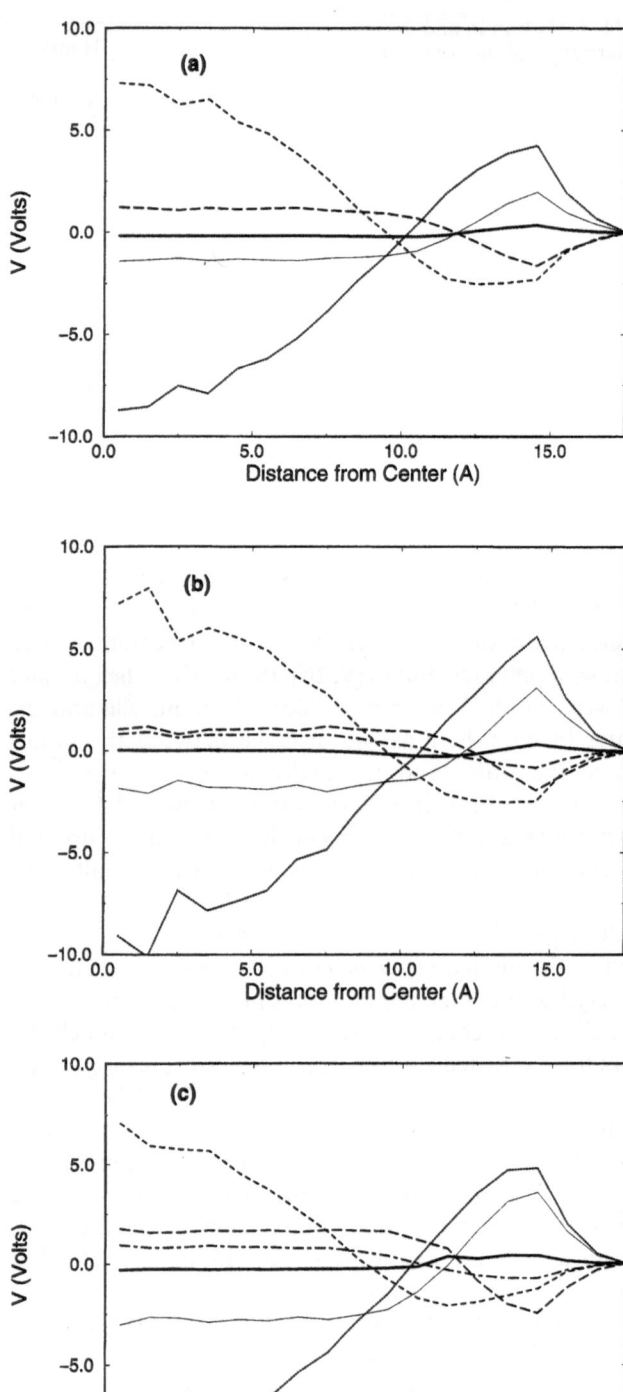

Fig. 4 The components of the electrostatic potential drop for the PW water model. The permanent, induced, quadrupolar and total contributions are depicted by the long-dashed, dashed, dot-dashed and solid lines, respectively

ions, yet it achieved solvent separation of ions. This is an illustration of one of the main reasons for developing and employing polarizable models. In principle, they should be more robust and predictive. They also provide some additional information concerning the nature of the electrostatic interactions within the system.

We have investigated the electrostatics of the micelle–water interface using Eq. (1) to calculate the contributions from the different types of electrostatic sites present. The electrostatics of the water molecules will be examined first. Figure 4 gives the contributions that the water molecules' permanent dipole, induced dipole and quadrupole moments make to the total electrostatic potential as a function of distance from the center of column of surfactant from the FP (PW water model) simulation. The induced dipole in the PW model makes a contribution that is similar to but somewhat smaller than that for the permanent dipole. This result is consistent with the observation that the induced dipoles of the water molecules tend to point along the direction of the permanent dipole in the water molecule and helps account for the similarity of the potential drops for the water in both the PW and SPC/E models (see Fig. 5). The contributions from the water molecule's quadrupole moment are small and noisy. This suggests that higher multipolar moments should not make significant contributions to the overall potential drop for this system. In the interfacial region, the potential due to the quadrupoles vaguely follows the pattern of the dipoles. Inside the micelle they give a contribution that is negative on average while that from the permanent dipoles is positive.

Figure 5 gives plots of the components making up the total electrostatic potentials. These electrostatic potentials

Fig. 5 The contributions to the electrostatic potential drops for the NP (a), PS (b), and FP (c) models. The Na$^+$ ions, carboxylate headgroups, total ionic (Na$^+$ plus carboxylate headgroups), water, and surfactant induction contributions to the electrostatic potential are plotted using dotted, dashed, narrow-solid, long-dashed and dot-dashed lines, respectively. The total potential drops are given by the thick solid lines

Progr Colloid Polym Sci (1997) 103:146–154
© Steinkopff Verlag 1997

have been measured relative to the potential in the corners of the hexagonal simulation cell. In all of these systems the Na$^+$ ions and carboxylate headgroups make consistent and by far the largest contributions (between 6 and 10 V) to the total potential drop. Since they tend to have opposite signs, the sum of these contributions is much smaller (between -1 and -3 V). Interestingly, the total electrostatic potential contribution from water is nearly identical but somewhat smaller than the sum of the electrostatic potential drops due to the permanent charges except that it has the opposite sign, as noted previously [25]. In the PS and FP models, contributions from the inductions in the surfactant ions from both the headgroups and the hydrocarbon chains follow a pattern similar to, but smaller than, that for the contribution from water. The overall tendency of the terms to cancel leads to a total electrostatic potential (see Fig. 6) that is dramatically smaller (extreme values of less than 0.5 V) than the various contributions. This is of course an illustration of dielectric screening, and our simulation clearly reveals the nature of such effects on small length scales. Our results are consistent with an effective dielectric constant between 3 and 9 for the interfacial region.

The total potential drop has the same general form for all of the models employed. Outside the micelle, in the region dominated by the Na$^+$ ions this potential is positive. As one enters the carboxylate headgroup region the potential becomes negative. After this the models differ somewhat. In the NP and FP models the potential remains constant while for the PS model potential rises back. The limiting values of the potentials within the micelles are -0.17, 0.04, and -0.29 V for the NP, PS, and FP systems, respectively. The corresponding potential drops

for globular micelles [11] for these three models are all roughly -0.2 V. The results for the PS model for the columnar micelles stand out from the other results because they are much smaller and positive. Why is the potential drop different for the simulation of the columnar micelle with the PS model? Unlike the globular micelles, the columnar micelles are very closely packed, so it is not possible to choose a test point that is *far away* from the micelle itself. As a result, the test point is in a region where the arrangement of the components of the system are likely sensitive to the nature of the model employed and thus the net potential drop may be sensitive to the details of the model. Excluding this particular result, it is suprising that the NP and FP models (and the PS model for the globular micelle) give results which are grossly similar despite the fact that the ions in the polarizable systems are further away from the headgroups implying that the electrostatic double layer will be significantly different. The answer likely rests on the idea that outside and perhaps to a limited extent inside the double layer, the net electrostatic potential is the result of an overall balance of all of the electrostatic interactions in the system and that this balance reflects some general features of the system. One side effect of this collective behavior is that it is hard to pick out particular aspects of the PS model that are giving rise to its atypical behavior in this case.

The terms $-\text{div}\,\mathbf{P}$ and $\text{div}(\text{div}\,\mathbf{Q})$ both appear in Eq. (1) in a manner equivalent to the charge densities which are easier to consider than the original vector and tensor quantities. In Fig. 7, the charge densities and these pseudo charge densities are plotted for the FP system. Note that the pseudo charge density for the water appears

Fig. 6 The total electrostatic potential drops for the NP (long-dashed), PS (short-dashed) and FP (solid) models

Fig. 7 The effective charge densities for the water (dashed), permanent charges (dotted), and surfactant induction sites (dot-dashed) for the FP model. The total effective charge density is given by the solid line

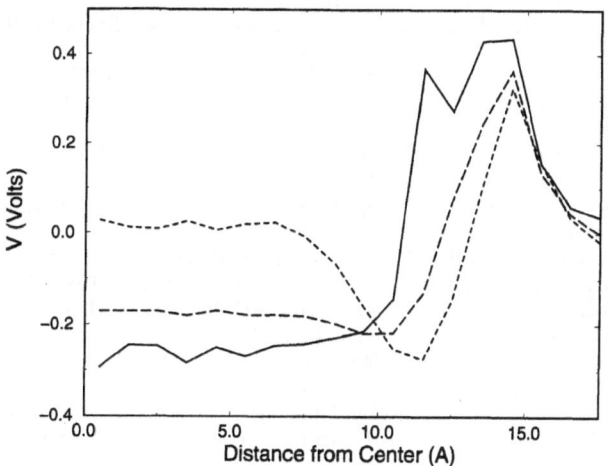

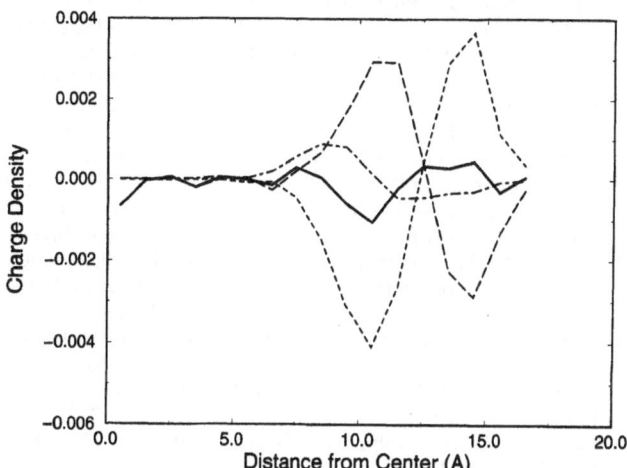

154
J.C. Shelley et al.
Structure and electrostatics of surfactant–water interface

in essentially the same location but with opposite sign as the actual charge density suggesting that the electrostatic response of the water to the presence of the ions is quite local. The contribution from the surfactant induction is similar to that from the water but is smaller and shifted into the micelle somewhat. The total effective charge density still displays the characteristic, yet much reduced, double layer pattern for these systems consisting of an outer positive and inner negative layer. As well, there is an additional, yet small, positive layer inside of the negative one due mainly to the surfactant induction. This layer appears in both of our simulations that employ polarizable models.

Conclusions

Our simulations show that the interfacial region has a narrow water–hydrocarbon interface about 4 Å thick, as has been found in previous simulation of micelles [9, 27]. The Na$^+$ ions and carboxylate head groups make up a characteristic electrostatic double layer around these micelles. In the polarizable models the Na$^+$ ions do not come in contact as much with the headgroups as in the original nonpolarizable model, in better agreement with experiment.

The analysis of the electrostatic interactions across the interface indicate that the individual contributions to the overall potential drop are large (some larger than 7 V). However these terms largely cancel to give a net potential drop across the interface with a magnitude of less than 0.3 V, a detailed illustration of dielectric screening at the molecular level. The most significant terms in this cancellation are those from the Na$^+$ ions and the carboxylate headgroups, as one might expect. This result contrasts with that for globular micelles where the most significant terms involved in the cancellation are those from the headgroups and the water [11]. Interestingly enough the overall pattern for the total electrostatic potential is the same for all of the models despite the different nature of the counter-ion association with the headgroups. The PS model does give a very small net positive potential drop through the interface. This is likely due in part to the danger involved in trying to estimate the potential drop within the double layer.

Finally, the use of the effective charge density is helpful. It clearly shows that the response of the water to the presence of the ions closely mirrors the net ionic density suggesting that the response is quite local. In addition, this treatment for the polarizable models suggests the presence of an additional, weak effective charge layer due to the inductions in the surfactant just inside of the normal double layer.

Acknowledgments We gratefully acknowledge the financial support of the National Institutes of Health, the National Science Foundation, the National Sciences and Engineering Council (Canada) and IBM. The majority of the computations were conducted using the resources of the Cornell Theory Center, which receives major funding from the National Science Foundation and IBM Corp, with additional support from the New York State Science and Technology Foundation and members of the Corporate Research Institute.

References

1. Naujok RR, Paul HJ, Corn RM (1996) J Phys Chem 100:10 497–10 507
2. Gragson DE, McCarty BM, Richmond GL (1996) J Phys Chem 100: 14 272–14 275
3. Benjamin I (1995) Acc Chem Res 28:233–239
4. Jonsson B, Bohmer MR, Teleman O (1986) J Chem Phys 85:2259–2271
5. Watanabe K, Ferrario M, Klein ML (1988) J Phys Chem 92:819–821
6. Watanabe K, Klein ML (1989) J Phys Chem 93:6897–6901
7. Shelley JC, Watanabe K, Klein ML (1991) Electrochemica Acta 36: 1729–1733
8. Watanabe K, Klein ML (1991) J Phys Chem 95:4158–4166
9. Shelley JC, Sprik M, Klein ML (1993) Langmuir 9:916–926
10. Laaksonen L, Rosenholm J (1993) Chem Phys Lett 216:429–434
11. Shelley JC, Sprik M, Klein ML (1996) in preparation
12. Berendsen HJC, Grigera JR, Straatsma TP (1987) J Phys Chem 91:6269–6271
13. Chandrasekhar J, Spellmeyer DC, Jorgensen WL (1984) J Am Chem Soc 106:903–910
14. Jorgensen WL (1981) J Am Chem Soc 103:335–340
15. Rychaert JP, Ciccotti G, Berendsen HJC (1977) J Comp Phys 23:327–341
16. van der Ploeg P, Berendsen HJC (1982) J Chem Phys 76:3271–3276
17. Hirshfelder JO, Curtis CF, Bird RB (1954) Molecular Theory of Gases and Liquids. Wiley, New York
18. Sprik M, Klein ML (1988) J Chem Phys 89:7556–7560
19. Sprik M (1991) J Phys Chem 95: 2283–2291
20. Sprik M (1991) J Chem Phys 95: 6762–6769
21. Sprik M (1991) Metals in Solution, In: Damay P, Leclercq F (eds) Journal de Physique IV, Colloque C5, Vol 1, pp 99–102
22. Ciccotti G, Ferrario M, Rychaert JP (1982) Mol Phys 47:1253–1264
23. Allen MP, Tildesley DJ (1987) Computer Simulations of Liquids. Clarendon, Oxford
24. Hoover WG (1985) Phys Rev A 31: 1695–1697
25. Sprik M, Shelley JC (1993) Tenside 4:243–246
26. Shelley JC, Watanabe K, Klein ML (1993) Int J Quant Chem 17:103–117
27. MacKerell AD Jr (1995) J Phys Chem 99:1846–1855
28. Wennerström H, Lindman B (1979) Phys Rep 52:1–86
29. Lindman B, Wennerström H (1980) Topics in Current Chemistry 87. Springer, New York

Progr Colloid Polym Sci (1997) 103:155–159
© Steinkopff Verlag 1997

F. Mavelli

Stochastic simulations
of surfactant aggregation kinetics

Received: 13 November 1996
Accepted: 1 December 1996

Dr. F. Mavelli (✉)
Dipartimento di Chimica
Campus Universitario
Via Orabona 4
70126 Bari, Italy

Abstract Results of stochastic simulations of micellization kinetics are presented. The algorithm used was derived from the general Monte Carlo method introduced by D.T. Gillespie (1976, J. Phys. Chem. 22:403–434) and applied to micelle formation according to a mechanism that allows association and dissociation among n-mers of whatever aggregation number. With a careful choice of thermodynamic and dissociation kinetic constants it was possible to reproduce both equilibrium and kinetic properties of a hypothetical surfactant solution. The results obtained by stochastic simulations are compared to experimental evidence available in the literature.

Key words Association – kinetics – Monte Carlo – micellization – simulation – surfactant

Introduction

In this paper, the results of a Monte Carlo method for the simulation of the stochastic time evolution of the micellization process are presented. The computational algorithm [1] used represents an optimization of a general procedure introduced by Gillespie some years ago [2]. It was applied to the case of surfactant reversible association according to the general mechanism reported in Fig. 1 that allows associations and dissociations among n-mers of whatever aggregation number.

Fig. 1 Reversible association scheme of surfactant molecules: all possible associations and dissociations are allowed

$$M_i + M_j \underset{k^B_{i,j}}{\overset{k^F_{i,j}}{\rightleftharpoons}} M_{i+j}$$

$i, j = 1, 2, \ldots$

This scheme can be considered suitable for the description of nonionic surfactant solutions and also to the case of charged amphiphilic molecules only if the kinetics of the ion absorption–desorption process from the micellar surface is very fast (high salt concentration). In the following section this algorithm will be briefly discussed without going into the mathematical details of the stochastic process theory [3], but providing the reader with essential skeleton of the simulation program. Then the problem of the kinetic constants definition will be dealt with, illustrating the semi-empirical approach used and, finally, the outcomes of the simulations will be shown and compared with the behaviour usually observed for the real surfactant solutions.

The simulation algorithm

In Gillespie's stochastic approach [2] to the chemical kinetics, the time evolution of a reacting system is seen as a sequence of time interval throughout which nothing occurs, called dead time τ, followed by an infinitesimal interval of time dt in which one of the possible reactions takes place. At time t, the density probability function that a dead time of length τ will occur is given by

$$P(\tau; t) = \exp[-A\tau] \tag{1}$$

where A represents the sum of the density probabilities of each reaction to occur in the infinitesimal interval dt. For the micellization process this sum remains defined [4] as follows:

$$\frac{A}{\Omega} = \sum_i^M \sum_j^{i-1} k_{i,j}^{\mathrm{F}} \frac{M_i}{\Omega} \frac{M_j}{\Omega} + \sum_i^M k_{i,i}^{\mathrm{F}} \frac{M_i}{\Omega} \frac{M_i - 1}{\Omega}$$

$$+ \sum_{i=1}^M \sum_{j=1}^{i\backslash 2} k_{i-j,j}^{\mathrm{B}} \frac{M_i}{\Omega}, \tag{2}$$

where $\Omega = (N_{\mathrm{A}} V)$, V being the system volume, while M is the highest aggregation number of n-mers present in the system and M_i ($i = 1, 2, \ldots, M$) are the n-mer numerical populations. The backslash "\" has been used to symbolize the integer quotient of a division between integer numbers, so that "$i\backslash 2$" means "$i/2$" if i is even or "$(i - 1)/2$" if i is odd. As shown by Eq. (2), A can be considered as the sum of the kinetic contributions for every association and dissociation processes according to the scheme in Fig. 1. After the dead time has occurred, the probability that a certain reaction among all the possible will take place in the system is given by the ratio of its own kinetic contribution over A. Therefore, this Monte Carlo algorithm simulating the time evolution of micellization consists of an initialization step and an iterative procedure. In the initialization step the starting value for the n-mer populations are set along with the kinetic constants and the volume V of the system. At each loop of the iterative procedure A is calculated and two pseudo-random numbers g_1 and g_2 uniformly distributed between 0 and 1 are drawn. They are used to simulate, respectively, the dead time, randomly distributed according Eq. (1), by the formula:

$$\tau = \frac{1}{A} \ln\left(\frac{1}{g_1}\right) \tag{3}$$

and to select the occurring reaction μ as follows:

$$\sum_{\rho}^{\mu-1} a_\rho < g_2 A \leq \sum_{\rho}^{\mu} a_\rho \tag{4}$$

having labelled all possible association and dissociation reactions in a sequential order and being a_ρ the kinetic contribution of the ρth reaction. Once the occurring reaction has been determined, the populations of involved n-mers have to be updated, according to Fig. 1, by summing $+1$ or -1 to M_i and M_j and -1 or $+1$ to M_{i+j} depending on whether μ is an association or a dissociation

reaction. So, by repeating the iterative procedure a suitable amount of times the kinetics of the micellization process can be entirely simulated.

The kinetic constants

It should be evident that, to run the simulations, the values of kinetics constants $k_{i,j}^{\mathrm{F}}$ and $k_{i,j}^{\mathrm{B}}$ are necessary. Regarding $k_{i,j}^{\mathrm{B}}$, we suppose that they can be described as follows:

$$k_{i,j}^{\mathrm{B}} = (i+j) \frac{i!\, j!}{(i+j-1)!} k_0^{\mathrm{B}} \tag{5}$$

that is, they linearly depend on the size of the partitioning n-mer times a combinatory term which makes relatively faster the dissociations towards a larger fragment and a smaller one rather than towards two fragments of comparable size. In order to achieve an expression for $k_{i,j}^{\mathrm{F}}$, the equilibrium distribution function of surfactant molecules over the n-mer size classes is introduced and supposed to be a Boltzman distribution function:

$$P(i) = \frac{i\, [M_i]_{\mathrm{eq}}}{[C]} = \frac{\exp\left[-\dfrac{\varepsilon_i}{k_{\mathrm{B}} T}\right]}{\sum_j \exp\left[-\dfrac{\varepsilon_j}{k_{\mathrm{B}} T}\right]}, \tag{6}$$

where ε_i can be defined as the work of transferring a surfactant molecule from a standard state to the ith size class and $[M]_{\mathrm{eq}}$ and $[C]$ represent the n-mer equilibrium concentrations and the overall surfactant concentration, respectively. ε_i will be a complex function both of the system macroscopic properties and of the surfactant molecular features. In a semi-empirical approach we assume that, at a certain concentration $[C]$, it can be expressed by the formula:

$$\varepsilon_i = i \left[\frac{i - n_{\max}}{n_{\max}}\right]^2 \varepsilon_0, \tag{7}$$

where ε_0 is a constant depending on the properties of the considered system and $n_{\max}$ represents the maximum in the function distribution. Under these assumptions the values for the thermodynamic constants can be easily obtained by using the mass-action law along with Eq. (6), according to Fig. 1:

$$K_{i,j} = \frac{ij}{(i+j)} \frac{1}{[C]} \frac{P(i+j)}{P(i) \cdot P(j)}. \tag{8}$$

Finally, the association kinetic constants can be easily determined:

$$k_{i,j}^{\mathrm{F}} = K_{i,j} k_{i,j}^{\mathrm{B}}. \tag{9}$$

Progr Colloid Polym Sci (1997) 103:155–159
© Steinkopff Verlag 1997

To calculate the numerical values of $k_{i,j}^F$ and $k_{i,j}^B$, we set the external parameter as simply as possible:

$$\frac{\varepsilon_0}{k_B T} = 1.0, \qquad k_0^B = 1.0, \qquad [C] = 1.0, \qquad n_{max} = 60$$

just imposing that at the unitary surfactant concentration the maximum in the distribution function (6) equals 60. It should be clear that not having defined the dimensions of ε_0, k_0^B and V the results obtained will be valid for hypothetical surfactant solutions.

The outcomes of the stochastic simulations

Different simulations were run, setting $\Omega = 2.0 \times 10^7$ and changing the value of $[C]$ in the range 0.0025–1.0, always

starting from a flat distribution of surfactant molecules in the dimensional range 1–100. In Figs. 2 and 3 the time evolution of the surfactant concentration $i[M_i]$ in the different size classes against time for $[C] = 1.0$ and $[C] = 0.025$, respectively are reported.

The collected results show the presence of a critical surfactant concentration below which no micellization is observed.

In Fig. 4 the final values of all simulations for the monomer concentration $[M_1]$ and the surfactant concentration of proper micelles $\sum_{i > 40} i[M_i]$ against $[C]$ are reported. As can be seen in Fig. 4, the trends are in agreement with experimental observation on surfactant solutions: the monomer concentration increases linearly with $[C]$ until the range 0.025–0.04, while above $[C] = 0.04$, it rapidly reaches the value ≈ 0.025. On the other hand, the

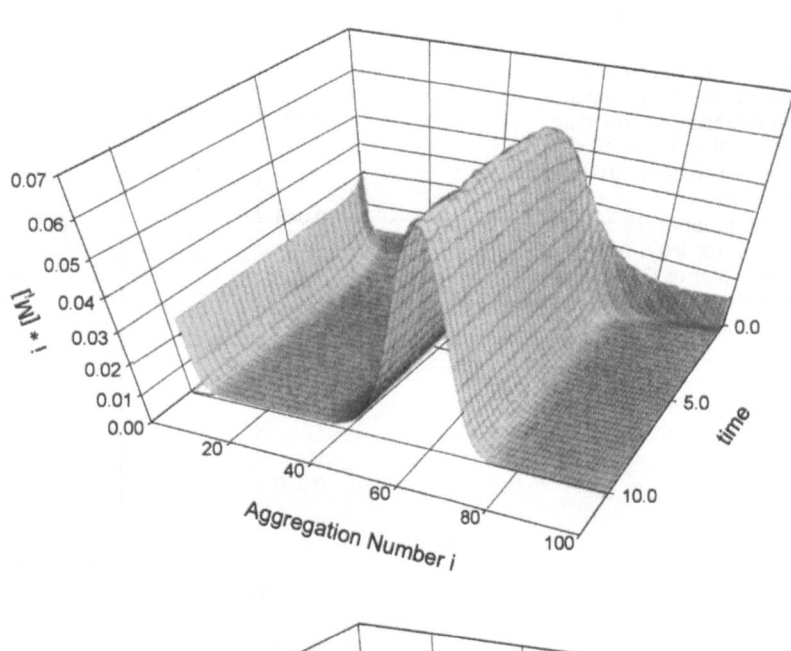

Fig. 2 Stochastic evolution of the surfactant concentration in the different size classes: results of a Monte Carlo simulation starting from a flat size distribution in the range 1–100 and setting $[C] = 1.0$ and $\Omega = 2 \times 10^7$

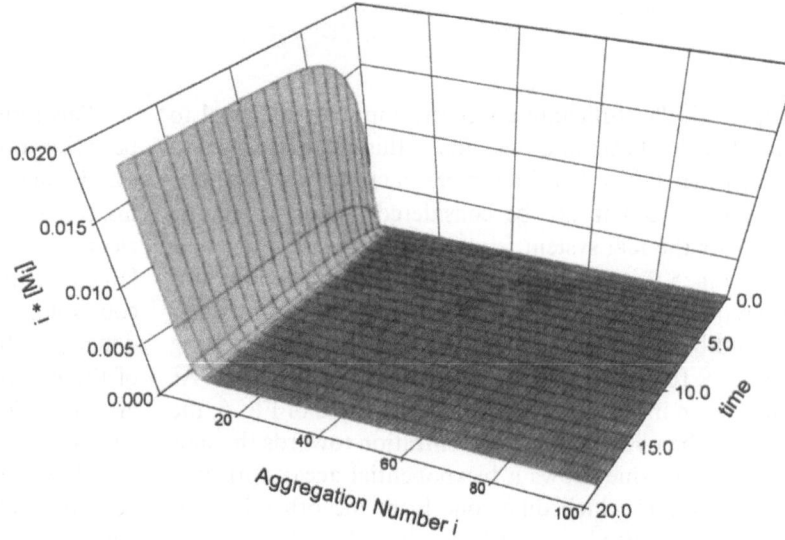

Fig. 3 Stochastic evolution of the surfactant concentration in the different size classes: results of a Monte Carlo simulation starting from a flat size distribution in the range 1–100 and setting $[C] = 0.025$ and $\Omega = 2 \times 10^7$

Fig. 4 Monomer concentration $[M_1]$ and surfactant concentration in the proper micelle size region ($i > 40$), against $[C]$. The results were obtained as final values of Monte Carlo simulations for different values of $[C]$ starting from flat size distributions in the range 1–100 and setting $\Omega = 2 \times 10^7$

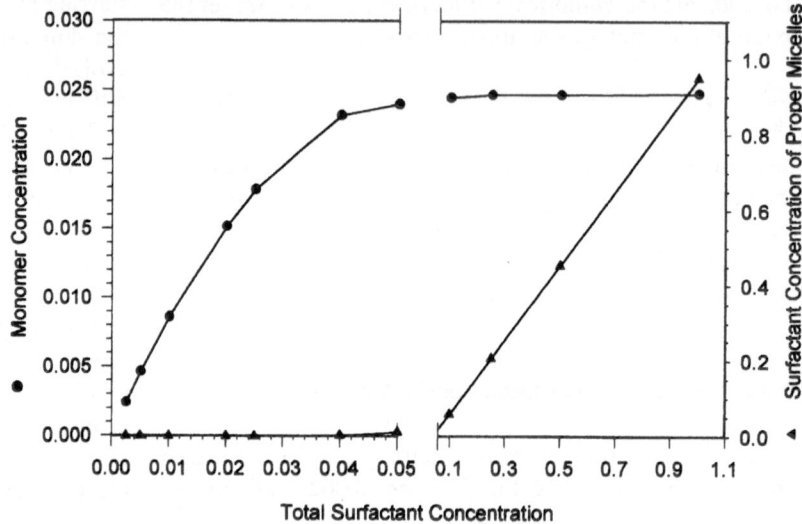

Fig. 5 Jump experiment simulation: overall n-mer concentration against time $[C] = 1.0$ and $\Omega = 2.0 \times 10^7$. The perturbation was obtained setting $\Omega = 2.2 \times 10^7$ at a certain point of the run. In the inset on the left a semi-logarithmic plot of the stochastic evolution after the perturbation is reported along with the results of a linear regression analysis

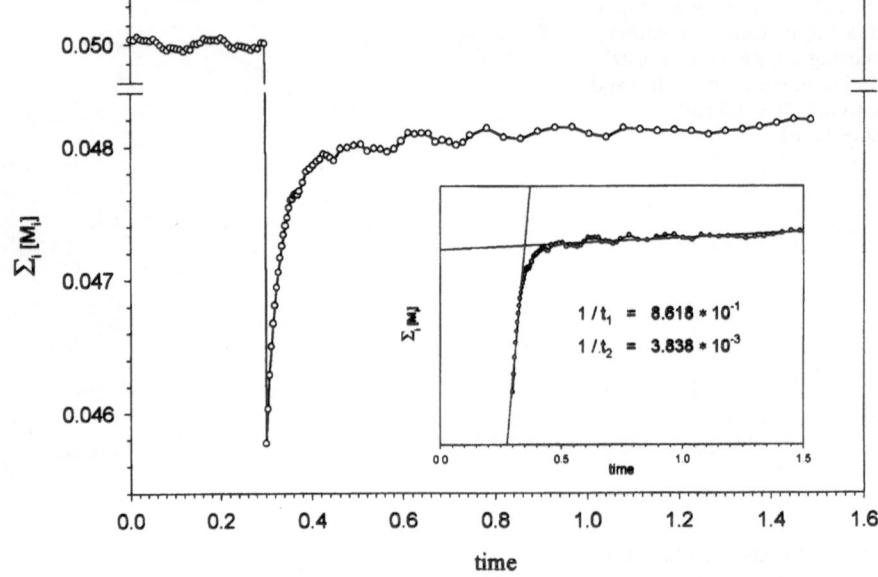

proper micelle surfactant concentration remains equal to 0 until $[C] = 0.025$ and only above this value it begins to increase. So, $[C] = 0.025$ represents a critical value for our simulations and it can be considered analogous to the CMC for the real system.

In Fig. 5 the result of a jump-experiment simulation is also reported. A system with $[C] = 1.0$, oscillating around its equilibrium size distribution, was perturbed by setting $\Omega = 2.2 \times 10^7$ at a certain point of the simulation run. As can be seen in the inset on the left-hand side of Fig. 5, the decay of the overall n-mer concentration towards the new equilibrium value shows a biexponential decay with two time constants which differ one from the other by two orders of magnitude: $t_1 = 1.16$ and $t_2 = 260.55$.

This result is in agreement with the surfactant solution behavior observed during jump experiments. The measured values for t_1 and t_2 are usually in the range of μs and ms, respectively. The presence of these relaxation times was clearly explained at first by the work of Aniansson and Wall [5] as a consequence of two different processes: a fast one, consisting of rapid exchange of small fragments (monomers, dimers, ...) by the proper micelle, and a slow one, consisting of the complete formation or dissociation of micellar aggregates. The simulations reproduce this feature well. In fact, as shown in Fig. 6, during the fast process the proper micelle surfactant concentration changes very rapidly (upper plot) while the concentration of the proper micelles remains practically constant (lower plot).

Progr Colloid Polym Sci (1997) 103:155–159
© Steinkopff Verlag 1997

Fig. 6 Jump experiment simulation: proper micelle concentration (upper plot) and surfactant concentration in the proper micelle size class (lower plot) against time, $[C] = 1.0$ and $\Omega = 2.0 \times 10^7$. The perturbation was obtained setting $\Omega = 2.2 \times 10^7$ at a certain point of the run

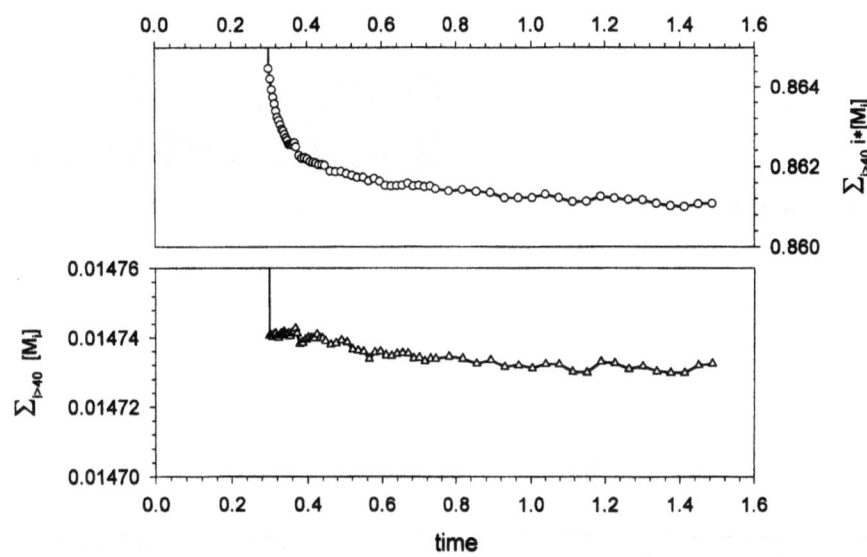

Conclusions

This Monte Carlo algorithm is able to reproduce both the equilibrium and kinetic properties of a generic surfactant solution in spite of the elementary definition of kinetic constants used. Of course, to get a more detailed description of micellization kinetics, a general theory to derive the kinetic constants as function of the surfactant molecular features and of the macroscopic properties of the system under consideration (temperature, pressure, ionic strength,...) has to be elaborated. Nevertheless, it should be stressed that the approach illustrated here can be considered as a useful tool to study those systems in which micelle formation is coupled with other processes as autopoietic micellar systems [6], hydrophobic compound solubilization in micelles [7], mixed micelle formations [8], and so on.

References

1. Mavelli F, Maestro M (to be submitted) J Phys Chem
2. Gillespie DT (1976) J Comput Phys 22:403–434 (1977) J Phys Chem 81: 2340–2369
3. Van Kampen NG (1981) Stochastic Processes in Physics and Chemistry, Chap 7. North-Holland, Amsterdam
4. Ginnel R (1979) Association Theory, Chap 2. Elsevier, Amsterdam
5. Aniansson EAG, Wall SN (1974) J Phys Chem 78:1024–1030 (1975) J Phys Chem 79:857–858
6. Luisi PL (1993) In: Stein W, Varela FJ (eds) Thinking about Biology. Addison-Wesley, New York, pp 3–25; Chizmadzhew YA, Maestro M, Mavelli F (1994) Chem Phys Lett 2267:56–62
7. Tachiya M (1987) In: Freeman GR (ed) Kinetics of Non-homogeneous Processes. Wiley, New York, pp 575–650
8. Wall S, Elvingson C (1985) J Phys Chem 89:2695–2705

Progr Colloid Polym Sci (1997) 103:160–169
© Steinkopff Verlag 1997

Cosurfactant facilitated transport in reverse microemulsions

J. Texter
B. Antalek
E. García
A.J. Williams

Received: 30 December 1996
Accepted: 20 January 1997

Dr. J. Texter (✉) · B. Antalek
E. García · A.J. Williams
Eastman Kodak Company
Rochester, New York 14650, USA

Abstract Faradaic electron transfer in reverse microemulsions of water, AOT, and toluene is strongly influenced by cosurfactants such as primary amides. Cosurfactant concentration, as a field variable, drives redox electron transfer processes from a low-flux to a high-flux state. Thresholds in this electron-transport phenomenon correlate with percolation thresholds in electrical conductivity in the same microemulsions and are inversely proportional to the interfacial activity of the cosurfactants. The critical exponents derived from the scaling analyses of low-frequency conductivity and dielectric spectra suggest that this percolation is close to static percolation limits, implying that percolative transport is along the extended fractal clusters of swollen micellar droplets. ^{1}H and ^{13}C NMR spectra show that surfactant packing transitions are also driven by changes in cosurfactant concentration. These packing transitions provide a physical basis for these electron transfer and conductivity percolation phenomena. Self-diffusion measurements derived from NMR pulsed gradient spin echo experiments show that water proton diffusion increases at the onset of electrical conductivity percolation and is transported along extended clusters. A dynamic partitioning model provides a direct measure of the volume fraction of these percolating clusters and an order parameter for quantifying water-in-oil droplet to percolating cluster microstructural transitions.

Key words Acrylamide – AOT – alkylamide – cluster formation – continuous transition – cosurfactant – order parameter – percolation – reverse microemulsion

Introduction

Transport in reverse microemulsions may be classified according to three categories: (1) ion and molecular diffusion through the (pseudo) continuous phase; (2) aggregate diffusion of reverse micelles; (3) ion and molecular exchange between reverse micelles. Processes occurring by ion and molecular diffusion through the pseudo-continuous phase are analogous to transport in simple solutions, with the provisos that the identification of the appropriate viscosity is not always straightforward and an obstruction factor must be taken into account because of tortuosity imposed by the discontinuous phase (swollen reverse micelles). Reverse micelles and substances solubilized therein diffuse at a rate characteristic of the effective micellar size and characteristically an order of magnitude more slowly than molecular diffusion. Viscosity and obstruction are factors that similarly affect such aggregate diffusion. The third category of transport, depending on exchange

Progr Colloid Polym Sci (1997) 103:160–169
© Steinkopff Verlag 1997

between swollen reverse micelles, may involve the convolution of the first two categories with binary micellar collision processes. These collisions have been pictured as hard sphere, and they have been modeled as the fusion of the colliding particles, followed by fission, keeping the long-term particle number density more or less constant [1–8].

This third category of transport has given rise to a broad spectrum of compartmentalized reaction chemistry. Reverse microemulsions may conveniently be formulated with various kinds of reagents in the water pools. When mixed, these microemulsions undergo transport according to category (3) above, and ensuing chemistry occurs when suitable reagents end up in the same swollen micellar water pool. Such chemistry has been greatly developed with respect to enzymatic transformations [9–11], organic latex formation in microemulsion polymerization [12, 13], inorganic nanoparticle precipitation [14], and organic–inorganic composite particle fabrication [15]. These chemistries result from intermicellar transport and solute exchange. Other transport processes such as electrical conductivity have been widely studied and found to be driven by various field variables such as disperse pseudophase volume fraction [16–23], temperature [24–27], and salinity [28]. Our focus in this review is an introductory look at such solute transport driven at constant temperature and at essentially constant aqueous volume (and generally decreasing volume fraction) by a different field variable, cosurfactant concentration.

Cosurfactant induced redox electron transfer

A cosurfactant-induced electron-transfer phenomenon is illustrated by the square wave voltammetry (SWV) data in Figs. 1 and 2 for the oxidation of $Fe(CN)_6^{4-}$ and for the reduction of $Ru(NH_3)_6^{3+}$ in a series of reverse water-AOT (sodium bis(2-ethylhexyl) sulfosuccinate)-toluene microemulsions. These data were obtained for a series of microemulsions containing varying amounts, ξ, of the cosurfactant acrylamide [29, 30]. Formulations are described in the caption to Fig. 1. The amount of water, AOT, and toluene is kept essentially constant in each series of microemulsions, and the amount of cosurfactant is varied. In such formulations, the mole fraction of cosurfactant (or equivalently weight %) is an effective field variable. In Fig. 2 it can be seen that the onset of facilitated $Fe(CN)_6^{4-}$ oxidation occurs at $\xi \approx 1.7\%$ and the onset of $Ru(NH_3)_6^{3+}$ reduction occurs at $\xi \approx 2.3\%$. Identical behavior to that illustrated in Fig. 2 was obtained for limiting currents in slow-scan cyclic voltammetry. It could then be concluded that this threshold phenomenon occurs under diffusion-limiting conditions and is not a simple manifes-

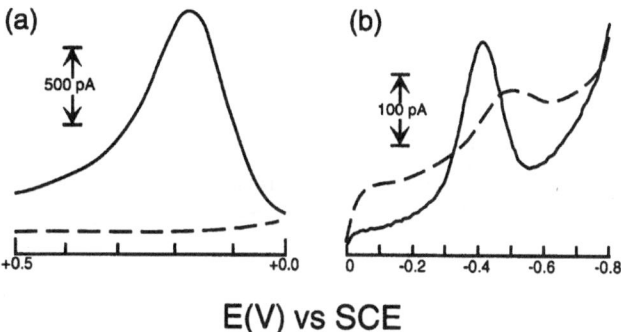

E(V) vs SCE

Fig. 1 SWV voltammograms (net currents) below (– – – –; $\xi = 0.2\%$, w/w) and above (———; $\xi = 5\%$, w/w) threshold for (a) $Fe(CN)_6^{4-}$ oxidation (at about $+ 30$ mV vs. SCE, saturated calomel electrode) and (b) $Ru(NH_3)_6^{3+}$ reduction (at about $- 400$ mV vs. SCE) at 10-μm diameter platinum ultramicroelectrode. Microemulsions were formulated with 1.84 g AOT, 7.32 g toluene, 0.833 g aqueous solution, and varying amounts of acrylamide (0–0.87 g) to cover the ξ-range of 0–5% (w/w). Aqueous $K_4Fe(CN)_6$ was 10.25 mM and aqueous $Ru(NH_3)_6Cl_3$ was 10.27 mM

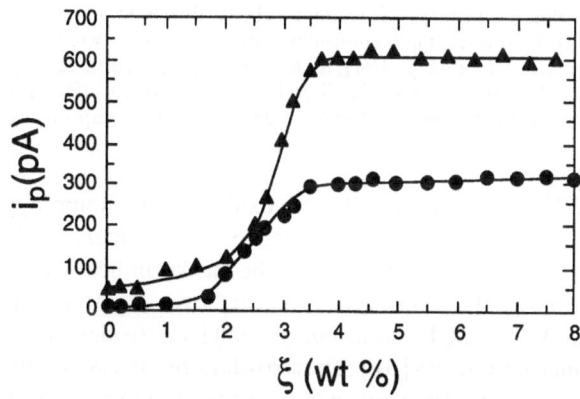

Fig. 2 SWV peak (net) currents, i_p, for $Fe(CN)_6^{4-}$ oxidation (●) and $Ru(NH_3)_6^{3+}$ reduction (▲) as a function of cosurfactant concentration, ξ (acrylamide weight percent). Microemulsions formulated as described in Fig. 1

tation of changes in electron transfer kinetics at the electrode surface. While various electrode kinetic models and hemimicellar hypotheses can be formulated to explain this phenomenon, considerations discussed below show these effects unequivocally to be bulk transport phenomena characteristic of these complex fluids.

Similar threshold phenomena have been reported for the reduction of persulfate in reverse microemulsion polymerization of acrylamide [12, 13], for ferrocyanide oxidation in similar microemulsions utilizing acetonitrile as cosurfactant [29], and for the autocatalytic oxidation of acrylamide and a variety of primary alkylamides where the respective amides served as cosurfactant [31].

162
J. Texter et al.
Cosurfactant facilitated transport in reverse microemulsions

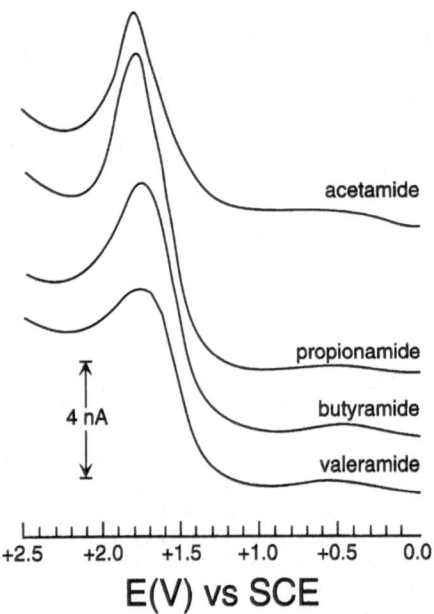

Fig. 3 Square-wave voltammograms (net currents, 24 °C, at 10 μm diameter platinum ultramicroelectrode) for irreversible oxidation of homologous primary alkylamides above threshold in reverse water–AOT–toluene microemulsions. Microemulsions were formulated with 1.77 g AOT, 7.04 g toluene, 0.785 g water, and varying amounts of amide, ξ = 8.4, 7.2, 7.3, and 7.9% (w/w) for cosurfactants acetamide, propionamide, butyramide, and valeramide, respectively

SWV voltammograms for the oxidation of a homologous series of primary alkylamides in reverse microemulsions are illustrated in Fig. 3. The peaks indicate that oxidation of the amide group occurs over the range of 1.7–1.8 V (vs. SCE, saturated calomel electrode) at Pt ultramicroelectrodes [31]. These oxidations are essentially irreversible since the reverse current voltammograms (not illustrated) are nearly featureless and nearly all the peak information is obtained in the forward current voltammograms. Measurement of oxidation currents at these potentials is normally impossible in aqueous solution, but the compartmentalization of water and electroactive amides in these reverse microemulsions provides an expanded accessible potential window. Such measurements in highly resistive solutions, where the continuous phase is basically toluene, are also impeded by large iR-drops, but ultramicroelectrode technology obviates these limitations due to the very low current fluxes involved.

This amide oxidation is an autocatalytic phenomenon. This autocatalysis is illustrated in Fig. 4, where peak oxidation currents are plotted as a function of cosurfactant (amide) concentration, ξ, for this series of homologous amides. Increasing the valeramide concentration further than illustrated in Fig. 4 leads to a two-phase system. Increasing ξ causes a dramatic increase from a low to a high-current (flux) state at a threshold concentration ξ_0.

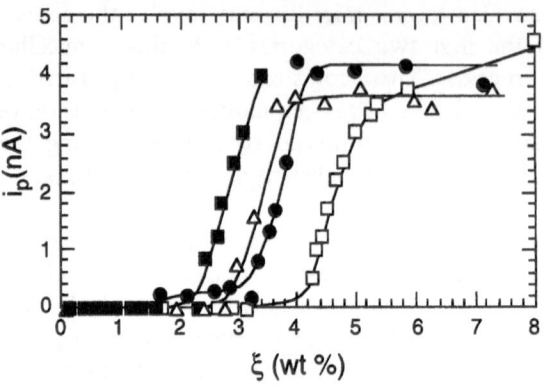

Fig. 4 SWV peak (net) currents for homologous amide oxidation as a function of amide (cosurfactant) concentration, ξ (weight %) in reverse water–AOT–toluene microemulsions formulated as described in the caption to Fig. 3: (◆) acetamide; (●) propionamide; (▲) butyramide; (■) valeramide

This ξ_0 is estimated by intersecting tangents to the initially slowly increasing current and to the steeply rising portion of the $i_p - \xi$ curve. It appears ξ_0 decreases as alkyl chain length of cosurfactant increases; this inverse proportionality is illustrated in Fig. 5. Extrapolation to $n = 0$ suggests the ξ_0 for formamide should occur in the range of 4.5–5.0%. This trend is consistent with the interfacial

Fig. 5 Cosurfactant threshold concentration, ξ_0, as a function of alkyl chain length, n, for the onset of autocatalytic cosurfactant oxidation in water-cosurfactant–AOT–toluene microemulsions described in Figs. 3 and 4

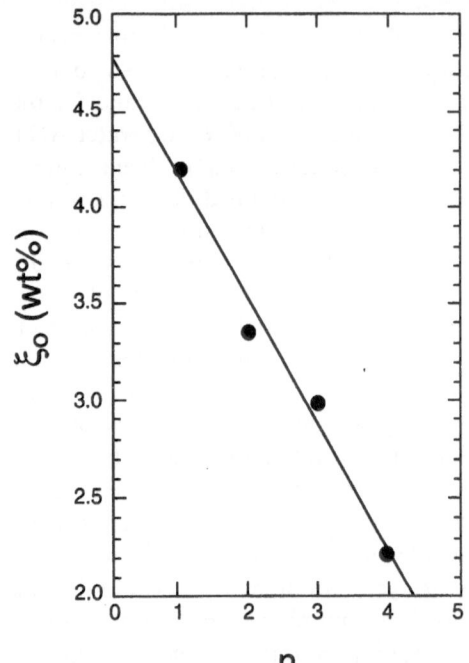

Progr Colloid Polym Sci (1997) 103:160–169
© Steinkopff Verlag 1997

activity and cosurfactancy of these amides; as alkyl chain length increases, water solubility decreases and propensity to adsorb at the oil–water interface increases [31].

The following mechanism was put forward [31] to explain this autocatalysis: (1) permeation by cosurfactant (amide) of the water–AOT–toluene interfacial regions as a result of partitioning equilibria with concomitant increase in polarity and dielectric constant in these regions; (2) diffusion of swollen micelle to proximity of electrode surface; (3) collision of swollen micelle with the electrode surface (de facto hemimicelle formation) or with a hemimicelle on the electrode surface and diffusion of amide through the AOT interfacial region within the electron transfer distance of the electrode; (4) irreversible oxidation of amide.

Association-colloid models must be invoked to explain aspects of mass transport since the electroactive species, electrolytes, as well as the cosurfactants have relatively low solubility in the toluene (pseudo) continuous phase. Since Faradaic redox currents in these microemulsions are diffusion limited, the complex fluid microstructure is expected to affect mass transport through the bulk to the electrode [32]. The sigmoidal current transition might possibly be assignable to a microstructural transition from a water-in-oil microemulsion to a bicontinuous microemulsion, where electroactive diffusion (and peak or limiting current) can increase by orders of magnitude. Phase studies [33] of the water–acrylamide–AOT–toluene system, however, indicate, that only a water-in-oil type system exists over the composition range investigated here. Another possible explanation for such a phenomenon could involve a dramatic decrease in the water pool droplet sizes with a concomitant increase in the diffusion coefficient of these droplets. Quasi-elastic light scattering of these microemulsions yielded droplet diffusion coefficients [32] that can be converted directly to droplet diameters with the aid of the Stokes-Einstein equation and independent viscosity measurements [34]. Results of quasielastic light scattering studies for various series of microemulsions show that the type of electrolyte has a negligible effect on the sizes, and the effect of added cosurfactant is a modest one [30].

Percolation

Percolation is a key phenomenon in signaling the onset of changes in transport (mechanisms) of charge and solutes confined principally to the disperse pseudo-phase of reverse microemulsions. Percolation is also a phenomenological indicator for morphological changes such as clustering and bicontinuity or sponge phase transformations. Cazabat et al. raised an important distinction when they asked which of these two morphologies best coincides

with percolation [18, 19]. Is percolation a process of conduction along paths defined by clusters and strings of reverse micelles [18, 21, 35–37] or along sponge-phase-microstructured volume elements bicontinuous in both water and oil [38, 39] wherein the interfacial surfactant film has low to zero mean-curvature? Both possibilities have been endorsed by various groups over the past decade, and the issue is complicated by ambiguity in how to define percolation and the onset of percolation. Percolation thresholds may be defined analytically without equivocation as an inflection point at ξ_p in an S-shaped curve of conductivity (log conductivity) versus field variable ξ [40, 41], and increases in conductivity may change 2–4 orders over a range of field variable. While percolation generally represents a regime in which transport changes from a low to high state, the precise definition of the *onset* of percolation at field variable $\xi = \xi_0$ does not appear to have an established basis.

In addition to percolation in reverse microemulsions being driven by field variables such as (disperse phase) volume fraction, temperature, and salinity, cosurfactant concentration appears also to suffice. Low-frequency electrical conductivity for the aqueous acrylamide–AOT–toluene reverse microemulsion system is illustrated in Fig. 6 [42]. The cosurfactant concentration $\xi_0 = 1.2\%$ (w/w) is annotated with the arrow (σ_p) in Fig. 6 and corresponds to the approximate *onset* of electrical conductivity percolation. The percolation *threshold* in conductivity occurs at $\xi_p = 3.09\%$.

A scaling analysis [42] obtained from dielectric spectroscopy studies of these aqueous acrylamide reverse microemulsions was done by examining the imaginary (ε'')

Fig. 6 Low-frequency conductivity (σ) of aqueous acrylamide–AOT–toluene reverse microemulsions at 25 °C as a function of (acrylamide) cosurfactant concentration, ξ. The σ_p and arrow at $\xi \approx 1.2\%$ (w/w) shows the onset of percolation in low frequency conductivity

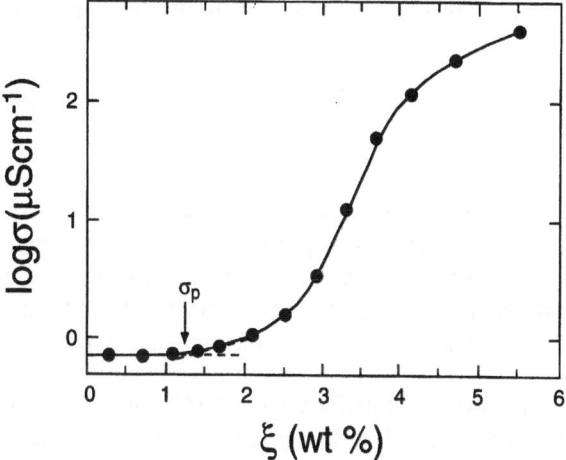

164

J. Texter et al.
Cosurfactant facilitated transport in reverse microemulsions

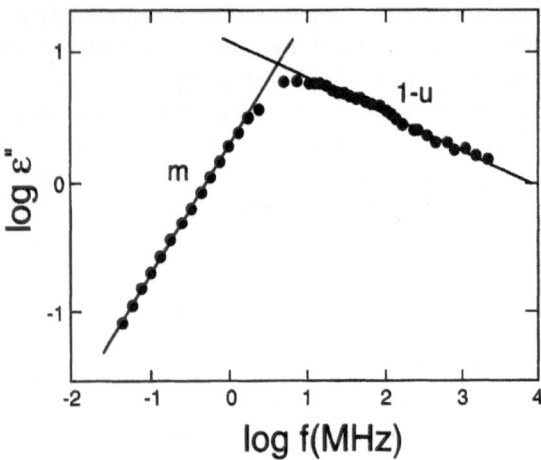

Fig. 7 Scaling analysis of the imaginary part of the dielectric constant as a function of frequency at 25 °C for an aqueous acrylamide–AOT–toluene reverse microemulsion with ξ = 4% (w/w)

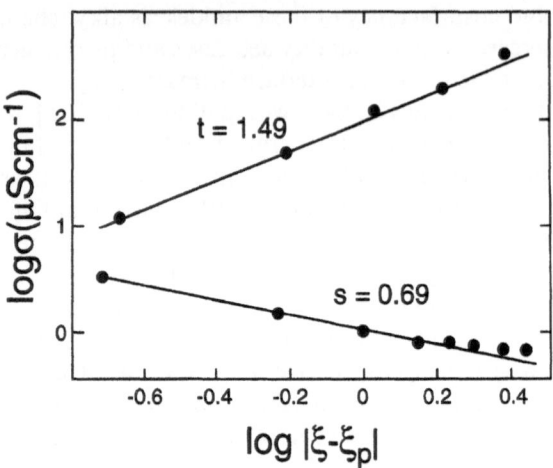

Fig. 8 Critical exponent analysis for low-frequency electrical conductivity in a neighborhood of the analytical percolation threshold (ξ_p = 3.09%) for aqueous acrylamide–AOT–toluene reverse microemulsions. The divergence below threshold yields the exponent s; the increase above threshold yields the exponent t

component of the dielectric permittivity to obtain two critical exponents, m and u [42, 43]. The derivation of these exponents from a log ε'' − log f plot is illustrated in Fig. 7 for a microemulsion with ξ = 4% (w/w) acrylamide [42]. The exponent u, derived [42] from both real and imaginary components of the complex permittivity $\varepsilon(\omega)$, appears invariant with increasing $\xi \geq \xi_p$; $\langle u \rangle = 0.73 \pm 0.05$.

The low-frequency conductivity data of Fig. 6 can be analyzed directly for another pair of exponents, s and t [44], where below threshold the conductivity diverges according to

$$\sigma \propto (\xi_p - \xi)^{-s} \tag{1}$$

and above threshold:

$$\sigma \propto (\xi - \xi_p)^t . \tag{2}$$

The derivation of these exponents is illustrated in Fig. 8 with $t = 1.49$ and $s = 0.69$. These exponents can be combined to provide another derivation of the exponent u [45],

$$u = \frac{t}{s + t} \tag{3}$$

and yield $u = 0.68$. This value is within one standard deviation of the 0.73 average derived from the frequency scaling of the imaginary component of the dielectric permittivity. The dynamic percolation limit typically has u exponents in the range of 0.59–0.61 with $t \sim 1.6$–1.9 and $s \sim 1.2$ [43]. The static percolation limit has exponents $t \sim 1.9$ and $s \sim 0.7$ [43], and yields an exponent u of 0.73. Clearly, the s exponent derived here of 0.69 appears closer to the 0.7 value typical of the static limit. The u exponent

values of 0.68–0.73 derived here appear significantly larger than the dynamic limit of 0.59–0.61, and very close to the static limit of 0.73. A consequence of such a correspondence is that the extended fractal clusters providing support for transport trajectories span the largest dimensions of the microemulsion and appear to have a lifetime competitive with the transport time. This dimensionality corresponds to a distance much much greater than a swollen micellar diameter. In dynamic percolation, clusters need only exist *locally* with respect to the (charge) carrier. Such clusters need to form in order to provide a path for carrier motion and for the next part of the trajectory, but those parts of the cluster already traveled may dissipate and dissociate.

Chemical shifts and relaxation

A carbon and chain labeling scheme [46, 47] for AOT is illustrated in Fig. 9, and proton NMR spectra illustrating assignments [48] of the various proton resonances of AOT are illustrated in Fig. 10. While most of the individual proton resonances are resolvable, several overlap. ^{1}H-^{13}C HETCOR experiments on microemulsions with $\xi = 0$ and 6% show changes in chemical shifts occur in both the carbon and proton dimensions as a result of adding cosurfactant (acrylamide). These HETCOR results showed that the terminal methyl protons experience a significant upfield shift with increasing ξ. In this case, H10 is affected to the greatest degree and H8 is affected the least. Protons H1 and H3 (and H1′ and H3′) showed no shift with acrylamide concentration.

Progr Colloid Polym Sci (1997) 103:160–169
© Steinkopff Verlag 1997

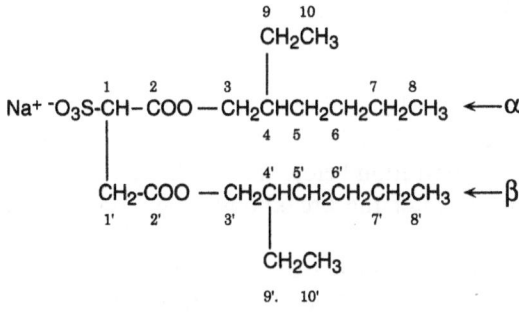

Fig. 9 Carbon and chain labeling for AOT

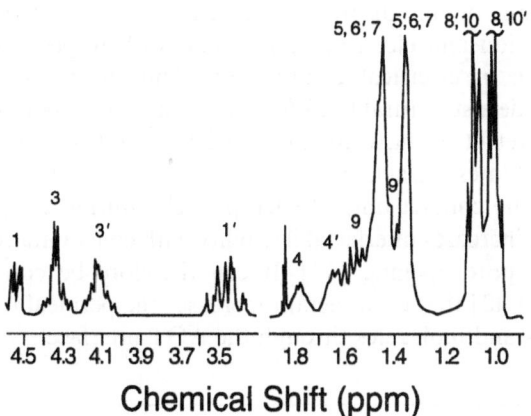

Fig. 10 Proton NMR spectra of reverse microemulsions at (acrylamide) $\xi = 5\%$ illustrating assignments of AOT protons. The numeric labels indicated for the resonances refer to the structure illustrated in Fig. 9

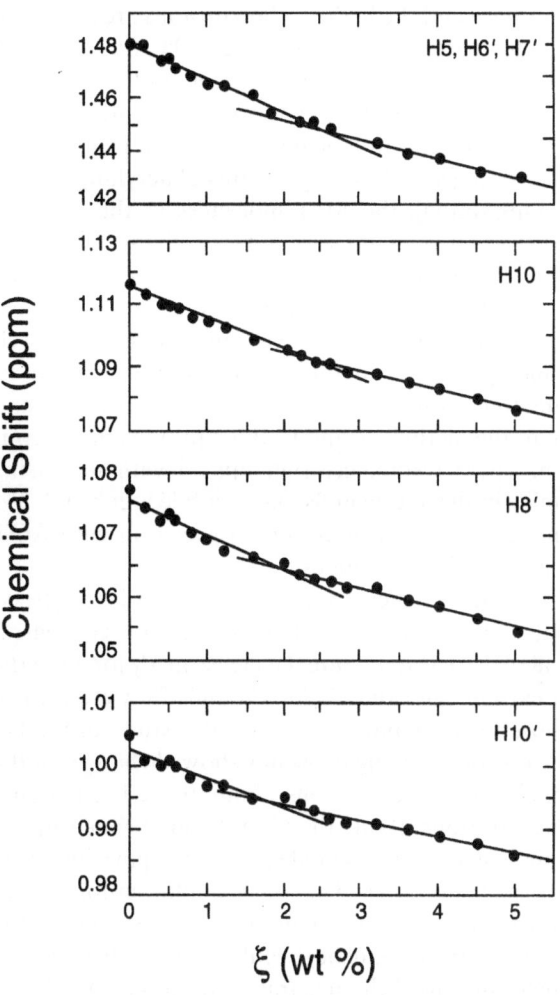

Fig. 11 AOT proton chemical shifts as a function of cosurfactant (acrylamide) concentration, ξ, for H5, H6′, and H7′ methylene protons; H10 methyl protons; H8′ methyl protons; and H10′ methyl protons

Not withstanding their similarity, the methyl groups from the α chain are distinguishable from the β chain [46]. Figure 11 shows proton chemical shifts as a function of ξ for the H5, H6′, and H7′ methylene, H10 methyl, H8′ methyl, and H10′ methyl groups. These resonances all exhibit breakpoints in the 2–3% interval, suggesting a change in interfacial AOT packing occurs during the onset of percolation [49]. Similar breakpoints are exhibited by the proton resonance trans to the amide group in acrylamide and by ^{13}C resonances of carbons C1, C1′, and C3 [48].

Only those carbons around the AOT headgroup exhibited significant chemical shift changes with the acrylamide level. Changes in proton chemical shifts are seen among the 2-ethylhexyl terminal methyl groups and generally along the 2-ethylhexyl chains. Since acrylamide, as a cosurfactant, tends to aggregate close to AOT and partitions among the AOT, the more concentrated the acrylamide is in the water pool the greater is the number of AOT molecules experiencing the presence of the acrylamide, and thus clear trends are observed in the shifts. It has

been concluded [50] that an increasing hydrophobic environment leads to an upfield shift for C2. However, no significant shift was observed for carbons C2′ and C3′, and it was suggested [48] that the acrylamide aligns preferentially nearer to the α chain. In the present system the downfield shifts for carbons C1 and C1′ can be attributed to a greater hydration of the headgroup area of AOT. Martin and Magid [50] reported a similar effect. This suggests that as the cosurfactant is incorporated, it disorders the AOT and facilitates water penetration further into the headgroup region.

Acrylamide proton shifts suggest that the region around the headgroup of AOT becomes more hydrated while the area around C2 and C3 of AOT becomes dehydrated [48, 49]. The acrylamide partitions at the interface in such a way so that the amide protons are within or near the water pool and the hydrocarbon protons are among or near the hydrocarbon chains of the AOT. Similar shifts with increasing acrylamide of the amide proton signals in these microemulsions and in aqueous acrylamide support this proposed orientation of amide groups in proximity to the water pool [49]. Two populations of acrylamide, that partitioning among the AOT molecules at the interface and in the water pool, must be in fast exchange with each other. This is because only one set of resonances for each species of acrylamide proton is observed; the resonances represent an average of these two populations. The acrylamide methylene and methine group protons exhibit an opposite trend with respect to the amide protons and relative to the methylene and methine protons in aqueous solution. These resonances shift upfield with increasing acrylamide in the microemulsions. A subtle break at 2–3% acrylamide is observed indicating association with AOT hydrocarbon chain dynamics.

^{13}C T_1 measurements were made for the 2-ethylhexyl terminal methyl carbons at four acrylamide concentrations [48]. There was a gradual increase in T_1 for all of the methyl carbons and the T_1 for C8 and C8' were greater than those for C10 and C10'. A detailed study of the H1 and H1' proton coupling constants showed changes in the rotational isomer populations about the C1–C1' bond. The three protons about the C1–C1' bond make up an ABX spin system that was modeled using a spin simulation software package [48]. The trans conformation corresponds to the α and β chains being trans across the C1–C1' bond. In the gauche-1 conformation, these chains are gauche and the β chain is trans with respect to H1. In the gauche-2 conformation the β chain is trans to the sulfonate group. The gauche-2 conformation dominates, which is consistent with the findings of others [46, 51, 52]. There is a slight increase in the gauche-2 conformation (from about 0.69 to 0.72) and a slight decrease in the trans conformation (from 0.11 to 0.08), suggesting a change in the packing of the chains. This packing modification suggests changes in average conformation. An increase in the trans conformation has previously been associated with chain ordering within the aggregated state as compared with the free amphiphile state [53, 54]. The converse indicates chain disordering. These data suggest a general disordering effect on the packing of the AOT due to the presence of the cosurfactant. De Gennes and Taupin have pointed out [39] that the cosurfactant was pictured as increasing the disorder of the interfacial film in the pioneering work of Schulman and Roberts in 1946. It was inferred [48] that the acrylamide resides close to C2 and C3, replacing water in the region along the α chain, because C2' and C3' reveal no shift with acrylamide concentration. The head group region is consequently disordered, allowing for greater hydration around C1 and C1'.

In addition to this disordering, the chemical shift data between 2% and 3.5% acrylamide exhibit breakpoints for many of the chain protons and for three of the carbons. The protons and carbons of the AOT molecule (and carbons of the acrylamide) react differently to changes in acrylamide concentration before the break than after. The shift dependence after the breakpoint is less sensitive to ξ compared to before the break. Such decreasing sensitivity is consistent with increased disorder and increased chain mobility. These breakpoints illustrating disordering may be likened to continuous or second-order transitions in interfacial structure, where compositional partitioning of cosurfactant in the interface appears concomitantly. The ramifications of this disordering appear significant with respect to ion and electron transfer and with respect to percolation in electrical conductivity. This disordering may provide a structural basis for the onset of percolation [12]. Carver et al. have hypothesized that added acrylamide facilitates attractive interactions between particles, increases the contact time of such particles during collisions, and increases interfacial flexibility with concomitant transitory pore opening [12]. It can therefore be concluded [30, 48] that cosurfactant increases the permeability of ions and molecules through the AOT interface.

Self-diffusion

Self-diffusion coefficients for all chemical components in microemulsions may be determined by NMR pulse gradient spin echo experiments as exemplified by Lindman and coworkers [55–57]. Self-diffusion coefficients for toluene, water, acrylamide, and AOT in this aqueous acrylamide-AOT-toluene system are illustrated in Fig. 12 as a function of ξ [58]. The self-diffusion coefficients for water, acrylamide, and AOT exhibit breakpoints near, but below the percolation threshold, in the neighborhood of 1.2–1.3% acrylamide. This breakpoint corresponds to the onset of percolation in electrical conductivity illustrated in Fig. 6 at ξ_0. The AOT diffusion coefficients run parallel to those for acrylamide, but are offset to lower values.

When this AOT diffusion is modeled in the context of a fast exchange (of AOT between the water-swollen droplets and monomeric AOT in the continuous toluene pseudo-phase) model, the observed AOT self-diffusion may be modeled by the following two-state model [58]:

$$D_{\text{obs}} = xD_{\text{c}} + (1 - x)D_{\text{mic}} . \tag{4}$$

D_{mic} corresponds to the swollen micelle self-diffusion coefficient and D_{c} represents the self-diffusion coefficient of monomeric AOT in the toluene phase. Nearly all of the AOT is in the disperse state, and $D_{\text{mic}} \approx D_{\text{AOT}}$ (D_{obs}) with relatively little error [58].

Progr Colloid Polym Sci (1997) 103:160–169
© Steinkopff Verlag 1997

167

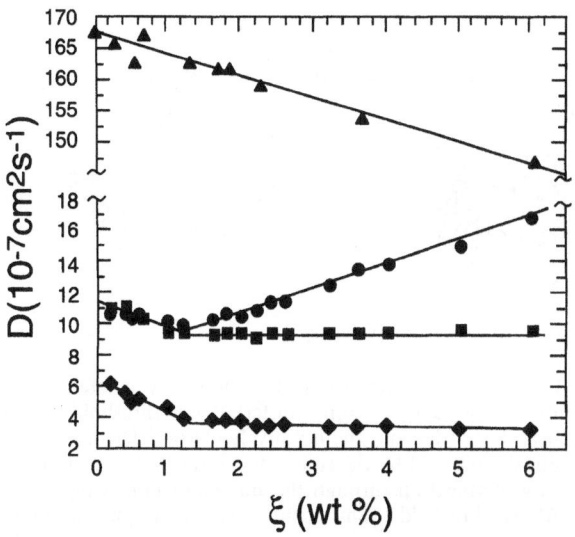

Fig. 12 Observed self-diffusion coefficients at 25 °C for toluene (▲), water (●), acrylamide (■), and AOT (◆) in reverse microemulsions (formulated as described in the caption to Fig. 1) as a function of acrylamide concentration, ξ

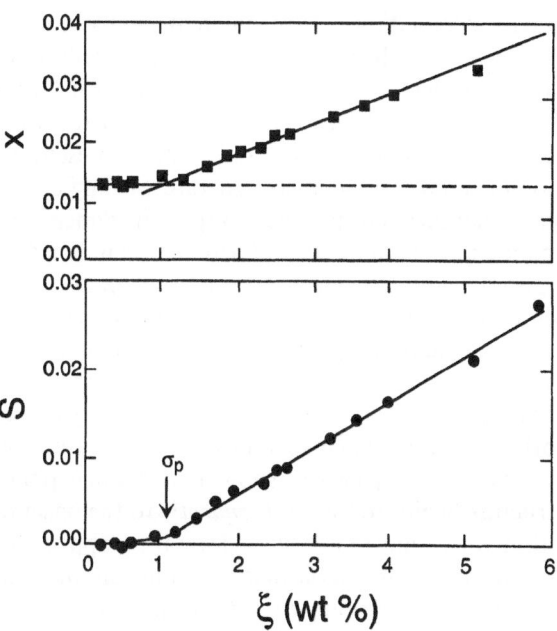

Fig. 13 (top) Mole fraction (x) water in the continuous (toluene) phase of the microemulsions as a function of acrylamide concentration, for aqueous acrylamide–AOT–toluene reverse microemulsions. (bottom) Order parameter for disperse pseudophase water derived from data of (a). The σ_p and arrow at $\xi \approx 1.2\%$ (w/w) acrylamide indicate the approximate onset of percolation in low-frequency electrical conductivity

Similarly water has been assumed to undergo rapid exchange between the continuous and disperse pseudophases. When the measured D_{obs} for water is considered in terms of the two-state model of Eq. (4), $D_{mic} \approx D_{AOT}$, and D_c is the diffusivity of dilute water measured in toluene, $5.41 \times 10^{-5} \, \text{cm}^2 \, \text{s}^{-1}$. The mole fraction of water in the respective pseudophases can then be calculated from Eq. (4), the observed diffusivity of water (Fig. 12), the observed diffusivity of AOT (D_{mic}), and the diffusivity of water in the continuous pseudo-phase, D_c. The apparent mole fraction results for water in the continuous phase, x, are illustrated in Fig. 13 (top). Below 1.2% acrylamide, the mole fraction of water in the continuous phase is relatively constant, at about $x = 0.013$. Above $\xi = 1.2\%$ the apparent mole fraction of water in the continuous phase steadily increases.

This differential increase in partitioning into the continuous phase over the subthreshold value of $x = 0.013$ suggests the formation of a third pseudo-phase assigned to percolating clusters of droplets. In this three-pseudo-phase model, if was assumed [58] that the molar ratio of water to AOT in the clusters is the same as that of the isolated swollen micelles, and it may then be concluded that this excess mole fraction derived for water in the continuous pseudophase represents the volume fraction of the percolating pseudo-phase (φ_{ex}). An order parameter (S) for the disperse pseudo-phase (clusters and isolated swollen micelles) was defined as [58, 59]

$$S = \frac{\varphi_{clstr}}{\varphi_{clstr} + \varphi_{mic}}, \tag{5}$$

where φ_{mic} is the volume fraction of the isolated swollen micelle pseudo-phase and φ_{clstr} is the excess volume fraction derived in Fig. 6 (top). Figure 6 (bottom) illustrates this order parameter. S increases from 0 to about 0.027 as cosurfactant concentration increases above ξ_0 at about $\xi = 1.2\%$.

Similar coincidences in the *onset* of electrical percolation with the *onset* of water–proton self-diffusion have been demonstrated in recent studies of temperature driven percolation [25] and of volume fraction driven percolation [60]. In each of these cases [42] ξ_0 describes this coincidence *and* the occurrence of a continuous transition in S (Eq. 4). Such correlations between the ξ_0 for electrical conductivity and water–proton self-diffusion may be found in the work of Geiger and Eicke [61] and in Jonströmer and coworkers [62] by carefully correlating the conductivity and self-diffusion data presented therein.

Summary

Percolation in water diffusion has been clearly resolved by NMR self-diffusion measurements. The onset of water self-diffusion increase as indicated by NMR measurements correlates with the onset of electrical conductivity percolation in five different microemulsion systems. These ξ_0 values also coincide with continuous transitions in order parameters defined by Eq. (5). These transition points ξ_0 indicate the formation of percolating clusters of water–swollen micelles and do not represent the formation of sponge phase microstructure. Scaling analysis suggests that these clusters tend to span the dimensions of the microemulsion in the case of percolation induced by acrylamide. Such lengthy clusters may be taken as a type of bicontinuity, in that it represents a connected network spanning the volume of the microemulsion. This cluster bicontinuity exists as a result of the *connectedness* of surfactant films, not as a result of sponge phase yielding a network of water channels spanning the microemulsion. Other studies [42, 60] clearly show that sponge phase microstructure begins to form *subsequently* to the onset of percolation at ξ_0, and sponge phase formation appears to coincide with increasing *surfactant* self-diffusion and distinctly different continuous transitions in order parameters such as given in Eq. (5). Cosurfactant concentration clearly can be used to drive percolation in Faradaic electron transport, electrical conductivity, and water self-diffusion. Distinct similarities with temperature and volume fraction driven percolation exist in that "onset" points ξ_0 tie electrical percolation together with water self-diffusion. This similarity is also connected to ξ_0 for Faradaic electron transport for a variety of electroactive solutes. Spin-lattice relaxation suggests surfactant packing *disorders* over the entire percolation regime. Chemical shift measurements also show that breakpoints in chemical shifts for key methyl protons and chain methylene protons and for headgroup carbons of AOT occur over the percolation regime in ξ, and some of these breakpoints coincide with ξ_0 obtained from electron transfer, electrical conductivity, and water self-diffusion. These observations show that cosurfactant-driven changes in AOT *packing* are key to the clustering attendant to the percolation effects

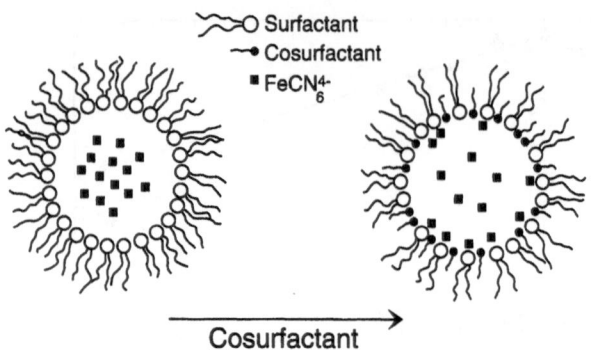

Fig. 14 Cartoon illustrating aspect of cosurfactant facilitated electron transport (Faradaic electron transfer). Below threshold (left) in the absence of cosurfactant, the negatively charged ferrocyanide is electrostatically repelled by the sulfonates of the AOT headgroups. Complex ion permeation through the surfactant film is highly activated. Above threshold (right) cosurfactant interdigitates among the surfactant headgroups, decreases charge density repulsion of the negatively charged complex ion, and modifies the AOT packing to produce a more flexible interfacial surfactant film of increased ionic permeability

observed and to the attractive interactions that lead to clustering and facilitated transport therein. A cartoon illustrating such interdigitated packing of cosurfactant in these microemulsions is illustrated in Fig. 14, where an electroactive solute such as ferrocyanide is dissolved in the water pools. Below ξ_0 these complex ions undergo some electrostatic repulsions from the AOT headgroups, and the AOT interfacial layer is relatively impermeable to ion transport. Similar effects are seen with uncharged and even with positively charged species, where AOT and ruthenium hexamine may even form an ion pair. Nevertheless, transport of such species across AOT interfacial layers is highly activated, until the AOT chain ordering is disrupted sufficiently to yield the high-flux transport observed in Faradaic electron transfer, electrical conductivity, and water proton self-diffusion.

References

1. Eicke HF, Shepherd JCW, Steinemann A (1976) J Colloid Interface Sci 56: 168–176
2. Fletcher PDI, Howe AM, Perrins NM, Robinson BH, Toprakcioglu C, Dore JC (1982) In: Mittal KL, Lindman B (eds) Surfactants in Solution, Vol. 3. Plenum, New York, pp 1745–1758
3. Zana R, Lang J (1987) In: Friberg SE, Bothorel P (eds) Microemulsions: Structure and Dynamics. CRC Press, Boca Raton, pp 153–172
4. Fletcher PDI, Howe AM, Robinson BH (1987) J Chem Soc Faraday Trans 1 83:985–1006

5. Howe AM, McDonald JA, Robinson BH (1987) J Chem Soc Faraday Trans 1 83:1007–1027
6. Jada A, Lang J, Zana R (1989) J Phys Chem 93:10–12
7. Jada A, Lang J, Zana R, Makhloufi R, Hirsch E, Candau SJ (1989) J Phys Chem 94:387–395
8. Lang J, Lalem N, Zana R (1991) J Phys Chem 95:9533–9541
9. Pileni MP (ed) (1989) Structure and Reactivity in Reverse Micelles. Elsevier, Amsterdam

10. Kabanov AV, Klyachko NL, Mametkin SN, Merker S, Zaroza AV, Bunik VI, Ivanov MV, Levashov AV (1991) Protein Eng 4:1009–1017
11. Stamatis H, Xenakis A, Menge U, Kolisis FN (1993) Biotech Bioeng 42:931–937
12. Carver MT, Hirsch E, Whittmann JC, Fitch RM, Candau F (1989) J Phys Chem 93: 4867–4873
13. García E, Texter J (1992) In: Mackay R, Texter J (eds) Electrochemistry in Colloids and Dispersions. VCH Publishers, New York, pp 257–270

14. Kizling J, Boutonnet-Kizling M, Stenius P, Touroude R, Maire G (1992) In: Mackay R, Texter J (eds) Electrochemistry in Colloids and Dispersions. VCH Publishers, New York, pp 333–344
15. Gan LM, Zhang LH, Chan HSO, Chew CH (1995) Mater Chem Phys 40:94–98
16. Laguës M, Ober R, Taupin C (1978) J Phys 39:L487–L491
17. Lagourette B, Peyrelasse J, Boned C, Clausse M (1979) Nature 281:60–62
18. Cazabat AM, Chatenay D, Langevin D, Meunnier J (1982) Faraday Discuss Chem Soc 76:291–303
19. Cazabat AM, Chatenay D, Guéring P, Langevin D, Lang J, Zana R (1984) In: Mittal KL, Lindman B (eds) Surfactants in Solution, Vol 3. Plenum Press, New York, pp 1737–1744
20. van Dijk MA (1985) Phys Rev Lett 55:1003–1005
21. Chatenay D, Urbach W, Cazabat AM, Langevin D (1985) Phys Rev Lett 54:2253–2256
22. van Dijk MA, Casteleijn G, Joosten JGH, Levine YK (1986) J Chem Phys 85:626–631
23. Peyrelasse J, Boned C (1990) Phys Rev A 41:938–953
24. Dutkiewicz E, Robinson BH (1988) Materials Sci Forum 25–26:389–392
25. Feldman Y, Kozlovich N, Nir I, Garti N, Archipov V, Idiyatullin Z, Zuev Y, Fedotov V (1996) J Phys Chem 100:3745–3748
26. Peyrelasse J, Boned C (1992) In: Chen SH, Huang JS, Tartaglia P (eds) Structure and Dynamics of Strongly Interacting Colloids and Supramolecular Aggregates in Solution, Kluwer Academic Publishers, Dordrecht, pp 801–806
27. Almgren M, Jóhannsson R (1993) J Phys IV 3:81–90
28. Guéring P, Lindman B (1985) Langmuir 1:464–468

29. García E, Texter J (1993) Proc Electrochem Soc 93–1:2166–2167
30. García E, Song S, Oppenheimer LE, Antalek B, Williams AJ, Texter J (1993) Langmuir 9:2782–2785
31. García E, Texter J (1994) J Colloid Interface Sci 162:262–264
32. García E, Song S, Oppenheimer LE, Texter J (1995) Colloid Surfs 94:131–136
33. Candau F, Leong YS, Pouyet G, Candau S (1984) J Colloid Interface Sci 101:167–183
34. Dayalan E, Qutubuddin S, Texter J (1992) In: Mackay R, Texter J (eds) Electrochemistry in Colloids and Dispersions. VCH Publishers, New York, pp 119–135
35. Borkovec M, Eicke HF, Hammerich H, Das Gupta B (1988) J Phys Chem 92:206–211
36. Moha-Ouchane M, Peyrelasse J, Boned C (1987) Phys Rev A 35:3027–3032
37. Jóhannsson R, Almgren M (1993) Langmuir 9:2879–2882
38. Kim MW, Dozier WD (1990) In: Chen SH, Rajagopalan R (eds) Micellar Solutions and Microemulsions – Structure, Dynamics and Statistical Thermodynamics. Springer, New York, pp 291–301
39. De Gennes PG, Taupin C (1982) J Phys Chem 86:2294–2304
40. Safran S, Webman I, Grest GS (1985) Phys Rev A 32:506–511
41. Cametti C, Codastefano P, Tartaglia P, Rouch J, Chen SH (1990) Phys Rev Lett 64:1461–1464
42. Antalek B, Williams AJ, Texter J, Feldman Y, Garti N (1997) Colloids Surf A Physicochem Eng Aspects 125:xxx–xxx
43. Feldman Y, Kozlovich N, Nir I, Garti N (1995) Phys Rev E 51:478–491

44. Laguës M (1979) J Phys 40:L331–L333
45. Peyrelasse J, Moha-Ouchane M, Boned C (1988) Phys Rev A 38:904–917
46. Yoshino A, Sugiyama N, Okabayashi H, Taga K, Yoshida T, Kamo O (1992) Colloids Surf 67:67–79
47. Olsson U, Wong TC, Soderman O (1990) J Phys Chem 94:5356–5361
48. Antalek B, Williams AJ, García E, Texter J (1994) Langmuir 10:4459–4467
49. Antalek B, Williams AJ, García, E Wall DH, Song S, Texter J (1996) In: Pillai V, Shah DO (eds) Dynamic Properties of Interfaces and Association Structures. AOCS Press, Champaign, IL, pp 183–196
50. Martin CA, Magid L (1981) J Phys Chem 85:3938–3944
51. Heatley F (1987) J Chem Soc Faraday Trans 83:517–526
52. Ueno M, Kishimoto H, Kyogoku Y (1976) Bull Chem Soc Jpn 49:1776–1779
53. Persson BO, Drakenberg T, Lindman B (1976) J Phys Chem 80:2124–2125
54. Cheney VB, Grant DM (1967) J Am Chem Soc 89:5319–5327
55. Lindman B, Kamenka N, Kathopoulis TM, Brun B, Nilsson PG (1980) J Phys Chem 84:2485–2490
56. Stilbs P, Lindman B (1984) Progr Colloid Polymer Sci 69:39–47
57. Lindman B, Olsson U (1996) Ber Bunsenges Phys Chem 100:344–363
58. Antalek B, Williams AJ, Texter J (1996) Phys Rev E 54:R5913–R5916
59. Antalek B, Williams AJ, Texter J (to be published) J Chem Phys
60. Texter J, Antalek B, Williams AJ (1997) J Chem Phys 106:xxx–xxx
61. Geiger S, Eicke HF (1986) J Colloid Interface Sci 110:181–187
62. Jonströmer M, Olsson U, Parker WO Jr (1995) Langmuir 11:61–69

Progr Colloid Polym Sci (1997) 103:170–180
© Steinkopff Verlag 1997

J.F. Rusling

Catalytically active, ordered films of proteins, surfactants and polyelectrolytes on electrodes

Received: 9 November 1996
Accepted: 16 November 1996

Abstract This paper reviews recent results on electrochemistry and electrochemical catalysis using stable, cast films containing proteins and surfactants. The first example of electrochemistry in such films involved in iron heme protein myoglobin (Mb), which gave greatly enhanced electron transfer rates in lamellar liquid crystal films of insoluble surfactants. Strongly adsorbed surfactant at the electrode–film interface inhibits adsorption of denatured proteins which otherwise block electron transfer. Enhanced electron transfer rates compared to bare electrodes were found with many surfactants. Formal potentials of the proteins depend on surfactant type and electrode material. Ferrodoxins and the metabolic enzyme cytochrome P450$_{cam}$ also gave reversible electron transfer in surfactant films. Spectroscopic and thermal studies showed that surfactants in the films are arranged in bilayers similar to biomembranes. Proteins are oriented and retain native conformations in films at medium pH. Films can be further stabilized by polyelectrolytes such as Nafion and by enzyme crosslinking.

Mb-surfactant films were used to reduce organohalides by electrochemical catalysis with enhanced rates compared to homogeneous reactions. Cyt P450$_{cam}$–lipid films were used to reduce trichloroacetic acid. Electrochemical reduction of cyt P450$_{cam}$ in the presence of oxygen mimics acceptance of electrons by cyt P450Fe(III) and cyt P450Fe(II)–O$_2$ during *in vivo* catalytic oxidations, by which pollutants are thought to be activated in carcinogenesis. Finally, films of heme proteins and the polyanion DNA also gave good electron transfer rates.

Key words Protein–surfactant films – protein electrochemistry – enzyme catalysis – protein orientation – protein–DNA films

Dr. J.F. Rusling (✉)
Department of Chemistry
Box U-60
University of Connecticut
Storrs, Connecticut 06269-4060, USA

Introduction

A little more than a decade ago, Kunitake and coworkers described ordered films of water-insoluble surfactants cast onto solid surfaces from organic solvents or aqueous vesicle dispersions [1]. Evaporation of the solvent after casting leaves thin films self-assembled into ordered stacks of bilayers. The same group later reported similarly ordered films stabilized by ionic polymers [2–4]. These techniques provide viable alternatives to the more tedious Langmuir–Blodgett film transfer method [5] to deposit multiple bilayers of surfactants on surfaces.

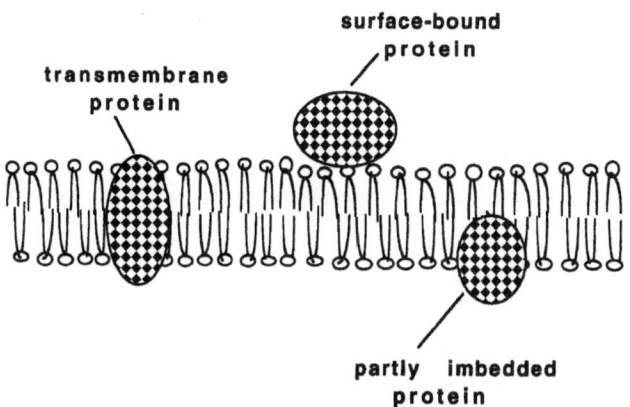

Fig. 1 Idealized drawing of a bilayer membrane showing three possible modes of protein binding

Bilayers of surfactants called phospholipids are integral structural units of biological membranes. In living organisms, membranes are made of roughly 40% lipid and 60% protein and generally exist in a partly fluid, selectively permeable state [6, 7]. The phospholipids are arranged in bilayers, and proteins can be adsorbed onto the surface or imbedded in the bilayer (Fig. 1).

Films of synthetic or natural bilayer-forming surfactants can be designed to give supramolecular aggregates with properties similar to biomembranes [1–5]. Properties of biological membranes such as catalysis, microporosity, selective permeability, and binding of molecules onto or within the membrane can be built into these films for specific chemical applications.

In the early 1990s, we [8–11] and Nakashima et al. [12, 13] began to explore the electrochemistry of small electroactive molecules imbedded in cast multi-bilayer surfactant films. In their liquid crystal states, these films exhibited excellent charge transport driven by injection of electrons from the electrode to the electroactive centers. Charge transport diffusion through these films showed a large phase-dependent increase as reflected in the electrochemical current for the reduction or oxidation of these centers as temperature is increased through the gel-to-liquid crystal-phase transition temperature (T_c). Electrical resistance of multi-bilayer films of DODAB/polystyrene sulfonate showed a large decrease when the temperature passed through the T_c region [14]. Surfactant films can also preconcentrate nonpolar organic reactants and reject multiply charged anions or cations [10, 11], providing selectivity that may be useful in specific applications.

Major difficulties have been encountered in the past in achieving direct electron transfer between electrodes and proteins. This problem is related to adsorption of the protein and macromolecular impurities on the electrode, and other possible reasons [15, 16]. Great progress

has been made in this area recently, mainly by electrode surface and protein modification [15–18]. However, specific methods must often be tailored for each protein of interest.

In 1993, we reported [19] reversible electron transfer between electrodes and the iron heme protein myoglobin imbedded in cast multi-lamellar liquid crystal films of didodecyldimethylammonium bromide (DDAB). Heretofore, reversible electron transfer from electrodes to myoglobin in solution had been accomplished only for highly purified myoglobin solutions on specially cleaned indium tin oxide electrodes [20, 21]. If enhanced electron transfer for proteins in surfactant or lipid films were to prove general, it might help solve longstanding problems in protein electrochemistry.

In this paper, we review recent progress in achieving direct electron transfer between electrodes and proteins by using surfactant films. We shall see that these films have been extended successfully to a number of redox proteins and enzymes. Specific applications, including films with catalytic activity, are also discussed. Such enzyme films have future applications as models for investigating the fundamental chemistry of normal and disease state processes, as biosensors, and as catalysts for fine chemical synthesis.

Direct electron transfer between electrodes and proteins in surfactant films

Molecular requirements

Surfactants are amphiphilic molecules with a charged or polar head group and one or more long hydrocarbon tails (Fig. 2). We rely on molecular features which cause surfactants to self-assemble spontaneously into bilayers. The molecular structure of the surfactant controls supramolecular architecture. Ability to form bilayers can be predicted by using the surfactant packing parameter $v/a_0 l_c$, where v is the volume of the hydrocarbon tail region per surfactant molecule, a_0 is the approximate headgroup area when the surfactant is in a bilayer, and l_c is the optimal chain length [22, 23]. While a_0 may be taken from X-ray diffraction or Langmuir film balance measurements, we can find the other two quantities from

$$l_c = 1.5 + 1.26n \text{ Å}, \tag{1}$$

$$v = 27.4 + 26.9n' \text{ Å}, \tag{2}$$

where n is the number of carbon atoms in the tail region, and n' is one less than the number of carbon atoms in the chain. Equation (2) must be applied for each hydrocarbon chain in the surfactant.

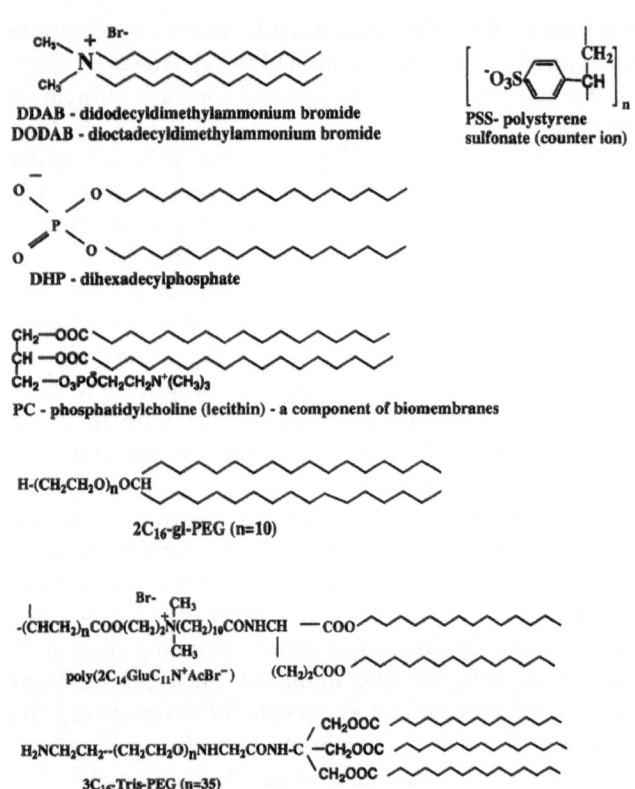

DDAB - didodecyldimethylammonium bromide
DODAB - dioctadecyldimethylammonium bromide

DHP - dihexadecylphosphate

PC - phosphatidylcholine (lecithin) - a component of biomembranes

$2C_{16}$-gl-PEG (n=10)

PSS - polystyrene sulfonate (counter ion)

poly($2C_{14}GluC_{11}N^+AcBr^-$)

$3C_{16}$-Tris-PEG (n=35)

Fig. 2 Bilayer-forming surfactants used to make ordered protein–surfactant films

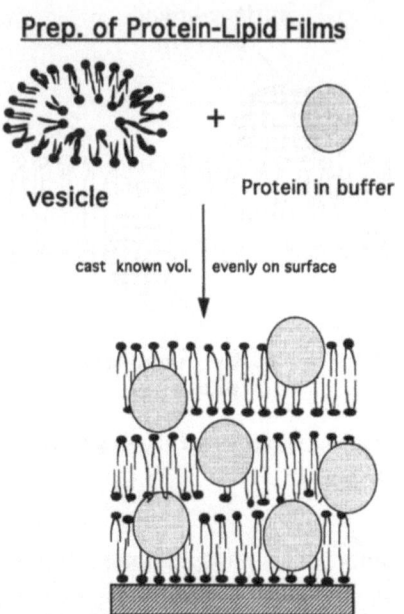

Prep. of Protein-Lipid Films

vesicle

Protein in buffer

cast known vol. evenly on surface

Fig. 3 Schematic representation of aqueous vesicle method of preparing surfactant films containing proteins, showing an idealized view of static structure

A surfactant packing parameter between $\frac{1}{2}$ and 1 is likely to lead to bilayer structures [22, 23]. These tend to be either lamellar bilayers or vesicles, which are closed spheroidal packages of surfactant [5]. A $v/a_0 l_c$ between $\frac{1}{2}$ and 1 is found in double-chain surfactants such as dialkyl-dimethylammonium salts or phosphatidylcholines of chain length 12–20 (Fig. 2). Such surfactants are insoluble in water and form bilayers when cast as films. Common single-chain surfactants such as sodium dodecylsulfate (SDS) and cetyltrimethylammonium bromide (CTAB) have $v/a_0 l_c < \frac{1}{3}$, are soluble in water, and form micelles.

Structures of surfactants suitable for insoluble films containing proteins are shown in Fig. 2. The films can be cast onto solid surfaces from aqueous vesicle dispersions prepared by sonication [24, 25] or from solutions in organic solvents [11, 19]. Films containing proteins have also been prepared from composites of bilayer-forming ionic surfactants with ionic polymers of opposite charge (Fig. 2). Examples include polystyrene sulfonate [24] or the ionomer Nafion with DDAB (cf. Fig. 2) [27].

Myoglobin in surfactant films

The iron heme protein myoglobin (Mb, MW 17 000) is essential for oxygen storage and transport in mammalian muscle. It can also be reduced and oxidized at the iron heme center, and catalyzes oxidations and reductions of organic molecules. Myoglobin in solution has been reduced electrochemically in a ligand-coupled reaction [28]. Under special conditions of protein purity and electrode preparation, Mb gives nearly reversible electrochemistry, but only on specially cleaned indium tin oxide (ITO) electrodes [20, 21].

As mentioned above, myoglobin (Mb) was the first protein for which we demonstrated that surfactant films enhanced electron exchange with electrodes. Films were initially prepared by casting films of DDAB from chloroform onto electrodes, then allowing the chloroform to evaporate [19]. These films were loaded by placing them into a Mb solution, from which the protein diffused into the film. More recently, we have prepared thin films by mixing a buffer solution containing protein with a vesicle dispersion of the surfactant, then casting this mixture onto an electrode and drying [25] (Fig. 3).

Electrochemical characterization of the films can be done by cyclic voltammetry [29–31]. In this technique, the potential difference of a working electrode in an electrochemical cell is scanned cyclically over a range of voltages. A cyclic voltammogram (CV) of a 20 μm Mb-DDAB film on a pyrolytic graphite (PG) electrode [19, 32] shows that a scan between 0.3 and − 0.5 V vs. SCE gives a distinct peak (Fig. 4c) for the MbFe(III)/MbFe(II) reduction. The reverse scan gives a negative peak characteristic of the

Progr Colloid Polym Sci (1997) 103:170–180
© Steinkopff Verlag 1997

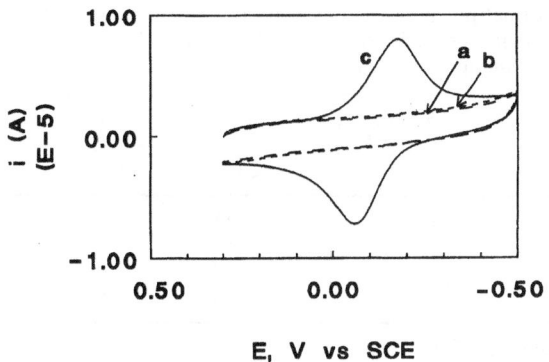

Fig. 4 Cyclic voltammograms at $100 \, \text{mV s}^{-1}$ and 25°C: (a) pH 5.5 buffer containing no protein on a bare PG electrode; (b) 25 μM Mb purified by ultrafiltration in buffer on bare PG; (c) ca. 20 μm Mb-DDAB film on PG in buffer, no Mb in solution (adapted from Ref. [32])

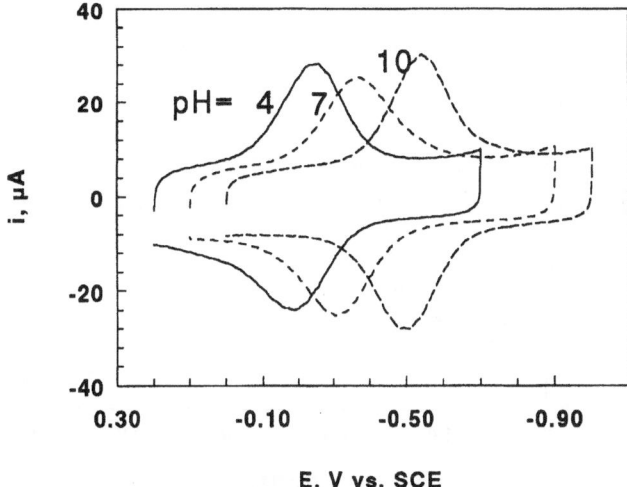

Fig. 5 Cyclic voltammograms of ca. 1 μm Mb-DLPC film on PG electrode at $2 \, \text{V s}^{-1}$ and 25°C in buffers of different pH $+0.1$ M NaBr (adapted from Ref. [33])

MbFe(II)/MbFe(III) oxidation. The electrode reaction is

$$MbFe(III) + e^- \rightleftharpoons MbFe(II) \,. \qquad (3)$$

CV peaks for Mb-DDAB films decreased less than 20% upon storage of the electrode in buffer for a month. A bare PG electrode placed into buffer or a myoglobin solution gives no peaks (Figs. 4a, b).

The shape of the CV for this rather thick Mb-DDAB film reflects diffusion-controlled electrochemistry. The peaks are not symmetric (Fig. 4c), and have a characteristic "diffusional tail". Furthermore, peak current is proportional to the square root of scan rate, as predicted by linear diffusion theory [29, 30].

Diffusion in this case refers to transport of charge through the film. The question arises as to the mechanism of this process. When a 20 μm film containing no protein was placed into a Mb solution, Mb reached the electrode in about 5 s, consistent with the diffusion coefficient of $5 \times 10^{-7} \, \text{cm}^2 \, \text{s}^{-1}$ estimated from CV [19]. This "breakthrough" experiment suggests that Mb diffuses physically through DDAB films. Similar experiments suggested that Mb may diffuse through films of phosphatidyl cholines, but not through films of dihexadecylphosphate [26].

CVs of Mb-phosphatidylcholine films (Fig. 5) have more symmetric peak shapes, and equal reduction and oxidation peak heights which were linear functions of scan rate. These characteristics indicate thin layer electrochemical behavior [29–31] in which all electroactive MbFe(III) in the films is converted to MbFe(II) on the forward CV scan. Unlike Mb-DDAB films, those made with phosphatidylcholines became thinner after soaking in buffer for about 1 h, but then remained stable. Thinning is a result of the slight solubility of phosphatidylcholines in water.

The final thickness of these films was estimated at 0.5–1 μm by scanning electron microscopy.

Reversible changes in voltammetric (Fig. 4) and spectroscopic [33] properties of Mb-surfactant films when varying pH showed that Mb within the films is subject to control by the acidity of the external solution. In control experiments, pH-independent voltammetry was found for ferrocene amphiphiles in films of DDAB or phosphatidylcholines, suggesting that pH-dependent voltammetry of Mb-surfactant films is characteristic of Mb, and not of the surfactant [33].

The pH dependence of electrochemical parameters of Mb-surfactant films has been investigated with respect to the detailed mechanism of electron transfer [33]. At pH 5–8, protonation of MbFe(III)–H_2O occurs prior to electron transfer. A protonated form of MbFe(III) is the actual electron acceptor, which may be a rapidly formed, partly unfolded conformational intermediate.

Thicker Mb-DDAB films (e.g. 20 μm) show the influence of gel-to-lamellar liquid-crystal-phase transitions in voltammetric currents obtained under diffusion control. Changes in film fluidity with temperature were observed by monitoring the limiting current obtained by normal pulse voltammetry (NPV). This limiting current depends on the square root of the charge transport diffusion coefficient, but not on the rate of the electron transfer reaction [29]. For example, Mb-DDAB films in pH 5.5 buffer [19] gave relatively small NPV limiting currents at T below the phase transition temperature (T_c), where they are in the rigid gel state. As temperature increases, a sigmoid shaped discontinuity appeared in plots of current vs. T (Fig. 6), signaling the conversion of the film to the liquid crystal

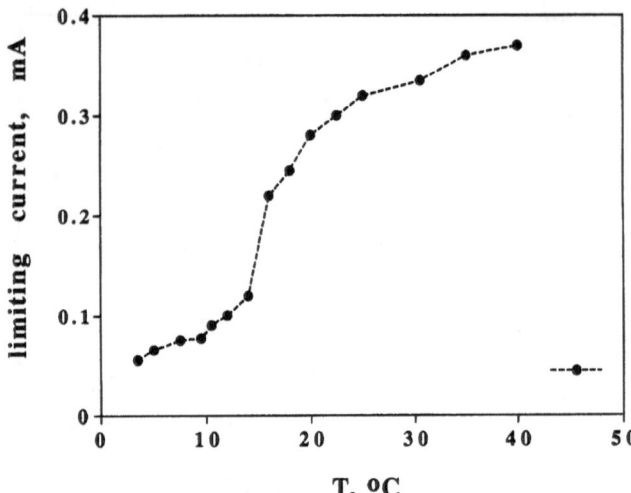

Fig. 6 Influence of temperature on limiting current from normal pulse voltammetry (pulse width 10 ms) for Mb-DDAB films on PG in pH 5.5 buffer (adapted from Ref. [19])

Table 1 Formal potentials and electron transfer rate constants for thin (0.5–1 μm) myoglobin–surfactant films on PG electrodes in pH 7 buffer containing no protein

Surfactant	Average k_s [s^{-1}] (from SWV)	Average $E^{0'}$ [mV] vs. NHE	
		CV	SWV
DDAB[a]	31	11	1
DHP[b]	90	−59	−68
DMPC[a]	60	−87	−102
DLPC[a]	60	−90	−102

[a] Data from Ref [33]
[b] Data from Ref [26] $E^{0'}$ extrapolated from pH 7.5 and pH dependence

Table 2 Formal potentials and electron transfer rate constants for 20 μm films of myoglobin and DDAB on different electrodes in pH 5.5 buffer containing no protein[a]

Electrode	$10^3 k^{0'}$ [cm s^{-1}]	$E^{0'}$ [mV] vs. NHE
Pt	11	120
PG	7	90
Au	3	50
ITO	3	−50

[a] Data from Ref. [32] obtained by CV

phase. The rising portion of this curve begins very near to the T_c value of about 12°C for Mb-DDAB films prepared with pH 5.5 buffers.

Cyclic and pulsed voltammetric studies of Mb-surfactant films have been used to obtain electrochemical parameters such as electron transfer rate constants and formal potentials ($E^{0'}$), i.e. apparent standard potentials under given experimental conditions. Recent work has shown that a Gaussian distribution model for protein molecules with slightly different $E^{0'}$-values fits voltammetric data in thin surfactant films [26, 33]. This model was used with nonlinear regression to extract average $E^{0'}$ and k_s values from square wave voltammograms (SWV). Simple models for voltammetry of single species confined to the electrode surface did not fit the data, but formal potentials could be estimated from the midpoints of CV cathodic and anodic peaks.

$E^{0'}$ depends rather strongly on the type of surfactant used. The $E^{0'}$ values for thin films of three different surfactants (Table 1) are all more negative than 50 mV vs, NHE found for Mb dissolved in pH 7 buffer [28]. The cationic surfactant DDAB gave the most positive $E^{0'}$, followed by anionic DHP, then the zwitterionic phosphatidylcholines. $E^{0'}$ also depended on electrode material (Table 2), in the order Pt > PG > Au > ITO. These data seem consistent with an electric double-layer influence on the electrode potential felt by the protein, dependent on electrode material and surfactant type.

Electron transfer rate constants for the proteins estimated from voltammograms of the films are difficult to interpret in a fundamental way. In particular, apparent kinetic constants may depend on various film properties

including fluidity, ion transport kinetics and energetics [34–36]. Nevertheless, the apparent rate costants k_s (in s^{-1}) for thin films and $k^{0'}$ (in cm s^{-1}) for thin films can be used comparatively to gain insight into the influence of surfactant types (Table 1) and electrode material (Table 2) on the efficiency of electron transfer in the films.

Comparisons must take into account that at the present level of analysis, the kinetic constants are reproducible to roughly ± 20%. In thin Mb films, values of k_s follow the trend DHP > DMPC > DDAB, but the dependence is not strong (Table 1). Similarly, the dependence of $k^{0'}$ on surfactant type was weak for thick liquid crystal films of most of the surfactants in Fig. 2 at pH 5.5 [24]. The dependence of $k^{0'}$ on electrode material measured for thick films was also weak, and we can conclude only that Pt, PG > Au, ITO (Table 2). Surfactant films on ITO are much less stable than on the other electrodes [32], so that metal and carbon electrodes are preferable except for spectroelectrochemistry.

Finally, Mb films of DDAB/PSS and 3C$_{16}$-Tris-PEG ($n = 35$) prepared by casting and by Langmuir–Blodgett film transfer were compared. In the liquid crystal phase, $E^{0'}$ values were 15–35 mV more, positive, and $k^{0'}$ was about 3-fold larger for LB films compared to cast films [24]. This suggests that the presumably higher order afforded by the LB method provides small improvements in electrochemical properties. For practical applications,

Progr Colloid Polym Sci (1997) 103:170–180
© Steinkopff Verlag 1997

these improvements do not seem large enough to justify the increased complexity and experimental difficulty of the LB method.

Cytochrome P450$_{cam}$ in surfactant films

The cytochromes P450 are a large family of iron heme enzymes that metabolize drugs and may activate pollutants toward carcinogenesis in a wide range of organisms [37]. One of the easiest of these enzymes to obtain is the bacterial cytochrome P450$_{cam}$ (cyt P450$_{cam}$).

We recently found that thin films of DMPC or DDAB containing cyt P450$_{cam}$ on PG electrodes gave reversible reduction–oxidation peaks by cyclic voltammetry (CV) in oxygen-free buffers (Fig. 7a1). Similar results were obtained in DDAB films, with formal potential ($E^{0'}$) from CVs about 100 mV more positive than in DMPC. Peaks were reproducible during a month's storage in buffer. In contrast, bare PG electrodes in cyt P450$_{cam}$ solutions gave no CV peaks, and were indicative of very slow electron transfer (Fig. 7a2).

CVs of cyt P450 films treated with excess carbon monoxide CO showed that the midpoint potential of the CVs (E_m) shifted +61 mV in DMPC films (Fig 7b2) and +45 mV in DDAB films. After removal of CO, a CV (Fig. 7b3) nearly identical to the reversible CV of Cyt P450$_{cam}$ (cf. Fig. 7b1) was found. Since the cyt P450Fe(II)-CO complex of rabbit liver cyt P450 has a formal potential about 150 mV positive of that of P450Fe(II) [39], the shift in the position of CVs of films in the presence of CO suggests the formation of Cyt P450$_{cam}$Fe(II)–CO.

Fig. 7 may be explained by the following equations:

$$\text{cyt P450Fe(III)} + e^- \rightleftharpoons \text{cyt P450Fe(II)}, \qquad (4)$$

$$\text{cyt P450Fe(II)} + CO \rightleftharpoons \text{cyt P450Fe(II)} - CO, \qquad (5)$$

$$\text{cyt P450Fe(II)} - CO \rightleftharpoons \text{cyt P450Fe(III)} + e^- + CO. \qquad (6)$$

Reversible electron transfer occurs (Figs. 7a1 and b1) in the absence of CO. With CO present, E_m shifted positive under the influence of the known rapid equilibrium with CO (Eq. (5) and Fig. 7b2). Thus, these results suggest that the initial electron transfer (Eq. (4)) involves the heme iron of cyt P450$_{cam}$ (Eq. (4)).

Films containing other proteins

Surfactant films facilitate electron transfer with electrodes for a number of proteins in addition to Mb and cyt P450$_{cam}$. The protein hemoglobin (Hb) was incorporated

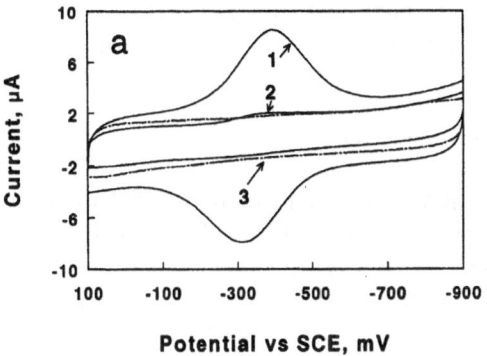

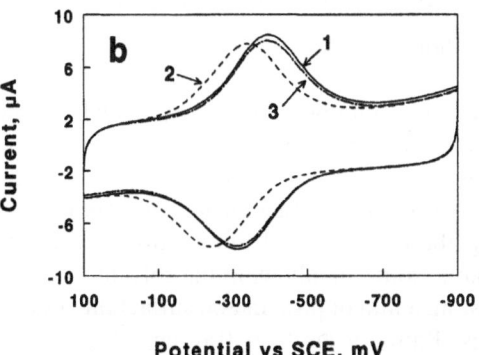

Fig. 7 Cyclic voltammograms on basal plane PG electrodes at 100 mV s^{-1} in pH 7 buffer +0.1 M KCl. (a) (a1) substrate-free cyt P450$_{cam}$–DMPC film in oxygen-free buffer containing no enzyme; (a2) bare electrode in oxygen-free buffer containing 40 μM cyt P450$_{cam}$; (a3) DMPC film in oxygen-free buffer containing no enzyme. (b) (b1) substrate-free cyt P450$_{cam}$–DMPC film in oxygen-free buffer containing no enzyme; (b2) same electrode as in (b1) after solution was purged with CO for 5 min; (b3) same electrode as in (b2) after solution was purged with N$_2$ for 20 min to remove CO. (adapted from Ref. [38])

into stable thin films on PG electrodes by the vesicle dispersion method (Fig. 3) using didodecyldimethylammonium bromide (DDAB). The heme Fe(III)/Fe(II) couple of Hb had a much greater electron-transfer rate in these liquid crystal films than on bare PG. Electrochemical diffusion studies suggested that diffusion of Hb is much slower than diffusion of Mb in DDAB films.

Bianco et al. examined the electrochemical properties of lipid films containing c-type of cytochromes [41–43], and investigated pH effects. Cytochrome c, cytochrome c$_3$, and cytochrome c$_{553}$ incorporated into phosphatidylcholine–cholesterol films doped with lauric acid gave nearly reversible voltammograms.

The ion–sulfur protein ferrdoxin from spinach in phosphatidylcholine–cholesterol films doped with dodecylamine or DODAB gave reversible CVs [44]. Chlorella ferredoxin gave reversible voltammograms in films of several of the synthetic surfactants in Fig. 2 [45, 46].

Films of proteins and Nafion

The polymer Nafion is an ionomer with $< 15\%$ ionizable sulfonate groups per monomer unit. It has a partly hydrophobic character and a very high affinity for hydrophobic cations [47, 48]. Nafion films feature segregation of hydrophobic and hydrophillic regions, and can be considered as insoluble "polymeric surfactants".

$$-(CF_2CF_2)_x-(CFCF_2)_y-$$
$$O-(CF_2)_3-O-CF_2CF_2-SO_3^- Na^+$$

Structure of Nafion

Stable, functional composite films were made from the ionomer Nafion, water-insoluble surfactants, and heme proteins or the enzyme alcohol dehydrogenase [27]. The films with the best electrochemical behavior were formed by first casting about $1-2$ μm of Nafion onto an electrode from an ethanolic solution, allowing the solvent to evaporate, then casting a film of protein and surfactant vesicles on top of this. Rates of electron transfer between PG electrodes and Mb and Hb and in these composite films were much larger than for Hb and Mb in solution on Nafion-coated or bare PG electrodes. The vesicle–protein dispersions apparently fill hydrophilic pockets in the porous Nafion film, allowing protein to gain proximity to the electrode for electron transfer reactions. DDAB is arranged in lamellar liquid crystal bilayers in these films, undoubtedly stabilized by hydrophobic and hydrophilic interactions with Nafion.

Films containing Nafion, DDAB, and hemoglobin or myoglobin retained 90% of their initial redox activity during 4 weeks storage at 4°C. These films were more stable than films of the proteins and DDAB alone. Films of Nafion–lecithin–Cyt c on PG also had reversible electrochemical properties, but were much less stable.

After crosslinking the enzyme alcohol dehydrogenase (ADH) with dilute glutaraldehyde, Nafion–lipid–ADH films stored for 2 days at room temperature retained twice the activity of the enzyme in a solution at 4°C. Crosslinked Nafion–lipid–ADH films had about 80% of their original activity for oxidation of ethanol after storage for 4 days [27].

Supramolecular film structures

A multitechnique approach was used to characterize the structures of protein–surfactant films. Important

questions to be answered about these films include: (i) How are the surfactants ordered? (ii) Do the proteins retain their native conformations? (iii) Are the proteins specifically oriented? The most detailed analysis of film structure has been done for Mb-surfactant films [19, 24, 25, 27, 33].

Phase tansitions

Evidence for surfactant aggregation into bilayers in protein–surfactant films was obtained by observing gel-to-liquid crystal-phase transitions by differential scanning calorimetry (DSC). This phase transition is related to the onset of fluidity of the hydrocarbon tails for surfactants arranged in bilayers [5–7]. Phase transition temperatures (T_c) of Mb-surfactant films [19, 24, 25] were observed for all the surfactants in Fig. 2. For a given surfactant, T_c values of films were within several °C of T_c values for the corresponding vesicle dispersions. These results indicate that all of the surfactants are arranged in bilayers in the films. The presence of the protein does not seem to influence T_c in any consistent manner.

Protein conformation

UV-Vis, ESR, and reflectance FT-IR spectroscopy were used to monitor protein conformation in the films. The Soret band for absorbance of visible light by the Fe(III)heme group of native Mb occurs at about 409 nm in solution, and at 410–414 nm for films of Mb alone. This band appears between 412 and 415 nm for films of the surfactants in Fig. 2 [19, 24]. Completely denatured Mb gives a band at ≤ 400 nm. Thus, results are consistent with native Mb in the surfactant films. ESR spectra are fully consistent with this view [24, 49], and also indicate that the heme iron is in the high spin form with water as the sixth axial ligand.

Shapes of amide infrared bands can be used to detect conformational changes in the polypeptide backbone of Mb [50–52]. The amide I band ($1700-1600$ cm^{-1}) is caused by C=O stretching of peptide linkages. Amide bands of proteins consist of many overlapped components, and resolution enhancement of spectra aids in detecting conformational differences. We illustrate studies of Mb conformation in films by using second derivative amide I reflectance spectra of Mb-DMPC films (Fig. 8).

The second derivative of a symmetric positive peak has a negative peak at the position of the original maximum, in between two smaller positive maxima. The strong negative peak at about 1659 cm^{-1} in the second derivative amide I spectra of pure Mb films without surfactant at pH 5–7

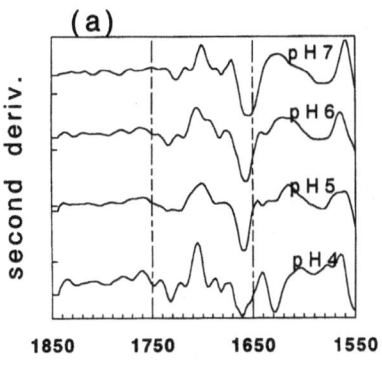

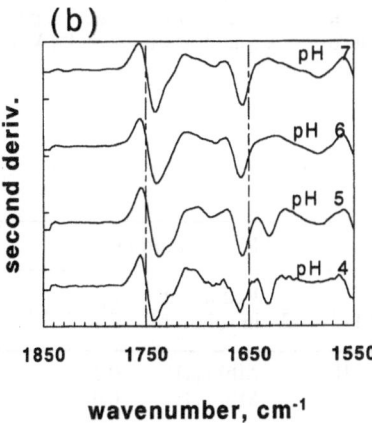

Fig. 8 Second derivative reflectance infrared spectra at source incidence angle 60° of thin films prepared using buffers from pH 4 to 7 and cast onto vapor deposited Al: (a) Mb film; (b) Mb-DMPC film (adapted from Ref. [33])

(Fig. 8a) is assigned to α-helix structures [51–53]. At pH 4, this peak becomes smaller and is accompanied by a new band close to 1630 cm^{-1}, assigned to extended disordered features of the polypeptide backbone.

Second derivative FT-IR spectra of Mb-DMPC films (Fig. 8b) at pH 6 and 7 are similar to those of Mb alone. The band at 1742 cm^{-1} is from DMPC ester carbonyl stretch. As pH decreases from 7 to 4, the α-helix band for Mb-DMPC films at 1657–1659 cm^{-1} decreases and is accompanied by a new band at 1630 cm^{-1} indicating increased protein disorder. This latter peak is clearly evident at pH 5, and is stronger at pH 4 (Fig. 8b). The new band suggests unfolding of helices at low pH compared to the native Mb structure. Second derivative spectra for Mb-DDAB also showed a band at 1630 cm^{-1} at pH 4, but not at pH 7.5, confirming increased extended chain features as pH decreased.

FT-IR spectra confirmed that between pH 5.5. and 7.5, Mb in the films has a secondary structure similar to the native state, with about 75% helix [33]. Partial unfolding

of the protein involving loss of helices is found at pH 4 in surfactant-free films (Fig. 8a), and at pH values up to 5.5 in DDAB and PC films (Fig. 8b). Roughly, 20% helical content is lost in going from pH 7.5 to 4 [33]. The surfactant films facilitate partial unfolding at pH values slightly higher than in the absence of surfactant.

Angular-dependent reflectance FT-IR analysis of the DDAB CH stretching and bending bands of Mb-DDAB films provided an estimate of the tilt angle of the hydrocarbon chains. A tilt angle of about 30° with respect to the film normal was found [19]. This is similar to tilt angles of phospholipid bilayers in crystals [7], and provides secondary evidence for surfactant ordering in the films.

Protein orientation

UV-Vis linear dichroism and ESR anisotropy provided information about the orientation of Mb in films [25]. Both experiments measure the dependence of spectroscopic features on the orientation of the film with respect to an instrumental frame of reference.

In linear dichroism (LD), the film is tilted with respect to the direction of light propagation, and LD is the difference in absorbance for the Mb Soret band when using perpendicular and parallel plane polarized light:

$$LD = A_{\parallel} - A_{\perp}. \tag{7}$$

Appropriate analysis of the data [25] provides the order parameter S:

$$S = (1 - 3\cos^2 \varphi)/2 \tag{8}$$

from which φ, the angle between the transition moment vector in the heme plane and the normal to the film plane, can be obtained. Thus, φ is the orientation of the Mb heme plane to the film normal.

In ESR, the film plane orientation with respect to the magnetic field is varied. Changing ratios of $g = 2$ and $g = 6$ peaks for different orientations of Mb surfactant films reflected the orientation of the heme plane [25]. The orientation angle of the heme was estimated by matching computer simulations to the experimental spectra. The disadvantage of the ESR method is that spectra must be taken at 5–10 K for acceptable signal to noise, and the data analysis is more complex than for LD.

ESR anisotropy at 10 K and Soret band LD at room temperature established the orientation of Mb in the films. The two techniques gave good agreement for the angle of orientation between the Mb heme plane and the film normal (Table 3). Almost no dependence of Mb orientation on the type of head group and phase state of these films was found. The charge on the head group did not

Table 3 Orientations of heme plane with respect to the film normal from linear dichroism (LD) and ESR anisotropy[a]

Surfactant	φ deg	
	LD	ESR
DDAB	59 ± 2	55
DHP	61 ± 3	60
DMPC	60 ± 3	60
DLPC		60

[a] Data from Ref. [25]

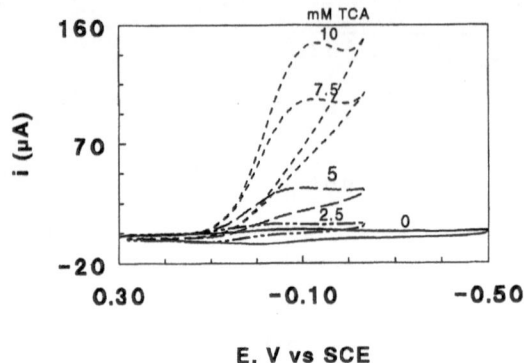

Fig. 10 Steady-state cyclic voltammograms at 100 mV s^{-1} in pH 5.5 buffer of Mb-DDAB films with trichloroacetic acid (TCA) at 0, 2.5, 5, 7.5, and 10 mM (adapted from Ref. [54]). The direct reduction peak for TCA on a DDAB-PG electrode is found at about -1.4 V vs. SCE (not shown)

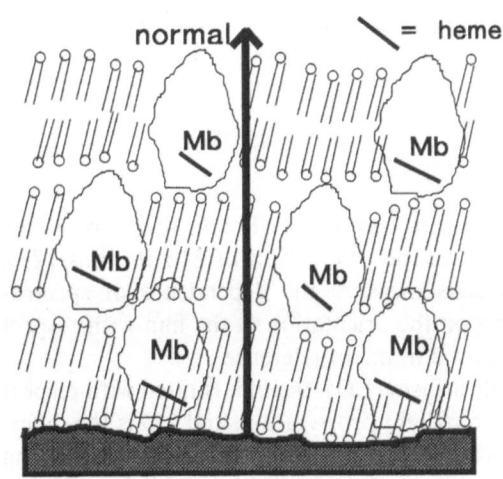

Fig. 9 Idealized model of several bilayers of a cast Mb-surfactant film, (adapted from Ref. [25]). The exact position of Mb within the bilayer regions is uncertain

Table 4 Comparison of reduction rates of organohalides by MbFe(II) at pH 5.5[a]

Mb-DDAB films			Mb in solution	
Reactant	$\Delta E_{1/2}^{\text{b)}}$ [mV]	k_1' [$M^{-1} s^{-1}$]	Reductant in solution	k_1 [$M^{-1} s^{-1}$]
Cl_3CCOOH	1300	8×10^3	MbFe(II)	0.2
$BrCH_2C_2CH_2Br$	1000	2×10^2	MbFe(II)	0.07

[a] Data from Ref. [54]
[b] Difference in potentials between catalyzed and uncatalyzed electrochemical reductions of the organohalides

influence orientation, suggesting that a significant fraction of the protein may be imbedded in hydrophobic surfactant bilayer regions [25]. An idealized structural model of Mb-surfactant films consistent with experimental findings is given in Fig. 9.

Catalysis with Mb and Cyt P450cam

Mb-DDAB films were used to catalyze reductions of several organohalides with significant lowering of activation free energy [54]. Myoglobin is made to act as a redox enzyme in these films. Ethylene dibromide and trichloroacetic acid were dehalogenated by MbFe(II) generated electrochemically in the films, with apparent reaction rates much larger than with MbFe(II) in solution.

A cyclic voltammogram of a Mb-DDAB film shows large increases in the peak current for reduction of MbFe(III) when increasing amounts of trichloroacetic acid are added to the solution (Fig. 10). Reduction occurs at potentials more than 1 V positive of the corresponding direct reductions at bare electrodes. These results are characteristic of catalytic reduction of trichloroacetic acid by the pathway [54]

$$MbFe(III) + e^- \leftrightharpoons MbFe(II) \quad k_1^{0'} \text{ (at electorde)}, \tag{9}$$

$$MbFe(II) + Cl_3CCOOH \rightarrow Cl_2\dot{C}COOH$$
$$+ MbFe(III) + Cl^- \quad (k_1, \text{rds}), \tag{10}$$

$$MbFe(II) + Cl_2\dot{C}COOH \rightarrow MbFe(III)$$
$$+ Cl_2^-CCOOH(\text{fast}), \tag{11}$$

$$Cl_2^-CCOOH + H^+ \rightarrow Cl_2CHCOOH \tag{12}$$

Voltammetric data were used to estimate rate constant k_1', the apparent k_1 in the films uncorrected for organohalide concentration (Table 4). Values of k_1' are much greater than the chemical rate constants for reactions with MbFe(II) in solution. Rates are enhanced by preconcentration of the organohalides in the films [54]. These ordered protein–surfactant films provide reaction environments resembling biomembranes.

A reduced form of MbFe(II) was also produced in these films at about -1.1 V vs. SCE, and used to dechlorinate tetra- and tri-chloroethylenes. Products of all the catalytic reductions were similar to those obtained from reductions by anaerobic bacteria or the enzyme cytochrome P450. Reactions proceeded for thousands of catalyst turnovers, and apparent rate constants were in the range $2 \times 10^2 - 10^4$ $M^{-1} s^{-1}$.

Very recent results demonstrated that Cyt P450$_{cam}$-surfactant films catalyzed electrochemically driven reduction of trichloroacetic acid [38]. Also, cyt P450Fe(II) formed at the electrode reacts with oxygen to give cyt P450Fe(II)–O$_2$, which appears to be reduced electrochemically. Thus, electrochemical reduction of cyt P450Fe(III) in the presence of oxygen mimics *in vivo* electron acceptance from enzymic redox partners by cyt P450Fe(III) and cyt P450Fe(II)-O$_2$ during catalytic oxidations.

Protein–DNA films

Recently, collaboration between the author, A-E. Nassar, and N. Nakashima resulted in the preparation of stable films of calf thymus double-stranded DNA and proteins on electrodes [55]. Direct electron transfer was achieved for myoglobin or hemoglobin in DNA films on pyrolytic graphite (PG) electrodes. As with the surfactant films, enhanced electron transfer rates were achieved compared to bare PG electrodes with proteins in solution. DNA films also extracted proteins from solution. Mb appears to diffuse through pure DNA films such faster than Hb. Conformational changes in both DNA and protein upon binding are likely within these films. DNA–protein films may find applications in electrochemical and spectroscopic studies of DNA–protein and DNA–enzyme–substrate interactions, and as biosensors for proteins.

Summary and future prospects

This review shows that stable, ordered surfactant films can facilitate electron exchange between proteins and electrodes. Heme proteins and ferredoxins exhibited enhanced electron transfer rates in these films, and it is likely that the method may prove to be somewhat general. Both surfactants and proteins have average specific orientations in the films. The protein environment is similar to that in biological membranes. These films have good electrochemical properties because of liquid crystal fluidity and strong adsorption of surfactant at the electrode–film interface, eliminating adsorption of proteins at the electrode which might otherwise block electron transfer.

Mb-surfactant films were used for the catalytic reduction of organohalide pollutants. Moreover, preliminary results in our laboratory show that oxidative catalysis should also be possible. Thus, films containing myoglobin, and the more difficult to obtain cyt P450$_{cam}$, provide models to study chemical process which may activate pollutants for carcinogenesis in the human liver. Cyt P450 enzymes thought to activate pollutants are membrane bound, so the lipid films provide simplified but realistic environments for chemical studies of this enzyme catalysis.

Finally, general applications of the films described herein should emerge. Unstable enzymes such as alcohol dehydrogenase can be stabilized by crosslinking within Nafion–lipid films [27], providing a possible strategy for biosensor design. Successful studies of the detailed reduction mechanism of Mb in surfactant films by combining electrochemical and spectroscopic probes [33] suggest that a variety of biochemical redox reactions would be amenable to similar studies in these films. For example, we are currently beginning studies of cyt P450-catalyzed pollutant oxidations relevant to carcinogenesis, as well as studies of the reaction of cyt P450 and its enzymic redox partners in the films. We feel that the future may reveal a wide variety of applications for ordered protein–surfactant films.

Acknowledgments The author's research described herein was supported by grant No. ES03154 from the National Institute of Environmental Health Sciences (NIEHS), NIH. Contents are solely the responsibility of the author and do not necessarily represent official views of NIEHS, NIH. The author is also grateful to students and colleagues named in joint publications whose valuable contributions and insight made possible much of the work described.

References

1. Nakashima N, Ando R, Kunitake T (1983) Chem Lett 1577–1580
2. Kunitake T, Shimomura M, Kajiyama T, Harada A, Okuyama K, Takayanagi M (1984) Thin Solid Films 121:L89–91
3. Kunitake T, Tsuge A, Nakashima N (1984) Chem Lett 1783–1786
4. Shimomura M, Kunitake T (1984) Polymer J 16:187.
5. Fendler JH (1982) Membrane Mimetic Chemistry. Wiley, New York
6. Kotyk A, Janacek K, Koryta J (1988) Biophysical Chemistry of Membrane Function. Wiley, Chichester, UK
7. Cevc G, Marsh D (1987) Phospholipid Bilayers, Wiley, New York
8. Rusling JF, Zhang H (1991) Langmuir 7:1791–1796
9. Hu N, Rusling JF (1991) Anal Chem 63:2163–2168

10. Zhang H, Rusling JF (1993) Talanta 40:741–747
11. Rusling JF (1994) Microporous Mater 3:1–16
12. Nakashima N, Masuyama M, Mochida M, Kunitake M, Manabe O (1991) J Electroanal Chem 319:355–359
13. Nakashima N, Wake S, Nishino T, Kunitake M, Manabe O (1992) J Electroanal Chem 333:345–351
14. Nakashima N, Eda H, Kunitake M, Manabe O, Nakano K (1990) J Chem Soc Chem Commun 443–444
15. Armstrong FA, Hill HAO, Walton NJ (1988) Acc Chem Res 21:407–413
16. Armstrong FA (1990) In: Bioinorganic Chemistry, Structure and Bonding, Vol 72. Springer, Berlin, pp 137–221
17. Heller A (1990) Acc Chem Res 23:128–134
18. Tarlov MJ, Bowden EF (1991) J Am Chem Soc 113:1847–1894
19. Rusling JF, Nassar A-EF (1993) J Am Chem Soc 115:11891–11897
20. Taniguchi I, Watanabe K, Tominaga M, Hawkridge FM (1992) J Electroanal Chem 333:331–338
21. Taniguchi I, Kurihara H, Yoshida K, Tominaga M, Hawkridge FM (1992) Denki Kagaku 60:1043–1049
22. Evans DF, Mitchell DJ, Ninham BW (1986) J Phys Chem 90:2817–2825
23. Isrealachvili J (1992) Intermolecular and Surface Forces, 2nd Ed. Academic Press, London
24. Nassar A-EF, Narikiyo Y, Sagara T, Nakashima N, Rusling JF(1995) J Chem Soc Faraday Trans 91:1775–1782
25. Nassar A-EF, Zhang Z, Chynwat V, Frank HA, Rusling JF, Suga K (1995) J Phys Chem 99:11013–11017
26. Zhang Z, Rusling JF (1996) Biophys Chem in press
27. Huang Q, Lu Z, Rusling JF (1996) Langmuir, 12:5472–5480
28. King BC, Hawkridge FM, Hoffman BM (1992) J Am Chem Soc 114:10603–10608
29. Bard AJ, Faulkner LR (1980) Electrochemical Methods. Wiley, New York
30. Murray RW (1984) In: Bard AJ (ed) Electroanalytical Chemistry, Vol 13. Marcel Dekker, New York, pp 191–368
31. Rusling JF, Suib SL (1994) Adv Mater 6:922–930
32. Nassar A-EF, Willis WS, Rusling JF (1995) Anal Chem 67:2386–2392
33. Nassar A-EF, Zhang Z, Hu N, Rusling JF, Kumosinski TF (1997) J Phys Chem in press
34. Majda M (1992) In: Murray RW (ed) Molecular Design of Electrode Surfaces, Techniques of Chem, Vol 22. Wiley, New York, pp 159–206
35. Andrieux CP, Saveant J-M (1992) In: Murray RW (ed) Molecular Design of Electrode Surfaces, Techniques of Chem., Vol 22. Wiley, New York, pp 207–270
36. Feldberg SW, Rubinstein I (1988) J Electroanal Chem 240:1–15
37. Schenkman JB, Greim H (eds) (1993) Cytochrome P450. Springer, Berlin
38. Zhang Z, Nassar A-EF, Lu Z, Schenkman JB, Rusling JF (1997) J Chem Soc, Faraday Trans. in press
39. Guengerich FP, Ballou DP, Coon MJ (1975) J Biol Chem 250:7405–7414
40. Lu Z, Huang Q, Rusling JF (1996) J Electroanal Chem in press
41. Bianco P, Haladjian J (1994) J Electroanal Chem 367:79–84
42. Bianco P, Haladjian J (1994) Electrochim. Acta 39:911–916
43. Hanzlik J, Bianco P, Haladjian J (1995) J Electroanal Chem 380:287–290
44. Bianco P, Haladjian J (1995) Electroanalysis 7:442–446
45. Tominaga M, Yanagimoto J, Nassar A-EF, Rusling JF, Nakashima N (1996) Chem Lett 523–524
46. Nassar A-EF, Rusling JF, Tominaga M, Yanagimoto J, Nakashima N (1997) J Electroanal Chem, in press
47. Martin CR, Freiser H (1981) Anal Chem 53:902
48. Szentirmay MN, Martin CR (1984) Anal Chem 56:1898
49. Rusling JF, Nassar A-EF, Kumosinski TF (1994) In: Kumosinski TF, Liebman MN (eds) Molecular Modelling. ACS Symposium Series 576. Amer Chem Soc, Washington, DC, pp 250–269
50. Susi H, Byler DM (1986) Methods Enzymol 130–290
51. Kumosinski TF, Unruh JJ (1996) Talanta 43:199–219
52. Rusling JF, Kumosinski TF (1996) Nonlinear Computer Modelling of Chemical and Biochemical Data. Academic Press, New York, pp 117–134
53. Dong A, Huang P, Caughey WS (1990) Biochemistry 29:3303–3308
54. Nassar A-EF, Bobbitt JM, Stuart JD, Rusling JF (1995) J Am Chem Soc 117:10986–10993
55. Nassar A-EF, Rusling JF, Nakashima N (1996) J Am Chem Soc 118:3043–3044

Progr Colloid Polym Sci (1997) 103:181–192
© Steinkopff-Verlag 1997

A. Tarazona
S. Kreisig
E. Koglin
M.J. Schwuger

Adsorption properties of two cationic surfactant classes on silver surfaces studied by means of SERS spectroscopy and ab initio calculations

Received: 25 November 1996
Accepted: 6 December 1996

Dr. A. Tarazona (✉) · S. Kreisig
E. Koglin · M.J. Schwuger
Institute of Applied Physical Chemistry
Research Center Juelich
52425 Juelich, Germany

Abstract Surface-enhanced Raman microprobe spectroscopy (micro-SERS) and near-infrared Fourier transform SERS spectroscopy (NIR-FT-SERS) are used to study, in situ, the adsorption process of alkylpyridinium bromide (C_nPyBr) and alkyltrimethylammonium bromide (C_nTAB) adsorbed on charged silver nanoparticle surfaces. Vibrational assignment was achieved by comparison of observed band position and intensity in the Raman spectra with wave numbers and intensities from ab initio LCAO-MO-SCF Hartree–Fock calculations at the 6-31G* level.

Information on the monolayer adsorption geometry at the charged surface was obtained on the basis of this assignment. The hydrocarbon tail length was varied from C_1 to C_8 and the chain vibrations were calculated at different conformational defects (from all tans (ttttt) to all gauche (ggggg)). The adsorption behaviour was investigated as a function of surfactant concentration and alkyl chain length.

The results from of ab initio calculations, alkyl chain length dependence, and the short-range effect of SERS suggest that the C_nPB surfactant molecules are adsorbed with the ionic headgroup towards the charged surface and the hydrophobic tail is directed away from the surface. In contrast, the C_nTAB surfactants are adsorbed with the headgroup pointing to the surface and the hydrocarbon chain is with the tail running parallel to the surface. An experimental model based on head-group model compounds was performed to reproduce the SERS spectra corresponding to both kind of surfactants.

Key words SERS microprobe spectroscopy – NIR-FT-SERS spectroscopy – monolayer composition – ab initio calculations

Introduction

The adsorption of surfactants at the interface of solid and liquid phase is one of the most important phenomena that have captured the interest of surfactant research [1]. In the case that the substrates are constrained to metallic surfaces, the surfactant adsorption on metal surfaces is extremely important in a variety of fields such as adhesion, lubrication, detergency and corrosion inhibition. The experimental evaluation of the adsorption of surfactants at the solid–liquid interface has been conventionally obtained by means of adsorption isotherms. The adsorption isotherms provide quantitative measures, but little information on the aggregate structures. In recent efforts to probe this structure fluorescence decay [2], neutron reflection [3]

182
A. Tarazona et al.
Adsorption properties of two cationic surfactant on silver surfaces

and atomic force microscopy (AFM) [4] were used. Another drawback from the isotherms is the lack of dynamical information on the adsorption process. Hence, it would be ideal to have a technique that yields in situ structural information, and at the same time provides information on the dynamics.

Surface enhanced Raman scattering (SERS) spectroscopy fulfils these requirements,. Eventhough, the use of SERS as an interfacial technique is limited to just a few metals, it nevertheless has great potential to provide information on the identification of the actual species adsorbed, its orientation, its rate of adsorption/desorption, and quantification that can be extended to other metal surfaces.

SERS was initially observed in 1974 [5], and in 1977 [6, 7] the role of the surface enhancement effect was discovered. Since then much efforts have been concentrated to explain the phenomenon, and it was not before the beginning of the 80s when people started to realize the analytical qualities of SERS for applied research. The SERS effect can be briefly described as up to a 10^6 enhancement of the Raman scattering cross section of adsorbates on the roughened metal surfaces. This enormous enhancement makes SERS to a very sensitive technique for detecting adsorbed species on metal surfaces, even in submonolayer coverage.

A great deal of efforts has been directed at trying to understand this effect, which is reflected in the number of review articles concerning this topic [8–12]. Although, there is still a strong controversy about the mechanism that causes the effect, it is widely accepted that there are two contributing mechanism: the classical electromagnetic enhancement and the "first-layer effect".

In the past few years, some work has focused on the adsorption of long alkyl-chain at the solid–liquid interface using vibrational spectroscopic techniques. For instance Ohtake et al. [13] have studied the adsorption of n-alkyltrichlorosilane on SiO_2 using FT IR spectroscopy. In this context Knoll et al. [14] have shown the equivalence between infrared and SERS measurements of monolayer assemblies of long alkyl-chained molecules on a variety of rough surfaces.

The first reported work using SERS as probe to investigate the adsorption of surfactants at the solid–liquid interface dated from the beginning of the 1980s. In their report, Heard et al. [15] have shown that SERS spectra could be obtained even at submonolayer converge from n-alkylpyridinium bromide adsorbed at Ag colloid surfaces. Dendramis et al. [16] have studied the adsorption of cetyltrimethylammonium bromide (CTAB) on copper electrode surfaces. They were able to determine the CTAB adsorption rates from aqueous solution and observed no changes in the spectra as the concentration was varied

between 1×10^{-4} and 5×10^{-3} M. By applying the "surface selection rules" [17], they have proposed that the surfactant is adsorbed with the head group attached to the copper surface and the hydrocarbon tail runs parallel to the surface. Wiesner et al. [18] have published a report on CTAB adsorption on colloidal particles and its effects on the colloid stabilizer using SERS. In this work partial information on the geometry of adsorption was obtained. The main information was obtained by analyzing the SERS bands corresponding to the head group. They have proposed that the head group is attached to the colloidalparticles while the alkyl-chain points towards the solution.

Adsorption of surfactants on roughened Ag electrode surfaces have also been subject of investigations using SERS. Sun et al. [19] study the potential dependence of the SERS spectra CTAB, cetylpyridinium chloride (CPC), sodium dodecylsulfate, polyoxylethylene (23)do-decanol, Brij-35 and Triton X-100. Only the results from CTAB and CPC were extensively presented. While for CTAB a "solid-like" structure of the tail with the head group attached to the electrode surface was proposed, for CPC it was postulated that the head group is attached to the surface with the alkyl-chain pointing away from the surface. They have based this statement on the observation that the SERS bands from the pyridinium head group showed larger intensities than the bands corresponding to the alkyl-chain. This interpretation would be valid if the head group and the alkyl-chain experience an enhancement that depends only on the distance to the surface. But as already known, the enhancement for the saturated hydrocarbons is generally smaller than for aromatic molecules [12]; so even if the chain and the head group are at the same distance from the surface, the head group modes experiences a larger enhancement.

The chemical methods applied by Koglin et al. [20] have led to a significant advance in the study of the adsorption geometry of CTAB at solid–liquid interface (of a silver electrode surface). By labelling the terminal methyl group of the cetyl chain with deuterium atoms, it was possible to show that CTAB adsorbs with the headgroup attached to and the tail running parallel to the surface. Nevertheless, no explanation on the possible reason for the observed behaviour was given.

Experiment

Materials

The single-chained surfactants C_nPB $(C_nH_{2n-1}N^+$ $C_5H_5Br^-$, $n = 10, 12, 16, 18)$ and C_nTAB $(C_nH_{2n-1}N^+$ $(CH_3)_3Br^-$, $n = 6, 10, 16)$ were prepared according to the formula from Shelton et al. [21], and Norcross et al. [22], respectively. The different forms of the pyridinium surfactant were synthesized by mixing pyridine with a molar amount of the appropriate n-alkyl bromide and holding the solution for about 16 h at 110–130°C. After than the yellowish product was recrystallized from acetone and petrol ether gaining a white crystalline powder.

The trimethylammonium surfactants were synthesized from the n-alkyl bromides and a double molar amount of a 40% aqueous trimethylammonium bromide solution. The mixture was diluted with ethanol, producing a 10% solution, and heated together for one hour at 60°C. The slightly yellow product was dissolved in ethanol and precipitated with diethyl ether. All other chemical reagents were of analytical quality from E. Merck (Darmstadt, FRG).

Silver colloids

The silver hydrosols used in the SERS measurements were prepared according to prescription from Lee and Meisel [23]: 45 mg $AgNO_3$ were dissolved in triple distilled water. After heating to 100°C a solution of 50 mg sodium citrate in 5 ml water was added dropwise. The mixture was boiled for at least one hour.

Electrochemical preparation

In order to obtain micro-SERS spectra from molecules adsorbed on charged silver surfaces it is necessary to subject the electrode surface to special cleaning and roughening procedure. The first step consists of mechanically polishing the silver surface with successively finer grades of alumina down to a 0.05 μm suspension (Buehler) leading to a scratch-free mirror-like surface. Ultrasonic treatment and hydrogen generation at − 1.7 V vs. SCE is used to clean the electrode surface of attached alumina. The surface roughness needed for acquisition of the SERS spectra is provided by an ex situ roughening procedure.

Several oxidation reduction cycles (ORCs) are performed in a 0.1 M KCl solution until the sum of the total coulometric charge reached a predefined limit (130 mC/cm^2). After thoroughly rinsing the yellowish silver surface with triple distilled water the SERS substrate is ready for use.

Opto-electrochemical equipment

The opto-electrochemical cell, with a capacity of about 1 ml, is a Teflon cylinder (⊘ 15 mm, depth = 15 mm) equipped with a polycrystalline silver working electrode (⊘ 2 mm) enclosed in a Teflon holder, a Pt counter electrode (⊘ 1 mm, length = 15 mm), and a saturated calomel reference electrode (SCE). Hence, all potentials mentioned in the text below are referred to SCE.

The electrochemical equipment used for roughening and controlling the surface potential of the silver electrode consists of a potentiostat (PAR, model 173), and a function generator (PAR, model 175) as programmer for the ORC. Furthermore, a digital coulometer is connected to measure the charge transfer during the ORC.

Microprobe set-up: MOLE S 3000

The SERS measurements in the visible spectral range were obtained using a triple monochromator unit MOLE S 3000 (Instruments S.A./Jobin Yvon). The spectrometer consists of a double monochromator DHR 320 (600 grooves/mm), a main monochromator HR 640 (600 or 1800 grooves/mm), a diode array detector (E-IRY 1024), and a microscope (Olympus BH 2) with CCD camera and monitor. The microscope focuses the radiation of the laser (488 and 514 nm: model 2020-03 Ar ion laser; 633 nm: model 127 HeNe laser, both Spectra-Physics) onto the sample giving a spatial resolution of 1 μm^2. The spectral solution is 8 1/cm.

NIR-FT Raman spectrometer

The FT-SERS spectra obtained in the near infrared spectral range were measured with a Bruker FT-Raman system (RSF-100) connected to a Raman microscope (Bruker R590B). Excitation radiation of 1064 nm is provided by a Nd–YAG laser. A spectral solution of 4 1/cm was obtained while the spatial resolution was 100 μm^2.

Calculation details

For the interpretation of vibrational spectra it is of crucial importance having a reliable normal mode assignment. Hence, a series of calculations, involving either the molecules studied experimentally or a model system of them, were performed varying basis sets, theory levels, and several molecular parameters, like conformational angles or length of alkyl-chains. Most of the ab initio calculations were done using Hartree–Fock theory and a 6-31g* basis set. All calculations were performed with a Gaussian set of programs [24] on a Cray Y MP-4 supercomputer. Geometries were optimized until the individual gradients were less than 10^{-5} Hartree/Bohr and the root-mean-force less than 10^{-7} Hartree/Bohr.

With the computational resources available at the present, it is still not possible using ab initio methods to obtain the vibrational spectrum, including both frequencies and intensities, at a satisfactory level of theory with a reasonable basis set from molecules larger than 30 heavy atoms. Consequently, rather large molecules have to be modelled by considering parts of them and studying the size effects of these oversimplifications. Since the class of molecules studied in this work were cationic surfactants a considerable amount of calculations were focused on obtaining the Raman spectra of the headgroup and the alkyl-chain, as well as from surfactants having a short alkyl-chain.

Results and discussion

Normal modes assingment in the CH spectral range

The dependence of the Raman bands in the CH spectral range on the hydrocarbon chain-length from the two different cationic surfactants is evaluated using ab initio calculations. Figs. 1a and b illustrate the calculated Raman spectra of methyl-, butyl- and heptyl-pyridininium and trimethylammonium, respectively.

For methylpyridinium there are three bands at 2939, 3025 and 3039 1/cm due to CH_3 modes and four modes corresponding to in-plane vibrations of the pyridinium ring at 3067, 3080, 3083 and 3100 1/cm. In the lowest spec-

trum of Fig. 1a only five bands can be resolved, since there are two ring modes having a separation of only 3 1/cm. For buthylpyridinium (middle spectrum in Fig. 1a) the bands due to CH_2 motions are clearly seen at 2880 (symmetric CH_2 stretching mode), 2891 (asymmetric CH_2 stretching mode) and 2904 1/cm (CH_2 symmetric stretching mode). Additionally, the methylene modes are remarkably affected due to the change in chemical environment, i.e. in the first case the methylene is bound to a nitrogen atom belonging to the pyridinium ring and for butylpyridinium it is attached to a propylene group. So, the two strong methyl bands from methylpyridinium observed at 3025 and 3039 1/cm converge to a single one at 3002 1/cm (asymmetric CH_3 vibration) with lower intensity while the symmetric CH_3 vibrational mode is shifted from 2939 to 2962 1/cm keeping roughly the same intensity. The calculated Raman spectrum from heptylpyridinium (Fig. 1 top) is dominated by the asymmetric CH_2 stretching mode at 2884 1/cm. It can be seen from the spectra that the three in plane vibrations of the ring are not affected by an increase in the alkyl chain length.

The Raman spectrum obtained from tetramethylammonium (TMA, bottom of Fig. 1b) shows the symmetric CH_3 mode at 2945 and the asymmetric mode at 3032 1/cm. Fig. 1b, middle, presents the corresponding calculated Raman spectrum of butyltrimethylammonium. The bands due to CH_2 motions appear at similar positions as the corresponding vibrations of the butylpyridinium ion, while the bands originating in CH_3 vibrations dominate the spectrum. By increasing the length of the hydrocarbon chain from 4 to 7, it can be seen that the bands due to

Fig. 1 Calculated Raman spectra from **A** alkylpyridinium and **B** alkyltrimethylammonium. The hydrocarbon tail-length was varied between C_1 and C_7 for each surfactant. For a better visualization, lorentzian envelopes with a fixed line-width of 8 1/cm for all Raman bands were used. The symmetric CH_2 stretching mode for both species is marked using a vertical line and labelled with the corresponding frequency, i.e. 2875 1/cm

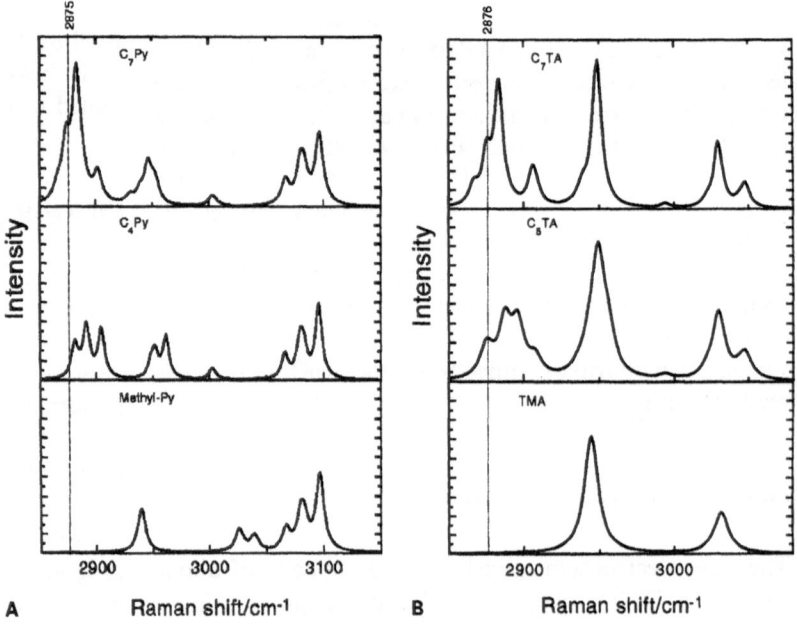

CH_2 vibrational motions gain intensity, while those coming from CH_3 keep their intensity constant. This tendency will later be used as the main feature required to interpret the SERS spectra.

All spectra presented in Fig. 1 were calculated assuming an all-trans conformation. In the experimental Raman spectra of both surfactants with hydrocarbon chains possessing more than 10 carbons, the band at 2884 1/cm (asymmetric CH_2) is the strongest band of all, followed by the one at 2875 1/cm (symmetric CH_2). The intensity ratio between these two bands can be used to study the conformation of the alkyl chain. Snyder et al. [25] have presented a very nice experimental treatment of this issue. To the best knowledge of the authors, up to now only force field calculations have been used to study this issue from a theoretical point of view [26, 27]. In this calculations the force field parameters were fitted to represent the observed frequencies.

In order to evaluate the conformational effects on the Raman bands by means of ab initio calculation, octane was used as a model for the hydrocarbon chain. Fig. 2 shows the Raman spectra from octane containing different amounts of gauche defects. The all-trans conformation spectrum shows the symmetric CH_2 mode at 2870 1/cm and the asymmetric one at 2880 1/cm. As the fraction of gauche defects increases it can be observed that the intensity of the symmetric and asymmetric bands changes and new features at lower frequencies of the asymmetric band appear (marked with an arrow in Fig. 2). These new features can unambigously be attributed to bands arising from gauche conformations. Comparing all calculated Raman spectra of octane, leads one to the conclusion that the position of the gauche defects in the molecules play a very important role in the behaviour of the Raman spectra. Raman spectra having one gauche defect (ttgtt and tgttt) show that the intensity of the symmetric band is stronger when the gauche defect is situated in the centre of the chain. Furthermore, the asymmetric band is also sensitive to the gauche position and loses more intensity in the case of ttgtt than in the case of tgttt. Furthermore, the asymmetric band is also sensitive to the gauche position and loses more intensity in the case of ttgtt than in the case of tgttt. The former shows simultaneously a larger increase of the feature related to the gauche defects. In the configurations possessing more gauche defects, the feature around the symmetric CH_2 vibration loses intensity for all conformations except when the gauche defects have substituted all the trans (ggggg). For this last configuration, the intensity of symmetric CH_2 band is a maximum. Therefore, for an all trans conformation (solid-like) the asymmetric band would be the most intense one and depending on the guache content, the symmetric band would gain intensity.

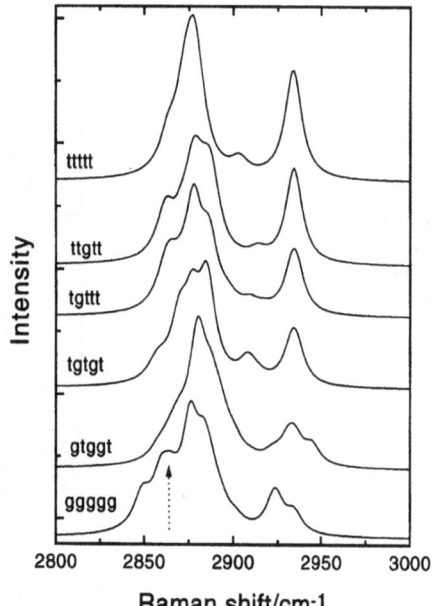

Fig. 2 Calculated Raman spectra from octane in the C–H stretching range. For all trans (ttttt) octane, octane with a single (ttgtt) gauche defects, octane with a single gauche defect (tgttt), octane with two (tgtgt) gauche defects, octane with three (gtggt) and octane with five gauche defects (ggggg). The envelopes were constructed using reasonables linewidth (8 1/cm). The arrow shows a spectral feature near the C–H symmetric stretching arising from the gauche defects

Comparing calculated and measured polycrystalline Raman

An example for the comparison between calculated and measured Raman spectra of surfactants having nearly the same alkylchain is shown in Fig. 3. The two spectra at the top of this figure are the results of an ab initio frequency/intensity calculation for heptylpyridinium and heptyltrimjethylammonium, respectively. The experimental Raman spectra of polycrystalline dodecylpyridinium and dodecyltrimethylammonium are printed at the bottom. Comparing the pyridinium spectra a good correspondence for the ring mode, at 3080 1/cm, and CH motions of the chain, occurring between 2900 and 3050 1/cm can easily be seen. The difference in the spectra can be explained by considering the occurrence of Fermi resonance of C–H modes and taking into account thermal and crystallinity interactions in real samples, which lead to line broadening.

The experimental Raman spectrum from polycrystalline dodecyltrimethylammonium is presented at the bottom of Fig. 3b, while the simulated spectrum is depicted at the top. In the latter one four features can be seen. The first one appears as a shoulder at 2866 1/cm, and has already been assigned to the CH_2 symmetric stretching mode. The

Fig. 3 Comparision between
lorentzian-envelop generated
Raman spectra, using a fixed line-
width of 8 1/cm from
heptylpyridinium and the Raman
spectrum from a polycrystalline
sample of dodecylpyridinium-
bromide. **A** shows the C–C
stretching spectral range and
B corresponds to the C–H
stretching range. The excitation
source was the 488 nm line from
an Ar ion laser, the scanning rate
was 2 1/cm/2s

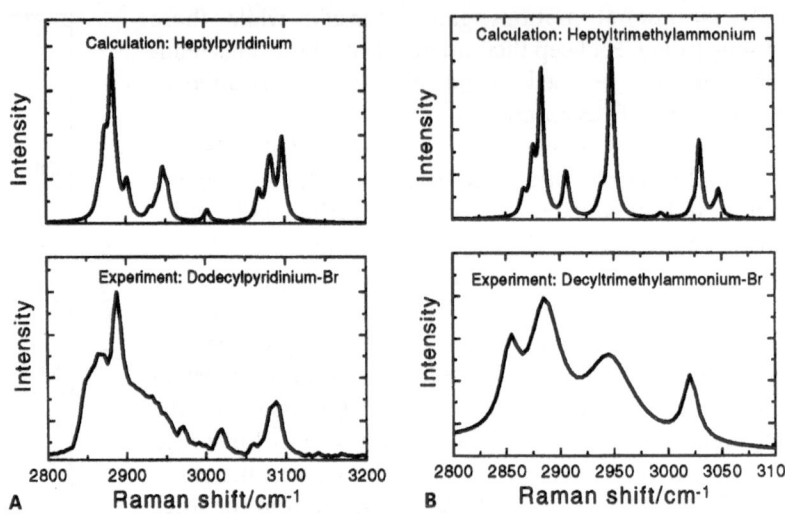

asymmetric stretching modes appear at 2883 and 2905 1/cm. Between the first shoulder and the intense band at 2883 1/cm, another should appears (2875 1/cm), produced by vibrations from terminal CH_3 of the chain. The band occurring at 2947 1/cm is due to the symmetric motions of CH_3 groups from the headgroup and the chain. The bands at 3030 and 3047 1/cm are produced by asymmetric stretching modes from the methyl-groups of the head.

After inspecting the lower part of Fig. 3b, it is quite easy to make the correspondence between the experiment and the calculations. Once again, the discrepancies are due to molecular interactions inside the crystal and anharmonic effects. The occurrence of overtones and Fermi resonances, which are not considered inside the harmonic approximation has already been discussed in the literature [28, 29].

SERS results

The adsorption of ionic surfactant on metalic surfaces at the liquid–solid interface is influenced and controlled by many factors, such as the surfactant concentration, counterion concentration and surface charge. The effect of these factors on the adsorption of alkylpyridinium on silver surfaces studied by means of the SERS spectroscopy are presented and discussed in this section.

Figure 4a shows a characteristic SERS spectrum from cetylpyridinium bromide ($C_{16}PyBr$). This measurement was recorded at -0.2 V vs. SCE, i.e. at a positively charged surface. With alkylpyridinium belonging to a class of cationic surfactant, the anions of the electrolyte solution may build a bridge to facilitate the adsorption of a positively charged molecule onto a positive surface.

The bands due to in-plane motions of the pyridine ring are marked with parallel bars ‖ and a band corresponding to out of plane motions of the pyridine ring is marked by means of ⊥. It can be clearly seen that almost all the enhanced bands from the pyridinium ring are due to in-plane motions of the ring and that their enhancement is stronger than the corresponding one of the alkyl-chain modes. This strongly point towards the hypothesis that the ring is bound to the surface.

The band at 1030 1/cm which is due to the ring breathing mode, is the most intense feature of the whole spectrum. The next strongest feature corresponds to an in-plane stretching mode at 1634 1/cm. The band at 3080 1/cm is due to in-plane motions of the pyridinium ring caused by three overlaping normal modes occurring around 3080 1/cm. The two bands appearing at 1172 and 1218 1/cm are due to in-plane motions of the ring.

Assuming, in accordance with the SERS selection rules [7], that the vibrations with atomic motion perpendicular to the surface couple more effectively to the surface electromagnetic waves, a perpendicular configuration of the pyridinium ring with respect to the surface is suggested. But after observing a band due to out of plane atomic motions at 870 1/cm is unlikely to be so. Although, this band at 870 1/cm has a relatively low intensity, in comparison to other bands from in plane motions, it experiences a large surface enhancement. This can be seen by comparing the relative intensities between the ring mode at 1030 1/cm and this out-of-plane band in both SERS and corresponding polycrystalline Raman spectra [19]. Hence, it seems more likely that the alkyl-pyridinium ring is adsorbed to the silver surface not in a completely vertical configuration.

The two arrows in Fig. 4b point to bands produced by CH_2 motions. The first one at 1457 1/cm corresponds to

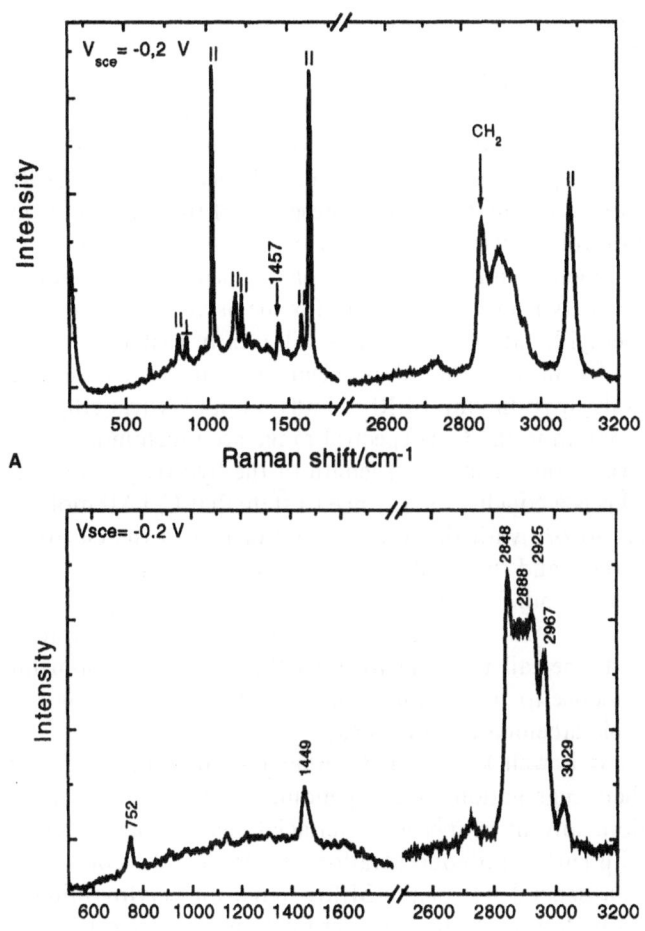

Fig. 4 SERS spectra from **A** cetylpyridiniumbromide (C_{16}PyBr) and **B** cetyltrimethylammonoumbromide (C_{16}TAB) on Ag electrode-surfaces; ex situ roughening; $E_s = -0.2$ V vs. SCE; concentration 10^{-3} M in 0.1 KCl; excitation wavelength at 488 nm; 10 mW; integration time 1.5 s; number of readings 60. $\perp$ pointed a band corresponding to a out of plane mode of the pyridine ring. $\parallel$ pointed the bands due to in plane motions of the pyridine ring. The frequencies of the main CH-bands is depicted over both spectra

facts suggest that most of the alkyl-chain has a liquid-like behaviour, implying that most of the alkyl-chain is neither bound to the surface nor even close to the surface. If this statement would be correct, it is to expected that by extending the length of the alkyl-chain the SERS signal should not be affected appreciably.

Fig. 4b shows the SERS spectrum of cetyltrimethylammonium bromide adsorbed on a silver electrode surface (-0.2 V vs. SCE). In this case, only two bands are clearly observed in the C–C stretching spectral range 500–1800 1/cm. The first one appears at 752 1/cm, and is due to C–N stretching mode at the headgroup. The second band is observed at 1450 1/cm and it has been assigned to the C–H scissoring mode of the alkyl chain.

In contrast to the two relative weak features described before, the C–H spectral range exhibits very strong bands, their shape is analogous to, that in the alkyl-pyridinium spectrum, but well resolved bands. The strongest band is at 2848 1/cm and is assigned to CH_2 symmetrical stretching. The next band, which is not well resolved, comes around 2888 1/cm. This band is due to the asymmetrical CH_2 atom motions. At 2925 1/cm, there is the band due to Fermi resonance produced by the overtone of the band at 1449 and the CH_2 symmetric stretching vibration. The bands at 2967 and 3020 1/cm are assigned to the CH_3 symmetric and asymmetric stretching modes, respectively.

In the C–H spectral range of C_{16}TAB all bands are better resolved and somewhat sharper than for C_{16}PyBr, indicating that the alkyl chains of the C_{16}TAB are somehow more rigid. Additionally, the bands corresponding to the methyl motions are quite strong, suggesting that the methyl groups are close to the electrode surface. This last statement implies an adsorption via the headgroup of the molecule. Considering that the headgroup as well as the surface is positively charged, the adsorption is facilitated in some way by anions in the solution.

In order to gain more information about the adsorption geometry, the influence of the alkyl chain length on the SERS spectra was investigated. Figure 5a shows the SERS dependence on the length of the alkyl chain for the pyridinium surfactant, when the number of carbon atoms was varied from 10 to 18. The strongest band observed in this figure corresponds to the symmetric CH_2 mode, which is observed at 2844 1/cm in each spectrum. The second strongest and well defined feature corresponds to the in-plane C–H stretching mode of the pyridinium ring. Between these two bands occur asymmetric CH mode and several Fermi resonances. It has already been pointed out that liquid alkanes show similar spectra in C–H range as that of cetylpyridinium bromide on the electrode. Furthermore all the spectra presented in Fig. 5a show exactly the same features and almost the same intensity relations, albeit a small dependence on the ratio between the ring

CH_2-scissoring vibration, showing a rather low intensity. On the basis of the short-range effect of the surface enhancement it can be stated that just a small fraction of the alkyl-chain is close enough to the electrode surface to be enhanced. The second marked band at 2844 1/cm is the CH_2 band due to the symmetric stretching of the alkyl chain. This band is the most intense feature of the alkyl-chain to be seen in this spectrum.

Moreover, the CH bands around 3000 1/cm (including those corresponding to the methyl group) resemble the bulk-liquid spectrum from an amphiphilic molecule having a long alkyl chain. A predominance of the asymmetric stretching at 2884 1/cm over all other bands is observed for ordered structures (polycrystalline Fig. 3). These two mode and the total integrated intensity of the C–H bands. In fact, the ring vibration at 3080 1/cm for C_{10}PyB has a higher relative intensity with respect to the band at 2844 1/cm than for C_{18}PyBr. Whereas this last observation would contradict the expected liquid-like behaviour exposed before, it is important to stress the fact that between C_{12}PyBr and C_{18}PyBr no difference is observed. Thus, it could be speculated that the surface enhancement is "felt" by bonding out to the C_{12} or C_{13}, methylene group and the C–H beyond this point do not contribute significantly to the Raman signal.

Following a similar sequence as for *n*-alkyl pyridinium, the dependence of the SERS intensities for alkyl trimethylammonium on SERS activated Ag electrode surfaces on the chain length is examined and some conclusions on the possible adsorption geometry are presented. Figure 5b shows the SERS spectra of alkyl-methylammonium for three different chain-lengths, cetyl-, decyl and tetra-methylammonium bromide (C_{16}TAB, C_{10}TAB and TMAB) from top to bottom, respectively. The SERS spectrum from tetramethylammonium bromide (TMAB) shows the bands due to the methyl-groups attached to the Ag-surface. By comparing the SERS with the calculated Raman spectrum, it can be seen that the symmetric CH_3 band at around 2950 1/cm appeared along with two new features indicating a strong interaction between at least one methyl-group with Ag-surface, with the consequent change in the force constants. The next SERS spectrum to be discussed corresponds to C_{10}TAB (Fig. 5b middle). First, the spectral features known from TMAB are also present in this spectrum. Moreover, the band due to the CH_2 symmetric stretching mode at 2848 1/cm (marked

by the vertical line) appears as well as the asymmetric one at 2888 1/cm. The intensity of the asymmetric vibration is stronger than the symmetric one. The stronger intensity of the methyl bands in comparison to the CH_2 related bands suggests that the methyl groups remain close to the surface, experiencing still a larger enhancement. The SERS spectrum from C_{16}TAB is presented at the top of Fig. 5b. For this spectrum it can clearly be seen that the CH_2 symmetric vibrational mode has gained intensity and its intensity is even larger than for the asymmetric one. The methyl band at 3020 1/cm is still clearly visible.

By now it should be clear that in comparision to *n*-alkylpyridinium SERS bands of *n*-alkyltrimethylammonium in the C–H spectral range are much more sensitive to the variation of length of the hydrocarbon chain. This is a conclusive evidence to state that C_nTAB molecules adsorb with the headgroup attached to the electrode surface and their tail running parallel to it. While on the other hand, C_nPyBr molecules adsorb with the pyridinium headgroup attached to the surface in a tilted configuration with the tail pointing towards the solution. Supporting evidence to these conclusions are obtained by using isotopic labelling methods [30].

It is well known that surfactants in solution change their aggregation state depending on the salt concentration, solvent properties, interfacial charge and especially with surfactant concentration [1]. Hence, the concentration dependence of the SERS signal for cetylpyridinium bromide was studied. The critical miceller concentration for C_{16}PyBr in 0.1 M KCl is around 4×10^{-4} M (25°C) [31]. The SERS spectra from C_{16}PyBr in 0.1 M KCl adsorbed on Ag were collected from solution having a

Fig. 5 SERS spectra obtained from **A** alkylpyridiniumbromide, corresponding to hydrocarbon length of 10, 12 and 18 carbon atoms are presented from bottom to the top, respectively; **B** alkyltrimethylammoniumbromide. In this case the hydrocarbon length was 1, 10 and 16. The used substrate was an ex situ electrochemically roughened Ag-electrode. The spectra were acquired at an applied potential of $E_s = -0.2$ V vs. SCE. All the solutions were 10^{-3} M in 0.1 KCl. The excitation wavelength was 632 nm

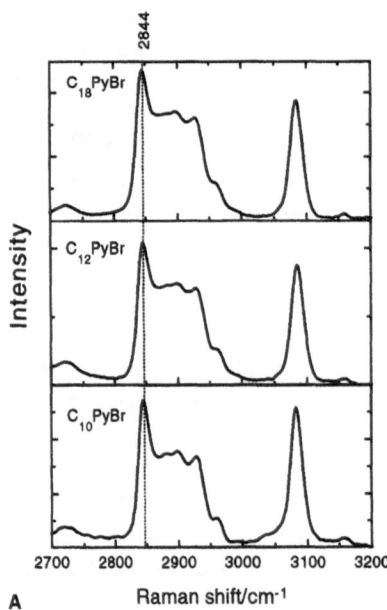
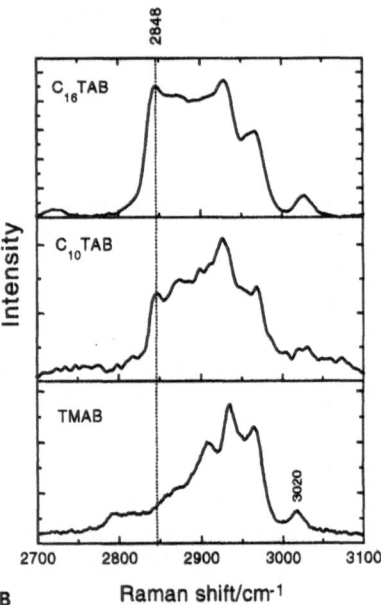

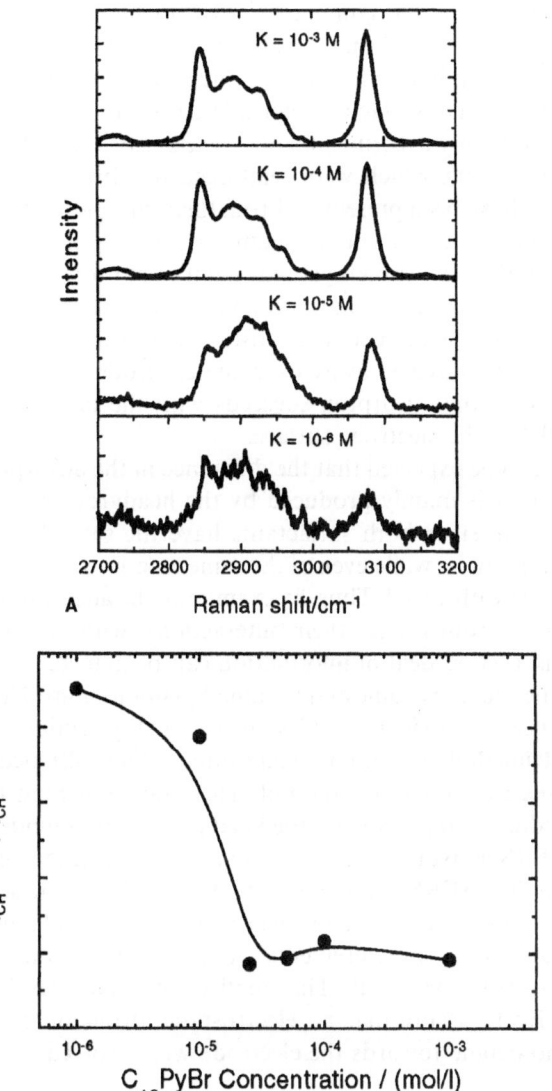

Fig. 6 Concentration dependence of the SERS intensity of cetylpyridinium bromide adsorbed on a Ag electrode; **A** C–H stretching modes spectral range **B** concentration dependence of the ratio between the integrated intensity of all the C–H modes belonging to the alkly chain centre at 2900 and the integrated intensity of the pyridinium ring at 3080; excitation wavelength was 514 nm; 10 mW; integration time 1.5 s; number of reading 60; ex situ roughening; $E_s = -0.2$ V vs. SCE

concentration ranging between 10^{-6} and 10^{-3} M. The exact concentrations were $10^{-3}, 10^{-4}, 5 \times 10^{-5}, 2.5 \times 10^{-5}, 10^{-5}$ and 10^{-6} M. Figure 6a shows the corresponding SERS spectra of the C–H stretching region. The merit of considering the C–H region results from being able to observe changes for even very low surfactant concentrations.

For concentrations larger than 10^{-5} M, the C–H stretching of the ring at 3080 1/cm exhibits the highest intensity. But for concentrations below 10^{-5} M, the 3080 1/cm band becomes weaker than all other C–H stretching bands. The band due to the C–H symmetric stretching at 2844 1/cm shows a clear dominance when the surfactant concentration is above 10^{-5} M, but at lower concentrations the Fermi resonance bands and the asymmetric band at 2884 1/cm dominate the spectrum. This variation can be seen as a transition from a liquid-like spectrum to a liquid crystalline spectrum [32].

Figure 6b shows the concentration dependence of the ratio between the integrated intensity from the alkyl chain bands and the ring stretching mode intensity. The alkyl band intensity was integrated between 2800 and 3015 1/cm and the ring band intensity was integrated between 3020 and 3115 1/cm. It can be clearly seen that the ratio remains practically constant without changes for surfactants concentrations above 10^{-5} M. However, at 10^{-5} M these ratio shows an abrupt jump. This abrupt change indicates a transition in the aggregation state of the cetylpyridinium at the surface. It is reasonable to think that for concentrations over 2×10^{-5} M a full monolayer is formed. While for lower concentrations there are not enough $C_{16}PyBr$ molecules to form such a monolayer, which means, that a part of the hydrophobic chain is able to get close to the surface. Note the similarities between the SERS spectra from $C_{16}PyBr$ at concentrations lower than 10^{-5} and the corresponding $C_{10}TAB$ and $C_{16}TAB$ spectra in Fig. 5b.

It is well known that the dependence of the SERS bands on the excitation energy can affect the relative intensity ratios, especially in cases where the "first-layer" effect plays an important role [12]. Consequently, it is crucial to corroborate the indepence of the surfactant SERS spectra from the excitation energy. Unfortunately, it is not possible to examine the CH bands around 3000 1/cm from the C_nPyBr by means of NIR FT-SERS spectroscopy [33]. As an alternative to overcome this drawback silver colloids were used. The laser beam was focused onto the sample cell and collected in backscattering configuration, using conventional optics instead of a microscope to improve the signal to noise ratio.

Figure 7 presents the concentration dependence of the SERS bands from $C_{16}PyBr$ in the C–H spectral range. In the spectra corresponding to the two lowest surfactant concentrations no SERS signals can be distinguished from the background level. At a surfactant concentration of 6×10^{-5} M, the SERS bands rise above background. Here, it can be observed that the band at 2850 1/cm, due to the symmetrical stretching mode of the CH_2 groups, appears in a similar fashion as in the case of C_nTAB or C_nPyBr (at low concentrations) when the SERS spectra were excited with visible light, i.e. its intensity is comparable to the other C–H bands. This supports the model, where the hydrocarbon chain runs parallel to the surface in both

190
A. Tarazona et al.
Adsorption properties of two cationic surfactant on silver surfaces

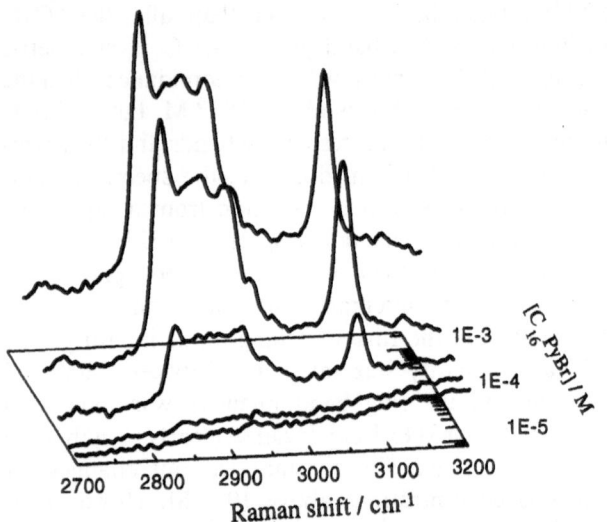

Fig. 7 Concentration dependence of the SERS bands from C$_{16}$PyBr in the C–H spectral range adsorbed on Lee-Meisel colloids (1 ml); the KBr concentration was 0.015 M; excitation wavelength 1064 nm; 350 mW; number of reading 100

cases. For higher concentrations, the band at 2850 1/cm dominates the SERS spectra and resembles the relative intensity ratio observed for liquid alkanes, suggesting that the hydrocarbon chain points away from the surface.

A similar behaviour, as the one from C$_n$PyBr, has already been reported for surfactants with a silane headgroup adsorbed on smooth silicon surfaces, using FT-IR [34] or vibrational sum-frequency generation [35]. After observing the same relative band ratios for excitation in the visible and in the near infrared spectral range, it can be

concluded that the comparison of Raman spectra from liquid alkanes and liquid crystals with SERS spectra can be used as a criterion for formulating an adsorption model for the cationic surfactants considered in this work.

So far, only experimental observations based on SERS measurements, which were reinforced by ab inito calculations, have been presented. From their interpretation the corresponding adsorption geometries for *n*-alkylpyridinium and *n*-alkyltrimethylammonium were proposed. But up to now, no reasons have been presented to explain why is one case the surfactant is adsorbed on the Ag-surface with its tail directed away from the electrode, and in the other case the adsorpton succeeds with the tail running parallel to the electrode surface.

It can be expected that the difference in the adsorption behaviour is mainly produced by the headgroup in each surfactant, since both surfactants have the same kind of aliphatic chain, while even of the same length strong differences were observed. Thus, by examining the adsorption of the headgroups and their interactions with aliphatic chains, a good deal of information can be gained.

Therefore, pyridine and tetramethylammonium (TMA) were used as model molecules for the alkylpyridinium and alkyltrimethylammonium headgroups, while 1-Br octane was used as model for the tail. The strategy behind this procedure is to pre-adsorb the head model compounds on the SERS activated electrode, and after acquiring the corresponding SERS spectrum, to add the 1-Br octane solution to look whether bands due to CH$_2$ vibrational modes appear. The measurements were carried out at $E_s = -0.8$ V vs. SCE. The applied potential was kept negative to assure that no electrostatic attraction of the bromo octane towards the electrode was involved.

Fig. 8 A SERS spectra pyridine before (bottom) and after the addition of 1-Br octane in a proportion of 1:5. **B** SERS spectra from Tetramethylammonium bromide before (bottom) and after the addition of 1-Br octane in a proportion of 1:5 (middle). For comparing purposes the SERS spectrum of C$_{16}$TAB (top). The used substrate was an ex situ electrochemically roughened Ag-electrode. The spectra were acquired at applied potential $E_s = -0.2$ V vs. SCE. All the solutions were 10^{-3} M in 0.1 KCl. The excitation was 488 nm

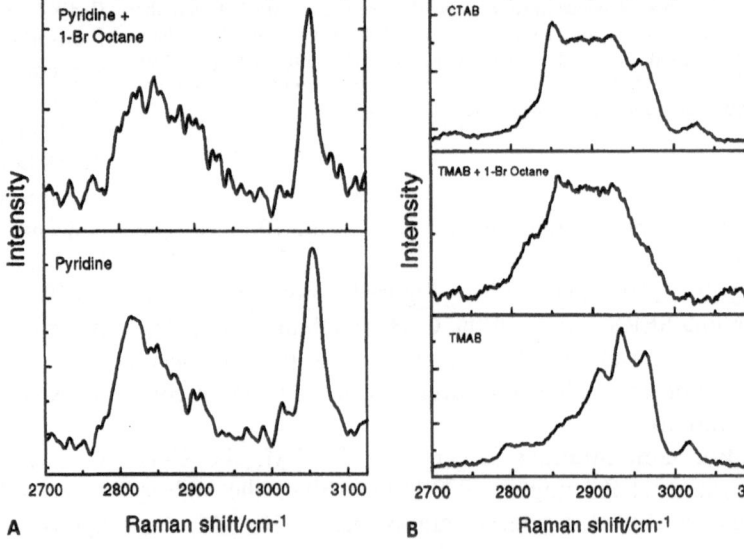

The results of pyridine are presented in Fig. 8a, with the SERS spectrum from pyridine at the bottom and the corresponding spectrum from the mixture (with pyridine and 1-Br octane in the volume ratio of 1 to 5) at the top. Comparing both spectra, no changes can be observed after the addition of the 1-Br octane solution. This may imply that the adsorbed pyridine keeps the 1-Br octane away from the electrode surface, blocking any enhancement of the Raman signal from 1-Br octane. The fact that pyridine shows some SERS bands between 2800 and 2900 1/cm (may suggest) the presence of linear hydrocarbon chains in the solution, but even the Raman spectrum from liquid pyridine shows a large number of weak bands in the spectral range just mentioned.

In Fig. 8b the SERS spectra from the simplified model system for alkyltrimethylammonium and for cetyl-trimethylammonium are presented. At the bottom of the figure the SERS spectrum of the 10^{-3} M tetramethylam-monium bromide (TMAB) solution is shown. As already discussed all observed bands correspond to those orig-inated in methyl motions. After recording the spectrum, 1-Br octane was added to complete a 1-Br octane: TMAB volume ratio of 5:1. The SERS spectrum corresponding to the mixture is depicted in the middle of Fig. 8b. It can clearly be seen that the spectrum is now dominated by a band at 2850 1/cm, which is due to the CH_2 symmetric stretching mode. Moreover, the relative band ratio re-sembles the Raman spectrum from liquid crystals in an ordered phase, as already discussed above. This result can be explained by considering that the alkyl chain is adsor-bed on top of the pre-adsorbed TMAB and orientated parallel to the electrode surface. After comparing this last spectrum with that of $C_{16}TAB$ (top) the self consistency of this model is evident. In addition the growth of the CH_2 bands can be controlled by amount of 1-Br octane added to the cell.

Conclusions

In this work, a good deal of information on the adsorption properties from two kinds of cationic surfactants (alkyl-pyridinium and alkyltrimethylammonium) have been ob-tained by combining SERS measurements and vibrational calculations based on Hartree–Fock methods.

The SERS spectra from alkylpyridinium surfactants exhibit a combination of an plane bands and one out-of-plane band, suggesting that the headgroup is adsorbed on the Ag surface in a tilted orientation. It must be pointed out that the criterion based on the relative SERS intensities between the headgroup and hydrocarbon chain bands cannot be applied as in previous works, because the bands experience different "first-layer effect" enhancements due to the presence of π-electrons of the headgroup. The independence of the SERS signals in C–H spectral range of the hydrocarbon chain and their similarities to the Raman spectra from liquid n-alkanes is a more reliable criterion on which to base the orienta-tion of the hydrocarbon chain. It has also been shown that variation of the excitation energy does not affect the observations or conclusions. For high concentrations the pyridinium rings, which are highly polar, cover the surface completely, blocking the metallic surface. This creates a hydrophobic surface that can be used as foundation to build a bilayer structure. For low concentra-tions, the SERS-spectra show a similar band relation in the C–H spectral range as the alkylmethylammonium surfac-tants.

The high sensitivity of the SERS signals from the alkyltrimethylammonium species to the hydrocarbon chainlength shows that the absorption takes place with the headgroup attached to the surface and the alkylchains running parallel to it, building a hydrophobic surface where the following layers stack on the previous ones. Additional support to this statement is given by the nar-rower width of the SERS bands in C–H spectral range from alkyltrimethylammonium in comparison to those from alkylpyridinium, which implies a more ordered structure for alkyltrimethylammonium than for alkyl-pyridinium. It was possible through an experimental model to emulate the headgroup behaviour and its interac-tions with the hydrocarbon chain, giving more support to the proposed adsorption orientation for both surfactant classes.

Acknowledgments One of the authors (A.T.) greatfully acknowledges the financial support giving by CONICIT (Venezuela). The authors are indebted to S. Schumann and H.D. Narres for technical support, continous encouragement, and for valuable dis-cussions.

References

1. Rosen MJ (1989) Surfactant and Inter-facial Phenomena, Chapter 2. Wiley, New York
2. Levitz P, Van Damme D, Keravis D (1984) J Phys Chem 88:2228.
3. McDermott DC, McCarney J, Thomas RK, Rennie AR (1994) J Colloid Inter-face Sci 162:304
4. Manne S, Gaub HE (1995) Science 270:1480
5. Fleischmann M, Hendra PJ, McQuillan AJ (1974) J Chem Phys Lett 26:163
6. Jeanmaire DL, Van Duyne RP (1977) J Electroanal Chem 84:1

7. Albrecht MG, Creighton JA (1977) J Am Chem Soc 99:5215
8. Chang RK, Furtak TE (eds) (1982) Surface-enhanced Raman Scattering: Plenum, New York
9. Furtak TE, Reyes (1980) J Surf Sci 93:351
10. Creighton JA (1980) Springer Series in Chemical Physics, Vol. 15. Springer, Berlin
11. Van Duyne, RP (1980) In: Moore CB (ed) Chemical and Biological Applications of Lasers, Vol. 4. Academic Press, New York
12. Otto A, Mrozek I, Grabhorn H, Akemann W (1992) J Phys Condens Matter 4:1143
13. Ohtake T, Mino N, Ogawa K (1992) Langmuir 8:2082
14. Knoll W, Philpott RM, Golden WG (1982) J Chem Phys 77:219
15. Heard MS, Grieser F, Barraclough (1983) Chem Phys Lett 95:154
16. Dendramis AL, Schwinn EW, Sperline RP (1983) Surf Sci 134:675
17. King FW, Van Duyne RP, Schatz GC (1978) J Chem Phys 69:4472
18. Wiesner J, Wokaun A, Hoffman H (1988) Prog Colloid Polym Sci 76:271
19. Sun S, Birke LR, Lombardi JR (1990) J Phys Chem 94:2005
20. Koglin K, Krug O (1992) In: Kiefer W, Cardona M, Schaak M, Schneider W, Schroetter HW (eds) Proc XIIIth Int Conf on Raman Spectroscopy. Wiley, Chichester
21. Shelton RS, Van Campen MG (1946) J Am Chem Soc 68:757
22. Norcross G, Openshaw HT (1949) J Chem Soc (London):1174
23. Lee PC, Meisel D (1982) J Phys Chem 86:3391
24. (a)Frisch MJ, Trucks GW, Head-Gordon M, Gill PMW, Wong MW, Foresman JB, Johnson BJ, Schlegel HB, Robb MA, Replogle ES, Gomperts R, Andres SL, Raghavachari K, Binkley JS, Stewart JJP, Pople JA (1992) In: Gaussian 92, Revision B. Gaussian, Pittsburgh PA (b) Frish MJ, Trucks GW, Schlegel HB, Gill PMW, Johnson BJ, Robb MA, Cheeseman JR, Keith T, Petersson GA, Montgomery JA, Ragavachari K, Al-Laham MA, Zakrzewiski VG, Ortiz JV, Foresman JB, Cioslowski J, Stefanov BB, Nanayakkara A, Challacombe M, Peng CY, Ayala PY, Chen W, Wong MW, Anf'dres JL, Replogle ES, Gomperts R, Martin RL, Fox DJ, Binkley JS, Defrees DJ, Barker J, Stewart JJP, Head-Gordon M, Gonzales C, Pople JA (1994) In: Gaussian 94 Revision C. Gaussian, Pittsburgh, PA
25. Snyder RG, Hsu SL, Krimm S (1978) Spectrochim Acta 34A:395
26. Snyder RG, Kim Y (1991) J Phys Chem 95:602
27. Snyder RG (1992) J Chem Soc Faraday Trans 88:1823
28. Snyder RG, Strauss HL, Elliger CA (1982) J Phys Chem 86:5145
29. MacPhall RA, Snyder RG, Strauss HL, J Chem Phys 77(3):1118
30. Tarazona A, Kresig SM, Koglin E, Schwuger MJ, In preparation
31. Fang M, Huang T, Gu T, Mo Y, Wang Z, Li X (1993) Spectrochimica Acta 49:1009
32. Van Winkle DH, Dierker SB, Clark NA (1989) J Chem Phys 91:5212
33. Kreisig SM, Tarazona A, Koglin E, Schwuger MJ (1996) Langmuir 12:5279
34. Ohtake T, Mino N, Ogawa K (1992) Langmuir 8:2082
35. Watanabe N, Yamamoto H, Wada A, Domen K, Hirose C, Ohtake T, Mino N (1994) Spectrochimica Acta 50A:1529
36. Albercht MG, Creighton JA (1977) J Am Chem Soc 99:5215
37. Barańska H, Labudzińka A, Terpiński J (1987) Laser Raman Spectrometry: Analytical Applications. Wiley, New York
38. Sandroff CJ, Garoff S, Leung KP (1983) Chem Phys Lett 96:547
39. Kreisig SM, Tarazona A, Koglin E, Schwuger MJ (1996) Langmuir 12:5279
40. Bryant MA, Pemberton JE (1991) J Am Chem Soc 113:3629
41. Bryant MA, Pemberton JE (1991) J Am Chem Soc 113:8284

Progr Colloid Polym Sci (1997) 103:193–200
© Steinkopff Verlag 1997

A.E. Kaifer

Electrodes derivatized with mono- and multilayer assemblies containing preformed binding sites

Received: 13 December 1996
Accepted: 17 December 1996

Dr. A.E. Kaifer (✉)
Chemistry Department
University of Miami
Coral Gables, Florida 33124-0431, USA

Abstract Recent work on the design, preparation, and characterization of thiolate-on-gold self-assembled mono-layers and multilayers containing preformed binding sites is reviewed. The use of these interfacial systems as molecular sensors in the vapor and solution phases is discussed.

Key words Self-assembled mono-layers – binding sites – receptors – host–guest interactions – amphi-philic structures – sensors – voltammetry – modified electrodes

Introduction

During the last decade the field of electrochemistry has witnessed a very fast progress on the modification of electrode surfaces. From the predominant use of random polymeric structures, prevalent in the electrode modifica-tion efforts of the late 70s and early 80s, electrochemists have learnt to control the molecular architecture of the electrode–solution interface to a degree that was clearly out of reach a decade ago. Many electrode modification methods developed recently rely on the use of thiolate self-assembled monolayers (SAMs) [1–3]. These systems offer unparalleled ease of preparation and levels of mo-lecular organization close to those that can be reached with Langmuir–Blodgett film methods. Therefore, elec-trodes derivatized with unfunctionalized or functionalized alkanethiolate monolayers have been the subject of exten-sive research work during the last few years [4, 5].

My group's fundamental research goal is directed to the use of SAM-related techniques for the preparation of electrodes derivatized with highly organized molecular assemblies containing preformed, active binding sites. Sev-eral years ago, we thought that we could draw from the numerous classes of receptors developed by organic chem-ists interested in molecular recognition phenomena to prepare electrodes with built-in selective properties. The fundamental idea is that a receptor or host that shows selective binding affinity for a substrate in homogeneous solution will maintain its binding properties when immo-bilized on an electrode surface. The resulting electrode will exhibit selective electrochemical reactivity for those sub-strates that can be bound to the receptor, while all other substrates will be rejected by the interfacial structure, thus preventing their electrochemical reactions at the underly-ing electrode surface. This idea is pictorially represented in Scheme 1.

Scheme 1 Idealized representation of an interfacial monolayer as-sembly containing "cylindrical" binding sites

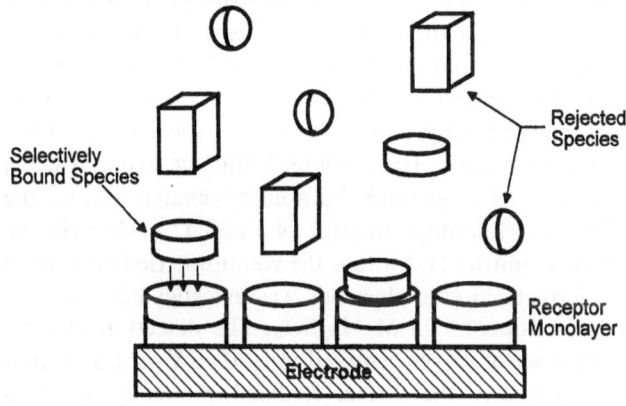

In this review, the efforts of several groups to realize the simple idea represented in Scheme 1 will be summarized. Particular emphasis will be placed on discussion of systems designed and studied in the author's laboratory.

Interfacial self-assembly of an inclusion complex

Among the many artificial receptors synthesized and studied during the last two decades, Stoddart's tetra-cationic cyclophane [6] 1^{4+} stands out for several reasons, such as (i) its rigid cavity in which the distance between the two bipyridinium groups is ideal for inclusion of an aromatic ring, (ii) the substantial π electron–acceptor character provided by the two bipyridinium subunits, (iii) the redox properties also afforded by the bipyridinium groups, and (iv) its tetracationic nature that provides an easy way to control solubility through the selection of appropriate counterions. Therefore, 1^{4+} is an excellent host for binding of substrates with π-donor moieties [7]. Stoddart and coworkers have utilized these properties to prepare a large number of rotaxanes, pseudorotaxanes, and catenanes including cyclophane 1^{4+} as a building block [8, 9].

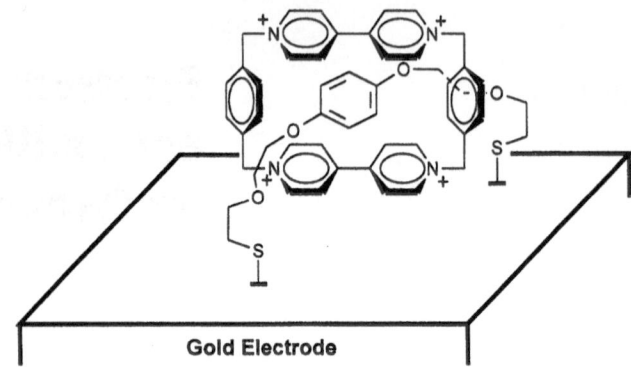

Scheme 2 A surface-attached inclusion complex

duction of the bipyridinium groups of the cyclophane host verified the surface anchoring of the inclusion complex [10] (see Scheme 2 for a pictorial representation of the surface-attached structure). Similar experiments with monothiol hydroquinol **3** (no bipyridinium reduction waves were detected in this case) confirmed that the surface immobilization of the inclusion complex takes place by a two-point anchoring mechanism. The surface attach-

We reasoned that these properties could also be used for the preparation of surface-attached molecular assemblies. For instance, host 1^{4+} forms an inclusion complex with the bis(thiol) hydroquinol derivative **2** due to the π-donor character of the aromatic nucleus of this guest. The equilibrium binding constant was determined to be $253 \pm 12\,M^{-1}$ at $25\,°C$ in acetonitrile [10]. When it is bound, the aromatic nucleus of **2** threads the cavity of the cyclophane so that the thiol-terminated side chains project out at different sides of the cavity. It is then possible to use the terminal thiol groups to anchor the inclusion complex to a gold surface. To accomplish this we exposed clean gold electrodes overnight to a deoxygenated acetonitrile solution containing a mixture of **2** and 1^{4+}. After rinsing with acetonitrile and water, the voltammetric behavior of the derivatized gold electrode was investigated in deoxygenated aqueous $0.1\,M\,Na_2SO_4$. The detection of a reversible set of waves centered at $-0.46\,V$ vs. SSCE that corresponds to the surface-confined, monoelectronic re-

ment of the $2 \cdot 1^{4+}$ inclusion complex is a low efficiency process. From the integration of the cathodic wave, we measured a surface coverage of $7\,(\pm 1) \times 10^{-12}\,mol/cm^2$, which represents about 8% of the maximum surface coverage expected for such species. This low efficiency is not surprising considering the complexity of the surface attachment process.

The successful trapping and subsequent immobilization of the $2 \cdot 1^{4+}$ inclusion complex on gold surfaces demonstrates that the ideas and concepts common to the fields of molecular recognition and supramolecular chemistry can also be utilized to assemble novel structures at the electrode–solution interface.

Interfacial self-assembly of thiolated receptors

In the previous section the surface attachment of an inclusion complex was discussed. Although the resulting

Progr Colloid Polym Sci (1997) 103:193–200
© Steinkopff Verlag 1997

molecular structure is highly unusual, it would be more interesting to prepare interfacial assemblies with free receptors having active binding sites. This would permit the investigation of the receptor's binding properties *in situ*, i.e., at the electrode–solution interface. The preparation of molecular assemblies with active binding sites is, of course, of much higher interest from the standpoint of developing new sensor technology.

In this context the already extensive work of Reinhoudt and coworkers on the preparation and characterization of self-assembled monolayers of resorcin[4]arenes [11–13], calixarenes [14], and even carceplexes [15] constitutes one of the most important efforts to develop this field. Self-assembled monolayers of resorcin[4]arenes exhibit selective adsorption properties for small chlorinated hydrocarbons in the gas phase [13], although serious questions have been raised as to how much the observed selectivity has to do with the presence of molecular cavities in the monolayer assembly [16]. Davis and Stirling have also reported on the self-assembly of similar compounds and demonstrated the formation of multilayers [17, 18]. The extensive and very systematic work of Crooks, Ricco and coworkers on the development of devices for the detection of volatile organic compounds deserves special mention [19]. Recently, Crooks and coworkers have taken a very interesting turn in their work, demonstrating that monolayer-attached dendrimeric structures exhibit remarkable selectivity for the sorption of gas-phase molecules [20, 21].

The author's group has focused its efforts on monolayers containing preformed cavities or binding sites designed to operate and respond when immersed in aqueous solutions. While numerous sensor applications can be found for aqueous and vapor phases alike, it can be argued that it should be easier to design systems based on strong host–guest interactions in an aqueous environment. The reasons for this are related to the importance that solvophobic interactions have in molecular recognition forces. Generally, a host tends to bind more strongly as the solvent polarity increases due to solvophobic effects [22]. Therefore, my group's research is directed to the preparation and characterization of self-assembled monolayers designed to exhibit effective binding sites in aqueous media. We were indeed encouraged by the pioneering work of Rubinstein and coworkers on the development of monolayers incorporating selective binding sites for transition metal ions [23, 24].

Prior to our work with immobilized receptors, we demonstrated interfacial binding in a monolayer possessing protruding ferrocene groups [25]. The ferrocene groups in this case constitute the guests and as such are bound by extremely low concentrations (0.05–2 μM) of an amphiphilic calix[6]arene host in the contacting aqueous solution. An important result from this work was the

observed lack of binding interactions with a non-amphiphilic analog of the calix[6]arene host. This finding points to amphiphilicity as a crucial property, which is required for a host to act as an effective interfacial receptor. Indeed, highly soluble, non-amphiphilic receptors are expected to be very well solvated in the bulk solution, thus missing the necessary driving force to approach and bind substrates located in interfacial environments.

Our group has recently reported two novel examples of interfacial molecular complexation in thiolate monolayers containing well-defined host binding sites [26, 27]. In both cases, the binding sites are provided by sulfur-containing monomeric receptors which can be co-assembled on Au surfaces to afford monolayers with molecular recognition properties.

Monolayers of thiolated β-CD

Synthetic conversion of all seven primary –OH groups of β-CD into –SH groups yields a receptor (**4**) that can be chemisorbed on gold surfaces via the formation of a maximum of seven thiolate-Au bonds per receptor molecule [26].

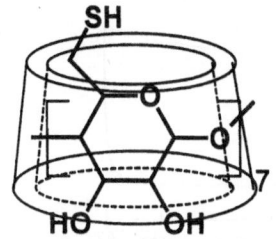

Receptor 4

The thiolated β-CD forms monolayers with substantial defect densities due to the lack of favorable lateral interactions between the CD molecules. In order to "patch the holes" in these monolayer assemblies without clogging the CD cavities, we devised procedures for the preparation of mixed monolayers containing both the thiolated β-CD receptors and pentanethiol molecules. These monolayers were characterized by reductive desorption and electrode capacitance measurements. The monolayer blocking properties were also investigated, i.e., the voltammetric response of solution redox probes such as $Fe(CN)_6^{4-}$ and $Ru(NH_3)_6^{2+}$ at the monolayer-covered electrodes were recorded. Our data supported the monolayer structure depicted in the Scheme 3 [26].

The well-defined hydrophobic cavities in a monolayer containing receptor **4** are ideally suited for molecular recognition at the electrode-solution interface. We selected ferrocene as an advantageous substrate for these studies because of its well documented and strong binding inside

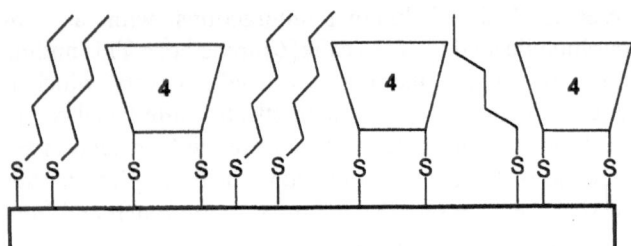

Scheme 3 Mixed monolayer formed by receptor **4** and pentanethiol

Fig. 1 Voltammetric response
of a Au/**4** + C$_5$SH electrodes at
0.5 V/s in 0.2 M Na$_2$SO$_4$ also
containing (A) 5 μM ferrocene,
(B) 5 μM ferrocene + 5 μM
MTA, and (C) 5 μM
ferrocene + 7.5 μM MTA

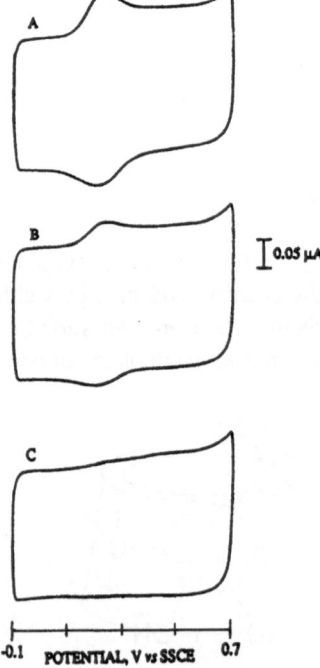

Furthermore, we determined the monolayer-bound ferrocene surface coverage as a function of ferrocene solution concentration and analyzed the data using the Langmuir isotherm model. The results allowed us to estimate both the surface density of active binding sites (Γ_{max}) and the effective binding constant (K) for the interfacial recognition interactions. The calculated value of Γ_{max} (2.5×10^{-11} mol/cm^2) represents about 60% of the surface coverage of **4** measured from the reductive desorption data, suggesting that a fraction of the CD cavities are deactivated probably during the pentanethiol sealing step [26]. The calculated K value (3.9×10^4 M^{-1}) is about an order of magnitude larger than typical binding constants measured for water-soluble ferrocene derivatives and β-CD in homogeneous solution [28].

Co-assembly of decanethiol and a "designer" interfacial receptor

The cyclophane receptor **1**$^{4+}$ exhibits interesting molecular recognition properties which we would like to express at the electrode–solution interface. In 1992, we reported the use of this host compound to detect catechol and indole using voltammetric techniques [29]. Our approach was then based on the immobilization of the tetracationic receptor inside a thin layer ($\sim$ 50 nm) of the polyelectrolyte Nafion. A more effective approach would rely on the immobilization of **1**$^{4+}$ in a monolayer directly attached to the electrode surface. To reach this goal, we targetted the disulfide receptor **5**$^{4+}$ which affords a binding cavity similar to that of **1**$^{4+}$, should chemisorb to Au surfaces through the formation of two thiolate-Au bonds, and appears reasonably accessible from a synthetic standpoint. My group has recently reported the preparation and characterization of **5**$^{4+}$ [27].

the β-CD cavity and because of its reversible electroactive character. Figure 1 shows the voltammetric response of a gold electrode covered with a mixed monolayer of **4** + C$_5$SH when immersed in a 5.0 μM solution of ferrocene (in aqueous Na$_2$SO$_4$). Control experiments clearly reveal that the observed voltammetric response results from the presence of the interfacial CD cavities. Binding of the ferrocene molecules in the monolayer cavities allow their cyclic voltammetric detection at the low micromolar levels utilized in this work. A very relevant experiment that demonstrates how the detection of ferrocene requires the presence of the CD cavities was performed by adding m-toluic acid (MTA) to the solution. MTA is also an excellent guest for β-CD; however, it is electroinactive in the potential range surveyed. As the solution concentration of MTA increases, the currents associated with the reversible oxidation of ferrocene decrease, revealing that MTA competes effectively with ferrocene for the available monolayer binding sites, gradually displacing it, and, thus, erasing the voltammetric ferrocene signal (see Fig. 1).

Receptor **5**$^{4+}$

Progr Colloid Polym Sci (1997) 103:193–200
© Steinkopff Verlag 1997

INDOLE BINDING ISOTHERM
Au/thiocyclophane/C10SH

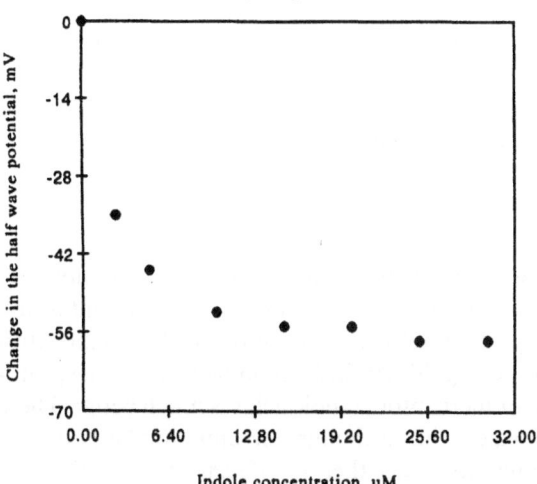

Fig. 2 Observed shifts in the formal potential for the monoelectronic reduction of the surface confined paraquat groups of Au/5^{4+} + $C_{10}H_{21}$SH electrodes immersed in phosphate buffer (pH = 7) solutions also containing variable concentrations of indole

This receptor and decanethiol co-assemble on Au surfaces to form mixed monolayers. Voltammetry of Au electrodes covered with 5^{4+} + C_{10}SH monolayers yields the response anticipated for the simultaneous monoelectronic reduction of the two bipyridinium groups of the receptor. The formal reduction potential value is very responsive to the presence of either catechol or indole in the contacting solution, while it is left unaltered by the addition of aromatic compounds with π-acceptor character, such as nitrobenzene and cyanobenzene [27]. A plot of formal reduction potential as a function of indole concentration is shown in Fig. 2. The shape of the plot strongly suggests that the π-donor indole molecules are bound by the π-acceptor sites afforded by the immobilized receptor molecules at the gold surface. Control experiments with model compounds demonstrated that the receptor must possess a cavity lined by two bipyridinium groups in order to exhibit the behavior shown in the graph. The potential-concentration data can be analyzed and fitted to a Langmuir isotherm. For the binding of indole we found $\Gamma_{max} = 3.2 \times 10^{-11}$ mol/cm^2 and $K = 2.7 \times 10^5$ M^{-1}; for the binding of catechol the values were $\Gamma_{max} = 2.9 \times 10^{-11}$ mol/cm^2 and $K = 1.0 \times 10^5$ M^{-1}. The two values for Γ_{max} (surface coverage of active binding sites) were identical within error margin ($\pm 10\%$), as expected since all these experiments were done with electrodes modified using identical procedures. The binding constant values obtained for both indole and catechol were again about an order of magnitude larger than those previously determined with similar host–guest systems in homogeneous aqueous solution.

These two recent reports by our group constitute novel examples of molecular recognition by supported monolayers, designed to contain well-defined binding sites. The approach is characterized by the use of known *molecular receptors*, capable of effective guest binding within monolayer systems adapted for voltammetric signaling while immersed in aqueous media. In these examples, the receptors were functionalized to effect their covalent immobilization on gold surfaces. While this approach is very reliable, the self-assembly of large molecules, such as these functionalized receptors, is very inefficient compared to the facile self-assembly of alkanethiols. Therefore, we had to use mixed monolayers containing both the receptor molecules, to provide binding properties, and alkanethiols to fill in the areas left uncovered by the receptors. Because of this, the preparation of these monolayers is complicated and experimentally demanding. In the next section an alternative approach currently under exploration in my group will be discussed in some detail.

Interfacial self-assembly of nonthiolated receptors

Is it necessary to use thiolated receptors to prepare monolayers with active binding sites? The answer is obviously no. An alternative and very promising approach would be based on the preparation of self-assembled monolayers from alkanethiols possessing appropriate terminal functionalities that could then be used, in a second step, to direct the self-assembly of the nonthiolated receptor molecules. For instance, the self-assembly of carboxylate-terminated alkanethiols is well known and yields carboxylate surfaces which have very interesting properties as substrates for the preparation of molecular assemblies. Carboxylate-terminated SAMs have been used as templates for the adsorption of cationic polyelectrolytes, metal ions, and other species [30–32]. Quite recently, the author's group has demonstrated the adsorption of viologen-containing polyelectrolytes on the surface of carboxylate-terminated monolayers [33]. Specifically, we investigated two polyelectrolytes having either butyl or undecyl linkages between the repeating viologen subunits (see Table 1 for structures). Both polyelectrolytes were found to adsorb at neutral pH on monolayers prepared by the self-assembly of either 3-mercaptopropionic or 8-mercaptooctanoic acids on gold. The extent of adsorption varies depending on the nature of the polyelectrolyte and the type of surface as reflected by the values of viologen surface coverages measured in voltammetric experiments and given in Table 1. The adsorption is optimized by using the more hydrophobic polyelectrolyte and the longer thiol. However, even under optimum conditions, the viologen polyelectrolyte is not tightly packed at the interface. The

Table 1 Viologen surface coverages (mol/cm^2) measured for the adsorption of viologen polyelectrolytes on three different surfaces

	Bare Au	Au/S(CH$_2$)$_2$COOH	Au/S(CH$_2$)$_7$COOH
$\left[\overset{+}{N} \bigcirc\!\!\bigcirc \overset{+}{N}\!\!-\!\!(CH_2)_4 \right]_n$	3.5×10^{-11}	3.2×10^{-11}	5.9×10^{-11}
$\left[\overset{+}{N} \bigcirc\!\!\bigcirc \overset{+}{N}\!\!-\!\!(CH_2)_{11} \right]_n$	4.6×10^{-11}	5.0×10^{-11}	7.2×10^{-11}

gradual deprotonation of the $-S(CH_2)_7COOH$ monolayer in the pH range 6–11 enhances the adsorption of the cationic polyelectrolyte [33]. From these data, we obtained an apparent pK_a of ~ 8 for the monolayer $-COOH$ groups, in agreement with previous reports [34–36].

This work reveals that carboxylate-terminated SAMs afford excellent surface properties for the assembly of cationic compounds, especially if they have amphiphilic character. With this idea in mind we decided to investigate the assembly of a series of novel amphiphilic cyclodextrin receptors, prepared by Coleman and coworkers [37] on this type of carboxylate surfaces. At neutral pH the amino groups on these cyclodextrin derivatives are extensively protonated and, thus, their structure is amphiphilic as the primary side harbors a crown of $-NH_3^+$ groups while the secondary side bears the aliphatic chains. We reasoned that these cyclodextrin hosts would be ideally suited to aggregate on a gold surface, especially if it is covered by a monolayer of carboxylate-terminated thiols, owing to the favorable electrostatic interactions between the negative charges on the monolayer surface and the positive charges on the cyclodextrin hosts.

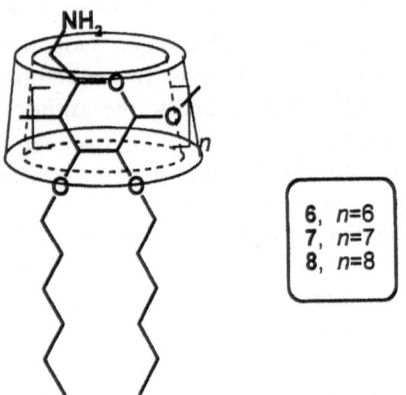

6, $n=6$
7, $n=7$
8, $n=8$

These expectations have been recently verified [38]. Cyclic voltammetric, contact angle, FT-IR, and quartz crystal microbalance (QCM) measurements indicate that all three receptors (**6–8**) aggregate effectively on gold surfaces covered by carboxylate-terminated SAMs. In particular, contact angle, FT-IR and QCM data indicate that

receptors **6–8** aggregate on bare gold, forming monolayers in which the cyclodextrins adopt random orientations. Conversely, the level of organization of the cyclodextrin aggregates on gold surfaces modified with mercaptopropionic or mercaptooctanoic acids is much better. These findings suggest that although the amphiphilic nature of **6–8** is enough to induce their interfacial accumulation, the electrostatic interactions between the monolayer's carboxylate groups and the cationic receptors constitute the primary forces leading to the formation of well-organized molecular assemblies. These investigations are currently underway in the author's laboratory, but from QCM data, we have clearly established that the amphiphilic cyclodextrin receptors form organized multilayer structures on carboxylate-terminated surfaces. The molecular recognition properties of these cyclodextrin assemblies are still under investigation. Preliminary results indicate that the receptors can actively bind ferrocene from a contacting aqueous solution. However, the presence of the alkyl chains extending from the cyclodextrin cavities modifies their selectivity and provides external, non-cavity binding sites for hydrophobic substrates in the multilayer structure. In any case, this work demonstrates that it is possible to prepare organized molecular assemblies containing active binding sites without relying on thiolate-based receptor immobilization reactions.

Non-sensing applications

The potential applications of these monolayer systems for the development of sensors has been made clear throughout this work. However, binding interactions at the electrode–solution interface can also be advantageously utilized with different goals, such as the design and preparation of novel materials. In collaboration with Professor Luis Echegoyen's group, at the University of Miami, we have recently completed some work which points in this direction. In this case, we modified a gold surface with cystamine, resulting in a surface-containing ammonium groups facing the solution. The surface coverage of the cystamine monolayer is rather low (about 35% of a well-packed monolayer) but this does not seem to be very

Progr Colloid Polym Sci (1997) 103:193–200
© Steinkopff Verlag 1997

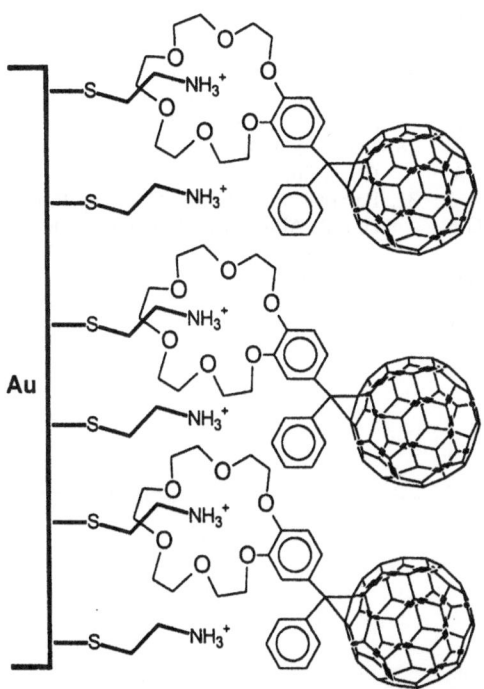

Scheme 4 Self-assembly of fullerene–crown derivatives

of a fullerene species. Our data suggest that lateral fullerene–fullerene contacts greatly assist the self-assembly of the monolayer of **9**, but control experiments strongly indicate that without the ammonium–crown ether interactions the self-assembly of the fullerene overlayer does not take place [39].

Conclusions and future work

The work described in this article represents an exciting research effort to utilize self-assembled monolayers for the design of novel sensor devices. Current efforts are addressing the development of gas-phase sensors for volatile compounds or sensors for the detection of soluble species in aqueous media. The selectivity of these monolayer-based devices does not yet approach the requirements for operation in complex biological or environmental samples. In the case of the systems designed for operation in aqueous solution, the sensitivity obtained so far is appropriate for some practical applications. In addition to these significant research avenues, schemes relying on interfacial molecular recognition events can also be used for the design of novel materials and films. Therefore, this field of research holds substantial promise and will probably see substantial growth in the next few years.

relevant. We then took advantage of the well-known interaction between $-NH_3^+$ groups and crown ethers to assemble a second layer over the cystamine monolayer (see Scheme 4) [39]. Exposure of the $Au/S(CH_2)_2NH_3^+$ electrode to a dichloromethane solution of the fullerene–crown ether derivative **9** (see structure in Scheme 4) leads to the fast and reversible formation of a reasonably well-packed monolayer of **9** molecules (1.4×10^{-10} mol/cm²). This constitutes a novel example of a monomolecular film

Acknowledgments The author is grateful to the National Science Foundation for the continuous support of this research. Several students have contributed substantially to the research in this area performed in the author's group, namely: Luis A. Godínez, María T. Rojas, and Litao Zhang. The author wishes to express his appreciation to Professors Luis Echegoyen, George W. Gokel, and J. Fraser Stoddart with whom he has collaborated in several aspects of the research work described here.

References

1. Whitesides GM, Laibinis PE (1990) Langmuir 6:87
2. Ulman A (1991) Ultrathin Organic Films, Chapter 3 Academic Press, New York
3. Dubois LH, Nuzzo RG (1992) Annu Rev Phys Chem 43:437
4. Zhong C-J, Porter MD (1995) Anal Chem 67:709A
5. Mandler D, Turyan I (1996) Electroanalysis 8:207
6. Odell B, Reddington MV, Slawin AMZ, Spencer N, Stoddart JF, Williams DJ (1988) Angew Chem Int Ed Engl 27:1547
7. See, for example: Goodnow TT, Reddington MV, Stoddart JF, Kaifer AE (1991) J Am Chem Soc 113:4335
8. Amabilino DB, Stoddart JF (1995) Chem Rev 95:2725
9. Philp D, Stoddart JF (1996) Angew Chem Int Ed Engl 35:1154
10. Lu T, Zhang L, Gokel GW, Kaifer AE (1993) J Am Chem Soc 115:2542
11. Thoden van Velzen EU, Engbersen JFJ, Reinhoudt DN (1994) J Am Chem Soc 116:3597
12. Thoden van Velzen EU, Engbersen JFJ, de Lange PJ, Mahy JWG, Reinhoudt DN (1995) J Am Chem Soc 117:6853
13. Schierbaum KD, Weiss T, Thoden van Velzen EU, Engbersen JFJ, Reinhoudt DN, Göpel W (1994) Science 265:1413
14. Huisman B-H, Thoden van Velzen EU, van Veggel FCJM, Engbersen JFJ, Reinhoudt DN (1995) Tetrahedron Lett 36:3273
15. Huisman B-H, Rudekevich DM, van Veggel FCJM, Reinhoudt DN (1996) J Am Chem Soc 118:3523
16. Grate JW, Patrash SJ, Abraham MH, Du CM (1996) Anal Chem 68:913
17. Adams H, Davis F, Stirling CJM (1994) J Chem Soc Chem Commun 2527
18. Davis F, Stirling CJM (1995) J Am Chem Soc 117:10385
19. For a recent example, see: Yang HC, Dermody DL, Xu C, Ricco AJ, Crooks RM (1996) Langmuir 12:726

20. Zhou Y, Bruening ML, Bergbreiter DE, Crooks RM, Wells M (1996) J Am Chem Soc 118:3773
21. Wells M, Crooks RM (1996) J Am Chem Soc 118:3988
22. Mirzoian A, Kaifer AE (1995) J Org Chem 60:8093
23. Rubinstein I, Steinberg S, Tor Y, Shanzer A, Sagiv J (1988) Nature 332:426
24. Steinberg S, Tor Y, Sabatini E, Rubinstein I (1991) J Am Chem Soc 113:5176
25. Zhang L, Godínez LA, Lu T, Gokel GW, Kaifer AE (1995) Angew Chem Int Ed Engl 34:235
26. Rojas MT, Königer, Stoddart JF, Kaifer AE (1995) J Am Chem Soc 117:336
27. Rojas MT, Kaifer AE (1995) J Am Chem Soc 117:5883
28. Godínez LA, Patel S, Criss CM, Kaifer AE (1995) J Phys Chem 99:17449
29. Bernardo AR, Stoddart JF, Kaifer AE (1992) J Am Chem Soc 114:10624
30. Li J, Liang KS, Scoles G, Ulman A (1995) Langmuir 11:4418
31. Bharathi S, Yegnaraman V, Rao GP (1995) Langmuir 11:666
32. Collinson M, Bowden EF (1992) Langmuir 8:1247
33. Godínez LA, Castro R, Kaifer AE (1996) Langmuir 12:5087
34. Bain CD, Whitesides GM (1989) Langmuir 5:1370
35. Troughton EB, Bain CD, Whitesides GM, Nuzzo RG, Allara DL, Porter MD (1988) Langmuir 4:365
36. Tarlov MJ, Bowden EF (1991) J Am Chem Soc 113:1847
37. Parrot-Lopez H, Ling C, Zhang P, Baszkin A, Albrecht G, Rango C, Coleman AW (1992) J Am Chem Soc 114:5479
38. Godínez LA, Kaifer AE, to be submitted
39. Arias F, Godínez LA, Kaifer AE, Echegoyen L (1996) J Am Chem Soc 114:6086

Progr Colloid Polym Sci (1997) 103:201–215
© Steinkopff Verlag 1997

D. Bizzotto
J. Lipkowski

Amphiphiles at electrified interfaces

Received: 19 December 1996
Accepted: 29 December 1996

Abstract The adsorption of insoluble surfactants, spread as a monolayer at the gas–solution interface (GS), onto an electrified metal–solution (MS) interface of a gold single-crystal electrode has been investigated. A Langmuir trough converted into an electrochemical cell was employed in these studies and the adsorption of insoluble surfactants on the Au electrode surface was measured using electrochemical techniques. The results of these experiments have shown that the transfer of insoluble surfactants from GS onto the MS interface is strongly affected by the electrode potential and the transfer ratio is 1:1 only at the potential of zero charge (pzc). UV-Vis and light scattering experiments were employed to demonstrate that insoluble surfactants may be desorbed from the electrode surface at very negative potentials. The desorbed surfactants are trapped in the subsurface region in the form of aggregates, most likely micelles. When the electrode potential is changed to a value close to pzc the micelles spread back onto the electrode surface. The potential induced adsorption of micelles proceeds through a hemimicelle stage. We have shown that the shape and size of the hemimicelle may be conveniently controlled by the electrode potential.

Key words Insoluble surfactants – adsorption at electrodes – Langmuir–Blodgett films – micelles – hemimicelles

D. Bizzotto · J. Lipkowski (✉)
Guelph-Waterloo Centre for Graduate
Work in Chemistry
Department of Chemistry and Biochemistry
University of Guelph
Guelph, Ontario, Canada N1G 2W1

Introduction

The adsorption of surfactants on solid–liquid interfaces is at the center of interest in colloid and surface science. The nature of interactions between the solid surface, surfactant and the solution phase need to be understood to control the adsorption of surfactants. Solid surfaces in contact with aqueous solutions, especially surfaces of clays or minerals, are frequently charged and this charge generates a potential at the solid–solution interface. The adsorption of surfactants on the solid–liquid interface quite often proceeds at an interface which is electrified. Unfortunately, the influence of the surface potential on adsorption of surfactants has been little studied.

Electrochemistry offers the best opportunity to control potential at the solid–solution interface and to study the behaviour of surfactant molecules in the presence of an electrostatic field. Miller [1, 2] was the first to show that a film of lipids spread at the air/solution interface of a Langmuir trough can be transferred onto a surface of a Hg electrode. He demonstrated that the field-induced changes of the film properties may be conveniently studied using electrochemical techniques. Miller's approach was applied later by Nelson and Guidelli for biomimicking research. They deposited films of phospholipids onto a Hg electrode surface [3–6] and investigated either the incorporation of a number of toxins [7, 8] or ion transport through lipid layers with and without incorporated ion channels [9–15]. Others have also pursued electrochemistry in a Langmuir trough [16–20].

The object of this review is to describe the spreading of insoluble surfactants onto an electrified metal solution interface. We will demonstrate that by converting a Langmuir trough into an electrochemical cell, one may determine the film pressure of surfactants transferred from the gas–solution (GS) interface of the Langmuir trough to the metal–solution (MS) interface of the electrode. We will measure the ratio of the film pressures at the MS and GS (the transfer ratio) and will demonstrate how this transfer ratio is affected by the electrode potential. We will also employ UV-Vis spectroscopies and light scattering experiments to demonstrate that insoluble surfactants may be desorbed from the electrode surface, when a large electrostatic field is applied to the electrode surface. We will show that the desorbed surfactants are trapped in the subsurface region in the form of aggregates, most likely as micelles. When the magnitude of the electrostatic field at the surface is reduced the micelles spread back onto the electrode surface. We will provide arguments that the potential induced adsorption of the micelles proceeds through a hemimicelle stage. We will show that the shape and size of the hemimicelle may be conveniently controlled by the electrode potential. These results are general in nature and may be helpful in understanding the behavior of a wide class of insoluble surfactants at solid–liquid interfaces.

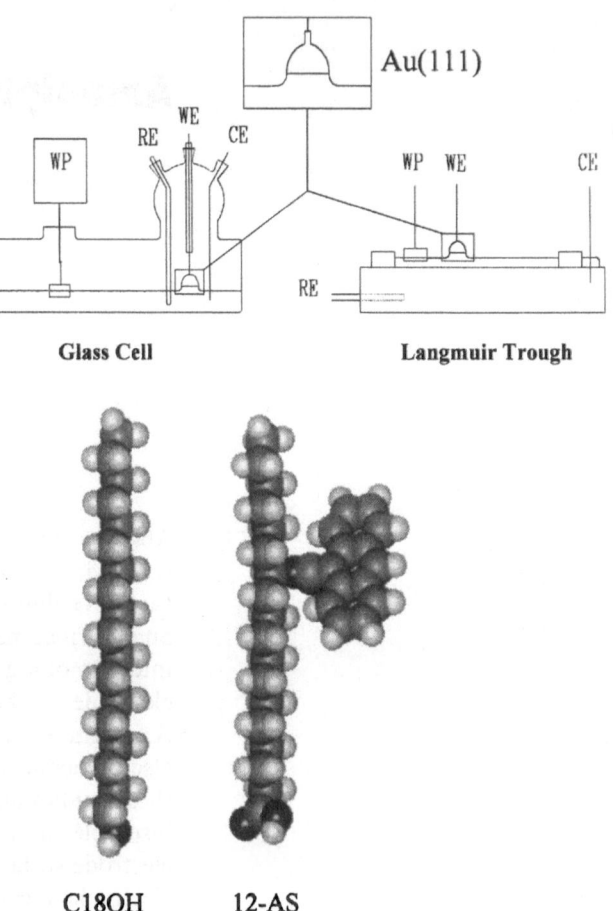

Fig. 1 Top section, schematics of the glass cell and the electrochemical Langmuir trough. Bottom section, space-filling models of surfactants used in our studies

Characterization methods

We have described experimental techniques and data-processing procedures in a series of recent publications [21–26]. Figure 1 shows the space-filling models of surfactants used in our studies. They included octadecanol, 12-(9-anthroloxy) stearic acid (12-AS) and 4-pentadecylpyridine. In addition, octadecanoic acid was used in some experiments [22]. A special glass cell or a home-made Langmuir trough converted into an electrochemical cell were used in our studies. Figure 1 is a diagram of the trough and the cell. The glass cell had a port through which a Wilhelmy plate (WP), attached to a microbalance, measured the film pressure of surfactant spread at the GS interface during the electrochemical experiment. The Langmuir trough was equipped with a reference electrode (RE) and a counterelectrode (CE) so that the electrochemical experiments could be performed on a monolayer compressed to a desired value of the film pressure at the GS interface. The surfactant was dissolved in chloroform and placed onto the surface of the electrolyte. When experiments were performed in the glass cell, enough surfactant was spread onto the surface to cover the surface twice over. This created a monolayer on the surface of the electrolyte whose film pressure was equal to the equilibrium spreading pressure. When the experiments were performed in the Langmuir trough, the pressure of the monolayer spread at GS was controlled by the position of the movable barrier. The films of surfactants were deposited onto the surface of a gold single-crystal electrode (working electrode (WE)) using the horizontal touching procedure. Only a few experiments were performed in which the film was deposited onto the surface of a hanging mercury electrode. The gold electrode was flame-annealed and cooled in Ar, then touched to the monolayer at the air/solution interface. The electrode was then slightly pulled away from the plane of the aqueous surface to form a hanging meniscus. This was termed a single touch layer. This hanging meniscus orientation ensured that only the polished surface of the electrode was exposed to the solution. Multiple touch layers could also be produced by immersing the electrode

Progr Colloid Polym Sci (1997) 103:201–215
© Steinkopff Verlag 1997

at potentials near the potential of zero charge (pzc), removing the small (typically) drop of electrolyte then retouching the monolayer covered GS interface as before resulting in the hanging meniscus again. The adsorbed surfactant was characterized using cyclic voltammetry, differential capacity and charge density measurements, details of which are described in our previous publications [21–23, 25, 27]. The chronocoulometric measurements combined with the electrocapillary equation allowed the film pressure of the adsorbed film to be determined and directly compared to the film pressure of surfactant at the GS interface measured using a Wilhelmy plate.

Electroreflectance, fluorescence and elastic light-scattering experiments used to study the mechanism by which surfactants spread at the MS interface are described in [25]. Monochromatic linearly polarized light was focused onto the electrode surface at a 45° angle. The specularly reflected light was collected in electroreflectance experiments. Fluorescence and elastically scattered light were collected at an angle of 90° with respect to the plane of incidence and 45° with respect to the electrode surface.

Transfer of an insoluble monolayer from GS onto MS interface

Gold and mercury electrodes display ideally polarizable behavior in a broad range of electrode potentials. Under these conditions, the metal solution interface behaves as a capacitor. The capacity of the electrode decreases when organic molecules are present at the interface [28, 29]. The measurement of the electrode capacity provides a convenient tool to study the spreading of an insoluble monolayer onto a metal electrode surface. The capacity may be measured by applying a linear voltage sweep to the electrode and recording the voltammetric current (CV-cyclic

voltammetry curves). The voltammetric current is equal to the product of the differential capacity of the electrode and the voltage sweep rate. The differential capacity can also be measured directly using an ac impedance technique [28].

Figure 2 shows the cyclic voltammetry and differential capacity curves determined for Au(1 1 1) and Hg electrodes horizontally touched to the GS interface that was film free (dotted lines) and covered by the film of octadecanol (solid or dashed lines). In order to compare the adsorption of insoluble surfactants at the two metals, all curves are plotted versus the rational potential defined as $E_{rational} = E - E_{pzc}$, where E is the electrode potential measured versus the saturated calomel electrode (SCE) and E_{pzc} is the potential of zero charge for a given metal ($E_{pzc} = 275$ mV (SCE) for Au(1 1 1) and $E_{pzc} = -450$ mV for Hg). The octadecanol film lowers the electrode capacity at potentials close to pzc indicating that surfactant molecules are adsorbed at the metal surface in this range of potentials. At the Hg electrode, the adsorption of octadecanol was investigated in a broad range of both negative and positive potentials. At the Au(1 1 1) electrode, oxidation of octadecanol took place at positive potentials and for this metal the studies were restricted to negative polarizations. Looking at the curves for the mercury electrode first, we note that adsorption of octadecanol is quite symmetric with respect to pzc and extends over a range of 300–400 mV in the negative and positive directions. The desorption of the film is observed at larger polarizations. The desorption is responsible for the appearance of the characteristic desorption peaks on the voltammetric curve and a sudden increase of a capacity on the differential capacity curve. At even larger polarization both the voltammetric current and the differential capacity curves recorded in the presence of the film merge with the

Fig. 2 Cyclic voltammetry (top panels) and differential capacity (bottom panels) recorded for Au(1 1 1) (left) and hanging mercury drop electrode (right) in contact with 0.05 M KClO$_4$ in the absence (dotted line) and in the presence (solid line-voltage scan in the positive direction, dashed line-voltage scan in the negative direction) of C$_{18}$OH spread at the GS interface. CV: 20 mV/s, capacity: 5 mV rms, 25 Hz for Au(1 1 1) or 253 Hz for Hg, 5 mV/s sweep rate, a series RC circuit was assumed in the calculation

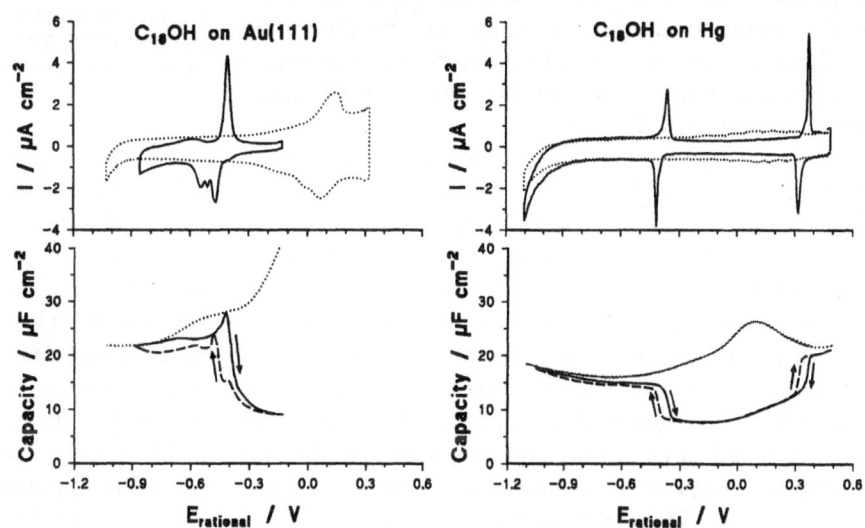

corresponding plots determined for the film free interface. When the direction of the voltage scan is reversed and E moves back towards the pzc, the film spreads again onto the interface at potentials about ± 300 mV with respect to pzc. There is an apparent hysteresis between the desorption and adsorption cycles which indicates that the film desorption and readsorption processes are slow. The slow kinetics of the desorption/adsorption phenomena are responsible for the absence of the absorption/desorption peaks on the differential capacity curves.

The curves determined for the Au(1 1 1) electrode show that, at negative potentials, the adsorption of octadecanol at gold displays a number of similarities to the adsorption at mercury. The capacity of the film covered electrode has a comparable value, the desorption of the film takes place at approximately these same potentials and similar adsorption hysteresis is observed at the two metals. The main difference between the behavior of the two electrodes is the presence of multiple desorption peaks on CV recorded for gold and a single desorption peak on CV determined for mercury. This behavior indicates that the adsorption/desorption kinetics are more complex on a solid than on a liquid metal. Tentatively, we explain these differences in terms of homogeneity of the electrode surface. The surface of the liquid electrode (mercury) is homogeneous. In contrast, the gold single-crystal surface is a mosaic of well ordered (1 1 1) domains separated by steps, domain boundaries, etc., and is not energetically homogeneous. Although the mercury electrode displays more ideal behavior than gold electrode, it is easier to perform spectroscopic experiments described in the next section on gold than on mercury. For this reason most of the studies described in this paper were performed on gold rather than on mercury. We note also that, due to a lower hydrogen overvoltage at the gold–water interface the CVs recorded at this electrode are somewhat distorted by the faradaic currents due to the hydrogen evolution reaction. Since the capacities measured using the ac impedance technique are less affected by the faradaic currents, we will use chiefly the differential capacity curves to characterize the adsorption of surfactants at gold electrode surfaces.

Since adsorption of an insoluble surfactant changes the electrode capacity it also changes the charge density at the metal surface. The charge density at the metal surface may be conveniently measured using a chronocoulometric technique described by us in earlier publications [21–23]. Figure 3a shows the charge density curves measured for the Au(1 1 1) electrode in the absence and presence of octadecanol. The curve marked as "desorption" corresponds to charge acquired by depositing the monolayer onto the metal surface at pzc and then desorbing the film by stepping the potential in the negative direction (desorbing the film). The curve marked "adsorption" corre-

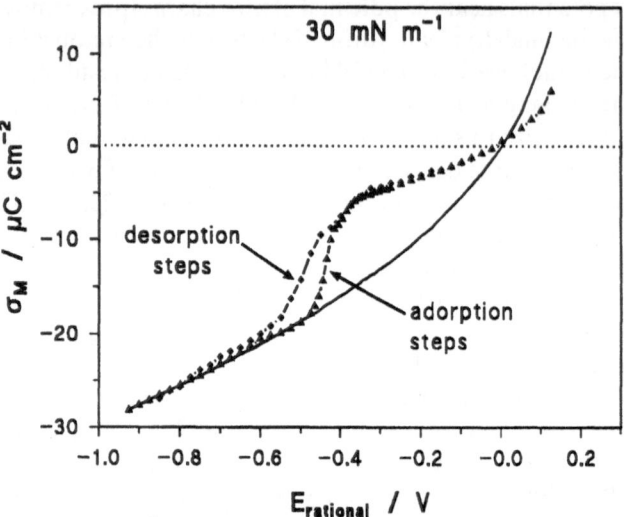

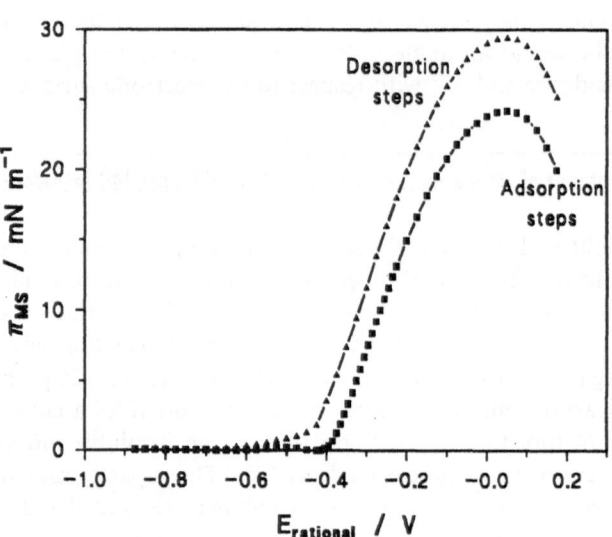

Fig. 3 Top panel, charge density versus electrode potential plots determined on Au(1 1 1) for 30 mN m^{-1} film pressure of C$_{18}$OH spread at the GS interface of the Langmuir trough. Bottom panel, the film pressure at MS calculated by integration of the charge density curves. Supporting electrolyte 0.05 M KClO$_4$

sponds to charge measured when the contact between the gold electrode at the GS interface was made at $E = -800$ mV, where the film does not adsorb on the electrode surface, and stepping the potential in the positive direction, to potentials where the insoluble surfactants adsorb onto the metal surface (adsorption of the film). The hysteresis in charge density corresponding to the film adsorption and desorption has a full analogy to the hysteresis observed previously on the cyclic voltammetry or the differential capacity curves. The adsorption of the insoluble surfactant apparently decreases the charge density at the electrode surface. Consequently, a characteristic step appears on the charge density curves in the presence of the film.

The product of charge and potential is equal to energy, therefore, the area contained between the charge density curves corresponding to the film free and the film covered electrode is equal to the change of the surface energy due to adsorption of the surfactant molecules. In fact, this energy change is equal to the pressure of the film at the MS (π_{MS}). Figure 3b shows the film pressures at the metal solution interface determined by integration of the two charge density plots. The curves are parabolic in shape with the maximum at the potential of zero charge. We note that, for a fixed value of the film pressure at the GS, the film pressure at the MS depends on the electrode potential. We also note that the charge densities corresponding to adsorption and desorption of the film give somewhat different film pressure values. We will show later that adsorption of the film is likely to be slower than desorption. For this reason we consider the film pressures determined from charge densities corresponding to desorption of the film as being closer to the equilibrium values.

The data in Fig. 3b show that electrochemistry provides a means to measure the pressure of the film that was transferred from the GS interface of a Langmuir trough onto the MS interface of a metal electrode. Using a trough converted into an electrochemical cell, we can measure simultaneously the pressure of the films at GS and MS interfaces. In this way the transfer of the film from GS onto MS may be studied quantitatively. Figure 4 shows the film pressure data determined in a series of experiments in which the film pressure of octadecanol spread at the GS was varied and for each value of the film pressure at GS the charge densities at the Au(1 1 1) electrode were measured. The film pressures were calculated only from the charge densities corresponding to the desorption of the film. The π_{MS} are plotted in a three-dimensional (3D) diagram in which the calculated values of π_{MS} are plotted on the vertical axis, while the film pressure at the GS interface (π_{GS}) and the electrode potential are plotted on the two axes of the basal plane. The 3D representation provides a condensed description of the dependence of π_{MS} on the two independent experimental variables π_{GS} and E. These data show clearly that spreading of the insoluble surfactants from GS interface onto the electrified MS interface depends strongly on the electrode potential. For a fixed value of π_{GS}, the π_{MS} may change from a maximum value at the potential of zero charge to zero at the most negative potentials. Apparently, the insoluble surfactant spreads easily from the GS onto the uncharged MS interface. The straight line joining the maxima on the π_{MS} versus E curves has a slope of unity. It corresponds to a 1:1 transfer of the insoluble monolayer from the GS onto the MS interface at the pzc. The transfer ratio drops below unity as the poten-

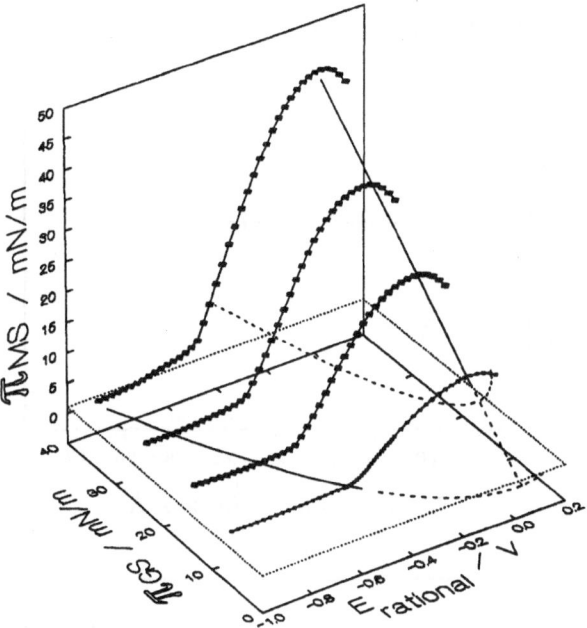

Fig. 4 Three-dimentional representation of the film pressure at MS interface as a function of the film pressure at the GS interface and of electrode potential

tial departs from the pzc and it eventually becomes equal to zero at very negative potentials where the charge density at the metal is sufficiently large.

An intersection of the curves in Fig. 4 with a plane parallel to the basal plane of the figure shows how π_{GS} has to change with potential in order to maintain a constant value of π_{MS}. The dotted line in Fig. 4, corresponding to π_{MS} equal to $10\,mN\,m^{-1}$ displays a quasi-parabolic dependence on E. The quasi-parabolic line lying in the basal plane ($\pi_{MS} = 0$) denotes the potential at the onset of the spreading of the insoluble monolayer from the GS onto the MS interface. At potentials located beyond this line the surfactant will not spread from the GS onto the MS interface regardless of the value of π_{GS}.

This result shows how the interaction of a surfactant molecule with the solid surface is influenced by the interfacial potential. It may be used to predict the behavior of neutral surfactants at various solid–liquid interfaces. It also shows that good quality films of insoluble surfactants can be deposited at solid surfaces only when the charge at the solid surface is close to zero. Our results show also that the adsorption of insoluble surfactants displays many similarities to the adsorption of small, soluble organic molecules. This point is illustrated by Fig. 5 where, the top panel (Fig. 5a), shows the film pressures at the pzc of the Au(1 1 1) electrode and at GS plotted against the mean molecular area (MMA) of $C_{18}OH$, at the surface of the Langmuir trough. The changes of π_{MS} track the changes of

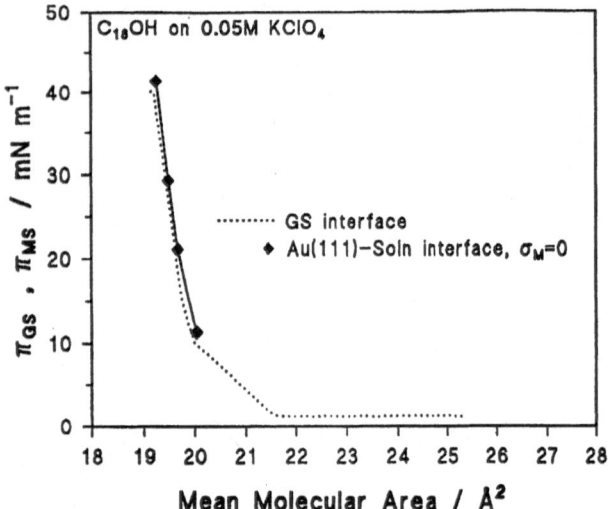

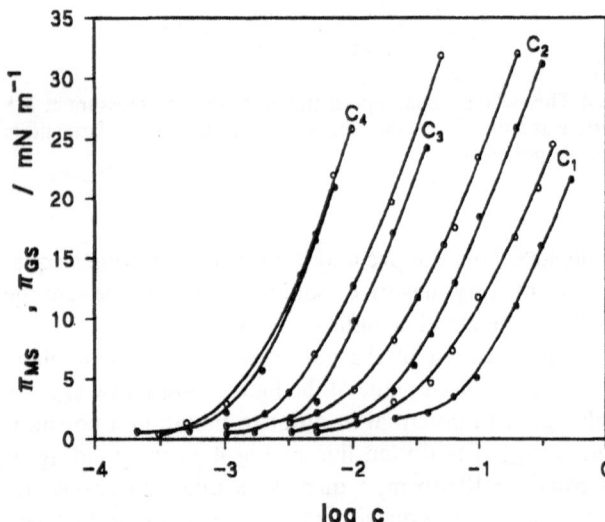

Fig. 5 Top panel, (dotted line) compression isotherms for $C_{18}OH$ on 0.05 M $KClO_4$ at (dotted line) the GS interface, (solid line) an uncharged MS interface. Bottom panel, (open points) plots of the film pressure at the MS of a Hg electrode (at the pzc) and (closed points) the film pressure at the air–solution interface, versus the logarithm of bulk concentration of normal, short chain, soluble, aliphatic alcohols. The number of carbon atoms in the molecule is indicated at each curve. Data taken from Ref. [30]

π_{GS} consistent with the 1:1 transfer from GS onto MS. The bottom panel, Fig. 5b, shows π_{MS} corresponding to adsorption of soluble, short-chain aliphatic alcohols at the pzc of a Hg electrode and the π_{GS} corresponding to the adsorption of these molecules at the air–solution interface, plotted against the logarithm of the bulk concentration of the alcohol. These data are taken from a paper by Frumkin and Damaskin [30]. They demonstrate that as the length of the n-alkyl alcohol increases, the difference be-

tween the interfacial tension and surface tension decreases and for hexanol the π_{MS} data essentially track the π_{GS} data in full agreement with the behavior of $C_{18}OH$ displayed in Fig. 5a.

The similarities between adsorption of insoluble surfactants and small organic molecules are illustrated further by quite a similar dependence of the adsorption of these two classes of compounds on the surface crystallography of the gold electrode. Figure 6a shows film pressures for adsorption of octadecanoic acid (ODA) at the three low index gold single-crystal electrodes. These data were acquired by keeping a constant pressure of the film of ODA at the surface of the Langmuir trough (15 mN m^{-1}) and recording the charge density curves for the three gold single-crystal electrodes. These results show that, for an identical film pressure at GS, the film pressure at the MS interface decreases by moving from the Au(1 1 1) to the Au(1 1 0) plane. The insoluble surfactant spreads less onto the more open (1 1 0) surface than onto the densely packed (1 1 1) surface. Figure 6b shows the film pressure curves for the adsorption of diethylether, a soluble organic molecule onto the same gold surfaces. These data were taken from Refs. [21, 31]. For this same bulk concentration of the diethylether, its film pressure at the MS displays similar dependence on the surface crystallography of gold to that observed for adsorption of ODA molecules in Fig. 6a. Apparently, the same dependence of adsorption on the surface crystallography is observed regardless whether the surfactant molecules arrive at the MS interface from the bulk of solution or from the GS interface. This fact sheds light on the mechanism of spreading of an insoluble surfactant from the GS to the MS interface. It is well established that the dependence of adsorption of diethylether on the crystallographic orientation of gold can be explained in terms of a solvent-substitution mechanism of adsorption from solution and variable hydrophilicity of different gold surfaces [21, 31]. The similarity between adsorption of an insoluble surfactant from GS interface and a soluble surfactant from the bulk of the solution, suggests that adsorption of the insoluble surfactant at gold electrodes is also a solvent-substitution reaction.

The strong dependence of the adsorption of an insoluble surfactant on the solid electrode surface crystallography explains also the differences between spreading of the insoluble film onto a liquid surface of a Hg and a solid surface of the Au(1 1 1) electrode discussed by us earlier. Defects inevitably present at the surface of a single crystal electrode may be treated as adsorption sites of different crystallographic orientations. Figure 6 shows that surfactant will spread differently at the (1 1 1) terraces and at the defect sites. Clearly, the film is likely to be less evenly spread on a single crystal surface of a solid than on a surface of a liquid electrode. The insoluble film will be

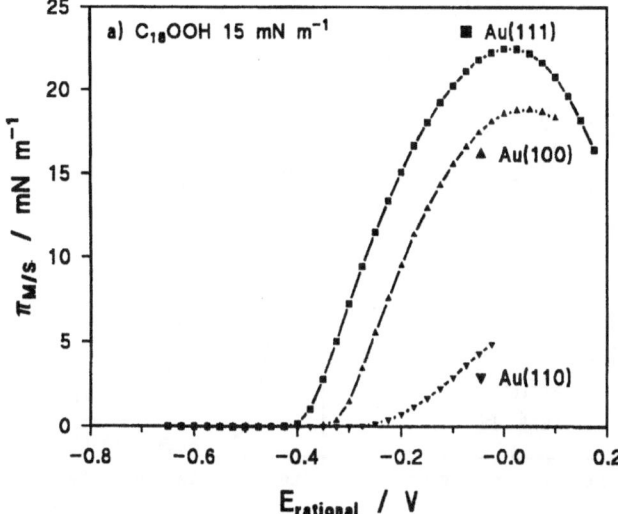

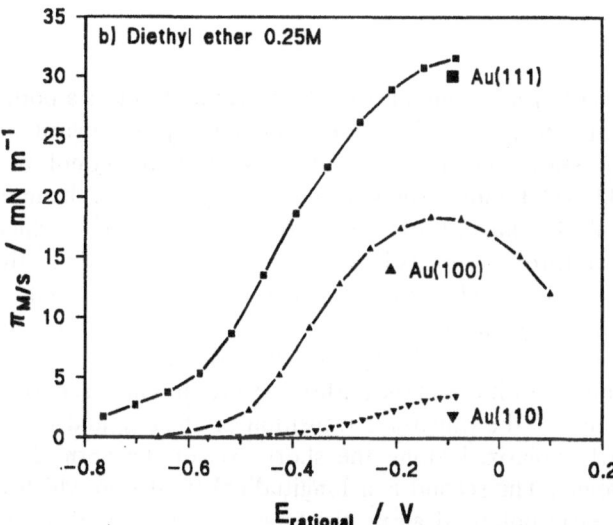

Fig. 6 Top panel, the pressure of the ODA film at the MS interface for Au(1 1 1), Au(1 0 0) and Au(1 1 0) electrodes when the ODA film pressure at the GS interface was equal to 15 mN m⁻¹. Bottom panel, the pressure of the diethylether film at the MS interface for Au(1 1 1), Au(1 0 0) and Au(1 1 0) electrodes in equilibrium with 0.25 M diethylether concentration in the bulk of the solution (data taken from Ref. [31])

much less evenly spread on a polycrystalline electrode consisting of micro- and macrofacets of different single crystal surfaces. Films deposited onto surfaces of polycrystalline materials are likely to have high concentrations of defects and so-called "pinholes".

Spectroscopic studies of spreading of insoluble surfactants at electrode surfaces

The experiments described in the preceding section led us to believe that surfactant molecules adsorbed at the MS

interface of an electrode are in equilibrium with surfactant spread at the GS interface. However, when the above experiments were completed an additional experiment was performed. A monolayer of $C_{18}OH$ was spread at the surface of the Langmuir trough and compressed to a film pressure of about 10 mN m⁻¹. The electrode was then brought in contact with the monolayer at the pzc. Next, the film at GS was decompressed to the film pressure of about 1 mN m⁻¹. The film pressure at MS measured for the electrode in contact with the film decompressed at GS was equal to 10 mN m⁻¹, exactly this same result as for the monolayer in the compressed state. This result suggested that an island of the monolayer initially spread at GS interface is trapped under the electrode surface and preserves its properties even when the rest of the monolayer at GS interface is decompressed. This island of the film could be repeatedly desorbed and later readsorbed onto the electrode surface without any loss of the initial film properties. We have also observed that a thicker film is deposited onto the metal surface, when the electrode initially covered by a monolayer of an insoluble surfactant is immersed at pzc and then retouched to the monolayer covered GS interface to form the hanging meniscus again. We believe a bilayer is formed using this procedure, however since we do not have sufficient evidence that this film has a bilayer structure we will refer to it as a double touch film. Similarly, for some surfactants, a triple touch film may be deposited onto the electrode surface by touching the monolayer covered GS three times. These multiple touch films may be desorbed from the electrode surface at sufficiently negative potentials and readsorbed back onto the metal surface when the potential returns to a value close to pzc. The desorption and adsorption processes are very repeatable and the film formed by readsorption of the surfactant molecules retains all its initial characteristics.

To understand the mechanism by which the surfactant is repeatedly desorbed and readsorbed, we have performed light-scattering experiments for the Au(1 1 1) electrode covered by the film of $C_{18}OH$ molecules. A beam of electromagnetic radiation was directed at the electrode surface at a 45° angle of incidence. The radiation scattered from the electrode was then detected at an angle of 45° with respect to the electrode surface and 90° with respect to the plane of incidence. Figure 7 shows the result of the combined electrochemistry–light-scattering experiment. A slow voltage sweep was applied to the electrode initially covered by a film produced by the single-touch and the double-touch procedure, the differential capacity of the electrode was measured and simultaneously the light elastically scattered from the electrode surface was recorded. The results show a very weak signal of scattered radiation measured for potentials at which the film was spread onto

Fig. 7 Capacity and intensity of elastically scattered light from a Au(1 1 1) electrode with C18OH adsorbed as a single (a) and double (b) touch layers. The incident light was 550 nm and s-polarized; (dotted line) differential capacity for the film free interface, (solid or dashed lines) differential capacities for the film covered electrode. The relative scales for the scattered light intensity are shown in the figure. Supporting electrolyte 0.05 M KClO₄

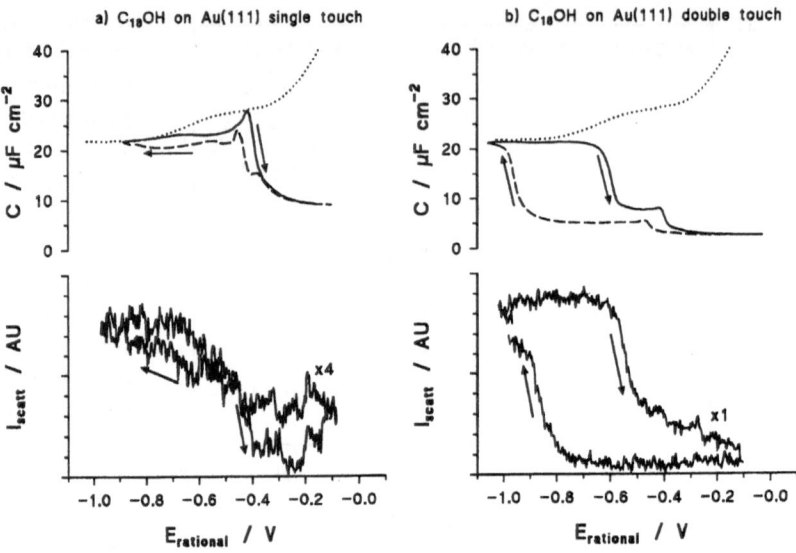

the electrode surface. However, at potentials at which the differential capacity shows that surfactant desorbs from the electrode surface, a dramatic increase in the amount of scattered radiation is observed. The scattered radiation reaches a plateau at the negative limit of potentials, where the film is totally desorbed from the electrode surface. When the direction of the voltage scan is reversed and we approach the pzc, the scattered radiation is quenched at potentials at which characteristic pit appears on the differential capacity curve, indicating that film is spread at the electrode surface. The changes of the intensity of the scattered radiation apparently follow the changes of the differential capacity. We note that the scattered radiation is about four times stronger for the film produced by the double-touch technique. In addition, the film produced by the double touch procedure is characterized by a lower capacity at potentials closer to pzc, it is apparently adsorbed in a broader range of the electrode potentials and displays a more pronounced hysteresis between the adsorption and the desorption cycle than the single-touch film. These features are consistent with a larger amount of surfactant being deposited onto the electrode surface by the double-touch procedure. The appearance of scattered radiation at potentials at which surfactant molecules are desorbed from the electrode surface suggests that the desorbed molecules form aggregates in subsurface region, which later spread back onto the metal surface at potentials close to pzc. We do not know the actual shape of these aggregates, however we will refer to them as micelles.

The potential-induced adsorption and desorption of insoluble surfactants was investigated further using 12-AS, an insoluble surfactant-dye molecule adsorbed onto the

Au(1 1 1) electrode surface. The 12-AS molecule is a popular fluorescent probe used to study fluidity and polarity of biological membranes and phospholipid monolayers [32–37]. Figure 8 shows a space-filling model of the molecule. It is an anthroloxy derivative of stearic acid in which an anthroloxy moiety is attached to the 12th carbon on the alkyl chain. The long axis of the anthroloxy group is collinear with the alkyl chain. The UV spectrum of this dye displays two absorption bands due to $(\pi* \leftarrow \pi)$ electronic transitions in the anthroloxy group. The first corresponds to the transverse transition at 360 nm, and is linearly polarized along the short axis of the anthroloxy moiety. The second is a longitudinal transition which is linearly polarized along the long axis of the anthroloxy group.

The adsorption of 12-AS on Au(1 1 1) was initially characterized using differential capacity. Figure 9 shows the differential capacity curves for the films produced by the single-, double- and the triple-touch procedures. The shape of the differential capacity curves apparently changes with the number of touches used for the film deposition. The capacity at potentials close to pzc becomes lower, the potential range in which the film is spread at the electrode surface becomes broader, and the hysteresis between the adsorption and desorption cycles becomes more pronounced with an increase in the number of touches. With increasing number of touches, the film becomes either more densely packed or thicker. We have performed electroreflectance, fluorescence and light-scattering experiments on so characterized films of 12-AS molecules. These studies were published recently [25], however, in order to make our review comprehensive we will present some of these results below.

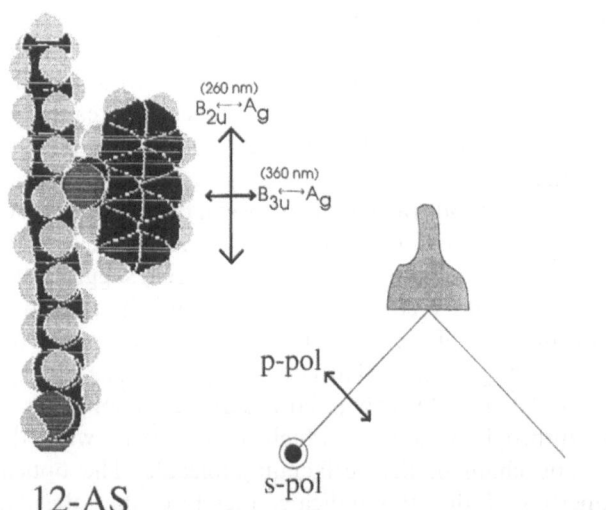

Fig. 8 Left side, space filling model of 12-AS molecule and the directions of the optical transitions. Right side, the geometry of the electroreflectance experiment

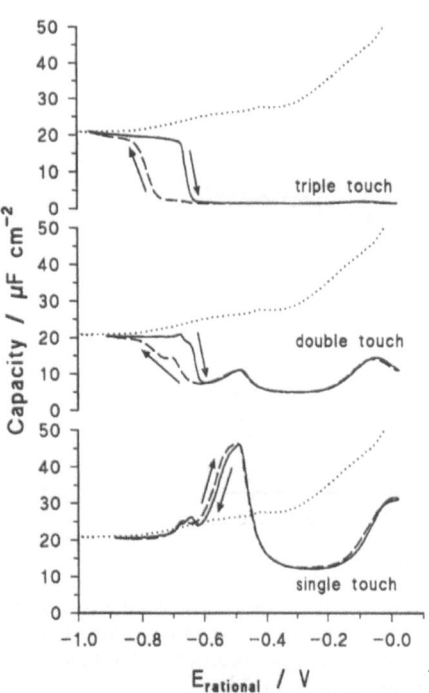

Fig. 9 Differential capacity of Au(1 1 1), without (dotted line) and with adsorbed 12-AS, single, double and triple touch film (as labeled). The positive (solid line) and negative (dashed line) directions of the voltage sweep are shown in each figure. The capacity was measured using 5 mV rms sine wave perturbation modulated at 25 Hz. A series RC circuit was assumed in determining the capacity

Due to complications caused by light scattering, the presentation of the electroreflectance spectra will be restricted to the single touch film for which only very weak elastic light scattering was observed at the desorption potential. The electroreflectance spectra were recorded by measuring the spectrum at $E_{ads} = +100$ mV/SCE, the potential where the monolayer is adsorbed onto the electrode, then stepping the potential to $E_{des} = -550$ mV/SCE at which the surfactant is desorbed (see Fig. 9) and recording the second spectrum. The two spectra were subtracted to give ΔR and then divided by the spectrum at the adsorption potential resulting in the calculated electroreflectance spectra $\Delta R/R$. Signal averaging was used to increase the signal to noise and typically four spectra were averaged before calculating ΔR. Both s and p linearly polarized incident light was used, spanning a wavelength range of 200–340 nm. The directions of the electric field of the photon with respect to the electrode surface are shown in Fig. 8. The electroreflectance $(\Delta R/R)$ spectra are presented in Fig. 10.

The changes in reflectivity $(\Delta R/R)$ for 12-AS on Au(1 1 1) can be described to a first approximation by

$$\frac{\Delta R}{R} = \frac{R(E_{ads}) - R(E_{des})}{R(E_{ads})}$$

$$= B\left[\cos^2 \theta(E_{des})\,\varepsilon(E_{des})\,\Gamma(E_{des})\right.$$

$$\left. - \cos^2 \theta(E_{ads})\,\varepsilon(E_{ads})\,\Gamma(E_{ads})\right], \qquad (1)$$

where R is the electrode reflectivity, θ is the angle between directions of the electric field of the photon and the transition dipole moment in the dye molecule, ε is the molar absorption coefficient, Γ is the surface concentration of the dye, B is a constant which incorporates all the other parameters of a specific model of the interface. The magnitude of $\Delta R/R$ depends on the mechanism by which the surfactant molecules are desorbed and readsorbed at the electrode surface. If the surfactant molecules are exchanged between MS and GS interfaces by moving out and onto the electrode surface, no dye molecules should be present in the optical path at potentials outside of the capacitive pit region at E_{des}. The first term in Eq. (1) should then be equal to zero ($\Gamma(E_{des}) = 0$) and the measured $\Delta R/R$ should be monopolar in character resulting in a typical absorption spectrum of the film of 12-AS molecules. According to the definition of $\Delta R/R$ given in Eq. (1) the spectrum should have a negative sign. In contrast, if the surfactant molecules are desorbed as micelles trapped under the electrode surface, the dye molecules would remain in the optical path even if they are desorbed from the electrode surface. In this case $\Delta R/R$ represents the difference between the absorption spectra of the dye molecules in the film at potentials within the pit and clustered as micelles at a potential outside the pit. The measured spectrum should then be different (bipolar) in character. The electroreflectance spectrum should be distinctly different

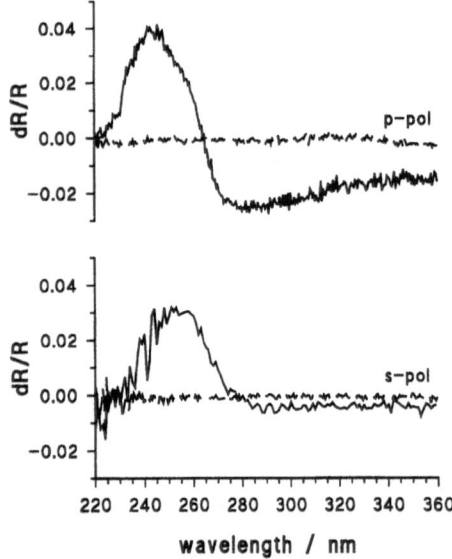

Fig. 10 Electroreflectance spectra (s and p polarized light); (solid line) for a single touch film of 12-AS adsorbed onto Au(1 1 1) and (dashed line) the background spectrum recorded without an adsorbed monolayer. Data taken from Ref. [25]

for each of the two mechanisms and hence the electro-reflectance experiments should lead to unambiguous identification of the surfactant desorption/readsorption mechanism.

The electroreflectance spectra, shown in Fig. 10, display absorption bands around 260 nm due to the longitudinal electronic transition in the anthroyloxy moiety of the surfactant molecules. The absorption band for p-polarized radiation is bipolar with the positive and negative sections having comparable amplitudes. The band for s-polarized light is predominantly unipolar; although the optical signal for this band changes sign at 280 nm, the negative portion of the spectrum is much weaker than its positive section. This band is dominated by the positive signal and hence we will refer to it as a unipolar band. In view of the above discussion the behavior of the p and s polarized bands may appear as contradictory. However, the predominant positive sign of the band recorded for s-polarized light indicates that the dye molecules are essentially optically inactive (or weakly active) at potentials in the capacitive pit (when they are spread as a film at the electrode surface), but become optically active at the potentials outside the film (when they are desorbed from the electrode surface). The positive sign of the unipolar band recorded for s-polarized light and the bipolar character of the band recorded using p-polarized radiation indicate that the desorbed surfactant molecules remain in the optical path, trapped in the electrolyte layer near the electrode surface probably in the form of micelles (or other

clusters such as flakes or vesicles). Therefore, the repeatable desorption and readsorption processes described above involve potential-induced micellization of the film at the negative potentials and spontaneous spreading of the micelles onto the electrode surface at more positive potentials.

The lack of optical activity (or weak optical activity) of the adsorbed dye molecules, when the electric field of the photon is parallel to the surface (s-polarized) suggests that the electric field vector of the photon and the transition dipole moment of the absorption band are orthogonal (or nearly orthogonal), so that $\cos^2 \theta(E_{ads}) \approx 0$. The absorption at 260 nm is linearly polarized along the long axis of the anthroyloxy moiety which is co-linear with the aliphatic chain of the surfactant molecule. The optical properties of the film indicate, therefore, that the dye molecules are packed into the film with the chains assuming a vertical orientation with respect to the surface. The characteristic bipolar behavior of the $\Delta R/R$ spectra determined using p-polarized radiation indicates that: (i) the adsorbed molecules are optically active ($\cos^2 \theta(E_{ads}) \approx 0$ for this polarization), (ii) the electronic transition for an adsorbed molecule takes place at approximately the same energies as for a molecule in the desorbed state and (iii) the spectrum for adsorbed molecules is significantly broadened and therefore the maximum band intensity was attenuated with respect to the spectrum of the solution species. These features of the spectrum of adsorbed molecules are consistent with the optical behavior of a monolayer of pyridine adsorbed at gold electrode surfaces [38] and with the optical spectra of other organic molecules adsorbed at group IB metals surfaces from the gas phase [39]. In these studies a significant broadening of the absorption bands for the adsorbed species was observed. In the present case, optical activity of adsorbed molecules interacting with p-polarized photons is expected for a vertical orientation. The lack of a band energy shift is consistent with an expected weak interaction between the optical tag and the metal surface when the surfactant molecule assumes the vertical orientation. The band broadening may be explained in terms of coupling of the dynamic dipole of the excited state of an adsorbed molecule to its image in the metal or to the electron–hole pair formed on the surface. Such a coupling can reduce the lifetime of the excited state by a factor of 10^5–10^6 resulting in a significant broadening of the optical transition [39].

The fluorescent properties of 12-AS have been widely used in membrane research. The fluorescence of the 12-AS molecules adsorbed at the gold surface should be quenched because of the reduced lifetime of the exited state. In contrast, the molecule should fluoresce if desorbed forming aggregates in the electrolyte layer near the electrode surface. These properties indicate that, if surfactant

Progr Colloid Polym Sci (1997) 103:201–215
© Steinkopff Verlag 1997

desorption/adsorption involves movement of the surfactant molecules parallel to the electrode surface (in and out of the MS interface), no fluorescence should be observed in the whole range of the investigated potentials. For potentials at which the surfactant is adsorbed the fluorescence should be quenched and at desorption potentials, surfactant molecules should not be present in the optical path. In contrast, if desorption involves the formation of aggregates in the subsurface region, the desorption of the surfactant molecules should be accompanied by an appearance of fluorescence.

Figure 11 shows the fluorescence spectra recorded for the film covered Au(1 1 1) electrode at a potential inside the capacitive pit, where the surfactant molecules are adsorbed, and at a potential outside the capacitive pit, where the surfactant is desorbed. The spectra were determined by changing the wavelength of the incident radiation between 220 and 440 nm and detecting the intensity of the fluorescence at a fixed wavelength of 470 nm. The top, middle and bottom panels show spectra determined for the film produced by the triple-, double- and single-touch procedure, respectively. The spectra were recorded using both s and p polarizations of the incident photon. Polarization dependence of fluorescence from agglomerates of dye molecules is complex and hence only the spectra acquired using p-polarized incident radiation are shown in Fig. 11 (the spectra for s-polarized light were included in our recent publication [25]). For the film-covered surface the spectra were recorded holding the electrode at $E = 0$ V (SCE). The spectra for desorbed surfactant molecules were acquired at potentials −590, −620 and −650 mV for the films produced by single, double and triple touching procedures, respectively. The fluorescence spectra have not been corrected for the intensity of the incident radiation. The spectra show two bands each characteristic for an anthroyloxy moiety. The band at 260 nm represented a transition along the long axis of the anthroyloxy moiety. The band at 360 nm represents a transition along the short axis of the fluorophore with associated vibrational fine structure. The difference between the intensity of the two bands arises from a combination of differences in quantum efficiency and changes in intensity of the incident radiation (output from the arc lamp) over the spectral ranges of the two bands.

The spectra clearly show that the dye molecules do not fluoresce when they are adsorbed at the electrode surface (at E_{ads}). In contrast, fluorescence was observed at the potential where the surfactant is desorbed from the surface. Therefore, the results of fluorescence experiments support the conclusion that the desorbed molecules reside in the subsurface region in the form of micelles (or other aggregates). The intensity of the fluorescence increased in a nonlinear fashion on going from the aggregates produc-

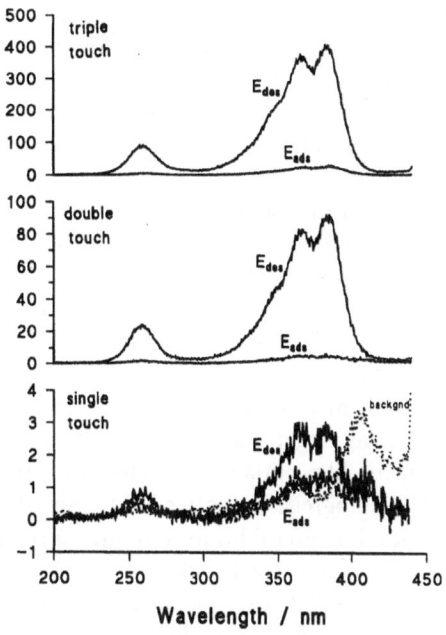

Fig. 11 Fluorescence excitation spectra (p-polarized incident light) at the adsorption potential (E_{ads}) and at the desorption potential (E_{des}) for single, double and triple touch layers of 12-AS adsorbed onto Au(1 1 1)

ed by desorption of the film deposited by single touch to the aggregates formed from the triple-touch film. The fluorescence by molecules desorbed from the single touch film is very weak and comparable with the background fluorescence. The fluorescence is, however, 100 times stronger in the case of molecules desorbed from the triple touch film.

The fluorescence for adsorbed molecules was apparently quenched due to the reduced lifetime of the excited state in a molecule adsorbed at the metal surface as discussed earlier. Quenching due to the energy transfer to the metal can also be observed in molecules that are desorbed but reside in the electrolyte layer near the electrode surface. According to the so-called Förster energy transfer mechanism, observed in the membrane studies [35, 40] and in Langmuir–Blodgett films [41], the change of the fluorescence intensity with separation of the fluorescent molecule from the quencher (metal) is described by the formula

$$\frac{I}{I_\infty} = \left(1 + \left(\frac{d_0}{d} \right)^4 \right)^{-1}, \tag{2}$$

where I/I_∞ is the fluorescence intensity normalized to the fluorescence intensity in the absence of the quencher, d_0 is the distance between fluorophore and the quencher where the fluorescence is reduced by one-half, and d is the separation between quencher and fluorophore. For

Langmuir–Blodgett films at gold the characteristic distance d_0 may be of the order of 100 Å [41]. Therefore, we can tentatively explain the dramatic increase in the amount of fluorescent radiation by moving from the aggregates desorbed from the single touch to the triple touch film in terms of an increasing distance between the majority of the fluorophore and the metal surface.

The optical measurements demonstrated that the potential-induced desorption of the film involves formation of aggregates in the electrolyte layer near the electrode surface at negative potentials, and readsorption of these aggregates at potentials close to pzc. The fluorescence and elastically scattered light measurements have shown that the apparent size of the aggregates depended on the number of monolayers in the film initially deposited onto the electrode surface. These aggregates scatter light, and stay in the subsurface region for several hours and later spread back onto the electrode surface without any loss of the material. This behavior suggests that the aggregates must be either macromolecules with a very low value of the diffusion coefficient and/or they must trap the charge of counterions from the diffuse part of the double layer and become attracted to the electrode surface by electrostatic forces. Although, the electroreflectance and the fluorescence measurements may be performed only using dye surfactants, the light-scattering experiments may be performed on surfactants without a chromophore group and they show that all insoluble films display the same behavior. Therefore, the mechanism of the potential-induced adsorption/desorption of a film of an insoluble surfactant must be general in character.

Mechanism of adsorption of insoluble surfactants at the solid–liquid interface

We have shown in the preceding section that changes of intensity of the scattered radiation correlate quite well with the changes of the differential capacity of the electrode. We note that the two signals probe quite a different physical state of surfactant molecules. The scattered light provides information about surfactant present in the form of a micelle in the subsurface region, in contrast differential capacity responds to the surfactant molecules which adsorb onto the electrode surface. In order to gain further insight into the mechanism of the potential-induced adsorption/desorption of insoluble surfactants we quantitatively correlated the changes of the scattered light intensity and the changes of the differential capacity. We have chosen a film of 12-AS molecules produced by the triple-touch procedure as an example, since the intensity of the scattered radiation is particularly intense in this case. The experiments were performed using a monochromatic light-

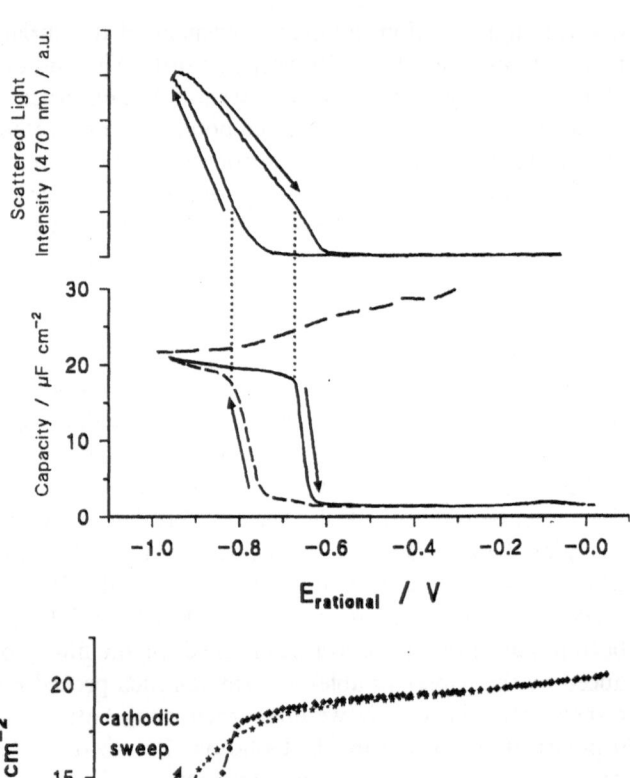

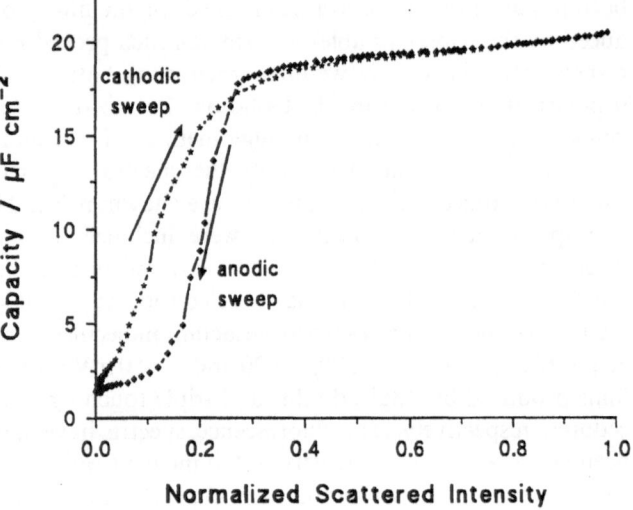

Fig. 12 Top panel, a comparison of the capacitive response and the changes in the intensity of elastically scattered 470 nm light with potential for a triple touch layer of 12-AS on Au(1 1 1). Bottom panel, the differential capacity plotted against the intensity of the elastically scattered light at 470 nm for a triple touch layer of 12-AS on Au(1 1 1)

of the wavelength 470 nm which is outside the range of the absorption bands for this molecule.

The two top panels in Fig. 12 show the scattered light intensity and differential capacity curves recorded when a slow (5 mV/s) voltage scan was applied to the electrode surface. In full analogy to the similar plots presented earlier in Fig. 7 for the films of $C_{18}OH$, we observe that the light scattering is quenched at potentials at which the film is spread onto the surface and that the light is scattered at potentials at which the film is desorbed. At first glance, the changes of the scattered light intensity correlate well with the changes of the electrode capacity. However, a close inspection of these curves reveal some important differences between the optical and capacitive signals. To assist further discussion, the two sets of curves were connected by dotted vertical lines at potentials corresponding to the onset of the capacitive pits on the differential capacity curve. These lines help us to note that a significant quenching of the scattered light takes place at potentials before the capacitive pits are formed and before the surfactant spreads onto the electrode surface. The difference between the two signals may be even better shown if the results of the optical and electrochemical measurements are combined and the capacity is plotted against the scattered light intensity. The bottom panel in Fig. 12 shows the plot of the differential capacity against the intensity of the scattered radiation. The intensity of the scattered radiation was normalized with respect to the minimum intensity observed at the most negative potential. In this graph, we can easily see that almost 80% of the scattered light intensity is quenched in the region where the differential capacity of the electrode decreases by about 10% and that only the remaining 20% of the scattered light intensity is quenched where 90% change of the differential capacity takes place and where the insoluble surfactant spreads onto the electrode surface. This behavior strongly suggests that adsorption/desorption of a film of insoluble surfactants proceedes through a hemimicelle stage. The formation of hemimicelles has long been postulated in the surface and colloid science literature; however, only recently AFM images of hemimicelles on mica [42] and neutron-scattering experiments [43] reveal the existence of these structures.

At that point we note that the desorption/spreading of insoluble films has similarities to the potential-dependent spreading of oil droplets onto the Hg electrode surface described recently by Zutic et al. [44, 45]. It appears that all these phenomena may be described by a common model. Zutic et al. demonstrated that spreading of the oil droplets onto the electrified interface is observed if the spreading coefficient defined as

$$S = \gamma_{SW} - \gamma_{SL} - \gamma_{LW} \qquad (4)$$

is larger than zero. In Eq. (4), γ_{SW}, γ_{SL}, γ_{LW}, are the interfacial tensions for the solid–water, solid–organic liquid and organic liquid–water interface, respectively. In fact, the same equation may be applied to describe spreading of micelles, with the symbols γ_{SW}, γ_{SL} and γ_{LW} representing the interfacial tension of the film free, film-covered electrode and the interfacial tension of a micelle, respectively. We note that the first two terms of Eq. (4) are equal to the film pressure at the metal solution interface $\pi_{MS} = \gamma_{SW} - \gamma_{SL}$. For insoluble surfactants, the spreading coefficient may be conveniently expressed as a difference between the film pressure and the interfacial tension of the micelle

$$S = \pi_{MS} - \gamma_{LW} . \qquad (5)$$

The thermodynamic condition for spreading of micelles onto the electrode surface is that the film pressure at the MS must be higher than the surface tension of the micelle $\pi_{MS} > \gamma_{LW}$. Figure 3 shows that the film pressure depends on the electrode potential and hence spreading of the insoluble film depends on the potential as well. Spreading of insoluble surfactants causes the appearance of the characteristic pit on the differential capacity curve. The potential at which the pit is formed should correspond to $\pi_{MS} \approx \gamma_{LW}$. For the film of 12-AS the capacitive pit is observed at film pressures of about 4 mN m^{-1}. We note that this value is consistent with the reported surface pressures of micelles [46]. At potentials outside the pit, where the film pressure $0 < \pi_{MS} < \gamma_{LW}$, the surfactant molecules are not yet desorbed from the surface. They must adhere there to the electrode surface in the form of droplets or hemimicelles. The contact angle for these droplets (hemimicelles) is described by Young's equation

$$\cos \theta = \frac{\gamma_{SW} - \gamma_{SL}}{\gamma_{LW}} = \frac{\pi_{MS}}{\gamma_{LW}} \qquad (6)$$

and apparently depends on the film pressure. Since the film pressure changes with potential, the contact angle varies with E as well. Its value should range from zero at potentials where $\pi_{MS} \approx \gamma_{LW}$ to 90° at more negative E where $\pi_{MS} \approx 0$. When the electrode potential becomes even more negative the contact angle should attain values higher than 90° and the droplets (micelles) should be detached from the electrode surface. The progressive change of the shape of the hemimicelle with potential is schematically shown in Fig. 13.

In Fig. 14 this thermodynamic model has been employed to explain how the properties of the film formed by 12-AS vary as a function of the electrode potential. At potentials near the pzc, the surfactant is strongly adsorbed in a highly oriented fashion. As the potential becomes negative, the film becomes less ordered and the organic layer is porated [24], a prelude to desorption which occurs at the most negative charge density measured. At this potential of desorption, the surfactant exists as an organized structure near the electrode surface. Interestingly, these organizations do not diffuse away from the interface and are easily readsorbed when the potential is made less negative. The respreading of the micellular structures occurs in a fashion similar to the spreading of organic droplets onto solid surfaces. Initially, the micelles adhere to the metal surface and form hemimicelles. The shape of the hemimicelles is controlled by the magnitude of the contact

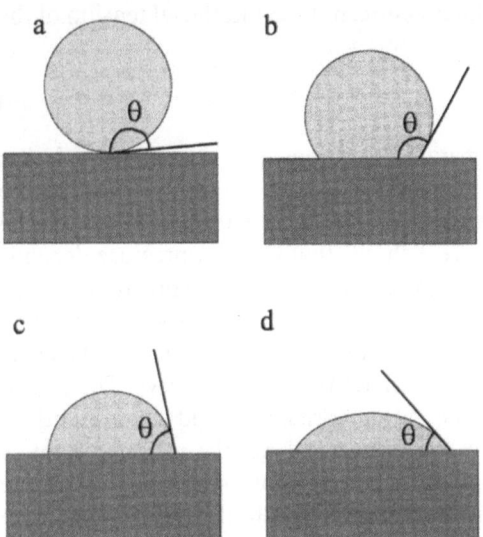

Fig. 13 A schematic representation of the changes in the contact angle between the solid surface, and adsorbed hemimicelle

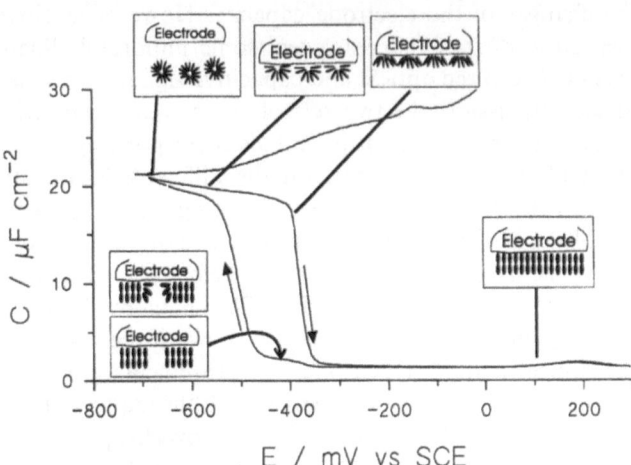

Fig. 14 A schematic of the proposed mechanism for the adsorption/desorption of a film of insoluble surfactants from/to the Au electrode

angle. The contact angle changes from 90 to 0° as the electrode potential becomes more positive and the film pressure increases. This progressive change of the contact angle explains a progressive quenching of light scattering and fluorescence within the potential range where hemimicelles are formed. When the contact angle is large, the hemimicelle covered electrode surface is rough and scatters electrode radiation. In addition, a significant fraction of the surfactant molecules contained in a hemimicelle is sufficiently far from the metal surface so that their fluorescence is not quenched. When the contact angle decreases, the surface becomes smoother and the surfactant molecules come closer to the electrode surface. The light scattering decreases and the fluorescence is quenched. At potentials where $\pi_{MS} > \gamma_{LW}$ the hemimicelles spread

onto the metal surface to form initially a loose and later the more ordered film.

To conclude, we would like to stress that formation of hemimicelles, as a first step in surfactant adsorption onto a solid–liquid interface, has long been postulated in the literature and has been proven recently by neutron scattering [43] and AFM imaging [42] experiments. The significance of our results is that they show that the shape and size of the hemimicelles formed at the metal–solution interface may be conveniently controlled by the electrode potential. This control offers a new possibility to study the mechanism by which surfactants adsorb at the solid–liquid interface.

Acknowledgments This work was supported by a grant from Natural Sciences and Engineering Research Council of Canada.

References

1. Miller IR, Rishpon J, Tennenbaum A (1976) Bioelectrochem Bioenerg 3:528
2. Miller IR (1981) Structural and Energetic Aspects of Charge Transport in Lipid Layers and in Biological Membranes. In: Milazzo G (ed) Topics in Biochemistry and Bioenergetics. Vol. 4. John Wiley and Sons, New York, pp 161–224
3. Nelson A, Benton A (1986) J Electroanal Chem 202:253
4. Nelson A, Auffret N (1988) J Electroanal Chem 244:99
5. Nelson A, Leermakers FAM (1990) J Electroanal Chem 278:73
6. Leermakers FAM, Nelson A (1990) J Electroanal Chem 278:53
7. Nelson A, Auffret N, Readman J (1988) Anal Chim Acta 207:47
8. Nelson A, Auffret N, Borlakoglu J (1990) Biochim Biophys Acta 1021:205
9. Nelson A (1991) J Electroanal Chem 303:221
10. Nelson A, van Leeuwen HP (1989) J Electroanal Chem 273:183
11. Nelson A, van Leeuwen HP (1989) J Electroanal Chem 273:201
12. Moncelli MR, Guidelli R (1992) J Electroanal Chem 326:331
13. Moncelli MR, Becucci L, Guidelli R (1994) Biophys J 66:1969
14. Moncelli MR, Becucci L, Herrero R, Guidelli R (1995) J Phys Chem 99:9940
15. Moncelli MR and Becucci L (1995) J Electroanal Chem 385:183
16. Fujihira M, Toshinari A (1986) Chem Lett 921
17. Zhang X, Bard A (1989) J Amer Chem Soc 111:8098
18. Charych D, Landau EM, Majda M (1991) J Amer Chem Soc 113:3340
19. Widrig C, Miller CJ, Majda M (1988) J Amer Chem Soc 110:2009

20. Belewicz R, Sawaguchi T, Chamberlin II RV, Majda M (1995) Lang 11:2256
21. Noel J, Bizzotto D, Lipkowski J (1993) J Electroanal Chem 344:343
22. Bizzotto D, Noël J, Lipkowski J (1993) Thin Solid Films 248:69
23. Bizzotto D, Noël JJ, Lipkowski J (1994) J Electroanal Chem 369:259
24. Bizzotto D, McAlees A, Lipkowski J, McCrindle R (1995) Lang 11:3243
25. Bizzotto D, Lipkowski J (1996) J Electroanal Chem 409:33
26. Sagara T, Zamlynny V, Bizzotto D, McAlees A, McCrindle R, Lipkowski J (in preparation)
27. Bizzotto D (1996) Characterization of the adsorption of insoluble surfactants onto an electrified interface. PhD Dissertation, University of Guelph, Guelph, ON. 238p
28. Damaskin B, Petrii OA, Batrakov VV (1968) Adsorption of organic Compounds of Electrodes. Nauka, Moscow
29. Damaskin B, Kazarinov VE (1980) The Adsorption of Organic Molecules. In: Bockris JO'M, Conway BE, Yeager E (eds) Comprehensive Treatise of Electrochemistry, Vol. 1. Plenum Press, New York, pp 353–396

30. Frumkin A, Damaskin B (1967) Pure Appl Chem 15:263
31. Lipkowski J, Nguyen van Houng C, Hinnen C, Parsons R, Chevalet J (1983) J Electroanal Chem 143:375
32. Cadenhead DA, Kellner BMJ, Muller-Landau F (1975) Biochim Biophys Acta 382:253
33. Cadenhead DA, Kellner BMJ, Jacobson K, Papahadjopoulos D (1977) Biochem 16:5386
34. Waggoner AS, Stryer L (1970) Proc Natl Acad Sci USA 67:579
35. Slavik J (1994) Fluorescent Probes in Celluar and Molecular Biology. CRC Press Inc., Boca Raton, FL, 295 pp
36. Haugland RP (1992) Molecular Probes. Handbook of Fluorescent Probes and Research Chemicals, 5th ed. Molecular Probes Inc., Eugene, OR, p 249
37. Kofman R, Garrigos R (1981) Thin Solid Films 82:73
38. Henglein F, Lipkowski J, Kolb D (1991) J Electroanal Chem 303:245
39. Avouris P, Demuth JE (1983) Surface Studies with Lasers. In: Aussenegg F, Leitner A, Lippitch ME (eds) Surface Studies with Lasers, Vol. 33. Springer, Berlin, pp 24–34

40. Sawyer WH (1988) Fluorescence spectroscopy in the study of membrane fluidity: model membrane systems. In: Methods for Studying Membrane Fluidity. Alan R Liss Inc, New York, pp 161–191
41. Kuhn H, Möbius D, Bücher H (1972) Spectroscopy of Monolayer Assemblies. In: Weissberger A, Rossiter BW (eds) Physical Methods of Chemistry. Part IIIB Optical, Spectroscopic, and Radioactivity Methods, Vol. 1. Wiley-Interscience, New York, pp 577–578
42. Manne S, Gaub HE (1995) Science 270:1480
43. McDermott DC, McCarney J, Thomas RK, Rennie AR (1994) J Colloid Int Sci 162:304
44. Zutic V, Kovac S, Tomaic J, Sveltlicic V (1993) J Electroanal Chem 349:173
45. Ivosevic N, Tomaic J, Zutic V (1994) Lang 10:2415
46. Tien HT (1967) J Phys Chem 71:3395

Progr Colloid Polym Sci (1997) 103:216–225
© Steinkopff Verlag 1997

R.K. Thomas

Neutron reflection from surfactants adsorbed at the solid/liquid interface

Received: 3 December 1996
Accepted: 6 December 1996

Dr R.K. Thomas (✉)
Physical and Theoretical Chemistry
Laboratory
South Parks Road
Oxford OX1 3QZ, United Kingdom

Abstract Neutron reflection has proved to be a very successful technique for investigating the structure of layers of surfactants adsorbed at the air/water interface, especially when isotopic substitution may be used to resolve the structure within the layers. Here the extension of the technique to surfactant layers adsorbed at the solid/liquid interface is discussed. The experimental conditions necessary for probing surface structure accurately are mainly determined by the possibilities available for varying the contrast through isotopic substitution. The simplest variation is of the H/D ratio in water, but definitive evidence for bilayer structures at the surface can usually only be obtained by varying the contrast within the surfactant layer. At the hydrophobic surface, where only monolayer adsorption is expected, the most sensitive contrast for observing a surfactant layer is when the hydrophobic layer on the surface is deuteriated. In comparison with other techniques that probe the solid/liquid interface the main strengths of neutron reflection are that it measures the absolute coverage on the surface and it determines the structure normal to the interface at a resolution down to 1–4 Å, depending on the level of isotopic substitution used.

Key words Surfactants – adsorption – neutron reflection

Introduction

Although adsorption at the solid/liquid interface plays an important role in many technological systems, in comparison with the solid/gas interface it has been relatively little studied by techniques capable of exploring molecular details. There are two main reasons for this. The powerful experimental techniques that have successfully explored the solid/gas interface are all vacuum techniques and therefore cannot be used to study a buried interface, and, in terms of the structure and dynamics of the interface, the gas/solid interface is dominated by energetic consider- ations whereas for the solid/liquid interface entropy is usually at least as important as energy, and this introduces considerable extra difficulty into any theoretical modelling. In the last decade a number of new experimental techniques have been developed which are capable of exploring buried interfaces and useful experimental results are gradually starting to emerge. Neutron reflection was one of the first of these techniques to be developed [1] and, although it has mainly been used for the study of the liquid/vapour interface [2] and thin polymeric films [3], it is increasingly being applied to the solid/liquid interface, which is what is now discussed in this paper.

Neutron reflection

The conceptual basis of the neutron reflection experiment is simple. A monochromatic and well collimated neutron beam is reflected off a large flat surface (typically 10 cm^2) and the intensity of the reflection measured as a function of the incident angle. For specular reflection, i.e. where incident and reflected angles are the same, the reflectivity as a function of angle is related to the neutron refractive index profile normal to the interface. For example, in the same way that reflection of light off a thin uniform film on a flat substrate leads to interference fringes, so neutron reflection also gives rise to interference fringes whose separation is directly related to the film thickness and whose amplitude depends on the relative refractive indices of the three media concerned. There are two key features that make neutron reflection much more powerful than reflection of light. One is that neutron wavelengths are two or three orders of magnitude shorter than for light and hence smaller dimensions can be probed. The other is that the neutron refractive index is not only simply related to composition but can be altered by isotopic substitution. Thus the reflectivity is easily related to composition and it is possible to alter the refractive index profile, and hence the reflectivity, by changing the isotopic composition. This offers the possibility of making the experiment sensitive only to the adsorbed layer, or even just to selected parts of the adsorbed layer [2].

For calculating the reflection of light it is convenient to use the refractive index and angle of incidence but for neutron reflectivity, it is customary to use scattering length density and momentum transfer. The momentum transfer κ is related to the grazing angle of incidence θ by

$$\kappa = \frac{4\pi \sin \theta}{\lambda}, \tag{1}$$

where λ is the wavelength, and the relation between the refractive index n and the mean scattering length density $\bar{\rho}$ is

$$n^2 = 1 - \frac{\lambda^2}{\pi} \bar{\rho}. \tag{2}$$

The mean scattering length density, as a function of the distance normal to the surface z is given by

$$\bar{\rho}(z) = \sum_i b_i n_i(z), \tag{3}$$

where b_i is the scattering length and n_i is the number density of atomic species i. The neutron scattering length is an empirically determined number and it varies erratically through the periodic table and from isotope to isotope. We shall make extensive use of the difference between the different scattering lengths of the isotopes of hydrogen for which $b_H = -3.74 \times 10^{-5}$ and $b_D = 6.67 \times 10^{-5}$ Å. Note that specular reflection is only sensitive to the *mean* scattering length density of each layer, i.e. the experiment does not detect any in-plane inhomogeneities. This point will be discussed further below.

For a given model of the structure normal to the interface, no matter how complex, it is possible to calculate the neutron reflectivity exactly using the same formulae, apart from the difference in the refractive index, as for light polarized at rightangles to the plane of reflection. For a multilayer structure the optical matrix method [4] can then be used, in which the interface is divided into as many layers as are required to describe it with adequate resolution. This method lends itself especially well to machine calculations and is therefore the most widely used method of analysing neutron reflectivity. However, it does not reveal the relatively simple relation between reflectivity and interfacial structure, which can be done more clearly using the kinematic approximation. In the kinematic approximation the reflectivity profile is given by [5, 6]

$$R = \frac{16\pi^2}{\kappa^2} |\rho(\kappa)|^2, \tag{4}$$

where $\rho(\kappa)$ is the one dimensional Fourier transform of $\bar{\rho}(z)$:

$$\rho(\kappa) = \int_{-\infty}^{\infty} \bar{\rho}(z) \exp(-i\kappa z)\, dz, \tag{5}$$

where we have simplified $\bar{\rho}$ to ρ. Two simple results that follow from eq. (4) are the reflectivity profiles for a perfectly smooth interface and for a uniform monolayer on a substrate. The density profile for a smooth surface is just a Heaviside function, the Fourier transform of which leads to the reflectivity profile

$$R = \frac{16\pi^2}{\kappa^4} \Delta\rho^2, \tag{6}$$

where $\Delta\rho$ is the difference in scattering length density between the two bulk phases. The kinematic reflectivity for a single uniform monolayer on a substrate is [7]

$$R = \frac{16\pi^2}{\kappa^4} \big[(\rho_1 - \rho_0)^2 + (\rho_2 - \rho_1)^2 $$
$$ + 2(\rho_1 - \rho_0)(\rho_2 - \rho_1) \cos \kappa\tau_1 \big], \tag{7}$$

where τ_1 is the thickness of the layer. An important case in neutron reflection experiments arises when the scattering length densities of the two bulk media on either side of the layer are equal, $\rho_0 = \rho_2 = \rho$, when

$$R = \frac{16\pi^2}{\kappa^4} \left[4(\rho_1 - \rho)^2 \sin^2 \left(\frac{\kappa\tau_1}{2} \right) \right] \tag{8}$$

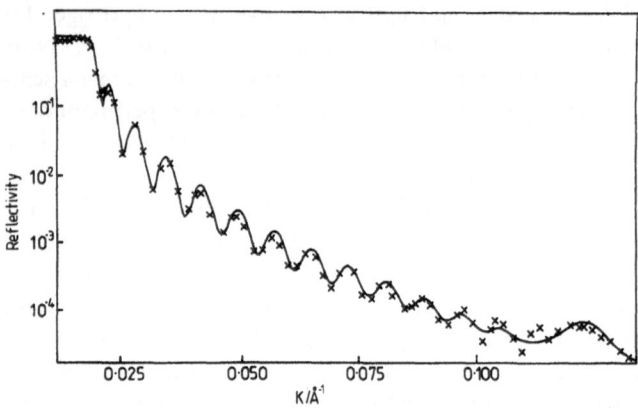

Fig. 1 Neutron reflectivity profile of an organic thin film on a glass substrate. The layer was a Langmuir–Blodgett film consisting of 29 layers of fully deuterated cadmium stearate. Although not strictly a uniform film the lower angle interference fringes are well described by Eq. (7). At higher angles the first order Bragg diffraction from the repeating structure within the layer can be observed

Table 1 Examples of scattering length densities of typical materials involved at the solid/liquid interface

Material	Scattering length density $(10^{-6}\,\text{\AA}^{-2})$
Si	2.1
Amorphous SiO_2	3.4
Crystal quartz	4.2
H_2O	-0.58
D_2O	6.35
Octane-h_{10}	-0.05
Octane-d_{10}	6.44
$C_{16}H_{33}N(CH_3)_3Br$	$-0.3f + (1-f)\rho^*_{\text{water}}$
$C_{16}D_{33}N(CD_3)_3Br$	$7.1f + (1-f)\rho^*_{\text{water}}$
$C_{18}H_{37}$-(nOTS fragment)	-0.37
$C_{18}D_{37}$-(dOTS fragment)	7.0

* f is the volume fraction of the layer occupied by the surfactant and ρ_{water} depends on the scattering length density of the water being used

Both Eqs. (7) and (8) give reflectivity profiles that consist of a series of interference fringes, whose separation $\Delta\kappa(=2\pi/\tau_1)$ gives the thickness of the layer directly, and whose amplitude is related to the scattering length densities of the monolayer and bulk phases and hence to their compositions through Eq. (3). Since, in general, the composition and scattering length density of the bulk phases will already be known it is then easy to derive the composition of the monolayer. Figure 1 shows an example of such fringes for a uniform layer on a substrate under conditions where Eq. (7) is appropriate.

The viability of neutron reflection from the solid/liquid interface depends on the interface being accessible to the neutron beam. Since the illuminated area of surface is large, access to the interface depends on either the liquid or solid being reasonably transparent. Apart from the special case of proton containing materials, which scatter neutrons incoherently to such an extent that they do not transmit sufficiently, the main scattering process is diffraction. Since the diffraction pattern of a liquid is a continuous function of angle it is impossible to avoid diffraction of the beam as it enters through the liquid phase. For the long path lengths required at grazing incidence, multiple diffraction then prevents the beam reaching the interface from the liquid side. For a solid, however, diffraction can be avoided either by suitable orientation of a single crystalline solid sample so as to avoid satisfying the Bragg condition, or by selecting a neutron wavelength above the Bragg cut-off for the solid (λ > twice the unit cell dimension). In principle, this should make a wide range of solids suitable, but, in practice, the pathlength through the crystal may be sufficiently long that coherent inelastic scattering, which is less restricted than implied by the Bragg

diffraction condition, leads to too great a loss of intensity through multiple scattering. Also, the size of the single crystals required (ideally not less than about $10 \times 5 \times 1$ cm^3) is so large that very few solid materials are available at an accessible price. However, three readily available materials are suitable for neutron reflection, and these are silicon, and crystalline and amorphous quartz. The convenience of these materials is that the silica surface is one of the most important in terms of practical applications. Other materials, e.g. sapphire, are probably also suitable but have not yet been tried. Unfortunately, it is unlikely that mica, which has been so well studied by other experimental techniques, can be used for neutron reflection studies of the solid/liquid interface except in quite unusual situations.

Because different nuclei may scatter neutrons with different amplitude, and, in the case of protons and deuterons, with opposite phase, the use of a combination of protonated and deuteriated materials can substantially change the reflectivity profile of a system while maintaining the same chemical structure. Thus it is possible, by adjustment of the H/D ratio, to prepare solvents that are matched to the solid to give zero contrast between the two bulk phases, and then the reflectivity profile arises only from the interfacial material. Such an experiment is surface specific. Table 1 compares the scattering length densities of silicon, and amorphous and crystalline quartz, with those of water and other possible solvents and with some possible monolayer structures. The composition of most proton containing solvents can be varied so that their scattering length densities can be made higher or lower than the three silicon containing substrates. Thus the "contrast match" condition can always be met. Figure 2 shows the effect of varying the scattering length density of water on the

Progr Colloid Polym Sci (1997) 103:216–225
© Steinkopff Verlag 1997

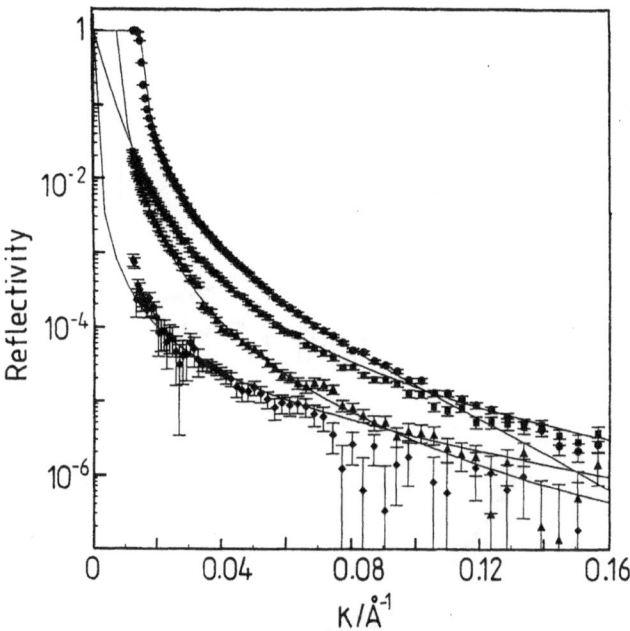

Fig. 2 Neutron reflectivity profiles from the Si/SiO$_2$/water interface at different isotopic compositions of water. The scattering length densities of the various water samples are (●) 6.35 (D$_2$O), (■) −0.58 (H$_2$O), (▲) 3.4, and (◆) 2.5 × 10^{-6} Å^{-2}. The continuous lines are the best fits of a 10 Å layer of SiO$_2$ with a roughness of 6 Å [8]

reflectivity from the silicon/water interface [8]. The scattering length density of D$_2$O is larger than that of Si and total reflection is observed for this contrast but that of H$_2$O is less and than there is no total reflection. The larger contrast is between Si and D$_2$O and this reflectivity is therefore the most intense (see Eq. (6)). When the scattering length density of the water is adjusted to 2.1 × 10^{-6} Å^{-2} it is exactly matched to Si, but the signal does not completely vanish because there is an oxide layer on the silicon which has a scattering length density of about 3.4 × 10^{-6} Å^{-2}. Thus at this contrast condition there is a weak signal just from the natural oxide layer. When the scattering length density of the water is adjusted to 3.4 × 10^{-6} Å^{-2} the silica layer is matched to the water and the reflection is then from the Si/SiO$_2$ interface. Thus, each water contrast is sensitive to different features of the interface and, with the assumption that the structure and composition of this interface are not affected by isotopic substitution, details such as the thickness and composition of the oxide layer, and the roughness of the Si/SiO$_2$ and SiO$_2$/H$_2$O interfaces, can be determined by fitting a suitable set of reflectivity profiles. For a typical polished Si block the thickness of the SiO$_2$ layer varies from 10–20 Å and consists of 70–100% SiO$_2$, any empty space being filled with water. It may also be necessary to include some roughness to account for the observed reflectivity profiles but it is difficult

to make a clear separation between the effects of roughness and incorporation of water into the oxide layer, both features being manifestations of roughness. As will be shown later the role of roughness in the behaviour of the silica surface is complex.

Self-assembled monolayers at the solid/liquid interface

A range of types of surface, suitable for neutron reflection experiments, can be prepared by grafting a molecular film on to the substrate. Most of the chemistry required to prepare a wide range of functionalized surfaces is well known and hence, in principle, any type of surface could be studied by neutron reflection using the silicon/silica solid as a universal substrate. The simplest and most widely studied example is the formation of a hydrophobic surface by self assembly of alkyl trichlorosilanes on silica [9]. Although the structure of such layers has been well established in the dry state, i.e. at the hydrophobic solid/air or vacuum interface, such layers have not been studied in the presence of liquid. Figure 3 shows the neutron reflectivities from protonated octadecyl trichlorosilane (OTS) layers deposited on the natural oxide surface of the (1 1 1) plane of silicon [10]. Given the discussion above about the usefulness of varying the water contrast in order to elucidate the structure more accurately, the reflectivities are shown for water matched to silicon (Fig. 3(a)), where the signal from the hydrophobic layer dominates the reflectivity, and for H$_2$O (Fig. 3(b)) when it can be seen from the comparison with the reflectivity in the absence of the hydrophobic layer, that the protonated hydrophobic layer becomes invisible. Figure 4(a) shows the reflectivity from a deuteriated OTS layer in water matched to silicon and Fig. 4(b) shows the same layer in D$_2$O, where one would expect that the layer should vanish. In fact, there is a significant signal from the layer in the latter conditions. This is partly because the match between an ideal deuteriated layer and D$_2$O is not as good as the match between the protonated layer and H$_2$O and partly because, in this particular case, the upper part of the OTS layer is slightly less dense than the lower part. This small difference could not be seen in the reflectivity from the "dry" OTS layer in air using neutron and X-ray reflectivity and it has been attributed to the effect of the underlying roughness of the silica [10]. Although it is generally thought that a self-assembled OTS layer generates a smoother surface than present initially, not all of the underlying roughness of the silica surface can be expected to be eliminated and this may manifest itself in defects in the outer part of the OTS layer.

In both Figs. 3(a) and 4(a) the main contribution to the reflectivity is from the OTS layer and eq. (8) should

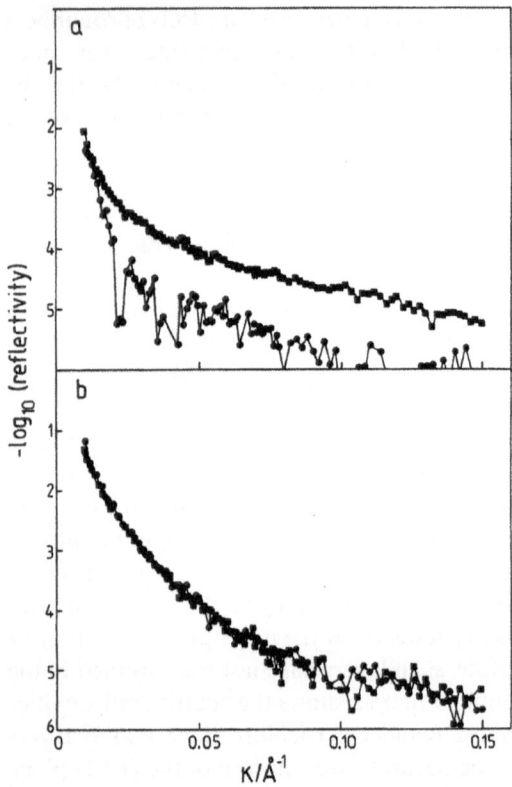

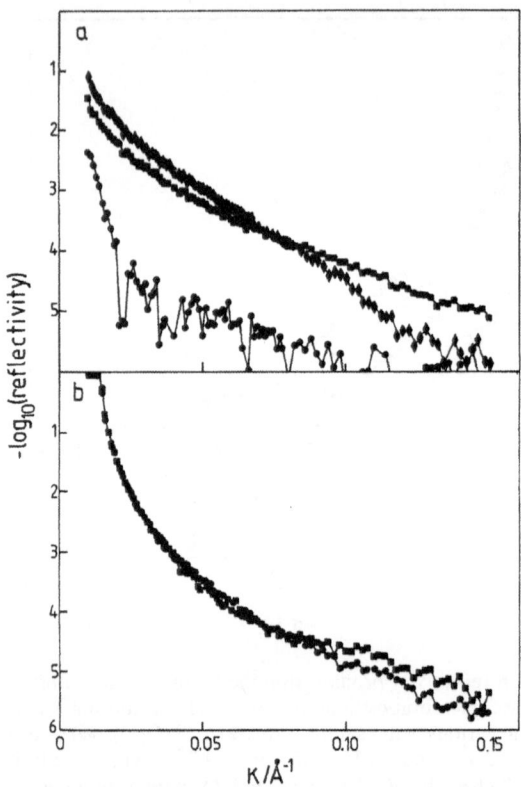

Fig. 3 Neutron reflectivity profiles from (●) the Si/SiO$_2$ surface, and (■) the same surface coated with fully protonated OTS. In (a) the scattering length density of the water $(2.5 \times 10^{-6} \text{ Å}^{-2})$ is approximately matched to silicon $(2.1 \times 10^{-6} \text{ Å}^{-2})$ and in (b) it is matched (pure H$_2$O) to that of the protonated OTS layer [10]

Fig. 4 Neutron reflectivity profiles from (●) the Si/SiO$_2$ surface, (■) the same surface coated with fully deuterated OTS, and (◆) the same surface coated with dOTS with an additional adsorbed layer of fully deuterated tetraethylene glycol monododecyl ether (C$_{12}$E$_4$). In (a) the scattering length density of water $(2.5 \times 10^{-6} \text{ Å}^{-2})$ is approximately matched to silicon $(2.1 \times 10^{-6} \text{ Å}^{-2})$ and in (b) it is approximately matched (pure D$_2$O) to that of the deuterated OTS layer [10]

describe the reflectivity profile approximately. The thickness of the OTS layer is about 29 Å and, according to Eq. (8), the first negative interference dip in the profile should occur at a momentum transfer of about $2\pi/\tau_1 \cong 0.22 \text{ Å}^{-1}$. Because of the κ^{-4} decay in the reflectivity and because the incoherent background level corresponds to a reflectivity of somewhat more than about 10^{-6}, this dip occurs just beyond the range of the experiment. Thus the ideal situation, suggested by the discussion of the fringes in Fig. 1, does not usually obtain for thin molecular layers, which means that it is not always obvious that it is possible to separate composition and thickness in the fitting process. The easiest way of resolving this ambiguity is to compare the intensities for the protonated (Fig. 3(a)) and deuteriated (Fig. 4(a)) OTS layers. Both profiles are described by Eq. (8) and their ratio is given by

$$\frac{R_D}{R_H} = \frac{(\rho_D - \rho)^2}{(\rho_H - \rho)^2} = \frac{(nb_D - \rho)^2}{(nb_H - \rho)^2} \tag{9}$$

and, apart from the ambiguity of sign of the square root, which is easily resolved, the number density of the molecu-

les in the layer can therefore be determined from the ratio of the reflectivities and the known values of b_D and b_H. In practice, the composition is determined by fitting models of the structure using the optical matrix method, but Eq. (9) demonstrates that isotopic substitution can ensure that the surface coverage is determined independently of the layer thickness, even though such a separation of composition and thickness determinations might not be possible from just one profile. When the system needs to be described in terms of more than one layer, ambiguities of interpretation may creep in and it is then essential to use isotopic substitution. Two other points are worth noting. The first concerns the sensitivity of the technique. Noting that the reflectivity is plotted on a logarithmic scale, it can be seen that the signal from the deuteriated OTS layer in Fig. 4(a) in the mid range of the reflectivity is three orders of magnitude more intense than the background and two orders of magnitude greater than the reflectivity from the SiO$_2$ layer, although the amount of OTS giving rise to the

Progr Colloid Polym Sci (1997) 103:216–225
© Steinkopff Verlag 1997

signal is only about 10^{-8} mol. Note also that this sensitivity is considerably reduced if only the protonated layer is studied because the factor $(\rho_1 - \rho)^2$ in Eq. (8) is much less. From the values in Table 1 the signal from the deuteriated layer in this contrast is nearly 4 times that from the protonated layer. If the substrate used had been quartz $(\rho = 4.2 \times 10^{-6} \text{ Å}^{-2})$ instead of silicon, the situation would have been reversed.

When the layer is made thicker by the adsorption of a monolayer of deuteriated surfactant on to the deuteriated OTS layer the dip in the reflectivity should move to lower κ and this can be seen in Fig. 4(a), where the position of the dip is very approximately 0.15 Å^{-1}, which gives a total thickness for the OTS + surfactant layer of about 42 Å. That such a conclusion does not need an elaborate fitting procedure illustrates the sensitivity of the neutron experiment to dimensions of the interface in the surface normal direction.

Surfactant layers at hydrophobic solid/liquid interfaces

On a hydrophobic surface surfactants are expected to adsorb from aqueous solution as monolayers with the hydrophobic group of the surfactant oriented towards the surface. A suitable choice of contrast for observing the thickness and composition of surfactant adsorption on to the hydrophobic OTS layer is for both layers to be deuteriated and the water to be matched to silicon. According to Eq. (8) the interference dip in the reflectivity profile should move towards a value of κ of $2\pi/(\tau_1 + \tau_2)$, where τ_2 is the thickness of the surfactant layer. Figure 4(a) shows the reflectivity when a layer of fully deuteriated surfactant tetraethylene glycol monododecyl ether, $C_{12}D_{25}(OC_2D_4)_4OH$, which we abbreviate to $dC_{12}dE_4$, is adsorbed on the deuteriated OTS layer [10]. The position of the dip is now within the range of the experiment at approximately 0.13 Å^{-1}, which gives a total thickness for the OTS + surfactant layer of about 48 Å. Given that the thickness of the OTS was already found to be 29 Å in a separate experiment, it is then possible to deduce that the thickness of the surfactant layer is approximately 19 Å. A different choice of contrast may be used to determine the orientation. For example, we have already shown in Fig. 4(b) that there is only a small signal from d-OTS in D_2O because there is only a small contrast difference between hydrophobic layer and solvent. If the fully protonated $hC_{12}hE_4$ is now adsorbed on the surface, it introduces a large change in contrast, the consequence of which is to cause a marked dip due to destructive interference at a κ value of about 0.045 Å^{-1}, as can be seen in Fig. 5(a). Consideration of the various interference terms that contribute to what is now quite a complicated structural

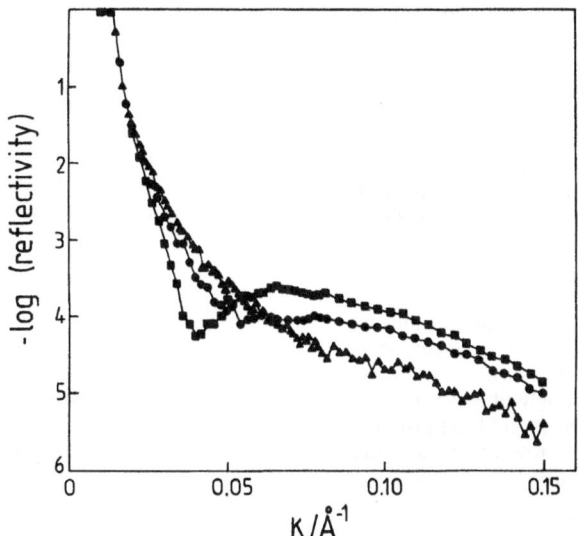

Fig. 5 Neutron reflectivity of (▲) the Si/SiO$_2$ surface coated with dOTS, and the same surface with an adsorbed layer of (■) fully protonated $C_{12}E_4$, and (●) $hC_{12}dE_4$. The scattering length density of water (D$_2$O) is approximately matched to that of the hydrophobic dOTS layer [10]

profile suggests that this negative interference should occur at $\pi/(\tau_{SiO_2} + \tau_{hphobe} + \tau_{surf})$. This is consistent with the separate determination of the layer thicknesses (respectively, 20, 29 and 19 Å) in that the sum of the thicknesses of the three layers is now approximately 69 Å ($2\pi/0.045$). To determine the orientation of the $C_{12}E_4$ layer we now use isotopic substitution to contrast the head and chain groups of the molecule. Still using dOTS and D$_2$O we adsorb $hC_{12}dE_4$. The deuteriated E$_4$ head group of the surfactant will now be almost the same scattering length as D$_2$O, especially since it is hydrated with D$_2$O (about 2 molecules per ethylene glycol group), and if the head group is on the D$_2$O side of the layer the thickness of the composite layer giving rise to the interference dip will be decreased by the thickness of the head group region of the layer. As can be seen in Fig. 5(a) the dip does indeed shift in the correct direction (to just above 0.05 Å^{-1}). This both confirms that the hydrophilic part of the surfactant points into the aqueous phase and gives an approximate value of the thickness of the head group region of 6–10 Å. Although one does not rely on such a crude analysis to determine the structure but rather on a full fitting of a wide range of contrasts, the analysis does demonstrate the basic sensitivity of neutron reflection to the detailed structure of the layer, provided that full advantage is taken of the possibilities of isotopic substitution.

The overall structure of the $C_{12}E_4$ layer adsorbed on a self-assembled OTS layer is shown schematically in Fig. 6, where it has been attempted to represent the main features

Fig. 6 Schematic structure of
the composite layer consisting
of a silicon oxide surface with
a self-assembled monolayer of
octadecyl trichlorosilane and
a layer of the surfactant
tetraethylene glycol
monododecyl ether adsorbed
from aqueous solution. The
diagram shows the dimensions
of the layers as deduced from
neutron reflection and
represents the proportions of
the various components in the
layers. The surfactant molecules
are strongly tilted with the
ethylene glycol head groups
pointing towards the aqueous
solution

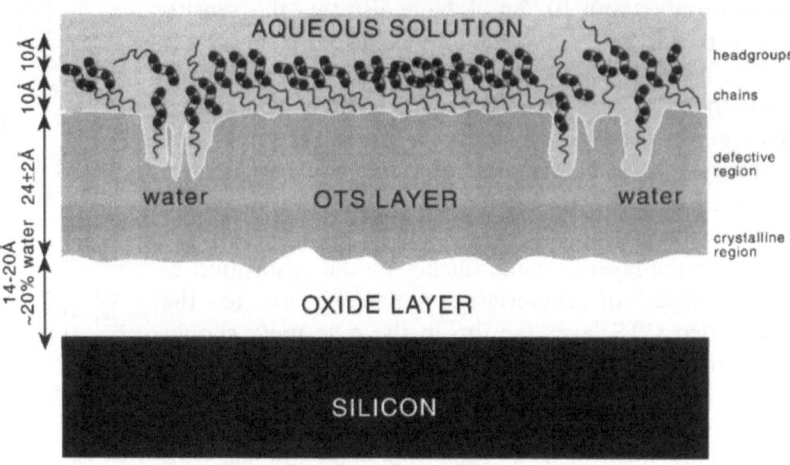

observed [10]. The first is that the defects in the outer half of the OTS layer occupy about 10% of the volume of this layer and this is filled partly by water and partly by surfactant chains. This may not be a general result but may be particular to the OTS layer studied, although it must be said that we have not yet been able to prepare an OTS layer where such penetration did not take place. There is some disorder in the surfactant layer in that some mixing of head groups and hydrocarbon chains is needed to fit all the data, and the layer is thin (19 Å) in relation to the fully extended length of the molecule of about 34 Å. This indicates that the molecule is tilted with a mean order parameter for the tilt away from the surface normal of $\langle \cos \theta \rangle \cong 0.60 \pm 0.05$. The maximum coverage of surfactant at the critical micelle concentration (cmc) corresponds to an area per molecule of 50 Å^2, which is close to the value at the air/liquid interface at the same concentration [11]. Interestingly, when the overall thickness of the $C_{12}E_4$ layer at the air/liquid interface is corrected by removing the contribution of capillary waves to the thickness, the thickness of the layer becomes almost the same as observed at the hydrophobic solid/liquid interface. The close resemblance of the two layers in both amount adsorbed and corrected thickness is maintained down to areas per molecule as low as 300 Å^2. Although this cannot be regarded as a general result, the adsorption behaviour of the anionic double chain surfactant aerosol OT was also found to be similar at hydrophobic solid/liquid and air/liquid interfaces, after capillary wave corrections had been made [12–14].

Finally, we consider a more complex example, the mixture of two surfactants adsorbed at the hydrophobic solid/liquid interface. It should be clear from the discussion so far that the composition of a mixed surfactant

layer can be determined by using combinations of isotopic species of the two surfactants [15]. The interest in such a measurement is that it opens up the possibility not only of using composition measurements to help formulate better models of surfactant mixing at interfaces, which is often far from ideal, but of using structural measurements of the vertical separation of two surfactants to assess the mechanism of interaction between the two compounds, for example, whether it is driven by chain–chain interactions or by head group interactions. Hines et al. [16] have studied mixtures of sodium dodecyl sulphate (SDS) and N, N', N''-dodecyldimethylaminoacetic acid (dodecyl betaine) at the OTS surface. SDS is negatively charged and is known to interact strongly with the zwitterionic betaine. Once again, the structure and composition of the mixed layer were found to be quite similar to those at the air/liquid interface.

Surfactant layers at hydrophilic solid/liquid interfaces

Experimentally the most accessible hydrophilic surface is the silica surface, although it is by no means simple. Except at very low concentrations, where there may be coulombic interactions between the charged surfactant and charged surface, adsorption of surfactants is generally expected to be dominated by the hydrophobic effect, which will cause the surfactant to adsorb in a bilayer or related structure. In principle, neutron reflection can identify a bilayer structure either from its overall thickness or from the surface coverage. In practice, for the two surfactants that have so far been most systematically studied by neutron reflection, the non-ionic surfactant $C_{12}E_6$ [17, 18] and the cationic surfactant hexadecyl trimethyl ammonium bromide,

Progr Colloid Polym Sci (1997) 103:216–225
© Steinkopff Verlag 1997

Fig. 7 Schematic representation of the adsorbed layer of $C_{12}E_6$ adsorbed at the hydrophilic silica/water interface [17]

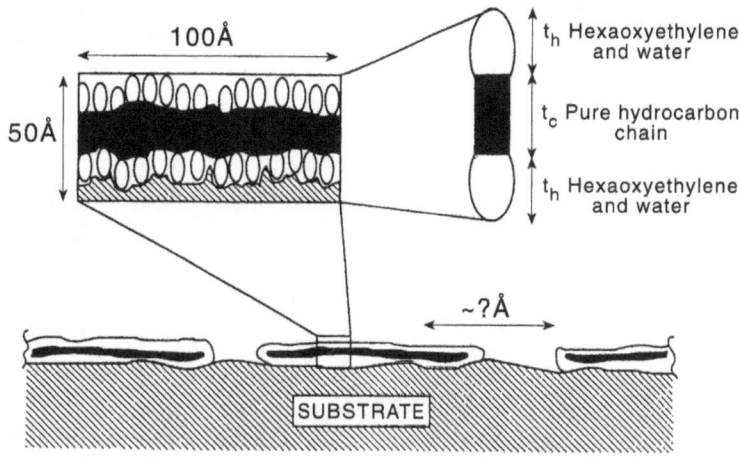

C$_{16}$H$_{33}$N(CH$_3$)$_3$Br, abbreviated to C$_{16}$TAB [19, 20], the coverage is not a completely reliable guide to the structure. It is always found to be considerably less than the 200% expected for a perfect bilayer, usually in the range 100–140%, which suggests that the structures formed are either defective bilayers or, possibly, flattened micelles. In all the ranges of concentration so far studied, the thickness, on the other hand, is always significantly larger than would be appropriate for a monolayer and is more consistent with bilayer aggregates. A typical structure, deduced from reflectivity data, is shown schematically for C$_{12}$E$_6$ adsorbed on hydrophilic silicon oxide in Fig. 7 [17].

The main weakness of specular neutron reflection is that it is not affected by the *in-plane* structure of the layer, although, in principle, such a determination is possible with off-specular reflection. Hence, the experiment cannot draw other than indirect conclusions about the in-plane heterogeneities. Recently, non-contact atomic force microscopy has shown unequivocally that surfactants absorb in aggregate structures on solid surfaces, often forming regular arrays of aggregates on the surface, which may have in-plane dimensions as small as those of spherical micelles [21, 22]. However, contrary to the neutron experiment, the AFM experiment is not sensitive to the *vertical* dimension of the aggregates. It seems reasonable to suppose that when the in-plane dimension of an adsorbed aggregate is typical of a spherical micelle then so will be its vertical dimension. It is therefore interesting to examine the thicknesses of adsorbed C$_{12}$E$_6$ and C$_{16}$TAB bilayers on silica.

Both the amount of C$_{16}$TAB and the thickness of the bilayer aggregates are sensitive to the condition of the underlying silica surface, which gives rise to a certain variability of the amount adsorbed on polished silica sam-

ples. In particular, Fragneto et al. [8] found that about 10% more C$_{16}$TAB was adsorbed on a smooth silica surface than on a rough one. Such sensitivity is not unexpected if it is aggregates that are adsorbed because one of the factors controlling adsorption would then be the distortion of the aggregate by the local surface structure and this could depend on the roughness of the surface. Fragneto et al. found that the thickness of the bilayer also depends on the roughness and that the thinnest, highest coverage, layers are formed on the smoothest surfaces. For C$_{16}$TAB bilayers the thickness has been found to vary between 32 and 40 Å on different silica surfaces. The fully extended length of the C$_{16}$TAB molecule is about 25 Å and the diameter of a spherical micelle should therefore be about 50 Å. The small value of the thickness of the bilayer aggregates is then more consistent with a defective bilayer rather than flattened micelles. On the other hand, the smallest thickness found for the C$_{12}$E$_6$ layer is (49 ± 4) Å [17, 18] to be compared with a fully extended length of about 38 Å. However, the value of 38 Å assumes that the ethylene glycol chain has its fully extended value of 21.6 Å, but experiments on the C$_{12}$E$_m$ series at the air/water interface indicate that this part of the layer will have a more disordered structure with a thickness, after allowing for the contribution of capillary wave roughness, of about 14 Å [23], which would make the effective maximum length of the whole molecule about 31 Å. It is therefore possible that C$_{12}$E$_6$ is in the form of flattened micelles.

Whereas all the neutron reflection studies of bilayer surfactant structures have relied on using the overall thickness of the layer to establish that it is a bilayer. Fragneto et al. were able to show this directly by using partially labelled C$_{16}$TAB molecules [8]. The series of isotopic species '0' C$_{16-m}$dC$_m$dTAB with m = 4, 8 and 12, where '0' indicates

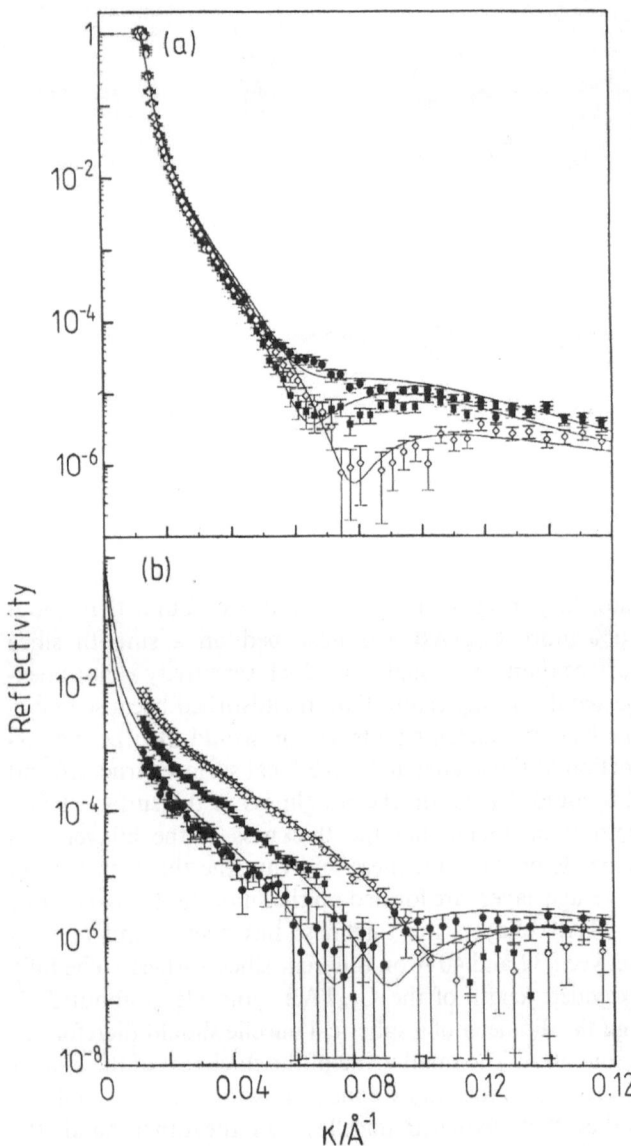

Fig. 8 Neutron reflectivity and calculated profiles at the Si/SiO$_2$/C$_{16}$TAB water interface for different isotopes of C$_{16}$TAB, (●) '0' C$_{12}$dC$_4$TAB, (■) '0' C$_8$dC$_8$dTAB, and (◇) '0' C$_4$dC$_{12}$dTAB, in (a) D$_2$O and (b) water (2.07×10^{-6} Å^{-2}) matched to silicon. The continuous lines are the best fits of a model of a surfactant bilayer to the data [8]

that the isotopic composition of the fragment indicated is such that its scattering length is zero, give quite different reflectivities, as shown in Fig. 8. With this range of different isotopic species of the surfactant together with the usual contrast variation of the solvent it was possible to show that the distribution of isotopic labels was such that it could only be consistent with a bilayer, i.e. a three layer model of the structure is the minimum that will account for

the set of reflectivity profiles. The dimensions of the models fitted for the different contrasts are shown in Fig. 9. Fragneto et al. were further able to show that the bilayer was symmetrical with respect to the surface normal when the surface was smooth, but was unsymmetrical on the rough surface with the outer, aqueous, side, of the bilayer being about 10% more dense than the inner.

Conclusions

Neutron reflection is clearly a very sensitive technique for determining structure and composition of surfactant layers adsorbed at the solid/liquid interface. There are, however, many other techniques that are becoming available and it is useful to compare the strengths and weaknesses of these techniques with neutron reflection.

We have already discussed the possible complementarity of non-contact atomic force microscopy and neutron reflection in that the former is primarily sensitive to in-plane structure, whereas this is the greatest limitation of neutron reflection. The more traditional technique of ellipsometry is extremely sensitive as a means of determining surface coverage and variation of thickness (see, for example, [26]) and provides a more rapid and accessible tool for such studies. In comparison, the power of neutron reflection lies in its ability to give an absolute thickness of the layer and to resolve the internal structure of the layer, although it must be emphasized that the latter may only be possible with systematic isotopic labelling. Spectroscopic techniques also examine the adsorbed layer directly and may be very sensitive to particular features of the structure [27], e.g. orientation of fragments of the layer, at a level of sensitivity much higher than conceivable with neutrons. However, the weakness of the spectroscopic techniques is generally that they cannot determine the absolute, and often not even the relative, coverage of the surface, and this may make it difficult to draw conclusions about, for example, the variation of orientation with coverage. Finally, there is the less direct method of the force balance, which can give extremely accurate measurements of the distance variation of the interaction and hence, with some assumptions, the thickness of adsorbed layers. For example, the thickness of 32 Å obtained from neutron reflection for the C$_{16}$TAB bilayer on a smooth surface is no more accurate than, and compares extremely well with, values of 32 and 33 Å obtained by Kekicheff [24] and Pashley [25], respectively, for C$_{16}$TAB on mica, the smoothest possible surface. Unfortunately, another deficiency of the neutron experiment is that it is restricted to a small number of solid substrates and mica is not one of

Progr Colloid Polym Sci (1997) 103:216–225
© Steinkopff Verlag 1997

225

Fig. 9 Schematic representation of the model used to fit the neutron reflectivity data from different isotopic species of $C_{16}TAB$ adsorbed at the Si/SiO$_2$/aqueous interface. The surfactant chains overlap and the shaded areas indicate the three layers which have to be used to account for the observed reflectivity for each different isotope [8]. CD$_2$ and CH$_2$ groups are respectively the filled and open "atoms" in the interdigitated hydrocarbon chains

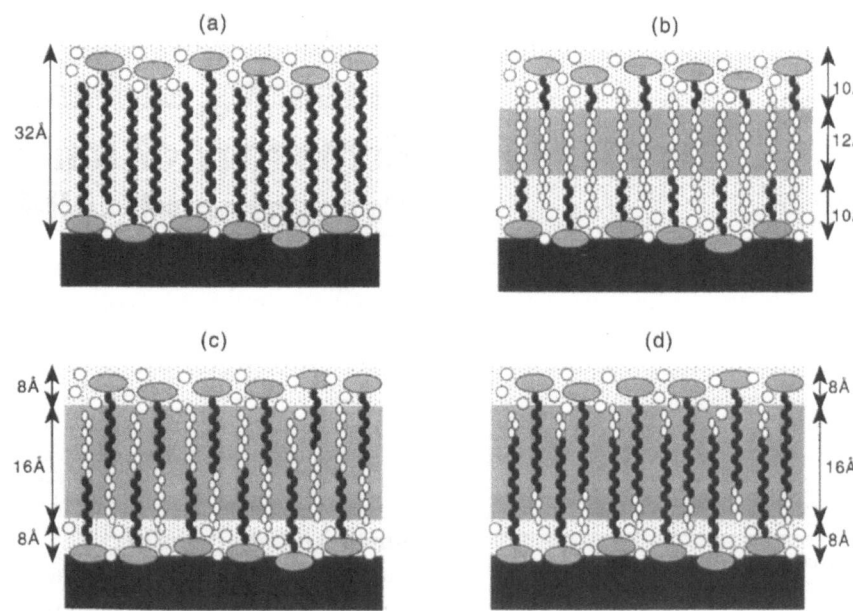

these. While this does not place much limitation on the type of surface that can be explored, since chemical modification of silica can mimic almost any type of surface, it does mean that no direct comparison can be made between surface force measurements and neutron reflection measurements.

References

1. Hayter JB, Highfield RR, Pullman BJ, Thomas RK, McMullen AI, Penfold J (1981) J Chem Soc Faraday Trans I 97:1437
2. Lu JR, Li ZX, Smallwood J, Thomas RK, Penfold J (1995) J Phys Chem 99:8233
3. Thomas RK, Penfold J (1996) Curr Opin Colloid Interface Sci 1:23
4. Heavens OS (1955) Optical Properties of Thin Films. Butterworths, London
5. Als-Nielsen J (1985) Z Phys B61:411
6. Crowley TL (1984) D Phil Thesis, University of Oxford
7. Lu JR, Lee EM, Thomas RK (1996) Acta Crystallogr. A52:11
8. Fragneto G, Thomas RK, Rennie AR, Penfold J (1996) Langmuir, 25:6036
9. Maoz R, Sagiv G (1987) Langmuir 3:1034
10. Fragneto G, McDermott DC, Lu JR, Thomas RK, Rennie AR, Satija SK, Gallagher D (1996) Langmuir 12:477
11. Lu JR, Li ZX, Su T, Thomas RK, Penfold J (1993) Langmuir 9:2408
12. Fragneto G, Li ZX, Thomas RK, Rennie AR, Penfold J (1996) J Colloid Interface Sci 178:531
13. Li ZX, Lu JR, Thomas RK (1996) Langmuir, in press
14. Li ZX, Lu JR, Thomas RK, Penfold J (1996) J Phys Chem, in press
15. McDermott DC, Kanelleas D, Thomas RK, Rennie AR, Satija SK, Majkrzak CF (1993) Langmuir 9:2404
16. Hines JD, Fragneto G, Thomas RK, Garrett PR, Rennie GK, Rennie AR (1996) J Colloid Interface Sci, in press
17. McDermott DC, Lu JR, Lee EM, Thomas RK, Rennie AR (1992) Langmuir 8:1204
18. Lee EM, Thomas RK, Cummins PG, Staples EJ, Penfold J, Rennie AR (1989) Chem Phys Lett 162:196
19. Rennie AR, Lee EM, Simister EA, Thomas RK (1990) Langmuir 6:1031
20. McDermott DC, McCarney J, Thomas RK, Rennie AR (1994) J Colloid Interface Sci 162:304
21. Manne S, Gaub HE (1995) Science 270:1480
22. Ducker W, Grant LM (1996) J Phys Chem 100:3207
23. Lu, JR, Li ZX, Thomas RK, Staples EJ, Thompson L, Tucker I, Penfold J (1994) J Phys Chem 98:6559
24. Kekicheff P, Christenson HK, Ninham BW (1989) Colloids and Surfaces 40:31
25. Pashely RM, McGuiggan PM, Horn RG, Ninham BW (1988) J Colloid Interface Sci 126:569
26. Tiberg F (1996) J Chem Soc Faraday Trans 92:551
27. Johal MS, Usadi EW, Davies PB (1996) J Chem Soc Faraday Trans 92:573

Progr Colloid Polym Sci (1997) 103:226–233
© Steinkopff Verlag 1997

S. Manne

Visualizing self-assembly: force microscopy of ionic surfactant aggregates at solid–liquid interfaces

Received: 10 December 1996
Accepted: 13 December 1996

Dr. S. Manne (✉)
Princeton Materials Institute
Princeton University
Princeton, New Jersey 08540, USA

Abstract The adsorption of surfactants from micellar solutions onto solid surfaces plays a crucial role in applications such as surface wetting, particulate detergency, and colloidal stabilization. Recent work (see below) has shown that the shape, size, and lateral organization of ionic surfactant aggregates at solid–liquid interfaces can be determined directly by atomic force microscopy (AFM). Imaging is performed using repulsive stabilization forces between surfactant layers adsorbed to both the tip and sample, and aggregate structures are determined by comparing AFM images with previous adsorption measurements. The observed interfacial structures for ionic surfactants result from a tradeoff between intermolecular interactions (i.e. geometric packing considerations) and molecule–surface interactions (i.e. the density, type, and crystalline order of adsorption sites). The hydrophobic, crystalline cleavage planes of graphite and MoS_2 adsorb and orient single-tail surfactants along the substrate symmetry axes, and this horizontal adsorption serves as a template for half-cylindrical aggregates. The hydrophilic, anionic surfaces of mica and silica interact with cationic headgroups, giving rise to spherical, cylindrical, or planar aggregates depending on the surfactants geometry and density of electrostatic binding sites.

Key words Force microscopy – self-assembly – double layer forces – micelles – interfaces

Introduction

The spontaneous aggregation of surfactants into micelles of finite size in aqueous solutions was first proposed by McBain in 1913 [1]. Micellization in solution has since been investigated using a variety of experimental techniques (e.g. X-ray scattering [2], fluorescence quenching [3] and cryo-transmission electron microscopy [4, 5]), and this process is now fairly well understood both in terms of the geometry [6] and thermodynamics [7] of aggregation. An analogous aggregation process at solid–liquid interfaces was first proposed by Fuerstenau in 1955 [8], based on the adsorption characteristics of anionic surfactants on alumina. However, an understanding of interfacial aggregation has been slower in coming, partly due to the experimental difficulties involved in detecting structure in nanometer-scale adsorbate films in a liquid environment. Surfactant adsorption has been quantified by solution-depletion methods [9], which measure the effective surface area per adsorbed molecule, and by surface force apparatus measurements [10–15], which measure the charge density (or surface potential) and thickness of the adsorbate layer; however, these measurements are often consistent with more than one adsorbate structure. Evidence for lateral structure in adsorbed surfactant films has come from neutron reflection [16], fluorescence quenching [17],

Progr Colloid Polym Sci (1997) 103:226–233
© Steinkopff Verlag 1997

surface force measurements [18] and calorimetry [19], but the size, shape and lateral organization of aggregates has proven difficult to quantify.

Atomic force microscopy (AFM) [20] has recently been used to image interfacial aggregates directly, *in situ* and at nanometer resolution [21, 22]. The key to this application lies in an unusual contrast mechanism, namely a pre-contact repulsive force ("colloidal stabilization force") between the adsorbed surfactant layers on the tip and sample. In contrast to previous adsorbate models of flat monolayers and bilayers, AFM images have shown a striking variety of interfacial aggregates – spheres, cylinders, half-cylinders and bilayers – depending on the surface chemistry and surfactant geometry. I review the AFM evidence for these structures and discuss the possible intermolecular and molecule–surface interactions involved.

Experimental

The AFM probe is simply a micromachined cantilever spring (typically $100 \times 20 \times 0.5 \, \mu m$) terminated by a small pyramidal tip ($\sim 4 \, \mu m$ in size) which is placed in close proximity to a sample surface. Forces applied by the sample surface on the tip cause proportional deflections of the cantilever, enabling the AFM probe to act as a force sensor. An AFM images a sample surface by sensing the cantilever deflection (i.e. tip–sample force) at each point of a raster scan and plotting these data on a color scale. Although the AFM traditionally uses contact forces to image sample topography, it has been recognized from the beginning [20] that *any* laterally varying force can give rise to imaging contrast; examples of noncontact imaging forces include electrostatic [23] and magnetic forces [24]. More recently, it has been demonstrated that repulsive electric double layer (e.d.l.) forces can also be used as a contrast mechanism in aqueous solutions [25], setting the stage for imaging ionic surfactant aggregates at solid–liquid interfaces.

The technique for imaging interfacial aggregates is schematically illustrated in Fig. 1 for cationic surfactant adsorption at either an anionic surface (such as mica) or a hydrophobic surface (such as graphite). An aqueous surfactant solution above the critical micelle concentration (cmc) is introduced into the AFM fluid cell. Spontaneous surfactant adsorption on the sample and on the AFM tip (whose surface is silica-like, and therefore anionic at neutral pH) charges both surfaces positively, setting up a precontact repulsive force (Fig. 1A). Using a force setpoint in this precontact region of the force curve allows the adsorbate topography to be imaged while "flying" the tip above the plane of hard contact (Fig. 1B). No instrument modification is necessary, and standard feedback electronics can be used for imaging.

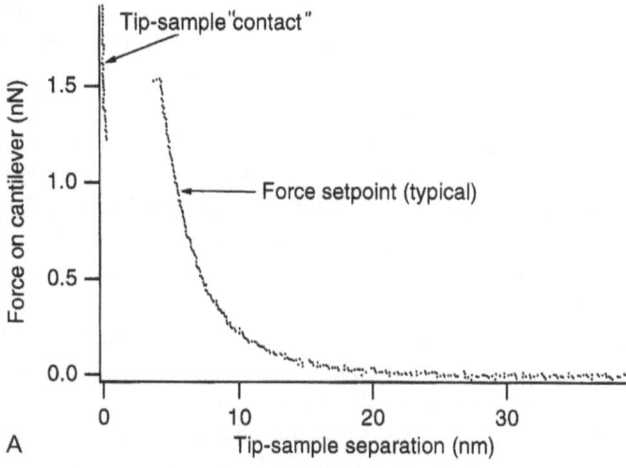

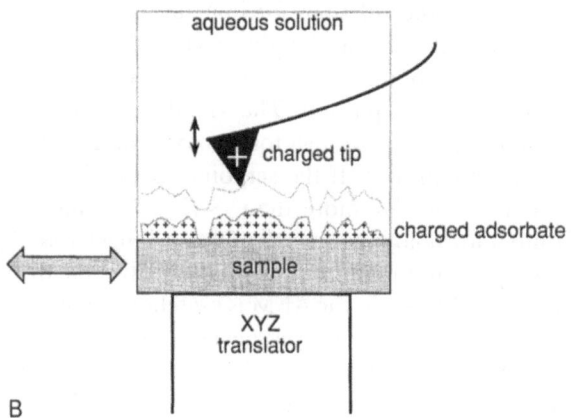

Fig. 1 (A) Typical data for force on the cantilever vs. tip–sample separation in ionic surfactant solution, in this case for a silica surface in 7 mM $C_{14}TABr$ ($2 \times cmc$). Here the precontact repulsion is detectable at a tip–sample separation ~ 20 nm and increases rapidly down to a separation of ~ 4 nm. The measured exponential decay length is 2.8 nm, compared to theoretical Debye lengths of 3.6 nm for a 1:1 electrolyte and ~ 2.9 nm when micelles are taken into account [44]. A typical force setpoint for imaging surface aggregates is shown. Below 4 nm, the surfactant layers of the tip and sample fuse together, causing an irreversible jump of the cantilever. The zero value for tip–sample separation (i.e. tip–sample "contact") is chosen to be at the onset of constant compliance [21]; it likely does *not* represent intimate contact between the tip and sample surfaces, but rather a contact between the hydrophobic groups of strongly adsorbed surfactant molecules on the two surfaces, as reported previously [13].) (B) Schematic of the technique for imaging surface aggregates by AFM. Adsorption of cationic surfactant on the tip and sample gives both surfaces a net positive charge, resulting in repulsive interactions ("colloidal stabilization forces") which are used to image the adsorbate structure in the same way that contact imaging is normally used. As the sample is raster scanned, the cantilever deflects in response to precontact repulsive forces and traces out the adsorbate morphology (shaded line). Although electric double-layer interactions are the most obvious source of this precontact repulsion, confinement effects also play an important role (see text)

I have used a commercial AFM (Nanoscope III, Digital Instruments) and commercial cantilevers made of either silicon (Park Scientific Instruments) or silicon nitride (Digital Instruments), cleaned by exposure to ultraviolet light for 15–60 min. Cantilever spring constants ranged from ~0.05 to 0.6 N/m. Stiffer cantilevers were less susceptible to thermal drift but gave poorer contrast. Introduction of the surfactant solution into the fluid cell was almost always accompanied by large changes in the free cantilever deflection, probably caused by thermal drift; immersion times of 5–60 min were necessary for the drift to equilibrate before imaging could begin. The surfactant solution was held in the fluid cell by surface tension alone; an O-ring was *not* used for containment, as it often led to unacceptable levels of mechanical creep due to improper seating and swelling.

A typical sequence of steps for imaging is as follows. After the cantilever is engaged in the usual way, a force vs. distance curve is obtained to check for the existence of a usable precontact repulsion. The viability of imaging depends on both the precontact force maximum F_{max} and on residual thermal drift. If the setpoint on the force vs. distance curve drifts by more than $\sim F_{max}/2$ during the time required for imaging (typically 60 s), a longer wait is usually necessary to minimize residual drift. After the drift is within acceptable levels, the AFM is switched to imaging mode and the setpoint initially set to a large negative value (say −10 V), thereby fully retracting the z translator and separating the tip from the sample. The lateral scan parameters are now entered, enabling the scan while the tip is still retracted from the sample. The feedback gains are set to moderate values (typically integral gain = 0.3, proportional gain = 0, and 2D gain = 0), and the surface aggregates are imaged in deflection mode. This is accomplished by increasing the force setpoint to a value just slightly (~0.03 V) positive of the free cantilever deflection, causing the feedback loop to activate and the sample to advance towards the tip, effecting force control in the precontact region. While monitoring the deflection image, the setpoint is carefully and incrementally increased until the surface aggregates become visible. If the setpoint is increased too far, the aggregates abruptly disappear as the cantilever goes "over the top" of the force curve and jumps into the sample surface. Then the setpoint is once again set to −10 V to retract the z translator and the procedure repeated as necessary.

Results and discussion

The first surfactant aggregate structure to be directly imaged by AFM was that of hexadecyltrimethylammonium bromide ($C_{16}TABr$) on graphite [21]. Since then interfacial structures have been determined [21, 22, 26] for several ionic surfactants on both hydrophilic and hydrophobic surfaces (Table 1). Mica and graphite are particularly useful as (respectively) model hydrophilic and hydrophobic substrates, since both are widely available as single crystals, and fresh atomically flat surfaces can be prepared simply by cleaving.

The shapes and lateral organizations of interfacial aggregates have been elucidated by comparing AFM images with published quantitative measures of adsorption (see references in Table 1, column 3). For ionic surfactant concentrations above the cmc, measured surface areas per adsorbed molecule are generally *consistent with* vertical

Table 1 Comparison of AFM results with previous models of interfacial surfactant aggregation from aqueous micellar solutions

Solid surface	Ionic surfactant	Previous aggregation models	Aggregate morphology by AFM
Mica (anionic cleavage plane)	Alkyltrimethlammonium halides $C_{12}H_{25}N^+(CH_3)_3Cl^-$, Br^- $C_{14}H_{29}N^+(CH_3)_3Br^-$ $C_{16}H_{33}N^+(CH_3)_3Cl^-$, Br^-	Uniform bilayer [10, 12–14]	Parallel flexible cylinders [22]
	$(C_{12}H_{25})_2N^+(CH_3)_2Br^-$	Uniform bilayer [11]	Uniform bilayer [22]
Silica (oxide layer on Si wafer)	$C_{14}H_{29}N^+(CH_3)_3Br^-$	Bilayer patches [18, 19, 36, 37] Spheres [38, 39]	Spheres and spheroids [22]
Graphite (hydrophobic cleavage plane)	Alkyltrimethylammonium halides $C_{12}H_{25}N^+(CH_3)_3Br^-$ $C_{14}H_{29}N^+(CH_3)_3Br^-$ $C_{16}H_{33}N^+(CH_3)_3Cl^-$, Br^-	Vertical monolayer [28]	Parallel half-cylinders [21, 22]
	$C_{12}H_{25}OSO_3^-Na^+$	Vertical monolayer [29] Hemispheres [30]	Parallel half-cylinders [26]
MoS_2 (hydrophobic cleavage plane)	$C_{14}H_{29}N^+(CH_3)_3Br^-$	(none)	Parallel half-cylinders [22]

Progr Colloid Polym Sci (1997) 103:226–233
© Steinkopff Verlag 1997

monolayers on hydrophobic surfaces and vertical bilayers on oppositely charged hydrophilic surfaces [9, 15]. AFM results generally show more complicated aggregate morphologies which are, however, also consistent with the adsorption data.

Hydrophobic substrates: Graphite and molybdenum disulfide

All single-chain surfactants investigated so far – whether cationic [21, 22], anionic [26], or zwitterionic [27] – have shown aggregates on graphite characterized by straight, parallel stripes above the cmc. Figure 2A shows a representative example, namely aggregates of the cationic surfactant hexadecyltrimethylammonium chloride ($C_{16}TACl$) on graphite. These stripes are (a) oriented perpendicular to one of the three symmetry axes of the underlying substrate (as determined by lattice resolution scans), and (b) spaced apart by a little over twice (between 2 and 3 times) the length of the surfactant molecules. (For sodium dodecylsulfate (SDS) on graphite, this spacing is observed to be somewhat dependent on surfactant concentration below the cmc but relatively constant above the cmc [26].)

These observations are not consistent either with flat vertical monolayers [28, 29] or with hemispherical aggregates [30], two earlier models for surfactant adsorption on graphite. Instead, these images are interpreted [21] as arising from half-cylindrical aggregates, as shown in Fig. 2B. This arrangement readily accounts for the observed striped motif and the measured spacing between stripes, which is set by the cylindrical diameter. It also accounts for adsorption isotherm data, since the occupational headgroup area of a half-cylindrical micelle, when projected onto a plane, does not differ much from the headgroup area of a vertical close-packed monolayer (see the appendix). Additionally, the half-cylindrical morphology is consistent with a number of adsorption studies [31–33] indicating that alkanes and simple alkyl derivatives adsorb horizontally on the graphite cleavage plane, with their alkane chains oriented along a graphite symmetry axis. A bottom row of surfactants adsorbed in this fashion (head-to-head and tail-to-tail) not only serves as a natural foundation for half-cylinders, but also fixes their orientation perpendicular to the graphite symmetry axes as observed.

The half-cylindrical morphology is probably favored for two reasons. First, the free energy change per molecule for horizontal adsorption on graphite is far greater than that for micellization [21], assuming that calorimetry results of SDS [29] are typical of ionic surfactants. Thus, molecules adsorbed horizontally in the first adsorption

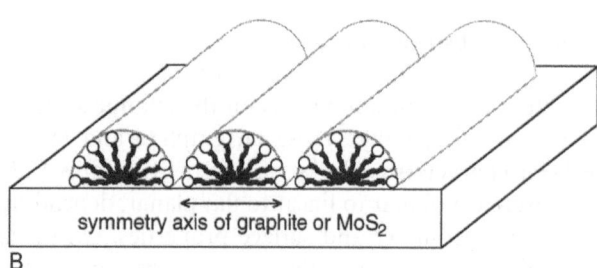

symmetry axis of graphite or MoS$_2$

Fig. 2 (A) 200 × 200 nm AFM image of $C_{16}TACl$ aggregates on graphite (2.6 mM solution, 2 × cmc), showing straight parallel stripes and two discrete spots in the Fourier transform (inset). The measured stripe spacing is 5.7 nm, and the stripes are oriented perpendicular to an underlying symmetry axis of the graphite substrate (not shown). For comparison, the extended chain length of C_{16} is 2.2 nm [45], and the extended length of the surfactant (assuming a headgroup diameter of ~0.6 nm) is therefore ~2.8 nm. Similar morphologies have been observed for $C_{12}TABr$ (S. Manne, unpublished results), $C_{14}TABr$ [22] and $C_{16}TABr$ [21] on graphite, and for $C_{14}TABr$ on MoS$_2$ [22]. (B) Proposed model of half-cylindrical aggregates at a crystalline hydrophobic substrate, with bottom row of molecules oriented along the substrate symmetry axes

step cannot be desorbed or "pulled up" by micelles in surrounding solution. Second, the presence of crystalline structure on the substrate introduces a directionality or anisotropy into its interactions with linear groups such as alkanes, making it difficult for surfactant tails to cross substrate symmetry axes. This would prevent hemispherical aggregation – which might otherwise be favored owing to the spherical "free curvature" – since hemispheres would require horizontal adsorption in roughly circular patches. The lattice of a crystalline hydrophobic substrate could thus control the shape and orientation of lyotropic liquid crystals, in much the same way that uniaxially textured polymers can control the interfacial phases of thermotropic liquid crystals [34].

Although an epitaxial relationship does exist between the surfactant tails and the graphite surface [31, 32] (the C–C spacing for an alkane chain differs from that of the graphite lattice by <2%), the surface control of aggregation is *not* dependent on this epitaxy. This is evidenced by the presence of similar striped structures, at similar spacings and relative orientations, on the hydrophobic cleavage plane of MoS_2 [22] – whose lattice has a different symmetry, periodicity, and elemental composition from the graphite lattice. The mere existence of a crystalline lattice may be enough to orient the tails, giving rise to linear aggregates at right angles to the substrates symmetry axes. Whether an *amorphous* hydrophobic substrate gives rise to hemispherical aggregates remains to be seen.

Hydrophilic substrates: mica and silica

Aggregates of cationic surfactants on the anionic surfaces of mica and silica, unlike those on graphite and MoS_2, have been characterized by a variety of structures [22] – from discrete globular to linear to flat planar, depending on surfactant geometry and surface properties. On mica, alkyltrimethylammonium surfactants give rise to parallel worm-like aggregates (Fig. 3A shows a representative example, $C_{12}TACl$ on mica), whereas dialkyldimethylammonium surfactants give rise to flat featureless images [22]. On silica, $C_{14}TABr$ (and probably other alkyltrimethylammonium surfactants as well) gives rise to aggregates that are discrete and globular in appearance (Fig. 4).

Surface force apparatus measurements [10–15] have indicated that adsorbed layers of quaternary ammonium surfactants on mica (above the cmc) have thicknesses of roughly twice the surfactant length. The standard interpretation of adsorption in the form of flat bilayers [10, 12–15] is, however, incompatible with AFM images for single-chain surfactants (Fig. 3A). This structure is instead interpreted as arising from flexible cylindrical aggregates lying on the substrate plane (Fig. 3B). Here the cylindrical diameter sets both the previously measured adsorbate thickness and the observed spacing between stripes on AFM images (between two and three times the surfactant length).

One can envision at least three related phenomenological approaches to understanding how this aggregate morphology arises. First, cylindrical surface aggregates can be regarded as a compromise between the curvature favored by intermolecular interactions (i.e. "free curvature', which is spherical for alkyltrimethylammonium surfactants [2])

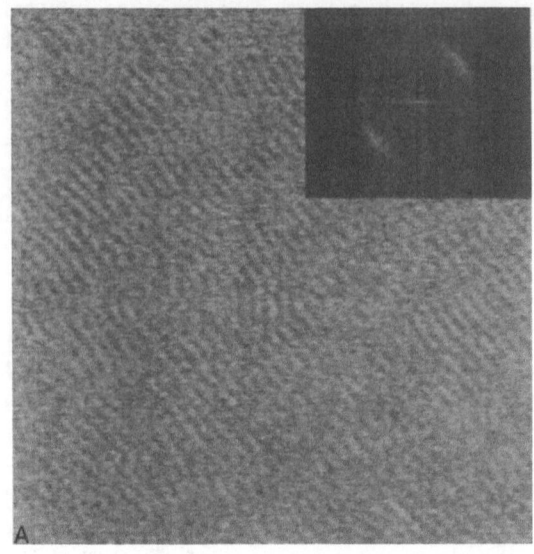

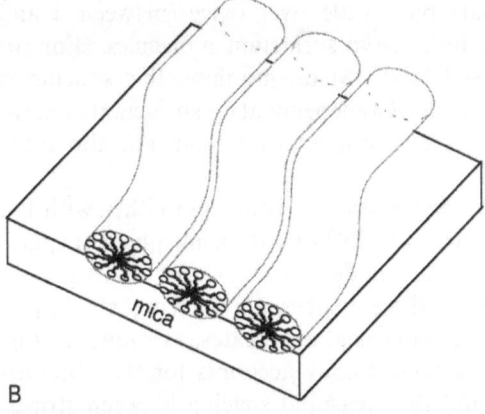

Fig. 3 (A) 200×200 nm AFM image of $C_{12}TACl$ aggregates on mica (38 mM solution, $2 \times$ cmc), showing parallel but meandering stripes as evidenced by a continuous arc in the Fourier transform (inset). The measured stripe spacing is 4.7 nm. For comparison, the extended chain length of C_{12} is 1.7 nm [45], and the extended surfactant length is ~2.3 nm. Similar morphologies have been observed for $C_{14}TABr$ [22], $C_{12}TABr$, $C_{16}TABr$ and $C_{16}TACl$ (S. Manne, unpublished results). (B) Proposed model of flexible cylindrical aggregates on mica

and that favored by headgroup–surface interactions (i.e. a flat bilayer). Second, the interfacial aggregates can be considered as the result of a surface-induced sphere-to-rod transition, wherein the mica acts as a highly charged "counterion" that causes a closer headgroup spacing and increases the dimensionless packing parameter. And third, the interfacial aggregates can be though of as simply a confined bulk lyotropic phase, caused by a locally high concentration of surfactant molecules attracted by the surface, e.g., parallel cylinders can be regarded as a hexagonal phase in two dimensions. It should be noted, however, that

Progr Colloid Polym Sci (1997) 103:226–233
© Steinkopff Verlag 1997

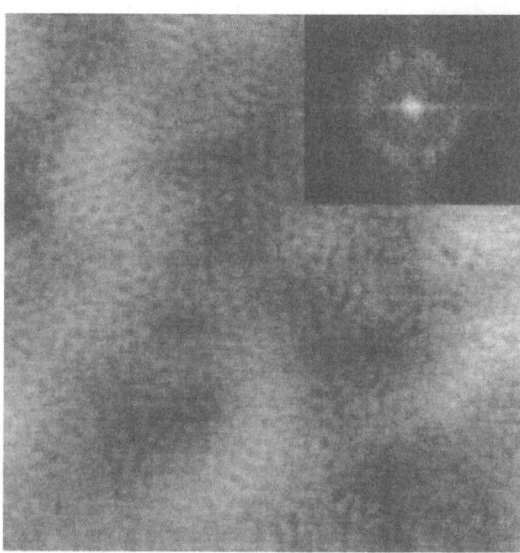

Fig. 4 200 × 200 nm AFM image of C_{14}TABr on silica (6.8 mM solution, 2 × cmc), showing a somewhat polydisperse aggregate structure consisting of spheres and spheroids (and perhaps short cylinders) in a random arrangement. The Fourier transform (inset) shows a continuous ring, indicating a well-defined spacing but little regular order. The measured nearest-neighbor spacing is ~5.8 nm. For comparison, the extended chain length of C_{14} is 1.9 nm [45], and the extended surfactant length is ~2.5 nm

this model is difficult to reconcile with the observation that *all* alkyltrimethylammonium surfactants in Table 1 show cylindrical aggregates on mica; C_{12}TACl is unique among the listed surfactants in exhibiting a pronounced micellar cubic phase [35] (at bulk concentrations that correspond to hexagonal phases in the other surfactants), yet AFM images of C_{12}TACl show cylinders rather than an "interfacial cubic phase" of spherical micelles on mica. A better understanding of interfacial aggregation on mica requires a more detailed investigation of aggregate geometry vs. dimensionless packing parameter, and such an investigation using gemini surfactants is currently under way.

Double-chain quaternary ammonium surfactants on mica show flat, featureless images [22] consistent with previous models of a uniform bilayer [11]. In this case no conflict exists between preferred curvature in solution and that at the surface, since dialkyl-dimethylammonium surfactants form bilayers (and vesicles) in solution.

Adsorption of alkyltrimethylammonium surfactants above the cmc on the surface of silica have been investi-gated using a variety of techniques [18, 19, 36–39], and the surface aggregates have been variously interpreted as bilayer patches [18, 19, 36, 37] and globular micelles

[38, 39]. AFM images of C_{14}TABr on silica (Fig. 4) show roughly spherical and spheroidal dots consistent with globular aggregates [22]. Although these aggregates are not arranged in a regular lattice, their nearest-neighbor distances are fairly well-defined for a given sample, as evidenced by the Fourier transform (inset). (However the interaggregate spacings have been observed to vary from one sample to the next [22], presumably due to uncontrol-led differences in surface chemistry; similar variations have also been observed in neutron reflection studies [16].)

Why do alkyltrimethylammonium surfactants form roughly spherical aggregates on silica but cylindrical ag-gregates on mica? The answer probably lies in the far greater density of electrostatic adsorption sites available on mica. Zeta potential measurements [40] in 1 mM KCl have shown that at pH 6 (a typical imaging condition), the surface potentials Ψ_0 of silica and mica are -80 and -120 mV, respectively. Using the Grahame equation

$$\sigma_0 = (0.731 e^- \, \text{nm}^{-2}) \times [\text{KCl}]^{1/2} \times \sinh[\Psi_0/(51.4 \, \text{mV})],$$

(1)

the calculated surface charge densities are $-0.05 e^- \, \text{nm}^{-2}$ for silica and $-0.12 e^- \, \text{nm}^{-2}$ for mica – more than a factor of two greater. The disparity is even more striking when we consider that the density of adsorption sites on mica is potentially much greater than the charge density, owing to ion exchange of the headgroups with the surface K^+ ions; the adsorption site density can therefore approach the charge density of the fully dissociated surface, $-2 e^- \, \text{nm}^{-2}$. Thus, a low density of adsorption sites on silica could lead to isolated adsorbates, whose hydropho-bic groups nucleate aggregates resembling bulk micelles [38], whereas a higher density of adsorption sites on mica leads to a higher-density form of interfacial aggregates, i.e. cylinders. On silica, the spacing between aggregates has been observed to increase with decreasing surface charge [22], giving further support to this model.

In addition to adsorption site density, one other im-portant difference between the two substrates deserves mention: The silica surface is amorphous whereas the mica is crystalline. While it is possible that this crystallinity plays a role in the formation of linear surface aggregates on mica (as with hydrophobic substrates), this effect is likely to be weak since the surface interacts with the symmetric headgroups rather than the highly asymmetric tails. The orientation of cylinders by the mica surface (Fig. 3), for instance, is distinctly weaker than the ori-entation of half-cylinders by graphite (Fig. 2). It is possible that the degree of surface order affects the degree of aggregate order (i.e. the length or "worminess' of the cylin-ders), but further experimentation is needed to test this hypothesis.

Imaging mechanism

Although the simple-minded schematic Fig. 1B implies that screened electrostatic forces alone are responsible for generating lateral contrast, the truth is probably much more complicated. At a typical force setpoint for precontact imaging (Fig. 1A), the cantilever position is just ~1 nm above the plane of hemifusion (where the irreversible jump takes place). At these small separations a host of non-e.d.l. repulsive interactions involving confinement effects come into play [41]. A *partial* list includes: finite ion size effects, caused by counterion confinement; hydration forces, caused by confinement of water molecules; and protrusion forces, caused by a suppression of (a) natural fluctuations of the surface aggregate layer and (b) molecular exchange between surface aggregates and solution. In addition, there are three other sources of uncertainty specific to AFM, namely (i) the shape of the imaging tip, (ii) the aggregate morphology on the highly curved tip surface, and (iii) the mechanical properties, especially the deformability, of aggregate layers at typical imaging forces.

While it will require many more investigations to untangle all of these effects and arrive at a good understanding of the contrast mechanism, comfort can be taken from the growing evidence that AFM images accurately reflect the actual aggregate morphologies at interfaces. This evidence comes from several directions. First, aggregate structures are observed to scale and rotate in the expected way when the AFM scan size and direction are changed (although scan direction-dependent aggregate *spacings* have been reported [26]), and the structures are observed to be independent of scan speed, ruling out scan-related dynamic effects such as "skid marks" [42]. Second, observed aggregate morphologies are independent of the cantilever used; cantilevers made of different materials, having different spring constants and tip shapes, give rise to similar aggregate images for a given surfactant and substrate. Conversely, a single cantilever can be used to image two or more different aggregate morphologies by changing the surfactant or substrate. Third, the observed correlation between aggregate orientations and graphite and MoS_2 symmetry axes [21, 22] is a strong indication that the aggregates are associated with the substrate and are not somehow created by surfactant confinement between the tip and sample. In addition, the same aggregate structure is generally recreated after deliberately rupturing the adsorbate layer at high force and re-imaging at low force. Fourth, observed aggregate curvature on mica varies with dimensionless packing parameter in the expected way, with double-tailed surfactants making flat aggregates and single-tailed surfactants making more curved (cylindrical) aggregates; consistent with this trend, preliminary results (not shown) with a *divalent* single-tailed surfactant (highly

repulsive headgroup) have revealed spherical aggregates on mica.

Conclusion

I have reviewed recent developments in direct imaging of surfactant aggregate morphology at solid–liquid interfaces using atomic force microscopy. Imaging is performed not in hard contact mode but by using precontact repulsion arising from electric double layer, hydration, and other confinement forces between the adsorbed surfactant layers on the tip and sample. These results reveal surface aggregates possessing a greater degree of curvature and complexity than previously supposed. Combining AFM images with previous measurements of absorbate thickness and density enables the identification of plausible aggregate morphologies for ionic surfactants (Table 1). For hydrophilic substrates, aggregates range in curvature from flat bilayers to parallel flexible cylinders to randomly arranged spheres, depending on surfactant geometry and surface chemistry (the density of binding sites and perhaps the degree of surface order or crystallinity). For crystalline hydrophobic substrates the aggregate morphology is much less variable, revealing parallel half-cylindrical aggregates, oriented perpendicular to the substrate symmetry axes, for each single-tailed ionic surfactant investigated; the responsible mechanism seems to be a horizontal adsorption of alkyl groups along symmetry axes, caused by a directional anisotropy in van der Waals and hydrophobic interactions between the tailgroup and surface lattice.

Acknowledgments I thank G.G. Warr for useful discussions, Digital Instruments for technical support, and the U.S. Army Research Office (grant DAAH04-95-1-0102) for financial support.

Appendix

For a given surfactant molecule, the dimensionless packing parameter [6] $g = v/(a_0 l)$, where v and l are the hydrocarbon chain volume and length and a_0 the optimal headgroup area. For linear, saturated, single-chain hydrocarbon tailgroups of $\gtrsim 10$ carbon atoms, v/l is approximately constant and has a value of 0.21 nm^2. For this special case (which encompasses most of the surfactants listed in Table 1), therefore $a_0 = 0.21 \text{ nm}^2/g$. A spherical aggregate curvature results when $g < \frac{1}{3}$ and a cylindrical curvature when $\frac{1}{3} < g < \frac{1}{2}$. For a half-cylindrical surface aggregate (Fig. 2B), we assume an effective g which lies just above the lower limit for cylindrical curvature, say $g = 0.35$. (This assumes that the surface exerts the smallest

Progr Colloid Polym Sci (1997) 103:226–233
© Steinkopff Verlag 1997

perturbation required to change the aggregate curvature from the spherical one favored in solution.) This gives $a_0 = 0.60 \text{ nm}^2$ for the occupational headgroup area in a half-cylindrical surface aggregate. The projected value of this area on the flat plane is then $a_0 \times [(2 \times \text{radius} \times \text{length})/(\pi \times \text{radius} \times \text{length})] = (2/\pi) \times a_0 = 0.38 \text{ nm}^2$. This value is quite close to the saturated occupational areas per molecule measured by solution depletion (0.42 nm^2 for SDS [43] and 0.40 nm^2 for $C_{16}TABr$ [28] on Graphon).

References

1. McBain JW (1913) Trans Faraday Soc 9:99
2. Reiss-Husson F, Luzzatti V (1964) J Phys Chem 68:3504–3511
3. Kunjappu JT, Somasundaran P, Turro NJ (1990) J Phys Chem 94:8464–8468
4. Talmon Y (1986) Colloids Surf 19:237–248
5. Vinson PK, Bellare JR, Davis HT, Miller WG, Scriven LE (1991) J Colloid Interface Sci 142:74–91
6. Israelachvili JN (1992) Intermolecular and Surface Forces, 2nd ed. Academic Press, London, pp 341–394
7. Rosen MJ (1978) Surfactants and Interfacial Phenomena. Wiley, New York, pp 83–122
8. Gaudin AM, Fuerstenau, DW (1955) Trans AIME 202:958
9. Hough DB, Rendall HM (1983) In: Parfitt GD, Rochester CH (eds) Adsorption from Solutions at Solid/Liquid Interfaces. Academic Press, London, pp 247–319
10. Pashley RM, Israelachvili JN (1981) Colloids Surf 2:169–187
11. Pashley RM, McGuiggan PM, Ninham BW, Brady J, Evans DF (1986) J Phys Chem 90:1637–1642
12. Kekicheff P, Christenson HK, Ninham BW (1989) Colloid Surf 40:31–41
13. Helm CA, Israelachvili JN, McGuiggan PM (1989) Science 246:919–922
14. Richetti P, Kekicheff P (1992) Phys Rev Lett 68:1951–1954
15. Parker JL (1994) Prog Surf Sci 4:205–271
16. McDermott DC, McCarney J, Thomas RK, Rennie AR (1994) J Colloid Interface Sci 162:304–310
17. Levitz P, Van Damme H, Keravis D (1984) J Phys Chem 88:2228–2235
18. Rutland MW, Parker JL (1994) Langmuir 10:1110–1121
19. Partyka S, Lindheimer M, Faucompre B (1993) Colloids Surf A 76:267–281
20. Binnig G, Quate CF, Gerber Ch (1986) Phys Rev Lett 56:930–933
21. Manne S, Cleveland JP, Gaub HE, Stucky GD, Hansma PK (1994) Langmuir 10:4409–4413
22. Manne S, Gaub HE (1995) Science 270:1480–1482
23. Terris BD, Stern JE, Rugar D, Mamin HJ (1989) Phys Rev Lett 63:2669–2672
24. Grütter P, Rugar D, Mamin HJ (1992) Ultramicroscopy 47:393–399
25. Senden TJ, Drummond CJ, Kekicheff P (1994) Langmuir 10:358–362
26. Wanless EJ, Ducker WA (1996) J Phys Chem 100:3207–3214
27. Ducker WA, Grant LM (1996) J Phys Chem 100:11507–11511
28. Saleeb FZ, Kitchener JA (1965) J Chem Soc 911–917
29. Zettlemoyer AC (1968) J Colloid Interface Sci 28:343–369
30. Koganovskii AM (1962) Colloid J USSR 24:702–708
31. Groszek AJ (1970) Proc Royal Soc London A 314:473–498
32. Yeo YH, Yackoboski K, McGonigal GC, Thompson DJ (1992) J Vac Sci Technol A10:600–602
33. Rabe JP, Buchholz S (1991) Sciences 253:424–427
34. Ishihara S, Wakemoto H, Nakazima K, Matsuo Y (1989) Liq Cryst 4:669–675
35. Balmbra RR, Clunie JS, Goodman JF (1969) Nature 222:1159–1160
36. Söderlind E, Stilbs P (1993) Langmuir 9:2024–2034
37. Yeskie MA, Harwell JH (1988) J Phys Chem 92:2346–2352
38. Leimbach J, Sigg J, Rupprecht H (1995) Colloids Surf A 94:1–11
39. Gu T, Huang Z (1989) Colloids Surf 40:71–76
40. Nishimura S, Tateyama H, Tsunematsu K, Jinnai K (1992) J Colloid Interface Sci 152:359–367
41. For a discussion see Israelachvili JN (1992) Intermolecular and Surface Forces, 2nd ed. Academic Press, London, pp 227–229 and 304–310
42. Leung OM, Goh MC (1992) Science 255:64–66
43. Greenwood FG, Parfitt GD, Picton NJ, Wharton DG (1968) In: Weber WJ, Matijevic E (eds) Adsorption from Aqueous Solutions. American Chemical Society, Washington, DC, pp 135–144
44. Pashley RM, Ninham BW (1987) J Phys Chem 91:2902–2904
45. Tanford C (1980) The Hydrophobic Effect, 2nd ed. Wiley, New York, PP 51–54

Progr Colloid Polym Sci (1997) 103:234–242
© Steinkopff Verlag 1997

A.C. Balazs
C. Singh
E. Zhulina
D. Gersappe
G. Pickett

Forming patterned films with tethered polymers

Received: 1 November 1996
Accepted: 23 November 1996

Dr. A.C. Balazs (✉) · C. Singh ·
E. Zhulina · D. Gersappe · G. Pickett
Department of Chemical and Petroleum
Engineering
1231 Benedum Hall
University of Pittsburgh
Pittsburgh, Pennsylvania 15261, USA

Abstract We use numerical and analytical models to investigate polymer films formed by tethering chains to flat surfaces and immersing the system in a poor solvent. Since the ends of the chains are immobilized on the surface, the polymers avoid the unfavorable solvent by clustering together into aggregates, or pinned micelles, on the surface. These micelles have a uniform size and spacing and form a distinct pattern. We demonstrate that more complex surface patterns can be generated by tethering homopolymers to two surfaces (so that the chains bridge the interfaces) or anchoring copolymers onto one surface. Our numerical calculations reveal the morphology of the layer and the scaling arguments indicate how the dimensions of the patterns depend on the characteristics of the chains and solvent. Finally, we investigate how free copolymers in solution can be exploited to modify the patterns in the film. These results provide further guidelines for controlling the structure of the tethered layer. Overall, the patterned films are useful in the fabrication of adhesives, lubricants and ordered colloidal arrays. The surfaces can also be used to regulate the flow of molecules in channels and for selective filtration systems.

Key words Tethered polymers – pinned micelles – patterned films – self-assembly

Introduction

Polymer films that contain well-defined patterns are a key component in a variety of novel applications. For example, if the films are composed of both hydrophilic and hydrophobic domains, the surface can be used as a template for growing biological cells with the desired shapes and sizes [1]. By depositing inorganic materials on the surface, the polymer films can also be used to fabricate organic/inorganic composites with tailored morphologies [2]. One means of forming patterned films is to anchor the ends of homopolymers onto a substrate (so that the ends are fixed and cannot move) and immerse the system in a poor solvent. The incompatibility between the polymer and solvent drives the system to phase-separate. With the chain ends immobilized on the surface, the polymers undergo a microsegregation by clustering with neighboring chains into distinct aggregates, or "pinned micelles" [3–9]. These micelles have a uniform size and spacing [10–14].

In this paper, we use both numerical and analytical models to indicate means of creating even more complex surface patterns with tethered chains in poor solvents. In the first example, we tether solvent-incompatible (solvophobic) homopolymers to *two* surfaces, so that the polymers form "bridges" between the interfaces. As demonstrated below, this system displays a novel phase transition as the surfaces are pulled apart: at a critical distance, the pinned micelles "jump" from one surface to the middle of the gap

Progr Colloid Polym Sci (1997) 103: 234–242
© Steinkopff Verlag 1997

between the two [15]. Thus, we expect that the system can be used as a sensor for detecting small changes in the separation between the plates. In the second example, we demonstrate that by introducing greater chemical diversity into the tethered chains, namely by grafting *copolymers*, we also drive the system to form complicated patterned layers [16–18]. These patterns provide a handle for engineering the interaction between polymer-coated substrates [19, 20]. Both the examples illustrate the dramatic effects that anchoring chains onto a substrate or confining polymer layers between two surfaces has on the morphology of the system. In the final study, we investigate how free copolymers in solution modify the patterns in the film. These results provide further guidelines for controlling the structure of the tethered layer.

Theoretical models

To carry out our investigations on tethered polymers, we use a two dimensional self-consistent mean field (SCF) theory [6, 15, 17–23] and scaling arguments. Through the 2D SCF theory, we determine both the vertical and the lateral density profiles, which are crucial for characterizing laterally inhomogeneous structures such as pinned micelles. Our SCF method is derived from the lattice theory of Scheutjens and Fleer [24]. In this treatment, the phase behavior of polymer systems is modeled by combining Markov chain statistics with a mean field approximation for the free energy. Given the probability of finding a single monomer at a particular site and the fact that all the monomers in a chain are connected, the statistics for chains of arbitrary length can be obtained through a series of recursion relations. These recursion relations involve the potential of mean-force acting at a site $\mathbf{r}$. This potential, in turn, is determined from the local distribution of all the components at the point $\mathbf{r}$, as well as the Flory–Huggins interaction parameters, or χ's, between the different components. Solving this series of equations numerically and self-consistently yields the equilibrium profiles for the polymers and solvent in the system [24]. These density profiles provide a picture of the local concentration gradients in the system. In the two dimensional SCF theory, the equations are written explicitly in terms of both the vertical (Z) and lateral (Y) directions [6, 21]. We assume that all the quantities are translationally invariant in the X direction.

We note that our assumption of translational invariance along X implies that the micelles in our system display a cylindrical morphology along the surface. The actual structures are expected to be more spherical in shape. Therefore, our SCF calculations indicate general

trends and yield qualitative predictions. To obtain more quantitative information, we employ scaling theory. In this model, each pinned micelle is composed of a spherical core, which contains the majority of the monomers, and strongly stretched tethers, or "legs", which anchor the core to the surface [3, 7]. The core is viewed as a densely packed system of thermal blobs, each of size $\xi \approx a/\tau$. Here, a is the monomer size, and given that T is the temperature $\tau = ((\theta - T)/\theta) > 0$ is the relative deviation from the θ-point. In addition, the micellar legs are considered to be strings of thermal blobs. The lateral size of the micelle is given by $D = (fs)^{1/2}$, where s is the area per grafted chain, and f is the number of chains within this aggregate. Further details of the scaling analysis are given in the appropriate sections.

Using these models, we isolated factors that control pattern formation in: homopolymers tethered to two surfaces, copolymer-coated surfaces and copolymer adsorption onto pinned micelles. Below, we describe each of these systems separately.

Homopolymers tethered to two surfaces

We examined the patterns formed by bridging homopolymers, where the end of each chain is tethered to one surface and the other end is tethered to a second parallel surface [15]. In the SCF calculations, we fix the length of the polymers at $N = 80$. The parameter ρ gives the grafting density (the area per chain is $s = 1/\rho$) and is fixed at $\rho = 0.0125$. (Pinned micelles only occur at relatively low grafting densities; at high grafting, the layer remains laterally uniform [5].) To model the poor solvent, the polymer–solvent energy, χ_{PS}, is set at 2. The polymer-surface energy χ_{PW}, equals zero and thus, is favourable relative to the polymer–solvent interactions. (The solvent–surface interaction is given by $\chi_{SW} = 0$.) Figure 1 shows the density profiles for the system as H, the vertical separation between the plates, is increased. When H is small ($\leq$ the height of the unperturbed micellar core), the system forms continuous bridging "bundles" that extend between the two plates (Fig. 1a). As the distance between the surfaces is increased, the bundles break up into pinned micelles that adsorb onto either of the surfaces (Fig. 1b). While the tethering segments on the opposite surface are relatively stretched, this conformation is favored because the adsorbed micelles avoid contact with the solvent. As H is increased further, the cost of stretching the chains increases and finally exceeds the gain in energy associated with having the micelles on the wall. At this point, there is a discontinuous "jump" of the micelles to the center of the gap. Figure 1c shows the conformation of this "centered" or "desorbed" micelle.

236
A.C. Balazs et al.
Forming patterned films with tethered polymers

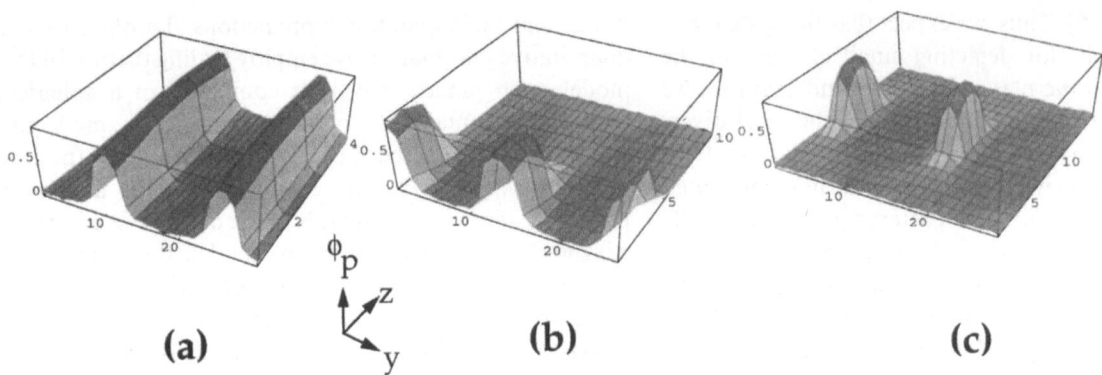

Fig. 1 SCF density profiles showing the effect of increasing the surface separation, H. The chain length is $N = 80$, $\rho = 0.0125$ and $\chi_{PS} = 2$. The chains are grafted in the XY plane and the two surfaces are separated along Z. The parameter ϕ_P is the polymer density. (a) The surfaces are highly compressed and thus, the micelles associate into "bundles" that extend from one surface to another. (b) Increasing H causes the micelles to absorb onto one of the surfaces. (c) Further increases in H drive the micelles to desorb and localize in the center of the gap

Further increases in H cause the micelles to grow in size and the distance between them to increase [25]. The aggregation of chains into larger micelles further reduces the unfavorable polymer–solvent contact and thus is energetically favored. When the separation is increased so that H is comparable to the chain length, the stretching energy dominates over the polymer–solvent interaction and the size of the micelles decreases. Finally, the micelles disappear and we obtain single, stretched chains with a uniform polymer density between the surfaces.

To analyze the transition further, we adopt a scaling model to determine the free energies of the pinned micelle for the "adsorbed" and "desorbed" (centered in the middle of the gap) states. In the adsorbed state, the core is deformed and partially spread on the grafting surface. This spreading is driven by the favorable interaction between the monomers and the surface. We let δ to be the adsorption energy per manomer (measured in thermal units). We consider the case where $|\delta| \ll \tau$, which allows us to assume that the average concentration of units within the core remains $\approx \tau$. The shape of the core is approximately a spherical cap with a contact angle θ, which is determined by Young's law:

$$\gamma_{sw} - \gamma_{cw} = \gamma \cos \theta . \qquad (1)$$

Here γ_{sw}, γ_{cw} and $\gamma \approx \xi^{-2} \approx \tau^2/a^2$ are the surface tensions at the solvent-wall, core-wall and core-solvent interfaces, respectively. The parameter $\Delta \equiv \cos \theta$ is related to the adsorption energy δ by [26]

$$\Delta = (\gamma_{sw} - \gamma_{cw})/\gamma \approx (\delta/\tau) , \qquad (2)$$

Where $\delta > 0$ (< 0) corresponds to an attractive (repulsive) interaction.

The equilibrium characteristics of the adsorbed state can be obtained by minimizing ΔF, the free energy per

chain in this micelle,

$$\Delta F = \Delta F_s + \Delta F_{el} \qquad (3)$$

with respect to f, the number of the chains in the deformed micelle. Here, ΔF_s is the surface free energy per chain of the partially spread core and ΔF_{el} is the free energy of stretching each leg (string of thermal blobs). Again, we measure ΔF in thermal units.

To determine the expression for ΔF_s, we note that the core of an adsorbed micelle consists of densely packed blobs; hence, the volume of the core is given by

$$V \approx (Nfa^3/\tau) . \qquad (4)$$

Given the height of the core, h, and Δ, simple geometric considerations show that the volume of the core is $V = (\pi h^3/3) ((2 + \Delta)/(1 - \Delta))$, and the area in contact with the wall is $A = \pi h^2((1 + \Delta)/(1 - \Delta))$. The surface area of the spherical cap is $S = 2\pi h^2/(1 - \Delta)$. Therefore, $\Delta F_s = (S\gamma - \gamma A\Delta)/f$. Using Eq. (4) and the latter definition of V, we can eliminate h from ΔF_s, and we find that for a partially spread globule, the equilibrium surface free energy per chain is

$$\Delta F_s \approx (\tau^{4/3} N^{2/3}/f^{1/3}) g(\Delta) . \qquad (5)$$

The term $g(\Delta) = (1 - \Delta)^{2/3}((2 + \Delta)/4))^{1/3} \leq 1$ accounts for the deformation of the core and is a convenient measure of the polymer–surface interaction.

The elastic stretching of the legs is

$$\Delta F_{el} \approx (\tau/2a) [(H^2 + fs)^{1/2} + (fs)^{1/2}] , \qquad (6)$$

where the two terms in Eq. (6) correspond to the legs attached to the upper and lower surfaces, respectively. Minimizing Eq. (3) with respect to f, with Eqs. (5) and (6) taken into account, one obtains the scaling dependence for

Progr Colloid Polym Sci (1997) 103:234–242
© Steinkopff Verlag 1997

the equilibrium number of chains, f_1, in the adsorbed pinned micelles,

$$f_1 \approx (\tau^{2/5} N^{4/5} / s^{3/5}) \, g(\Delta)^{6/5} . \tag{7}$$

It follows from Eqs. (7) that for higher values of δ, f_1 becomes smaller, i.e., the adsorbed, spread ($g < 1$) pinned micelles are smaller than the non-adsorbed ($g = 1$) micelles. While neglected from the above equation, minimizing Eq. (3) also shows that f_1 displays a weak dependence on H, increasing with an increase in H.

The second state of a pinned micelle is that in the middle of the gap between the two surfaces. Here, the core is undeformed, i.e., it is a sphere of radius R_0 and the legs emanating from both surfaces are stretched to the same extent [25]. For a centered micelle containing f chains, the volume of the core satisfies $4\pi R_0^3/3 \approx (Nfa^3/\tau)$. The surface and the elastic components of the free energy are given by

$$\Delta F_s = (4\pi\gamma R_0^2/f) \approx (\tau^{4/3} N^{2/3} / f^{1/3}) , \tag{8}$$

$$\Delta F_{el} \approx (\tau/a) \, (H^2/4 + fs)^{1/2} . \tag{9}$$

Minimizing the total free energy, $\Delta F_s + \Delta F_{el}$, one obtains the equilibrium number of chains, f_2, in a "desorbed" micelle [25]; asymptotically,

$$f_2 \approx \begin{cases} (\tau^{2/5} N^{4/5}/s^{3/5}), & H \ll \tau^{1/5} N^{2/5} s^{1/5} . \quad (10a) \\ (\tau^{1/4} N^{1/2} H^{3/4}/s^{3/4}), & H \gg \tau^{1/5} N^{2/5} s^{1/5} . \quad (10b) \end{cases}$$

Thus, in the second or "desorbed" state, the number of chains in the micelle increases noticeably with increasing H. These predictions agree with observations from the SCF calculations noted above.

By substituting the exact values of f_1 into Eqs. (5) and (6) and the complete expression for f_2 into Eqs. (8) and (9), we can readily compare the equilibrium free energies. This is shown graphically in Fig. 2, for various polymer–surface interaction energies, g. At a critical value, $H^*(g)$, the free energy curves cross, and any increase in H causes the cores of the micelles to jump abruptly into the middle of the gap. With increasingly attractive polymer-surface interactions (higher values of g), the micelles are more spread on the surface and higher values of H are necessary to "desorb" the micelles [15].

The discontinuous change in both the position and size of the cores leads to abrupt changes in the force vs. separation profile, which could be observed by the chemical forces apparatus [27]. The changes in polymer density with surface separation also lead to changes in the index of refraction or viscosity of the fluid in the gap. Thus, the system can be used to modulate the propogation of light in the medium or control the diffusion rate of particles through slits and channels. Though some hysteresis may be inevitable in actual systems, one can cycle back and

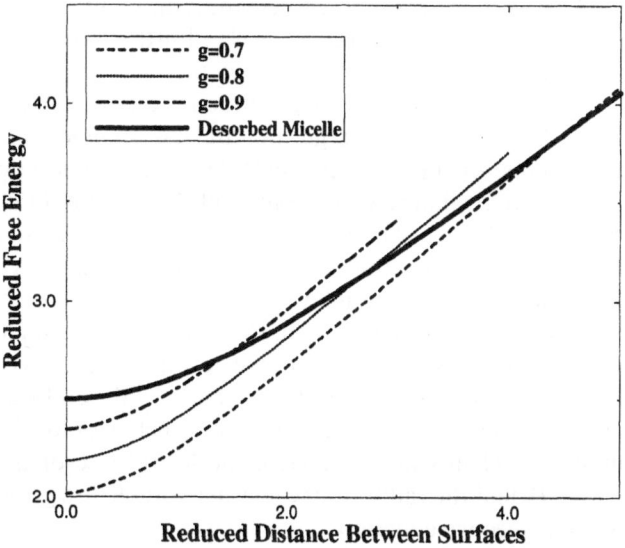

Fig. 2 The reduced free energy, $\Delta F/(\tau^{6/5} s^{1/5} N^{2/5})$, versus the reduced distance between the surfaces, $H/(\tau^{1/5} s^{1/5} N^{2/5})$, for the adsorbed (dashed lines) and desorbed (solid line) states of the pinned micelles. Here, ΔF is scaled by the free energy per chain for non-bridging micelles and H is scaled by the lateral size of the pinned micelle formed by non-bridging chains. For the adsorbed state, the curves are for different values of g, or polymer–surface interactions. A higher value of g corresponds to a greater polymer–surface attraction

forth between these states. Due to this distinctive change in morphology, the entire assembly could be used as a force sensor, altering the nature of a signal that passes through the layer when the confining surfaces change their separation.

The results also reveal that altering the separation between the substrates provides a mechanical means of controlling the microstructure of the film. As we describe below, by compressing two *copolymer-coated* surfaces, we can again tailor the morphology of the film between the substrates. To understand this behavior, we first describe the structure of a copolymer film on a single surface.

Copolymer-coated surfaces and their interactions

We examined the behavior of AB diblock copolymers that are sparsely grafted onto a flat surface [17, 18]. We assume that the blocks are relatively compatible so that the Flory–Huggins parameter between the A and B monomers, χ_{AB}, is small. We further assume that the B block is highly solvophobic; to model this interaction, the B-solvent interaction parameters, χ_{BS}, is set at 2. We varied the A-solvent interaction from poor to good by systematically decreasing χ_{AS} from 1 to 0. The interaction of the blocks with the surface is the same as that with the solvent. The grafting density is again fixed at $\rho = 0.0125$.

We first consider the case where the A and B blocks are comparable in length, or $N_A \approx N_B$. Furthermore, the chains are tethered by the ends of the B blocks, and $\chi_{AS} = 1$. Thus, the surrounding solution is a poor solvent for both blocks, but B segments are less soluble than A segments. In this case, the diblocks self-assemble into pinned micelles that resemble "onion-like" structures [17], as shown in Fig. 3a. The B blocks form the inner core and the A blocks form a thin outer layer. The encircling A's shield the B's from the unfavorable solvent and thereby lower the surface tensions within the system. Using simple scaling arguments, we can determine how the properties of these micelles depend on the characteristics of the diblocks and solvent [17]. In particular, the scaling behavior for the number of chains in a micelle, f, the lateral size of the micelle D, and the radius of the micellar core, R, are given by the respective equations:

$$f \sim \tau_B^{2/5} N_B^{4/5}/s^{3/5} , \tag{11}$$

$$D \sim (fs)^{1/2} \sim a(\tau_B s)^{1/5} N_B^{2/5} , \tag{12}$$

$$R \sim a(fN_B/\tau_B)^{1/3} \sim a(\tau_B s)^{-1/5} N_B^{3/5} . \tag{13}$$

Fig. 3 SCF density profiles. The plots reveal the effect of increasing N_A, while keeping N_B fixed at 80. The diblock copolymers are grafted by the less soluble B component. The parameter N_A has the following values: (a) $N_A = 120$, and (b) $N_A = 320$. Here, $\rho = 0.0125$, $\chi_{BS} = 2$, $\chi_{AS} = 1$, and $\chi_{AB} = 0$. The chains are grafted in the XY plane and ϕ_P denotes the polymer density. The plots marked "B" show the polymer density of the B blocks, while the plots marked "A" show the density of the A blocks. In (a), the A layer coats the more solvophobic B core and the micelle resembles an onion-like structure. In (b), the long A blocks shield multiple B cores and the structure resembles a garlic

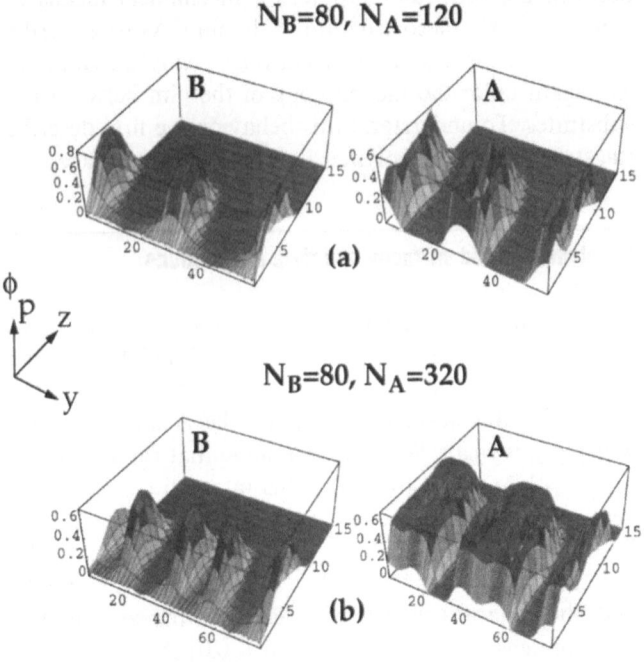

These relationships also characterize pinned micelles formed by pure B homopolymers. Thus, when the A segment is less than or comparable to the length of the B block, the micelles are predominantly governed by the properties of the B block: the solubility parameter τ_B, and the molecular weight N_B. The micelles are affected by N_A only through a weak, non-power law dependence.

When the length of the A block is increased, the B blocks are shielded more effectively, and the dominant losses arise from unfavourable contacts between the outer A layer and the solvent. Eventually, the losses at the boundaries of the A layer become sufficiently large that two adjacent A shells are driven to merge and thereby, minimize their contact with the solvent. Now multiple B cores are simultaneously shielded by a single, common A shell, and the micelle resembles a "garlic-like" structure (see Fig. 3b) [17]. The number of chains in the A shell and individual B cores, f_A and f_B, respectively, scale as

$$f_A \sim \tau_A^{2/5} N_A^{4/5}/s^{3/5} , \tag{14}$$

$$f_B \sim \tau_B^{2/5} N_B^{4/5}/s^{3/5} , \tag{15}$$

and thus, one finds $f_A/f_B > 1$ B cores per each A shell. Correspondingly, the lateral dimensions of the aggregates scale as $D_i = (f_i s)^{1/2}$, where i equals A or B. In addition, the size of cores, R_i, scales as $(f_i N_i/\tau_i)^{1/3}$.

If χ_{AS} is decreased to 0 (while the other conditions above are held fixed), the diblocks form a "flower-like" structure; the solvent-incompatible B blocks form a dense micellar core and the soluble A blocks form an extended corona of "petals" around the B micelles [18]. This A coating reduces the extent of B-solvent contact and stabilizes the system. Such flower-like micelles can be seen on each surface in Fig. 4a. The characteristics of the micelles are again given by eqs. (11)–(13). However, the thickness of the spherical outer shell, H, is given by [18]

$$H = a\tau_A^{1/5} N_A^{3/5} f^{1/5} \tag{16}$$

in good solvent conditions, and by

$$H = a N_A^{1/2} f^{1/4} \tag{17}$$

in theta solvent conditions.

If two such sparsely coated layers are brought into close contact, the pinned micelles from each surface interact and form novel structures. Little is known of the interactions in systems that involve both solvophobic *and* solvophilic (solvent-compatible) chains at low grafting densities [19, 20, 28]. Probing the structure and energies between these layers can yield insight into the nature of the adhesive forces between polymer-coated substrates and provide guidelines for tailoring the interactions between the interfaces.

Progr Colloid Polym Sci (1997) 103:234–242
© Steinkopff Verlag 1997

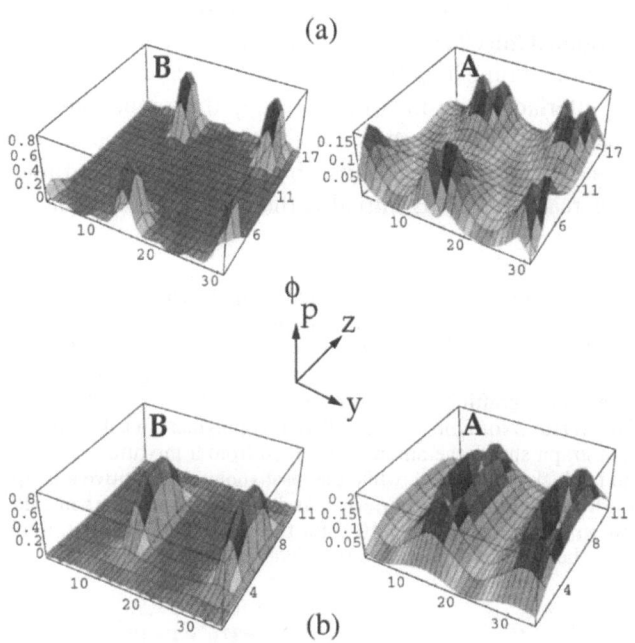

(a)

(b)

Fig. 4 SCF density profiles showing the effect of decreasing the surface separation. H, for symmetric AB diblocks grafted by the end of the less soluble B blocks. The respective block lengths are $N_A = N_B = 40$. The grafting density is $\rho = 0.0125$, $\chi_{BS} = 2$, $\chi_{AS} = 0$ and $\chi_{AB} = 0$. In (a), $H = 16$ and in (b), $H = 10$. The chains are grafted in the XY plane and the two surfaces are separated along Z. The parameter ϕ_P is the polymer density. The plots marked "B" show the polymer density of the B blocks, while the plots marked "A" show the density of the A blocks. In (a), the flower-like structures are clearly visible on each surface. Here, the soluble A's form an extended corona of "petals" around the B core

To investigate these interactions, we determined the effect of compressing two surfaces that are sparsely covered with end-grafted AB diblocks [19]. We consider symmetric diblocks where the length of the A and B blocks are equal. There are two cases: (1) the chains are tethered by the B end, and (2) they are grafted by the A end. In both cases, the interactions between the surfaces display an attractive region, even when the blocks are incompatible.

The basis for this behavior is illustrated in Fig. 4, where the chains are tethered by the end of the B block and solution is a good solvent for A. The chains form flower-like structures on each surface, with the A's forming an extended layer around the B core (Fig. 4a). With the outer layer extending into the solvent and the B blocks near the surface, one anticipates that the entropic losses associated with compressing the swollen A petals give rise to a purely repulsive interaction. However, as the surfaces are brought into close contact, the B cores from the opposite sides merge, thus providing mutual protection from the unfavorable solvent. Due to the solvophobic interactions, the micelles merge at a distance that is significantly greater than the vertical extent of an isolated micelle [19]. The

A blocks from both sides now form a layer that surrounds the central B core (Fig. 4b). In effect, the "flowers" from the two surfaces have merged to yield a larger flower that is located half way between the substrates.

To quantify these associations, in Fig. 5a we plot the energy of interaction, F_{int}, versus surface separation, H, for three values of χ_{AB}. When $F_{int} < 0$, the interaction is attractive, while for $F_{int} > 0$, it is repulsive. Intially, the free energy of interaction displays a small repulsive part, which arises from contact between the stretched, solvophilic A coronas. With further compression, however, the curves exhibit a distinct attractive interaction; this attraction arises from the merging of the micelles seen in Fig. 4b. At small surface separations, the entropic losses associated with further compression of the corona dominate the interactions and F_{int} again becomes repulsive. Note that all the curves display a sharp minimum, which indicates that the adsorbed layers provide an optimal separation between the substrates. If these diblocks are tethered to the surface of colloidal particles, the chains would drive the particles to sit at a well-defined distance from each other and thus, could cause the particles to form an ordered array.

When the diblocks are attached by the A end, the B cores lie further from the surface. Consequently, these

Fig. 5 (a) Free energy of interaction, F_{int}, as a function of surface separation. H, for symmetric AB diblocks. The respective block lengths are $N_A = N_B = 40$. The grafting density is $\rho = 0.0125$, $\chi_{BS} = 2$, $\chi_{AS} = 0$. The curves are for three different values of χ_{AB}:0, 0.5 and 1.0. In (a), the chains are grafted by the end of the less soluble B blocks. In (b), the diblocks are grafted by the end of the more soluble A blocks

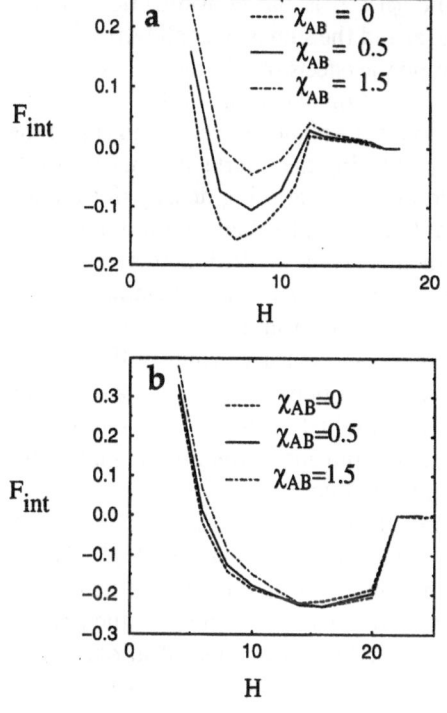

segments undergo less stretching and the merging of the micelles happens at greater surface separations than in the above example. As shown in Fig. 5b, the plot of F_{int} versus distance now shows a broad, flat attractive region [19]. Thus, the optimal distance between particles coated in this manner can fluctuate over a finite range and the colloids would not organize in a highly regular manner. These results indicate that the interactions between colloids can be tailored by tethering the appropriate polymers on the surface of the particles [19, 20].

Copolymer adsorption onto pinned micelles

Another means of systematically controlling the surface patterns is to introduce free chains in the solution that adsorb onto the substrate and thereby enhance or alter the size and shape of the structures [29]. To probe this behavior, we determined how different AB copolymers interact with a layer of pinned micelles, which are formed from B homopolymers grafted by one end onto a single surface. The various copolymers have the same length and chemical composition, but vary in sequence distribution, or the arrangement of the co-monomers along the chain. We set $\chi_{BS} = 2$ and $\chi_{AS} = \chi_{AB} = 0$; thus, the A's do not distinguish between the solvent and the B monomers. The respective monomer–surface interactions are given by $\chi_{AW} = \chi_{BW} = 0$. We fix ϕ^b, the concentration of the chains in solution, and determine the amount of copolymer adsorbed at the surface, which we define by the quantity $\theta^{ex} = (1/L) \sum_{y,z}(\phi(y,z) - \phi^b)$. Here, L is the length of the lattice along Y and $\phi(y,z)$ is the concentration of free polymer in layer (y, z) and the sum is over all lattice layers.

We first determine the effects of adding AB alternating copolymers into the solution above the grafted layer. The copolymers adsorb onto the pinned micelles and shield the B domains, thereby reducing the interfacial tension. Plotting θ^{ex} as a function of ρ, the grafting density (for a fixed value ϕ^b), we find that the adsorbed amount increases monotonically with increasing ρ. Within this range, increasing ρ increases the size of the pinned micelles and thus, the increase in θ^{ex} results from the larger surface area available for adsorption. The density profile of the system, however, shows that the copolymers do not perturb the morphology of the grafted layer. The alternating copolymer just coats the outside of the micelle and is not present within the micellar core.

We also determined the effects of placing AB diblock copolymers in the solution. At low values of ϕ^b, the diblocks behave in the same way as the alternating chains. That is, there is a monotonic increase in θ^{ex} as the gratifying density is increased. At higher concentrations of ϕ^b, however, the adsorption curve shows a maximum (see

Fig. 6a). At low grafting densities, the B blocks penetrate the pinned micelle and the A blocks form a corona around the core, diminishing the B-solvent contacts and reducing the interfacial tension. As a result, the pinned micelle expands in both the vertical and lateral directions. As ρ is increased, more diblocks adsorb and the micelle continues to increase in size. Eventually, the A coronas of adjacent

Fig. 6 (a) Plot of θ^{ex}, the amount of free copolymer adsorbed on the pinned micelles, as a function of ρ, the grafting density on the surface. The plots are for diblock copolymers at three values of ϕ^b, the bulk concentration. A maximum in the amount of diblock adsorbed as a function of grafting density is observed for $\phi^b = 1 \times 10^{-6}$. (b) Plot of θ^{ex} versus ρ for a mixture of different copolymers in solution. The main graph shows the amount adsorbed from a mixture of diblock and triblock copolymers, while the inset shows competitive adsorption from a mixture of alternating and diblock copolymers in solution. In both cases $\phi^b = 1 \times 10^{-6}$ for the different copolymers. In all cases, the adsorbed amount is the highest for the diblocks

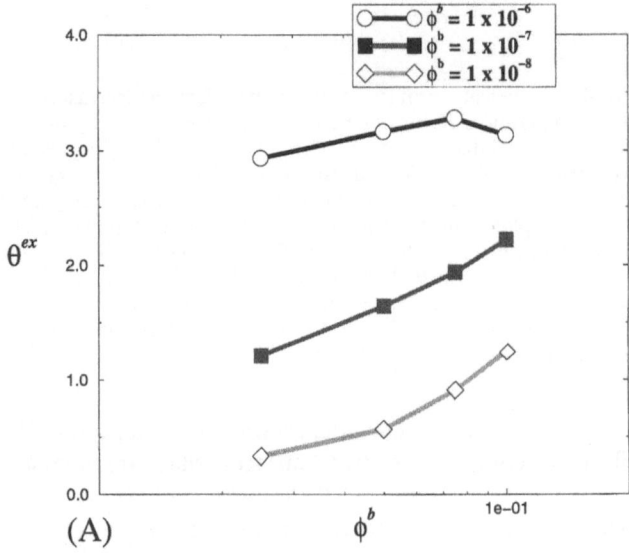

(A)

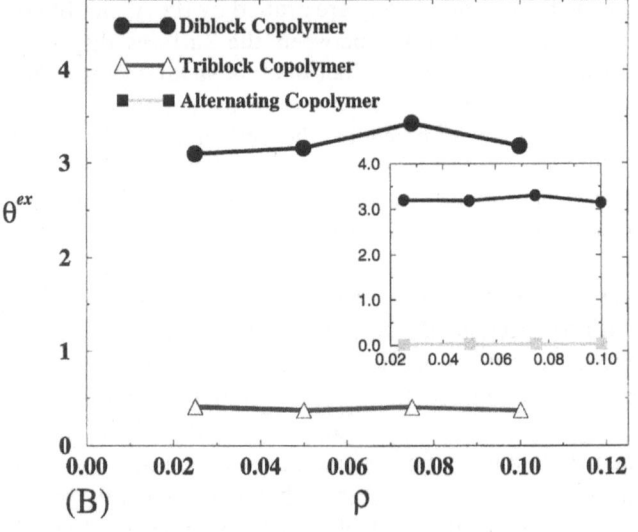

(B)

Progr Colloid Polym Sci (1997) 103:234–242
© Steinkopff Verlag 1997

micelles begin to overlap, resulting in a repulsion between the neighboring micelles. As the grafting density is increased further, this repulsion increases and in effect, decreases the total amount of adsorbed diblock, causing the maximum seen in Fig. 6a. Finally, the repulsion becomes large enough to destroy the structure of the micelles and the tethered chains form a laterally homogeneous layer.

The above calculations also revealed that the adsorbed amount is significantly higher for the diblocks than the alternating copolymer [30]. This finding gives rise to an intriguing question: If the solution contained a *mixture* of various copolymers (of fixed molecular weight but different sequence distributions) would the pinned micelles "distinguish" and preferentially adsorb the copolymer that is the most effective at reducing the interfacial tension (γ)? It is possible that the pinned micelles can "recognize" the most favorable component, namely the diblocks [31]. To test this hypothesis, we performed calculations in which the solution contained an equal mixture of alternating and diblock copolymers. We also performed a second set of calculations in which the mixture contained an equal mixture of triblock and diblock copolymers. Figure 6b shows the adsorbed amount versus ρ for these cases and reveals that the adsorbed amount is significantly higher for the diblock than the other architectures. If in fact the solution contained a greater variety of sequence distributions, the micelles should again preferentially adsorb the diblocks since they are the most effective at reducing γ, and therefore minimizing the free energy of the system.

It is of particular interest to determine whether this surface would display even greater selectivity. Namely, if the solution also contained a variety of diblocks, would the surface preferentially adsorb the diblocks that not only maximize the reduction in γ, but also reduce the steric repulsion between neighboring micelles. We are currently in the process of carrying out these calculations. The results will provide guidelines for fabricating systems for separation processes, and yield insight into the more general issue of how adsorbing chains interact with soft, responsive surfaces.

Conclusions

Tethered copolymers form a rich variety of patterns in poor solvents. The dimensions and morphology of the patterns can be tailored by varying the relative solubility of each block, and the composition of the copolymer. The patterns, in turn, effect the interactions between copolymer-coated surfaces and can be exploited in the fabrication of adhesives, lubricants and ordered colloidal arrays (which are useful as diffraction gratings or optical devices).

The morphology of the patterned film can also be significantly altered by compressing or stretching the tethered layer(s). This behavior can be utilized in regulating the flow of molecules in a poor solvent. Consider the two polymer-coated surfaces as the walls of a channel. At relatively large surface separations, the pinned micelles coating the surfaces will not significantly impede the diffusion of the particles. As the surfaces are compressed, however, the micelles or bundles that span the confining walls would greatly inhibit their motion. A subsequent increase in L would again allow unimpeded diffusion. Note that the changes in L need only be relatively small (from sufficiently greater than to less than the height of the micelles) to greatly alter, and thereby control, the flow behavior.

Copolymers added to the solution above the tethered layer can also be used to modify the surface pattern. In particular, the dimensions of the pattern are enhanced by the addition of alternating copolymers, which coat the layer, or a small volume fraction of diblocks, which penetrate and increase the size of the micelles. These and the above results highlight an advantage of using tethered polymers to create patterned surfaces. Namely, the dimensions of the pattern can be readily varied and thus, the films can be tailored for the desired specific applications.

Acknowledgments The authors gratefully acknowledge financial support from ONR through grant N00014-91-J-1363, from DOE through grant DE-FG02-90ER45438, from the NSF through grant DMR-9407100. C.S. gratefully acknowledges financial support from NSF grant DMR-92-17935 to David Jasnow.

References

1. Singhvi R (1994) Science 264:696
2. Aksay IA (1996) Science 273:892
3. Klushin LI (1992) unpublished work
4. Lai P, Binder K (1992) J Chem Phys 97:586
5. Yeung C, Balazs AC, Jasnow D (1993) Macromolecules 26:1914
6. Huang K, Balazs AC (1993) Macromolecules 26:4736
7. Williams DRM (1993) J Physique II 3:1313
8. Grest GS, Murat M (1993) Macromolecules 26:3108
9. Soga KG, Guo H, Zuckermann MJ (1995) Europhys Lett 29:531
10. The size of the micelles is determined by a balance between the entropic losses due to the stretching of the chains to form these aggregates and the energetic gain due to the mutual shielding of the chains from the poor solvent.
11. Siqueira DF, Kohler K, Stamm M (1995) Langmuir 11:3092
12. Zhao W, Krausch G, Rafailovich M, Sokolov J (1994) Macromolecules 27:2933

13. O'Shea SJ, Welland ME, Rayment T (1993) Langmuir 9:1826
14. Stamouli A, Pelletier E, Koutsos V, Vegte E. van der, Hadziiozannou G (1996) Langmuir 12:3221
15. Singh C, Zhulina E, Gersappe D, Pickett G, Balazs AC (1996) Macromolecules 29:7636
16. Zhulina E, Balazs AC (1996) Macromolecules 29:2667
17. Zhulina E, Singh C, Balazs AC (1996) Macromolecules 29:6338
18. Zhulina E, Singh C, Balazs AC (1996) Macromolecules, 29:8254
19. Singh C, Balazs AC (1996) Macromolecules, 29:8904
20. Singh C, Pickett G, Balazs AC (1996) Macromolecules 29:7559
21. Huang K, Balazs AC (1991) Phys Rev Lett 66:620
22. Israels R, Gersappe D, Fasolka M, Roberts VA, Balazs AC (1994) Macromolecules 27:6679
23. Singh C, Balazs AC (1996) J Chem Phys 105:706
24. Fleer G, Cohen-Stuart MA, Scheutjens JMHM, Cosgrove T, Vincent B (1993) Polymers at Interfaces. Chapman and Hall, London
25. Zhulina EB, Birshtein TM, Priamitsyn VA, Klushin LI (1995) Macromolecules 28:8612
26. Johner A, Joanny JF (1991) J Phys. II 1:181
27. Hadziioannou G (1996) private communication.
28. Chen C, Dan N, Dhoot S, Tirrell M, Mays J, Watanabe W (1995) Israel J Chem 35:41
29. Gersappe D, Balazs AC (1997, submitted, J. Chem. Phys.) Interactions Between Free Chains and Pinned Micelles.
30. In the case of a planar, penetrable interface, the amount of adsorbed diblock is also significantly greater than that for the alternating copolymer (see ref. [31])
31. Gersappe D, Balazs, AC (1995) Phys Rev E 52:5061; Lyatskaya Y, Gersappe D, Gross N, Balazs AC (1996) J Phys Chem 100:149

Progr Colloid Polym Sci (1997) 103:243–250
© Steinkopff Verlag 1997

D.W. Grainger

Synthetic polymer ultrathin films for modifying surface properties

Received: 10 January 1997
Accepted: 17 January 1997

Dr. D.W. Grainger (✉)
Department of Chemistry
Colorado State University
Fort Collins, Colorado 80523-1872, USA

Abstract A review of recent strategies to fabricate polymer thin film architectures on supports is presented. Polymer layers fabricated by (1) chemisorption of functionalized polymers to solid surfaces, and (2) films of stratified polyelectrolyte layers formed by electrostatic attraction are described. More traditional Langmuir–Blodgett methods for polymer thin film deposition are not discussed.

Key words Self assembly – surface modification – ultrathin films – polymers

Introduction

Interest in polymer thin film fabrication methods extends to several decades. This prolonged and extensive interest stems from a number of technologies that would benefit greatly from well-controlled, durable organic material architectures as overlayers or as stratified films. Applications in optics, sensors, tribology and biomedical devices are often cited [1–4]. Newer needs in biotechnology include combinatorial chemistry involving surface-immobilized species – polysaccharides, peptides, genetic material or libraries of immobilized molecules in spatially organized domains. Nearly all of these applications require well-defined and robust surface chemistry and/or functional groups. Many others demand spatial control, either laterally or vertically, of molecular dimensionality (nm scale) within the film architecture. Lastly, the processing advantages and other properties of polymers make the use of these materials in thin films and coatings attractive. In the absence of facile "molecular tweezers" to "place" molecules with the desired properties in precise locations, proximities and geometries, fabrication methods to build films with complex or anisotropic architectures must rely on self-organization and molecular recognition events that result in the so-called "supramolecular assembly" – material structures where the function of the entire assembled aggregate transcends the organization of individual molecules. Aspects of supramolecular chemistry that provide defined organic thin films have focused traditionally on Langmuir–Blodgett methods and their products. These approaches have been the subject of several comprehensive reviews [4–8] and are not covered herein. Another method to produce stratified organic thin films include casting of preformed colloidal amphiphile aggregates (micelles, vesicles) from liquid dispersions [9, 10] to yield layered films in desiccated states. Here, lateral control of molecule positional order is difficult as is interlayer alignment. Instead, polymer thin film fabrication methodologies that utilize solution absorption strategies will be reviewed. Layered copolymer structures have been recently produced by solvent casting [11] then subsequent thermal annealing [12] or electric field poling [13]. Again, molecular positional correlation between strata on molecular scales is difficult, although field effects [13] may be able to control segregation to dimensions approximating segment lengths in block copolymers.

This review attempts to describe other, perhaps newer routes reported to control structure in thin polymer films as coatings over solid supports. Again, combinations of chemistry and physical chemistry have been exploited in various attempts to manipulate film structure and interfacial properties. In general, the approaches described below are capable of producing stratified organic thin films where chemistry can be controlled to some extent along a vertical direction to the surface plane. The most significant challenge seems to be developing facile and convenient methods to tailor surface chemistry in the plane of the film surface – to control the *lateral* microstructure within the polymer film at the outermost surface [13].

Chemisorbed polymer ultrathin films

Diacetylenic alkyl thiols and disulfides

A virtual revolution in the development of organic coatings has been the formation of the so-called "self-assembled monolayers" (SAMs) of alkyl-thiols or disulfides on various metal surfaces (namely: gold, copper, silver and platinum) [14, 15]. Despite their noted advantages of high organization, convenient, innocuous fabrication conditions and monolayer thicknesses (approx. 20 Å), these films have some significant disadvantages for various applications, including their tendency to desorb, oxidize and remodel. An older, equally well-studied yet more chemically complex chemisorbed adlayer-former, the alkylsilane, exhibits the potential for true covalent coupling to its underlying substrate (solid-oxide surfaces) providing a more robust organic monolayer film than its alkylthiol cousin [16, 17]. However, newer evidence suggests that many long-standing assumptions surrounding the formation of true chemisorbed alkylsilane *monolayers* are poorly understood and that precipitated alkylsilane networks *physisorbed* to solid surfaces are often mistaken for the otherwise intended well-organized chemisorbed monolayer [18, 19]. In fact, it appears evident that true alkylsilane monolayer formation is the exception rather than the rule. Therefore, while the popular alkyltrichlorosilanes, in particular, are capable of both substrate reaction as well as two-dimensional covalent network formation, yielding an attractive bound polysiloxane overlayer, consistent yields of this desired ultrathin film are very difficult and routine in few labs.

A recent alternative approach is the use of diacetylene-containing alkylthiol monomer units that, when attached to metals via the thiolate bond, form planar arrays that are photopolymerized [20–22]. Such a strategy, shown in Fig. 1, combines the advantages of SAM formation with the improvements of a more robust polymer end-product. Previous work with L–B films of diacetylenic amphiphiles has demonstrated in-plane requirements for topo-chemically controlled polymerization [23]. "Templating" of the diacetylene topochemistry via the thiol "docking" to metal surface lattice sites spaced according to the metal

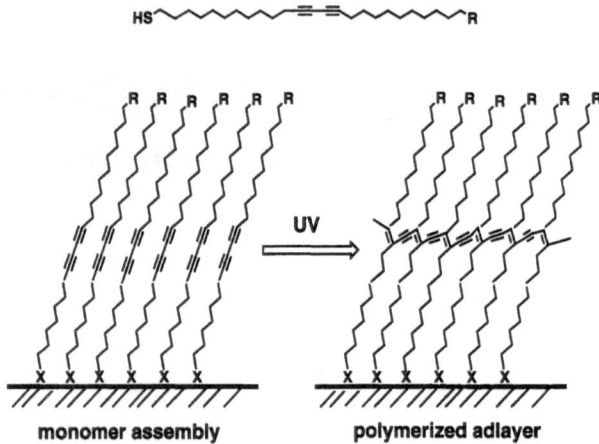

Fig. 1 Self-assembled monolayers of polymerizable diacetylene monomers on gold (see refs. [20–22])

surface lattice dimensions is fortuitous as the gold three-fold hollow site density helps produce a diacetylene adlayer with the optimal intermolecular distances and tilt angles for photopolymerization [20]. These materials can be patterned on surfaces and used as photoresists [22] and polymerized into chemically resistant ultrathin films [20, 21]. A significant additional feature is the capability to build multilayer structures in well-organized step-by-step strata [24] by fabrication layers with functional exposed terminal groups in the first adlayer that react with successive adlayers [25]. Defects known to occur in films of L–B diacetylenic analogues, likely along polycrystalline grain boundaries, have not been well characterized to date.

A similar although not topochemically governed approach to polymerize thiol-bearing monomers has been recently reported using mercaptomethyl styrene monomer [26]. Assembly of the monomer into dense-packed monolayers on gold is followed by free radical- or photo-induced polymerization in situ. Presence of little residual monomer or oligomer by mass spectrometry supports other spectroscopic evidence for monolayer polymerization.

Polymeric thiols and disulfides as chemisorbed ultrathin films

Polymeric L–B films fabricated by spreading various polymers at the air/water interface are well-known [6–8], indicating the capability to take a three-dimensional polymer coil in dilute solution and form a quasi-two-dimensional array at the air–water interface. By analogy, the adsorption of polymers bearing chemically reactive side chains (e.g., thiols, disulfides or silanes) specific for certain interfacial sites, could take advantage of the same organizational principles to yield multipoint attached polymer

Progr Colloid Polym Sci (1997) 103:243–250
© Steinkopff Verlag 1997

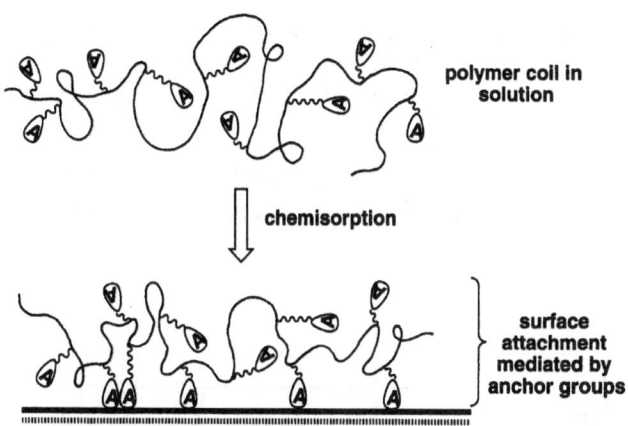

polymer coil in
solution

chemisorption

surface
attachment
mediated by
anchor groups

Fig. 2 Conceptual perspective for fabricating chemisorbed ultrathin organic films from preformed polymers bearing reactive anchoring groups (see refs. [27–30, 33–41])

films on solid supports, shown in Fig. 2. This approach has been experimentally described using a number of different polymer systems to date. The primary objectives have been to (1) fabricate well-defined polymer ultrathin films by chemical attachment to surfaces, and (2) provide a more robust alternative to monomeric alkylthiols by multi-point anchoring. Both objectives aim to improve upon problems mentioned above for both alkysilane and alkylthiol SAMs.

A survey of chemistries used for spontaneously anchoring polymer thin films is shown in Fig. 3. Early work by Stouffer and McCarthy [27] in this area showed that styrene polymers bearing only one terminal thiol group or styrene–propylene sulfide block copolymers could both form stable films via attachment to gold surfaces. Since those initial studies, several different polymer chemistries and architectures have been characterized as chemisorbed monolayers on gold, extending this concept.

For example, Schlenoff's group has more recently reported gold surface adsorption behavior for polystyrene and poly(styrene sulfonate) copolymers bearing (mercaptomethyl) styrene units [28]. Polymers of hydrophobic as well as charged hydrophilic (polyelectrolytes) were studied to examine the ability of chemisorptive interactions (thiol-gold) to overcome barriers blocking adsorption, in this case entropic losses of polymer surface immobilization or repulsive polymer charge/surface charge interactions. Introduction of the thiol groups along the polymer chain was shown to lead to irreversible polymer adsorption: higher percentages of polymer-bound thiol anchors exhibit reduced steady-state polymer surface adsorption values. Negatively charged poly(styrene sulfonate) bearing mercaptomethyl styrene units could also overcome repulsive anionic surface charge, adsorbing to charged gold surfaces

via thiolate bond formation. Polymer adsorption did not occur if thiol groups were not present.

Work with various acrylate copolymers containing sulfur-bearing side chains has shown that these also readily form monolayer films irreversibly bound to gold surfaces. Lenk et al. have shown that a copolymer of methyl methacrylate with 2-(methylthio)ethyl methacrylate binds readily to gold surfaces via polymer thiolate chemisorption [28]. Film thickness is a function of both solution concentration and density of sulfur anchoring groups. Additionally, "chaser" studies using a second small molecule fluorinated thiol after polymer assembly showed that not only could monomer thiols penetrate the pre-adsorbed polyacrylate monolayer but that this exposure could also displace some fraction of polymer from the surface. This displacement depended again on sulfur concentration along the polymer backbone. All polymer films appear to be porous and relatively amorphous – not consolidated, crystalline or defect-free.

Methacrylic acid polymers having two long alkyl chains terminated in ethyl disulfide groups at each monomer unit not only irreversibly adsorb to gold surfaces but are also capable of changing film thickness/swelling in response to local pH both reversibly and without detachment [30]. The authors used electrochemistry and ellipsometry to assert that polymer chain length and conformational state (mediated by protonated versus deprotonated methacrylic acid groups) either in solution prior to adsorption or after surface anchoring govern electron transfer behavior and film thicknesses.

Long, dialkyl chain monomers have also been reported in another polymer monolayer system on gold [31], but the difference lies in the terpolymer system that separates the methacrylate ethyl methyl asymmetric disulfide anchor monomer from the hydrophobic dialkyl methacrylate monomer (nonanchoring) using a third hydrophilic hydroxyethyl acrylate spacer monomer along the backbone. These architectural differences were deliberately introduced in an attempt to "decouple" the hydrophobic side chains from the anchoring side chains. Such a design has been effective in assisting organization in polymerized amphiphiles in lyotropic systems [8, 32]. In fact, the polymer design in this case is proposed as a surface "tethered", swellable and hydrophobically grafted polymer film that preserves an aqueous environment near an interface but might also be used to passively-tether-amphiphilic vesicle bilayers via association with the film's exposed, hydrophobic alkylated monomer units [31, 33]. Surface plasmon spectroscopy and cyclic voltammetry have been used to show that water permeates the bound polymer monolayer freely but that hydrophobic monomers associate to present a barrier to redox chemistry across the film.

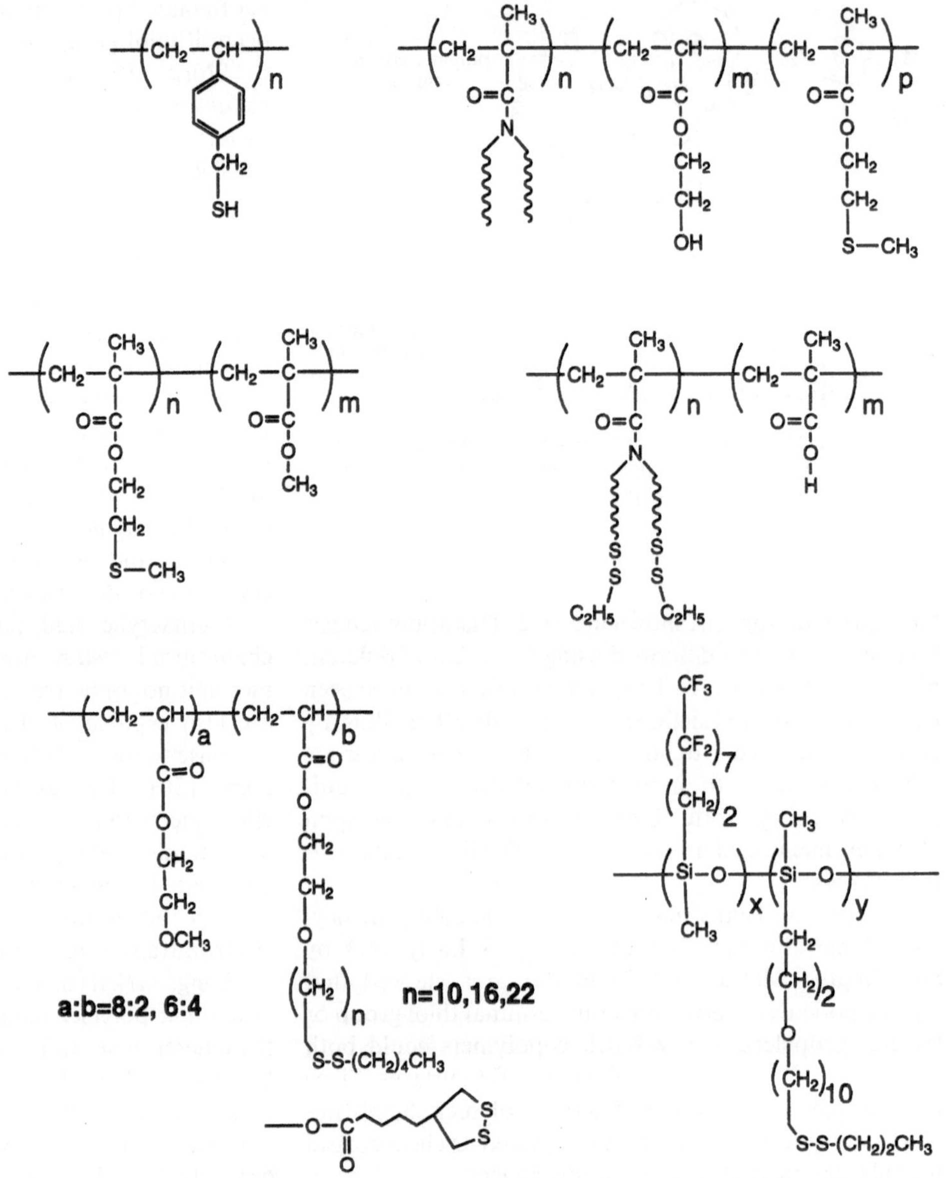

Fig. 3 Survey of chemistries
used for spontaneously
anchoring polymer thin films

One other acrylate polymer system bearing sulfur anchors has been extensively published. Again, the chemistry for this polymer monolayer system [34] is based upon analogous polymer chemistry known to produce ordered polymeric L–B films [8, 32]. Poly(hydroxyethyl acrylate)-co-methoxyethyl acrylate was esterified with alkyl disulfides of varying chain lengths [34]. These polymers spontaneously chemisorb to gold to yield bound monolayers (~20 Å thick depending on side chain length (density). A combination of surface analytical tools was used to extensively characterize these polymer thin films [35–37]. One important feature discovered was that increasing alkyl side chain length does *not* improve monolayer organization or lateral packing as might be expected from stud-ies of monomeric SAM components on gold. Polarized FTIR shows little evidence for side chain structural anisotropy in these samples as a function of chain length [36]. A second point is that stability is enhanced in these films: conditions that readily remove monomeric SAMs (e.g., octadecane thiol reflux in ethanol or chloroform) do not remove anchored sulfur-containing polymers from the gold surface [35]. Third, cyclic voltammetry indicates considerable disorder and defects in these films as monolayers, especially if polymers are anchored by long (>C10) alkyl side chains [36, 37]. Fourth, these materials were the first to exhibit two distinct sulfur binding environments using high-resolution XPS work [35]. The more sulfur-bearing side chains are present along the polymer backbone, the

Progr Colloid Polym Sci (1997) 103:243–250
© Steinkopff Verlag 1997

fewer (percentage-wise or proportionally) will actually anchor to the substrate via thiolate bonds. The sulfur binding assignments have since been a subject of some controversy [38] but are important to understanding the film anchoring, stability and structure. Several other polymer monolayer examples [28, 29] show analogous behavior manifested in inverse film thicknesses versus anchor group contents. Lastly, these polymers show surface microstructural adsorption specificity: they selectively chemisorb to submicron gold islands from solution, mediated by thiolate bond formation [36].

Polysiloxanes have also been modified with sulfur side chains to mediate their spontaneous adsorption to gold surfaces [39–41]. Like all of the polyacrylates described above, incorporation of thiol or disulfide along the polysiloxane backbone facilitates rapid and irreversible chemisorption to gold surfaces to yield ultrathin, stable films 20–34 Å thick. Detailed XPS work has shown again that two distinct sulfur populations are present in films of polymers bearing a high density of sulfur functionality. These have been assigned to thiolate (bound to gold) and free sulfur (unbound) [38, 39]. In one study the free, unbound thiol in the film was exploited to further derivatize the bound polysiloxane monolayer films in situ with further chemistries using photochemistry [39]. Two other recent studies have focused on the use of perfluorinated side chains on polysiloxanes to control film structure and the outermost, exposed surface chemistry [40, 41]. The highly flexible, mobile nature of the polysiloxane backbone permits polymer reorganization during the surface adsorption process. Since perfluorinated substituents have lower interfacial energies than and limited miscibilities with polydimethylsiloxane, surface rearrangements favoring the final placement of perfluorinated components at the film/ambient air interface are expected. Experimentally it was shown that perfluoroalkyl side chains attached to polysiloxane terpolymers also bearing alkyldisulfide anchoring chains do enrich the outer 15 Å of the 32 Å-thick bound monolayer film. XPS depth profiling combined with ToF-SIMS work has detailed the fluorine compositional gradient as a function of film depth for a series of polymers having different perfluoroalkyl chain contents [40]. Polarized FTIR analysis strongly suggests that the perfluoroalkyl chains in films of densely grafted systems are oriented roughly parallel to the surface normal [40]. This strongly suggests that within the 34 Å film thickness that perfluoroalkyl side chains occupy the outer 15 Å, the polysiloxane backbone lies below this, and the alkyl anchoring chains are located adjacent to the gold surface as shown in Fig. 4. Despite side chain structuring, electrochemical studies indicate that these films have defects that compromise their barrier properties at the molecular size limit.

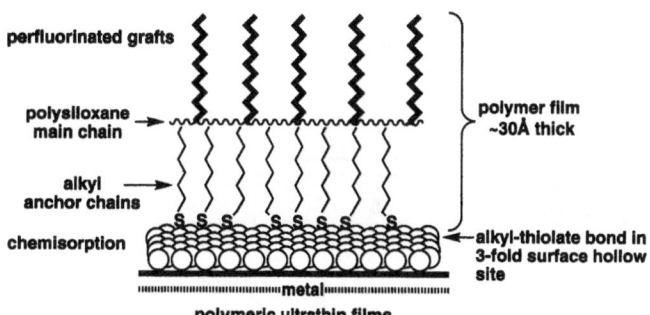

Fig. 4 Schematic of surface enrichment of grafted perfluoroalkyl side chains in anchored polydimethyl siloxane ultrathin films on gold (see ref. [40])

By contrast, analogous ultrathin polysiloxane films grafted with perfluropolyethers instead of the perfluoroalkyl side chains exhibit completely different structural behavior [41]. XPS surface enirchment of perfluorinated substituents occurs but only weakly for the perfluroether case. Reduced film stratification is observed. Additionally, no FTIR evidence for side chain organization is observed. This concurs with NEXAFS work that shows no chain structural anisotropy in these bound polymer films [41].

Polyelectrolyte multilayer films by electrostatically mediated adsorption

A very active area of research directed at fabrication of multilayer assemblies is based on consecutively alternating adsorption of anionic and cationic bipolar amphiphiles and/or polyelectrolytes. This method has been rapidly developed by Decher's group (see [42, 43] and references therein) and has been expanded by others to many types of polymers, inorganic and layered colloidal materials [44–50]. The strategy, shown in Figs. 5 and 6, uses counterion electrostatic attraction, often in combination with simple step-wise aqueous immersion, to build layered materials. Chemical reactivity is replaced by more convenient and reliable electrostatic forces. As depicted in Figs. 5 and 6, a solid substrate (of most any geometry) is chemically functionalized with a cationic surface charge (most commonly an aminosilanizing agent or plasma). From this step forward, all multilayer fabrication occurs by adsorption of polyelectrolytes from aqueous solutions. An anionic polyelectrolyte adsorbed to the cationic substrate in step A at high enough concentration to leave excess anionic surface charge. After an aqueous rinsing, a cationic polyelectrolyte is subsequently adsorbed at high enough concentration to reverse the surface charge again in step B.

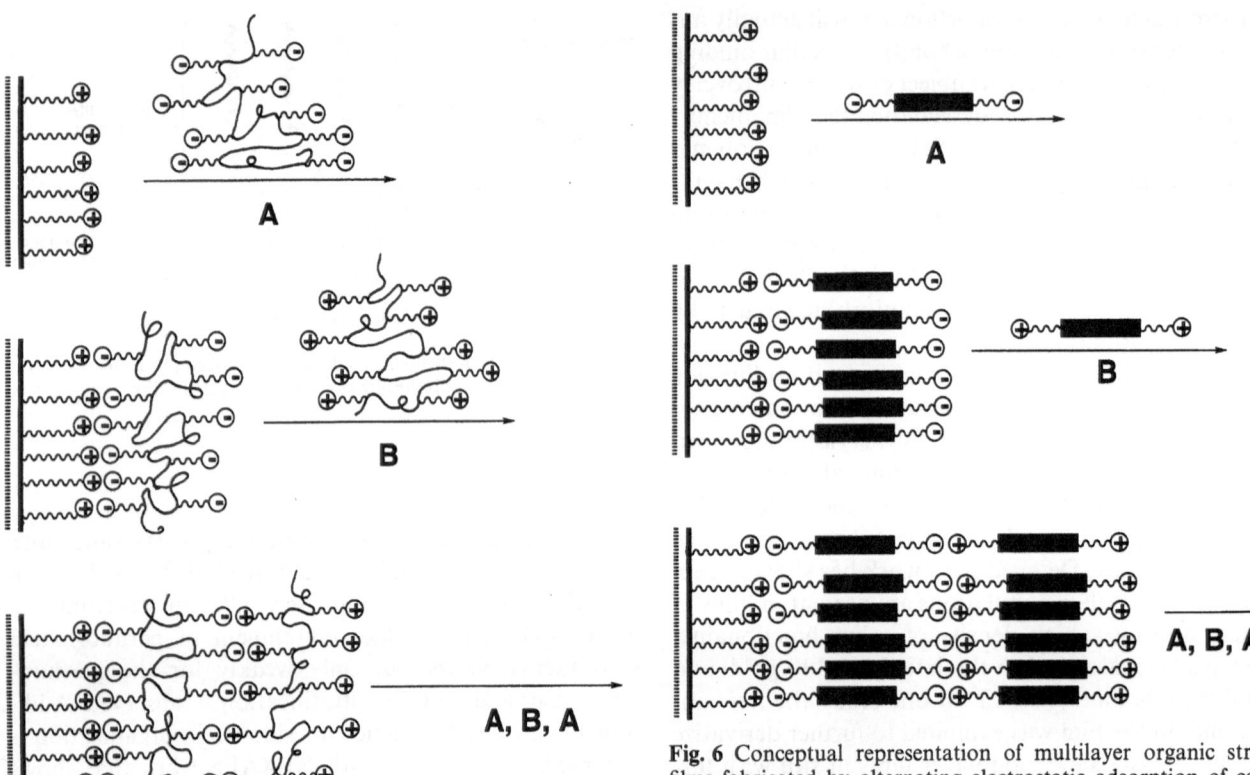

Fig. 5 Conceptual representation of multilayer polymer formation using electrostatic adsorption of polyelectrolytes (see refs. [42, 43])

Fig. 6 Conceptual representation of multilayer organic stratified films fabricated by alternating electrostatic adsorption of cationic and anionic bipolar amphiphiles (see ref. [53])

This process is repeated as many times and with as many different polyelectrolytes as desired, resulting in multilayer materials of many combinations. Film thickness scales with the number of layers, and in many cases x-ray reflectivity studies show evidence of substantial layering.

Decher's group has reported multilayer films of alternating poly(styrene sulfonate) and poly(allylamine) [51], poly(vinyl sulfate) and poly(allylamine), and these same polyanions with either poly-L-lysine or poly-4-vinylbenzyl-(N, N-diethyl-N-methyl-)-ammonium iodide [53]. Additionally, they have reported multilayered materials comprising polymer layers including DNA [54] and proteins [55, 56].

Related work has been done by Rubner's group for materials potentially useful to thin-film electronic and optical devices. Conjugated polymers in neutral and conducting chemistries [57–60], molecular dyes [61] and fullerenes [62] have all been stacked in multilayer arrays. Functional light-emitting diodes have been fabricated from multilayers of poly(p-phenylenevinylene) (PPV) and two different polyanions on active electrodes [63]. These devices have absorption and emission bands and photo-

luminescence of PPV that are dependent upon the polyanion used and can be deliberately shifted. The authors indicate that multilayer heterostructure devices with dramatically enhanced efficiencies are possible [63]. Other work in this area is focused on the same goal using PPV and similar fabrication methods [50, 64–67].

Incorporation of metal particles, inorganic oxides and other types of colloidal materials into these layered assemblies has also been pursued [44, 45, 48, 68–73]. Colloidal layers have provided photochemical, structural, catalytic, conductive or semi-conducting functions in these composite type structures.

A modification of the sequential polyelectrolyte multilayer fabrication strategy uses small molecule bipolar amphiphiles in place of polymers [53, 74–76]. Many tens of layers of small molecules can be stacked by alternate adsorption, approaching total layer thicknesses between 150 and 200 nm. Organization in these multilayers is reduced over that observed in polyelectrolyte multilayers. This order as well as stability can be improved by hybrid structures comprising alternating bipolar amphiphile and polyelectrolyte charged layers [53]. Bipolar diacetylenes can also be utilized [77, 78], providing interlayer stabilization via photopolymerization after multilayer assembly.

Conclusions

A menu of new fabrication tools and strategies is contributing to growth and interest in alternative methods to fabricate organized polymer thin films and coatings other than with the traditional L–B approach. Both direct chemical immobilization of polymers and sequentially directed adsorption methods have been shown to produce stable new and potentially useful ultrathin films and layered structures on solid supports. Applications for such materials are often touted, yet only a few technologically promising, prototypical devices have been reported. Challenges include (1) extension of this work to demonstrate the utility of these creative methods and polymer architectures to stimulate new technological developments, and (2) development of further strategies to control molecular positional order and density of functional molecular guests, lateral domains and clusters within film layers to create local nanoenvironments and to attempt lateral spatial organization of functional groups.

References

1. (1991) Adv Mater, Special Issue: Organic Thin Films 3
2. Swalen JD, Allara DL, Andrade JD, Chendross EA, Garoff S, Israelachvilli J, McCarthy TJ, Murray R, Pease RF, Rabolt JF, Wynne KJ, Yu H (1987) Langmuir 3:932
3. Frank CW, Rao V, Despotapoulou MM, Pease RFW, Hinsberg WD, Miller RD, Rabolt JF (1996) Science 273:912
4. Albrecht O, Sakai K, Takimoto K, Matsuda H, Eguchi K, Nakagiri T (1994) In: Birge R (ed) Molecular and Biomolecular Electronics. Adv Chem Ser 240 ACS Press, Washington, DC, pp 341–371
5. Peterson IR (1990) J Phys D: Appl Phys 23:379
6. Roberts GG (1990) Langmuir–Blodgett Films Plenum Press, NY
7. Miyashita T (1993) Prog Polym Sci 18:263
8. Ringsdorf H, Schlarb B, Venzmer J (1988) Angew Chem Intl Ed Engl 22:113
9. Kunitake T, Shimomura M, Kajiyama T, Harada A, Okayama K, Takayanagi M (1984) Thin Solid Films 121:L89
10. Kuo T, O'Brien DF (1991) Langmuir 7:584
11. Grainger DW, Okano T, Kim SW, Castner DG, Ratner BD Briggs D, Sung YK (1990) J Biomed Mater Res 24:547
12. Coulon G, Russell TP, Deline VR, Green PF (1989) Macromolecules 22:2581
13. Morkved TL, Lu M, Urbas AM, Ehrichs EE, Jaeger HM, Mansky P, Russell TP (1996) Science 273:931
14. Dubois LH, Nuzzo RG, (1992) Ann Rev Phys Chem 43:437
15. Ulman A (1991) Introduction to Ultrathin Organic Films. Academic Press, NY
16. Wasserman SR, Tao YT, Whitesides GM (1989) Langmuir 5:1074
17. Pomerantz M, Segmüller A, Netzer L, Segiv J (1985) Thin Solid Films 132:153; Sagiv J (1980) J Am Chem Soc 102:92
18. Sagiv J, Degenhardt D, Möhwald H, Quint P (1995) Supramolec Sci 2:9
19. Tripp CP, Hair ML (1995) Langmuir 11:149; Tripp CP, Hair ML (1992) Langmuir 8:1961
20. Batchelder DN, Evans SD, Freeman TL, Häußling L, Ringsdorf H, Wolf H (1994) J Am Chem Soc 116:1050
21. Kim T, Crooks RM (1994) Tet Lett 35:9501
22. Chen KC, Kim T, Schoer JK, Crooks RM (1995) J Am Chem Soc 117:5877
23. Tieke B, Wegner G, Naegle D, Ringsdorf H (1976) Angew Chem Int Ed Engl 15:764
24. Kim T, Crooks RM, Tsen M, Sun L (1995) J Am Chem Soc 117:3963
25. Kim T, Ye Q, Sun L, Kwok CC, Crooks RM (1996) Langmuir 12:6065
26. Ford JF, Vickers TJ, Schlenoff JB (1996) Langmuir 12:1944
27. Stouffer J, McCarthy TJ (1988) Macromolecules 21:1204
28. Schlenoff JB, Dharia JR, Xu H, Wen LQ, Li M (1995) Macromolecules 28:4290
29. Lenk TJ, Hallmark VM, Rabolt JF, Häußling L, Ringsdorf H (1993) Macromolecules 26:1230
30. Niwa M, Mori T, Higashi N (1993) Macromolecules 26:1936
31. Erdelen C, Häußling L, Naumann R, Ringsdorf H, Wolf H, Yang J, Liley M, Spinke J, Knoll W (1994) Langmuir 10:1246
32. Laschewsky A, Ringsdorf H, Schmidt G, Schneider J (1987) J Am Chem Soc 109:788
33. Häußling L, Knoll W, Ringsdorf H, Schmitt F-J, Yang J (1991) Makromol Chem Makromol Symp 46:145
34. Sun F, Grainger DW (1993) J Polym Sci A, Polym Chem 31:1729
35. Sun F, Grainger DW, Castner DG (1993) Langmuir 9:3200
36. Sun F, Grainger DW, Castner DG, Leach-Scampavia D (1994) J Vac Sci Technol 12:2499
37. Sun F, Lei Y, Grainger DW (1994) Colloids Surf 93:191
38. Castner DG, Hinds K, Grainger DW (1996) Langmuir 12:5083
39. Sun F, Grainger DW, Castner DG, Leach-Scampavia D (1994) Macromolecules 27:3053
40. Sun F, Castner DG, Mao G, Wang W, McKeown P, Grainger DW (1996) J Am Chem Soc 118:1856
41. Wang W, Castner DG, Grainger DW (in press) Supramolec Sci
42. Decher G (1996) In: Sauvage J-P, Hosseini MW (eds) Comprehensive Supramolecular Chemistry. Pergamon Press, Oxford, Vol 9, pp 507–528
43. Decher G (1996) In: Salamone JC (ed) The Polymeric Materials Encyclopedia: Synthesis, Properties, and Applications. CRC Press Inc, Boca Raton, Vol 6, pp 4540–4546
44. Gao M, Zhang X, Yang B, Shen JA (1994) J Chem Soc Chem Commun 2229
45. Keller SW, Kim HN, Mallouk TE (1994) J Am Chem Soc 116:8817
46. Kleinfeld ER, Ferguson GS (1994) Science 265:370
47. Lvov Y, Ariga K, Kunitake T (1994) Chem Lett 2323
48. Fendler JH, Meldrum FC (1995) Adv Mater 7:607
49. Fou AC, Rubner MF (1995) Macromolecules 28:7115
50. Tian J, Wu CC, Thompson ME, Sturm JC, Register RA, Marsella MJ, Swager TM (1995) Adv Mater 7:395
51. Decher G, Schmitt J (1992) Progr Colloid Polym Sci 89:160
52. Lvov Y, Decher G, Möhwald H (1993) Langmuir 9:481
53. Decher G, Hong JD (1991) Ber Bunsenges Phys Chem 95:1430
54. Lvov Y, Decher G, Sukhorukov G (1993) Macromolecules 26:5396
55. Hong J-D, Lowack K, Schmitt J, Decher G (1993) Progr Colloid Polym Sci 93:98
56. Decher G, Lehr B, Lowack K, Lvov Y, Schmitt J (1994) Biosens Bioelect 9:677

57. Cheung JH, Fou AC, Ferreira M, Rubner MF (1994) Thin Solid Films 244:806
58. Ferreira M, Rubner MF (1995) Macromolecules 28:7107
59 Cheung JH, Fou AC, Rubner MF (1994) Thin Solid Films 244:985
60. Fou AC, Rubner MF (1995) Macromolecules 28:7115
61. Yoo D, Lee J-k, Rubner MF (1996) MRS Symp Proc 413:395
62. Ferreira M, Rubner MF, Hsieh BR (1994) MRS Symp Proc 328:119
63. Fou AC, Onitsuka O, Ferreira M, Rubner MF, Hsieh BR (1996) J Appl Phys 79:7501
64. Tian J, Thompson ME, Wu CC, Sturm JC, Register RA, Marsella MJ, Swager TM (1994) Polym Prepr 35:761

65. Lehr B, Seufert M, Wenz G, Decher G (1996) Supramolec Sci 2:199
66. Hong H, Davidov D, Avny Y, Chayet H, Faraggi EZ, Neumann R (1995) Adv Mater 7:846
67. Onoda M, Yoshino K (1995) Jpn J Appl Phys 34:L260
68. Schmitt J, Decher G, Dressik WJ, Brandow SL, Geer RE, Shashidhar R, Calvert JM (in press) Adv Mater
69. Ishinose I, Ohno S, Kunitake T, Lvov Y, Ariga K (1995) Polym Prepr, Jpn 44:2544
70. Kotov NA, Dékány I, Fendler JH (1995) J Phys Chem 99:13065
71. Feldheim DL, Grabar KC, Natan MJ, Mallouk TE (1996) J Am Chem Soc 118:7640

72. Lvov Y, Ariga K, Ichinose I, Kunitake T (1996) Langmuir 12:3038
73. Mallouk TE (1996) In: Alberti G, Bein T (eds) Comprehensive Supramolecular Chemistry. 7:189
74. Decher G, Hong JD (1991) Makromol Chem, Makromol Symp 46:321
75. Sellergren B, Swietlow A, Arnebrandt T, Unger K (1996) Anal Chem 68:402
76. Mao G, Tsao Y-H, Tirrell M, Davis HT, Hessel V, Ringsdorf H (1995) Langmuir 11:942
77. Saremi F, Maassen E, Tieke B, Jordan G, Rammensee W (1995) Langmuir 11:1068
78. Saremi F, Tieke B (1995) Adv Mater 7:378

Progr Colloid Polym Sci (1997) 103:251–260
© Steinkopff Verlag 1997

A.A. Khan
Y. Shnidman

Molecular mean-field models of capillary dynamics

Received: 8 January 1997
Accepted: 17 January 1997

A.A. Khan
Department of Physics and Astronomy
University of Rochester
Rochester, New York 14627, USA

Dr. Y. Shnidman (✉)
Imaging Research
and Advanced Development
Eastman Kodak Company
Rochester, New York 14650–2216, USA

Abstract A novel lattice-gas approach has been developed to model the effect of molecular interactions on dynamic interfacial structure and flows of liquid–vapor and liquid–liquid systems in micro-capillaries. Within a mean-field approximation, discrete time evolution of species and momentum densities consists of alternating convective and diffusive steps subject to local conservation laws. Stick boundary conditions imposed during the convective step cause momentum transfer to lattice particles in contact with sheared solid walls. However, low densities, and diffusive relaxation of interfaces towards equilibrium give rise to velocity slip at the walls. We demonstrate applications of the single- and two-species isothermal versions of the model to capillary and wetting dynamics. The connections of the new approach to existing models of interfacial dynamics in two-phase flows are discussed, as well as its possible generalizations.

Key words Dynamics – interfaces – wetting – shear – lattice-gas

Existing approaches for modeling interfacial structure and dynamics

Lattice-gas (LG) models have been widely used to model a variety of static interfacial phenomena on the microscopic scale. Using a mean-field (MF) approximation, it is straightforward to calculate the equilibrium interfacial structure in such models. This is done by minimizing an inhomogeneous variational free energy with respect to the relevant local order parameters (e.g., see Ref. [1]). The resulting system of coupled, nonlinear difference equations can be solved at a relatively modest computational cost. This enables an efficient exploration of the dependence of the static interfacial structure and macroscopic properties (such as surface tensions and contact angles) on microscopic interactions. The continuum, coarse-grained limit of a lattice-gas MF free energy results in a Landau–Ginzburg free-energy functional (occasionally, it is useful to coarse-grain it further into a free energy that depends

only on collective interfacial degrees of freedom). Minimization of such functionals leads to an equivalent system of Euler–Lagrange equations with macroscopic phenomenological parameters governing the mesoscale interfacial structure and properties. Far from critical points, where fluctuations predominate, the resulting interfacial structure approximates that arising from computationally more intensive Monte Carlo (MC) and molecular dynamics (MD) simulations. This has been amply demonstrated for interfaces of simple liquids coexisting with their own vapor, or with another liquid, in the vicinity of solid substrates of varying wettability [2–4], as well as for such interfacial phenomena in more complex fluids containing surfactants [5] and polymers [4].

Modeling the dynamics of interfaces driven out of thermodynamic equilibrium is more challenging. Their time evolution must be governed by the same local laws of conservation of species, momenta and energies that form the foundation of mechanics, hydrodynamics, and irreversible thermodynamics. The latter two theoretical approaches were first formulated for homogeneous macroscopic phases. As in equilibrium, extensions to multiphase systems exist that model dynamic interfaces by introducing Gibbsian surfaces of zero thickness carrying excess properties [6]. For an interesting application of such an approach to the problem of dynamic wetting, the reader is referred to the recent work of Shikhmurzaev [7]. In phenomenological approaches of this type, the connection of the interfacial structure and dynamics to microscopic interactions and processes is lost. Instead, one has to rely on postulated constitutive relations and phenomenological parameters, which tend to proliferate in the case of dynamic interfacial phenomena. MD simulations, based on integrating the equations of Newtonian mechanics for a large system of particles with microscopic interactions, indicate that this connection is crucial for understanding and modeling such phenomena as the dynamics of wetting [8, 9]. Unfortunately, because of fundamental limitations on the size of the timestep, MD trajectories of interfacial systems are limited to the nanosecond range, while many dynamic interfacial phenomena occur on a much slower timescale. Recently, hydrodynamic lattice-gas [10] and lattice-Boltzmann [11] approaches have been applied to interfacial systems. Originally developed as tools for simulating systems governed by macroscopic hydrodynamic equations, these consist of deterministic or stochastic rules that are devised to simulate the hydrodynamic equations by alternating convective and dissipative time steps. The dissipative step mimics collisions of molecules in a gas in a manner that recovers separation into coexisting multiple phases, as well as macroscopic hydrodynamic behavior. However, this leaves much arbitrariness in the choice of the rules. As such rules do not depend explicitly on the microscopic interactions, the effect of the latter on the interfacial structure and dynamics is lost in this approach.

Convective–diffusive lattice-gas model: physical picture

To make possible efficient explorations of the dependence of interfacial dynamics on microscopic interactions, we have recently developed a novel, dynamic MF approach, which we call the convective–diffusive lattice-gas (CDLG) model. Lattice-gas representations of microscopic interactions and kinetic energies are used as the starting point. As in the old theory of transport in liquids developed by Eyring [12] and Frenkel [13] (hence referred to as EF theory), we assume that in a fluid, molecules are surrounded by cages of neighbors, but some cages are vacant.

On a short timescale, molecules undergo internal motion within a cage, but on a longer timescale they may hop from that cage to a neighboring vacant one. In the context of interfacial dynamics, the EF theory served as the basis for the molecular-kinetic (MK) theory of dynamic wetting [14–16]. If Monte Carlo simulations or mean-field approximations are used to solve the master equation governing the local hopping dynamics under an imposed profile of driving forces, the EF theory becomes a variety of a driven-diffusive lattice-gas [17]. Though interfacial dynamics in such models is of intrinsic interest, its applicability to real liquids is questionable. Unlike in driven-diffusive lattice gases, in real liquids the diffusive hopping between cages is augmented by convective motion of the cages between hops, and the driving forces, instead of conforming to a fixed profile, evolve subject to a local momentum conservation law.

Our convective–diffusive lattice-gas model, defined below, reflects these characteristics of real liquids. As in lattice-Boltzmann methods, discrete time evolution of local species and momentum densities within a mean-field approximation consists of alternating convective and dissipative steps subject to local conservation laws. However, in contrast with the lattice-Boltzmann methods, where the dissipative step is implemented by lattice collision rules without explicit dependence on microscopic interactions, in our CDLG model the dissipation is diffusive in nature, arising from a Markov chain representation of local hops from an occupied site to an empty site. The hopping transition rates have an explicit dependence on microscopic interactions, in a manner assuring relaxation to the same equilibrium state as exhibited by the corresponding static MF model at the same level of approximation. Stick boundary conditions imposed during the convective step cause momentum transfer to lattice particles in contact with sheared solid walls. Nevertheless, low densities, and diffusive relaxation of interfaces towards equilibrium, give rise to velocity slip at the wall, as in the MK theory of dynamic wetting [14–16]. Unlike the latter, however, our model incorporates convective flows and momentum conservation, similar to classical hydrodynamic models of dynamic wetting, though in our case the interfacial properties vary smoothly over a zone of finite width, rather than being discontinuous or singular at a sharp (infinitesimally thin) interface. This is similar in spirit to "volume-of-fluid" hydrodynamic methods (e.g., Ref. [18]), where an interfacial region of finite thickness interpolating between bulk densities is introduced as a mathematical convenience. In our case, the interface thickness is physical, and is determined by a correlation length that is a function of the microscopic interactions, the temperature, and external forces.

Lattice-gas Hamiltonian and approximate probability distribution

For simplicity, we present here a two-dimensional, isothermal version of the convective–diffusive lattice-gas model, appropriate for liquid and vapor phases of a single-species system in a microcapillary. Consider a rectangular slab Λ of $N = L_x \times L_y$ sites $\mathbf{r} = (x, y)$ on a square lattice, with lattice constant a and unit vectors $\hat{\mathbf{e}}_{1,2} = (\pm 1, 0)a$, and $\hat{\mathbf{e}}_{3,4} = (0, \pm 1)a$. The boundary layers B_1 and B_2 at $y = 0$ and $y = (L_y - 1)a$ are adjacent to solid walls W_1 and W_2 at $y = -a$ and $y = L_y a$, respectively. We will assume periodic boundary conditions in the $\hat{\mathbf{y}}$ direction. Sites are assigned spin variables $S_{\mathbf{r}} = \pm 1$, representing occupancy by a single particle species of mass μ, or a vacancy at site $\mathbf{r}$, respectively. Furthermore, assume that this closed microcapillary system between the two walls contains a fixed number of particles, and that the system is isothermal, as if each site were in contact with a heat reservoir at temperature T. Let us define a fundamental timestep $\Delta t = \tau$. At any given time, we will assume a velocity $c\mathbf{u}_{\mathbf{r}}$, defined at each site, where $c = a/\tau$ is a unit velocity, and $\mathbf{u}_{\mathbf{r}}$ is a dimensionless velocity field measured in fractions of the unit velocity. For the remainder of this paper, length and time will be expressed in units of a and τ, respectively.

Assuming symmetric square-well potentials for fluid–fluid and fluid–solid interactions, the Hamiltonian of such a system has the form:

$$H = -\sum_{\mathbf{r} \in \Lambda} \left[\frac{1}{2} \sum_{k=1}^{4} J S_{\mathbf{r}} S_{\mathbf{r} + \hat{e}_k} + h_{\mathbf{r}} S_{\mathbf{r}} - \frac{1}{4}(1 + S_{\mathbf{r}})\alpha \mathbf{u}_{\mathbf{r}}^2 \right]. \quad (1)$$

The last term (with $\alpha = \mu c^2$) represents the kinetic energy contribution to the Hamiltonian, while the remaining terms represent microscopic interaction energies of the Ising form. Here $(1 + s_{\mathbf{r}})/2$ is a projection operator assuring a site contribution to the kinetic energy only if that site is occupied. J models the attraction between a nearest-neighbor pair of LG particles, and $h_{\mathbf{r}}$ represents the effect of the interactions of these particles with the solid walls. For simplicity, both bulk and surface interactions are assumed to be short range, and thus $h_{\mathbf{r}}$ vanish everywhere except at the boundary layers B_1 and B_2.

Similarly to lattice-Boltzmann (LB) methods, we assume that the time evolution is modeled by alternating convective and dissipative timesteps. The physical picture is that, between hops, molecular cages are being convected by the mean velocity field. Such a mechanical convection step may drive interfacial configurations out of equilibrium. This is followed by a dissipative relaxation of the interfaces towards equilibrium by diffusive hops of molecules from an occupied cage to a neighboring vacant cage.

Within the MF approximation, we assume factorization of the N-particle probability distribution

$$P(\{S_{\mathbf{r}}\}, \{u_{\mathbf{r}}\}) = \prod_{\mathbf{r}} p_{\mathbf{r}}(S_{\mathbf{r}}) q_{\mathbf{r}}(u_{\mathbf{r}}). \quad (2)$$

We define a phase separation order parameter $m_{\mathbf{r}} = \langle S_{\mathbf{r}} \rangle$, related to the mean site density by $n_{\mathbf{r}} = (1 + m_{\mathbf{r}})/2$. In conjunction with the normalization condition, this leads to $p_{\mathbf{r}}(\pm 1) = (1_{\mathbf{r}} \pm m_{\mathbf{r}})/2$. We also assume that, between the hops, the particles have enough time to achieve local thermal equilibrium with the surrounding cage. This implies a local Maxwell–Boltzmann distribution of site velocities

$$q_{\mathbf{r}}(\mathbf{u}_{\mathbf{r}}) \propto \exp\left[-\alpha(\mathbf{u}_{\mathbf{r}} - \mathbf{v}_{\mathbf{r}})^2 / 2k_B T \right], \quad (3)$$

where $\mathbf{v}_{\mathbf{r}} = \langle \mathbf{u}_{\mathbf{r}} \rangle$.

Convective–diffusive dynamics from local conservation laws

Let a generic site density $\rho_{\mathbf{r}}$ represent either the (scalar) number density $n_{\mathbf{r}}$ or the components of the (vector) momentum density $\mu c n_{\mathbf{r}} \mathbf{v}_{\mathbf{r}}$. Similarly to lattice-Boltzmann (LB) methods, we evolve the system in time by alternating discrete convective and dissipative timesteps. The physical picture is that molecular cages are convected according to the mean velocity field at a site, between diffusive hops of molecules from an occupied site (cage) to adjacent empty sites (cages).

An initial density $\rho_{\mathbf{r}}$ evolves into a density $\rho_{\mathbf{r}}^C$ after a convective timestep of duration τ. The time evolution is assumed to obey a local conservation law. This means that the change in the number, or momentum, densities within a timestep is accounted by balancing all the conjugate density currents across the bonds into and out of this site, and, in the case of momentum density, a force that is either of external origin, or arising from internal stresses. Thus,

$$\rho_{\mathbf{r}}^C - \rho_{\mathbf{r}} + \tau \sum_{i=1}^{4} j_{\mathbf{r}, \mathbf{r} + \hat{\mathbf{e}}_i}^{\rho C} = \tau F_{\mathbf{r}}^{\rho}, \quad (4)$$

where

$$j_{\mathbf{r}, \mathbf{r} + \hat{\mathbf{e}}_i}^{\rho C} = \rho_{\mathbf{r}} (\mathbf{v}_{\mathbf{r}} \cdot \hat{\mathbf{e}}_i) - \rho_{\mathbf{r} + \hat{\mathbf{e}}_i} (\mathbf{v}_{\mathbf{r} + \hat{\mathbf{e}}_i} \cdot \hat{\mathbf{e}}_i) \quad (5)$$

are the net convective currents into a site from its nearest neighbors, and $F_{\mathbf{r}}^{\rho} = 0$ for $\rho = n$. In the case of two solid walls being sheared at opposite constant velocities $\pm v_{\mathbf{w}}$, we assume that if a particle occupies an adjacent boundary layer site during the convective step, the mean velocity at this site becomes identical to the wall velocity. Thus, we use the stick (no-slip) boundary conditions: $\mathbf{v}_{\mathbf{r}}^C = n_{\mathbf{r}}(\pm v_w, 0)$ if $\mathbf{r} \in B_1$ or $\mathbf{r} \in B_2$, respectively (the prefactor $n_{\mathbf{r}}$ reflects the probability that this site is indeed occupied).

The convective step is followed by a diffusive step, for simplicity assumed here to be of the same duration. As it is also subject to local conservation laws, we have

$$\rho_{\mathbf{r}}^{D} - \rho_{\mathbf{r}}^{C} + \tau \sum_{i=1}^{4} j_{\mathbf{r},\mathbf{r}+\hat{\mathbf{e}}_i}^{\rho D} = 0 \ . \tag{6}$$

The net diffusive currents $j_{\mathbf{r},\mathbf{r}+\hat{\mathbf{e}}_i}^{\rho D} = j_{\mathbf{r},\mathbf{r}+\hat{\mathbf{e}}_i}^{(\rho,\,0)} - j_{\mathbf{r},\mathbf{r}+\hat{\mathbf{e}}_i}^{(0,\,\rho)}$ are obtained by balancing forward and reverse diffusive fluxes into a site from its neighbors. We assume that the local diffusive hops constitute a Markov chain. The transition rates of the corresponding master equation are given by $\Gamma^{\rho}\varphi(\Delta H/k_{\mathrm{B}}T)$, where Γ^{ρ} is a microscopic friction coefficient setting the diffusive process timescale, and ΔH is the energy difference resulting from the hop. Hence, within the mean-field approximation of Eq. (2), the diffusive currents assume the form

$$j_{\mathbf{r},\mathbf{r}+\hat{\mathbf{e}}_i}^{(\rho,\,0)} = \Gamma^{\rho}\rho_{\mathbf{r}}^{C}(1 - n_{\mathbf{r},\mathbf{r}+\hat{\mathbf{e}}_i}^{C})\,\varphi(\langle\Delta H\rangle/k_{\mathrm{B}}T) \ . \tag{7}$$

The assumption of local detailed balance $\varphi(\lambda) = e^{-\lambda}\varphi(-\lambda)$ assures dissipative relaxation of the system towards equilibrium during the diffusive step. However, the short duration of the diffusive step is typically insufficient to achieve equilibrium before the start of next convective step. The new input for the convective step is

$$n_{\mathbf{r}}' = n_{\mathbf{r}}^{D} \ ,$$

$$\mathbf{v}_{\mathbf{r}}' \cdot \hat{\mathbf{e}}_i = \mathbf{v}_{\mathbf{r}}^{D} \cdot \hat{\mathbf{e}}_i + (j_{\mathbf{r},\mathbf{r}+\mathbf{e}_i}^{nD} - j_{\mathbf{r},\mathbf{r}-\hat{\mathbf{e}}_i}^{nD})/n_{\mathbf{r}}^{C} \ . \tag{8}$$

Here we alternate single convective steps with single diffusive steps. In general, the time sequence of convective and diffusive steps could follow any regular or random distribution. Different choices of the convective–diffusive sequence, as well as different values of the microscopic friction coefficients Γ^{ρ}, lead to variations in dynamic be-

havior, and in the system's transport coefficients, such as diffusivities and viscosities.

Applications: dynamic wetting and capillary dynamics

We originally developed the CDLG model as a tool for studying dynamic wetting phenomena. The CDLG MF time evolution equations presented above were applied to a variety of liquid–vapor systems in a microcapillary, under different boundary and initial conditions. As a first example consider a single-species lattice-gas governed by Hamiltonian (1) in a channel between two planar walls defined by $L_x = 128$ and $L_y = 32$, that is periodic along the x-direction. The initial density distribution was a bimodal distribution, $n_{\mathbf{r}}^{0} = 0.05$ or 0.95, antisymmetrically distributed about $x = 64.5$, and the initial velocities were $\mathbf{v}_{\mathbf{r}}^{0} = 0$. The temperature and the interaction parameters were fixed at $J/k_{\mathrm{B}}T = 0.3$ and $h_{\mathbf{r}} = 0$. We have first calculated the local equilibrium densities $\{n_{\mathbf{r}}^{\mathrm{eq}}\}$ by minimization of the variational mean-field free energy $F(\{n_{\mathbf{r}}\}) = U - TS$, where $U = \langle H\rangle$, $S = \langle k_{\mathrm{B}}\log P\rangle$, and the averaging is with respect to the MF approximation (2). Interpolated density contours resulting from this equilibrium calculation are shown in Fig. 1a. Note the coexistence of a liquid (high density) phase with a vapor (low density) phase separated by two planar interfaces centered about the contour line defined by $n_{\mathbf{r}} = 0.5$, which meet the walls at a static contact angle $\theta_0 = 90°$. Using the calculated values of $n_{\mathbf{r}}^{\mathrm{eq}}$, and $\mathbf{v}_{\mathbf{r}}^{\mathrm{eq}} = 0$ as the initial conditions, we then verified that the equilibrium solution is also a steady state solution of the CDLG MF time evolution

Fig. 1 Interpolated density contours at $n_{\mathbf{r}} = 0.2, 0.5,$ and 0.8 in a microcapillary between two solid walls. (a) Static walls ($u_{\mathrm{w}} = 0$). (b) Sheared walls ($u_{\mathrm{w}} = 0.02$). Upper (lower) wall moves to the right (left), respectively

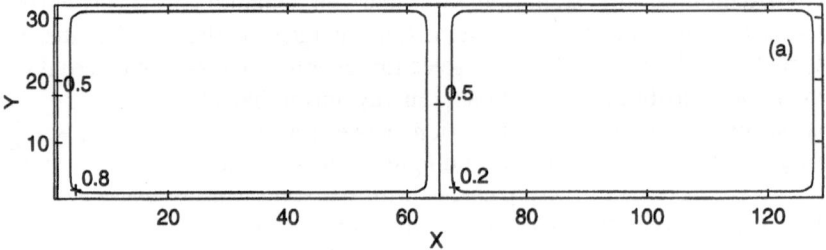

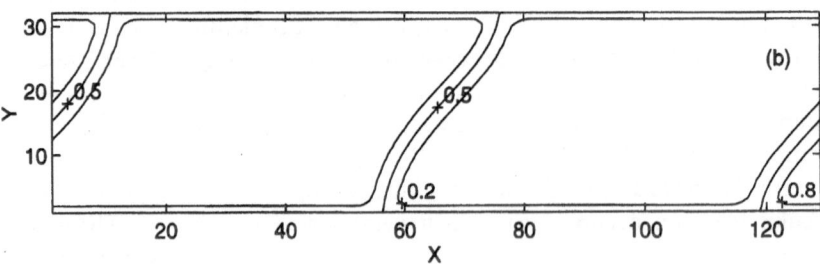

equations at vanishing wall velocities $u_w = 0$, within a common numerical accuracy. If, on the other hand, the boundary conditions at the wall are abruptly changed to $u_w \neq 0$, a nonzero velocity profile starts propagating from the walls in a diffusive manner, building up shear stresses that distort the interfaces from their equilibrium configuration and increasing the contact angle from its static value. At long times a new, nonequilibrium steady state (NESS) is approached. The density and velocity distributions in a NESS within a channel sheared at $u_w = 0.02$ are shown in Fig. 1b. This figure exhibits a significant distortion of the liquid–vapor interfaces from the planar equilibrium configuration. The interfaces now meet the walls at a dynamic contact angle $\theta > \theta_0$. The arrow plot of the mean velocity distribution, superimposed on a density contour plot, is shown for a region around the central interface in Fig. 2. Note the differences between the velocity distributions on the liquid and vapor phases. In the interfacial region, a rolling, "tank-tread" velocity profile tangential to the interface is established on the liquid side, while outgoing and incoming "jets" are observed at the advancing and receding contact zones on the vapor side. Far from the liquid–vapor interfaces, the velocity profile approaches a linear Couette-like profile with $v_y = 0$ and a constant $\partial v_x / \partial y$. This is seen very clearly in Fig. 3, showing an interpolated mesh plot of the x-component of the velocity profile v_x. Note that the velocity slip at the wall is small on the liquid side of the interface, while on the vapor side it is large. This leads to different values of $\partial v_x / \partial y$, and thus to different viscosities, in the two bulk

phases. By exploring examples such as these we have established that the bulk viscosities are quite sensitive to the values of the temperature, microscopic interactions, friction coefficients, and the sequence of convective and diffusive timesteps chosen for the CDLG model.

It is instructive to study the interplay of the effects of shear stresses and surface interactions on a liquid drop partially wetting one of the channel walls. For different values of the ratio $b = h_r / J$ at the wall, we first equilibrated a drop in contact with the bottom wall, and then generated a series of steady-state drop profiles (defined by the interpolated density contour $n_r = 0.5$) at different wall shear velocities u_w. Figure 4 displays the drop profiles resulting from calculations using $J/k_B T = 0.4$, and the same channel geometry as in the examples presented above. To initiate the CDLG time evolution equations, we used $v_r^0 = 0$ and a bimodal density distribution defined by a single square of sites fixed at a high density $n_r^0 = 0.95$, attached to the lower wall and surrounded by sites fixed at a low density of $n_r^0 = 0.05$. For fixed microscopic interactions and temperature, the equilibrium drop profiles in Fig. 4a were obtained from the steady-state limit of the time-evolution equations with the static boundary conditions $u_w = 0$. The steady-shear profiles in Fig. 4b were obtained as steady-state solutions of the CDLG evolution equations with sheared-channel boundary condition $u_w = 0.02$, using $v_r = 0$ and the equilibrium density distribution $\{n_r^{eq}\}$ as the initial condition. Note the asymmetry in the drop shape that develops with increasing u_w, and in particular the asymmetry between the advancing and receding contact

Fig. 2 Distribution of mean velocities about the central interface in Fig. 1b. For clarity, only site velocities of odd values of y are shown

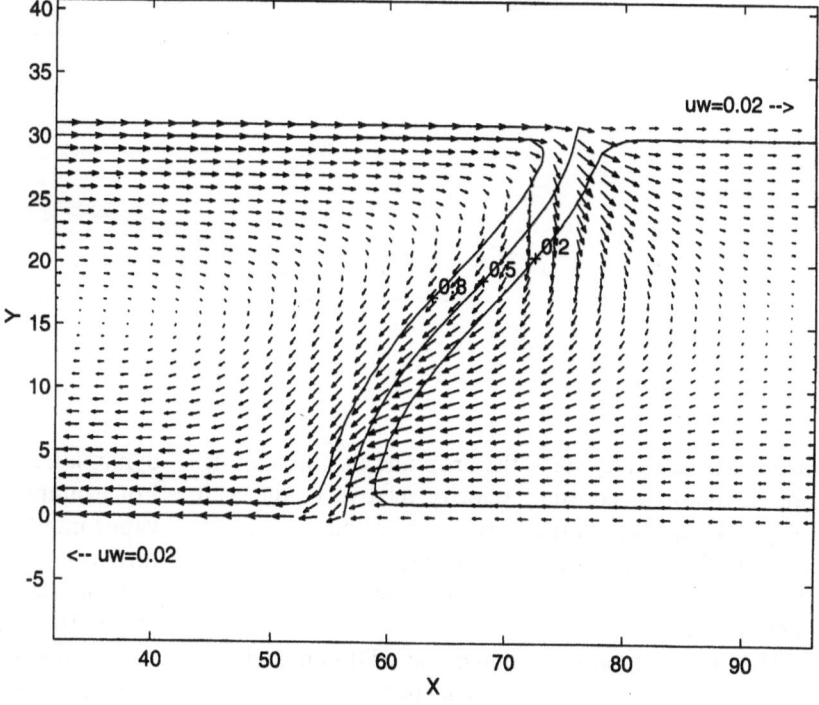

Fig. 3 Interpolated mesh plot of v_x about the central interface in Fig. 1b

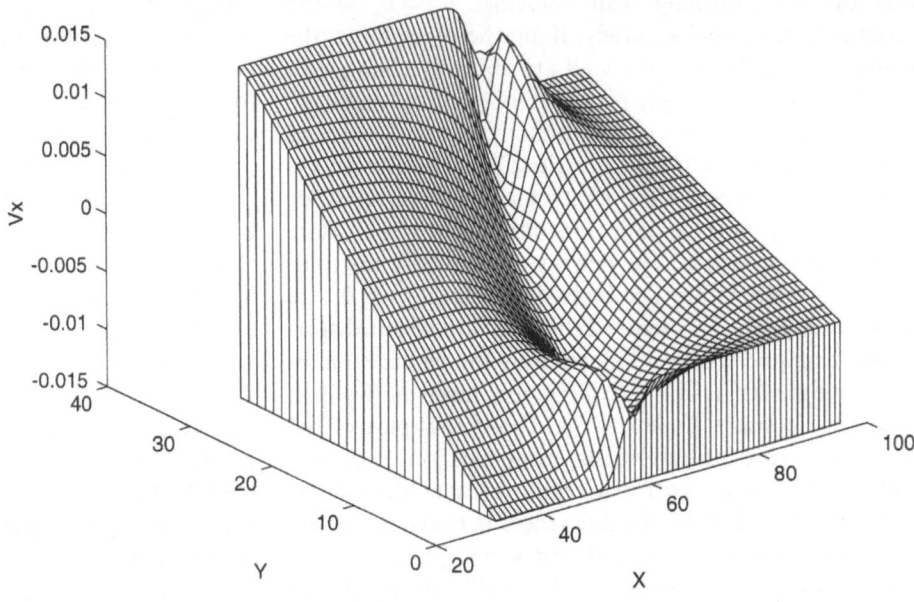

Fig. 4 Nonequilibrium steady state interface profiles (density contours at $n_r = 0.5$), for liquid drops in sheared microcapillaries. (a) $u_w = 0.02$. (b) $u_w = 0.06$. Symbols $+$, $\circ$, $\times$, and $*$ denote $b = 0.6, 0, -0.6$ and -1.2, respectively, where $b = h_r/J$. Bottom wall moves to the left

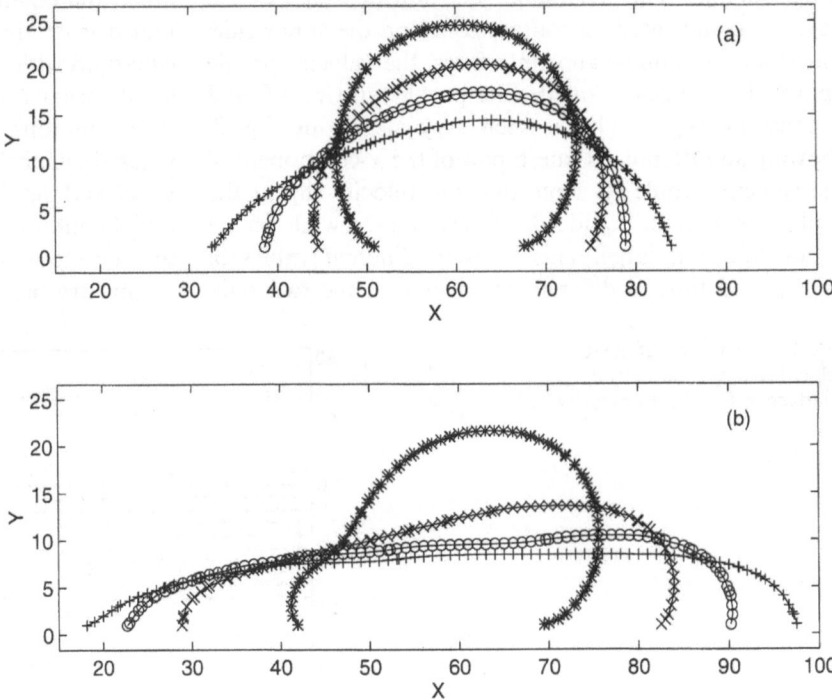

angles. At the steady state, the drop and its contact lines slide at almost a constant velocity relative to the wall, except for small oscillations arising from the discreteness of the lattice approximation. We have verified that the advancing contact angle is a monotonically increasing function of this velocity.

The time evolution equation of our CDLG models can also be used to study time dependence of the interfacial dynamics, as well as steady-state structure and velocities. For example, consider the time evolution of the liquid–vapor interface profile $y(x, t)$, starting at $t = 0$ with a perturbation centered at x_0, that is imposed on a steady-state interfacial profile (Fig. 5a). The steady state profile at $y(x) = 16.5$ separates a bottom liquid phase from a top vapor phase between two static solid walls in a system characterized by $L_x = 65$, $L_y = 32$, $J/k_B T = 0.4$

Progr Colloid Polym Sci (1997) 103:251–260
© Steinkopff Verlag 1997

Fig. 5 (a) The perturbed interface profile (density contour at $n_r = 0.5$) of $y(x,t)$ at $t = 0$. (b) Interfacial profiles $y(x,t)$ at three different times $t = 100$ (solid line), $t = 200$, (dotted line) and $t = 300$ (dashed-dotted line). (b) Oscillations of the interface $y(x,t)$ at $x = x_0$ for three time intervals of 200 timesteps starting at: $t = 100$ (dotted line), $t = 300$ (dashed-dotted line) and $t = 500$ (solid line)

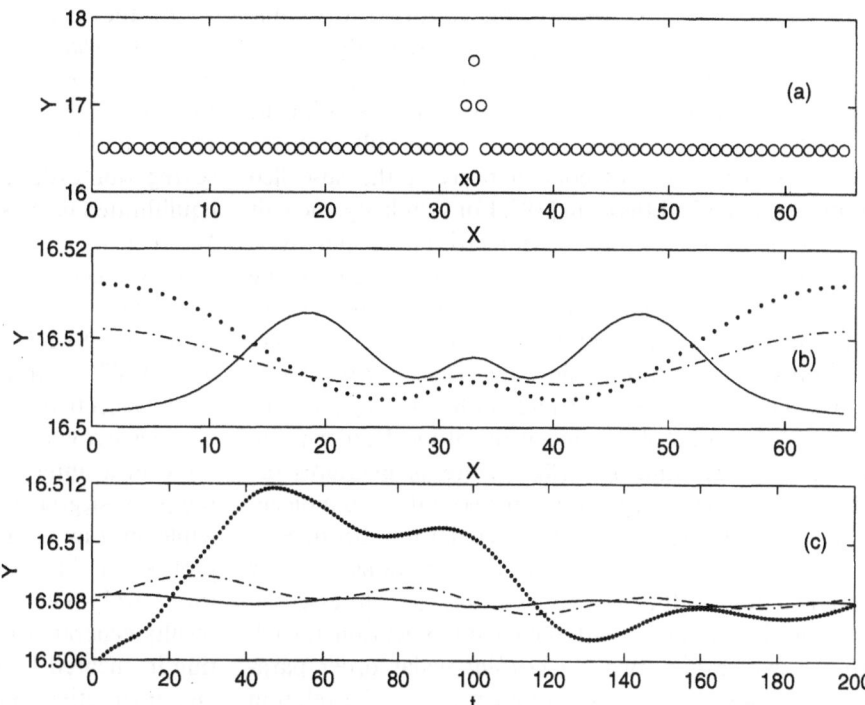

and $h_r = 0$. Subsequent iteration of the CDLG MF time evolution equations exhibits propagation and damping of a wave excited by this perturbation. This is demonstrated by Fig. 5b, showing the spatial variation of the profile $y(x, t)$ at three different times, and by Fig. 5c, displaying the oscillations of $y(x_0, t)$ with time. Analysis of such signals can provide quantitative information about wave dispersion relations, their dependence on microscopic interactions and the link to macroscopic parameters such as surface tensions and viscosities. The liquid–vapor surface tension can be found independently from static data, for example by plotting the difference between interior and exterior pressure across the interface of spherical liquid drops coexisting with vapor, versus drop curvature. As expected, we have found that such plots are almost linear as provided by the Young–Laplace relation, with a small systematic deviation from linearity at high drop curvatures.

Multiple species and surfactants

The simplest CDLG MF model described above can be used as a starting point for generalizations along many different directions. One obvious direction is towards CDLG models of systems with more than one species. A general projection operator formalism exists for generating static lattice-gas Hamiltonians of such systems. Their

dynamics can be modeled by adding kinetic energy terms to the Hamiltonian and generalizing the convective–diffusive time evolution to accommodate additional species.

For example, we have constructed a CDLG model for a binary liquid containing two molecular species. Molecular dynamics simulations of dynamic wetting in a sheared channel, using two-species Lennard–Jones binary liquids, provided a convincing demonstration of the existence of a molecular-scale velocity slip region about the ideal contact line [8, 9]. We will show here how CDLG can be generalized to study the same phenomenon in binary-liquids with short-range interactions. For comparison with the MD simulation, we will define the generalized CDLG model on a cubic lattice in a three dimensional $L_x \times L_y \times L_z$ channel between two solid walls perpendicular to $\hat{z}$ that is translationally invariant along $\hat{y}$, and periodic in $\hat{x}$. In this model we assign to each site a three-state degree of freedom $S_r = \pm 1$ or 0, where the value of ± 1 corresponds to a site occupied by one molecular species or the other, and the value of 0 represents a vacant site. The dynamic Hamiltonian now assumes the form

$$H = -\sum_{r \in \Lambda} \left[\frac{1}{2} \sum_{k=1}^{6} (J S_r S_{r+\hat{e}_k} + K S_r^2 S_{r+\hat{e}_k}^2) + h_r S_r \right.$$
$$\left. + \Delta_r S_r^2 - \frac{1}{2} S_r^2 \alpha u_r^2 \right], \tag{9}$$

where $\{\hat{\mathbf{e}}_k\}$ denote now the six unit vectors of the cubic lattice. The last term (with $\alpha = \mu c^2$) represents the kinetic energy contribution to the Hamiltonian. For simplicity, we consider here the case of identical molecular masses for the two species. The remaining terms represent bulk and surface interaction energies contributions in the so-called Blume–Emery–Griffiths form [19]. For simplicity, the pair interactions were chosen to be symmetric under spin inversion. Here S_r^2 is a projection operator assuring a site contribution to the kinetic energy only if that site is occupied, and h_r and Δ_r represent the effect of the interactions of the fluid species with the solid walls. The site MF probabilities $p_r(\pm 1) = (n_r \pm m_r)/2$ and $p_r(0) = 1 - n_r$ are now linear combination of two independent local order parameters $m_r = \langle S_r \rangle$ and $n_r = \langle S_r^2 \rangle$. Here m_r measures the extent of local phase segregation between the two molecular species, while n_r is the total molecular number density at a site. The CDLG time evolution equations are very similar to those of the one-species lattice gas. The essential difference is that, along with the local momentum at each site, we evolve now two *independent* scalar order parameters n_r, and m_r, as they no longer satisfy the relation $n_r = (1 + m_r)/2$, that was valid for the one-species system. Equations (4)–(8), written as they are in a generic form, as well as the wall boundary conditions, are still valid, except that the generic density ρ_r stands now for m_r, n_r, and the momentum components $\mu c n_r \mathbf{v}_r$, and a new iteration closure $m_r' = m_r^D$ is used.

In the example presented here, we chose $L_x = 60$, $L_z = 21$, $J/k_B T = 0.3$, $K/k_B T = -0.15$, and $\Delta_r/k_B T = 0.15$. To assure a static contact angle of $\theta_0 = 90°$, we also took $h_r = 0$. The initial configuration had mean velocities $\mathbf{v}_r^0 = 0$, a uniform density of $n_r^0 = 0.8$, and a bimodal phase segregation order parameter $m_r^0 = \pm 0.9$. These were first equilibrated using static boundary conditions with $u_w = 0$. We then used the equilibrium order parameters and velocities to generate a nonequilibrium steady state. Figure 6 shows the corresponding phase segregation order parameter in a NESS generated using the boundary condition $u_w = 0.0035$, with upper (lower) walls moving to the right (left), respectively.

The arrow plot of the local velocity distribution about the central interface, superimposed on a contour plot of the phase-segregation order parameter, is shown in Fig. 6, while in Fig. 7 the x-component the same velocity is plotted for the first four molecular layers adjacent to the lower wall. The resulting flows are very similar to the results generated from MD simulations of dynamic wetting in sheared Lennard–Jones binary liquids [8, 9]. It is also instructive to compare these figures with the analogous Fig. 2 and Fig. 3 presented above for liquid–vapor wetting in a sheared microcapillary. Unlike Fig. 2, in Fig. 6 the velocities are distributed symmetrically between the two liquid phases, including tangential, rolling "tank-tread" velocity profiles along opposite directions on the two sides of the liquid–liquid interface. Because of

Fig. 6 Steady liquid–liquid wetting: distribution of mean velocities (arrows) around the central liquid–liquid (indicated by contours of m_r at the three levels 0 and ± 0.6)

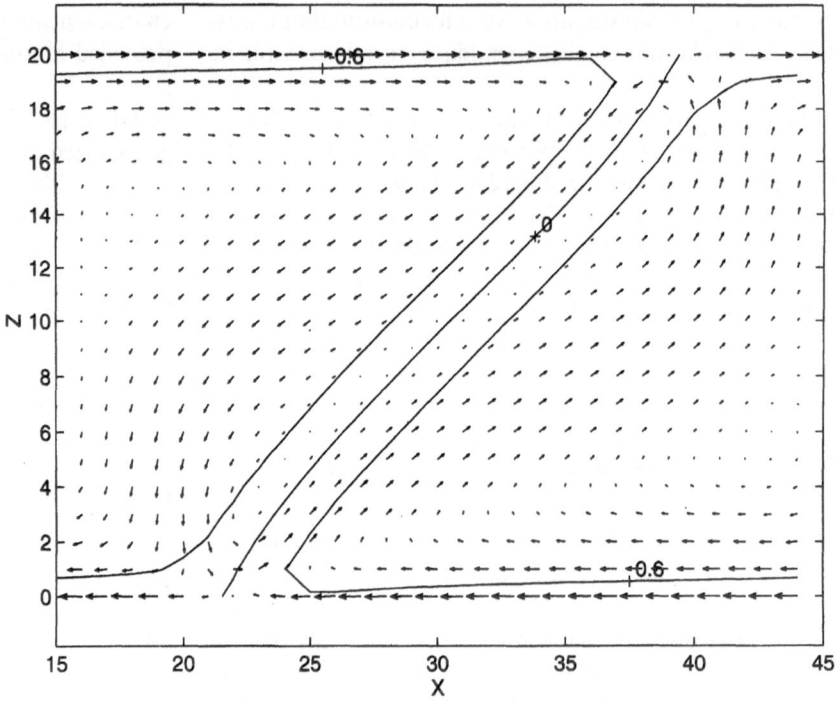

Fig. 7 Plots of v_x versus x for LG layers separated from the bottom wall by one (○), two (×), three (+) and four (∗) lattice constants, superimposed on a contour plot of m_r, for the same system as in Fig. 6

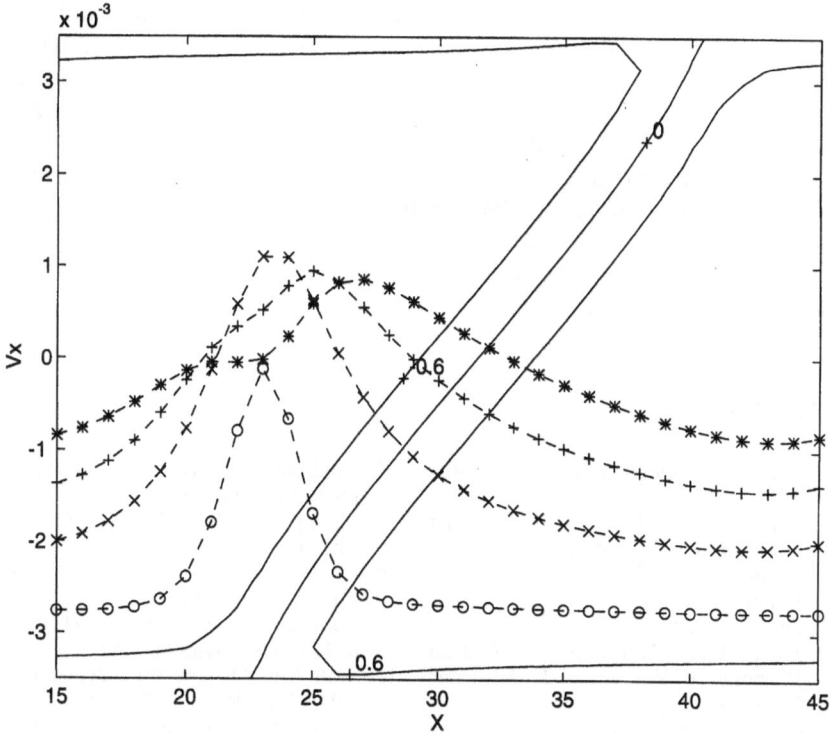

the simplifying assumptions used in this example, the viscosities of the two bulk phases are now identical. Figure 7 shows how velocity slip builds up at the contact zone, but decays over a few lattice spacings, as has been reported in MD simulations of binary Lennard–Jones liquids [8, 9].

Obviously, the simplifying symmetries used above are not essential for modeling a two species system, and can be removed. In fact, sometimes an asymmetry between the two species is essential for capturing the relevant physics and chemistry. For example, consider the case when one of the species represents water, while the other a (nonionic) surfactant. We have devised a three-state model to study the effect of surfactants on the dynamic wetting of an aqueous droplet coexisting with vapor. In this model, $S_r = 1$ represents occupancy by water, while $S_r = 0$ corresponds to a site occupied by a surfactant molecule, and $S_r = -1$ stands for a vacancy at that site. Since the vapor phase can be considered hydrophobic relative to the liquid aqueous phase, at equilibrium such a system can be modeled by the same Hamiltonian that has been suggested by Gompper and Shick as a model for microemulsions [5]. Their model is a generalized Blume–Emery–Griffiths Hamiltonian that includes "three-in-a-line" interaction terms of the form $LS_{r - \hat{e}_k}(1 - S_r^2)S_{r + \hat{e}_k}$, thus lowering the energy of the system when surfactants are at the interface between the hydrophilic and hydrophobic phases. Site contributions to the kinetic energy of water and sur-

factants can be expressed as $\alpha_+(S_r^2 + S_r)\alpha_+ \mathbf{u}_r^2/4$ and $\alpha_0 S_r^2 \alpha_0 \mathbf{u}_r^2/2$, respectively. However, since there is an order of magnitude discrepancy between the molecular weights of surfactant and water, one must have $\alpha_0 \gg \alpha_+$, as well as very distinct microscopic friction coefficients $\Gamma^0 \gg \Gamma^+$ in the hopping transition rates for surfactant and water. We are now in the process of studying the dynamics of such surfactant-containing CDLG models.

Summary and outlook

We have reviewed here the simplest, isothermal version of CDLG models for two-phase fluid dynamics on the microscopic scale. Applications of these models for studying interfacial dynamics in liquid–vapor and liquid–liquid systems in microcapillaries were discussed. The main advantage of our approach is that it models the explicit dependence of the interfacial structure and dynamics on molecular interactions, including surfactant effects. However, an off-lattice model of microscopic MF dynamics may be required for incorporating viscoelastic and chain-connectivity effects in complex fluids. Isothermal CDLG MF dynamics is based on the same local conservation laws for species and momenta that serve as a foundation for mechanics, hydrodynamics and irreversible thermodynamics. As in hydrodynamics and irreversible thermodynamics, the isothermal version of CDLG model can be generalized by incorporating local conservation of energy, which we have already done for single-species liquid–vapor systems. The nonisothermal CDLG allows self-consistent evolution of temperature alongside the local order parameters and the momenta. This can be very useful for studying the dependence of thermocapillary dynamics on microscopic interactions. Similarly to the static case, one can expect that an appropriate coarse graining of CDLG will result in a time-dependent Landau–Ginzburg (TDLG) theory subject to the same conservation laws (e.g., a vis-cous Cahn–Hilliard model (Model H), or its generalizations [20]). Indeed, Fig. 2 seems to be consistent with the order parameter contour plot and stream lines obtained by Seppecher [21] in the inner region of a contact zone, using a numerical solution of an isothermal, viscous Cahn–Hilliard fluid model at a steady state.

Acknowledgment The authors would like to thank Dr. Terry Blake (Kodak Ltd.) for encouraging this work and for many illuminating discussions.

References

1. Varea C, Robledo A (1992) Phys Rev A 45:2645
2. Binder K (1983) In: Domb C, Lebowitz J (eds) Phase Transitions and Critical Phenomena, Vol 8, p 1. Academic Press, London
3. Dietrich S (1988) In: Domb C, Lebowitz J (eds) Phase Transitions and Critical Phenomena, Vol 12, p 1. Academic Press, London
4. de Gennes PG (1985) Rev Mod Phys 57:827
5. Gompper G, Schick M (1994) Self-Assembling Amphiphilic Systems. Academic Press, London
6. Edwards DA, Brenner H, Wasan DT (1991) Interfacial Transport Processes and Rheology. Butterworth–Heinemann, Boston
7. Shikhmurzaev YD (1993) Int J Multiphase Flow 19:589
8. Thompson PA, Robbins MO (1989) Phys Rev Lett 63:766
9. Koplik J, Banavar JR (1995) Ann Rev Fluid Mech 27:257
10. Rothman DH, Zaleski D (1994) Review of Modern Physics 66:1417
11. Swift MR, Osborn WR, Yeomans JM (1995) Phys Rev Lett 75:830
12. Eyring H (1936) J Chem Phys 4:283
13. Frenkel JI (1946) Kinetic Theory of Liquids. Oxford University Press, Oxford
14. Blake TD, Haynes JM (1969) J Colloid Interface Sci 30:421
15. Blake TD (1993) In: Berg JC (ed) Wettability (Surfactant science series), Vol 49, p 251. Marcel Dekker, New York
16. Ruckenstein E, Dunn CS (1977) J Colloid Interface Sci 59:135
17. Schmittmann B, Zia RKP (1995) Statistical Mechanics of Driven Diffusive Systems. Academic Press, London
18. Brackbill JU, Kothe DB, Zemach C (1992) J Comp Phys 100:335
19. Blume M, Emery V, Griffiths RB (1971) Phys Rev A 4:1071
20. Chaikin PM, Lubensky TC (1995) Principles of Condensed Matter Physics. Cambridge University Press, Cambridge
21. Seppecher P (1996) Int J Eng Sci 34:977

Progr Colloid Polym Sci (1997) 103:261–267
© Steinkopff Verlag 1997

H. Bianco
Y. Cohen
M. Narkis

Probing the structure of inhomogeneous colloidal particles by small-angle x-ray scattering

Received: 18 November 1996
Accepted: 2 December 1996

Abstract Small angle X-ray scattering (SAXS) is well suited for a detailed study of both the shape and the inner structure of suspended colloidal particles. A "trial-and-error" modeling was used in a SAXS study of two-stage latices (TSL), composed of polystyrene (PS) and polytribromostyrene (PTBrS, 15 wt%). The TSL particles were found to have a concentric core–shell structure. When a PTBrS latex was used as a seed, its particles were overcoated with a PS shell during the second-stage polymerization. However, only a small portion of the seed particles were overcoated with a PTBrS shell when using a PS seed. The size distributions of the TSL and the PTBrS latex particles were determined from the scattering curves, using the method of indirect Fourier transformation. The resulting average radii were in good agreement with the values obtained from TEM observations. The contrast variation method, designed to separate the information on the particle as a whole from that of its inhomogeneities, was employed in a SAXS study of a model copolymer latex, composed of styrene and pentabromobenzyl acrylate (PBBA, 40 wt%). The separation of the homogeneous function allows direct calculation of the size distribution of the spherical particles (volume average diameter, (26.7 ± 1.3) nm). The SAXS analysis reveals a particle's inner structure described as a continuous copolymer phase, of composition being slightly richer in PBBA, within which domains of PS are randomly distributed. The volume fraction of the PS domains was estimated as 8 vol%, and their characteristic length as 5.1 nm.

Key words Emulsion polymerization – contrast variation – small-angle X-ray scattering – inhomogeneous particles

H. Bianco · Y. Cohen (✉) · M. Narkis
Department of Chemical Engineering
Technion-Israel Institute of Technology
Haifa 32000, Israel

Introduction

Structural studies of heterogeneous colloidal particles are very important from both fundamental and applied aspects in many fields, such as aggregates in surfactant solutions, dissolved natural or synthetic macromolecules and polymer latices. When describing a particle, one should first consider the overall properties: The average chemical composition, the shape of the particle's envelope, its size and size distribution. Additional information is related to the inner structure: The number of phases, their volume fractions and the particle's morphology.

Small angle scattering of neutrons (SANS), X-rays (SAXS) and light (SALS) are well suited for a detailed study of both the shape and the inner structure of suspended colloidal particles. These techniques possess the advantage of having good statistics and thus allow quantitative and accurate determination of the structural parameters. On the other hand, the interpretation of small-angle scattering data is often model-dependent. One has to find a model particle whose scattering pattern agrees with the experimental curve, but this model might not be a unique solution. If several models can be offered, on the basis of a priori information obtained using transmission electron microscopy, for example, a comparison of their predicted scattering curves with those obtained experimentaly allows selection of the most appropriate model and thus determine its structural parameters. In many cases, however, no a priori information is available. In homogeneous systems, a model can be selected using a "trial-and-error" procedure, as the particle's shape can be frequently described by a simple three-axial body (for example a sphere, an ellipsoid or a rod), at least at low resolution. This approach cannot be applied for heterogeneous particles, since even in the simplest case, of a two-phase particle, the number of possible arrangements is enormous. One of the most useful tools under these circumstances is the contrast variation method, described in the next section.

The contrast variation method

The basic idea of the contrast variation method [1–4] is to separate the information on the particle as a whole from that of its inhomogeneities, thus allowing better determination of the particle's shape and some quantitative measures about its inner structure. The experimental procedure involves recording of several scattering curves, each having a different scattering density of the dispersing medium. As a first step, a straightforward indication of the particle's heterogeneity can be obtained directly from the experimental data, without any further calculation. The scattering curves of homogeneous systems are essentially independent of contrast, simply scaling as the contrast is changed. If, however, any inner structure exists in the colloidal particles, changes in contrast will alter both the intensity and the angular dependence of the scattered intensity. For a quantitative analysis of these changes, the scattering intensity $i(h)$ can be split into three contrast-independent "basic functions" [1–4]: The homogeneous part $i_h(h)$, the heterogeneous part $i_i(h)$ and the cross-term $i_{ih}(h)$

$$i(h) = (\Delta\rho)^2 i_h(h) + (\Delta\rho) i_{ih}(h) + i_i(h), \qquad (1)$$

where $h = 4\pi \sin\theta/\lambda$ is the scattering vector (2θ and λ are the scattering angle and the wave length, respectively), the contrast is defined as $\Delta\rho = \bar{\rho} - \rho_m$, $\bar{\rho}$ is the particle's average scattering density (i.e. electron density, coherent scattering length or refractive index in SAXS, SANS and SALS, respectively) and ρ_m is the scattering density of the dispersing medium.

The homogeneous basic function is mathematically identical to the scattering curve that would be obtained for a hypothetical sample, containing homogeneous particles having the same shape as those in the studied sample. Thus, conventional methods developed for homogeneous bodies, i.e. determination of the radius of gyration, the volume and the specific surface, as well as "trail-and-error" modeling, can be applied for a separate analysis of the particle shape using the homogeneous function. The typical features of the inner structure are more explicitly apparent in the heterogeneous basic function. The initial slope of this function is an estimate of the distance between the center of mass of the shape and that of the inhomogeneities, vanishing when the sub-structures have a common center of mass [4]. An important structural parameter, the mean-squared density fluctuation of the inner structure, can be obtained from the invariant (i.e. integrated intensity) of the heterogeneous function. The sign of the cross-term at the smallest angles reflects the relative arrangement of the higher and lower density regions with respect to the center of mass. If the denser phase is closer to the periphery of the particle, the cross-term will have negative values at the low-angle region, and vice versa [4].

Theoretically, if three scattering curves at different contrasts are recorded, the data for each measured scattering vector h_i yield a set of three linear equations from which the values of the basic functions related to h_i can be evaluated. In practice, however, one has to determine the average density before this procedure can be carried out. Each experimental curve is extrapolated to zero angle using a Guinier plot. Since $i_{ih}(0) = i_i(0) = 0$ [1], it follows that

$$i(0, \Delta\rho) = (\Delta\rho)^2 i_h(0), \qquad (2)$$

In monodispersed systems, a plot of $\sqrt{i(0, \Delta\rho)}$ vs. the scattering density of the dispersing medium results in a straight line. The average density can be evaluated by extrapolating this line to the matching point, where $\bar{\rho} = \rho_m$ and therefore the forward scattering vanishes. However, in polydispersed systems a non-vanishing minimum forward scattering appears [3]. If no a priori information is available, many curves have to be recorded in order to evaluate the average density using Eq. (2), to verify the linearity of the plot or the location of the minimum.

An alternative procedure to evaluate the average density can be performed using three ρ_m-independent functions $I_h(h)$, $I_{ih}(h)$ and $I_i(h)$ [4]. These functions, defined as

$$I_h(h) = i_h(h), \tag{3a}$$

$$I_{ih}(h) = -2\bar{\rho}\,i_h(h) - i_{ih}(h), \tag{3b}$$

$$I_i(h) = \bar{\rho}^2 i_h(h) + \bar{\rho}\,i_{ih}(h) + i_i(h). \tag{3c}$$

are related to the scattering intensity by the following expression:

$$i(h, \rho_m) = \rho_m^2 I_h(h) + \rho_m I_{ih}(h) + I_i(h). \tag{4}$$

Extrapolation of these functions to $h = 0$ yields two measures of the average density:

$$\bar{\rho} = \frac{-I_{ih}(0)}{2 I_h(0)}, \qquad \bar{\rho}^2 = \frac{I_i(0)}{I_h(0)}. \tag{5}$$

The second method can also be used in the general case, namely, a polydispersed system in which the average density varies with particle size, and requires only three scattering curves. The ratio $\langle \bar{\rho}(R) \rangle / \sqrt{\langle \bar{\rho}^2(R) \rangle}$ can be used as an estimate of the width of the average density distribution.

Modeling of the scattering intensity

As previously mentioned, a priori information obtained using other experimental methods or the general features of the structure obtained from the contrast variation method can be used as a basis for more detailed analysis, i.e. selecting a structural model. In most cases, the structural model is described by a function, connecting the local scattering density at any point within the particle to the coordinates of that point. In simple cases this function can be used for analytical calculation of the theoretical scattering curve. For example, the scattering density of a homogeneous sphere, having constant electron density ρ and radius R, is given by

$$\rho(r) = \begin{cases} \rho & \text{at } r \le R, \\ 0 & \text{at } r > R. \end{cases} \tag{6}$$

The total scattering intensity per unit volume of a dilute sample containing homogeneous spheres of varying dimensions, normalized to the scattering intensity of a single electron, is given by [5, 6]

$$i(h) = \frac{4\pi}{3} \Delta\rho^2 \phi \int_0^\infty f_v(R) R^3 \Phi^2(hR)\,dR, \tag{7}$$

where ϕ is the volume fraction of the particles in the sample, $f_v(R)\,dR$ is the volume of particles having a radius between R and $R + dR$ per total volume of particles and

$\Phi(hR)$ is the shape factor of a spherical particle:

$$\Phi(hR) = 3\frac{\sin(hR) - hR\cos(hR)}{(hR)^3}. \tag{8}$$

Another simple example is a concentric core–shell structure, in which the outer radius R is proportional to the core radius R_c: $R = \alpha R_c$. The scattering density is given by

$$\rho(r) = \begin{cases} \rho_c & \text{at } r \le R_c, \\ \rho_s & \text{at } R_c < r \le R_s, \\ 0 & \text{at } r > R, \end{cases} \tag{9}$$

and the scattering intensity can be calculated using the volume fraction of the shell ϕ_s [6],

$$i(h) = \frac{4\pi}{3} \frac{\phi_s \alpha^3}{\alpha^3 - 1} \int_0^\infty f_v(R) R^3$$

$$\times \left[\Delta\rho_s \Phi(hR) - \frac{\Delta\rho_s}{\alpha^3} \Phi\left(\frac{hR}{\alpha}\right) + \frac{\Delta\rho_c}{\alpha^3} \Phi\left(\frac{hR}{\alpha}\right) \right]^2 dR, \tag{10}$$

where ρ_s and ρ_c are the shell and the core scattering densities, respectively, $\Delta\rho_s = \rho_s - \rho_m$ and $\Delta\rho_c = \rho_c - \rho_m$.

A special class of a particle heterogeneity is found in particles having irregular inner structures, namely multicomponent particles in which the phases are randomly distributed [4]. Practical examples are porous particles, inner structures arising from phase separation, polymer latex particles having a "domain" morphology and so on. As in the analogous situation of matter with random distribution occupying the whole irradiated volume [7, 8], this type of inner structure is best described by its characteristic length ξ, rather than by a well-defined function of $\rho(r)$. In the simple case of a spherical particle of radius R, composed of two phases, the scattering intensity is given by [4]

$$i(h) = (\Delta\rho)^2 \left(\frac{4\pi}{3} R^3 \right)^2 \Phi^2(hR)$$

$$+ \frac{32\pi^2}{3} (\xi R)^3 \frac{\phi_1(1 - \phi_1)(\rho_1 - \rho_2)^2}{(1 + h^2\xi^2)^2}, \tag{11}$$

where ϕ_1 is the volume fraction of phase "1" with respect to the particle volume, ρ_1 and ρ_2 are the scattering densities of phases "1" and "2", respectively.

Equation (11) shows two unique features of irregular structures. First, the scattering intensity has only two contributions: The homogeneous part resulting from the particle shape, and the heterogeneous part resulting from the inner structure. The heterogeneous part is similar to the one obtained for randomly distributed matter occupying the whole irradiated volume; the only difference is the proportionality to $(\xi R)^{3/2}$, which can be regarded as a characteristic volume of the heterogeneity. There is no cross-term, since the average amplitude is zero for any h, due to the random distribution. Furthermore, it is clearly seen that the heterogeneous part has a significant contribution to the forward scattering. The relative contribution of the two parts mostly depends on the ratio of the contrast and the "inner contrast" $(\rho_1 - \rho_2)$.

Practical examples: SAXS studies of latex particles

Core–shell morphology in two-stage latices

Polymer latices are colloidal dispersions of polymer particles in an aqueous medium. The preparation of two stage latices involves a second stage of emulsion polymerization, in the presence of a seed latex. Several methods have been used to investigate the morphology of the resulting two-phase particles, such as transmission electron microscopy [9, 10], fluorescence technique [11], SANS [12–14] and SAXS [15–20]. In a recent paper [6], a SAXS study of two stage latices (TSL), composed of polystyrene (PS) and polytribromostyrene (PTBrS), was presented. A unique feature of tribromostyrene is its high reactivity and ability for thermal initiation. Thus the reaction kinetics leading to structure formation during emulsion polymerization differs significantly from the previously studied systems. However, on the basis of these studies it was possible to offer several structural models, that were used as a basis for the SAXS analysis.

Two TSL samples were prepared, each containing 15 wt% PTBrS. The one, denoted as PS/PTBrS, was obtained using PS latex as a seed, and the other, denoted as PTBrS/PS was obtained using a PTBrS seed latex. For modeling of the SAXS intensities it was assumed that the latex sample is a dilute system of particles of identical shape and varying dimensions. Using the method of indirect Fourier transformation, developed by Glatter [21, 22], the correction for the dimensions of the incident beam (length and width) and calculation of the size distribution were carried out as a single step procedure. Figure 1 displays the desmeared SAXS intensities of the bromine-containing latices. The calculated intensities, shown as solid lines, were obtained from the size distribution displayed in Fig. 2.

The scattering curve of the PTBrS latex (Fig. 1) was fit by a model of homogeneous spheres (Eq. (7)). The number average radius resulting from the size distribution is 3.72 nm (TEM: 3.66 nm). Such a small mean radius is very unusual in emulsion polymerization. It was suggested that this arises from the high reactivity of TBrS and its ability to undergo thermal initiation [23–25]. A combined initiation mechanism, namely, entry of a water-borne radical to

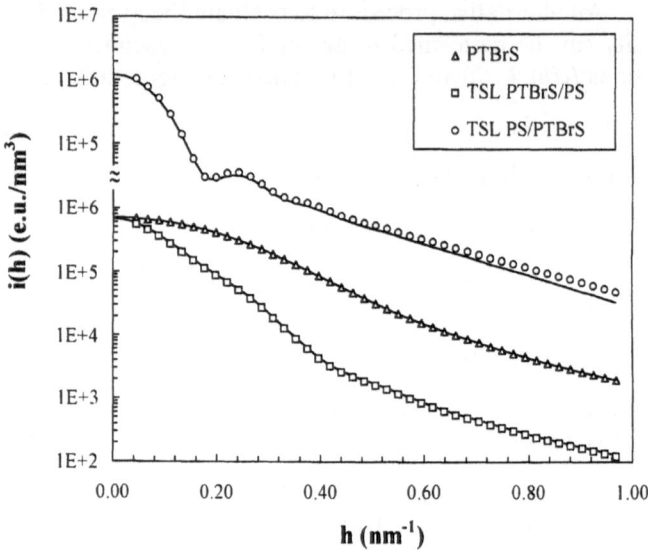

Fig. 1 Desmeared SAXS scattering intensities of the bromine-containing latices. Solid curves represent the theoretical intensities, calculated from the size distributions displayed in Fig. 2

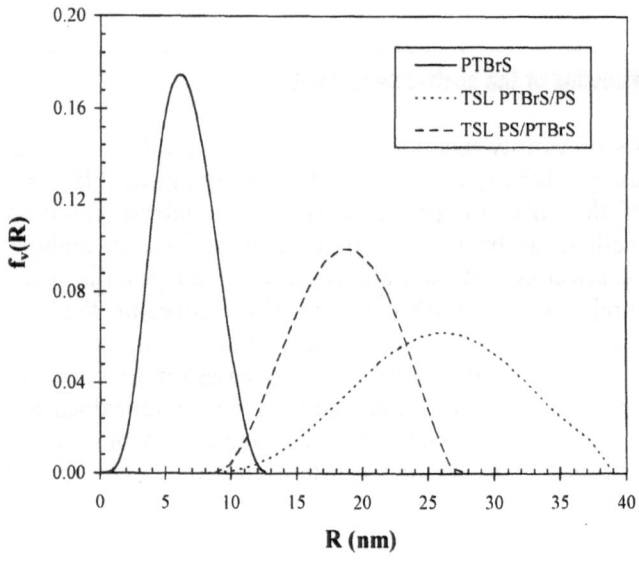

Fig. 2 Volume distribution of the latex particles, calculated using ITP

a monomer-swollen micelle (as in an ordinary emulsion polymerization process) and thermal initiation within the micelles, increases the number of nucleated particles, thus leading to a decrease of their size.

The analysis of the scattering curves from the TSL samples leads to the conclusion that the particles have a concentric core–shell structure, in which a core of the first-stage polymer is enclosed in a shell of the

Progr Colloid Polym Sci (1997) 103:261–267
© Steinkopff Verlag 1997

Fig. 3 TEM micrographs of the
bromine-containing latex
particles: (a) PTBrS, (b) TSL
PTBrS/PS, and (c) TSL
PS/PTBrS. Bar indicates
100 nm

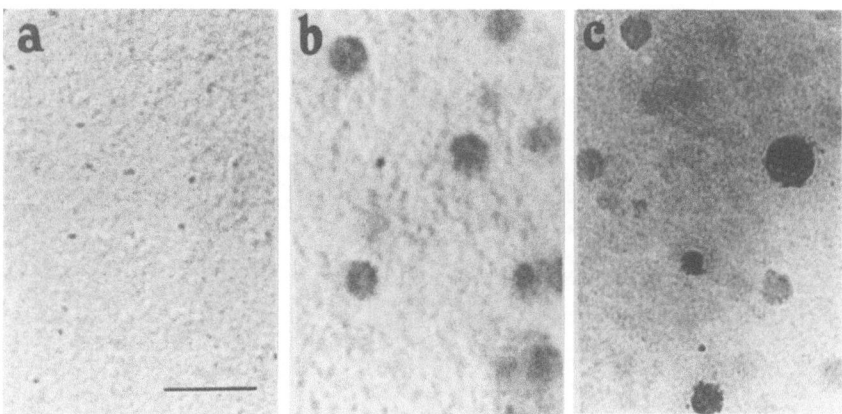

second-stage polymer. The fitting of the scattering curve of
TSL PTBrS/PS to this model (Eq. (10)) yields an average
content of 93 vol% PS in a core–shell particle, close to the
one calculated from mass balance (92 vol%). This suggests
that all the seed particles were overcoated with a PS shell
during the second-stage polymerization. However, fitting
of the scattering curve of the TSL PS/PTBrS to Eq. (10)
yields an average PTBrS content in the core–shell particles
of 56 vol%, compared to 8 vol% calculated from mass
balance. This indicates that only a minority of the PS seed
particles are overcoated with the second-stage polymer,
possibly due to the rapid polymerization kinetics of TBrS.
In semi-batch polymerization (i.e. a dropwise addition of
the monomer in the second polymerization stage), two
competing reactions exist simultaneously, namely nucleation of new chains on a seed particle and propagation of
these chains. As TBrS is highly reactive, propagation
should be much faster than nucleation, causing a situation
where most of the second stage monomers will lengthen
chains that were already initiated. TEM micrographs (Fig. 3c)
of the TSL PS/PTBrS supports the suggestion that
only part of the particles contain PTBrS. These particles
appear dark, indicating that they have a higher density
than the others. The micrographs also show a small
number of tiny dark particles, that indicates some secondary nucleation of PTBrS particles. Modification of Eq. (10)
to account for these particles gives a better fit to
the experimental results, indicating that 1.65 vol% of
the PTBrS is forming the secondary nucleated particles.
A good agreement in the outer radii obtained by SAXS
and TEM was found: PTBrS/PS: 21.0 nm SAXS, 21.6 nm
TEM; PS/PTBrS: 16.3 nm SAXS, 15.2 nm TEM.

Inhomogeneity in copolymer latex particles

Emulsion copolymerization has been widely used in the
industry for production of polymeric materials with

tailored properties. It is well known that quite often heterogeneity on the molecular level exists in these
copolymers [26–32]. Yet, remarkably little quantitative
information is available on heterogeneity on the colloidal
scale, determining the morphology of a latex particle.
Since no a priori information related to the particle's inner
structure was available, a recent SAXS study [33] of
a model copolymer latex, composed of styrene and pentabromobenzyl acrylate (PBBA, 40 wt%), employed the contrast variation method.

The electron density of the dispersing medium was
varied by adding sucrose to the latex sample. The smeared
SAXS intensities measured at different contrasts are
displayed in Fig. 4. As can be seen, changes in the contrast
affect both the intensity and the angular dependence of
the scattered radiation, indicating inhomogeneity of
the latex particles. The TEM micrographs (Fig. 5) reveal
the spherical shape and the size polydispersity of the
latex particles, but do not show any further structural
details. Thus, additional information related to the
particle's inner structure was obtained from the basic
functions.

The average density was determined by extrapolation
to $h = 0$ of the functions $I_i(h)$, $I_{ih}(h)$ and $I_h(h)$. From Eq.
(5), values of $\langle \bar{\rho}(R) \rangle = 403.4$ e.u./nm^3 and $\sqrt{\langle \bar{\rho}^2(R) \rangle} =
427.9$ e.u./nm^3 were obtained, i.e. about 6% difference be-
tween the two measures of the average density. The
smeared basis functions were calculated using Eq. (1). For
each experimental point h_i, a multi-variable linear regression procedure was employed, setting $\Delta\rho_i$ and $\Delta\rho_i^2$ as the
independent parameters. Each basic function was then
desmeared separately.

The homogeneous (desmeared) basic function $i_h(h)$ is
displayed in the lower curve of Fig. 6. Since the shape is
known to be spherical, the correction for the dimensions of
the primary beam (length and width) and the calculation of
size distribution could be carried out as a single step
using the program ITP. This yields a size distribution

266
H. Bianco et al.
SAXS studies of Inhomogeneous particles

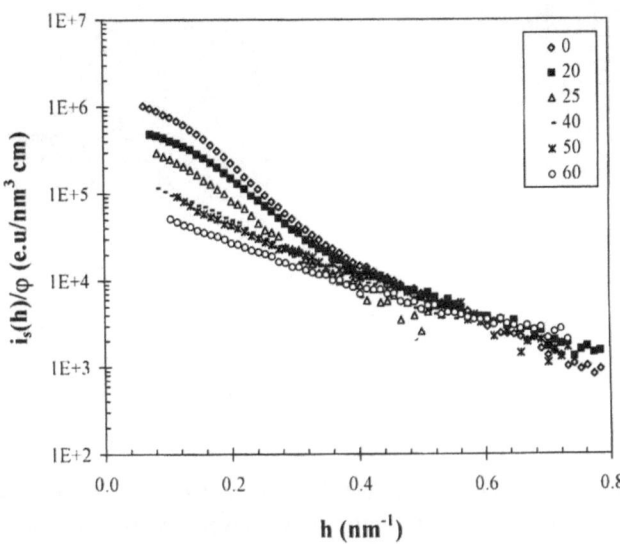

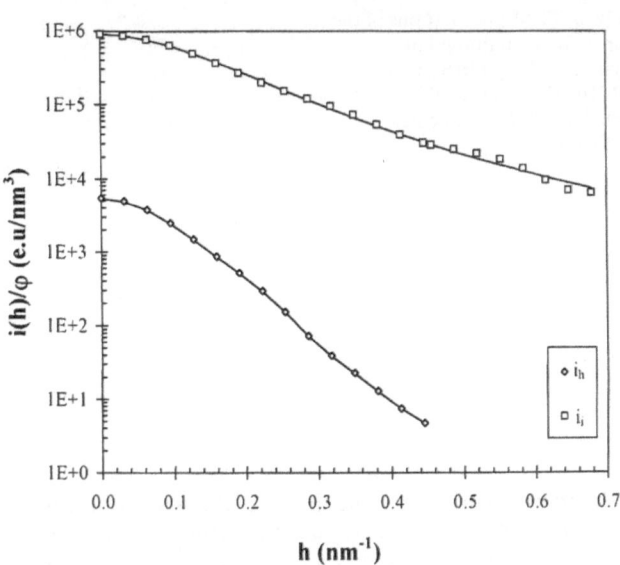

Fig. 4 Smeared SAXS scattering intensities obtained from 1.5 vol% copolymer latex at different contrasts. Legends indicate the weight percentage of sucrose in the solution.

Fig. 6 Desmeared homogeneous and heterogeneous basis functions. Solid lines represent the theoretical intensities.

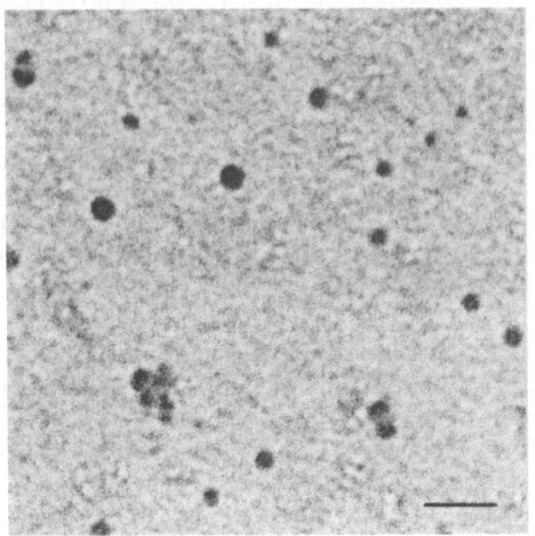

Fig. 5 TEM micrograph of the styrene/PBBA copolymer latex particles. Bar indicates 100 nm

for which the volume average diameter is 26.7 ± 1.3 nm and the number average radius is 22.6 ± 1.1 nm.

The desmeared heterogeneous function $i_i(h)$ is displayed in the upper curve of Fig. 6. Although a direct measurement at $h = 0$ is impossible, the trend of the angular dependence in the lowest measured angles suggests that the $i_i(h)$ function does not vanish at zero angle. This is an indication to the presence of irregular inner structure,

namely two phases that are randomly distributed within the particle. Moreover, this function could be easily fit by the theoretical form expected for such a structure, as shown by the solid line. The fitting yields values of $\xi = 4.68$ nm for the characteristic length of the inner structure, and mean-square density fluctuation was found to be $\bar{\eta}^2 \equiv \phi_1(1 - \phi_1)(\rho_1 - \rho_2)^2 = 352.3$ e.u./nm^6. However, although the value of the average density $\bar{\rho} = \phi_1\rho_1 + (1 - \phi_1)\rho_2$ is known, the SAXS analysis cannot give a unique solution ρ_1, ρ_2 and ϕ_1 unless one of the parameters can be determined by other means.

Due to the statistical nature of the random copolymerization reaction, each copolymer chain consists of sequences of various lengths of the repeating units, where the mean sequence length of each component is determined by the overall copolymer composition. Small angle scattering techniques are not sensitive to heterogeneity on the molecular level [2]. Yet, if long sequences of one monomer are formed, they might phase-separate and form a discrete phase, having a different electron density than the surrounding copolymer. Thus, phase separation might be an inherent property of the copolymer, and if this is the case, it is likely that the discrete phase mainly contains styrene segments, which is the major component in the studied system. Another possibility is that long sequences are formed as a result of composition drifts: the more reactive monomer disappears faster, leaving a higher concentration of the less reactive monomer which will polymerize later during the polymerization cycle. The composition of the formed copolymer obviously changes in the opposite

Progr Colloid Polym Sci (1997) 103:261–267
© Steinkopff Verlag 1997

direction. The reactivity ratios of PBBA and styrene are not known, but the kinetic data measured by Yuan et al. [34] suggest that PBBA is more reactive. Thus, composition drifts are also expected to induce longer styrene sequences at high conversions.

Although it is hard to decide which of the two possible mechanisms dictates the phase separation, they both lead to the same description: A major phase of PBBA/styrene copolymer, and a minor phase of polystyrene. The SAXS analysis results were re-examined by setting a value of $\rho_1 = 339.8$ e/nm^3 for the polystyrene phase, which further allowed direct calculation of the other parameters. This yields values of $\phi_1 = 0.08$ and $\rho_2 = 408.9$ e/nm^3. Namely, the polystyrene phase is indeed the minor one, occupying 8 vol% of the particle, and the major phase has a slightly higher electron density than the average value. The characteristic length of the minor phase was found to be $\xi_1 = \xi/(1 - \phi_1) = 5.1$ nm and the characteristic length of the major phase was $\xi_2 = \xi/\phi_1 = 58.4$ nm. This suggests that the inner structure of the copolymer can be described as a continuous copolymer phase, its composition being slightly richer in PBBA, in which domains of pure polystyrene are randomly distributed. The characteristic length of the domains is significantly smaller than the radius of gyration of a polymer chain (about 15 nm), supporting the suggestion of phase separation of long sequences, while the characteristic length of the copolymer phase roughly corresponds to the particle's diameter.

References

1. Balta-Calleja FJ, Vonk CG (1989) In: X-ray Scattering of Synthetic Polymers. Elsevier, New York
2. Glatter O, Kratky O (1982) In: Small Angle X-ray Scattering. Academic Press, London
3. Stuhrmann HB (1974) J Appl Crystallogr, 7:173
4. Bianco H, Narkis M, Cohen Y (1997) J Appl Crystallogr, submitted for publication
5. Guinier A, Fournet G (1955) In: Small-Angle scattering of X-rays. Wiley, New York
6. Bianco H, Narkis M, Cohen Y (1996) J Polymer Sci Polymer Phys, 34:2775
7. Debye P, Anderson HR, Brumberger H (1957) J Appl Phys 28:679
8. Porod G (1951) Kolloid-Z 124:83
9. Narkis M, Talmon Y, Silverstein M (1985) Polymer 26:1359
10. Jönsson JE, Hassander H, Jansson LH, Törnell B (1991) Macromolecules 24:126
11. Winnik MA (1984) Polymer Eng Sci 24:87
12. Yang SI, Klein A, Sperling LH (1989) J Polymer Sci Part B: Polymer Phys 27:1649
13. Mills MF, Gilbert RG, Donald HN, Rennie AR, Ottewill RH (1993) Macromolecules 26:3553
14. Dabdub D, Klein A, Sperling LH (1992) J Polymer Sci Part B: Polymer Phys 30:787
15. Hergeth WD, Schmutzler K, Wartewig S (1990) Makromol Chem, Macromol Symp 31:123
16. Beyer D, Lebek W, Hergeth WD, Schmutzler K (1990) Colloid Polymer Sci 268:744
17. Grunder R, Kim YS, Ballauff M, Kranz D, Müller HG (1991) Angew Chem Int Engl 30:1650
18. Dingenouts N, Ballauff M (1993) Acta Polymer 44:178
19. Grunder R, Urban G, Ballauff M (1993) Colloid Polymer Sci 271:563
20. Dingenouts N, Pulina T, Ballauff M (1994) Macromolecules 27:6133
21. Glatter O (1977) Acta Phys Austr 43:307
22. Glatter O (1977) J App Crystallogr 10:415
23. Cubbon RCP, Smith JDB (1969) Polymer 10:479
24. Cubbon RCP, Smith JDB (1969) Polymer 10:489
25. Bianco H, Cohen Y, Narkis M (1996) Polym Adv Tech 7:809
26. Poehlein GW (1988) In: Encyclopedia of Polymer Science and Engineering, Vol. 6, p. 1. Wiley, New York
27. Omi S, Kushibiki K, Iso M (1987) Polymer Eng. Sci 27:470
28. Dougher EP (1986) J Appl Polymer Sci 32:3051
29. Lee KC, El-Aasser MS, Vanderhoff JW (1992) J Appl Polymer Sci 45:2207
30. Lee KC, El-Aasser MS, Vanderhoff JW (1992) J Appl Polymer Sci 45:2221
31. Šňupárek J Jr, Krška F (1977) J Appl Polymer Sci 21:2253
32. van Doremaele JHJ, Schoonbrood HAS, Kurja J, German AL (1992) J Appl Polymer Sci 45:957
33. Bianco H, Narkis M, Cohen Y (1996) Macromolecules submitted for publication
34. Yuan Y, Siegmann A, Narkis M (1996) J Appl Polymer Sci 60:1475

Progr Colloid Polym Sci (1997) 103:268–279
© Steinkopff Verlag 1997

M.M. Lipp
K.Y.C. Lee
J.A. Zasadzinski
A.J. Waring

Protein and lipid interactions in lung surfactant monolayers

Received: 12 December 1996
Accepted: 13 December 1996

M.M. Lipp · K.Y.C. Lee ·
Dr. J.A. Zasadzinski (✉)
Department of Chemical Engineering
University of California
Santa Barbara, California 93106-5090, USA

A.J. Waring
Department of Pediatrics
King/Drew University Medical Center and
Perinatal Laboratories
Harbor-UCLA School of Medicine
Los Angeles, California 90059, USA

Abstract Human lung surfactant protein SP-B and its amino terminus ($SP-B_{1-25}$) alter the phase behavior of palmitic acid (PA) monolayers by inhibiting the formation of condensed phases and creating a new fluid PA-protein phase. This fluid phase increases the compressibility of the monolayers by forming a network that separates condensed phase domains at coexistence and persists to high surface pressures. The network changes the monolayer collapse nucleation from a heterogeneous to a more homogeneous process through isolating individual condensed phase domains. This results in higher surface pressures at collapse, and monolayers easier to respread on expansion, factors essential to the in vivo function of lung surfactant. The network is stabilized by low line tension between the coexisting phases as confirmed by the formation of extended linear domains or "stripe" phases. Similar stripes are found in monolayers of fluorescein-labeled $SP-B_{1-25}$, suggesting that the reduction in line tension is due to the protein. Comparison of isotherm data and observed morphologies of monolayers containing $SP-B_{1-25}$ with those containing the full SP-B protein shows that the peptide retains most of the native activity of the protein, which may lead to cheaper and more effective synthetic replacement formulations.

Key words Langmuir trough – isotherms – fluorescence microscopy – Brewster angle microscopy

Introduction

Basic interfacial science and engineering principles can be applied to fundamental problems in biology to bring a new perspective to identifying the mechanisms of disease. Work in our laboratory has involved the study of lung surfactant monolayers. Lung surfactant (LS) is a complex mixture of lipids and proteins that lines the alveoli and allows the lungs to function properly [1]. LS consists primarily of the saturated phospholipid dipalmitoylphosphatidylcholine (DPPC), along with significant amounts of unsaturated and anionic phospholipids such as phosphatidylglycerols and lesser amounts of neutral lipids such as palmitic acid (PA) and cholesterol. LS also contains four proteins; two of these, termed SP-B and SP-C, are hydrophobic and surface active. LS works by both lowering the surface tension inside the lungs to reduce the work it takes to inhale and by stabilizing the lungs through varying the surface tension as a function of lung volume. To accomplish this, the LS mixture must adsorb rapidly from the type II cells lining the alveoli where LS is produced to the air/fluid interface of the alveoli. Once at the interface, LS must form a monolayer that can both achieve low surface tensions upon compression and vary the surface tension as a function of the alveolar radius. This monolayer must also be capable of respreading rapidly if it collapses due to over-compression during exhalation. A failure of this system, either due to insufficient levels of LS or inactivation of existing LS can lead to respiratory distress syndrome (RDS), a potentially lethal disease in both adults and premature infants [2].

When RDS occurs, the lack of a properly functioning LS system results in a progressive failure of the lungs, which is manifested clinically by atelectasis (collapsed alveoli), decreased lung compliance (stiff lungs), decreased functional residual capacity (FRC, a measure of the amount of air left in the lungs after exhalation), systemic hypoxia (oxygen starvation), and lung edema (bleeding in the lungs) [3]. Neonatal RDS (nRDS) is known to be caused by a lack of sufficient surfactant levels (surfactant-deficient infants typically have less than 5 mg/kg of LS in their lungs, while typical healthy newborns have approximately 100 mg/kg). There are over 40 000 cases of neonatal RDS diagnosed in the United States annually, resulting in thousands of deaths [3]. Additionally, nRDS has traditionally been very difficult and costly to treat, as affected infants require intensive care for extended periods. Up until 1989, the primary treatment method available for neonatal RDS has been forced lung ventilation [4]. However, the large ventilation pressures needed to overcome the high alveolar surface tension of nRDS infants often complicates matters by causing mechanical barotrauma. This ventilation-induced trauma can result in lung leakage and hemorrhaging, rupture and necrosis of lung tissues, injury of the epithelial cells lining the alveoli, and permanent damage to the lungs such as bronchopulmonary dysplasia (BPD, a chronic lung disease most likely due to ventilation-induced barotrauma and is characterized by excess CO_2 levels in the blood and hypoxia, as well as an abnormal lung morphology that persists long after birth) if the infant survives [4]. Lung hemorrhaging is particularly problematic, as soluble blood serum proteins such as albumin are believed to inactivate any LS present in the alveoli; this can result in a downward cycle wherein ventilation actually worsens the condition.

In 1989, the FDA approved two formulations for the treatment of neonatal RDS by surfactant replacement therapy (SRT), in which exogenous surfactant mixtures are administered directly into the lungs of affected infants [5]. The first of these, called Survanta, consists of an organic extract of bovine LS lipids and hydrophobic proteins supplemented with synthetic palmitic acid (PA) and tripalmitin. The second formulation, called Exosurf, contains DPPC combined with synthetic, nonbiological emulsifying lipids with no relation to natural LS. Administration of mixtures such as these to affected infants has been proven to be a very effective treatment method for nRDS [6]. Benefits of treatment include: improvement in systemic oxygenation, reduced need for ventilation, more uniformly inflated lungs, increased lung compliance, increased

stability during deflation, and an increased FRC. The initiation of replacement therapies has coincided with a significant decline in the mortality rate of nRDS infants. SRT has been shown to reduce mortality rates by 30-50% for nRDS infants, and it has been estimated that 80% of the decline in the infant mortality rate of the United States between 1989 and 1990 could be attributed solely to the use of surfactant [6, 7]. SRT has also resulted in a significant savings for the treatment of nRDS infants. In a recent study examining the clinical effectiveness and financial ramifications of SRT, it was shown that inflation-adjusted charges per survivor declined by 10%, whereas the cost of care for each infant who died declined by over 30% after initiation of SRT, resulting in an estimated net saving of approximately $90 million per year [6].

However, further improvements in SRT formulations have been hindered by the lack of a fundamental understanding of how the LS system works and what the specific roles of each of the individual components are. Proposed replacement formulations can be divided into four classes: (i) natural LS extracts, (ii) modified LS extracts (natural extracts supplemented with synthetic lipids), (iii) synthetic formulations modeled on natural LS, and (iv) synthetic formulations with no relation to natural LS. Natural LS extracts have been shown to be effective both in vitro and in vivo, but human sources are limited, while animal sources are difficult and expensive to purify and pose the risk of containing viral or proteinaceous contaminants. For the case of modified LS extracts the same concerns exist, and although supplementing LS extracts with PA typically results in an improvement of activities the mechanism behind this improvement is unknown.

As with any natural product, both natural and modified LS extracts can have a wide variability in composition from batch to batch. These concerns have led to the study of synthetic LS formulations for which the composition can be better defined. Mixtures containing synthetic LS lipids with highly purified natural proteins such as SP-B and SP-C have been shown to be of comparable or better activity than many natural extracts; however a fundamental rationale for choosing the ratio of lipids to proteins for such mixtures besides their overall apparent activities in vivo is still lacking. Furthermore, LS proteins are difficult and expensive to obtain in a highly-purified form, and there does not yet exist a suitable host-vector system for the production of these proteins on a large scale through genetic engineering techniques (which is a common problem for surface-active proteins).

An ideal replacement formulation would be a mixture of synthetic lipids, in a ratio based on a good understanding of their individual functions in LS, combined with simple peptide sequences, produced via solid-state synthesis techniques, which capture the full activity of the native LS mixture. Such a mixture could be easily and cheaply produced, and the composition could be tailored to optimize the properties of the mixture for the treatment of specific cases, ranging from rapid distribution and spreading for nRDS infants to an increased resistance to inactivation by serum proteins in barotrauma cases. Currently, SRT is not yet widely available in medical hospitals and centers around the United States [6]. Although the mortality rate for nRDS infants has been declining since the advent of successful surfactant replacement strategies, the incidence of low birth weight infants has been steadily rising [7]. The availability of standardized replacement formulations with a fully quantified mechanism of action could lead to a more widespread use of SRT to help save the lives of both infants and adults. The key to designing effective replacement formulations is a thorough understanding of the function and activity of each of the LS components.

Although it was known as early as 1929 that the surface tension in normal lungs was low [8], LS itself was not identified as a substance until 1959 by Avery and Mead [9]. Using a Langmuir trough, they demonstrated that DPPC was the key component for the surface tension lowering effect. It was also discovered that nRDS infants had significantly lower amounts of LS than normal infants. However, the administration of aerosolized DPPC directly into the lungs of nRDS infants was shown to have no beneficial effect whatsoever in most cases (due to its rigidity at physiological temperatures, DPPC does not spread well at the air/water interface; it also does not adsorb well to the air/water interface from solution), which indicated the importance of the non-DPPC components of LS [10]. Subsequent studies showed that whole LS obtained from mature animals could be used as an effective LS replacement in both premature animals and infants [11,12]. Several in vitro and in vivo studies on both natural LS extracts and synthetic lipid mixtures demonstrated that the unsaturated and anionic lipids in LS as well as the proteins act to fluidize the DPPC-rich mixture, allowing it to adsorb rapidly to the air/fluid interface of the alveoli [13–16]. The fact that these fluidizing components are individually incapable of forming monolayers that can attain low surface tensions has led to the hypothesis that these components are "squeezed-out" of the LS monolayer, leaving it enriched in DPPC [17–19]. However, removal of all of these fluidizing components would prevent the DPPC-enriched LS monolayer from respreading rapidly upon the event of collapse. Moreover, it has been shown that the presence of LS proteins in pre-formed monolayers of DPPC greatly enhances the rate of adsorption of new materials to the interface [20]. A systematic study of monolayers of unsaturated or anionic LS lipids containing surface-active LS proteins can help in the

understanding of the LS system, as well as determine the mechanism by which additives such as PA enhance the activity of LS extracts.

Our recent work has centered on the elucidation of the roles of both PA and SP-B protein in both the natural LS system and in replacement surfactant formulations. As a large percentage of the fluidizing LS lipids are anionic, while the surface-active LS proteins such as SP-B have a net positive charge, it is plausible that these components may interact synergistically in LS monolayers to increase their surface activity. Although PA has been shown to greatly improve the activity of replacement formulations, the mechanism of action of this component is still not understood [14,21]. While the phase behavior of pure PA is well known, it does not indicate how PA "improves" the properties of LS mixtures. It has been postulated that PA may interact selectively with SP-B in LS monolayers, resulting in the retention of both components in the monolayer up to collapse. This would possibly allow these components to perform such functions as enhancing the respreading rate of material from the collapsed phase, as well as increasing the rate of adsorption and incorporation of material from the subphase. The study of mixed PA/SP-B monolayers will reveal the presence of any synergistic interactions between these components and may help explain how these components function in the whole LS monolayer.

Protein engineering can also be used to relate the amino acid sequence of SP-B protein to its activity in LS. Due to the current lack of a suitable expression system for the production of SP-B and the significant cost of synthesizing the native protein on the scales needed to treat RDS, work in our lab has focused on attempting to figure out what is the key functional moiety of the protein in the hopes of developing simpler sequences for replacement formulations. We have found that a 25 amino acid long sequence based on the amino-terminus of native SP-B possesses similar activity as that of the native protein both in vitro and in vivo [22,23]. It appears that the key to the activity of both the full-length and the shortened sequence is the presence of an amphipathic alpha helix with the positively charged residues aligned on one side of the helix and the hydrophobic residues on the other [24].

The goal of our work is to determine the specific mechanism of action of both PA and SP-B in LS. Using the information obtained from the study of simple binary mixtures, we hope to form a basis for the study of a complete synthetic LS mixture. While this is a common approach to studying multicomponent phase diagrams in chemical engineering and materials science, it is surprisingly uncommon in biology and medicine. Quite often, the number of components in a biological system is so large that this approach is impossible. However, LS is

Progr Colloid Polym Sci (1997) 103:268–279
© Steinkopff Verlag 1997

a relatively simple mixture and lends itself to a systematic approach. We have utilized direct monolayer imaging techniques such as fluorescence, polarized fluorescence, and Brewster angle microscopies to study the phase behavior and collapse mechanics of mixed PA and SP-B monolayers. We have discovered that synergistic interactions between SP-B and PA remove the driving force for squeeze-out of either component from LS monolayers, and explain how respreading of the post-collapse monolayer is facilitated. We have also manipulated the protein through the systematic introduction of point mutations to examine the effects of altering the hydrophobic/charge balance of the protein. This information can hopefully aid the rational design of synthetic LS formulations and the creation of simpler and cheaper peptide sequences to replace native LS proteins.

Experimental: visualizing lipid/protein monolayers

To study model LS monolayers, we have designed an integrated fluorescence, polarized fluorescence, and Brewster angle microscope/Langmuir trough assembly (an overview of this system is displayed in Fig. 1). Langmuir troughs have been used for the past 70 years to obtain surface tension vs. area isotherms of surfactant monolayers at the air/water interface. Surfactant monolayers can exist in several 2D phases as a function of the lateral density; at very high areas per molecule monolayers exist in a gas-like state, while monolayers with low areas per molecule exist in liquid-crystalline and solid phases. Unfortunately, isotherms can only provide macroscopic information concerning the phase state of the monolayer, the details of the nucleation and coexistence phenomena of the phase transitions cannot be determined (however, just as in 3D gases and liquids, a good understanding of the phase diagram is a necessary first step).

Several techniques have been developed over the last 15 years to visually probe the morphology of surfactant monolayers at the air–water interface. In fluorescence microscopy, a small amount of fluorescently labeled surfactant molecules is added to a monolayer; due to steric effects these tagged molecules tend to partition into less-ordered phases, which results in a visual contrast between coexisting phases [25–29]. Fluorescence microscopy has been used to determine domain sizes and shapes during phase transitions [25, 28, 30]. Polarized fluorescence microscopy (PFM) provides additional information on the lipid hydrocarbon chain ordering within condensed monolayers, especially in areas where the lipid hydrocarbon chains are tilted with respect to the surface normal [31, 32]. The interaction of the electric field vector of the polarized light with the absorption dipole moment of the

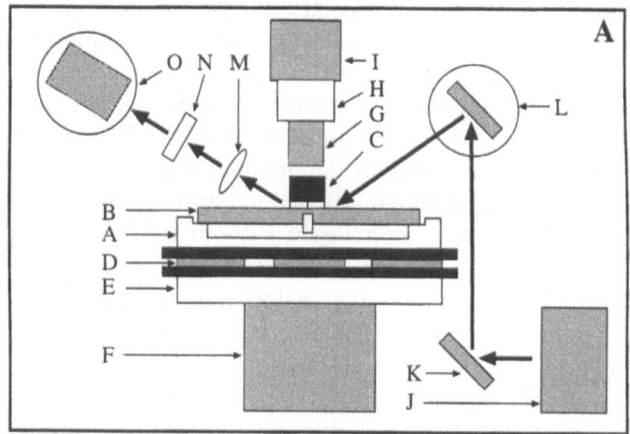

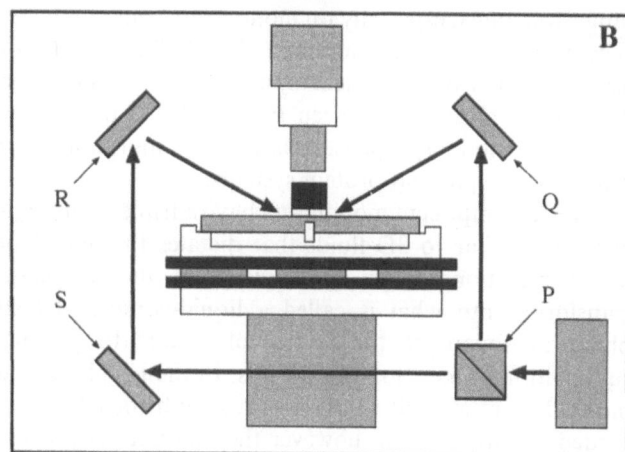

Fig. 1 A schematic of the FM/PFM/BAM assembly. **(A)** Configuration for dual FM/BAM operation. The labeled parts are: (A) Langmuir trough, (B) barrier, (C) surface pressure sensor, (D) thermoelectric elements, (E) water bath, (F) motorized *xyz* translation stage, (G) microscope objective, (H) mercury lamp/fluorescence filter cube assembly, (I) SIT camera, (J) laser, (K) mirror, (L) rotatable mirror, (M) lens, (N) polarizer cube (analyzer), (O) CCD camera (mounted on a rotatable stage). **(B)** Configuration for PFM operation. The parts specific to the PFM are (including the laser): (P) beam-splitter cube, (Q,R,S) mirrors. The bold arrows in both drawings denote the light path of the laser beam in the BAM and PFM modes. For the PFM mode, switching between the two beams shown in **(B)** from the beam-splitter rotates the plane of incidence of the beam with the surface by 180° and results in a reversal of contrast for regions of differing tilt directions in monolayers

fluorophore is dependent on the relative orientation of the molecule, hence using a polarized light source incident on the monolayer at an oblique angle as the excitation source allows for the detection of regions of different tilt directions.

A third optical microscopy technique, called Brewster angle microscopy (BAM), has only recently been developed [33, 34]. The benefits of BAM are that is provides information similar to fluorescence and polarized fluorescence microscopies without requiring the addition of fluorescent probes. A p-polarized light beam is weakly reflected by a monolayer when the surface is illuminated at the Brewster angle (about 53° from the surface normal for the air/water interface). The intensity of the reflected light is a function of the local state of the monolayer; variations in the refractive index of the monolayer result in image contrast due to different reflected intensities. Allowing for the study of these monolayers without the presence of fluorescently labeled molecules, BAM confirms that the fluorescent probe molecules at low mole percentages do not influence the phase behavior seen via FM and PFM, and provides additional information on local tilt and other anisotropic orientational effects. However, little fluorescence or Brewster angle microscopy has been done on lung surfactants, especially on monolayers containing LS-specific proteins. The combined FM, PFM, and BAM assembly provides an ideal system for examining the mechanism of action between PA and SP-B protein, as well as the monolayer phase behavior and morphology of other mixed lipid and protein systems.

Single-component monolayers have a triple point temperature similar to 3D fluids that dictates the nature of phase transitions. Below the triple point, the gas phase transforms into what is called a liquid-condensed (l.c.) phase, analogous to the reverse of a solid to gaseous sublimation process for 3D systems. In the l.c. phase, the molecules are aligned at the interface with their tails extended towards the air, however they do not yet possess any long-range positional order and the hydrocarbon chains are relatively disordered (although they can possess long-range orientational order, forming tilted phases with a uniform tilt direction over macroscopic dimensions). The terminus of the gas–l.c. transition is exhibited in the isotherm by a sharp increase in surface pressure upon compression, termed the lift-off point, which corresponds to the disappearance of the gas phase. Above the triple point, the gas phase transforms into a 2D liquid-like phase, termed a liquid-expanded (l.e.) or fluid phase. In this "fluid-like" phase, the molecules are randomly arrayed and free to diffuse, and the hydrocarbon chains are in the liquid state. This fluid phase has a much higher compressibility relative change in area per change in surface pressure (i.e. $-(-1/A)(\partial A/\partial \Pi)_T$) than the l.c. phase, as evidenced by a decreased absolute value of the slope in the isotherm. Under further compression, this fluid phase undergoes a first-order transition into the l.c. phase, seen by the appearance of a plateau in the isotherm (there is an equivalent of the Gibbs phase rule in two-dimensions which states that first-order transitions in single-component monolayers should occur at a constant surface pressure, although this postulate has been greatly debated).

At elevated pressures both below and above the triple point, the l.c. phase undergoes a second-order phase transition (indicated by a kink point in the isotherm) into a solid-like phase, referred to as either solid-condensed (s.c.) or crystalline depending on the degree of order of the hydrocarbon chains. In these phases, the molecules are close-packed and aligned in a lattice, possessing quasi-long range positional order. Compressing the monolayer below the minimum area per surfactant molecule in the crystalline state results in the collapse of the monolayer and formation of a 3D phase either above or below the monolayer. Although the formation of 3D phases may be favored under equilibrium conditions for monolayers at relatively low pressures, there typically exists a substantial energy barrier for the collapse process to occur. This makes collapse a kinetically driven process that requires the input of an activation energy, and depends on such factors as the rate of compression of the monolayer. The ability to resist collapse determines the maximum pressure (or minimum surface tension) a monolayer may achieve. Monolayers which require higher activation energies to initiate collapse can attain and maintain higher surface pressures than those requiring lower activation energies. This leads to lower ultimate surface tensions, and these monolayers are "better" lung surfactants.

Considering the reverse process, the structure and location of the collapsed phase and the reversibility of the collapse process determines whether the collapsed phase will reincorporate itself upon re-expansion of the monolayer. Monolayers that are more resistant to collapse (and can achieve high surface pressures) usually do not respread well; collapse is an irreversible process in these systems and the collapsed phase remains in a bulk 3D state at the interface or is lost into the subphase. For proper functioning, ideal LS monolayers should resist collapse (and thus achieve low surface tensions) yet respread rapidly and completely (which seems to be a dichotomy in behavior and is incompatible with single-component monolayers).

We have found that the addition of SP-B protein results in a drastic alteration of the phase behavior of PA monolayers [35]. The presence of protein fluidizes PA monolayers under all experimental conditions. Although Langmuir isotherms show changes in phase behavior of PA due to the presence of protein, they do not reveal the mechanisms by which the protein effects these changes. Similar changes in both the compressibility and collapse pressure of monolayers have been observed in mixed fatty acid and polycationic polymer systems [36]. This suggests that these effects may be more general in nature, and provides additional motivation for studying simpler peptide sequences for replacement formulations. Direct visualization of the phase transitions and collapse processes occurring in mixed PA/SP-B monolayers allows us to relate monolayer morphology to isotherm data which

helps us to elucidate the mechanisms behind this phenomenon.

Fluorescence microscopy of PA/SP-B$_{1-25}$ monolayers

Upon deposition of pure PA at an area per molecule of
approximately 60 Å^2 on a pure water subphase at 16°C,
the monolayer exists in a gas–l.c. coexistence region. As
shown in Fig. 2A, this is seen as a coexistence of circular
light gray l.c. domains in a dark gaseous background (the
probe used in this study, NBD-HDA, quenches when it
comes into contact with the subphase, making the gas
phase appear dark). At the liftoff point, the monolayer is
entirely in the l.c. phase, as seen by a lack of contrast in the
monolayer (Fig. 2B). Upon further compression, the
monolayer transforms from the l.c. phase into a solid
phase of homogeneous contrast. When 20 wt% SP-B$_{1-25}$
is added to these PA monolayers, a new bright fluid phase
can be seen to form (in addition to the gas and l.c. phases)
at high areas per molecule. This fluid phase persists past
the liftoff point, forming a network that partitions the
PA-rich l.c. phase into small, circular domains (Fig. 2C).
The network persists through the solid phase transition,
segregating the solid domains prior to collapse.

We observed a similar trend at temperatures above
the triple point of PA on a pure water subphase. At
such temperatures, pure PA monolayers exist as a
homogeneous bright fluid phase at the liftoff point. Upon

compression, the monolayers proceed through a first-order fluid to l.c. transition, evidenced by the formation of
a plateau region in the isotherm. During this transition,
the l.c. phase can be seen to nucleate from the bright fluid
phase background and grow into large, dark, circular l.c.
domains (Fig. 2D). These domains eventually pack together at high pressures, leading to the disappearance of
the l.e. phase and the formation of a homogeneous sheet of
l.c. phase. This l.c. phase again transforms into a homogeneous solid phase at low areas per molecule (Fig. 2E).
The addition of SP-B can be seen to greatly affect the fluid
to l.c. phase transition. The protein decreases the size and
increases the nucleation density of the l.c. phase (Fig. 2F).
This again results in a partitioning of the l.c. phase into
small, circular domains as seen at lower temperatures. In
the presence of protein, the bright fluid phase remains
upon transition of the l.c. phase into the solid phase and
persists to high surface pressures.

We have discovered that this partitioning of the condensed phase domains by the protein-induced network has
a drastic effect on the collapse processes occurring in these
monolayers. For pure PA monolayers on a pure water
subphase, compressing past the limiting area per molecule
in the crystalline state results in the collapse of the monolayer and the formation of 3D bulk phases. As seen in
Figs. 3A and C, both below and above the triple point,
collapse occurs heterogeneously at isolated points across
the monolayer and results in the formation of large, rigid,
dendritic-like collapsed-phase domains. This process is

Fig. 2 Fluorescence images of PA and PA/SP- B$_{1-25}$ films containing 0.5 mol% NBD-HDA on a pure water subphase. (A–C) Images taken at
16°C of (A) a pure PA film in the l.c.-g coexistence region (the l.c. domains are light gray, the gas phase is black), (B) the same film at the liftoff
point, consisting entirely of l.c. phase, and (C) a PA/20 wt% SP-B$_{1-25}$ at the liftoff point, showing the existence of a new, bright fluid phase
which breaks up the l.c. domains. (D–F) Images taken at 28°C of (D) a pure PA film in the fluid-condensed coexistence region (area per PA
molecule 35 Å^2), (E) upon further compression of the same film into the solid phase, and (F) a PA/20 wt% SP-B$_{1-25}$ film in the fluid-condensed
coexistence region (area per PA molecule 35 Å^2), showing a decreased size and increased nucleation density of the condensed domains

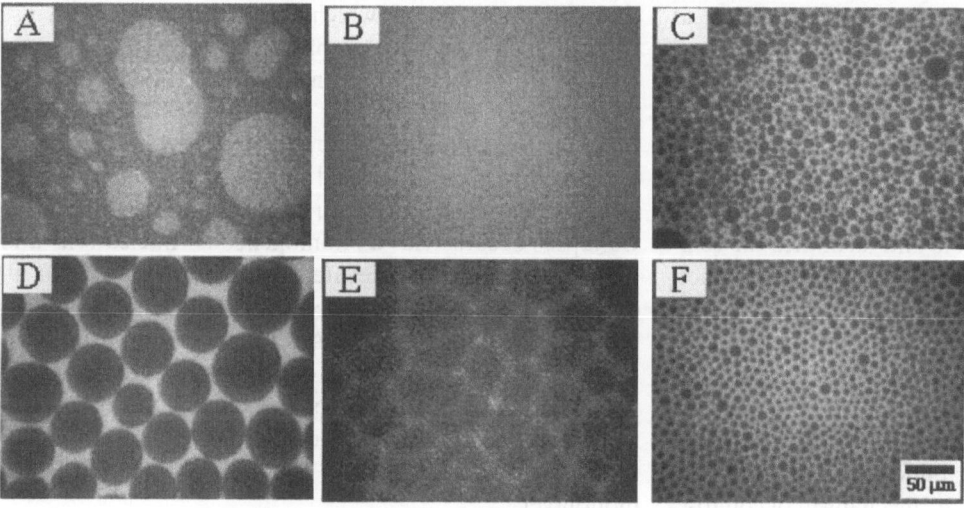

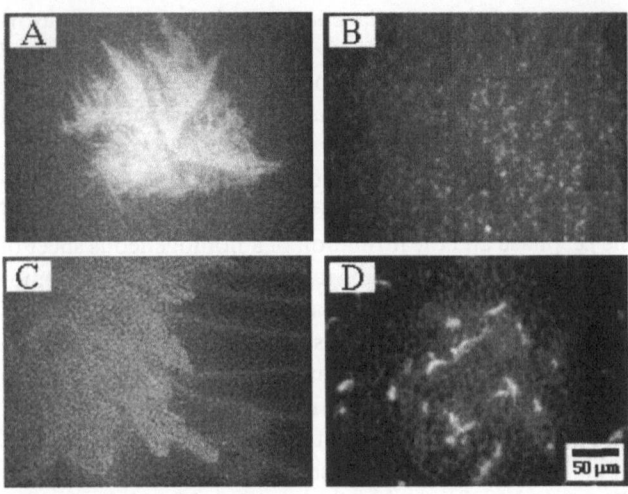

Fig. 3 Fluorescence images of the collapse behavior of PA and PA/SP-B$_{1-25}$ monolayers containing 0.5 mol% NBD-HDA on a pure water subphase. (A, B) Images taken at 16°C of the collapsed phase domains for (A) pure PA and (B) PA/20 wt% SP-B$_{1-25}$. (C, D) Images taken at 28°C of the collapsed phase domains for (C) pure PA monolayer and (D) PA/20 wt% SP-B$_{1-25}$

very irreversible, with subsequent expansion and recompression of the monolayer resulting in a large hysteresis, which implies a loss of material from the surface. The presence of the protein network prevents the heterogeneous nucleation process from occurring; each partitioned domain must now nucleate collapse independently and homogeneously. This requires a larger activation energy, which results in a rise in the collapse pressure over that of pure PA. This phenomenon is the two dimensional analog of the classic experiments of Turnbull, who showed that many simple metallic liquids could be undercooled far below their thermodynamic freezing points [37]. By subdividing the liquid into micron-size droplets, Turnbull was able to reduce the likelihood of heterogeneous nuclei in a given droplet, leading to homogeneous nucleation at large undercooling. Similar effects have been observed for supercooling water in emulsion droplets, polymer gels, or porous media [38, 39]. Additionally, since collapse must occur independently in each domain, the resulting collapsed phase domains are significantly smaller and more evenly distributed (Fig. 3B and D), which makes the collapse event more reversible and facilitates the reincorporation of this phase into the monolayer upon re-expansion.

Although the phase behavior of PA is altered for monolayers on physiological, buffered saline subphases (pH = 6.9, 0.15 M NaCl), the presence of SP-B protein has similar effects to those seen on a pure water subphase. For pure PA monolayers, the buffered saline subphase conditions result in a change in the mechanism of collapse.

The ionization of the headgroups inhibits the formation of bulk collapsed phases above the monolayer, where the electrostatic repulsion would be high. As a result, PA monolayers require a higher pressure to nucleate collapse, and collapse occurs via a bulk fracturing mechanism in which the monolayer cracks cooperatively over large length scales (Fig. 4A). This process is also irreversible, with the fractures persisting upon expansion. Below the triple point, the addition of SP-B protein again leads to the formation of a bright fluid phase which breaks up and partitions the condensed phase domains (Fig. 4B). In the presence of protein, the collapse mechanism shifts to a more homogeneous process. This network prevents the fracture event from occurring and shifts the mechanism to a nucleation and growth process, resulting in the appearance of a uniform distribution of small collapsed phase domains across the film (Fig. 4C).

Above the triple point on a buffered saline subphase, PA again undergoes a first-order fluid to l.c. transition, although the l.c. domains are smaller and more numerous due to the higher electrostatic repulsion within the l.c. phase nuclei (Fig. 4D). Domain shapes at a particular point of compression of a monolayer usually result from a balance between electrostatic repulsion and line tension; electrostatic repulsion favors the elongation of domains of the more condensed phase due to their higher lateral density, while line tension favors circular domains which minimize the perimeter. This balance most likely also has an influence on the size of the nuclei in monolayers undergoing a phase transition; the high electrostatic density in charged monolayers results in a smaller critical radius for the nuclei (with a corresponding increase in the required supersaturation pressure, assuming the line tension remains unchanged). However, upon further compression, the l.c. domains are actually seen to fuse together (Fig. 4E), eventually forming a contiguous sheet of l.c. phase prior to the solid transition which fractures upon collapse (Fig. 4F). The presence of protein reduces the size and increases the nucleation density of the l.c. domains at the fluid to l.c. transition (Fig. 4G). Upon further compression, the l.c. domains undergo limited fusion (Fig. 4H), however the size of the fused domains remains small and a bright phase network again persists right up to collapse. Collapse occurs via a homogeneous nucleation and growth mechanism at elevated pressures (Fig. 4I). For all cases, the increase in collapse pressure and the ease of the respreadability of the collapsed phase has important implications on the functioning of LS monolayers, effectively removing the driving force for the squeeze-out of components with low individual collapse pressures from LS monolayers and facilitating the respreading of the LS monolayer.

Progr Colloid Polym Sci (1997) 103:268–279
© Steinkopff Verlag 1997

Fig. 4 Fluorescence images of PA/SP-B$_{1-25}$ films containing 0.5 mol% NBD-HDA on a buffered saline subphase. (A–C) Fluorescence images at 16°C of (A) bulk fracturing collapse of a pure PA monolayer, (B) existence of a protein-rich bright phase network at elevated pressures of a PA/20 wt% SP-B$_{1-25}$ monolayer, and (C) collapse of the same monolayer via a homogeneous nucleation and growth mechanism. (D–F) Images of a pure PA film at 25°C in (D) the fluid-condensed coexistence region, (E) after fusion of individual domains into a sheet of condensed phase upon further compression, and (F) the post-collapse fracturing of the film. (G–I) Images of a PA/20 wt% SP-B$_{1-25}$ film at 25°C (G) in the fluid-condensed coexistence region (area per PA molecule 35 Å^2) (H) after limited fusion of the condensed domains, and (I) post-collapse, showing the nucleation and growth of small collapsed phase domains

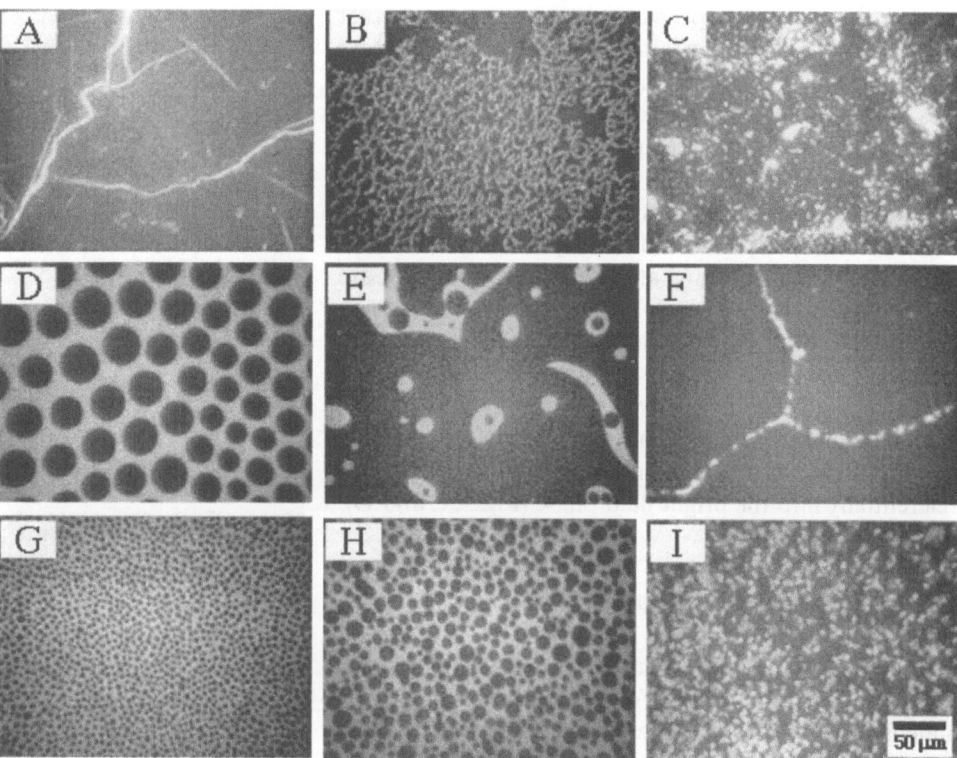

The use of fluorescently labeled protein allows us to further investigate the specific mechanisms by which the protein alters the phase behavior of PA. In all cases shown so far, the formation of a bright phase network surrounding the small condensed phase domains results in the existence of a large amount of line interface between the phases. In analogy to the interfacial tension that exists between two bulk 3D phases, there exists a line tension between coexisting phases in two dimensions. If the line tension between two phases is high, there would be a large energetic cost upon increasing the perimeter between the phases while maintaining a constant area ratio between the phases. However, for this system it appears that the increase in perimeter between the fluid and condensed phases is allowed to occur due to the fact that the protein acts to lower the line tension between these phases. Direct evidence of this is obtained on buffered saline subphases at low temperatures and high areas per molecule, in which case stripe phases are formed in monolayers of PA/20 wt% SP-B (Fig. 5A). The formation of stripe phases in the presence of protein indicates a low line tension between the coexisting phases in these films (for coexisting domains in Langmuir monolayers, the characteristic width of a domain possessing an increased electrostatic density with respect to its surroundings scales as $\exp(\lambda/\mu^2)$, where λ is the line tension and μ is the difference in dipole density

Fig. 5 Fluorescence images of stripe phases and partitioning characteristics of SP-B$_{1-25}$ and F-SP-B$_{1-25}$ in PA monolayers. (A) A monolayer of PA/20 wt% SP-B$_{1-25}$ on a buffered saline subphase at an area per PA molecule of 63 Å^2 showing the existence of stripe domains. (B) An image of a fluorescein-SP-B$_{1-25}$ (F-SP-B$_{1-25}$) monolayer on a pure water subphase at 23°C showing the formation of a "stripe" phase at low surface pressures. (C, D) Images of a monolayer of PA/20 wt% F-SP-B$_{1-25}$ also containing 0.5 mol% Texas-red DPPE on a pure water subphase at 28°C showing the fluorescence from (C) the lipid probe and (D) the fluorescently labeled protein

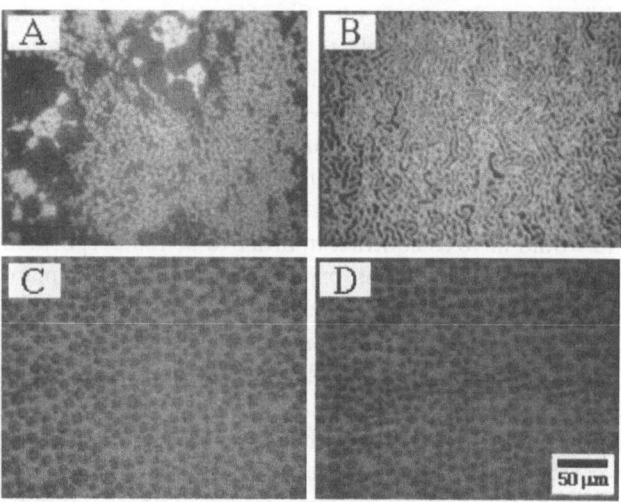

between the phases; stripe phases can occur when either λ is very small or μ is very large). Similar stripe phases are seen in pure films of fluorescein-labeled SP-B, which indicates that the low line tension is a direct effect of the presence of the protein (Fig. 5B). Using fluorescently labeled SP-B also allows us to pinpoint the location of the protein in the mixed films. A dual-probe experiment, in which both a lipid probe and SP-B protein labeled with a fluorescent tag emitting at different wavelengths are used, allows us to see which phase the protein tends to partition into during phase transitions. Rapidly switching between two filter cubes specific to each of the fluorophores in the film allows us to see the fluorescence from both the lipid phases and the protein at the same location. In all cases, the protein is seen to partition preferentially into the bright fluid phase (Figs. 5C and D).

Polarized fluorescence microscopy of PA/SP-B₁₋₂₅ monolayers

PFM gives us additional information on the collapse behavior of these monolayers. As previously described, for condensed lipid phases in which the molecules are tilted in a specific direction with respect to the surface normal, the use of polarized light as the fluorescence excitation source results in contrast between regions of different tilt directions. PFM allows us to determine the influence of lipid molecule tilt on both the nature of the nucleation of collapse and the lack of reincorporation of the collapsed phase upon re-expansion. For pure PA monolayers at 16°C, the l.c. phase consists of domains of differing tilt directions bordering at defect lines and points (Fig. 6A). If these defect points are not allowed to anneal out when the film transforms to the untilted solid phase prior to collapse, they may act as heterogeneous nucleation sites and allow collapse to occur at lower pressures. In the presence of protein, the condensed phases are individually either untilted or of uniform tilt (Fig. 6C). This removes the influence of any tilt defects on the collapse nucleation process and requires each domain to nucleate collapse homogeneously. Furthermore, PFM reveals that the collapsed phase itself is tilted, while the underlying monolayer is untilted (Fig. 6B). This means that the collapsed phase is at a lower packing density with respect to the monolayer, and would have to contract in order to be reincorporated into the monolayer upon expansion, which may explain why these domains do not readily reincorporate into the monolayer upon re-expansion. For the protein-containing films, the collapsed phase domains themselves do not appear to be tilted (Fig. 6D), which in combination with their smaller sizes may ease their reincorporation into the monolayer. A similar process occurs at temperatures above the triple point; the large collapsed phase domains

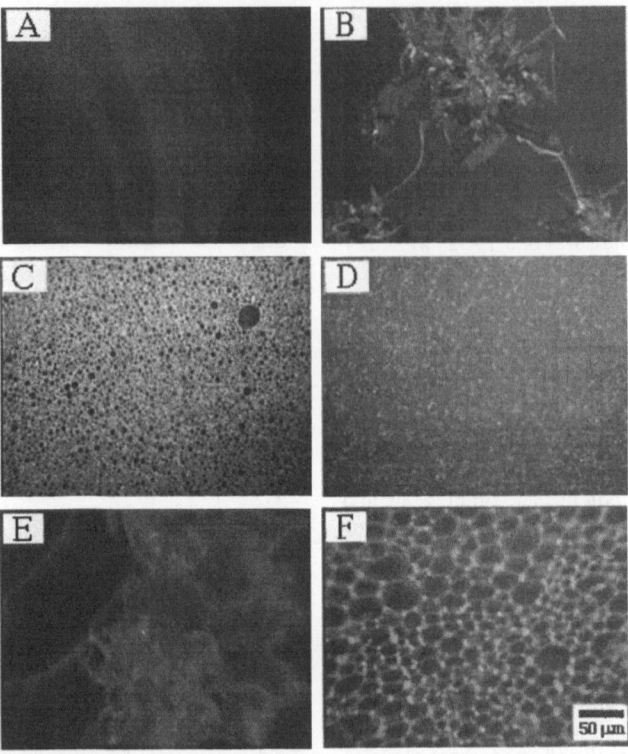

Fig. 6 Polarized fluorescence images of the collapse behavior of PA and PA/SP-B₁₋₂₅ monolayers containing 0.5 mol% NBD-HDA. (A, B) Polarized fluorescence images of a PA film on a pure water subphase at 16°C (A) at a pressure of 15 mN/m in the liquid-condensed phase (with contrast resulting from domains of differing tilt direction), and (B) after collapse of the monolayer, showing the growth of a tilted collapsed phase domain. (C, D) Polarized fluorescence images of a PA/20 wt% SP-B₁₋₂₅ film on a pure water subphase at 16°C (C) at a pressure of 15 mN/m (showing a lack of tilt contrast within the small condensed domains), and (D) after collapse of the monolayer, showing the homogeneous distribution of small collapsed phase domains. (E, F) Polarized fluorescence images of collapsed phase domains for films on a pure water subphase at 28°C for (E) pure PA and (F) PA/20 wt% SP-B₁₋₂₅

for pure PA are again tilted (Fig. 6E), while in the presence of 20 wt% SP-B₁₋₂₅ the collapsed phase domains are smaller and do not grow past the boundaries of the original solid phase domains (Fig. 6F).

Brewster angle microscopy of PA/SP-B₁₋₂₅ monolayers

A significant concern with the use of fluorescence and polarized fluorescence microscopies for the study of surfactant monolayers has been the possibility of artifacts due to the addition of foreign probe molecules to the system. BAM provides equivalent information to PFM without requiring the addition of a foreign probe molecule. Thus, we have used BAM to confirm that the probe molecules, present at low concentrations, do not influence the

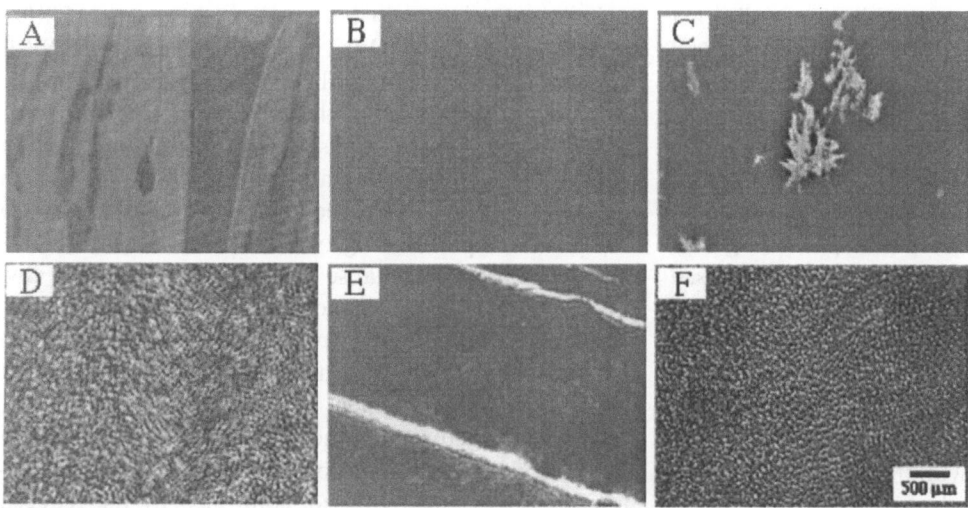

Fig. 7 Brewster angle microscope images of the collapse behavior of PA and PA/SP-B$_{1-25}$ monolayers. (A–C) Brewster angle microscope images of a film of pure PA on a pure water subphase at 16°C (A) in the l.c. phase (with contrast arising from the existence of domains of differing tilt direction), (B) upon transition to the solid phase (showing the loss of tilt contrast), and (C) post-collapse (showing the presence of tilt contrast within the collapsed phase domain). (D) Brewster angle microscope image of a PA/20 wt% SP-B$_{1-25}$ film on a pure water subphase at 16°C showing the growth of the uniform collapsed phase. (E, F) Images from monolayers on buffered saline subphases at 16°C showing (E) bulk fracture of a PA monolayer and (F) nucleation and growth of a collapsed phase for a PA/20 wt% SP-B$_{1-25}$ film

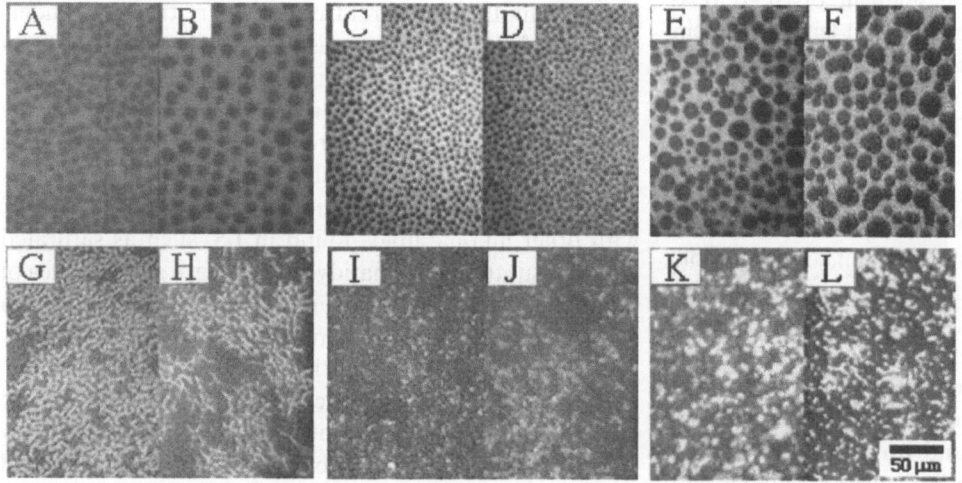

Fig. 8 Comparison of PA films containing SP-B$_{1-25}$ and SP-B$_{1-78}$ (also containing 0.5 mol% NBD-HDA) under various experimental conditions. (A, B) Fluorescence images of PA/SP-B films on a pure water subphase at 28°C in the l.e.–l.c. coexistence region containing (A) 20 wt% SP-B$_{1-25}$ and (B) 20 wt% SP-B$_{1-78}$. (C, D) Fluorescence images on a buffered saline (0.15 M NaCl, pH = 6.9) subphase at 25°C in the fluid-condensed coexistence region containing (C) 20 wt% SP-B$_{1-25}$ and (D) 20 wt% SP-B$_{1-78}$. (E, F) Fluorescence images of limited fusion of lc domains PA/SP-B films on a buffered saline (0.15 M NaCl, pH = 6.9) subphase at 25°C in the fluid-condensed coexistence region containing (E) 20 wt% SP-B$_{1-25}$ and (F) 20 wt% SP-B$_{1-78}$. (G, H) Fluorescence images of stripe phases occurring in PA/SP-B films on a buffered saline subphase at 16°C containing (G) 20 wt% SP-B$_{1-25}$ and (H) 20 wt% SP-B$_{1-78}$. (I, J) Fluorescence images of collapsed phase domains occurring in PA/SP-B films on a pure water subphase at 16°C containing (I) 20 wt% SP-B$_{1-25}$ and (J) 20 wt% SP-B$_{1-78}$. (K, L) Images of PA/SP-B films post-collapse on a buffered saline subphase at 25°C containing (K) 20 wt% SP-B$_{1-25}$ and (L) 20 wt% SP-B$_{1-78}$

morphologies observed in lipid/protein films via fluorescence and polarized fluorescence microscopies, and also to obtain images at larger length scales. In all cases so far in our laboratory, BAM has provided completely analogous images of monolayers of various components of LS to those obtained with fluorescence and PFM. This confirmation is of particular importance for the study of collapse of PA/SP-B$_{1-25}$ films; sites of accumulation of

278
M.M. Lipp et al.
Protein and lipid interactions in lung surfactant monolayers

fluorescently labeled molecules could potentially act as heterogeneous nucleation sites. However, at all experimental conditions, BAM images of collapse show a similar shift in collapse mechanism as seen with fluorescence and polarized fluorescence microscopies. On pure water subphases, BAM images show the presence of tilt in the l.c. phase (Fig. 7A), the disappearance of this tilt contrast on transition to the solid phase (Fig. 7B), and the growth of large, isolated, tilted collapsed phase domains for pure PA films (Fig. 7C). On buffered saline subphases, BAM images further reveal the macroscopic dimensions of the fracture cracks in collapsed films of pure PA (Fig. 7E). For both subphases, in the presence of SP-B$_{1-25}$, BAM images show the homogeneous and uniform nature of the growth of the collapsed phase domains at a much higher nucleation density (Figs. 7D and F).

Comparison of SP-B$_{1-25}$ to full-length SP-B protein

Earlier isotherm data obtained in our laboratory indicated that the amino-terminal segment of SP-B possesses most of the activity of the full-length protein [40]; our fluorescence images also confirm this fact [35]. Under all experimental conditions, similar images are obtained using SP-B$_{1-25}$ and full-length SP-B (Fig. 8). This result has important implications on the design of more cost-effective replacement surfactant formulations. Currently, sources of natural SP-B from human or animal sources are limited and expensive to obtain, and a suitable expression system for the production of genetically engineered SP-B on a practical scale does not exist. Short peptide sequences are easy and cheap to synthesize and can facilitate the large scale production of replacement formulations. The elucidation of the minimal functional requirement of SP-B protein and how it relates to the amino acid sequence may allow us to design simpler and less expensive sequences which perform the same functions as SP-B.

Conclusions

An ideal replacement surfactant consists of a mixture of synthetic lipids based on those found in natural LS and synthetic protein sequences based on the LS proteins with a fully quantified mechanism of action and in vivo effect. This replacement mixture would (1) reduce the mortality rate and occurrence of BPD in nRDS infants, (2) reduce the treatment costs, and (3) provide a replacement mixture to be used in other cases of RDS such as adult RDS. We have discovered that SP-B interacts synergistically with anionic lipids such as PA to make the properties of the mixture more suitable for LS. The protein accomplishes this by acting like an emulsifying agent in the monolayer which breaks up and divides the condensed phase domains prior to collapse. This partitioning of the condensed phase shifts the mechanism of collapse from a heterogeneous, low pressure process to a homogeneous, high pressure process, and facilitates the reincorporation of collapsed phase material back into the monolayer. We have found that the amino-terminal segment of full-length SP-B, SP-B$_{1-25}$, captures the activity of the native protein, producing similar isotherm and morphological changes. This interaction is electrostatically based, confirmed by the reduced activity of a mutant form of SP-B$_{1-25}$ with neutral serine residues replacing the four positively-charged residues (data not shown). The presence of the protein network may also enhance the adsorption rate of material to the interface by providing docking sites for the attachment of vesicles from the subphase.

Current and future work in our laboratory is based on studying LS on several fronts. We are trying to understand the functional mechanism of action of SP-B protein by inducing various point mutations to study the effect of varying the charge to hydrophobic balance of the protein. We are also currently engaged in studying simpler peptide sequences that capture the full activity of native SP-B. The interactions of both SP-B and SP-C with other anionic lipids found in LS, such as saturated and unsaturated phosphatidylglycerols are another area of our focus. We hope to eventually use the information gained from binary and ternary model LS systems to study a complete mixture of synthetic LS containing DPPC, anionic lipids, and synthetic protein sequences. We are also examining the mechanism of inhibition of LS by serum proteins such as albumin, an understanding of which would allow us to engineer an enhanced resistance to inhibition into these model replacement mixtures. The role chemical engineers play in surfactant research is to provide clinicians and medical researchers with novel replacement mixtures engineered for optimum performance at low cost, and to provide a complete and detailed picture of the phase behavior and properties of the mixture. Clinicians could then use this information as a basis for both designing and analyzing tests of the surfactant in animal model systems. Instead of having an extract of unknown composition with unspecified properties, clinicians would have a fully quantified system of known properties and phase behavior.

References

1. Shapiro DL, Notter RH, (eds) (1989) Surfactant Replacement Therapy. Liss, New York
2. Soll R (1992) Res Staff Phys 38:19–23
3. Kopelman AE, Mathew OP (1995) Ped Rev 16:209–217
4. Bancalari E, Sosenko I (1990) Ped Pulmon 8:109–116
5. Jobe AH (1993) New Eng J Med 328:861–868
6. Schwartz R, Anastasia M, Luby M, Scanlon J, Kellogg R (1994) New Eng J Med 330:1476–1480
7. Singh G, Yu S (1995) Am. J. Public Health 85:957–964
8. Pattle RE (1955) Nature 175:125–1126
9. Avery ME, Mead J (1959) Am J Dis Child 97:517–523
10. Robillard E, Alarie Y, Dagenais-Perusse P, Baril E, Guilbeault A (1964) Can Med Assoc J 90:55–57
11. Enhorning G, Robertson B (1972) Pediatrics 50:55–66
12. Fujiwara T, Tanaka Y, Takei T (1980) Lancet 1:55–59
13. Possmayer F, Yu S, Weber J, Harding P (1984) Can J Biochem Cell Biol 62:1121–1131
14. Tanaka Y, Tsunetomo T, Toshimitsu A, Masuda K, Akira K, Fujiwara T (1986) J Lip Res 27:475–485
15. van Golde L, Batenburg J, Robertson B (1988) Phys Rev 68:374–455
16. Fujiwara T, Robertson B, Robertson B, Van Golde, L, Batenburg J (eds) (1992) Pharmacology of Exogenous Surfactant Elsevier, Amsterdam, pp 561–592
17. Egberts J, Sloot H, Mazure A (1989) Biochim Biophys Acta 1002:109–113
18. Yu S, Possmayer F (1992) Biochim Biophys Acta 1126:26–34
19. Pastrana-Rios B, Flach C, Brauner J, Mautone A, Mendelsohn R (1994) Biochemistry 33:5121–5127
20. Oosterlaken-Dijksterhuis M, Haagsman H, van Golde L, Demel R (1991) Biochemistry 30:8276–8281
21. Cockshutt A, Absolom D, Possmayer F (1991) Biochem Biophys Acta 1085:248–256
22. Waring, A, Taeusch W, Bruni R, Amirkhanian J, Fan B, Stevens R, Young J (1989) Pep Res 2:308–313
23. Gordon LM, Horvath S, Longo M, Zasadzinski JA, Taeusch HW, Faull K, Leung C, Waring AJ (1996) Prot Sci 5:1662–1675
24. Waring, A, Gordon L, Taeusch H, Bruni R, In: Epand R (ed) The Amphipathic Helix. CRC Press, Boca Raton, pp. 143–171
25. McConnell HM (1991) Ann Rev Phys Chem 42:171–195
26. Mobius D, Mohwald H (1991) Adv Mater 3:19–24
27. Weis R (1991) Chem Phys Lipids 57:227–239
28. Knobler CM, Desai R (1992) Ann Rev Phys Chem 43:207–264
29. Mohwald H (1993) Rep Prog Phys 56:653–685
30. Mohwald H (1990) Ann Rev Phys Chem 41:441–476
31. Riviere S, Henon S, Meunier J, Schwartz DK, Tsao MW, Knobler CM (1994) J Chem Phys 101:10045–10051
32. Schwartz D, Tsao M, Knobler C (1994) J Chem Phys 101:8258–8261
33. Henon S, Meunier J (1991) Rev Sci Instrum 62:936–939
34. Honig D, Mobius D (1991) J Phys Chem 95:4590–4592
35. Lipp MM, Lee KYC, Zasadzinski JA, Waring AJ (1996) Science 273:1196–1199
36. Chi L, Johnston R, Ringsdorf H (1991) Langmuir 7:2323–2329
37. Turnbull D (1952) J Chem Phys 20:411–432
38. Tanaka T, Ishiwata S, Ishimoto C (1977) Phys Rev Lett 38:771–774
39. Bruggeller P, Mayer E (1980) Nature 288:569–570
40. Longo ML, Bisagno A, Zasadzinski JA, Bruni R, Waring AJ (1993) Sci. 261:453–456

Progr Colloid Polym Sci (1997) 103:280–285
© Steinkopff Verlag 1997

J.I. Siepmann

Monte Carlo calculations for vapor–liquid phase equilibria in Langmuir monolayers

Received: 10 December 1996
Accepted: 13 December 1996

Presented at the 210th ACS National
Meeting, Symposium on "Interfacial
Structure: Amphiphiles at Vapor–Liquid
Interfaces", Chicago, Illinois, 1995

Dr. J.I. Siepmann (✉)
Department of Chemistry
University of Minnesota
207 Pleasant St. SE
Minneapolis, Minnesota 55455-0431, USA

Abstract Configurational-bias
Monte Carlo simulations in the Gibbs
ensemble have been carried out to
determine the vapor–liquid
coexistence curve for a pentadecanoic
acid Langmuir monolayer. Two
different force fields were studied: (i)
the original monolayer model of
Karaborni and Toxvaerd including
anisotropic interactions between alkyl
tails, and (ii) a modified version of this
model which uses an isotropic united-
atom description for the methylene
and methyl groups and includes
dispersive interactions between the
tail segments and the water surface.

The calculated phase diagram for the
Karaborni and Toxvaerd force field
deviates significantly from the
experimental observations, but the
modified model gives a more
quantitative description. In both
cases, the shape of the coexistence
curve can be fitted using the critical
exponent of the three-dimensional
Ising model. The two force fields yield
qualitatively different structural
properties.

Key words Molecular simulation –
Langmuir monolayer – Gibbs
ensemble

Introduction

Knowledge of the phase behavior of Langmuir monolayers is of importance in many technological applications
and can enhance our understanding of the properties of
the biological membrane [1]. Langmuir monolayers form
spontaneously when certain amphiphilic molecules are
spread on the air/water interface [2]. Monolayers consisting of pure fatty acids, alcohols, esters, or phospholipids
are very sensitive to changes in the surface area per molecule and/or temperature and display a rich phase diagram
with a variety of ordered low-temperature, high-density
structures and three distinct fluid phases (for excellent
reviews, see refs. [3–7]). At sufficiently low densities, Langmuir monolayers exist in a gaseous state (G). A two-phase
coexistence regime marks the onset of condensation to the
liquid-expanded (LE) phase. At higher surface densities,

many mesomorphous phases (liquid-condensed, LC) analogous to those of smectic liquid crystals are found. Further
compression results in transition to crystalline poly-
morphs and eventually to the collapse of the monolayer.

Whereas the experimental determination of the vapor–liquid coexistence curves for bulk fluids is a routine
task, the determination of the phase boundaries and of the
critical points for the two-dimensional Langmuir monolayers remains a great challenge [3]. Measurements of the
surface-pressure/area (π–A) isotherms of monolayers of
pentadecanoic acid (PDA) have been carried out as early
as 1926 [8]. In the 1970s, Kim and Cannell [9, 10] showed
that the LE and G phases are connected by horizontal
isotherms inferring a first-order transition. From the coexistence curve they concluded that the two-phase region
ends in a critical point at around 26°C. In the 1980s, Pallas
and Pethica [11] undertook another set of isotherm
and surface potential experiments using extremely pure

Progr Colloid Polym Sci (1997) 103: 280–285
© Steinkopff Verlag 1997

systems. They confirmed the first-order nature of the LE-G transition but found markedly different coexistence densities and argued that the critical temperature should be higher than 50°C. From the isotherm measurements, it is also not clear whether the shape of the coexistence curve can be better described using the mean-field or the two-dimensional Ising critical exponent [3]. The controversy between the two isotherm studies could only be resolved using a new technique, fluorescence microscopy, which allows direct observation of phase coexistence due to different solubilities of a fluorescence impurity in each phase [12–14]. Comparing isotherm and fluorescence data, Moore et al. [15] confirmed the coexistence densities reported by Pallas and Pethica and showed that the LE-G and LE-LC coexistence lines merge in a triple point. Using the fluorescence technique, LE-G phase coexistence was observed at temperatures approaching 70°C [15]. Close to critical points, however, the results of fluorescence studies should be viewed with caution, since the effect of the fluorescence probe on the stabilities of the coexisting phases is unknown. In contrast, Brewster angle microscopy [16, 17] allows to visualize the different monolayer phases without the need for a specific probe molecule. This technique has recently been successfully applied to determine the surface-pressure vs. temperature (π–T) phase diagrams of Langmuir monolayers of long-chain fatty acids [18–20]. To the author's knowledge, this technique has not yet been employed to obtain the π–A phase diagram and to study the coexistence regimes close to the critical points. Buontempo et al. [21–23] have observed very slow relaxation phenomena during the compression cycle in isotherm measurements on long-chain fatty acid monolayers. They conclude that nonequilibrium monolayer structures might have been present in some of the earlier isotherm studies.

Computer simulation may be a good alternative in the investigation of the phase transitions of Langmuir monolayers and the coexistence regions in microscopic detail. However, until recently it was considered impossible to calculate the critical properties of such complex systems using conventional techniques [24]. Even if a reliable force field for the aqueous monolayer would be available, direct simulations in the canonical or microcanonical ensemble cannot yield reliable data on phase coexistence and phase boundaries, since phase segregation and stability are dramatically perturbed by the finite size of the systems ($\mathcal{O}(10^2)$) in constant-area simulations [25–27]. Simulation of two coexisting phases by conventional techniques would require at least thousands of chain molecules, thus far exceeding the current limits of even the most powerful supercomputers. Our recent results [28, 29], however, have indicated that the limitations of the conventional techniques can be successfully overcome using a combination of the Gibbs-ensemble Monte Carlo approach and the

configurational-bias Monte Carlo method. In the next section, we will briefly review this new simulation methodology and highlight the differences between the two force fields used in this study for the fatty acid Langmuir monolayer. Thereafter, we present results for the phase diagrams and discuss the structural properties of the monolayers.

Simulation details

The Gibbs-ensemble Monte Carlo (GEMC) technique introduced by Panagiotopoulos [30–32] utilizes two *separate* simulation boxes. Besides the conventional random displacement of particles, additional types of Monte Carlo moves are applied to ensure that the two boxes are in equilibrium, i.e. have equal temperature, pressure and chemical potential of each species [30]. The particular advantage of simulations in the Gibbs ensemble is that if conditions are such that the system wants to phase separate the simulation yields a vapor phase in one box and a liquid phase in the other without having to create a "real" vapor–liquid interface. As a result, the coexistence properties can be determined *directly* with a surprisingly small number of particles. The rate-determining step in a GEMC simulation is the equilibration of the particles between the two boxes, i.e. the swapping of a molecule from the vapor to the liquid box and vice versa. While this step is relatively easy for systems containing atoms or small molecules, it is virtually impossible to insert chain molecules into their liquid phase using the conventional techniques. To enhance the sampling of chain insertions we have used the configurational-bias Monte Carlo (CBMC) algorithm in which a chain molecule is inserted atom by atom such that conformations with favorable energies are preferred [33–35]. The resulting bias of the particle-swap step in the Gibbs ensemble is removed by special acceptance rules [36]. Besides allowing simulations of fluids with articulated structure in the Gibbs ensemble, the CBMC approach has been shown to be very efficient to relax the conformational degrees of freedom which are believed to be the rate-limiting step for equilibration of the high-density phases of monolayers [23, 27]. Over the last years, the combination of GEMC and GBMC techniques has been utilized to calculate the bulk vapor–liquid coexistence curves for many flexible chain molecules, such as bead-spring polymers [36], the homologous series of n-alkanes [28, 38], and some alkanols [39].

Karaborni and Toxvaerd model

In the fatty acid Langmuir monolayer model of Karaborni and Toxvaerd [40–42], a pentadecanoic acid (PDA) molecule consists of 15 pseudoatoms, each representing either the carboxylate headgroup, an internal methylene group, or the terminal methyl group. The interactions of tail segments are modeled by Toxvaerd's anisotropic united-atom model which accounts implicitly for the hydrogen atoms bonded to the carbon backbone [43]. Amphiphile–water interactions are represented by external potentials which define a nonzero width interface [40]. The interactions of the headgroups are governed by a combination of excluded volume and dipolar repulsion in the form of a purely repulsive 12 + 3 potential [40]. For long-chain fatty acid monolayers at surface areas ranging from 18.5 to 40 $Å^2$ per chain, Karaborni et al. [40, 41] and Karaborni [42] reported very satisfactory agreement between simulation and experiment for a variety of properties including indirect confirmation of some of the phase transitions at higher densities, and it was concluded that this model yields an adequate description of the phase behavior of Langmuir monolayers of fatty acids. However, our earlier Gibbs-ensemble simulations [29] have shown that the Karaborni and Toxvaerd model fails to yield an adequate description of the vapor–liquid phase envelope of PDA (see Results and Discussions).

Model II

Model II is the first attempt to derive an improved force field for the fatty acid Langmuir monolayers starting from the Karaborni and Toxvaerd model. From calculations of the phase diagrams of normal alkanes [38], we have learned that the anisotropic potential of Toxvaerd [43], which governs the tailgroup interactions in the Karaborni and Toxvaerd model, is not an ideal choice with the original parameterization and underestimates the critical temperatures for the longer alkanes. Therefore, this part of the force field is replaced by an isotropic united-atom alkane model [28], which yields good agreement with experiment for longer alkanes. A second shortcoming of the Karaborni and Toxvaerd model is that the attractive forces due to dispersion interactions between tail segments and the water surface are neglected. Omission of these interactions in combination with the external potential (the energy penalty for "solvating" methylene beads (40)) manifests itself by relatively erect molecules (small tilt angles) in the vapor phase [29], i.e. missing the tendency of the amphiphilic molecules to lie flat on the water surface in the vapor phase. Peters et al. [44] have reported that the surface tension of the original Karaborni–Toxvaerd model is too low and have also attributed this failure to the missing dispersive tail–water interactions. Model II mimics the dispersion forces in the external potential by including the attractive part of 9–3 potential which can be

obtained by integrating the interactions of the tail segments over a half-sphere of the supporting water phase [45]. In addition, the diameter (σ_{COOH} parameter) used for the repulsive 12 + 3 headgroup interactions has been adjusted. The value of $\sigma_{COOH} = 4.22$ Å suggested by Karaborni and Toxvaerd [40] was taken straight from earlier monolayer simulations of van der Ploeg and Berendsen [46] in which a standard 12–6 Lennard–Jones potential was used. Since the Karaborni and Toxvaerd model yields substantially too high liquid densities, an enlarged value of $\sigma_{COOH} = 6.5$ Å has been used in model II.

Nine simulations were carried out for the Karaborni and Toxvaerd model covering the temperature range from 298 top 373 K [29], and six simulations (308 to 358 K) were performed for model II. In both cases, a total of 60 PDA molecules were used which were initially equally distributed between the two simulation boxes of the Gibbs ensemble. Equilibration periods of 30 000 to 60 000 Monte Carlo cycles were used at each temperature, and the production periods lasted for a least 30 000 cycles. Further details of the simulations can be found in ref. [29].

Results and discussion

The calculated phase diagram for pentadecanoic acid using the Karaborni and Toxvaerd model has been reported in a previous publication [29]. The results can be summarized as follows: (i) the model yields coexistence between a liquid and a vapor phase; (ii) the liquid phase of the model monolayer is substantially denser than the LE phase of the "real" monolayer, and the critical point seems to be shifted to higher temperatures; and (iii) the coexistence curve obtained from simulations can best be fitted with a scaling exponent of $\beta \approx 0.32$, supporting a three-dimensional character of the finite-size model system. Thus the Karaborni and Toxvaerd model yields merely qualitative agreement for the G–LE coexistence with experiments on the same systems.

Figure 1 shows a comparison of the simulated phase diagrams for the original Karaborni and Toxvaerd model [29] and for model II. In the latter case, the liquid densities and the critical temperature are reduced in comparison to the Karaborni and Toxvaerd model, and the overall agreement with the experimental data has improved. It should be noted here that the lowering of the critical temperature is at first hand surprising, since use of the isotropic alkane force field in model II should raise the critical temperature with respect to Toxvaerd's anisotropic model. However, the additional 9–3 term in the external potential and the enlarged head group effectively lower the heat of vaporization and overall lead to a decrease of the critical temperature. The coexistence curves for both models can best be

Progr Colloid Polym Sci (1997) 103:280–285
© Steinkopff Verlag 1997

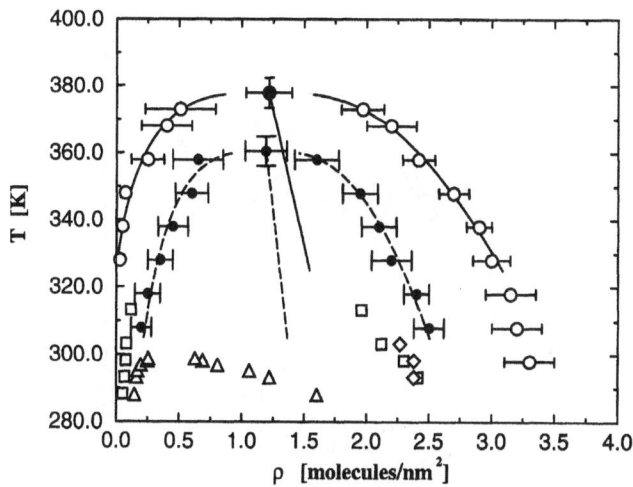

Fig. 1 Vapor–liquid coexistence densities for pentadecanoic acid monolayers. The results of the isotherm measurements of Kim and Cannell [9] and Pallas and Pethica [11] and the fluorescence data of Moore et al. [15] are shown as open triangles, squares, and diamonds, respectively. The simulation data are depicted by open and filled circles for the original Karaborni and Toxvaerd model [29] and an improved force field (this work). The solid and dashed lines are fits of the simulation data using the scaling law and the law of rectilinear diameters

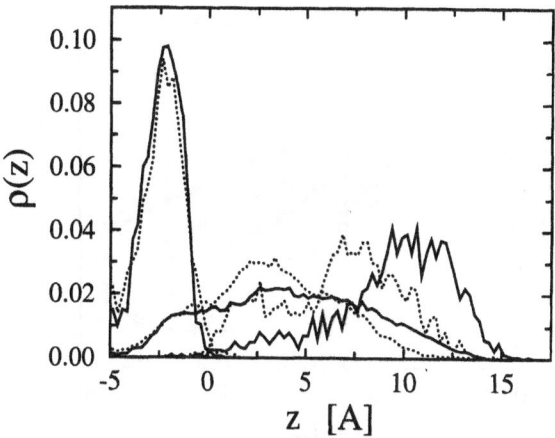

Fig. 2 Normalized bead density profiles in the direction perpendicular to the air/water interface at $T = 338$ K. Results for the original Karaborni and Toxvaerd model and the improved force field are shown as solid and dotted lines, respectively. The left-hand pair of bead density profiles with a peak at $z \approx -2.4$ Å are only for the carboxylic headgroups, the right-hand pair of curves are the bead density profiles for the terminal methyl groups, and the shallow curves depict the total density profiles summing over all beads, including the head and tail groups

described using a critical exponent β close to 0.3. Considering that in our simulations the fluctuations perpendicular to the interface which arise from the flexibility of the amphiphilic tails together with the finite width provided by the external potentials approach the magnitude of the fluctuations allowed by the periodic boundaries used for the interface plane, it is not surprising that we find an exponent close to the one expected for three-dimensional systems. Due to limitations in computer speed, GEMC simulations for much larger Langmuir monolayer systems are at present not feasible.

The simulations of the G-LE phase envelopes also yield a wealth of information on the microscopic structures of the two phases. Figure 2 shows the bead density profiles in the direction perpendicular to the air/water interface for the liquid phases at $T = 338$ K. The distributions of the carboxylic head groups are very similar for both force fields. Whereas the denisty profiles for the terminal methyl groups are strikingly different. For the Karaborni and Toxvaerd model, we find most of the tail segments in the region 7.5 Å $< z <$ 15 Å, where z is the distance to the air/water interface. For model II, the distribution of the methyl groups is shifted to much lower values and a secondary peak appears at $z \approx 2.5$ Å. Correspondingly, the molecular tilt angle is much larger for model II ($\theta_m = 52°$) than for the Karaborni and Toxvaerd model ($\theta_m = 35°$). Surprisingly, the molecular tilt angles found in the vapor phase are identical (within the statistical accuracy) to those of the coexisting liquid phases.

The differences in the density profiles and the molecular tilt angles for the two force fields can be mainly attributed to the attractive surface interactions in model II. The alkyl chains respond to this potential by aproaching the surface. In addition to the increase in tilt angle, the alkyl chains for model II have an enhanced tendency to the formation of loops which is evident in the secondary peak found in the methyl bead denisty profile. These structural changes are accompanied by a large decrease in the liquid densities for model II, which are approximately 30% lower than those for the Karaborni and Toxvaerd model at the same temperatures.

The difference in tilt angle and the formation of loops is also manifested in changes in the S_{CD} order parameter along the chain backbone (see Fig. 3). The alkyl chains of model II are significantly more disordered than for the Karaborni and Toxvaerd model. In addition, negative S_{CD} values are found for model II at carbons 10 to 12 which is indicative of a tendency for the C–H bonds at these carbons to be perpendicular to the interface as would be required at the top of a loop. The S_{CD} order parameters can also be used to distinguish between the vapor and liquid phases. For the Karaborni and Toxvaerd model, in particular, alkyl tails in the vapor phases are distinctively more disordered than those in the liquid phases. As expected, the differences between the coexisting phases diminish when the critical temperature is approached. There is much less difference in the order of the coexisting phases for model II. The dispersive interaction seems to

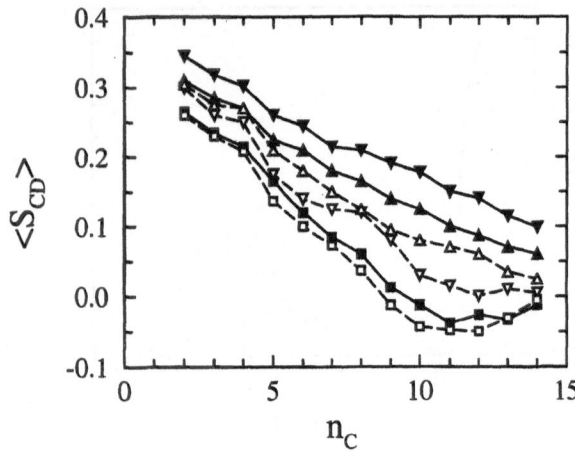

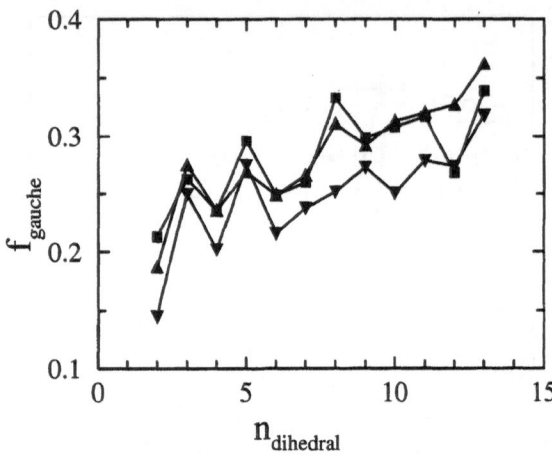

Fig. 3 S_{CD} bond order vs. carbon segment number. The upwards and downwards pointing triangles show the results for the original Karaborni and Toxvaerd model at $T = 338$ K and 368 K, respectively. The squares are used for the data obtained with the improved model at $T = 338$ K. Filled symbols (solid lines) and open symbols (dashed lines) are used for liquid and vapor phases, respectively

Fig. 4 Fraction of *gauche* dihedral angles vs. the bond number. The different symbols represent the same simulations as in Fig. 3

be sufficiently strong to force the alkyl tails into similar conformations in both phases. The fraction of dihedral angles in *gauche* conformations is shown in Fig. 4. Again, more defects and conformational disorder is found for model II.

muir monolayers, the available force fields lack the accuracy required for quantitative predictions. The model II proposed in this work shows moderate improvements, but the large σ_{COOH} parameter would greatly destabilize any phase with surface areas lower than approximately 30 Å^2 per chain and would thus preclude the study of the high-density phases. The two force fields yield distinctively different structural properties. Thus with the help of new experimental data probing the structures of the Langmuir monolayers in more detail, it might be feasible to find the right ingredients for a more satisfactory force field for these systems.

Conclusions

The combination of Gibbs-ensemble and configurational-bias Monte Carlo is a viable technique to calculate the fluid phase diagrams of monolayer systems (in addition to bulk systems for which the technique has been previously used). However, for the relatively complex fatty acid Lang-

Acknowledgments I am very grateful to Mike Klein and Sami Karaborni for their contributions to the work presented here. I would also like to acknowledge many stimulating discussions with Doug Tobias and Richard Pastor. This work was in part supported through a Camille and Henry Dreyfus New Faculty Award and through a Grant-In-Aid from the Graduate School, University of Minnesota.

References

1. Gaines Jr GL (1996) Insoluable Monolayers at Liquid Gas Interfaces. Wiley, New York
2. Langmuir I (1933) J Chem Phys 1:756
3. Knobler CM (1990) Adv Chem Phys 77:397
4. Möhwald H (1989) In:Rinte T, Sherrington D (eds) Phase Transitions in Soft Condensed Matter. Plenum Press, New York, pp 145–159
5. Möhwald H (1990) Annu Rev Phys Chem 41:441
6. McConnell HM (1991) Annu Rev Phys Chem 42:171
7. Knobler CM, Desai RC, (1992) Annu Rev Phys Chem 43:207
8. Adam NK, Jessop G (1926) Proc Roy Soc London A 110:423
9. Kim MW, Cannell DS (1975) Phys Rev Lett 35:889
10. Kim MW, Cannell DS (1976) Phys Rev A 13:411
11. Pallas NR, Pethica BA (1987) J Chem Soc Faraday Trans 83:585
12. von Scharner V, McConnell HM (1981) Biophys J 36:409
13. Lösche M, Sackmann E, Möhwald H (1983) Ber Bunsen-Ges Phys Chem 87:848
14. Knobler CM (1990) Science 249:870
15. Moore BG, Knobler CM, Akamatsu S, Rondelez F (1990) J Phys Chem 94:4588
16. Hénon S, Meunier (1991) Rev Sci instrum 62:936
17. Hönig D, Möbius D (1991) J Phys Chem 95:4590
18. Overbeck GA, Möbius D (1993) J Phys Chem 97:7999
19. Rivière S, Hénon S, Meunier J, Schwartz DK, Tsao M-W, Knobler CM (1994) J chem Phys 101:10 045
20. Tsao M-W, Fischer TM, Knobler CM (1995) Langmuir 11:3184

21. Buontempo JT, Rice SA (1993) J Chem Phys 98:5825
22. Buontempo JT, Rice SA (1993) Chem Phys 99:7030
23. Buontempo JT, Rice SA, Karaborni S, Siepmann JI (1993) Langmuir 9:1604
24. Allen MP, Tildesley DJ (1987) Computer Simulation of Liquids. Oxford University Press, Oxford
25. Bareman JP, Cardini G, Klein ML (1988) Phys Rev Lett 60:2151
26. Siepmann JI, McDonald IR (1993) Langmuir 9:2351
27. Karaborni S, Siepmann JI (1994) Mol Phys 83:345
28. Siepmann JI, Karaborni S, Smit B (1993) Nature 365:330
29. Siepmann JI, Karaborni S, Klein ML (1994) J Phys Chem 98:6675
30. Panagiotopoulos AZ (1987) Mol Phys 61:813
31. Panagiotopoulos AZ, Quirke N, Stapleton M, Tildesley DJ (1988) Mol Phys 63:527
32. Panagiotopoulos AZ (1992) Fluid Phase Eq. 76:97
33. Siepmann JI (1990) Mol Phys 70:1145
34. Siepmann JI, Frenkel D (1992) Mol Phys 75:59
35. Siepmann JI (1993) In: van Gunsteren WF, Weiner PK, Wilkinson AJ (eds) Escom, Leiden
36. Mooij GCAM, Frenkel D, Smit B (1992) J Phys Cond Matt 4:L255
37. Siepmann JI, McDonald IR (1993) Mol Phys 79:457
38. Smit B, Karaborni S, Siepmann JI (1995) J Chem Phys 102:2126
39. van Leeuwen ME (1996) Molec Phys 87:87
40. Karaborni S, Toxvaerd S (1992) J Chem Phys 96:5505
41. Karaborni S, Toxvaerd S, Olsen OH (1992) J Phys Chem 96:4965
42. Karaborni S (1993) Langmuir 9:1334
43. Toxvaerd S (1990) J Chem Phys 93:4290
44. Peters GH, Toxvaerd S, Svendsen A, Olsen OH (1994) J Chem Phys 100:5996
45. Steele WA (1974) The interaction of Gases with Solid Surfaces. Pergamon Press, Oxford
46. van der Ploeg P, Berendsen HJC (1982) J Chem Phys 76:3271

Progr Colloid Polym Sci (1997) 103:286–293
© Steinkopff Verlag 1997

W.A. Goedel

Hydrophobic polymers tethered to the water surface

Received: 2 December 1996
Accepted: 11 December 1996

Dr. W. A. Goedel (✉)
Max-Planck-Institut für Kolloid-
und Grenzflächenforschung
Haus 9.9
Rudower Chaussee 5
12489 Berlin, Germany

Abstract Properties of monolayers of polymers tethered to the water surface are reviewed with emphasis on non-glassy hydrophobic polymers with polar head groups. These monolayers offer a convenient method to investigate "polymer melt brushes". They are especially suited to probe the statistical thermodynamics of the polymer coils which are stretched away from the interface upon increasing tethering densities. They can be used to generate films of controlled thickness in the nanometer range and nanometer thin freely suspended elastic membranes.

Key words Tethered polymer – polymer monolayer – polymer brush – crosslinked monolayer – rubber elastic membrane – Langmuir-Blodgett technique – telechelics – macroions

Introduction

Polymers differ from low molecular weight compounds in two important aspects:
(i) the *enthalpy* of interaction with surrounding molecules is basically multiplied by the number of repeat units. Therefore polymers usually have low vapour pressure, low solubility and may adsorb strongly to attractive interfaces.
(ii) The polymer chain can assume a huge number of conformations, thus the *entropy* often is dominated by conformational contributions. The structure of polymeric adsorbate layers often results from balancing enthalpic and entropic contributions to the Gibbs free energy. Since the entropic contributions to the free energy in polymers often is dominated by conformational terms, polymeric adsorbate layers often drastically differ from the low molecular weight counterparts.

In addition, polymers can be varied by copolymerization. Most important in this context is the modification of polymers by attaching a short second chain or a small number of single groups which drastically differ in their properties from the repeat units of the polymer chain.

These so-called telechelic polymers can be regarded as large amphiphiles. (In analogy to the low molecular weight case, it is appropriate to use the term "head" for the shorter chain or single groups and "tail" for the main chain. The term "end" will be reserved for terminal groups of the polymer chains, which do not significantly differ in their properties from the other repeat units.) One thus can classify polymers according to their solubility (soluble/insoluble) and surface activity ((i) non-surface active; (ii) surface active repeat units; (iii) surface active head groups).

This paper focuses on water insoluble polymers which are bound to the interface with a polar head group. These polymers are often called "tethered polymers" or "polymer brushes". Because of the high surface energy of the water surface most polymers have at least slightly surface active repeat units and form monolayers via adsorbtion of the repeat units. Non-surface active polymers like polydienes are the exception, rather than the rule. However if the head group adsorbs more strongly than the repeat unit, all significant effects of a "polymer brush" can be observed at "high" surface concentration of the polymers.

Progr Colloid Polym Sci (1997) 103:286–293
© Steinkopff Verlag 1997

"Swollen brush" versus "melt brush"

Depending on whether the main chain is soluble or insoluble, it is convenient to distinguish between a "swollen brush" and a "melt brush". In both systems the interactions between neighbouring polymer chains lead to a distortion or stretching of the random coils away from the interface. The "swollen brush" (see Fig. 1b) is composed of up to 90% vol of solvent, the concentration of polymer segments gradually decreases from close to the interface towards the bulk solution [1–3]. Because of the high degree of swelling, the polymer chains usually are mobile enough to assume their equilibrium conformations. If the bulk solution is removed and the "wet brush" is dried, the polymer coils collapse and the properties drastically change.

The "melt brush" (see Fig. 1a), on the other hand, is free of solvent. Thus, its properties depend significantly less on the presence of a solvent. The "melt brush" has its main importance in the context of phase separated block copolymers and the often complex morphologies of these systems (for example, lamellar [see Fig. 1c], hexagonal, cubic) are a result of the balance between the interfacial tension and the interactions of the closely packed polymer chains [4–6]. While there is considerable interest in these bulk systems, it has been very advantageous to study polymer brushes as *monolayers at flat surfaces*. In such a monolayer it is relatively easy to determine and tune the surface concentration of the head groups and to give the system a preferred orientation in space.

Fig. 1 Schematic comparison of (a) monolayers of tethered polymers that are free of solvent, (b) monolayers of tethered polymers swollen with solvent, and (c) one lamella of phase separated blockcopolymers

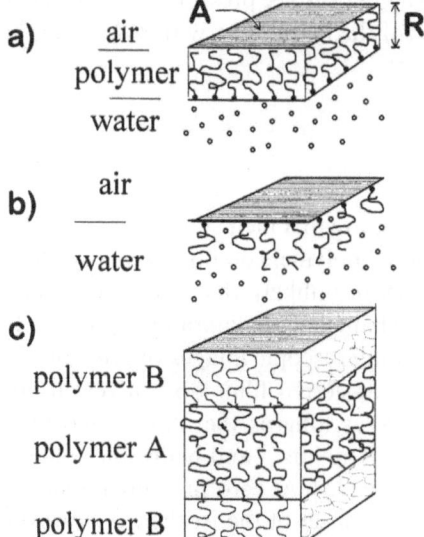

"Melt brush" versus glassy polymer

If a glassy, insoluble and non-surface active polymer is applied to a water surface, it forms hard objects. For example, if polystyrene is spread from benzene solutions onto a water surface it forms hard disk-like objects. Depending on the concentration of the spreading solution these objects can even be monomolecular. Once the benzene is evaporated, the objects vitrify. They do not change shape if the monolayer is laterally compressed. They do not fuse to form a continuous film and the properties are determined by the non-equilibrium geometry and arrangement of the hard objects and the friction between them [7, 8].

In this case, the polymer chains have no conformational freedom. For the general properties of these films it is, therefore, unimportant, whether the objects are made out of polymer at all. Films made out of inorganic colloidal particles have the same general features. Because the properties are determined by friction between hard objects, attaching a hydrophilic head group to the polymer does not significantly change the properties [9, 10].

If, however, the hydrophobic polymer is investigated above its glass transition, the polymer chains can rearrange, flow and fuse to form a continuous film. In the case of non-glassy non-surface active chains *without* a hydrophilic head group, the polymer will not form a thin continuous layer, but retract into small droplets. If, however, a polar head group is attached to these polymer chains, the polymer is tethered to the interface via these head groups and forms a "melt brush".

Synthesis

Hydrophobic polymers with well-defined head groups usually can be made via anionic polymerization followed by suitable termination of the living chain end. Like any other chemical reaction, the termination reaction often is incomplete or gives rise to side reactions. If a quantitative interpretation of the isotherms is intended, the degree of head group functionalization has to be determined and the polymer has to be chromatographically purified [11].

Isotherms

Christy, Petty and Roberts published a paper on monolayers of polybutadiene-1,2 with a quaternized ammonium head group [12]. The polymer has a chain length of approximately 25 carbon atoms (molecular weight 600–700 g/mol). Thus, the material is at the borderline

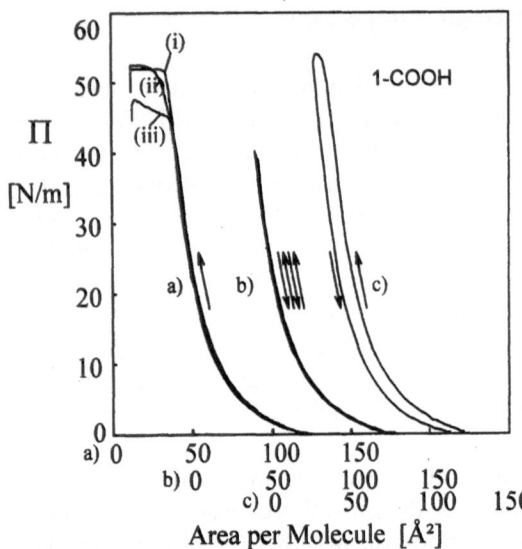

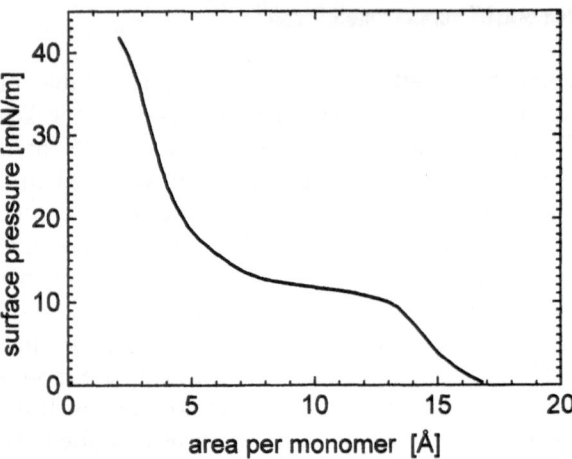

Fig. 2 Isotherm and hysteresis measurement of Perfluoropolyethers with carboxylic head groups. For clarity the data are horizontally shifted with respect to each other. (a) variation of compression speed: (i) $1 \text{ Å}^2 \text{ s}^{-1}(\text{repeat unit})^{-1}$, (ii) $10^{-1} \text{ Å}^2 \text{ s}^{-1} (\text{repeat unit})^{-1}$, (iii) $10^{-2} \text{ Å}^2 \text{ s}^{-1}(\text{repeat unit})^{-1}$, (b) Two consecutive cycles of compression and expansion up to a maximum pressure of 40 mN/m; (c) Compression down to 30 Å^2 and subsequent re-expansion

Fig. 3 Isotherms of Polydimethylsiloxane (1 kg/mol) with hydrophilic head groups [15]. The first plateau at zero surface pressure indicates the transition from a gaseous state to a monolayer with respect to the repeat units. The second plateau at roughly 10 mN/m indicates the transition from a monolayer with respect to the repeat units to a "melt brush" Data taken from ref. [15]

between long chain polymers and low molecular weight substances. The isotherm is very similar to expanded isotherms of liquid substances like oleic acid (chain length = 18 carbon atoms). The authors chose to compare it to isotherms of polymers with surface active repeat units. They do not consider this polymer as being qualitatively different from polymers without hydrophilic head groups.

Kim and Chung point out the qualitative difference between pure polybutadiene, which does not form a monolayer on the water surface, and polybutadienes with hydroxyl side groups. This contribution mainly focuses on polymers with a large number of hydroxyl groups per chain. The lowest number of hydrophilic groups is on average one hydroxyl group in every ten repeat units and the authors discuss the system in the context of polymers with surface active repeat units [13].

Isotherms of perfluoropolyethers with chain lengths of approximately one hundred atoms with various hydrophilic head groups have been reported [14]. As in the case of polybutadiene the water insoluble polymer backbone does not form a monolayer by itself. Hydrophilic head groups, however, make the polymer surface active, and isotherms of an expanded type can be recorded. (see Fig. 2).

Lenk, Lee and Koberstein showed that even the surface active polydimethylsiloxane (PDMS) can be transformed into a polymer brush by attaching strongly hydrophilic head groups. Methyl terminated PDMS forms

a monolayer with respect to the repeat unit, which collapses at 10 mN/m. Further compression of this film, yields a plateau in the isotherm, which signifies the coexistence of the monolayer and excess material. If the PDMS chain, however, is terminated with an amine or carboxylic group, an additional rise of the surface pressure upon compression can be recorded [15]. This part of the isotherm is similar to the expanded isotherms recorded in the case of the above-mentioned polydienes and perfluoropolyethers with non-surface active repeat units. The isotherms just seem to be shifted to higher surface pressure by the value of the collapse pressure of the monolayer with respect to the repeat unit (see Fig. 3).

Lateral structure

The hydrophobic chains are in a melt state. The expanded isotherms of the tethered polymers do not show any phase transitions. It is therefore unlikely that films of a single polymer of uniform chain length laterally segregate to form domains. Indeed microscopic images of films on the water surface and of transferred films do not reveal any sign of lateral structure as long as the film is imaged at surface pressures above 0 mN m^{-1} and below the collapse pressure (see Fig. 4h) [19, 25]. At zero pressure (areas per molecule larger than the onset of the isotherm), however, the film breaks up into patches of two-dimensional foam,

Progr Colloid Polym Sci (1997) 103:286–293
© Steinkopff Verlag 1997

which is in coexistence with the "bare" water surface (see Fig. 4a–g). This observation is a general feature of any insoluble monolayer [16] and can be interpreted as a two-dimensional liquid/gas coexistence.

Atomic force microscopy (AFM) images of films transferred to solid substrates reveal a smooth surface

of the film [17]. The film has occasional holes of μm diameter, that are smaller than the resolution limit of the light microscopic techniques. It is not clear yet whether these holes are created during the transfer, or whether they are already present on the water surface.

Fig. 4 Monolayers of polyisoprene with a sulfonate head group and $N = 140$ repeat units imaged with Brewster angle microscopy at various areas per head group. The pictures cover an area of approximately 0.75 mm × 0.75 mm. a)–g): A/n decreasing from initially 220 Å^2 in picture a) to 185 Å^2 in picture g); h): $A/n = 130$ Å^2, i: $A/n = 85$ Å^2. The corresponding isotherm is included in Fig. 7 left side

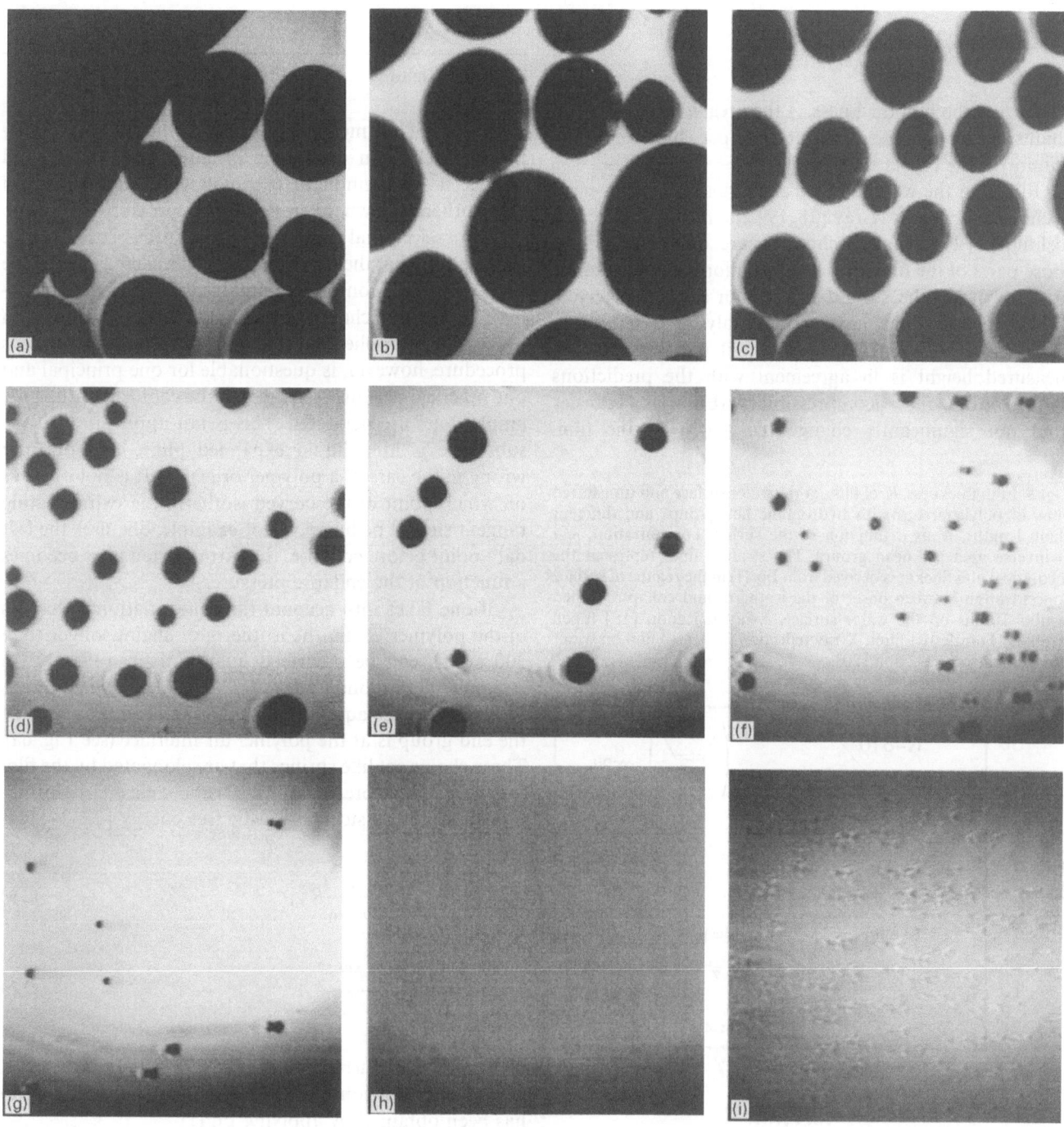

Vertical structure

The hydrophobic region of the film does not take up significant amounts of water. It is therefore reasonable to assume that the density of polymer segments is very close to the bulk melt of the pure polymer and nearly constant throughout the film. This conservation of the polymer density implies that the film thickness should be proportional to the chain length and inverse proportional to the area per head group.

$$RA = nNv \Leftrightarrow R = Nv\left(\frac{A}{n}\right)^{-1}, \tag{1.}$$

where R is the film thickness, A the area, n the number of chains, N the number of repeat units per chain and v the volume of the repeat unit.

Films on the water surface and transferred films have been investigated by X-ray [18, 19] and neutron reflection technique [20]. These techniques are only sensitive to those parts of the film, that have a uniform height. Bumps or droplets on the surface of the film are not detected. The reflection data have been analysed assuming a homogeneous film of constant height (see Fig. 5). The measured height is in agreement with the predictions derived from the incompressibility (Eq. (1)). Transfer does not significantly change the height of the film.

Fig. 5 Film thickness, R, of films at the water surface and transferred films of polyisoprenes with hydrophilic headgroups and different chain lengths, N as a function of the surface concentration, n/A (=inverse area per head group). The straight lines represent the theoretical film thickness derived from Eq. (1) in the regime of surface concentration between onset of the isotherm and collapse. (Filled symbols) films on the water surface, X-ray reflection [18] (open symbols) Transferred films, X-ray reflection [19], (+) film on water surface neutron reflection [20]

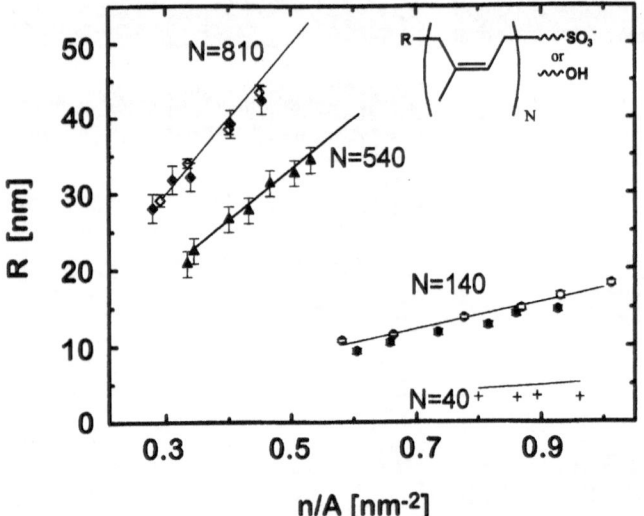

The above-mentioned AFM investigations did allow to measure the depth of the occasional holes. The thus measured step height satisfactorily agrees with the theoretical thickness as well as with the thickness obtained via X-ray reflectivity studies.

Thus, in principle, one can fine tune the film thickness of a polymer monolayer, just by selecting the appropriate area per molecule at transfer. This dependency of the film thickness on the surface concentration might be of great value for the application of transferred films.

Thermodynamics

In the case of low molecular weight amphiphiles, especially those which form crystalline or liquid crystalline solid phases, it is common to extrapolate the linear part of the isotherm down to zero surface pressure to obtain the "cross-sectional area". This procedure is based on the assumption, that the film is laterally compacted against the hard core repulsion of the amphiphiles, while the conformations do not change. Occasionally, the extrapolation procedure is applied to expanded isotherms [12]. This procedure, however, is questionable for one principal and one practical reason: (i) changes in the conformation of the amphiphiles are neglected. This is not appropriate, if the substance is in a liquid expanded phase and outright wrong in the case of a polymer brush. (ii) It is not obvious on which point of the curved isotherm the extrapolating tangent should be based. If, for example, one uses the last data point before collapse, the extrapolated area becomes a function of the collapse pressure.

If one takes into account the conformational changes of the polymer chain, the isotherm of chains longer than 300 atoms can be described quantitatively [19, 21]. In a simple picture one can assume that all polymer chains are bound to the aqueous phase by the head group, while the end group is at the polymer air interface (see Fig. 6a). These chains act like springs that are elongated by the film thickness. Therefore, due to this "rubber elastic" deformation these chains store an elastic free energy [22–24]:

$$F_{\text{elastic}} = nk_{\text{B}}T\frac{3}{2}\frac{1}{\langle r^2 \rangle_0}R^2, \tag{2.a}$$

$$F_{\text{elastic}} = k_B T\frac{3}{2}\frac{n^3 N^2 v^2}{\langle r^2 \rangle_0}A^{-2}, \tag{2.b}$$

where the undisturbed mean square end-to-end distance, $\langle r^2 \rangle_0$, is proportional to the chain length. Equation (2b) has been obtained by applying Eq (1).

a)

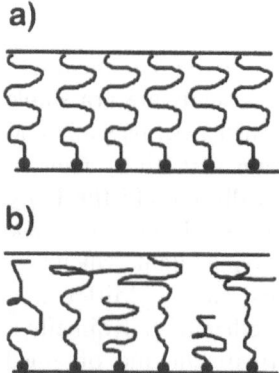

b)

Fig. 6 Schematic comparison between (a) a uniformly stretched brush, and (b) a non-uniformly stretched polymer brush similar to the Semenov scenario

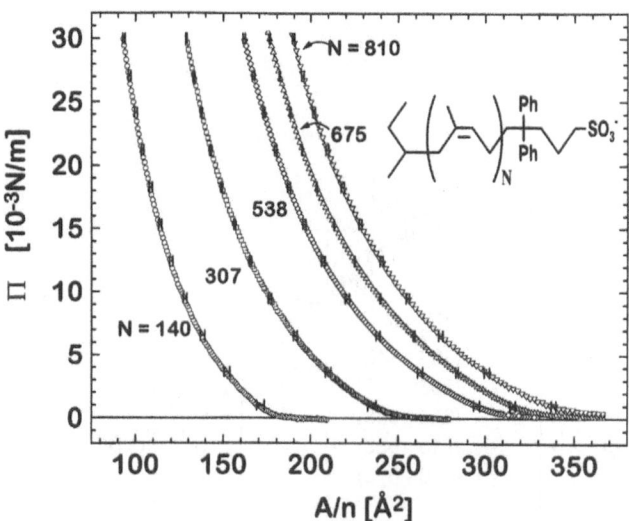

Fig. 7 Isotherms of polyisoprenes with sulphonate headgroups and different chain lengths, N is the number of repeat units (average of at least 5 measurements)

The first derivative of this elastic energy is given by

$$\frac{\partial F_{\text{elastic}}}{\partial A} = -3k_{B}T\alpha^{2}N\left(\frac{A}{n}\right)^{-3}. \qquad (3.)$$

In the case of a freely jointed chain, the constant $\alpha = v/\sqrt{\langle r^{2}\rangle_{0}/}$ can be interpreted as the cross-sectional area of a chain segment. The surface pressure depends on this first derivative.

It can be shown that all other components to the surface pressure are independent of the chain length N:

$$\Pi = f\left(\frac{A}{n}, \text{not } N\right) + ck_{B}T\alpha^{2}N\left(\frac{A}{n}\right)^{-3}, \qquad (4.)$$

with $c = 3$. Two features of this description are important: (i) at a given area per head group, A/n, the "elastic part" of the surface pressure depends linearly on the chain length; (ii) it is inversely proportional to the third power of the area per head group. Therefore, the surface pressure data of different chain lengths but the same area per head group should give a straight line if plotted against the chain length. The slope of this line is independent of the first term in Eq. (4) and should be inversely proportional to the third power of the area per head group. If we rescale by the "cross-sectional area of the chain", α, we obtain a dimensionless representation:

$$\frac{\Delta\Pi}{\Delta N}\frac{\alpha}{k_{B}T} = c\left(\frac{A}{n}\cdot\frac{1}{\alpha}\right)^{-3} \Leftrightarrow \qquad (5.a)$$

$$\log\left\{\frac{\Delta\Pi}{\Delta N}\cdot\frac{\alpha}{k_{B}T}\right\} = \log c - 3\log\left\{\frac{A}{n}\cdot\frac{1}{\alpha}\right\}. \qquad (5.b)$$

The double logarithmic version of Eq. (5.b) predicts a straight line. The slope of that line is -3 and the intercept is given by the prefactor c. In this rescaled plot all pairs of

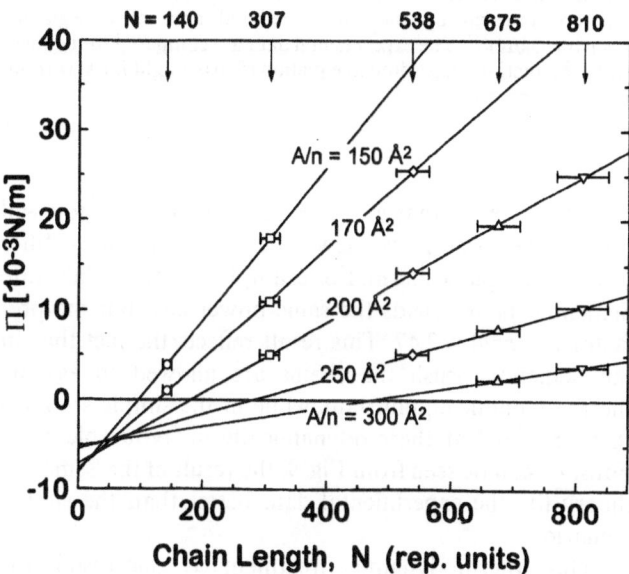

Fig. 8 Surface pressure of Polyisoprene-SO₃ films as a function of chain length at constant area per head group (same data as in Fig. 7, for clarity only the data for only a limited number of areas per head group are included)

polymers will be represented by a single line, even if we compare chains of different flexibility. The predictions of eqs. (4) and (5) have been confirmed using a set of sulphonate terminated polyisoprenes (see Figs. 7–9).

The assumption that the free ends are located at the "upper" surface of the film seems to be quite artificial. A scenario as depicted in Fig. 6b is more likely. It can be

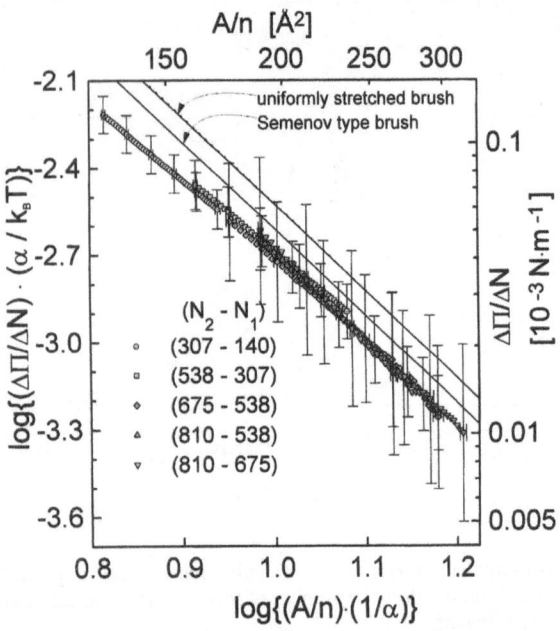

Fig. 9 Double logarithmic plot of pressure difference over chain length difference, versus area per head group. The left and bottom axis are rescaled by the "cross-sectional area of a segment" $\alpha = v/(\langle r^2 \rangle_0/N)^{1/2}$. The experimental data are compared to the theoretical predictions according to equation (5.b) (straight lines) without any fitting procedures

Cross linking

Until now, the films considered are room temperature liquids. Thus, the transferred films usually are not stable [12, 14, 25]. They can easily be washed away by water and can creep away from the substrate with time. Thus, it might be advantageous to chemically modify the films. Christy, Petty and Roberts [12] did stabilize polyisoprene oligomers via irradiation with ultraviolet light, which presumably cross links the polymer chains. Mirley and Koberstein degraded films of tethered polydimethylsiloxanes with oxidative ozone treatment and obtained thin films of mechanically stable inorganic coatings [26]. These films were treated after transfer to the substrates, thus the transferred films might substitute films that were applied to the surface via spin coating, adsorbtion or evaporation.

The Langmuir Blodgett technique offers the unique advantage to cover holes in perforated substrates. It has been shown, that films of tethered polyisoprenes with photoreactive side groups can be cross linked on the water surface via irradiation [27]. These cross-linked films of 40 nm thickness can be transferred to grids and perforated substrates and can span holes of up to 300 μm in diameter. Since these freely suspended films are made out of a cross-linked polymer melt, they show rubber elastic behaviour.

shown, however, that any type of affine deformation will lead to the same power law. The prefactor c may differ from the simple scenario. For example, the more elaborate Semenov theory yields the same power law, but the prefactor $c = \pi^2/4 \approx 2.47$. This result reflects the fact that in the Semenov brush the chains are allowed to assume thermodynamic equilibrium, while in the simple scenario they are fixed at thermodynamically unfavourable positions. As can be seen from Fig. 9, the result of the Semenov theory fits the experimental data better than the simple scenario.

This thermodynamic treatment is successful for polymers longer than 300 atoms. Shorter chains deviate from that theory. For chain lengths of approximately 100 atoms, the elastic contribution to the isotherm seems to be unimportant [25].

Conclusions

Monolayers of hydrophobic polymers, tethered to the water surface are suitable model systems for the investigation of polymer "melt brushes". Especially they offer a convenient way to vary the area per tethered head group and to give the brush a well defined orientation and make it accessible to other experimental techniques. Simple measurements of isotherms already yield valuable information on the thermodynamics of these brushes.

From the applied point of view they offer a convenient method to prepare nanometer thick coatings and thus might be an alternative to other coating techniques. Especially promising is the possibility of generating nanometer thick rubber elastic membranes, that can span holes of the size of one-third of a millimeter.

References

1. Kent MS, Lee L-T, Farnoux B, Rondelez F (1992) Macromolecules 25:6231
2. Milner ST (1988) Europhys Lett 7:659
3. Review article: Halperin A, Tirrell M, Lodge TP (1992) Adv Polymer Sci 100:31
4. Semenov AN (1985) Sov Phys JETP 61:733
5. Semenov AN (1993) Macromolecules 26:6617
6. Review article: Bates FS, Fredrickson GH (1990) Ann Rev Phys Chem 41:525
7. Kumaki J (1986) Macromolecules 19:2258
8. Kumaki J (1988) Macromolecules 21:749
9. Niwa M, Hayashi T, Higashi N (1990) Langmuir 6:263

Progr Colloid Polym Sci (1997) 103:286–293
© Steinkopff Verlag 1997

10. Yoshikawa M, Worsfold DJ, Matsuura T, Kimura A, Shimidzu T (1990) Polymer Commun. 31:414
11. Czichocki G, Much HR, Heger WA Goedel (submitted) J Chromatogr
12. Christie P, Petty MC, Roberts GG (1985) Thin Solid Films 134:75
13. Kim MW, Chung TC (1988) J Collid Interface Sci 124:365
14. Goedel WA, Xu C, Frank CW (1993) Langmuir 9:1184
15. Lenk TJ, Lee DHT, Koberstein JT (1994) Langmuir 10:1857
16. Mann EK, Henon S, Langevin D, Meunier J (1992) J Phys II France 2:1683

17. Boehnke UC (in preparation)
18. Baltes H, Schwendler M, Helm C, Heger R, Goedel WA (submitted) Macromolecules
19. Heger R, Goedel WA, Macromolecules (1996) 29:8912
20. Gentle IR, Saville PM, White JW, Penfold J (1993) Langmuir 9:646
21. See also for a more concise version of the theory, Goedel WA, Heger R, J Supramolecular Science (Proc '96 European Conference on Organised Films, to be published 1997)
22. Treolar LRG (1975) The Physics of Rubber Elasticity. Clarendon Press, Oxford, p.56/57

23. deGennes PG (1979) Scaling Concepts in Polymer Physics. Cornell University Press, Ithaca & London p 31
24. Mark JE, Erman B (1988) Rubberlike Elasticity a Molecular Primer. Wiley, New York, p 30
25. Goedel WA, Wu H, Friedenberg MC, Fuller GG, Foster M, Frank CW (1994) Langmuir 10:4209
26. Mirley CL, Koberstein JT (1995) Langmuir 11:1049
27. Heger R, Goedel WA, J Supramolecular Science (Proc '96 European Conference on Organised Films, to be published 1997)

Progr Colloid Polym Sci (1997) 103:294–299
© Steinkopff Verlag 1997

V. Rosilio
A. Kasselouri
G. Albrecht
A. Baszkin

The effect of the chemical nature of grafted chains on the interfacial miscibility of amphiphiles

Received: 29 October 1996
Accepted: 17 November 1996

Dr. V. Rosilio (✉) · G. Albrecht · A. Baszkin
Laboratoire de Physico-Chimie des Surfaces
URA CNRS 1218
Université Paris-Sud
5, rue J.B. Clément
92296 Chatenay-Malabry, France

A. Kasselouri
UPR CNRS 180
Université Paris-Sud
5, rue J.B. Clément
92296 Chatenay-Malabry, France

Abstract The miscibility of two phospholipids dipalmitoylphosphatidylcholine (DPPC) and dimyristoylphosphatidylcholine (DMPC) possessing both a choline head group, with per-(6-dodecanoylamino-6-deoxy) β-cyclodextrin ($C_{11}CONH$-β-CD) and poly(ethylene oxide)-bearing lipid (PEO-lipid), respectively, has been assessed by surface pressure measurements of binary monolayers under dynamic conditions. Although the four studied amphiphiles had similar hydrophobic moieties constituted of hydrocarbon units with the number of carbons ranging from 12 to 16, PEO-lipid markedly differed from other amphiphiles due to its bulky poly(ethylene oxide) chain containing 13 ethylene oxide units totally immersed in the aqueous subphase. The additivity rule applied to these binary mixtures clearly showed that molecular areas for both systems deviated from linearity. For DMPC/PEO-lipid mixtures this deviation was attributed to the presence of the poly(ethylene oxide) chain which hindered the establishment of a close contact between the film forming molecules. For DPPC/$C_{11}CONH$-β-CD mixtures the deviation from linearity was surmised to be due to an interaction at the level of hydrophobic chains protruding into the air phase. The thermodynamic relationship given by Joos and Demel (BBA 183: 447–457) analysing miscibilities in monolayers according to their collapse pressures, confirmed the deviation from ideality for both systems and indicated that the effect was much more pronounced for DPPC/$C_{11}CONH$-β-CD system.

Key words Spread monolayers – amphiphilic cyclodextrin – poly (ethylene oxide) lipid – dynamic surface pressure – interfacial miscibility

Introduction

The investigation of the miscibility of lipids and biocomponents in monolayers, which are particularly suitable models of biological membranes, may be considered as one of the essential factors in the understanding of physical properties and biological functions of the latter [1, 2]. The understanding of intermolecular forces between hydrocarbon chains at aqueous interfaces is also fundamental for theories of interactions in monolayers and two-dimensional phase transitions. The available information from the experimental data with insoluble lipid monolayers at the air/water interface show that chain-chain interaction and the role of the water in the pair interactions is the major factor in defining the two-dimensional properties of these films [3–7].

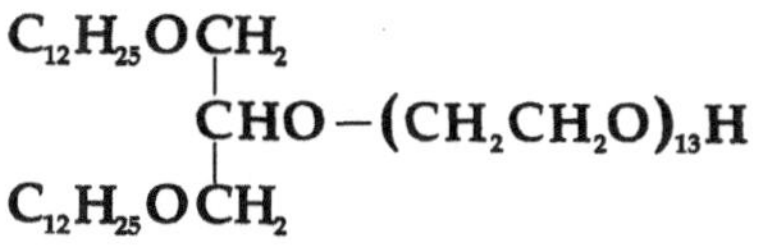

Progr Colloid Polym Sci (1997) 103: 294–299
© Steinkopff Verlag 1997

C$_{12}$H$_{25}$OCH$_2$
|
CHO $-$ (CH$_2$CH$_2$O)$_{13}$H
|
C$_{12}$H$_{25}$OCH$_2$

Fig. 1 Structure of PEO-lipid (12, 13)

C$_{11}$CONH-β-CD

Hydrophilic secondary face

Fig. 2 Schematic representation of the amphiphilic β-cyclodextrin

We are interested in the construction of nanomeric structures based on mixed amphiphilic biocomponents for which the mixing behavior from almost ideal mixing to complete immiscibility would strongly depend on hydrocarbon chain interaction between two components and the differences in their headgroup structure [8].

These criteria prompted us to review and to compare two systems recently studied by us in which mixed monolayers were formed with either amphiphilic cyclodextrins or poly(ethylene oxide)-bearing lipids and phospholipids [9–12]. The phospholipids simultaneously deposited with either per-(6-dodecanoylamino-6-deoxy) β-cyclodextrin (C$_{11}$CONH-β-CD) or poly (ethylene oxide)-bearing lipid (PEO-lipid) to form mixed monolayers were dipalmitoylphosphatidylcholine (DPPC) and dimyristoylphosphatidylcholine (DMPC), respectively.

The synthesis of a PEO-lipid containing a double-aliphatic C12 chain and a hydrophilic chain consisting of 13 ethylene oxide residues to which the C12 chains were bound was realized for the first time by Kuwamura [13] and then by Sunamoto's group [14]. The chemical structure of this PEO-lipid is shown in Fig. 1.

Cyclodextrins which are cyclic oligosaccharides built up of six, seven or eight 1-4-linked glucopyranose units (α-CD, β-CD and γ-CD, respectively) may be modified by lipophilic substitution on one of their two hydrophilic faces surrounding a hydrophobic cavity to form amphiphilic molecules. A schematic model of a modified β-CD with 7 grafted C12 chains at the primary face and with 14 hydroxilic groups at the secondary face is illustrated in Fig. 2. The synthesis route of such a molecule starting from per-(6-amino-6-deoxy) β-cyclodextrin is depicted in Fig. 3.

Compressional behavior of mixed monolayers of phospholipids and amphiphiles containing hydrophobic and hydrophilic grafted chains

It is generally admitted that if the experimental surface pressure (π)–area (A) isotherm for a mixture of two amphiphiles coincides with the calculated average curve, no mixing or interaction occurs. The shift of experimental data to the left with respect to the calculated average will indicate a condensation effect or the existence of an interaction between film-forming components. An interference in the packing of mixed molecules would correspond to a displacement of the data to the right of the calculated average curve [15].

However, the above considerations concern the systems in which none of the two film forming molecules contain any bulky chains that extend deeply into the aqueous phase.

Let us consider first the mixing behavior of PEO-lipid (12, 13) with DMPC. For such a system both an ideal or a non ideal mixing may occur. The occurrence of an interaction may arise at the level of hydrophobic moieties of two molecules mixed in a monolayer which have a comparable chain length (12 carbons for PEO-lipid and 16 carbons for DPPC) or at the level of their hydrophilic

Fig. 3 Synthetic route for the preparation of amphiphilic C$_{11}$CONH-β-CD

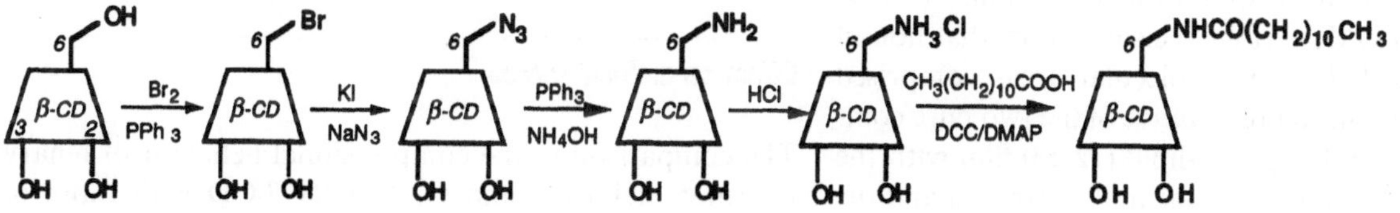

296
V. Rosilio et al.
Interfacial miscibility of amphiphiles

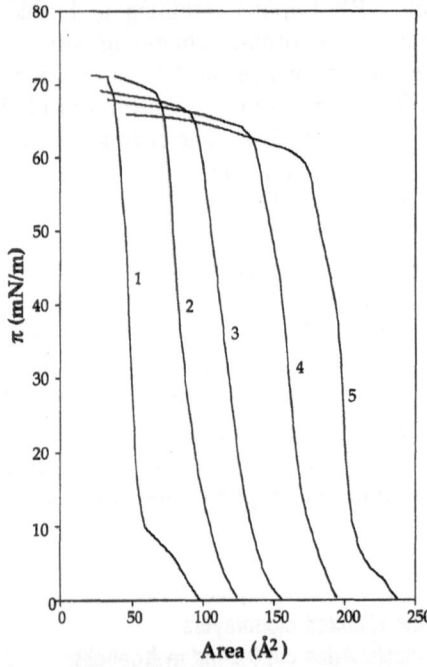

Fig. 4 Surface pressure–area isotherms of DPPC (1), $C_{11}CONH$-β-CD (5), and their mixtures: 7:3 (2), 1:1 (3) 2:8 (4)

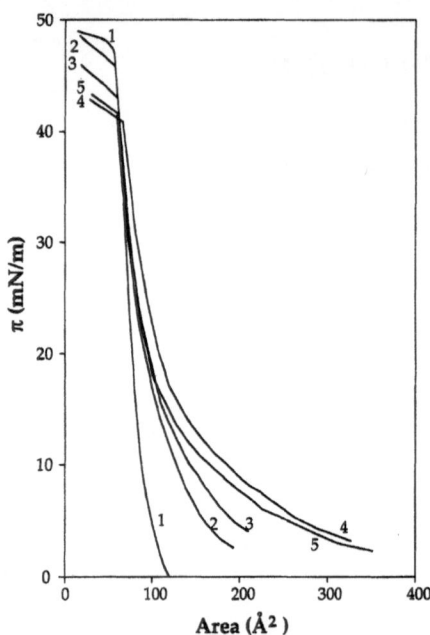

Fig. 5 Surface pressure–area isotherms of DMPC (1), PEO-lipid (12, 13) (5), and their mixtures, 7:3 (2), 1:1 (3) and 2:8 (4)

moieties immerged in the aqueous subphase. An interaction at both levels may also be envisaged. The hydrophilic moiety of the PEO-lipid, which is more bulky than the choline group of DMPC, would considerably affect mixing behavior of these two molecules. Conversely, for the modified β-cyclodextrin (β-CD) one would expect the second possibility to be unlikely.

The surface pressure (π)–area (A) isotherms for pure $C_{11}CONH$-β-CD and its mixed monolayers with DPPC are illustrated in Fig. 4, and those for pure PEO-lipid (12, 13) and its monolayers with DMPC are shown in Fig. 5. From Fig. 4 it is apparent that while at all studied surface concentrations of $C_{11}CONH$-β-CD the two components behave independently of each other, as indicated by the location of all isotherms of mixed films falling between those corresponding to pure components, their collapse pressures gradually increased with the increase in the DPPC surface concentration. From Fig. 4 it is also apparent that the presence of $C_{11}CONH$-β-CD led to the disappearance of the liquid-expanded (LE) to the liquid-condensed (LC) phase transition characteristic of pure DPPC.

The π–A profiles recorded for the mixed films of PEO-lipid (12, 13) and DMPC (Fig. 5) clearly show that not all of the isotherms fall to the molecular areas comprised between the isotherms representative of the two pure components. For the DMPC/PEO-lipid (12, 13) film with the molar ratio (2:8) both the average molecular area and the

area at the film collapse are higher than the corresponding areas of pure PEO-lipid (12, 13). Also its collapse pressure is lower than that of pure PEO-lipid (12, 13).

Evidently, binary mixtures of $C_{11}CONH$-β-CD and DPPC differ markedly from those of PEO-lipid (12, 13) and DMPC. If for the former system the addition of the cyclodextrin to a monolayer of DPPC shifted its molecular area toward higher values (but lower than the mean molecular area of pure $C_{11}CONH$-β-CD), and would suggest that an ideal mixing occurred in monolayers of the two components, for the mixtures of PEO-lipid and DMPC the molecular characteristics were consistent with a nonideal mixing. Two arguments may support such a prediction. First, not all of the π–A isotherms corresponding to binary systems are comprised between the π–A isotherms of the pure components, and second, despite that all mixtures exhibited molecular areas at the collapse pressure virtually the same as that of pure DMPC, the increase in the PEO-lipid content in the mixture brought about the diminution of collapse pressure. The second argument would indirectly indicate that DMPC was expelled from binary monolayers at higher PEO-lipid ratios.

Effect on molecular area

The comparison of the compressional behavior of binary mixtures of DPPC and $C_{11}CONH$-β-CD with those of

Progr Colloid Polym Sci (1997) 103:294–299
© Steinkopff Verlag 1997

Fig. 6 Schematic model of the organization of $C_{11}CONH$-β-CD molecules in mixed DPPC/$C_{11}CONH$-β-CD monolayers at low and high surface pressures

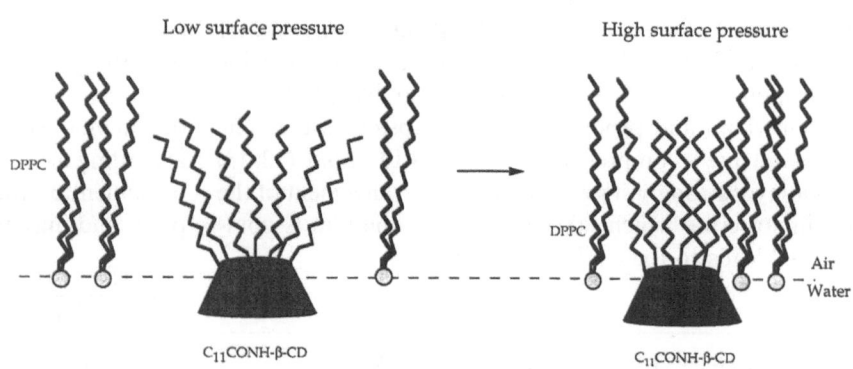

DMPC and PEO-lipid (12,13) shows that whereas for the former the addition of an amphiphile with grafted hydrophobic chains ($C_{11}CONH$-β-CD) resulted in the gradual increase of molecular areas throughout the whole compression cycle, for the latter this increase in molecular area was limited to low pressures. The compression of binary DMPC/PEO-lipid (12,13) films beyond about 40 mN/m caused gradual expulsion of DMPC from the interface which resulted in the diminution of collapse pressures of the mixtures and their displacement toward the value of collapse pressure of pure PEO-lipid (12,13).

Figures 6 and 7 schematically illustrate possible arrangements of studied amphiphiles under compression. For a monolayer constituted of molecules such as $C_{11}CONH$-β-CD, DMPC or DPPC, possessing hydrophobic chains oriented toward the air phase and polar groups located at the level of the air/water interface both the molecular area and surface pressure are determined by the packing of molecules in plane of the monolayer. The compression of such monolayers would solely bring about a change in the orientation of hydrocarbon chains and their confinement to an upright position (Fig. 6). This is reflected by a steep rise of the surface pressure observed with $C_{11}CONH$-β-CD, DMPC and DPPC molecules in the range of molecular areas immediately beyond their respective LE–LC transitions (Figs. 4 and 5).

The PEO-lipid (12,13) represents another type of amphiphiles. Spread molecules of PEO-lipid (12,13) form stable monolayers with hydrophobic C12 chains protruding into the air phase and the hydrophilic poly(ethylene oxide) chain immersed in the aqueous subphase. The compressional behavior of such a molecule strongly depends upon the hydrophilic chain length [10]. While an amphiphile with a short PEO chain (5 ethylene oxide monomers) was shown to form condensed-type monolayers characteristically displaying a steep increase in surface pressure, a moderate slope, characteristic of an expanded film behavior was observed for longer PEO chain lipid

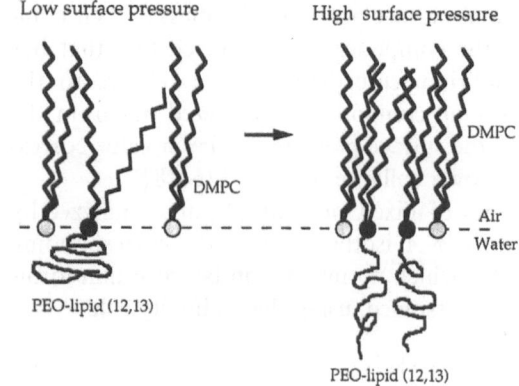

Fig. 7 Schematic model of the arrangement of PEO-lipid (12,13) molecules in mixed DMPC/PEO-lipid (12,13) monolayers at low and high surface pressures

derivatives (12,13 or 12,31) [9,10]. Lateral compressibility $(1/A)(\mathrm{d}A/\mathrm{d}\pi)$ of the PEO-lipid (12,13) against its surface density plots showed a minimum at 2.6×10^{-2} m/mN which was superior to that of DMPC (1.3×10^{-2} m/mN) and would suggest that the PEO chain immersed in the water subphase controls the packing of the entire molecule at the interface. Indeed, as indicated by the high value of its limiting area (55 Å^2) the PEO moiety occupies a large area and prevents the establishment of a close contact between hydrophobic chains of film-forming molecules. Compared to DMPC which has a hydrophobic chain of similar length, the PEO-lipid (12,13) displays a higher compressibility and a lower collapse pressure. Based on these considerations compressional arrangements of PEO-lipid (12,13) at the interface may resemble those schematically depicted in Fig. 7.

The deployment of the PEO chain in water may take different forms. If the chain were in the form of a random coil, then on compression one would expect a large molecular area and rather a low collapse pressure. Conversely,

298

V. Rosilio et al.
Interfacial miscibility of amphiphiles

an unfolded elongated form of the PEO chain in water would reduce its molecular area and yield a limiting value corresponding to a cross-section of the double C12 hydrocarbon chain and produce a high collapse pressure. The observed high compressibility of the PEO-lipid (12, 13) monolayer strongly suggests that the rearrangement of the chain structure which takes place when the film is compressed leads to the chain elongation.

Ideality–non-ideality of the systems. Occurrence of an interaction between two film-forming components

When two film-forming components are immiscible, then according to the phase rule, their mixed monolayers collapse at the same surface pressure regardless of their composition; e.g. the component of the mixed film that has a lower equilibrium spreading pressure relative to the other film-forming component is squeezed out from the monolayer as the surface pressure reaches a value corresponding to its own collapse pressure [16–18].

The behavior of mixed monolayers can be analyzed by comparing their π–A isotherms with the corresponding ideal curves, for which no interaction between film-forming molecules is assumed, using the additivity rule

$$A_{\text{mix}} = xA_1 + (1 - x)A_2,\tag{1}$$

where A_{mix} corresponds to the mean molecular area per molecule of the mixture, x and $(1 - x)$, and A_1 and A_2 are the mole fractions and molecular area at a given surface pressure of components 1 and 2, respectively. In the case of ideal miscibility or complete immiscibility, A_1 and A_2 are constant and equal to the values for the pure components.

According to the theory developed by Joos and Demel [3], the collapse pressure of a mixed insoluble monolayer, Π_{cm}, with completely miscible components is given by the relation

$$1 = X_1^s \gamma_1 \exp\left(\frac{\Pi_{\text{cm}} - \Pi_{\text{c1}}}{kT} A_1\right)$$
$$+ X_2^s \gamma_2 \exp\left(\frac{\Pi_{\text{cm}} - \Pi_{\text{c2}}}{kT} A_2\right),\tag{2}$$

where X_1^s and X_2^s are the mole fractions at the surface, A_1 and A_2 are the molecular areas at the corresponding Π_{c1} and Π_{c2} collapse pressures of components 1 and 2, respectively; γ_1 and γ_2 represent the surface activity coefficients at the collapse point of component 1 and 2, respectively, k is the Boltzmann constant and T the absolute temperature. In the case where no interaction between two components occurs, surface activity coefficients will be

equal to unity ($\gamma_1 = \gamma_2 = 1$), and the theoretical curve describing the collapse pressure of the mixed film, according to Eq. (2), should coincide with the experimental curve drawn on the basis of measured points with that calculated according to Eq. (2). For a non-ideal mixture, γ_1 and γ_2 are dependent on the interaction parameter, ξ, at the collapse point and may be calculated from

$$\xi = \frac{\Pi_{\text{c1}} - \Pi_{\text{cm}}}{kTX_2^s} A_1.\tag{3}$$

The interaction energy ΔE may then be calculated using Eq. (4), and assuming a hexagonal lattice formation in closely packed mixed monolayers, where each molecule is surrounded by 6 neighbors ($ZL = 6$):

$$\Delta E = \frac{\xi RT}{ZL}.\tag{4}$$

Figure 8 clearly shows that the mixing behavior of binary $C_{11}CONH$-β-CD/DPPC monolayers displayed an important deviation of the experimental curve from the ideal curve and that this deviation was observed as well when the Joos and Demel approach was applied [3, 11] as from the data of the molar area versus $C_{11}CONH$-β-CD molar fraction relationship [11]. It should also be noted that an existence of the interaction for this system is particularly remarkable from the analysis of the collapse

Fig. 8 Experimental and ideal collapse pressures (π_c) calculated from the Joos and Demel approach, and experimental and ideal average molecular areas determined by the additivity rule for the DPPC/$C_{11}CONH$-β-CD system at collapse pressure, versus $C_{11}CONH$-β-CD molar fraction

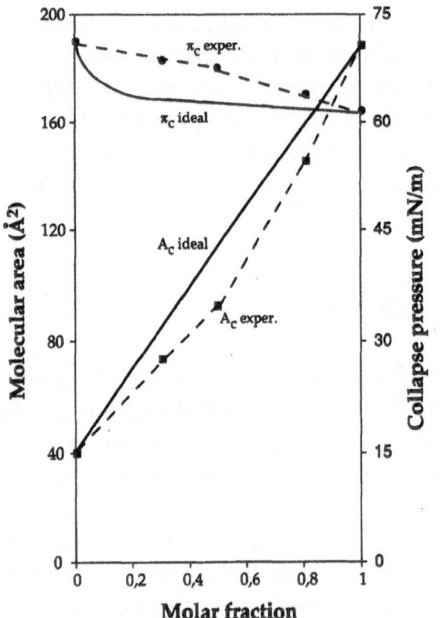

Progr Colloid Polym Sci (1997) 103:294–299
© Steinkopff Verlag 1997

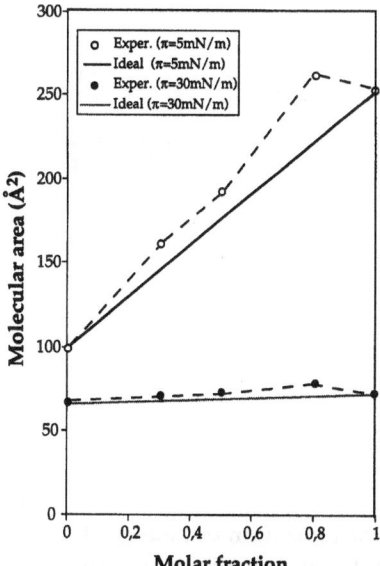

Fig. 9 Experimental and ideal average molecular areas determined by the additivity rule for the DMPC/PEO-lipid (12, 13) system at low (5 mN/m) and high (30 mN/m) surface pressures versus PEO-lipid molar fraction

pressure versus molar fraction relationship (Fig. 8). Negative values of both the interaction parameter and the interaction energy, respectively, equal to -5.5 and -540 cal/mol, are quite significant and would indicate the occurrence of an interaction between the hydrophobic chains grafted on the β-CD and DPPC. Since no deviation from ideality was observed for binary mixtures of DMPC with amphiphiles of β-CDs bearing shorter than C12 hydrocarbon chains, it has been inferred that the interaction between these two film-forming components may only

occur when hydrophobic chains on both molecules have a comparable length [11].

Similarly for the $C_{11}CONH$-β-CD/DPPC system, application of the additivity rule to binary DMPC/PEO-lipid (12, 13) mixtures shows that the miscibility of these two components in monolayers is far from ideal (Fig. 9). In this case the observed deviation from linearity may be either due to the steric hindrance caused by the hydrophilic PEO moiety of the molecule, or due to the interaction between these two film-forming molecules at the level of their hydrophobic chains. The exact location of the level at which this interaction takes place appears to be difficult to establish since both the hydrophobic and hydrophilic moieties of PEO-lipid molecules may contribute to its occurrence. Previously published results on the miscibility of DMPC with PEO-lipid-bearing PEO moieties with varying chain length have clearly demonstrated that hydration of hydrophilic moieties essentially affects the miscibility of the system. The hydration effect should account for the possible existence of two zones of water, one directly bound by hydrogen bonds to a poly(ethylene oxide) chain and water in the vicinity of polymer chain in which there has been significant restructuring [9]. The latter effect would be all the more important as the length of the poly(ethylene oxide) chain increased and for the PEO-lipid (12, 13), as suggested by the surface potential data, the hydrophilic part of the molecule would profoundly influence the overall behavior of its mixed monolayer with DMPC.

An additional support which would reinforce the above conclusion is that the calculated values of the interaction parameter ($\xi = -1$) and interaction energy ($\Delta E = -98.7$ cal/mol) for DMPC/PEO-lipid (12, 13) are much lower compared to the DPPC/$C_{11}CONH$-β-CD system.

References

1. Birdi KS (1989) Lipid and Biopolymer Monolayers at Liquid Interfaces. Interscience, New York
2. Inoue T, Tasaka T, Shimozawa R (1992) Chem Phys Lipids 63:203–212
3. Joos P, Demel RA (1966) Biochim Biophys Acta 183:447–457
4. Pagano RE, Gershfeld NL (1972) J Colloid Interface Sci 41:311–317
5. Smaby JM, Brockman HJ (1991) Langmuir 7:1031–1034
6. Smaby JM, Brockman HJ (1992) Langmuir 8:563–570
7. Hasmonay D, Billoudet F, Badiali JP, Dupeyrat M (1994) J Colloid Interface Sci 165:480–490
8. Marsh D (1990) Handbook of Lipid Bilayers, CRC Press, Boston, pp 229–263
9. Rosilio V, Albrecht G, Okumura Y, Sunamoto J, Baszkin A (1996) Langmuir 12:2544–2550
10. Rosilio V, Albrecht G, Baszkin A, Okumura Y, Sunamoto J (1996) Chem Lett 657–658
11. Kasselouri A, Coleman AW, Baszkin A (1996) J Colloid Interface Sci 180: 384–397
12. Kasselouri A, Coleman AW, Albrecht G, Baszkin A (1996) J Colloid Interface Sci 180:398–404
13. Kuwamura T (1961) Kogyoukagakuzasshi 64:1958–1969
14. Ries HE Jr, Swift H (1989) Colloids Surf 40:145–165
15. Gaines GL (1966) Insoluble Monolayers at the Liquid–Gas Interfaces. Interscience, New York, pp 208–300
16. Minones J, Yebra-Pimentel E, Conde O, Iribarnegaray E, Casas-Parada M, Vila Jato JL, Seijo B (1995) J Pharm Sci 84: 508–511
17. Minones J, Yebra-Pimentel E, Conde O, Iribarnegaray E, Casas M, Vila Jato JL, Seijo B (1994) Langmuir 10:1888–1893

Progr Colloid Polym Sci (1997) 103:300–306
© Steinkopff Verlag 1997

N.L. Abbott

Active control of interfacial properties of aqueous solutions using ferrocenyl surfactants

Received: 2 December 1996
Accepted: 3 December 1996

Abstract This paper reviews our recent progress towards the development of principles for active control of interfacial properties of aqueous solutions. Three mechanisms are identified by which oxidation of ferrocene to ferrocenium within surfactants with the structure $Fc(CH_2)_n N^+(CH_3)_3$ (Fc = ferrocene; $n = 8, 11, 15$) can cause changes in the surface tension of aqueous solutions: the changes in surface tension are large (up to 23 mN/m) and can be made reversible by using electrochemical methods to repeatedly oxidize and reduce the ferrocene. We also report principles for delivery of surfactants at controlled rates to spatially defined locations in a solution (with resolution of micrometers). These principles can, we believe, form the basis of methods for the localized release of solubilizates, permeabilization of biological membranes, and control of the stability of thin films of liquid. Finally, we report that lifetimes of surface-active states of ferrocenyl surfactants formed electrochemically can be manipulated from $\sim 10^{-2}$ to 10^3 s by changing the concentration of an oxidizing agent (Fe^{3+}) dissolved within the bulk of the aqueous solution; the Marangoni flow of fluid away from an electrode protruding from the surface of a solution is found to result from a balance in the rates of electrochemical creation and chemical removal of $Fc(CH_2)_{11}N^+(CH_3)_3$ at the surface of the solution.

Key words Active control – surfactants – surface tension – ferrocene – electrochemical

Dr. N.L. Abbott (✉)
Department of Chemical Engineering
and Materials Science
University of California at Davis
Davis, California 95616, USA

Introduction

This paper reviews our recent progress towards the development of principles for active control of interfacial properties of aqueous solutions [1–4]. The goal of this research is to develop methods and materials that will (i) make possible *in situ* transformations in the surface activity of water-soluble molecules, (ii) permit these transformations to be reversed, thereby allowing interfacial properties of surfactant systems to be cycled over time, and (iii) to provide the capability to direct these changes in surface activity to localized regions in space, thus providing control of gradients in interfacial properties of aqueous systems.

The approach we describe herein for active control of surface properties is based on the use of ferrocenyl surfactants and electrochemical methods. Because ferrocene is chemically stable under a wide range of conditions, and because the electrochemical oxidation of ferrocene to the ferrocenium cation is reversible in aqueous solutions, amphiphilic molecules that host ferrocene have been the subject of a number of past studies. These studies include the use of water-insoluble amphiphiles such as $Fc(CH_2)_{16}$ COOH (Fc = ferrocene) for studies of electron transfer across Langmuir–Blodgett films deposited on electrodes [5]; use of water-soluble surfactants such as $N^+(CH_3)_3$ $(CH_2)_{11}Fc$ for manipulation of the stability of dispersions of water-insoluble dyes in solution [6]; and use of amphiphilic molecules such as $HS(CH_2)_{10}COFc$ chemisorbed on surfaces of gold for *in situ* control of the wetting of fluids on surfaces [7].

Here we describe the use of surfactants I–IV (Fig. 1) in a study of principles for active control of interfacial properties of aqueous solutions. Whereas Tajima and coworkers have reported oxidation of $FcCH_2N^+(CH_3)_2$ $(CH_2)_nCH_3$ ($n = 8, 10, 12$) to lead to small changes (< 5 mN/m) in the surface tension of aqueous solutions [8], and whereas Anton and coworkers have observed surfactants with the structure $FcCH_2N^+(CH_3)_2(CH_2)_n$ CH_3 to be unstable in their oxidized states (causing redox reactions to be irreversible) [9], we report that surfactants I–IV permit large (up to 23 mN/m) and reversible changes in the surface tension of aqueous solutions over a wide range of concentrations of surfactant.

The remaining sections of this overview are organized as follows. First, we briefly describe electrochemical characteristics of ferrocenyl surfactants. Second, we report measurements of the surface tension of aqueous solutions of the surfactants, and in doing so, identify three mechanisms by which redox transformations of ferrocenyl surfactants can lead to changes in surface tension. Third, we describe the use of ferrocenyl surfactants and electrochemical methods to cycle properties of surfactant solutions, to direct changes in the properties of surfactant solutions with spatial resolutions of micrometers and,

finally, to tune the lifetime of surface active states of molecules from milliseconds to minutes.

Synthesis and redox characteristics of ferrocenyl surfactants

Methods for the synthesis of surfactants I–IV have been reported previously [2, 4, 6]. Gram quantities of each surfactant can be readily prepared in a few days. Purification of the surfactants can be achieved by repeated recrystallization from solutions of ethyl acetate/acetone until the surface tensions do not change with further recrystallization.

Surfactants I–III each undergo a one-electron oxidation to form a dication ($N^+(CH_3)_3(CH_2)_nFc^+$). The half-wave potential for oxidation of a millimolar solution of II (0.1 M Li_2SO_4, pH 2) is 0.17 V (vs. SCE). We have measured the half-wave potential of the series I–III to increase as a function of chain length. Solutions of I–III, when left exposed to air, oxidize over hours and days: the rate of oxidation decreases with increasing chain length of the surfactant. Electrochemical characterization of IV demonstrates that the two ferrocene groups oxidize independently at a half-wave potential of 0.15 V (vs. SCE) [4].

Surfactants I–IV can also be oxidized or reduced by addition of iron (III) sulfate or sodium hydrosulfite, respectively. Equilibrium surface tensions (see below) were measured to be indistinguishable when using chemical or electrochemical methods to manipulate the oxidation states of the surfactants.

Surface tensions

Figure 2 shows the equilibrium surface tensions (Wilhelmy plate method) of aqueous solutions (0.1 M Li_2SO_4, pH 2) of surfactants I–IV in their reduced and oxidized states [2, 4]. Below we discuss features of these plots that demonstrate the existence of three different mechanisms that lead to changes in surface tension upon oxidation of ferrocene to ferrocenium.

First, for all reduced surfactants dissolved in solution at concentrations near their critical micelle concentrations (CMC), oxidation leads to an increase in the surface tension of the solution (Fig. 1). In the case of surfactants I and II, oxidation returns the surface tension of the solution to a value that is similar to the surfactant-free solution of electrolyte (approx. 72 mN/m). The excess surface concentration of surfactant, estimated using the Gibbs adsorption equation, decreases in the case of surfactant II from 2.0×10^{-6} to $< 0.1 \times 10^{-6}$ mol/m^2 upon oxidation. Clearly, oxidation drives the desorption of surfactant from the surface of the solution. The increase in surface tension of

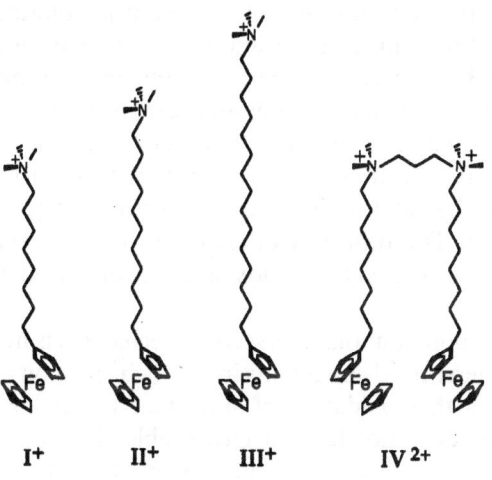

Fig. 1 Structures of ferrocenyl surfactants I–IV

I^+ II^+ III^+ IV^{2+}

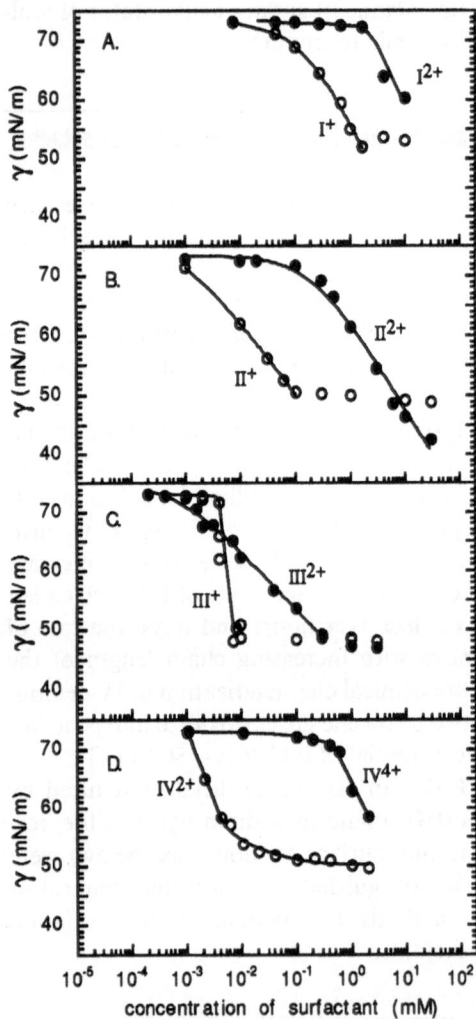

Fig. 2 Equilibrium surface tensions of aqueous solutions (0.1 M Li$_2$SO$_4$, pH 2, 25°C) of ferrocenyl surfactants: A. (O) I^+ (●) I^{2+}; B. (O) II^+ (●) II^{2+}; C. (O) III^+ (●) III^{2+}; D. (O) IV^{2+} (●) IV^{4+}

the solution can be as large as 23 mN/m and can be reversed by chemical or electrochemical reduction of the surfactant.

Although *in situ* control of the extent of adsorption of surfactant to a surface can be used to effect large changes in surface tension (as described above), inspection of Figs 2A–C shows that the maximal change in surface tension is limited to a narrow range of concentrations around the CMC of each of the reduced surfactants I–III. Surfactant **IV** – a dimeric ferrocenyl surfactant – was synthesized with the goal of increasing the range of concentrations of surfactant over which maximal changes in surface tension can be obtained by desorption. Because the extent of adsorption of surfactant to the surface of an aqueous solution is determined by the standard free energy of adsorption, and because the change in the standard free

energy of adsorption that accompanies oxidation of IV^{2+} to IV^{4+} will be roughly twice that of the corresponding monomeric surfactant (I^+), we hypothesized that control of the oxidation state of dimeric ferrocenyl surfactants would provide the capability to control surface tensions over a wider range of concentrations than is possible with monomeric surfactants. The results presented in Fig. 1D demonstrate this hypothesis to be correct: the range of concentrations of surfactant leading to a maximal change in surface tension can be extended to almost two orders of magnitude by using a dimeric surfactant [4].

The second feature of the surface tension plots that we wish to highlight is found in Fig. 2B. At concentrations of II^+ greater than 8 mM, oxidation of II^+ to II^{2+} results in a *decrease* in surface tension, in contrast to the *increase* in surface tension observed at lower concentrations. Estimates of the excess surface concentrations of surfactant indicate that the limiting values for II^+ and II^{2+} are similar: 2.0×10^{-6} mol/m^2. That is, at concentrations of II^+ much greater than its CMC, oxidation to II^{2+} does not result in measurable desorption of surfactant. The limiting excess concentration of this surfactant at the surface of its solution does not appear to be controlled by the electrostatic charge on the surfactant (which is, of course, changed by oxidation). The electrostatic charge on the surfactant does, however, appear to influence the surface pressure exerted by the monolayer: oxidation leads to an increase in the density of charge within the monolayer of surfactant and thus to an increase in the electrostatic contribution to the surface pressure (and decrease in surface tension).

This second mechanism leading to changes in surface tension, namely *in situ* change in the density of charge within a monolayer of surfactant adsorbed at the surface of a solution, has resulted in changes in surface tension of only 6 mN/m or less over the range of concentrations studied to date. We believe, however, that this second mechanism is a promising one because (i) it should become insensitive to the concentration of surfactant in solution provided a threshold concentration is exceeded and (ii) it leads to low surface tensions. We note, in addition, that this second mechanism is one that can, in principle, lead to rapid changes in surface tension because it does not require the transport of surfactant to or from the surface of the solution in order to effect a change in surface tension. The transport of charge to or from the surfactant can be effected by a redox mediator dissolved in solution.

Finally, we point out that this second mechanism is not general to all ferrocenyl surfactants in Fig. 1. Inspection of Fig. 2C shows that oxidation of surfactant **III** at high concentrations does not lead to measurable changes in surface tension. As discussed below, in contrast to **II**,

Progr Colloid Polym Sci (1997) 103:300–306
© Steinkopff Verlag 1997

oxidation of III^+ to III^{2+} does lead to substantial change in the surface excess at high concentrations of III.

Inspection of Fig. 2C reveals the third mechanism by which changes in surface tension can be effected by changes in the oxidation state of ferrocenyl surfactants. At concentrations less than the CMC of III^+, oxidation of III^+ to III^{2+} leads to a *decrease* in the surface tension of the solution. This behavior contrasts to that of surfactants I, II and IV, where oxidation leads to an increase in surface tension at low concentrations (via desorption of surfactant from the interface).

Although we do not yet fully understand this third mechanism leading to changes in surface tension, below we point out some interesting differences in the behavior of III and the other ferrocenyl surfactants, and use these observation to propose a possible explanation for the cause of the decrease in surface tension upon oxidation of III^+ to III^{2+} (as seen in Fig. 2C).

First, we note that the line shape of the plot of surface tension for III^+ in Fig. 2C differs from that of I^+ and II^+. The surface tension of aqueous solutions of III^+ decreases abruptly at a concentration of 0.003 mM whereas the surface tensions of solutions of I^+ and II^+ decrease gradually with increasing concentration. The adsorption isotherm of III^+ cannot be described by the Langmuir model of adsorption. The limiting area occupied by III^+ (< 28 Å^2/molec) is also substantially less than either I^+ (55 Å^2/molec) or II^+ (85 Å^2/molec).

Second, we observe the behavior of sulfur dust on the surface of solutions III^+ to differ from solutions of II^+ and water: sulfur dust sprinkled on the surface of solutions of III^+ moves together, whereas the behavior of dust on the surface of solutions II^+ resembles that on water (relative motion of particles of sulfur is observed). These observations suggest that films formed by III^+ on the surface of aqueous solutions are more rigid than films formed by II^+.

Third, inspection of Fig. 2A–C demonstrates that the threshold concentrations of I^+, II^+, and III^+ required to reduce the surface tension below that of the aqueous solution of electrolyte are 0.02, 0.001 and 0.003 mM, respectively. It is surprising that the threshold concentration of III^+ is greater than II^+ because, typically, the threshold concentration for measurable surface activity decreases with increasing chain length of a surfactant.

Fourth, whereas the maximum excess surface concentrations of I and II change little upon oxidation, a substantial change in the surface excess accompanies oxidation of III at high concentrations. Oxidation of solutions of III causes the limiting surface excess to decrease from $> 6 \times 10^{-6}$ to 1.5×10^{-6} mol/m^2. These values correspond to surface areas of < 28 and 110 Å^2/molec, respectively.

The observations above, when combined, support the view that III^+ forms condensed monolayers at the surface of aqueous solutions. Because the limiting area of III^+ was found to be < 28 Å^2/molec, we infer that III^+ assumes an extended conformation within these monolayers with ferrocene located at the outer surface (air side) of the monolayer. We do not expect ionic surfactants to form multilayers at the surface of an aqueous solution. Upon oxidation of ferrocene to ferrocenium, however, both the ferrocenium and trimethylammonium cations will strongly associate with the aqueous subphase (to decrease the self-energy of their charges). Monolayers of III^+ would, therefore, be expected to undergo extensive reorganization at the surface of the solution upon oxidation. The reorganization, driven by hydration of the ferrocenium cation, appears to be reflected in the substantial increase in the limiting area occupied by III^+ upon oxidation to III^{2+} (increase from < 28 to 110 Å^2/molec).

We hypothesize that reorganization of the surfactant-laden surface, as described above, can account for the decrease in surface tension measured upon oxidation of III^+ at concentrations less than 0.004 mM. We propose that the change in conformation upon oxidation of III^+ to III^{2+} causes the state of the monolayer to change from a condensed one to an expanded one. The abrupt change of surface tension at 0.004 mM is consistent with the formation of condensed domains of III^+ on the surface of the solutions with bulk concentrations less than 0.004 mM. A decrease in the surface tension of an aqueous solution of III^+ requires interaction of the condensed domains of III^+; molecules of III^{2+}, in contrast, are dispersed homogeneously on the surface of the aqueous solutions and, therefore, plausibly reduce the surface tension of these solutions at concentrations lower than that of III^+. Ongoing experiments based on X-ray reflectivity, Brewster angle microscopy and reflection IR spectroscopy will, we believe, permit the above proposed physical pictures of ferrocenyl surfactant-laden interfaces to be tested. Finally, we note that the absence of change in surface tension upon oxidation of III^+ at high concentration appears to result from the opposing effects of a decrease in the excess concentration of surfactant and an increase in the charge per molecule of surfactant (as discussed in Ref. [2]).

In summary, measurements of surface tension presented in Fig. 2 demonstrate that three distinct mechanisms cause changes in surface tension upon oxidation of ferrocenyl surfactants: (i) desorption of surfactant from the surface of an aqueous solution; (ii) ionization of a surfactant without significant desorption from the surface of the solution; and (iii) change in the phase state of a spontaneously adsorbed monolayer of surfactant, possibly driven by a change in the conformation of the surfactant within the monolayer.

Reversible changes in surface tension

The changes in surface tension described above were driven by oxidation of the electrically neutral ferrocene to the ferrocenium cation. These changes can be reversed by reduction of the ferrocenium cation back to ferrocene [1]. The surface tensions of aqueous solutions of **II** can be cycled (> 14 cycles) on time-scales of minutes by using electrochemical methods to sequentially oxidize **II**$^+$ and reduce **II**$^{2+}$. Figure 3 shows 14 cycles ($\Delta\gamma \sim 16$ mN/m) of the dynamic surface tension measured using a maximum bubble pressure tensiometer. The dynamic surface tensions of aqueous solutions of **II**$^+$ are higher (~ 8 mN/m) than the equilibrium values measured using a Wilhelmy plate (see Fig. 2). The response time of the maximum bubble pressure tensiometer was approximately one minute. The very small loss of surface activity (~ 0.1 mN/m per cycle) that accompanied each cycle was caused by unknown processes taking place during the electrochemical transformation; the loss of surface activity was not due to chemical instability of **II**$^+$ or **II**$^{2+}$ in solution.

Spatially-directed delivery of surfactant

Gradients in surface (or interfacial) tension can accelerate the spreading of fluids, enhance the stability of surfactant-laden films of liquid, emulsions, and foams, and increase rates of mass transport across interfaces. The motion of fluid driven by a gradient in surface tension is referred to as a "Marangoni flow". We have demonstrated that electrochemical reduction of **II**$^{2+}$ to **II**$^+$ at an electrode that

protrudes from the surface of the solution can cause the Marangoni flow of an aqueous solution [3]. Because the surface tensions of aqueous solutions of **II**$^{2+}$ differ from **II**$^+$, the localized transformation of **II**$^{2+}$ to **II**$^+$ results in a gradient in the surface tension of an aqueous solution. Temporal control of the potential applied to an electrode permits active control of both the location and the rate of delivery of surfactant to an aqueous solution and thus the strength of the Marangoni flow.

The Marangoni phenomenon is shown in Fig. 4. Sulfur powder was used to visualize the motion of fluid on the

Fig. 4 Top view of displacement of fluid on the surface of an aqueous solution of 0.1 mM **II**$^{2+}$; (100 mM Li$_2$SO$_4$, pH 2.0) induced by application of -0.3 V to a Pt working electrode protruding from the surface of the solution. The surface of the solution was dusted with sulfur powder. Panel a is a schematic illustration of the arrangement of the electrodes. The reference electrode (RE) was a SCE and the counterelectrode (CE) was a platinum flag. Panels b–d show the distribution of sulfur powder on the surface of the solution before application of the reducing potential (b) and 30 s (c) and 60 s (d) after application of the reducing potential to the WE

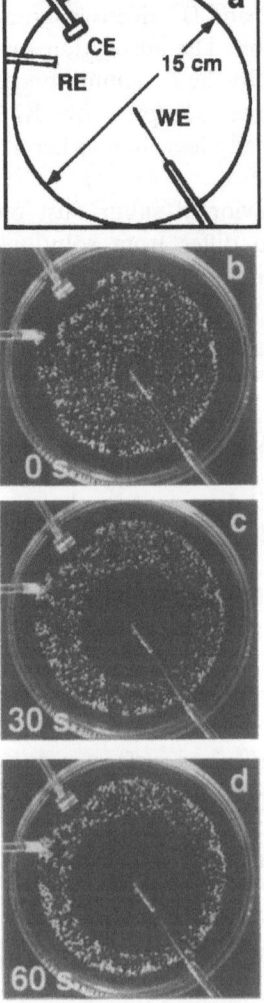

Fig. 3 Dynamic surface tension of an aqueous solution of 0.3 mM **II**$^+$/**II**$^{2+}$ (0.1 M Li$_2$SO$_4$, pH 2, 20°C) measured during the repeated cycling of the surfactant between oxidation states. The high values of surface tension correspond to a solution of **II**$^{2+}$ and the low values correspond to a solution of **II**$^+$

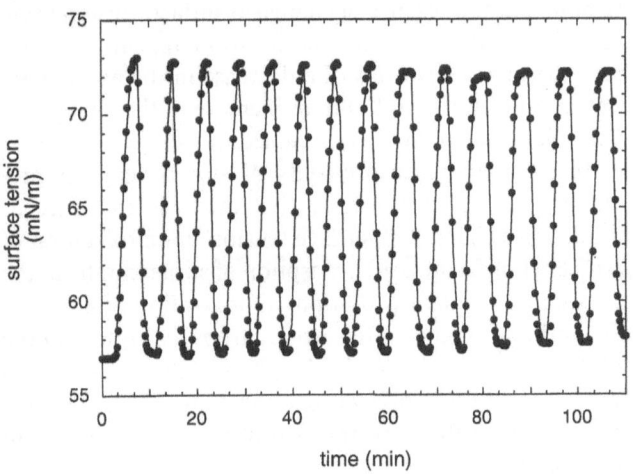

Progr Colloid Polym Sci (1997) 103:300–306
© Steinkopff Verlag 1997

surface of a solution of II^{2+}. The surface of each aqueous solution of II^{2+} was placed in motion by application of a reducing potential to a platinum working electrode. Application of -0.3 V to a Pt electrode immersed in a solution of 0.1 mM II^{2+} caused the surface of the fluid to move radially outward from the Pt working electrode toward the edge of the petri dish. Sulfur powder supported on the surface of the solution was displaced from the center of the dish toward the meniscus of the fluid at the vertical walls of the petri dish. The motion of fluid was sustained for times greater than 10 min at -0.3 V.

The effect of the potential applied to the electrode on velocity of fluid motion was investigated by using a solution of 0.1 mM II^{2+} prepared by chemical oxidation of a solution of II^+ with ca. 1.5 equivalents of Fe^{3+}. The onset of the motion of fluid occurred between 0.2 and 0.1 V, which spans the half-wave potential measured for oxidation of II^+ to II^{2+} (0.17 V vs. SCE). The velocity of fluid motion increased with increasingly negative (reducing) potentials up to ca. -0.2 V where the velocity reached a plateau. Above -0.2 V, mass transport of the surfactant to the working electrode plausibly limits the rate of generation of surfactant and thereby limits the velocity of fluid motion.

Control of life-time of surface-active states of molecules

Surfactants form the basis of procedures for the extraction of proteins from biological membranes and for the preparation of liposomes. The exclusion or removal of low concentrations of surfactants from solutions following their use can, however, be tedious and difficult (e.g. dialysis). Low concentrations of surfactants can also denature proteins when extracting biological membranes. By combining ferrocenyl surfactants and electrochemical methods, we have established principles for the creation of surfactant molecules with lifetimes of 10^{-2}–10^3 s in localized regions of solutions, thereby providing new principles for the use of surfactants in aqueous systems in which their sustained or widespread presence is detrimental [3].

The principle we report is based on a competition between the rate of electrochemical generation of a surface-active species and its chemical removal through the presence of a redox-active species dissolved in solution. Surface-active II^+ was generated electrochemically by reduction of II^{2+} at an electrode immersed in a solution of II^{2+} and Fe^{3+}. The lifetime of II^+ so-formed was controlled by its rate of formation at the electrode and its rate of chemical oxidation by Fe^{3+}. Manipulation of the concentration of Fe^{3+} can be used to control the average lifetime of the surface-active state of the ferrocenyl surfactant.

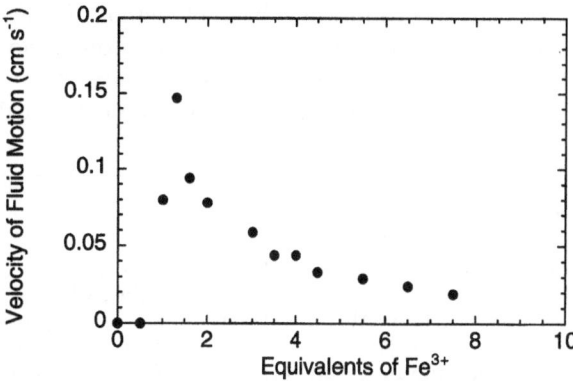

Fig. 5 Dependence of the velocity of fluid motion at the surface of an aqueous solution (100 mM Li_2SO_4, pH 2.0) of 0.1 mM II^{2+} on the number of equivalents of Fe^{3+} added to the solution. The fluid motion was induced at the surface of the solution by application of -0.3 V to a Pt electrode protruding from the surface of the solution

The Marangoni phenomenon at the surface of an aqueous solution of II^{2+} (see above) can be used to demonstrate the above principle for the formation of surfactant with lifetimes that range from 10^{-2} to 10^3 s. We observed the velocity of fluid motion near an electrode to reflect a balance between rates of formation and removal of II^+ from the surface of a solution (Fig. 5). At low concentration of Fe^{3+} (0.01 molar excess equivalents of Fe^{3+}), the steady-state velocity of fluid motion was low because surfactants generated at the electrode were long-lived and quickly saturated the surface region of the solution. By addition of 0.5 molar excess equivalent of Fe^{3+}, and thus by tuning the lifetime of the ferrocenyl surfactant to ca. 1 s, steady-state gradients in surface tension could be maintained on the surfaces of aqueous solutions for tens of minutes. The gradients caused substantial surface flows. At concentrations of Fe^{3+} greater than ca. 1 mM, however, II^+ is consumed by Fe^{3+} over times-scales that are short (tens of ms) compared to time-scales associated with the spreading of the surfactant on the surface of the solution: the velocity of fluid motion is observed to decrease rapidly.

Conclusion

This paper summarizes three mechanisms by which oxidation of ferrocene to ferrocenium within ferrocenyl surfactants can drive changes in the surface tension: the changes are large (up to 23 mN/m) and can be made reversible by using electrochemical methods. In addition, the principles we report for delivery of surfactants at controlled rates to defined locations in a solution can, we believe, be extended to form the basis of methods for the localized release of solubilizates, permeabilization of biological membranes, and control of the stability of thin films of liquid. The development of photochemical methods for control of the oxidation states of ferrocenyl surfactants will, we believe, expand possible uses of ferrocenyl surfactants in schemes for active control of interfacial properties.

References

1. Gallardo B, Hwa M, Abbott N (1995) Langmuir 11:4209–4212
2. Gallardo B, Metcalfe K, Abbott N (1996) Langmuir 12:4116–4124
3. Bennett D, Gallardo B, Abbott N (1996) J Am Chem Soc 118:6499–6505
4. Gallardo B, Abbott N (1997) Langmuir 13:203–208
5. Guo L, Facci J, McLendon G (1995) J Phys Chem 99:4106–4112
6. Saji T, Hoshino K, Ishii Y, Goto M (1991) J Am Chem Soc 113:450–456
7. Abbott N, Whitesides G (1994) Langmuir 10:1493–1497
8. Tajima K, Huxur T, Imai Y, Motoyama I, Nakamura A, Koshinuma M (1995) Colloids Surf A 94:243–251
9. Anton P, Heinze J, Laschewsky A (1993) Langmuir 9:77–85

Progr Colloid Polym Sci (1997) 103:307–317
© Steinkopff Verlag 1997

A.R. Pitt

The efficiency of dynamic surface tension reductions within homologous series of surfactants in aqueous gelatin solution

Received: 18 December 1996
Accepted: 31 December 1996

A.R. Pitt (✉)
Kodak European R&D
Kodak Ltd.
Headstone Drive
Harrow, Middlesex HA1 4TY
United Kingdom

Abstract Here we consider the dynamic surface tension (DST) reduction at approximately 0.1 s surface age of three homologous series of surfactants in aqueous solution containing 7% deionised type IV bone gelatin (Kodak type TCG-II). The three homologous series represent three different classes of surfactant, anionic, nonionic and nonionic–anionic. Whereas the static or equilibrium values of surface tension at the critical aggregation concentration tend to decrease with increasing hydrophobic tail length and reach limiting values with the more hydrophobic homologous, the DST reduction always shows a region of maximum efficiency as a homologous series is ascended. This region of maximum efficiency covers approximately one log unit of concentration (2×10^{-4} to 2×10^{-3} M). The maximum arises as the consequence of competition between increasing surface activity and decreasing monomer flux to the surface as the tail length increases. The decrease in

monomer flux is believed to be due to two processes: first, the formation of micelles which diffuse more slowly in gelatin solution and have increased stability towards breakdown to monomer; secondly, an increase in surfactant–gelatin interactions with increasing tail length. To a large extent, the most efficient homologues display their maximum efficiency at sub-micellar concentration. Higher homologues, i.e. those more hydrophobic than the most efficient homologues, show marked decreases in their efficiency of DST reduction with increasing tail length. This decreasing efficiency is clearly related to an increasing stability of their micelles. Overall, the most efficient members of the anionic surfactant series are 2 × more efficient than the corresponding members of either the nonionic or nonionic–anionic series which are of similar efficiency.

Key words Dynamic surface tension – homologues – surfactants–gelatin interactions

Introduction

Surfactants find applications in a wide variety of industries. Whilst equilibrium properties can often be a key to their use, there are many processes in which dynamic properties are foremost in importance. Examples of such

processes are coating, foaming, emulsification, dispersion and lubrication. The first two of these processes relate to the liquid/air interface, which is the area of focus for this paper. In the field of simultaneous multi-layer coating, Valenti et al. [1] have reported that a dewetting defect, known as the "receding edge effect", correlates with the dynamic surface tension properties of the coating solutions and not their static surface tension properties. A study of foaming carried out on a range of alkyl benzene sulphonates by Garret and Moore [2] has shown that foam stability correlates with dynamic surface tension lowering and not static or equilibrium values.

The question we wish to explore is the part played by the structure of a surfactant in determining its dynamic behaviour, particularly its "efficiency" at lowering surface tension. "Efficiency" here is used in the spirit of Rosen's definition [3] ($-\log[\text{concentration}/M]$ of surfactant required to lower surface tension by $20 \, \text{mN} \, \text{m}^{-1}$), but is defined alternatively [4] as the $\log[\text{concentration}/M]$ required to reach a surface tension midway between the value of the solvent and the value at the critical micelle concentration (CMC). Note, the alternative definition allows consideration of systems in which the interfacial tension lowering is less than $20 \, \text{mN} \, \text{m}^{-1}$. Another method of measuring efficiency is introduced later in the paper, which effectively considers the relative reduction in DST (dynamic surface tension) at a specific molar concentration. However, this definition is more arbitrary and the concentration employed is particularly dependent on surface age. Overall, efficiency is an interesting concept inasmuch as it is largely independent of the ultimate surface tension lowering, which can be varied, within limits, by simply changing the chemistry and geometry of any particular surfactant type [4]. Of particular interest, is how the efficiency of dynamic surface tension (DST) changes on ascending a homologous series of surfactants. To our knowledge there are three previous reports of how dynamic effects change on ascending a homologous series, all of which relate to foaming.

The first involves a study by Malysa et al. [5] of the surface elasticity and dynamic stability of wet foams for a homologous series n-alcohols (C_4–C_{10}). On ascending the series, the authors found that foam stability passed through a maximum with the C_6–C_8 alcohols. This effect was found to be unrelated to the Marangoni dilational modulus. However, the authors also determined a parameter they called the "effective elasticity" which depended on the kinetics of adsorption. They found that at short time scales (0.05–0.10 s), the effective elasticity correlated well with foam stability, reaching a maximum value with the C_7 alcohol.

The second report [6] involves a similar study with n-alkyl fatty acids (C_5–C_{10}). Here Malysa et al. show that the fatty acids follow exactly the same trends as the n-alcohols. There is a maximum in foam stability that corresponds with a maximum in effective elasticity as the homologous series is ascended (for short time scales, 0.02–0.64 s). The authors conclude from the first two studies that foamability does not relate simply to the

surfactant's surface activity, but must be dependent on a dynamic parameter such as effective elasticity.

The third report of dynamic effects within a homologous series, by Garrett and Moore [2], involves the foaming and dynamic properties of a range of alkyl benzene sulphonates. Like Malysa et al., they found a maximum in foam stability on ascending their homologous series in the face of a monotonic increase in surface activity. On investigating a parameter dependent on the kinetics of adsorption, in their case dynamic surface tension (DST) at 0.1 or 0.5 s surface age, they obtained a good correlation with their foam stability measurements. So all three studies effectively found the same phenomenon, namely that surfactant adsorption at short time scales (~ 0.05–0.5 s) goes through a pronounced maximum on ascending a homologous series, which correlates with maximum foam stability.

Surfactants are used as coating aids in the coating of photographic materials, the main medium for which is aqueous gelatin solution. In consequence, there is a particular interest in the dynamic behaviour surfactants in such media [1]. The purpose of this study is to investigate the dynamic performance of homologous series of surfactants in aqueous gelatin solution with a view to modelling their resulting behaviour in the near future. Before describing the experimental procedures and materials in detail it is helpful to outline the general approach.

A concentration of 7% gelatin w/w in water was chosen as the standard for the investigation. This is reasonably commensurate with its level of use in photographic systems and is often used as a standard concentration for evaluating other physical characteristics, such as bloom strength. Three different classes of surfactants were selected to seek differences or common factors in their dynamic behaviour: anionic, nonionic and anionic/nonionic (hydrophile comprising both anionic group and nonionic block).

The sodium trialkyl sulphotricarballylate (STS) class was selected for the anionic series [7]. Their tri-tail structure results in a relatively large jump in CMC for each extra carbon added to the tail, and hence fewer homologous are required to cover a broad range of CMC. With anionic surfactants in aqueous gelatin solution there is a cooperative association between micelles and gelatin, which occurs at much lower concentrations than the CMC, in water. To avoid confusion, the cooperative association in gelatin solution has been termed the critical aggregation concentration (CAC) [8,4]. This term is adopted hereafter.

For the nonionic study, a well-defined series of sugar-based surfactants was selected. These materials, termed dialkyl bis-gluconamides, were chosen on the basis of previous work which had shown them to be of good purity and to possess classic nonionic surfactant behaviour in terms of change of CMC with increasing hydrophobic tail length [9, 10].

For the anionic/nonionic study, the sulphated polyethoxylate class was selected. To obtain clean, single compounds, a range of pure, single-component alkyl pentaethoxylates were procured and sulphated. Details of this class of surfactant are given together with details of the other two classes in the Materials section.

The DST at ~0.1 s surface age was chosen as the dynamic parameter for this study. Use of a single point parameter is justified by the previous studies [2, 5, 6] insofar as they all show that the comparative dynamics of simple homologous series of surfactants are relatively unaffected by surface age in the range 0.02–0.50 s. Furthermore, this permits a relaxation in the choice of DST method in terms of the precision of defining its surface age. Hence, we have been able to use a relatively simple, rapid method of DST measurement, the overflowing circular weir [11–13] which gives an approximate surface age of 0.1 s (i.e. ± 0.05 s) under the conditions used. Using this technique, DST measurements were made on all 3 classes of surfactant as a function of homologue and concentration.

Dynamic surface tension is an important property of a surfactant solution and a recent book [14] provides an excellent review of the area. However, the effects that surfactant type and structure, micelle formation, CMC, and equilibrium tension have on the decay of surface tension with time ($\gamma(t)$) are still unclear. The Ward and Tordai equation [15] describes the diffusion-controlled adsorption of surface-active molecules at a fresh interface, but does not offer an analytical solution for surface tension decay $\gamma(t)$. Instead, limiting laws tend to be used which only apply at short times where the surface tension is close to that for the pure solvent, or at long times where the surface tension is close to the limiting value, i.e. in the vicinity of the CMC. Examples of such approximations for neutral molecules have been published by Miller et al. [16]. Hence, there is a problem describing the complete lifetime of surface tension decay, particularly at intermediate times. This issue has recently been revisited by Eastoe et al. [17] who studied the dynamics of homologous series of two types of nonionic surfactant, dialkyl bis-monosaccharides and CiEj. Because of these problems, we are endeavouring to develop an improved model that fits the dynamic behaviour in a more precise way. However, this is the subject of a further paper. The first step is to determine if there are any pattern of behaviour that can be related to surfactant dynamics.

Experimental

Apparatus

Measurements of static surface tension (SST) and DST using the Wilhelmy plate and overflowing circular weir methods have been described in detail in a previous publication [4]. The set of conditions specified for DST measurements yields a surface age of 0.1 ± 0.05 s with low-viscosity solutions.

Reagents

ROOC.CH$_2$ (I) Sodium trialkyl sulphotricarballylates (STSs), where

ROOC.CH

ROOC.CH.SO$_3$Na

$R = n\text{-}C_4H_9, n\text{-}C_5H_{11}\ n\text{-}C_6H_{13}\ n\text{-}C_7H_{15}$
$(CH_3)_3.C.CH_2, (CH_3)_2.CH.(CH_2)_2, (CH_3)_3.C.(CH_2)_2,$
$(C_2H_5)_2.CH.CH_2, (CH_3)_2.CH.(CH_2)_3, n\text{-}C_4H_9.CH(C_2H_5).CH_2$

The synthesis of the sulphotricarballylate surfactants (I) used in this study has been described elsewhere [18]:

R CH$_2$.NH.CO.(CHOH)$_4$.CH$_2$OH   (II) Dialkyl bis-gluconamides (DBGs)

\ /

C $R = n\text{-}C_5H_{11}, n\text{-}C_6H_{13}, n\text{-}C_7H_{15}$ and $n\text{-}C_8H_{17}$

/ \

R CH$_2$.NH.CO.(CHOH)$_4$.CH$_2$OH

The synthesis of the dialkyl bis-gluconamides has also been described previously [19]:

R–(OCH$_2$CH$_2$)$_5$OSO3.Na (III), Alkyl pentaethyleneoxide sulphates (sodium salt) where $R = n\text{-}C_{10}H_{11}, n\text{-}C_{12}H_{25},$ $n\text{-}C_{14}H_{29}, n\text{-}C_{16}H_{33}$

The alkyl pentaethyleneoxide sulphates were made by sulphation of the alkyl pentaethyleneoxides from NIKKO Chemicals which are single component nonionic compounds (~98%). The sulphation procedure is described by Gilbert [20].

The gelatin used in this study, was Eastman Kodak TCG-II deionized gelatin. A fuller description of its properties is given elsewhere [4]. The water used was purified by passage through a Millipore system comprising a carbon filtration unit, an RO unit and a MilliQ SP unit; the resultant water had a resistivity of 18.2 MΩ and a surface tension of 72.7 mN m^{-1} at 20°C. The surface tension data of the 7% w/w solution of gelatin in water were SST = 60 (± 1) mN m^{-1}; DST = 61.5 (± 0.5) mN m^{-1}.

Results

Anionic series

Figure 1 shows the combined DST results for the tri-n-C$_2$, tri-n-C$_4$, tri-n-C$_5$, tri-n-C$_6$, tri-n-C$_7$ and tri-2-ethyl-hexyl STS series in aqueous solution containing 7% deionized gelatin. Like Garrett's [2] study, these results show that the efficiency of lowering DST reaches a maximum as a homologous series is ascended, in this case with the C$_5$ homologue. Where curves cross, such as with the C$_4$ and C$_6$ homologues, the lower, more hydrophilic of the two describes the shallower curve. The tri-n-C$_7$ and tri-2-ethyl-hexyl compounds behave very similarly, which is to be expected since branching reduces the hydrophobicity [21]. Table 1 shows the CACs and values of SST at the CAC for

the same series. In Garrett's study [2] of pure aqueous solution, the CMCs and static surface tensions of the homologous series show a monotonic reduction with increasing tail length. In the present study, similar results were obtained, though the presence of the gelatin appears to cause limiting values of both the CAC and the surface tension in the plateau region beyond the CAC, for the more hydrophobic members of the series. The reason for these limiting values is likely be due to surfactant–gelatin interactions. Interestingly, the next most hydrophobic compound in the n-alkyl series, n-octyl STS, shows the first signs of insolubility in the STS/aqueous gelatin system as indicated by a cloudy solution.

Table 1 CAC and SSTs at CAC for anionic, nonionic and nonionic-anionic series

Surfactant	Alkyl group	CAC	SST at CAC [mN m^{-1}]
Sulphotricarballylates	n-C$_4$H$_9$	0.0094	33
	n-C$_5$H$_{11}$	0.00045	29
	n-C$_6$H$_{13}$	0.00018	26
	n-C$_7$H$_{15}$	0.00018	26
	2-et-hexyl	0.00016	26
Di-alkyl-bis-mono-saccharides	n-C$_5$H$_{11}$	0.017	33
	n-C$_6$H$_{13}$	0.0015	32
	n-C$_7$H$_{15}$	0.00025	28
	n-C$_8$H$_{17}$	0.00018	27
Sodium-alkyl-pentha-ethyleneoxide sulphates	n-C$_{10}$H$_{21}$	0.0021	39
	n-C$_{12}$H$_{25}$	0.00035	38
	n-C$_{14}$H$_{29}$	0.00083	36
	n-C$_{16}$H$_{33}$	0.00029	41

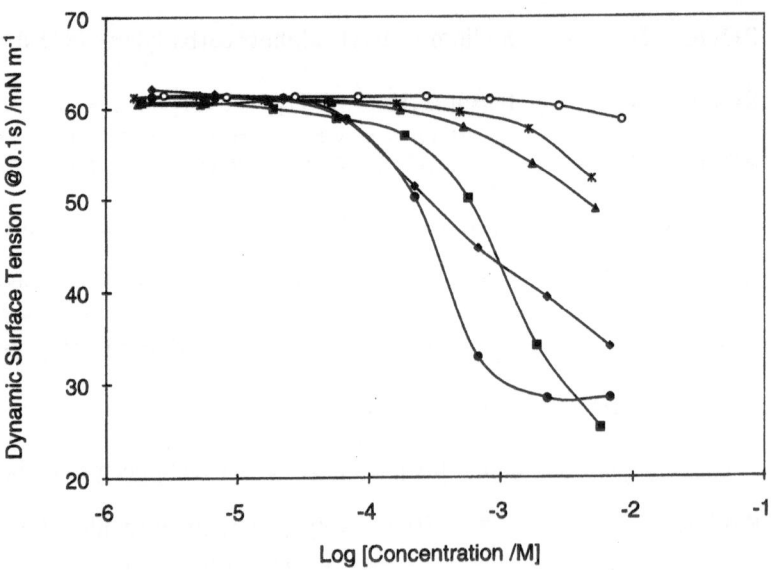

Fig. 1 Dynamic surface tension (at 0.1 s surface age) of a series of sodium tri-alkyl sulphotricarballylates, CH$_2$(CO$_2$R)CH(CO$_2$R) CH(CO$_2$R)SO$_3$Na, in aqueous solution containing 7% deionized gelatin. R groups: ○ n-C$_2$H$_5$; ♦ n-C$_4$H$_9$; ● n-C$_5$H$_{11}$; ■ n-C$_6$H$_{13}$; ▲ n-C$_7$H$_{15}$; ∗ 2-ethyl-hexyl

Nonionics

Figure 2 shows the DST results for the homologous series of nonionic dialkyl bis-gluconamide surfactants. Like the anionic series above, the nonionic series also shows that the efficiency of lowering DST reaches a maximum as the homologous series is ascended, in this case with the di-C_6 homologue. Hence, nonionic surfactants are no different from anionics in this respect. This suggests that, for a given surface age, there will be a maximum efficiency of DST reduction irrespective of surfactant type when a homologous series is ascended. The CACs and values of SST at the CAC are given in Table 1. The nonionic series shows exactly the same trends as the anionic series above. At first there is a monotonic decrease in CAC and SST at the CAC with increasing tail length, then with the more hydrophobic homologues, the CAC and SST at the CAC tend to reach limiting values.

Anionic/nonionics

This class involves the combination of an anionic group (sulphate) with a nonionic pentaethyleneoxide block as a complex hydrophilic headgroup. Figure 3 shows the DST results for the n-C_{10}, n-C_{12}, n-C_{14} and n-C_{16} homologues. Like the anionic and nonionic classes, this class also shows a maximum efficiency in the reduction of DST as the homologous series is ascended and similar trends in CACs and SSTs at the CAC (Table 1).

Efficiency of DST reduction

The combined evidence of all three class types of surfactant suggest a strong degree of common behaviour in their dynamics in aqueous gelatin solution: they all exhibit a maximum efficiency in DST reduction as their homologous series are ascended. Efficiency can be defined in several ways. Here, efficiency is represented in two different ways, both of which are designed to be independent of absolute values of surface tension lowering. The results for the three classes can then be combined and compared.

The first definition of efficiency used here, termed $E(\gamma_{1/2}^{dyn})$, is as described in the Introduction, i.e. the log[concentration/M] required to reach the surface tension midway between the value of the solvent and the value at the CMC, or, in this case the CAC. Hence, the lower $E(\gamma_{1/2}^{dyn})$, the better the efficiency. This concept has been applied to the DST data and is plotted as a function of CAC in Fig. 4. In the case of the STS series, data for a number of branched chain analogues have been included to broaden the range of CAC. The figure shows a number of features:

1. The maximum efficiency is not a sharp function of CAC; it extends over almost one log unit of concentration, i.e. $\sim 2 \times 10^{-4}$ to $\sim 2 \times 10^{-3}$ M.
2. At a CAC of approximately 2×10^{-4} M, the efficiency of DST reduction falls suddenly.
3. The same pattern of behaviour occurs at more or less the same molarity for all three surfactant classes.

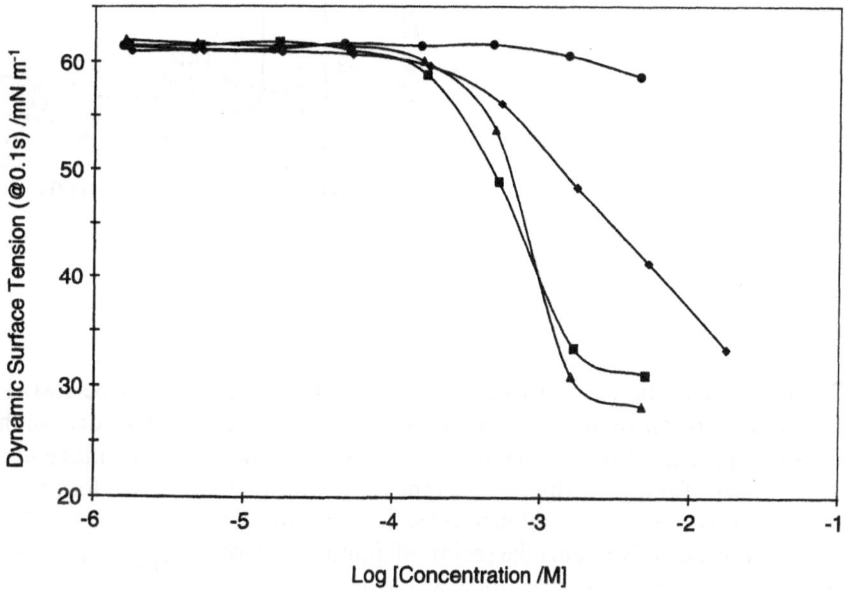

Fig. 2 Dynamic surface tension (at 0.1 s surface age) of a series of di-n-alkyl bis-gluconamides, $CR_2[CH_2NHCO(CHOH)_4CH_2OH]_2$, in aqueous solution containing 7% deionized gelatin. R groups: ◆ n-C_5H_{11}; ■ n-C_6H_{13}; ▲ n-C_7H_{15}; ● n-C_8H_{17}

Fig. 3 Dynamic surface tension (at 0.1 s surface age) of a series of sodium alkyl-pentaethyleneoxide-sulphates $(R.[O.CH_2CH_2]_5.OSO_3.Na)$ in aqueous solution containing 7% deionized gelatin. R groups: ♦ n-$C_{10}H_{21}$; ■ n-$C_{12}H_{25}$; ▲ n-$C_{14}H_{29}$; ● n-$C_{16}H_{33}$

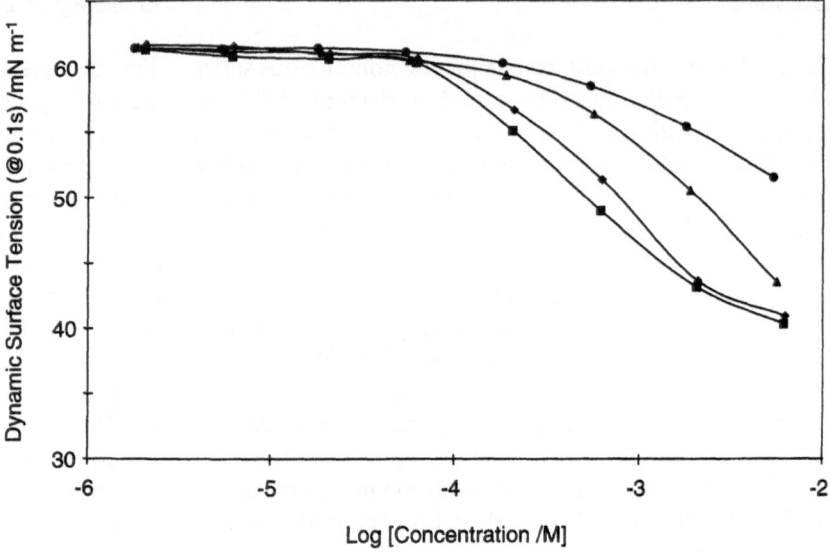

Fig. 4 Efficiency of reducing dynamic surface tension at 0.1 s surface age in terms of the log concentration of surfactant required to reach the mid-point of the DST curve as function of CAC: ♦ sodium tri-alkyl sulphotricarballylates (anionic); ■ di-alkyl bis-gluconamides (nonionic); ▲ alkyl-pentaethyl-eneoxide-sulphates (nonionic-anionic)

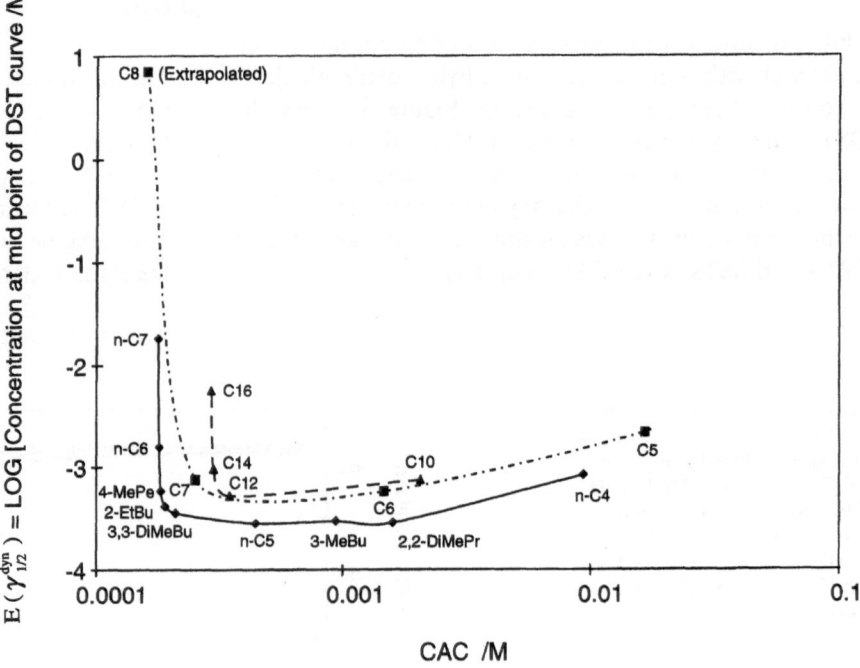

The second definition of efficiency used here is based on the surface tension reduction at a specific concentration, chosen to give maximum differentiation between homologous series. Taking all three surfactant classes into account, the obvious concentration to select for this purpose is 10^{-3} M. It adroitly avoids the region of limiting surface tension, the so-called "plateau" zone, where differentiation would be lost. However, to make the second definition independent of the absolute reduction in surface tension, it is advantageous to define the efficiency, $E(10^{-3}$ M$)$, in relative terms as follows:

$$E(10^{-3}\ M) = \frac{\gamma_{0.001M}^{dyn} - \gamma_{CAC}}{\gamma_{SOL}^{dyn} - \gamma_{CAC}}, \tag{1}$$

where $\gamma^{\mathrm{dyn}}_{0.001\,\mathrm{M}}$ is the dynamic surface tension at a surfactant concentration of 10^{-3} M, $\gamma^{\mathrm{dyn}}_{\mathrm{SOL}}$ is the dynamic surface tension of the solvent (7% w/w gelatin in water), and γ_{CAC} is the static surface tension at the CAC. $E(10^{-3}$ M$) = 1.0$, when the DST at 0.001 M is still at the dynamic value of the solvent ($\gamma^{\mathrm{dyn}}_{0.001\,\mathrm{M}} = \gamma^{\mathrm{dyn}}_{\mathrm{SOL}}$), and $E(10^{-3}$ M$) = 0.0$, if the DST at 0.001 M reaches the value attained at its CAC ($\gamma^{\mathrm{dyn}}_{0.001\,\mathrm{M}} = \gamma_{\mathrm{CAC}}$).

Figure 5 shows $E(10^{-3}$ M$)$ as function of CAC for all three surfactant classes, in a manner similar to $E(\gamma^{\mathrm{dyn}}_{1/2})$ in Fig. 4. The efficiency of DST reduction at 0.001 M shows essentially the same picture of efficiency as $E(\gamma^{\mathrm{dyn}}_{1/2})$ insofar as

1. The maximum efficiency zone is again demonstrated to extend over the same one log unit of concentration, i.e. $\sim 2 \times 10^{-4}$ to $\sim 2 \times 10^{-3}$ M.
2. The efficiency of DST reduction again drops off suddenly at CAC values of 2–3×10^{-4} M.

Figures 4 and 5 also appear to give the same overall picture insofar as they both show the nonionic and nonionic–anionic classes to be less efficient at lowering DST than the anionic class, though differentials appear to be greater in Fig. 5. However, the two definitions of efficiency are measuring slightly different things. $E(\gamma^{\mathrm{dyn}}_{1/2})$ demonstrates that the most efficient compounds from the anionic series are 0.27–0.31 of log unit in concentration more efficient than the corresponding compounds from either the nonionic or nonionic–anionic series. This effectively says that twice as much nonionic or nonionic–anionic is required relative to the anionic to reach the

mid-point of their DST curves for the most efficient compounds. On the other hand, $E(10^{-3}$ M$)$ simply demonstrates that at 10^{-3} M, the DST of the most efficient anionic is 96% of the way towards its surface tension at its CAC, whereas the best case nonionic and nonionic–anionic are 73% and 64%, respectively. Being limited to a specific concentration, albeit a very careful choice, $E(10^{-3}$ M$)$ is not quite as fundamental a parameter as $E(\gamma^{\mathrm{dyn}}_{1/2})$. The concept of $E(\gamma^{\mathrm{dyn}}_{1/2})$ can be applied to DSTs at any surface age, whereas the choice of a concentration for differentiating DST curves, such as 10^{-3} M, will need to be changed according to surface age.

Rationale behind maximum in efficiency of DST reduction

The cause of the maximum in efficiency of reducing DST (~ 0.1 s surface age) on ascending a homologous series of surfactants in dilute aqueous solution must lie in the existence of at least two opposing processes. Garrett and Moore [2] suggest that for their systems it is competition between decreasing equilibrium surface tension and increasing gradients of DST with reciprocal time. The increasing gradients of DST with reciprocal time are correlated with rate of micellar breakdown, the longer the hydrophobic tails of the surfactant the slower the micellar breakdown rate. Their other way of stating this is increasing surface activity versus decreasing rate of supply of monomer from micelles. In our dilute surfactant systems in aqueous gelatin, we believe that there is a related but slightly different reason for the maximum efficiency in

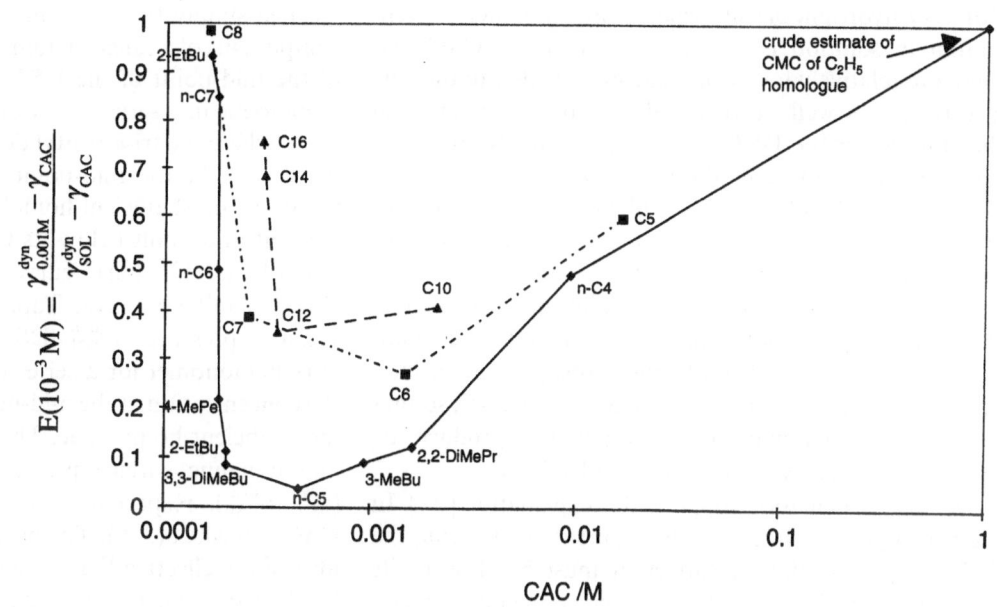

Fig. 5 Efficiency of reducing dynamic surface tension at 0.1 s surface age in terms of the relative reduction in surface tension at 10^{-3} M concentration as a function of CAC: ♦ sodium tri-alkyl sulphotricarballylates (anionic); ■ di-alkyl bis-gluconamides (nonionic); ▲ alkyl-pentaethyleneoxide-sulphates (nonionic-anionic)

DST reduction on ascending a homologous series, though we concur with the importance of increasing surface activity.

In our systems, we believe that the rate of micellar breakdown is not the important factor at the dilute concentration where the most efficient homologue exhibits its maximum efficiency in DST reduction. However, we believe that micellar breakdown rates become important for the more hydrophobic homologues at higher concentrations. We believe the reason behind the maximum is one of competition between increasing surface activity (increasing with increasing tail length) and decreasing monomer flux to the surface (decreasing with increasing tail length, i.e. with decreasing CAC). In other words, at the concentration of the mid-point of the DST curve of the most efficient homologue (which is below its CAC), it is proposed that the micellar content of higher homologues plays little role in surface tension lowering on the 0.1 s time scale. In effect, the micelles remove monomer from solution, diffuse more slowly than monomer, and so reduce monomer flux to the surface.

It is postulated that at short times the DST is determined by the diffusion of surfactant from the solution to the surface, i.e., adsorption of surfactant at the interface is rapid. Surfactant can exist in gelatin solution in several forms: monomer, micelles bound to gelatin [8], free micelles and possibly monomer bound to gelatin for the most hydrophobic species. For simple water-soluble surfactants in dilute solution, only the first two forms need to be considered. In dilute gelatin solution, monomer diffusion is relatively rapid (essentially that of a small molecule in water), but micelles diffuse at a slower rate. At concentrations near the CAC therefore, the flux of surfactant to the surface is controlled by the monomer concentration. For very hydrophobic surfactants, the monomer concentration in solution is essentially that of the CAC and therefore relatively low. Consequently, the flux to the surface is low as well. A very hydrophobic surfactant will therefore lower the DST only relatively slowly, even if it eventually gives low equilibrium or static surface tension. On the other hand, very hydrophilic surfactants do not adsorb strongly to the air–water interface, so that very high concentrations are necessary in order to achieve low values of surface tension, even at equilibrium. The surfactant exhibiting maximum efficiency under dynamic conditions is therefore one that adsorbs strongly to the surface, but at the same time does not have too low a CAC, so that there is a high enough flux of surfactant to produce the necessary surface excess to lower the DST.

A rough calculation supports this explanation [22]. In order to have a substantial effect on surface tension, the surface concentration of surfactant must be close to its saturation value, which for anionic surfactants corre-

sponds to about $2 \mu M m^{-2}$. If we assume that the contribution of monomer from micellar surfactant is negligible on the time-scale of the DST measurement for very dilute solutions very close to their CAC, then this amount of surfactant must be supplied by monomer from a slab of solution whose thickness is roughly the average diffusion distance of a monomer. This thickness is given approximately by $\sqrt{Dt}$, where D is the diffusion coefficient and t the time scale. Values of D from the literature for anionic surfactant monomers are: $6.6 \times 10^{-10} m^2 s^{-1}$ (sodium octyl sulphate [23]) and $8 \times 10^{-10} m^2 s^{-1}$ (sodium dodecyl sulphate, [24]). Hence, variations of the hydrophobic group of a surfactant have little impact on D. On the other hand, values of D for micellar anionic systems are an order of magnitude larger, e.g. $8 \times 10^{-11} m^2 s^{-1}$ (sodium alkyl benzene sulphonates [2]). For our calculations, we use a value of $4 \times 10^{-10} m^2 s^{-1}$ for D (monomer) to take into account the small retarding effect of gelatin. With $t = 0.1$ s, the average diffusion distance is about 6 mm. The necessary concentration of surfactant monomer in this 6 mm slab to complete a monolayer is then of the order of 0.3 mM. This is in the window of CAC where we see the maximum efficiency of DST reduction (0.2–2 mM). A surfactant whose CAC is close to 0.3 mM can deliver a maximal surface concentration in the given time; if the CAC is lower, less surfactant will get to the surface, while if the CAC is higher, the surface binding (which is determined by the same balance of forces as the CAC) will be less, and the surface excess of surfactant will be less. This rough calculation is consistent with the simplest possible model of surfactant binding, in which diffusion to the surface is rate determining.

Further evidence that monomer flux to the surface is a key issue in the maximum efficiency phenomenon can be gleaned simply by examining the monomer and micellar components of a range of homologues at the concentration of the mid-point of the DST curve of the most efficient homologue in a series. For example, take the anionic STS series where the tri-n-pentyl derivative is the most efficient derivative. At the mid-point, the efficiency determining point, the most efficient homologue is actually 0.2 of a log concentration unit below its CAC. Being on the steepest part of the γ^{DST}–log C curve, presumably the surface is close to a full monolayer. Table 2 shows the DST, dynamic surface pressure (γ_{SOL}^{dyn}–DST) and molar ratio of micelle/monomer for a series of STS at 0.00028(5) M, i.e. the concentration at the mid-point of the DST curve of the most efficient homologue. The table also shows the DST and dynamic surface pressure at the CAC, γ_{CAC}^{dyn} and (γ_{SOL}^{dyn}–γ_{CAC}^{dyn}), respectively, for those homologues whose CAC < 0.00028(5) M. The parameters at the CAC give an idea of the effective DST reduction that would be caused by their monomer component alone, assuming that their

Progr Colloid Polym Sci (1997) 103:307–317
© Steinkopff Verlag 1997

Table 2 DST, dynamic surface pressure (γ_{SOL}^{dyn}–DST) and molar ratio of micelle/monomer for the STS series at 0.00028(5) M (the concentration of the mid-point of the DST curve of the most efficient homologue, n-C_5) plus their DST and dynamic surface pressure at their CAC where their CAC < 0.000285 M

STS alkyl tail R	DST [mN m^{-1}]	γ_{SOL}^{dyn}–DST [mN m^{-1}]	DST at CAC (γ_{CAC}^{dyn}) [mN m^{-1}] (if CAC < 0.000285 M)	γ_{SOL}^{dyn}–γ_{CAC}^{dyn} [mN m^{-1}] (if CAC < 0.000285 M)	Molar ratio micelle monomer at 0.000285 M
n-C_2	61.3	0.2	—	—	—
n-C_4	49.9	11.6	—	—	—
n-C_5	47.0	14.5	—	—	—
n-C_6	55.2	6.3	57.1	4.4	0.58
n-C_7	59.1	2.4	59.8	1.7	0.58
2-et-hexyl	60.1	1.4	60.5	1.0	0.78

monomer concentration remains constant above their CAC.

The micelle/monomer ratio of the n-C_6 homologue at 0.000285 M is 0.58, i.e. (0.000285–CAC)/CAC)), assuming the n-C_6 monomer concentration to remain constant above its CAC. The monomer ratio of n-C_6 to n-C_5 homologue at the concentration of the mid-point of the n-C_5 DST curve is 0.63 (CAC/0.000285). Hence, the maximum efficiency seen with the n-C_5 homologue appears to be due to the fact that at the concentration of the mid-point of its DST curve, the n-C_6 has only 63% monomer relative to the n-C_5 (which is 100% monomer), the remaining 37% being tied up in micelles which, although prone to breakdown, are slow to diffuse and too few in number of influence surface tensions at short time scales like 0.1 s. Referring back to Table 2, if micellar breakdown is considered to contribute the difference between the actual DST at 0.000285 M and the DST at its CAC, then it only contributes a further 1.9 to the 4.4 mN m^{-1} dynamic surface pressure coming from monomer (6.3 mN m^{-1} overall). Hence, in a sense, it is the presence of micelles at 0.000285 M that reduces the monomer flux to the interface for the more hydrophobic homologues and so brings about the maximum in efficiency of DST reduction. In tracking the mid-point concentration of the n-C_5 up to the n-C_6 DST curve, the expectation would be that the latter DST curve should display a shallower $d\gamma/d \log C$ gradient as it can only achieve 63% of the coverage of the n-C_5 homologue (Fig. 1); this is exactly what is seen.

The next most hydrophobic member of the series, the n-C_7 derivative, shows the same CAC as the n-C_6, but shows less DST reduction. This is due to a decrease in activity from both the monomer concentration and micelle breakdown. Using the same arguments as before, this corresponds to a 1.7 mN m^{-1} reduction from monomer and a further 0.7 mN m^{-1} due to monomer from micelle breakdown (2.4 mN m^{-1} overall). Hence, the largest contribution to the reducing dynamic activity with increasing tail length comes from a decreasing activity of monomer rather than micelle breakdown. This is most likely due to increasing surfactant gelatin interactions with increasing

tail length. Increasing micellar stability does contribute a little to the decreasing dynamic activity at 0.000285 M concentration, but never more than 2 mN m^{-1} in STS series studies. We are currently developing a more sophisticated mathematical model for the dynamic behaviour of surfactant systems, which will be published shortly. However it should be noted that Diamant and Andelman have published such a model very recently [25, 26].

Rationale behind reducing efficiency of DST reduction with the higher homologues

Although we have proposed that micellar breakdown per se only has a small effect on the origin of the maximum in efficiency other than that the breakdown is a slow process, it is clear that it influences the efficiency of the more hydrophobic homologues whose DST curves show a progressive march to higher concentrations with increasing hydrophobicity (see Figs. 1–3). The maximum in efficiency occurs where the mid-point of the pertinent DST curve occurs at a concentration below the CAC, whereas with the higher homologues, the mid-point of their DST curve occurs at concentrations significantly above the CAC. In fact, the DST reduction of these higher homologues at their CAC is only a few mN m^{-1} at best (see Table 2 for anionic STS homologous series). In other words, by the time the higher homologues reach the mid-point of their DST curves, the extra surface tension reduction can only be coming from micellar breakdown. Table 3 shows that as the hydrophobicity of their tails increases, the ratio of micelle to monomer at their DST mid-points increases dramatically, despite virtually constant CAC (see Table 1). Thus in changing from the n-C_6 to n-C_7 homologue, which possess the same CAC, 19 times as much surfactant is required to reach the mid-point of the DST curve. Hence the decreasing efficiency of DST reduction with increasing tail length with the more hydrophobic homologues must be directly related to increasing stability of the micelles (assuming D varies relatively little from homologue to homologue).

Table 3 Dynamic surface pressure (γ_{SOL}^{dyn}-DST) and molar ratio of micelle/monomer for the STS series at the concentration of the mid-point of the DST curve, including monomer and micelle-breakdown contributions, assuming the monomer contribution = (γ_{SOL}^{dyn}-γ_{CAC}^{dyn}), i.e. is the same as at their CAC

STS alkyl tail R	Molar ratio: micelle monomer	γ_{SOL}^{dyn}-DST (at DST mid-point) [mN m^{-1}]	Monomer contribution (γ_{SOL}^{dyn}-DST) or (γ_{SOL}^{dyn}-γ_{CAC}^{dyn})*	Micelle-breakdown contribution (γ_{SOL}^{dyn}-DST-γ_{CAC}^{dyn}) [mN m^{-1}]
n-C$_4$	—	14.3	14.3	0
n-C$_5$	—	14.6	14.6	0
n-C$_6$	5.3	18.0	4.4*	13.6
n-C$_7$	103	18.0	1.7*	16.3
2-et-hexyl	168	18.0	1.1*	16.9

Conclusions

1. All three class types of surfactant studied exhibit a common behaviour in their dynamic surface tension properties in aqueous gelatin solution: all show a maximum efficiency in DST reduction at a surface age of 0.1 s as their homologous series are ascended.

2. The maximum is believed to be due to competition at the surface between reducing monomer flux that decreases with decreasing CAC and increasing surface activity that increases with decreasing CAC. The decrease in monomer flux is believed to result from mainly two processes: first, from the formation of micelles which diffuse more slowly in gelatin solution and release relatively little monomer as a result of breakdown during the 0.1 s time scale; and secondly from an increase in surfactant-gelatin interactions with increasing tail length. There is also a small effect due to increasing micelle stability with increasing tail length.

3. Higher homologues, i.e. those more hydrophobic than the most efficient homologues, show marked decreases in the efficiency of DST reduction as the tail length increases. This appears to be due to increasing micelle stability (decreasing rates of micelle breakdown).

4. There is a common zone of maximum efficiency in DST reduction for all three surfactant classes studied, which extends over a broad range of concentration ($\sim 2 \times 10^{-4}$ to $\sim 2 \times 10^{-3}$ M, i.e. approximately one log unit).

5. There appears to be a shallow approach to the maximum efficiency zone on the hydrophilic side of a homologous series (high CAC) and an abrupt drop where efficiency is suddenly lost on the hydrophobic side, at approximately 2×10^{-4} M. This sudden loss is believed to be due to an insufficient concentration of monomer within an average diffusion distance from the surface to form a monolayer capable of lowering surface tension (for the 0.1 s time scale).

6. Although all three classes of surfactant show similar dynamic behaviour, the most efficient anionic derivatives appear to be twice as efficient as the nonionic or nonionic-anionic series.

7. The static properties of the three surfactant classes studied show tendencies to reach limiting values as homologous series are ascended: CACs initially decrease with increasing tail length, but appear to reach a limiting value around 0.0002 M or just under; and surface tension values at the CAC decrease with increasing tail length, but reach tend to reach a limiting value with the higher homologues.

Acknowledgment The author is greatly indebted to the following colleagues within the Kodak organisation: Dr. C.B.A. Briggs, Dr. I.M. Newington, and Dr. T.J. Wear for the synthesis of the surfactants described in this study; Dr. T.D. Blake, Dr. A.M. Howe, and Dr. E.A. Simister for useful discussions on surfactant dynamics; and especially to Dr. T. Whitesides for discussions on surfactant dynamics and his calculations relating to the amount of surfactant within an average diffusion length of the surface in a given time scale.

References

1. Valentini JE, Thomas WR, Sevenhuysen P, Jiang TS, Lee HO, Liu Y, Yen S-C (1991) Ind Eng Chem Res 30:453
2. Garret PR, Moore PR (1993) J Colloid Interface Sci 159:214–225
3. Rosen MJ (1989) In: Surfactants and Interfacial Phenomena; 2nd Ed. Wiley New York, p 84
4. Pitt AR, Morley SD, Burbidge NJ, Quickenden EL (1996) Colloids Surfaces A 114:321–335
5. Malysa K, Lunkenheimer K, Miller R (1985) Colloids Surfaces 16:9–20
6. Malysa K, Lunkenheimer K, Miller R (1991) Colloids Surfaces 53:47–62
7. Linfield WM (1976) In: Anionic Surfactants: Part II; Surfactant Science Series, Vol. 7, Ch. 12 Dekker: New York
8. Whitesides TH, Miller DD (1994) Langmuir 10:2899

Progr Colloid Polym Sci (1997) 103:307–317
© Steinkopff Verlag 1997

9. Briggs CBA, Newington IM, Pitt AR (1995) J Chem Soc Chem Commun 379–380
10. Eastoe J, Rogueda P, Harrison WJ, Howe AM, Pitt AR (1994) Langmuir 10:4429–4433
11. Padday JF (1957) 2nd Int Congress of Surface Activity, Vol. I. Butterworths, London. p 1
12. Bergink-Martens DJM, Bos HJ, Prins A, Schulte BC (1990) J Coll Int Sci 138:1–9
13. Bergink-Martens DJM, Bos HJ, Prins A (1994) J Coll Int Sci 165:221–228
14. Dukhin SS, Kretzschmar G, Miller R (1995) In: Möbius D, Miller R (eds), Dynamics of Adsorption at Liquid In terfaces, Theory, Experiment, Application. Studies in Interface Sciences Series, Vol I. Elsevier, Amsterdam
15. Ward AF, Tordai L (1946) J Chem Phys 14:453
16. Fainerman VD, Makievski AV, Miller R (1994) Coll Sur A 87:61
17. Eastoe J, Dalton JS, Rogueda P, Lodhi AN, Crooks ER, Pitt AR, Simister EA (in press) J Coll Int Sci
18. US Patent No. 4,988,610 (1991)
19. US Patent 4,892,806 (1990)
20. Gilbert EE (1969) Synthesis Int J Methods Synthetic Organic Chem (1):3
21. Rosen MJ (1989) In: Surfactants and Interfacial Phenomena, 2nd ed Wiley, New York, 120–121
22. Whitesides TH, Eastman Kodak Co, private communication
23. Lindman B, Puyal MC, Kamenka N, Rymden R, Stilbs P (1984) J Phys Chem 88:5048
24. Leaist DG (1986) J Coll Int Sci 111:230
25. Diamant H, Andelman D (1996) J Phys Chem 100:13732–13742
26. Diamant H, Andelman D (1996) Europhys Lett 34(8):575–580

Progr Colloid Polym Sci (1997) 103:318–326
© Steinkopff Verlag 1997

S. Boussaad
R.M. Leblanc

Pure and mixed chlorophyll $\underline{a}$ Langmuir and Langmuir–Blodgett films. Structure, electrical and optical properties

Received: 12 February 1997
Accepted: 18 February 1997

S. Boussaad · Dr. R.M. Leblanc (✉)
Chemistry Department
Cox Science Building
Room 315
University of Miami
Coral Gables, Florida 33124, USA

Abstract Photosynthetic pigments, which undergo electron transfer reactions, are specifically associated with the D_1–D_2-Cyt_{b559} protein complex. A particular aggregated state of *in vivo* chlorophyll $\underline{a}$ (Chl $\underline{a}$) is responsible for the high-energy conversion efficiency. The Langmuir–Blodgett (L–B) technique allows preparation of systems that potentially mimic the packing of Chl $\underline{a}$ *in vivo*. Water has a pronounced effect on the optical and electrical properties of chl $\underline{a}$ L–B films, and is a key element in the formation of dimers with (anti) parallel transition moments of the constituent monomers. The incorporation of lipid makes the transport of photogenerated charge carriers more difficult. In the case of Chl $\underline{a}$-Cytochrome c mixtures, keto groups of chl $\underline{a}$ play a major role in the pigment–protein association.

Key words Photosynthetic pigments – chlorophyll $\underline{a}$ – Langmuir–Blodgett films

Introduction

Photosynthetic organisms are able to oxidize water, upon absorption of sunlight, to produce oxygen. In higher plants this process takes place in the thylakoid membranes located inside the chloroplast [1]. These membranes have two key reaction centers, P700 and P680, both of which operate in two distinct photosystems, photosystem I (PS I) [2] and photosystem II (PS II) [3] and consist of chlorophyll $\underline{a}$ (Chl $\underline{a}$). P700 and its electron acceptors make up PS I, whereas P680 and its electron carriers make up PS II. The latter is the site where the primary reaction takes place. Both Chl $\underline{a}$ and Pheophytin $\underline{a}$ (Phe $\underline{a}$) are pigments known to be present in this reaction center [4] where they act as the primary electron donor (P680) and acceptor, respectively. Also present in the PS II reaction center are the plastoquinones [5] (Q_A and Q_B) which always act in series with the carotenoid molecules [6].

The pigments which undergo electron transfer reaction are bound to the D_1–D_2-Cyt_{b559} protein complex [7, 8],

an integral part of the PS II core complex. The reaction center of PS II probably contains four molecules of Chl $\underline{a}$, two molecules of Phe $\underline{a}$ and two quinones [9, 10]. Compared to the PS II reaction center, the PS I reaction center is more complicated [11]. Its Chl $\underline{a}$ content varies between 100 and 200 molecules [12], most of which are not photochemically active, but, serve as part of the antenna system that absorbs light and transfers energy to P700 [13]. Excitation of PS II drives the transfer of an electron from P680 to a Phe $\underline{a}$ molecule which immediately reduces Q_A. After being transferred to Q_B, the electron flows to the oxidized Chl $\underline{a}$ in PS I (P700$^+$) through many electron carriers, including cytochrome f (Cyt f) and plastocyanin. The oxidized Chl $\underline{a}$ complex (P680$^+$) provokes the splitting of H_2O to O_2 through four Mn atoms which are probably bound to the proteins that hold P680. This splitting is followed by a release of an electron used to reduce P680$^+$.

Chl $\underline{a}$ and Chl $\underline{b}$ organized in other pigment–protein complexes serve as antennas [13]. Light-harvesting complexes [14, 15], LHC I and LHC II, are the outer antennas of the membrane and they transfer energy to the reaction

Progr Colloid Polym Sci (1997) 103:318–326
© Steinkopff Verlag 1997

centers. This function is achieved through the absorption of light and formation of electronic excited states useful for energy transfer [16, 17]. All Chl b in the thylakoid membrane is associated with proteins of the LHC I and LHC II. The ratio of Chl a/Chl b is regulated by the adaptation of chloroplast to the prevailing light conditions.

Chl a has a broad absorption spectrum, and aggregation through self-assembly typically leads to changes in its optical properties [18]. Red shifts are commonly observed in in vitro Chl a systems, such as thin films, monolayers and colloidal dispersions, used as models for the in vivo system [19, 20]. One such system based on the Langmuir–Blodgett (L–B) technique [21] for forming ordered thin films, allows the orientation of molecules in a molecular monolayer at an air/water interface and the subsequent transfer onto solid substrates. The transferred L–B film is well organized, and the use of this technique allows preparation of systems that potentially mimic the packing of Chl a in the in vivo system.

Methodology

The Langmuir and Langmuir–Blodgett techniques

The Langmuir technique

The amphiphilic molecules are first suspended in a solution where they are soluble, usually chloroform, glycerol, or a hydrocarbon, such as benzene. A small amount of the solution is placed onto the surface of very pure water sitting within the Langmuir trough, and, while the spreading solvent evaporates, the amphiphiles will gather at the air/water interface. Amphiphiles have both a hydrophobic and hydrophilic end so the hydrophobic end will stay in the air and the hydrophilic end will be in contact with the water. Furthermore, the amphiphiles will spread throughout the entire surface area of water in the trough. A movable barrier is used to constrain the amphiphiles to a smaller area of the water surface and pushes them closer together until the molecules form a monolayer.

While the barrier confines the amphiphiles to a smaller area, the force exerted by the monolayer is continuously measured and a surface pressure (Π)–area (A) isotherm can be drawn at a constant temperature. This curve plots the surface pressure (force per unit length) versus the mean molecular area occupied by the amphiphiles at the air/water interface. Usually, a Π–A isotherm shows four interesting regions [21]. An initial horizontal region where the mean molecular area is large and the interaction between molecules is small so the surface pressure is approximately constant. The first linear region deviates from the

horizontal where the monolayer compressibility is approximately constant. The second linear region occurs after the slope suddenly increases, indicating a change in phase of the monolayer. Within this region of the curve the monolayer has adopted an ordered structure and the compressibility still remains approximately constant but reduced by an order of magnitude. If the second linear region is extrapolated to zero, the x-intercept gives an area value known as the limiting area. In the final region, a maximum is reached in the curve and the monolayer collapses with the compressibility rapidly approaching infinity. During collapse, the molecules begin to pass over each other to form disordered multilayers or may dissolve in the subphase.

The Langmuir–Blodgett technique

The details of the deposition procedure depend upon the type of substrate and method used for the transfer of the monolayers [21–23]. If the substrate has a hydrophilic surface (mica, Au(111), glass) then as it is lowered vertically through the monolayer covering the water surface it will not be coated with amphiphilic molecules present there because the hydrophobic ends are sticking up into the air. However, when the slide is withdrawn from the water, the hydrophilic ends submerged in water will adhere to the substrate surface forming one monolayer L–B film on both sides of the substrate. By repeating this procedure, multilayers L–B film can be built up on the substrate. If the substrate has a hydrophobic surface (graphite, bare Si), then as it is lowered through the water surface it will become covered with a monolayer of molecules. In case where the substrate is withdrawn, a bilayer L–B film will form on both sides of it. The L–B films formed with either hydrophilic or hydrophobic substrate are called Y-type films.

The scanning probe microscopy technique

Principles of scanning tunneling and atomic force microscopes

Both scanning tunneling microscopy (STM) [24] and atomic force microscopy (AFM) [25] are able to image a material's surface by scanning a sharp probe very close to it. For STM, the probe searches for a specific value of current which passes between it and the sample. The probe comes very close to the sample's surface (usually less than 1 nm away) but, in theory, should not touch. This means that electrons must pass from the tip to the sample or vice versa by tunneling, a quantum mechanical phenomenon which occurs when subatomic particles are incident upon a potential barrier which is thin and short. The tunneling current, I, will be governed by the expression

$$I \propto \frac{V}{z} \exp\{ - Cz\sqrt{\bar{\phi}} \} , \qquad (2.1)$$

where V is the voltage bias, z is the tip–sample separation distance, C is a constant with the value $10.25 \, [\mathrm{nm} \times (\mathrm{eV})^{1/2}]^{-1}$, and $\bar{\phi}$ is the average of the tip and sample work functions. This expression is called Simmons' formula, and was derived for a plane, parallel tunnel junction [26]. Due to the value of C in the argument of the exponential for the tunneling current, the STM is able to achieve atomic resolution (the current will change by nearly a factor of 3 for a change in distance as small as 0.1 nm).

On the other hand, AFM exploits the forces which exist between atoms and molecules. The force exerted upon a tip mounted onto a cantilever with a known spring constant is monitored as the tip passes over the surface. The cantilever deflects during the scan and from the measurement of these an image of the surface topography is obtained. All types of materials experience these forces so the AFM is not limited to conductors. There are two regimes of force which can be felt by the probing tip. If the tip is scanned extremely close to the surface, the force $F(r)$ will be expressed as

$$F(r) = \frac{12A}{r^{13}} - \frac{6B}{r^7} , \qquad (2.2)$$

where A and B are constants which depend upon the material of both the tip and sample, and r is the separation distance between atoms of the tip and sample [27]. For small separation distances that are usual for contact scanning, the first term (the repulsive term) will dominate implying that only the tip atoms nearest to the sample surface will contribute most of the information about its topography. The resolution is best with the tip in contact with the surface due to the more rapid decay of the repulsive force term (fewer atoms are participating in the imaging).

Pure chlorophyll _a_ films

Visible and infrared spectroscopies of chlorophyll _a_ Langmuir–Blodgett films

The properties of Chl _a in vitro_ greatly depend on its environment and state of aggregation, properties which have been studied in solution [28, 29], in vesicles [30, 31], in micelles [32, 33] and in monolayers [34–42]. The maximum absorbances of various systems of Chl _a_ recorded in the blue and red regions of the visible spectrum are summarized in Table 1. The bathochromic shifts noticed especially in the red region are clear evidence of the transformations that Chl _a_ undergoes. Considering the importance of Chl _a_ in photosynthesis, it is vital to understand how Chl _a_ interacts with various chloroplast components and with itself. The Langmuir technique has proven to be valuable in the study of Chl _a_ interactions because it is easier to control the specific interaction in a two-dimensional layer as opposed to a bulk three-dimensional system.

The absorption spectrum of a pure Chl _a_ monolayer at a nitrogen/quartz interface exhibits a strong Soret band at 440 nm, and a red band centered at 680 nm [35, 43, 44]. The excitation spectrum of the monolayer, with the

Table 1 List of maximum absorbancies of various systems of Chl _a_ recorded in the blue and red regions of the visible spectrum. (s): shoulder; (n/a): data not available; (––): no band

System	λ [nm]								Ref.
	blue				red				
Dry Chl _a_ in _n_-octane	n/a	n/a	n/a	n/a	662.0	678.0^s	–	–	68
Dry Chl _a_ in cyclohexane	–	429	–	–	662.0	673.0^s	–	–	69
Dry Chl _a_ in benzene	n/a	n/a	n/a	n/a	662.0	678.0^s			68
Wet Chl _a_ in _n_-hexane	–	–	–	448.0	–	–	–	745.0	29
Chl _a_ in dry ethanol	–	–	430.0	–	665.0	–	–	–	64
Chl _a_ in wet ethanol	n/a	n/a	n/a	n/a	–	675.0	–	700.0^s	64
Chl _a_ in dioxane	–	–	433.1	–	662.0	–	–	–	37
Chl _a_ in vesicles	416.0	–	434.0	–	669.0	–	–	–	30
Chl _a_ monolayer	–	–	–	440.0	–	–	680.0	–	35
Chl _a_ monolayer in dioxane vapors	–	–	–	446.0	–	–	688.0	–	37
Chl _a_ monolayer in water vapors	–	–	–	440.0	–	–	–	742.0	37
Wet Chl _a_ monolayer	–	–	–	440.0	–	678.0	–	745.0	29
Wet Chl _a_ thin film	–	–	–	448.0	–	–	–	746.0	29

Progr Colloid Polym Sci (1997) 103:318–326
© Steinkopff Verlag 1997

emission monochromator set at 700 nm, has the same shape as the corresponding absorption spectrum. The main peaks are located at 386, 416 and 440 nm. The absorption and fluorescence spectra of Chl $\underline{a}$ L–B films were found independent of the number of monolayers in the range of 2–12 [45]. The examination of these optical properties under wet (water-saturated nitrogen atmosphere) and dry (nitrogen atmosphere) conditions showed the importance of water and the water content of the sample [45]. Lowering the temperature of hydrated Chl $\underline{a}$ L–B films does not drastically influence the absorption band profiles and half-bandwidths. The main band maximum in the red shifts slightly to 678 nm and the Soret band position remains practically unchanged. However, the fluorescence maximum shifts from 740 nm at room temperature to 760 nm on cooling to 85 K.

Removal of water has a pronounced effect on both absorption and fluorescence of Chl $\underline{a}$ L–B films. Changes in fluorescence spectra were noticed upon prolonged exposure (~ 1 h) of the L–B films to a stream of dry nitrogen. The result is a gradual decrease of emission intensity at 740 nm accompanied by the appearance of a growing component at 698 nm. However, lowering the temperature leads to a decrease of the 698 nm component and to the appearance of a long-wavelength component, the position of which depends on temperature. The absorption maxima in spectra of dried L–B films are at 438 and 678 nm at room temperature. The drying process brings the disappearance of the long-wavelength wing and an increase of absorption at 678 nm. This wing, recognizable as a wide band at 710 nm in hydrated L–B films at 85 K, is attributed to Chl $\underline{a}$-water aggregates that contain more water molecules per Chl $\underline{a}$ than those absorbing at 678 nm. It should be noted that both hydrated and dried L–B films are composed of the same two kinds of Chl $\underline{a}$ aggregates and differ only in their relative content.

The nature of the interactions between water molecules and monolayers or multilayers of Chl $\underline{a}$, obtained by the L–B technique, has been examined by infrared spectroscopy (IR) [40]. Following deposition of the monolayer of Chl $\underline{a}$, repetitive scans showed some modifications in the IR spectra which are interpreted as a reorganization of the molecules as some water molecules leave the array. Drying the sample further modifies the IR spectra, which indicates the departure of more tenacious water molecules. However, putting the sample in a moist atmosphere does not restore the original spectrum. This is an indication of a nonreversible reorganization in the Chl $\underline{a}$ array.

The IR spectra of Chl $\underline{a}$ monolayers are much more complicated than those of the multilayer, owing to a nonintegrating effect of the monolayer, which reveals the perturbing effect of the different dielectric environment on each functional group. To understand the influence of

water molecules on Chl $\underline{a}$ monolayers, we refer to a model of aggregation of Chl $\underline{a}$ into a multilayer array [39]. This multilayer model consists of a polymer of Chl $\underline{a}$ molecules linked through C=O---Mg coordinate bonds, and by water hydrogen bonding bridges. With a freshly prepared multilayer, the number of polymeric links is greatest so that the free ketone band (1695 cm^{-1}) would be absent immediately after the multilayer has been prepared. As the more labile molecules of water leave the multilayer aggregate, the multilayer collapses, breaking many polymeric links as evidenced by the decrease in the ketone C=O---Mg band at 1662 cm^{-1}.

On the basis of the analysis of the IR spectra and the information gathered from the surface pressure isotherms, a model (see Fig. 1a) is proposed for the monolayer arrangement at the air/water interface [40] which implies two sets of dimers of water per molecule of Chl $\underline{a}$. One pair of dimers constitutes the water of the first kind and is composed of vapor-like dimers. The two water molecules are held together by a hydrogen bond, and on one side by a weak coordinate bond with the Mg atom of one Chl $\underline{a}$ molecule and on the other side by a weak hydrogen bonds with the π-electron network of another Chl $\underline{a}$ molecule. This kind of water is situated between the porphyrin planes of Chl $\underline{a}$ molecules and is easily removed from the monolayer. The second pair of dimers is composed of water of the second and third kinds situated between the

Fig. 1 Schematic representation of the structure and orientation of the Chl $\underline{a}$ in a Langmuir film A (see Ref. [40]), within the dimer imaged by scanning tunneling microscopy B (see Ref. [47])

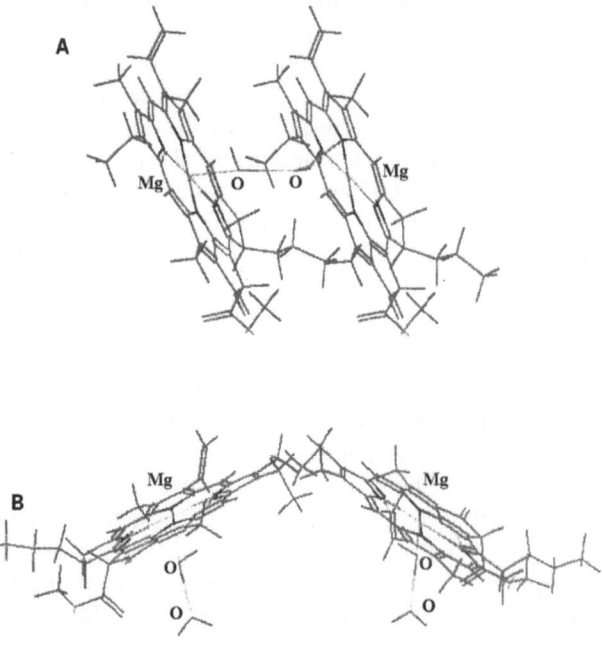

Mg atom of the Chl _a_ molecules and the water of the subphase. The second kind of water is closest to the Mg atom and is the most difficult one to remove. The third kind of water is closest to the water surface and its mobility is intermediate between that of water of the first kind and that of water of the second kind.

Scanning tunneling microscopy of chlorophyll _a_ Langmuir–Blodgett films

The organization of Chl _a_ molecules at the air/water interface is important for the determination of the L–B film structure. Chapados and Leblanc [40] have proposed a model describing the organization of Chl _a_ macrocycles in a monolayer compressed at 20 mN/m. They assume that Chl _a_ macrocycles are tilted by 62.5° with respect to the water surface and the distance between the Mg centers is about 0.70 nm. The value of this angle corresponds to the one estimated using polarized visible reflection of Chl _a_ monolayers compressed to the same surface pressure [46]. This orientation is supposedly maintained by water molecules that form bridges between the Mg centers and the water surface [40]. Considering the area of the porphyrin plane (1.987 nm^2), one can easily see the correlation between the molecular area (0.980 nm^2 at 20 mN/m) and this orientation of the Chl _a_ macrocycle.

The analysis of the STM images [47], shown in Fig. 2, suggest a different structure than in the above model [40]. The differences are related to the formation of a complex larger than one Chl _a_ molecule and the absence of the 0.70 nm gap between the Mg centers. We assume that each corrugation represents a dimer of Chl _a_ with (anti) parallel transition moments of the constituent monomers. In this dimer, the macrocycles are tilted relative to each other by 30° with respect to the substrate surface (see Fig. 1b). This angle of orientation corresponds to the angle calculated using the orientational properties of the reaction center triplet of PS II [48]. However, it is much smaller than the angle proposed by Chapados and Leblanc [40] and Okamura et al. [46]. The change in the orientation angle (from 62.5° to 30°) has probably occurred during the transfer of the monolayer. This new arrangement of the Chl _a_ macrocycles correspond with the diameter (3.00 ± 0.15 nm) and the heights (0.55 ± 0.05 nm) of the corrugations. The average distance estimated between 2 maxima (3.20 ± 0.15 nm) is in the range of the distance center to center (3.00 nm) calculated using transient absorption difference spectroscopy of the reaction center of PS II [49]. We should mention that the 3.00 nm distance corresponds to the proposed distance between the Chl _a_ accessory and the P680. On the other hand, the distance between the Mg centers (1.20 ± 0.15 nm) within the dimer

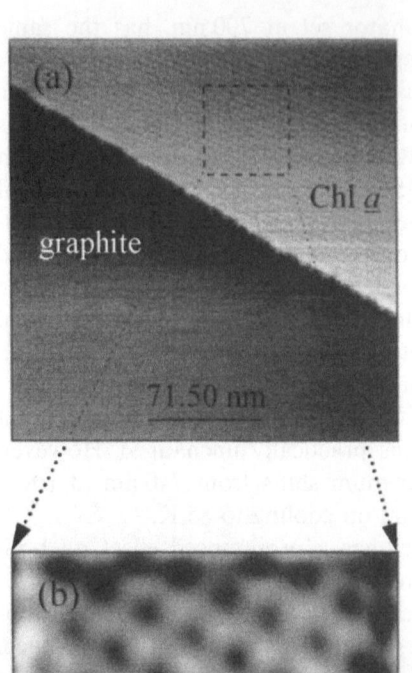

Chl _a_

graphite

71.50 nm

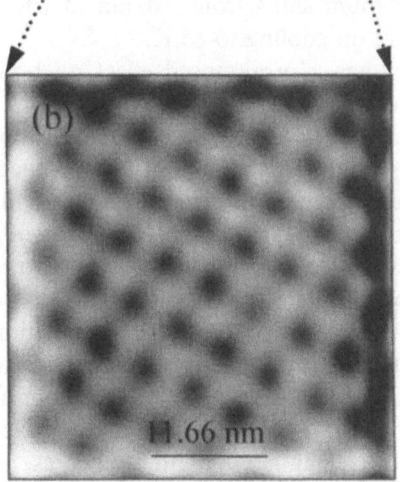

(b)

11.66 nm

Fig. 2 STM images of one monolayer of Chl _a_ ([**a**]: 214.5 × 214.5 nm, [**b**]: 35 × 35 nm) transferred at a surface pressure of 20 mN/m onto graphite. The imaging was in air and the tunneling current, the bias voltage were set to 0.1 nA and −300 mV, respectively

agrees with the calculated distance (∼1.00 nm) between centers of the Chl _a_ that makes the special pair identified as the P680 nm [50]. It is possible that the distance between the Mg centers, in the compressed monolayer at the air/water interface, is smaller than 1.20 ± 0.15 nm. Knowing the orientation angle estimated at the air/water interface (62.5°), the distance calculated between the Mg centers (0.70 ± 0.05 nm) corresponds with the gap estimated between the macrocycles [40].

Compared to the special pair of the purple bacteria (_Sp. viridis_), Chl _a_ molecules in the reaction center of PS II are weakly coupled to each other [51]. Therefore, the presence of a certain distance (1.20 nm), between the Mg centers in the dimer can be considered a reasonable argument for weak coupling. Using the exciton formula of McRae and Kasha [52] and assuming a distance of 1.20 nm between the Mg centers, this dimer would be expected to absorb at

684 nm [53]. We consider that only a few water molecules are holding the stretched dimer together, and their departure could easily alter the monolayer structure. This may result in a complete reorganization of the macrocycles. The proposed orientation resembles the proposed arrangement for the Chl a accessory molecules which are most likely bound to the histidines 118 of the D_1 and D_2 proteins [54]. This similarity in organization is probably one of the reasons for the resemblance of the optical properties of the monolayer and the PS II. In the light of these considerations, P680 is most likely a dimer with a geometry other than C_2-symmetry [55].

Mixed chlorophyll a films

Interfacial and electrical properties of chlorophyll a-lipid mixtures

The thylakoid membrane of chloroplast contains, in addition to chlorophylls, a large proportion of glycolipids. The major lipid components are the uncharged monogalactosyldiacylglycerol (MGDG) and digalactosyldiacylglycerol (DGDG), which combined amount to about 60–70 mol% of the total lipids present in the membrane. Charged lipids such as sulfoquinovosyldiacylglycerol (SQDG) form only 13% of the total amount.

Properties of Chl a-DGDG monolayers at the nitrogen/water interface were investigated using surface pressure (Π), surface potential (ΔV) and ellipsometric-area (A) isotherms [56]. Π–A isotherms show a pronounced negative deviation with respect to ideality, whereas ΔV-A isotherms follow the additivity law for mixed monolayers. As the mole fraction of Chl a increases from 0.1 to 0.5, each of the corresponding ellipsometric isotherm is shifted towards more negative values in the liquid-expanded region. For a mole fraction above 0.5, the ellipsometric $\delta\Delta$ values become more positive, with this trend continuing throughout the 0.5–0.9 range. The combined effects of anisotropy in both the refractive indices and the absorption coefficient can explain the turnover observed in the ellipsometric $\delta\Delta$ isotherms as the mole fraction of Chl a becomes larger than 0.5. In contrast to the Chl a-DGDG mixtures, the 0.8 (Chl a molar fraction) mixture of Chl a-SQDG displays almost ideal behavior, while that of 0.4 (Chl a molar fraction) is analogous to that shown by the Chl a-SQDG system. Furthermore, the negative charge of the polar head-group of the sulfolipid does not seem to influence the overall ellipsometric behavior of the mixed films which exhibit analogies with neutral galactolipid monolayers. Hydrophilic interactions have been ruled out on the basis of the agreement with the additivity law, which is observed for surface potentials of the mixed monolayers [57]. The observed deviation exhibited in the Π-A isotherms is interpreted in terms of the intermolecular cavity effect.

Sandwiches of the type Al/mixed monolayers of Chl a-SQDG/Ag have been fabricated in order to examine the effects of charged lipid on the photoelectric properties of

Table 2 List of various photovoltaic parameters of Chl a, Chl a-SQDG (see Ref. 58) and Chl a-Cantha (see refs. [59] and [60]) cells. In the case of Chl a and Chl a-SQDG cells, the incident power light is 21 μW/cm^2, whereas for Chl a-Cantha cells it is 13 μW/cm^2. (n/a): data not available

Type of cells	λ [nm]	I_{sc} [nA/cm^2]	V_{oc} [mV]	FF	η (%)
Al/Chl a/Ag Monolayers: 44	672	24	400	0.29	2.2×10^{-2}
Al/Chl a-SQDG/Ag Monolayers: 44 Molar ratio: 1, 0.025	672	20	450	0.27	1.1×10^{-2}
Al/Chl a-SQDG/Ag Monolayers: 44 Molar ratio: 1, 0.250	672	18	40	0.26	2×10^{-3}
Al/Cantha/Chl a/Ag Monolayers:10, 34	678 500	0 0	0 0	n/a n/a	0 0
Al/Chl a-Cantha/Ag Monolayers: 44 Molar ratio: 0.5, 0.5	678 500	4 2	10 7	n/a n/a	10^{-5}–10^{-4} 10^{-5}–10^{-3}
Al/Chl a/Chl a-Cantha/ Chl a/Ag Monolayers: 20, 14, 10 Molar ratio: 0.7, 0.3	678 500	15 5	500 300	n/a n/a	$6 \times (10^{-3}$–$10^{-2})$ 10^{-3}–10^{-2}

Chl a cells [58]. Various parameters of the cells, i.e. short circuit photocurrent (I_{sc}), open-circuit photovoltage (V_{oc}), fill factor (FF) and power conversion efficiency (η), were measured at 672 nm (the maximum absorption of Chl a in the red region of the optical spectrum). The results presented in Table 2 were obtained for an incident light power of 21 μW/cm^2 and molar ratios of Chl a:SQDG of 1:0.025 and 1:0.25. These parameters indicate that the introduction of an increased quantity of SQDG in Chl a decreases the cell efficiency. This is possibly due to the intervening effect of SQDG which makes the transport of photogenerated charge carriers across the multilayers of Chl a-SQDG more difficult. The introduction of recombination sites by SQDG may also be a reason for the lower efficiency of cells with increased quantities of SQDG.

A comparative study was carried out using a pigment instead of the lipid. Carotenoid canthaxanthin (Cantha) was added to Chl a in order to improve the efficiency of Al/Chl a monolayer/Ag solar cells, [59]. It was observed that the cells Al/mixed monolayers of Chl a and Cantha (0.5:0.5)/Ag showed a small photocurrent, however, the cells of the type Al/Chl a monolayers/mixed monolayers of Chl a and Cantha (0.7:0.3)/Chl a monolayers/Ag showed both photocurrent and photovoltage (see Table 2). The power conversion efficiencies of these cells calculated at 680 nm, the maximum of the red absorption band of Chl a, for incident light power of $\sim$13 μW/cm^2 are $\sim$0.05% and are comparable to those of Chl a cells [60]. Although the efficiencies of the cells did not show any improvement over Chl a and Chl a-SQDG cells, a large photocurrent at 500 nm (absent in Chl a cells) was observed. This was explained as due to the spectral sensitization of Chl a by Cantha. Further, the similarity of the action spectrum, when the cells were illuminated through an Al electrode, with the absorption spectrum of the pigment suggested the presence of a Schottky barrier at the Al/pigment interface.

Complexation of chlorophyll a and cytochrome c in Langmuir films

In photosynthetic membranes it is known that Chl a is associated with membrane proteins [61, 62]. This molecular association can only be extracted and purified with the use of detergents, which can reasonably alter the protein integrity. Therefore, a synthetic approach is considered in order to reconstitute, gradually, the biological membrane functions. The Langmuir technique is used to investigate films of mixed compounds at several molar ratios [63, 64]. Despite such advantages, it is not easy to investigate Chl a-protein complexes using the Langmuir technique, since the integral proteins are highly hydrophobic and spreading them at the air–water interface may induce some conformational changes.

The absorption spectrum in the red of Chl a dissolved in ethanol [64] is characterized by a low-energy band, whose maximum is at 665 nm. This corresponds to the electronic transition polarized along the Y-axis of the Π-electron conjugated system of the porphyrin ring [65]. When a small amount of ethanolic solution of Chl a is injected into water in order to obtain an ethanol:water ratio of 3:97, v/v ($\sim$10^{-5} M of Chl a), the QY transition of Chl a decreases in energy giving rise to a broad absorption band with a maximum at 675 nm and a shoulder at 700 nm. These changes are assigned to the formation of a new form of Chl a caused by an aggregation [45, 66, 67]. On the other hand, when the same experiment is repeated using the Cyt c solution ($\sim$10^{-7} M) instead of pure water, a single absorption band appears in the spectrum at 672 nm. This feature is attributed to the Chl a-Cyt c complexation, since the removal of aggregated excess of Chl a by centrifugation does not affect the maximum of absorption.

The binding of cytochrome c (Cyt c) to a monolayer of Chl a is studied using surface pressure (Π) vs area (A), surface potential (ΔV) vs area (A) and (^{14}C)Cyt c surface-concentration (Γ) vs area (A) isotherms [62]. The protein is incorporated into a Chl a monolayer compressed at a surface pressure of 20 mN/m. On expansion, the quantity of Cyt c incorporated into the monolayer gradually increases. This is supported by the increase of Π and ΔV after the protein is being injected in the subphase. The subsequent cycles of compression and expansions result in similar isotherms which are distinct from that measured at the first expansion. In addition to the latter observation, the surface radioactivity of (^{14}C)Cyt c indicates the irreversibility of protein incorporation into the Chl a monolayer. In fact, the surface properties of the binary film are completely different from those of either pure compounds. As a result, calculated values of ΔV for mixed monolayers assuming additivity deviate from experiment.

On the other hand, the incorporation of Cyt c in the monolayer of Chl a produces a blue shift of the entire spectrum. The Soret band is displaced from 439 to 436 nm and the red band from 680 to 673 nm. In addition, the red band is 3 nm narrower than that of pure Chl a monolayer. These spectroscopic characteristics of Cyt c-Chl a monolayers are comparable to those of mixed films of Chl a-MGDG in a molar ratio of 1:100. Therefore, it is assumed that in mixed Chl a-protein films, Chl a is diluted by Cyt c as it is in mixed Chl a-lipid films. However, the fluorescence lifetime of Cyt c-Chl a films is shorter than 0.2 ns and has the characteristics of a concentrated Chl a system. This apparent contradiction can be lifted out by taking into consideration energy transfer occurring

Progr Colloid Polym Sci (1997) 103:318–326
© Steinkopff Verlag 1997

between the monomer and aggregated forms of Chl $\underline{a}$. In this case, the role of Cyt c as a matrix can be that of incorporating different forms of Chl $\underline{a}$.

Langmuir–Blodgett films of the Chl $\underline{a}$-protein mixtures were analyzed using absorption, fluorescent and Fourier transform infrared spectroscopic techniques [64]. The IR spectroscopy showed that at protein molar fractions higher than 0.1, Cyt c molecules undergo drastic conformational changes from α-helix to β-sheet and turn structures. Such conformational changes are interpreted in terms of protein aggregation and denaturation. Spectroscopic evidence has also indicated the participation of the keto group in the Chl $\underline{a}$-protein interaction and the presence of Chl $\underline{a}$ molecules in an aggregated form in the complexes.

The Chl $\underline{a}$-Cyt c system does not directly mimic the specific functional complexes found *in vivo*. However, the findings discussed here can help to develop an understanding of the specific mechanisms involved in the Chl $\underline{a}$-protein interaction. One of these results is the fact that protein does not work as solvent, dispersing Chl $\underline{a}$ molecules in the whole range of molar fractions. The real mixture appears only at a defined molar fraction and is combined with relatively strong interactions. This behavior can be interpreted as an expression of the stoichiometric complex formation rather than the mixing processes. The interaction in the lipid–protein system should take into consideration the native conformation of the protein since the latter seems to undergo denaturation when its critical point is exceeded. Furthermore, the keto groups of Chl $\underline{a}$ play an important role in the Chl $\underline{a}$-Cyt c complexation.

Conclusion

The Langmuir–Blodgett technique is a powerful tool for building molecular-organized assemblies. Its use can bring complementary understanding of the pigments organization pattern in the native membranes. The study of photophysical and photochemical properties of Chl $\underline{a}$ assemblies in Langmuir and Langmuir–Blodgett films demonstrated the pronounced effect of water. This key element determines the assembly structure through a network of hydrogen bonds and oxygen coordination. In the models which describe the packing of Chl $\underline{a}$ in the monolayer and the L–B film, dimers of water maintain an array of oriented macrocycles. However, departure of some of these water molecules can cause a nonreversible reorganization of the array.

The study of the pigment-protein interaction is of great importance because *in vivo* Chl $\underline{a}$ is associated to a specific protein complex of the photosynthetic membrane. The use of the L–B technique combined with visible and infrared spectroscopy has demonstrated once more a great ability for bringing a complementary understanding of such complex systems. In the case of Chl $\underline{a}$-Cyt c, our finding indicated that the keto groups of the Chl $\underline{a}$ played a major role in the pigment–protein complexation. Also, the protein does not act as a dispersing solvent in the whole range of mole fractions. The real mixture, considered as an expression of the stoichiometric complex, appears only at a defined mole fraction where relatively strong interactions take place.

References

1. Hoober JK (1984) Chloroplasts. Plenum Press, New York
2. Margulies MM (1989) Plant Sci 64:1
3. Rögner M, Boekema EJ, Barber J (1995) TIBS 21:44
4. Parson WW (1991) In: Scheer H (ed) Reaction centers in Chlorophylls. CRC Press, Boca Raton, FL, p 1159
5. Parson WW (1991) In: Scheer H (ed) Reaction centers in Chlorophylls. CRC Press, Boca Raton, FL, p 1167
6. Goodwin TW (1980) In: Biochemistry of the Carotenoids, Vol 1, Chapman & Hall, New York, p 77
7. Nanba O, Satoh K (1987) Proc Natl Acad Sci USA 84:109
8. Ghanotakis DF, De Paula JC, Demetriou DM, Bowlby NR, Petersen J, Babcock GT, Yocum CF (1989) Biochim Biophys Acta 974:44
9. Danielius RV, Satoh K, VanKan PJM, Plijter JJ, Nuijs AM, Vangorkom HJ (1987) FEBS Lett 213:241
10. Rochaix JD, Erickson J (1988) Trends Biochem Sci 13:56
11. Golbeck JH (1992) Annu Rev Plant Physiol Plant Biol 43:293
12. Ikeuchi M (1992) Plant Cell Physiol 33:669
13. Thornber JP, Alberte RS (1977) In: Trebst A, Avron M (eds) The Organization of Chlorophyll *in vivo*, Encyclopedia of Plants Physiology. Springer, Berlin, p 574
14. Kühlbrandt W (1984) Nature 307:478
15. Kühlbrandt W (1987) J Mol Biol 194:757
16. Knox R, Lin S (1988) In: Scheer H, Schneider S (eds) Photosynthetic Light-harvesting Systems. Walter de Gruyter, Berlin, p 567
17. Knox JP, Dodge AD (1985) Phytochemistry 24:889
18. Frackowiak D, Zelent B, Malak H, Planner A, Cegielski R, Munger G, Leblanc RM (1994) J Photochem Photobiol A 78:49
19. Katz JJ, Bowman MK, Michalski TJ, Worcester DL (1991) In: Scheer H (ed) Chlorophyll Aggregation: Chlorophyll/Water Micelles as Models for *in vivo* Long-Wavelength Chlorophyll in Chlorophylls. CRC Press, Boca Raton, p 212
20. Katz JJ (1994) The Spectrum 7:1
21. Hann RA (1990) In: Roberts G (ed) Molecular Structure and Monolayer Properties in Langmuir–Blodgett Films. Plenum Press, New York, p 17
22. Munger G, Lorrain L, Gagné G, Leblanc RM (1987) Rev Sci Instrum 58:285
23. Petty MC, Barlow WA (1990) In: Roberts GG (ed) Langmuir–Blodgett Films. Plenum, New York, p 93
24. Zasadzinski JAN (1989) BioTechniques 7:174; Ogletree F, Salmeron M (1990) Prog Sol Stat Chem 20:235

25. Bustamante C, Dorothy AE, Keller D (1994) Curr Opin Struct Biol 4:750; Hansma HG, Hoh JH (1994) Annu Rev Biophys Biomol Struct 23:115; De Rose JD, Leblanc RM (1995) Surf Sci Rep 22:73

26. Simmons JG (1963) J Appl Phys 34:1793

27. Burns G (1985) In: Solid State Physics. Academic Press, San Francisco

28. Brody SS, Broyde SB (1968) Biophys J 8:1511

29. Volkov AG, Gugeshashvili MI, Munger G, Leblanc RM (1993) Bioelectr Bioener 29:305; Zelent B, Galant J, Volkov AG, Gugeshashvili MI, Munger G, Tajmir-Riahi H-A, Leblanc RM (1993) J Mol Struct 297:1; Munger G, Leblanc RM, Zelent B, Volkov AG, Gugeshashvili MI, Gallant J, TajmirRiahi H-A, Aghion J (1992) Thin Solid Films 210/211:739

30. Carpentier R, Dijkmans H, Leblanc RM, Aghion J (1983) Photochem Photobiol 5:245

31. Dijkmans H, Leblanc RM, Cogniaux F, Aghion J (1979) Photochem Photobiol 29:367

32. Worcester DL, Michalski TJ, Katz JJ (1986) Proc Natl Acad Sci USA 83:3791

33. Bowman MK, Michalski TJ, Tyson RL, Worcester DL, Katz JJ (1988) Proc Natl Acad Sci USA 85:1498

34. Tweet AG, Bellamy WD, Gaines Jr GL (1964) J Chem Phys 41:2068

35. de B Costa SM, Froines JR, Harris JM, Leblanc RM, Orger BH, Porter G (1972) Proc Roy Soc A 326:503

36. Leblanc RM, Galinier G, Tessier A, Lemieux L (1974) Can J Chem 52:3723

37. Leblanc RM, Chapados C (1977) Biophys Chem 6:77

38. Chapados C, Leblanc RM (1977) Chem Phys Lett 49:180

39. Chapados C, Germain D, Leblanc RM (1980) Biophys Chem 12:189

40. Chapados C, Leblanc RM (1983) Biophys Chem 17:211

41. Tancrède P, Munger G, Leblanc RM (1982) Biochim Biophys Acta 689:45

42. Picard G, Munger G, Leblanc RM, Le-Sage R, Sharma D, Siemiarczuk A, Bolton R (1986) Chem Phys Lett 129:41

43. Bellamy WD, Gaines Jr GL, Tweet AG (1963) J Chem Phys 39:2528

44. Désormeaux A, Leblanc RM (1985) Thin Solid Films 132:91

45. Krawczyk S, Leblanc RM, Marcotte L (1988) J Chim Phys 85:1073

46. Okamura E, Hasegawa T, Umemura J (1995) Biophys J 69:1142

47. Boussaad S, Tazi A, Leblanc RM (1996) Proc Natl Acad Sci USA, in press

48. Van Micghem FJE, Satoh K, Rutherford AW (1991) Biochim Biophys Acta 1058:379

49. Schelvis JPM, Van Noort PI, Aartsma TJ, Van Gorkom HJ (1994) Biochim Biophys Acta 1184:242

50. Braun P, Greenberg BM, Scherz A (1990) Biochem 29:10376

51. Kwa SLS, Eijckelhoff C, van Grondelle R, Dekker JP (1994) J Phys Chem 98:7702; Michel H, Deisenhofer J (1988) Biochem 27:2

52. McRae EG, Kasha M (1964) Fundamental Processes in Radiation Biology. Academic Press, New York, p 23

53. Kreutz W (1968) Z Naturforsch 23:520

54. Kobayashi M, Maeda H, Watanabe T, Nakane H, Satoh K (1990) FEBS Lett 260:138

55. Shipman LL, Cotton TM, Norris JR, Katz JJ (1976) Proc Natl Acad Sci USA 73:1791

56. Ducharme D, Shibata O, Munger G, Leblanc RM (1991) Can J Chem 69:1475

57. Ducharme D, Shibata O, Munger G, Leblanc RM (1989) Thin Solid Films 180:135

58. Nsengiyumva S, Hotchandani S, Leblanc RM (1994) Thin Solid Films 243:505

59. Diarra A, Hotchandani S, Kassi H, Leblanc RM (1991) Appl Surf Sci 48/49:567

60. Diarra A, Hotchandani S, Max J-J, Leblanc RM (1986) J Chem Soc Faraday Trans II 82:2217

61. Markwell JP, Thornber JP, Boggs RT (1979) Proc Natl Acad Sci USA 76:1233

62. Brecht E (1986) Photochem Photobiol 12:37

63. Lamarche F, Picard G, Téchy F, Aghion J, Leblanc RM (1991) Eur J Biochem 197:529

64. Es-Sounni A, Gruszecki WI, Tajmir-Riahi H-A, Zelent B, Wang G, Leblanc RM (1995) J Colloid Interface Sci 171:134

65. Sauer K (1975) In: Bioenergetics of Photosynthesis. Academic Press, New York, p 115

66. Inamura I, Ochiai H, Toki K, Watanabe S, Hikiro S, Araki T (1983) Photochem Photobiol 38:37

67. Shipman LL, Norris JK, Katz JJ (1976) J Phys Chem 80:877

68. Norris JR, Scheer H, Katz JJ (1975) Ann New York Acad Sci 244:260

69. Katz JJ, Ballschmiter K, Garcia-Morin M, Strain HH, Uphaus RA (1968) Proc Natl Acad Sci USA 60:100

Progr Colloid Polym Sci (1997) 103:327–328
© Steinkopff Verlag 1997

AUTHOR INDEX SUBJECT INDEX